Lecture Notes in

Volume 107

The series “Lecture Notes in Networks and Systems” publishes the latest developments in Networks and Systems—quickly, informally and with high quality. Original research reported in proceedings and post-proceedings represents the core of LNNS.

Volumes published in LNNS embrace all aspects and subfields of, as well as new challenges in, Networks and Systems.

The series contains proceedings and edited volumes in systems and networks, spanning the areas of Cyber-Physical Systems, Autonomous Systems, Sensor Networks, Control Systems, Energy Systems, Automotive Systems, Biological Systems, Vehicular Networking and Connected Vehicles, Aerospace Systems, Automation, Manufacturing, Smart Grids, Nonlinear Systems, Power Systems, Robotics, Social Systems, Economic Systems and other. Of particular value to both the contributors and the readership are the short publication timeframe and the world-wide distribution and exposure which enable both a wide and rapid dissemination of research output.

The series covers the theory, applications, and perspectives on the state of the art and future developments relevant to systems and networks, decision making, control, complex processes and related areas, as embedded in the fields of interdisciplinary and applied sciences, engineering, computer science, physics, economics, social, and life sciences, as well as the paradigms and methodologies behind them.

**** Indexing: The books of this series are submitted to ISI Proceedings, SCOPUS, Google Scholar and Springerlink ****

More information about this series at http://www.springer.com/series/15179

H. S. Saini · R. K. Singh ·
Mirza Tariq Beg · J. S. Sahambi
Editors

Innovations in Electronics and Communication Engineering

Proceedings of the 8th ICIECE 2019

Editors
H. S. Saini
Guru Nanak Institutions
Hyderabad, Telangana, India

R. K. Singh
Guru Nanak Institutions Technical Campus
Hyderabad, Telangana, India

Mirza Tariq Beg
Department of Electronics and
Communication Engineering
Jamia Millia Islamia
New Delhi, Delhi, India

J. S. Sahambi
Indian Institute of Technology Ropar
Rupnagar, Punjab, India

ISSN 2367-3370 ISSN 2367-3389 (electronic)
Lecture Notes in Networks and Systems
ISBN 978-981-15-3174-3 ISBN 978-981-15-3172-9 (eBook)
https://doi.org/10.1007/978-981-15-3172-9

This Springer imprint is published by the registered company Springer Nature Singapore Pte Ltd.
The registered company address is: 152 Beach Road, #21-01/04 Gateway East, Singapore 189721, Singapore

Committees

Editorial Board Members

Committee Members

Dr. S. P. Yadav, HOD1, ECE, GNITC
Dr. K. Santhi, HOD, EEE, GNITC
Dr. R. Prabhakar, Professor, GNITC
Prof. Anithaswamidas, GNITC
Prof. A. Mohan, GNITC
Dr. Binod Kumar, Associate Professor, GNITC
Dr. T. Vijayakumar, Professor, GNIT
Dr. Khushboo Pachori, GNITC
Dr. Harpreet Kaur, Associate Professor, GNITC
Prof. V. Bhagya Raju, GNITC
Mr. B. Sridhar, Associate Professor, GNITC
Mr. Sandeep Patil, Associate Professor, GNITC
Mr. Md. Shoukat Ali, Associate Professor, GNITC
Mr. D. Surendra Rao, Associate Professor, GNITC
Mr. K. Shashidhar, Associate Professor, GNITC
Mr. N. Srinivas, Associate Professor, GNIT
Mr. Sharath Chandra, Associate Professor, GNIT

Conference Committee Members

Department of Electronics and Communication Engineering (ICIECE 2019)

Budget

Dr. R. K. Singh
Mr. Sandeep Patil

Pre-conference Tutorials

Mr. Md. Shoukath Ali
Mr. V. Sai Babu

Receiving Papers and Acknowledgment, Attending Queries of Authors

Dr. Rashid Mahmood
Mrs. C. Sailaja
Mr. Srinivas Nanda
Mrs. N. Ramya Teja
Mr. I. Sharath Chandra (GNIT)

Conversion of Papers

Prof. A. Mohan

VIP Committee

Dr. Rashid Mahmood
Prof. B. Kedarnath

Invitation Preparation and Distribution

Dr. Rashid Mahmood

Registration (Online/Spot)

Mr. Ch. Raja Rao

Reception

Prof. Anitha Swamidas
Mythili Devi

Conference Office

Dr. Binod Kumar Prasad

VC, MD, JNTUH Messages

Prof. A. Mohan

Banners

Mr. K. Krishna Kumar

Proceedings

Dr. Binod Kumar Prasad
Mr. N. V. S. Murthy
Mr. L. Shiva

Transport Committee

Mr. D. Surendra Rao (VIP)

Program Committee

Dr. S. J. Sugumar
Dr. Rashid Mahmood
Dr. Harpreet Kaur
Mr. K. Shashidhar

Purchase Committee

Prof. S. Maheswara Reddy
Mr. A. V. Rameswara Rao

Certificates

Dr. R. Prabhakar
Mr. L. Shiva

Food Committee

Dr. R. Prabhakar
Mrs. B. Anitha
Mr. R. Gopinath
Dr. Kushboo Pachori
Mr. Chinna Narasimhulu
Mr. K. Raju

Inauguration and Valedictory Function

Dr. R. K. Singh
Dr. Shatrughna Prasad Yadav
Prof. S. Maheswar Reddy
Mr. Sandeep Patil

Photographs and Videos

Mr. O. Ravinder

Press Release and Media

Prof. A. Mohan

Conference Report

Dr. S. J. Sugumar

After Conference Attending Queries of Authors

Dr. Rashid Mahmood
Dr. Kushboo Pachori
Mrs. C. Sailaja

Keynote Speech Arrangements

Ms. Simarpreet Kaur

VIP Accompany

Dr. S. J. Sugumar
Ms. B. Aruna

Cultural Events

Mrs. Preethi

Reviewers List

Internal Reviewers

Prof. R. K. Singh, Associate Director
Dr. S. V. Ranganayakulu, Dean/R&D, GNITC
Dr. Shatrugna Prasad Yadav, HOD/ECE, GNITC
Prof. S. Maheswara Reddy, HOD/ECE, GNITC
Dr. K. Santhi, HOD/EEE, GNITC
Dr. S. J. S. Sugumar, Professor/ECE
Prof. Anita Swamidas, Professor/ECE
Prof. A. Mohan, Professor/ECE
Mr. V. Bhagya Raju, Professor/ECE
Dr. Rashid Mahmood, Professor/ECE
Dr. Binod Kumar, Associate Professor/ECE
Dr. Prabhakar, Associate Professor/ECE
Dr. Khushboo Pachori, Associate Professor/ECE
Dr. Harpreet Kaur, Associate Professor/ECE
Mr. K. Shashidhar, Associate Professor/ECE
Mr. N. V. S. Murthy, Associate Professor/ECE
Mr. B. Sridhar, Associate Professor/ECE
Mrs. G. Kiran Maye, Associate Professor/ECE

External Reviewers

Dr. Sandeep Kumar, Sreyas Institute of Engineering & Technology, Hyderabad
Dr. Ravindra Kumar Yadav, Director, Skyline Institute of Engineering and Technology, Greater Noida, G. B. Nagar (U.P.)
Dr. Maneesh Kumar Singh, Department of ECE, National Institute of Technology Delhi, India
Dr. Rohit Raja, Sreyas College of Engineering & Technology, Hyderabad
Imran Ahmed Khan, Galgotias College of Engineering and Technology, Greater Noida, Uttar Pradesh
Prof. (Dr.) Shamimul Qamar, King Khalid University, Abha, Saudi Arabia
Dr. Jugul Kishor, National Institute of Technology Delhi
Dr. M. Nasim Faruque, University Polytechnic, Faculty of Engineering and Technology, Jamia Millia Islamia, New Delhi-110025
Dr. Neeta Awasthy, Noida International University, Greater Noida
M. Lakshmanan, Department of ECE, Galgotias College of Engineering and Technology
Dr. Ashish Gupta, ITS Engineering College, Greater Noida, Affiliated to Dr. A.P. J. Abdul Kalam Technical University, Lucknow
Dr. Deepak Batra, Thapar University, Patiala, Punjab, India
Dr. B. Thiyaneswaran, Sona College of Technology, Salem-636005, Tamil Nadu
Dr. N. Malmurugan, Principal, Mahendra College of Engineering, Salem, Tamil Nadu
Dr. R. Maheswar, Sri Krishna College of Technology, Coimbatore
Korlapati Keerti Kumar, Department of ECE, Vaageswari College of Engineering, Karimnagar
Dr. Prabha Selvaraj, Malla Reddy Institute of Engineering & Technology, Campus 3, Maisammaguda, Dulapally
Dr. D. Jackuline Moni, Professor, ECE Department, Karunya Institute of Technology and Sciences, Coimbatore, Tamil Nadu
Dr. Suresh Merugu, Associate Dean, R&D Centre, CMR College of Engineering & Technology, Hyderabad, Telangana, India
Chaitanya Duggineni, Professor, GRIET, Bachupally, Kukatpally, Hyderabad.
Dr. Pushpa Mala S., Department of ECE, School of Engineering, Dayananda Sagar University, Kudlu Gate, Bengaluru-560068
Dr. K. V. Ramprasad, Professor, Department of ECE, Kallam Haranadhareddy Institute of Technology, Guntur-522019, A.P.
Dr. Jithin Kumar M. V., Kerala Technological University, Kerala
M. Aravind Kumar, JNTUK, GVIT Engineering College, Bhimavaram
M. Nizamuddin, Jamia Millia Islamia
Dr. S. Arul Jothi, Sri Ramakrishna Engineering College, Coimbatore
Sathish Kumar Nagarajan, Professor/ECE, Sri Ramakrishna Engineering College
Deepika Ghai, Assistant Professor, School of Electronics and Electrical Engineering, Phagwara
Dr. Kirti Rawal, Lovely Professional University, Phagwara, Punjab, India

Dr. V. A. Sankar Ponnapalli, Department of Electronics and Communication Engineering, Sreyas Institute of Engineering & Technology, Hyderabad

Hemlata Dalmia, Associate Professor, Sreyas Institute of Engineering & Technology

Rahul Hooda, Government College, Jind-126102

Dr. Anuj Singal, GJUS&T, Hisar

Dr. P. Venkateswara Rao, Vignana Bharathi Institute of Technology, Aushapur, Ghatkesar, Hyderabad-501301

Dr. D. Jayanthi, Professor, Department of ECE, Gokaraju Rangaraju Institute of Engineering & Technology, Hyderabad-500090, Telangana

Agha Asim Husain, ITS Engineering College, Plot No. 46, Knowledge Park-3, Greater Noida

Dr. Srinivas Bachu, Associate Professor, Marri Laxman Reddy Institute of Technology and Management, Dundigal, Hyderabad

Shilpa Rani, Flat-101, Geetika Towers, Kundanbagh, Begumpet, Hyderabad

Md Ehtesham, Electrical Department, Jamia Millia Islamia, New Delhi

Dr. Deepika Vodnala, B V Raju Institute of Technology, Vishnupur, Narsapur, Medak, Telangana

Abhiruchi Passi, Manav Rachna International Institute of Research and Studies, Sector 43, Delhi Surajkund Road

Dr. Sanjay Dubey, Professor and Associate HOD, ECE Department, BVRIT, Narsapur, Medak, Telangana

Dr. B. Anil Kumar, Associate Professor, ECE Department, VBIT Hyderabad

Prabhakar Sharma, CDAC Mohali

Dr. Sanjeev Kumar, H. No. 277E/1, Shastri Colony, Faridabad, Haryana

Dr. Leena Arya, Professor, ITS Engineering College, Greater Noida, U.P.

Dr. Jaishanker Keshari, ABES Engineering College, Ghaziabad

Madan Kumar Sharma, Galgotias Institutions

Manish Kumar Singh, Galgotias College of Engineering and Technology, Greater Noida

Dr. Caffiyar Mohamed Yousuff, C. Abdul Hakeem College of Engineering & Technology, Vellore

Arvind Kumar, Accurate Institute of Management and Technology, Greater Noida

Dr. Sreenivasa Rao Ijjada, ECE Department, GITAM, Rushikonda

Dr. Srilatha Indira Dutt Vemuri, Department of ECE, GITAM, Visakhapatnam

Shilpi Birla, Manipal University Jaipur

Dr. Neeraj Kanwar, Manipal University Jaipur

Shatrughna Prasad Yadav, Symbiosis University of Applied Sciences, Indore, M.P.

Syed Jalal Ahmad, Bukhari House, Malik Market, Dr. A. Q. Salaria Lane, Narwal, Jammu

Dr. Tarun Kumar Dubey, Department of Electronics and Communication Engineering, Manipal University Jaipur
Dr. S. P. Singh, Professor and Head, ECE Department, MGIT
Sikander Hans, Thapar University, Patiala
Dr. Munish Kumar, DCRUST, Murthal, Sonipat, Haryana
D. Elizabath Rani, GITAM (Deemed to be University), Visakhapatnam
Dr. Monika Bhatnagar, AKTU, Lucknow, U.P.
Dr. Vijayalaxmi Biradar, Sai Ram Residency, Flat No. 304, Backside Vishnu Theatre, Prashanth Nagar, Vanasthalipuram, Hyderabad-500070
Dr. Biswajeet Mukherjee, IIITDM, Dumna Airport Road, P.O.-Khamaria, Jabalpur-482005
Usha Sharma, IET, Sec. F, Jankipuram, Lucknow
J. Beatrice Seventline, Professor, GIT, GITAM (Deemed to be University), Visakhapatnam
Shilpa Rani, Department of ECE, GITAM (Deemed to be University), Visakhapatnam
I. Sharath Chandra, Guru Nanak Institute of Technology
Dr. Sandeep Jaiswal, Mody University of Science and Technology, Lakshmangarh, Sikar, Rajasthan
Deependra Sinha, E 46 Gamma 1, Greater Noida, U.P.
Dr. Amrita Rai, G. L. Bajaj Institute of Technology & Management
Sridevi Katamaneni, Associate Professor, Department of ECE, GIT, GITAM, Visakhapatnam
Amit Sehgal, G. L. Bajaj Institute of Technology & Management
Keshav Patidar, Associate Professor, School of Automobile and Manufacturing Engineering, Symbiosis University of Applied Sciences, Indore
Dr. Jitender, Plasma Research Centre, Gandhinagar
Pramod Kumar, IPEC Ghaziabad
Naqui Anwer, TERI School of Advanced Studies
Abrar Ahmad, Jamia Millia Islamia, New Delhi

National and International Advisory Board List

International Advisory Board

Prof. Mohammed H. Bataineh, Yarmouk University, Irbid, Jordan
Saman Halgamuge, The Australian National University, Canberra
Prof. Akhtar Kalam, Victoria University, Australia
Dr. Alfredo Vaccaro, University of Sannio, Benevento, Italy

National Advisory Board

Preface

The 8th International Conference on “Innovations in Electronics and Communication Engineering” (ICIECE 2019) is organized by the Department of Electronics and Communication Engineering, Guru Nanak Institutions Technical Campus, Hyderabad, India, during August 2–3, 2019. More than 509 papers were received from India and across the globe including Sweden, Italy, Iraq, Saudi Arabia, Australia, Malaysia, Bangladesh, Oman, Ethiopia, etc. Seventy-five papers have been selected by reviewers for publication in Springer “Lecture Notes in Networks and Systems.” The research contributions cover a wide range in the domain of electronics and communication engineering, which includes five tracks: communication engineering, signal/image processing, embedded system, VLSI and miscellaneous.

Distinguished professors and scientist from India and abroad joined as keynote speakers and shared their valuable ideas for innovation and integration in the field of electronics and communication engineering. We acknowledge keynote speakers from Malaysia—Prof. Mohd Rizal Bin Arshad and Dr. Goh Kam Meng.

The papers selected were presented by the authors during the conference, in front of session chairs. Parallel sessions were conducted for each track to accommodate all the authors and to give them ample time to discuss their ideas. The conference has grown exponentially over the years and has become a platform for scientists, researchers and academicians to present their ideas and share their cutting-edge research in various fields of electronics and communication engineering. The focus for this year conference was “innovation and integration in the field of electronics and communication engineering.”

This conference was funded by All India Council for Technical Education (AICTE), Delhi; Council of Scientific and Industrial Research (CSIR), Delhi; and Defence Research and Development Organisation–Naval Science and Technological Laboratory (DRDO-NSTL), Visakhapatnam.

We would like to thank all the keynote speakers, participants, session chairs, committee members, reviewers and international and national board members, Guru Nanak Institutions Management and all the people who have directly or indirectly contributed to the success of this conference. We would also like to thank Springer Editorial Team for their support and for publishing the papers as part of the "Lecture Notes in Networks and Systems" series continuously since last four years.

Hyderabad, India H. S. Saini
Hyderabad, India R. K. Singh
New Delhi, India Mirza Tariq Beg
Rupnagar, India J. S. Sahambi

About This Book

The objective of this book is to disseminate the new ideas and research submitted by the authors for the 8th International Conference on Innovations in Electronics and Communication Engineering (ICIECE 2019). This is a continuous process of technology upgradation. The latest inputs especially in the areas of signal and image processing, communication engineering, radar signal processing, antenna/microwave, embedded systems, VLSI design, biomedical electronics, IOT, virtual reality, Digital India, Smart Cities, etc., have been incorporated.

This book aims at bringing the research in latest fields of communication at one place for the readers to update themselves and start their pursuit of excellence in the field of electronics and communication. Innovate is the buzz word which is highlighted to succeed in life and make it more comfortable.

The readers can formulate their own field of research and innovation with this as the base and pick up ideas and technology which appeals them. This book is a very good reading for young researchers, the technocrats and academicians who seek to acquaint with the latest in the field of electronics and communication engineering.

Contents

About the Editors

H. S. Saini Managing Director of Guru Nanak Institutions, obtained his Ph.D. in the field of Computer Science. He has over 22 years of experience at university/college level in teaching UG/PG students and has guided several B.Tech., M.Tech. and Ph.D. projects. He has published/presented more than 30 high-quality research papers in international, national journals and proceedings of international conferences. He is the editor for Journal of Innovations in Electronics and Communication Engineering (JIECE) published by Guru Nanak Publishers. He has two books to his credit. Dr. Saini is a lover of innovation and is an advisor for NBA/NAAC accreditation process to many Institutions in India and abroad.

R. K. Singh Associate Director Guru Nanak Institutions Technical Campus, is an alumina of REC (now MNIT Jaipur) and did his M.Tech. from IIT Bombay, in the field of Communication Engineering. He has completed Ph.D. on Radar Signal Processing from GITAM Deemed to be University. He has served Indian Army in the core of Electronics and Mechanical Engineering for 20 years before hanging his uniform as Lt Col. He has rich industrial experience as Army Officer managing workshops and has been teaching faculty for more than six years while in services. He started his career as a teaching Assistant at MNIT Jaipur for one year before joining army. As a Professor, he has served for more than eleven years after premature retirement from the army services. He has served as HOD, Vice Principal and Principal of engineering college before being approved as Associate Director of this institute. The Professor had hands-on experience on high-tech electronic equipments and has done many courses on radars and simulators. He has published many papers on microstrip antennas, VLSI and radar signal processing in national and international conferences.

Mirza Tariq Beg is a Professor and Head Department of Electronics and Communication Engineering, Faculty of Engineering and Technology, Jamia Millia Islamia, New Delhi. He received Ph.D. degree from Jamia Millia Islamia New Delhi in the year 2003, M. Tech. from Delhi University Delhi in the year 1987 and B.Tech. from Aligarh Muslim University Aligarh in 1985. He started his career as an Assistant Professor in the Department of Electronics and Communication Engineering from Jamia Millia Islamia New Delhi in 1987. Now, he is working as a Professor since 2003 in the same organization. He was also Director of Centre for Distance & Open Learning (CDOL), Jamia Millia Islamia, New Delhi. His research area includes microwave and communication engineering. He has guided several Ph.D. students and authored and co-authored more than 50 research papers in peer-reviewed, international journals.

J. S. Sahambi received the graduation degree in electrical engineering from Guru Nanak Engineering College, Ludhiana, India, M.Tech. degree in computer technology from the Electrical Engineering Department and the Ph.D. degree in the area of signal processing, in 1998, both from the Indian Institute of Technology (IIT) Delhi, India. In June 1999, he joined Electronics & Communication Engineering Department, IIT Guwahati, and moved to the Department of Electrical Engineering, IIT Ropar, since 2010. His research interests include signal processing, image processing, wavelets, DSP embedded systems and biomedical systems.

Communications

Fault Analysis for Lightweight Block Cipher and Security Analysis in Wireless Sensor Network for Internet of Things

Shamimul Qamar, Nawsher Khan, Naim Ahmad, Mohammed Rashid Hussain, Arshi Naim, Noorulhasan Naveed Quadri, Mohd Israil, Mohammed Salman Arafath and Ashraf A. El Rahman

Abstract The integration of wireless sensor nodes (WSNs) in the Internet of things (IoT) may generate new security challenges for establishing secure channels between low-power sensor nodes and Internet hosts. It includes a bunch of challenges right from redefining the existing ones from scratch of legacy codes to the current trend of proposing and designing new key establishment and authentication protocols. This paper is the successful attempt in resource constraint environment like wireless sensor network (WSN) to integrate cipher authentication and lightweight key management solutions in one bundle so that it can be deployed in IoT domains WSN is also a part of IoT. The cipher should be authenticated, lightweight, and should be flexible to use in the cryptanalysis. The cipher gives the better authentication techniques and performance but limited to the flexibility. Lightweight block cipher (LBC) uses less

S. Qamar (✉) · N. Khan · N. Ahmad · M. R. Hussain · A. Naim · N. N. Quadri · A. A. El Rahman
College of Computer Science, King Khalid University, Abha 62529, Kingdom of Saudi Arabia
e-mail: sqamar@kku.edu.sa

N. Khan
e-mail: nawsher@kku.edu.sa

N. Ahmad
e-mail: nagqadir@kku.edu.sa

M. R. Hussain
e-mail: humohammad@kku.edu.sa

A. Naim
e-mail: arshi@kku.edu.sa

N. N. Quadri
e-mail: qnaveed@kku.edu.sa

A. A. El Rahman
e-mail: rezaalh@kku.edu.sa

M. S. Arafath
College of Engineering, King Khalid University, Abha, Kingdom of Saudi Arabia
e-mail: salman@kku.edu.sa

M. Israil
College of Science, Al Jouf University, Al Jouf, Sakakah, Kingdom of Saudi Arabia
e-mail: misrail@ju.edu.sa

H. S. Saini et al. (eds.), *Innovations in Electronics and Communication Engineering*, Lecture Notes in Networks and Systems 107,
https://doi.org/10.1007/978-981-15-3172-9_1

computation and secures the data with low computational cost and it is used for many purposes like radio-frequency identification (RFID) tags and sensor network nodes. Advanced Encryption Standard is applied in many devices for the encryption, but it cannot be applied to the smaller embedded devices like wearable devices, etc. LBC works for the many devices with low resources, and research on the LBC has been increased. In this research, Data Encryption Standard (DES) and analyzed algorithm are used to increase the efficiency in terms of solving time and Improved Fault Analysis (IAFA) is used to increases the security of the LBC in wireless sensor node for IoT.

Keywords Advanced encryption standard · Improved algebraic fault analysis · Lightweight block cipher · PRESENT · Radio-frequency identification

1 Introduction

The wireless sensor network (WSN) constructed by many large number of sensor nodes placed in order in a required area and in an organized manner to collect data. The WSN has no central processing control node in the wireless network, and the source to destination information transmission would be attained by the intermediate nodes in a multihop forwarding of information in many ways [1]. The wireless sensor network characteristics are the flexible, distributed, and dynamic are applied in wide range of applications, such as in military use for war in the battle, calamity relief, exploration of minerals, gas and oil, environmental risk innovation, and other areas of interest [2]. The sensor nodes, however, often grieve from a mixture of attacks and other exterior devastation since WSN is regularly used in harsh environments. Besides this, the cost of production of wireless sensor node depends on minimum manufacturing cost and restricted resources and minimum power and the large area of radio coverage. The accuracy of the captured data depends on large number of factors so the factors such as capturing, monitoring storing, and transmitting data must be accurate. Therefore, the fault detection in any parameter of the network is essential for measuring the accuracy of monitoring results of the collected data. The fault detection algorithms for sensing and collecting of data of using sensor nodes in WSN have been categorized into centralized fault detection which has central control over all other helping nodes and distributed fault detection in which nodes are distributed and according to different data processing methods can be applied to collect the required information [3]. The lightweight block cipher (LBC) algorithm is designed for the fast-developing technology with low-resource devices and it is applicable for wireless sensor networks, radio-frequency identification, and Internet of things [1, 4]. The most used LBC are PRESENT and DES. These ciphers are designed and tested for under 3000 logic gate equivalent and complication of hardware and items increases with gate equivalent [2]. As, these devices are used to transfer the sensitive or private data, adequate level of data security is a basic requirement. Use of RFID

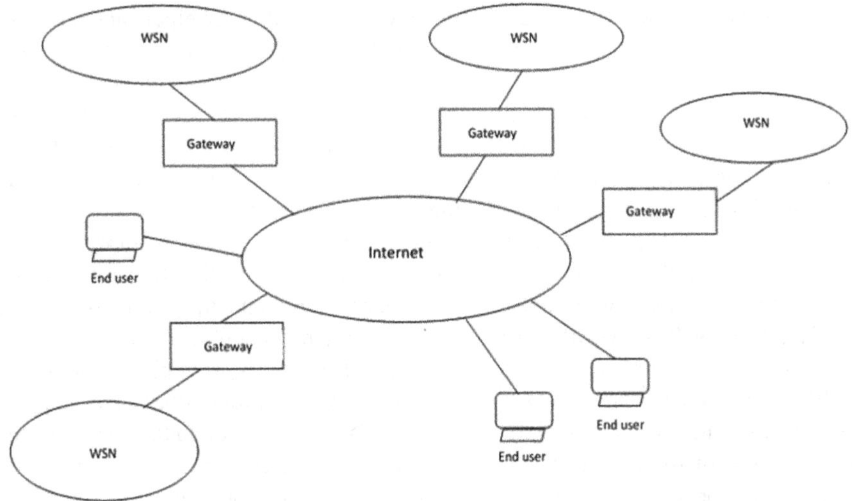

Fig. 1 IoT-based WSN

has increased recently and it cannot support the advanced encryption standard and standard public-key cryptosystems [3, 5].

LBC is used in RFID because it operates at low resources and it provides efficient and power-saving solution [6, 7]. The attack on the cipher is increasingly reported and system security becomes more vulnerable. The intended fault attacker can use the secret key searching method and possibly may recover the secret key by using the parameter of this fault and injecting fault parameters in the system and fault be able to be generated by the power supply, frequency of the electronic synchronized external clock, exposing laser in the circuit [8, 9]. The major attack is side-channel attacks, which illegally analyze the secrete keys using physical information. The IoT based on WSN is given in Fig. 1.

Later, fault attack has been combined with differential cryptanalysis is called differential fault analysis.

The remaining work of the paper is structured as follows. Section 2 contains the literature review of recent year paper in the literature. Section 3 contains the proposed system. Section 4 contains the algebraic fault algorithm. Section 5 discusses the result, and Sect. 6 contains the conclusion.

2 Literature Review

Li et al. [10] proposed QTL cipher technique. The author had described a new ultra-lightweight block cipher and also explains the importance of QTL supports 64- and 128-bit keys and the use of key size. The traditional Feistel-type structures have the limitation of the slow diffusion and it can convert only half of the block message in an iterative round. The Feistel network structure has been generalized and is applied in the design aim of QTL, which helps to overcome the disadvantages of the traditional Feistel-type structure. This design aim of QTL improves the speed of diffusion of the parameters of the Substitution–Permutation Network structures. These diffusion improved parameters increase the security in the Feistel network structure. The key size of this design is larger and it consumes more time to process.

Mohd et al. [11] objective was to implement the lightweight cipher in the hardware and optimize the design metrics. They measured the design metric by advanced design flow with implementing ciphers and conducting simulations. The KATAN/KTANTAN algorithm was used to implement and optimized the hardware. The program was processed in the hardware of Field Programming Gate Array (FPGA). The measuring was done with designs with various block size, number of implementation rounds and key scheduling. This research found that key scheduling involves in the consumption of resources, power, and energy. However, the lightweight concept is also used in software to achieve performance [12].

Lara-Nino et al. [13] designed and propose a standardized lightweight cipher in designing of hardware of cipher to defeat the security issues and problems in tremendously controlled environments. Several methods were analyzed and two novel designs were made. They used three state-of-the art designs of the cipher that have been calculated and compared by area, performance in terms of the security, low level of energy, and efficiency. The first design produces the best result and the second design produces the poor measure in terms of performance and energy consumption. Arafath et al. [14] have done a wonderful successful attempt to achieve security in WSN in software using the concept of three different cryptographic techniques.

Zhang et al. [15] presented the formal fault mode was examined to face the latest fault attacks. Algebraic fault analysis is used to inject faults in lightweight block cipher called LBlock to evaluate the security performance of the method. It involves in three processes, namely finding the fault and then to encrypt the fault parameters, to key scheduling of cipher, and modifying the round number in cipher or counter for the numbers. The best result of Fan Zhang fault mode shows that a single fault addition recovers the master key of LBlock in the complex scenario.

The proposed method is verified by applying algebraic fault analysis to three types of other block ciphers, i.e., first is Data Encryption, second is Standard, and third is PRESENT, and Twofish. This method has some limitations in algebraic fault analysis mainly it was a highly complex cipher such as Twofish.

Kotel et al. [16] selected the low-resources embedded system such as RFID tags. In this research, the analysis had been made in the programming software and execution of lightweight block cipher using fair estimation of lightweight cryptographic system used in low-resources embedded system. Then, the assessment and calculation of execution time, RAM power consumption and ROM memory consumption on the 8-bit AVR microcontroller is done using 16 bit MSP microcontroller. The 16-bit MSP microcontroller is applied and tested. The performance of software is decreased while increases in the security.

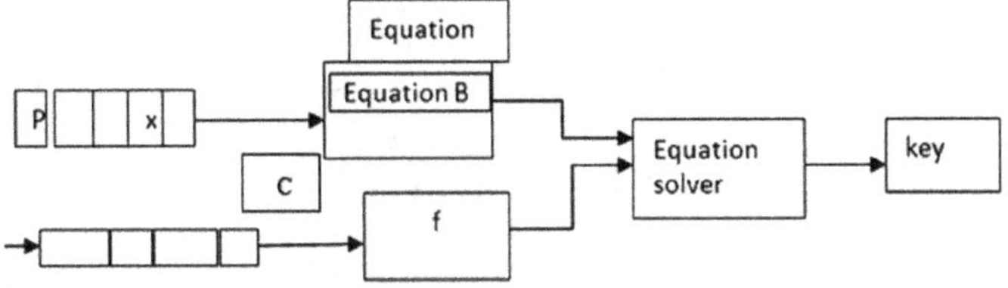

Fig. 2 Architecture of AFA

3 Proposed System: Algebraic Fault Analyzer

The analysis of security in the LBC by using fault attack has been made in this section and method used is AFA. It has four methods: Cipher descriptor, equation builder, machine solver and cipher of LBlock. Then, the analysis is made with the help of AFA method, which helps to solve the fault in the system.

3.1 Cipher Descriptor

Cipher descriptor has two modules, namely design and implementation, the module contains a **Descriptor** class which describes a cipher, and a **Cipher** class for using a cipher (e.g., keys, IVs, etc.). The design is used to solve the problem in cryptography.

3.2 Equation Builder

The cipher was designed as per fault analyzer and the designed cipher equation has been infused the cipher system for the proper operation. The equation builder gives the procedure for building the equation. This will automatically transfer operation into cipher descriptor into algebraic equations. The AFA has equation builder and equation solver to build equation set for the operation in the cipher. As shown in Fig. 2.

3.3 Machine Solver

Machine solver is a block which has an input equation and process to solve it by using the mathematical formula automatically by software, and output is applied as an input into the sat solver. The CryptoMiniSat techniques have been applied to solve the problems in cipher system based on the state variables, and it has to identify and remove the problem due to fault automatically.

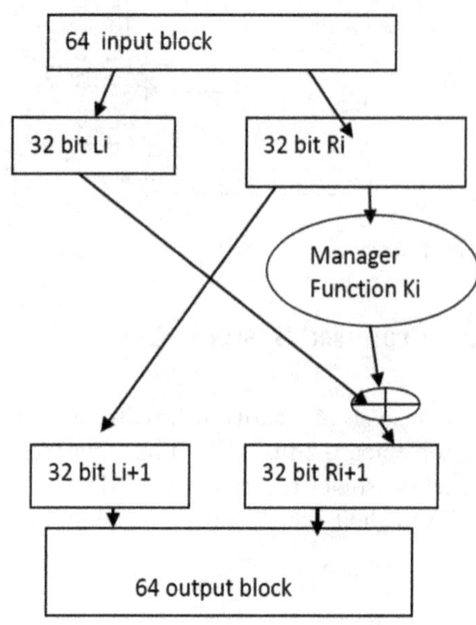

Fig. 3 Specification for lightweight block

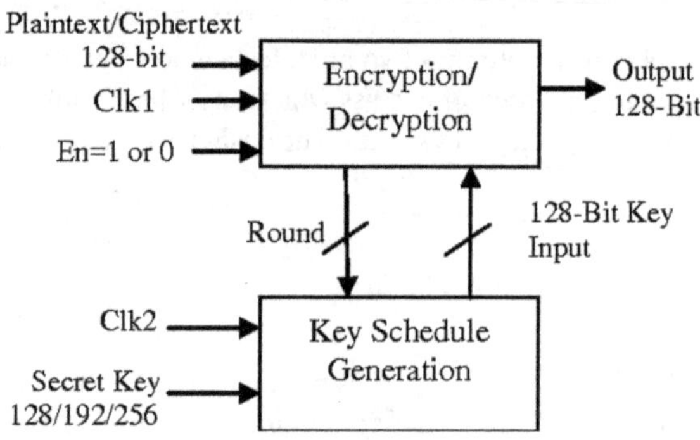

Fig. 4 Block diagram of encryption/decryption process

The process of LBlock is shown in Fig. 3 and input is given as plain text and it obtained as ciphertext. Decryption of the process is same as that of the encryption in the inverse order and key is formed by key scheduling. As shown in Fig. 4.

LBlock consists of three blocks, which are (1) encryption (2) key scheduling (3) decryption.

4 Algebraic Fault Analysis

Algebraic fault analysis has been redesign and applied to build the equation for the cipher system and the equation for the cipher's given below.

```
RandomPT(P);Input data Plan Text
Key: K1 = KS1 (K1,L);Key.
FOR rch = 1; rch<; rc++ do.
To generate:GenKSRdES(rc, ); //#1 Generate Key.
END
The iteration: FOR k = 0; k < N; i++ do.
Ci = Enc(Pi,K);
FOR rch = r − 1; rch<; rch++ do
To generate: GenEnRdES ( , , );//#2
END
The Input: GenInputES();
Ck * = InjectFault(Enc( , K),);Fault Input
FOR rch = r − 1; rc<; rc++ do
To End: GenEnRdES ( , , );//#3
END
To Generate Input: GenInputES();
The procedure for fault Insertion
Input: GenFaultyES (f = +);// #4: 18. END.
Generate RandomPT();
Cv = Enc( , K);
Iteration: FOR rch = 0; rch<; rch++ do
Generate: GenEnRdES ( , , );//#5:.
END:.
Generate: GenInputES( , );:.
Execution: Tsol = RunAFA();
```

In above algorithm, the value of P is for plan text data and K denote keys used in the cipher for the round effectively in the AFAI algorithm. Ks is used as key scheduling and Enc represents the encryption function used in the cipher. The result analysis is carried out for the problem in the system by the AFA. This algorithm generates and assigns the key for each plain text Pi for the correct encryption and fault is injected.

5 Results and Discussion

The analysis of the proposed system has been made and the result is presented in this section, and AFA is implemented in different scenarios. The cipher is applied in the LBlock and it contains 32 rounds of the Feistel network structure and analysis done above the 32 rounds. The r denotes the total executed number of rounds in AFA system of the networks (Fig. 5).

A fault is injected into r in the LBlock. This fault can be accessed at the commencement of each time instance. The effect of fault may increase or decrease the performance in terms of e total number of rounds in ciphers. In this case $r \geq 32$,

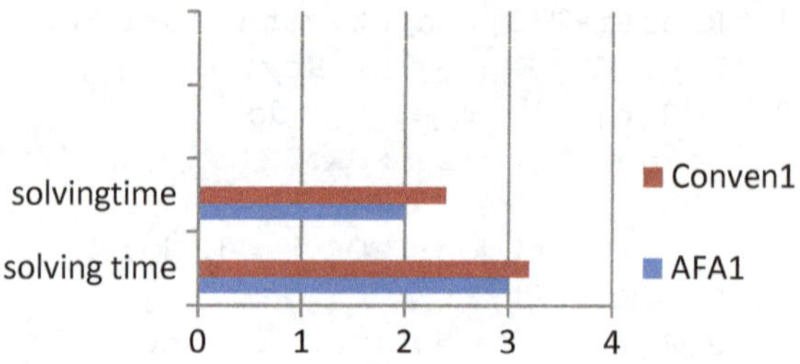

Fig. 5 Solving time and comparison in conventional and AFA1

LBlock produces the additional round taken place after the normal encryption and it increases the time. In this case $r \leq 32$, LBlock does not produce the first r_{max} round and skip the remaining rounds and causes the key recovery to reduce. The solving time for algebraic fault analysis is less as compared to conventional method and hence more efficient.

6 Conclusion

The encryption standard is effective encryption system, but it cannot run in the low-resource system below block size of 128 bit. The LBC provides security that works in the low-resource system like wearable devices, RFID, and sensor nodes, etc. The fault attack can recover the secret key by injecting the fault in the system. In this research, the AFA uses the different algorithms including LBlock, and DES and advanced encryption standard and fault is injected into the system. This LB cipher research analyzes the various faults in the system by the IAFA algorithm in various scenarios.

Acknowledgements We thank the Deanship of Scientific Research, College of Computer Science, Computer Engineering Department, Guraiger campus, King Khalid University, Abha KSA, colleagues, family, and friends for their support.

References

1. E.M. do Nascimento, J.A.M. Xexeo, A flexible authenticated lightweight cipher using Even-Mansour construction, in *IEEE International Conference on Communications (ICC)* (2017), pp. 1–6
2. S. Sadeghi, N. Bagheri, M.A. Abdelraheem, Cryptanalysis of reduced QTL block cipher. Microprocess. Microsyst. (2017)
3. B.J. Mohd, T. Hayajneh, A.V. Vasilakos, A survey on lightweight block ciphers for low-resource devices: comparative study and open issues. J. Netw. Comput. Appl. **58**, 73–93 (2015)
4. M.S. Arafath, K.U.R. Khan, K.V.N Sunitha, Security in opportunistic sensor network and IoT having sensors using light weight key generation and cryptographic algorithm, in *International*

Conference on Recent Innovations in Electrical, Electronics & Communication Engineering—(ICRIEECE) July 27th & 28th 2018 (School of Electrical Engineering, Kalinga Institute of Industrial Technology (KIIT), Bhubaneswar, India)

5. M.S. Arafath, K.U.R. Khan, K.V.N Sunitha, Pithy review on routing protocols in wireless sensor networks and least routing time opportunistic technique in WSN, in *10th International Conference on Computer and Electrical Engineering 11–13 Oct 2017, Journal of Physics: Conference Series* (Edmonton, Canada)
6. M. Yoshikawa, Y. Nozaki, Electromagnetic analysis attack for a lightweight cipher PRINCE, in *IEEE International Conference on Cybercrime and Computer Forensic (ICCCF)* (2016), pp. 1–6
7. M.S. Arafath, K.U.R. Khan, Opportunistic sensor networks: a survey on privacy and secure routing, in *2017 2nd IEEE International Conference on Anti-Cyber Crimes (ICACC)*, 26–27 Mar 2017, pp. 41–46
8. Y. Wei, Y. Rong, X.A. Wang, New differential fault attack on lightweight cipher LBlock, in *International Conference on Intelligent Networking and Collaborative Systems (INCoS)* (2016), pp. 285–288
9. M. Garcia-Bosque, C. Sánchez-Azqueta, G. Royo, S. Celma, Lightweight ciphers based on chaotic map-LFSR architectures, in *12th Conference on IEEE Ph.D. Research in Microelectronics and Electronics (PRIME)* (2016), pp. 1–4
10. L. Li, B. Liu, H. Wang, QTL: a new ultra-lightweight block cipher. Microprocess. Microsyst. **45**, 45–55 (2016)
11. B.J. Mohd, T. Hayajneh, K.M.A. Yousef, Z.A. Khalaf, M.Z.A. Bhuiyan, Hardware design and modeling of lightweight block ciphers for secure communications. Fut. Gener. Comput. Syst. (2017)
12. M.S. Arafath, K.U.R. Khan, K.V.N Sunitha, Incorporating security in opportunistic routing and traffic management in opportunistic sensor network. Int. J. Adv. Intell. Paradigms (IJAIP), ISSN online: 1755-0394, ISSN print: 1755-0386. http://www.inderscience.com/info/ingeneral/forthcoming.php?jcode=IJAIP
13. C.A. Lara-Nino, A. Diaz-Perez, M. Morales-Sandoval, Lightweight hardware architectures for the PRESENT cipher in FPGA. IEEE Trans. Circuits Syst. I Regul. Pap. (2017)
14. M.S. Arafath, K.U.R. Khan, K.V.N. Sunitha, Incorporating privacy and security in military application based on opportunistic sensor network. Int. J. Internet Technol. Sec. Trans. (IJITST). (ISSN: (Print), ISSN: 1748–5703 (Online)) (2017). https://doi.org/10.1504/IJITST.2017.091514 (Published online 8 May 2018)
15. F. Zhang, S. Guo, X. Zhao, T. Wang, J. Yang, F.X. Standaert, D. Gu, A framework for the analysis and evaluation of algebraic fault attacks on lightweight block ciphers. IEEE Trans. Inf. Forensics Secur. **11**(5), 1039–1054 (2016)
16. S. Kotel, F. Sbiaa, M. Zeghid, M. Machhout, A. Baganne, R. Tourki, Performance evaluation and design considerations of lightweight block cipher for low-cost embedded devices, IEEE (2016)

A Literature Review on Quantum Experiments at Space Scale—QUESS Satellite

C. S. N. Koushik, Shruti Bhargava Choubey, Abhishek Choubey and Khushboo Pachori

Abstract QUESS/Micius (Quantum Experiments at Space Scale) satellite is a sun-synchronous orbit satellite which is a joint Chinese-Austrian satellite mission operated under Chinese Academy of Sciences (CAS) and with the University of Vienna and the Austrian Academy of Sciences (AAS). It has base stations in Austria and in China at a distance of separation of 7600 km, at a height of 300 km. QUESS is designed to do quantum experiments at space scale, which uses a laser with unknown polarization leading to the quantum states, in space. It is used over long distances for the development of quantum encryption and quantum teleportation technology. It provides a secure communication of data with the least delay of data transmission along with the help of teleportation achieved by quantum entanglement along with usage of the principle of superposition. It works on the principle of Sagnac effect. Any eavesdropping attempt if done, it causes the change in the quantum states of the photons, such that it can be detected with the help of other qubits in the same pair. Quantum key distribution is used to generate a random key for every quantum entangled pair for quantum cryptography of the pair, which can be opened only at the receiver, maintaining a high level of secrecy. These codes are processed by quantum processors in order to open the lock that is placed on them by the random key, with the help of quantum computers at the receiving ends. It is also used to understand the quantum teleportation.

Keywords Quantum entanglement · Quantum teleportation · Quantum key distribution (QKD) · Sagnac effect · Quantum processing

C. S. N. Koushik · S. B. Choubey · A. Choubey · K. Pachori (✉)
Sreenidhi Institute of Science and Technology (SNIST), Yamnanmpet, Hyderabad 501301, India

S. B. Choubey
e-mail: shurtibhargava@sreenidhi.edu.in

A. Choubey
e-mail: abhishek@sreenidhi.edu.in

Guru Nanak Institutions Technical Campus (GNITC), Ibrahimpatnam, Hyderabad, India

H. S. Saini et al. (eds.), *Innovations in Electronics and Communication Engineering*, Lecture Notes in Networks and Systems 107,
https://doi.org/10.1007/978-981-15-3172-9_2

1 Introduction

QUESS is a Chinese-Austrian sun-synchronous satellite mission controlled by Chinese Academy of Sciences (CAS) since the year 2017 but it was launched in 2016. Its orbit is at an altitude of about 500 km, with an inclination of 97.37. QUESS facilitates quantum optics experiments for the development of quantum communications and quantum teleportation technology. It is also called as "Micius" after a fifth-century B.C. Chinese scientist named Mozi (Micius in Latin). This satellite was developed to have the most secret and secure way of communication by generating a random key to each photon pairs with varying polarization, which can only be unlocked at the receiver end only; hence, it could find the applications mostly in defense sector and for future internet connectivity and for space exploration. It was first proposed by Jianwei Pan (PI) of CAS in 2003. It can be used up to a range of 300–7500 km via optical fiber cable between the ground stations in Austria and China [1]. It has two deployable solar arrays designed to be operational for two years. It is a revolutionary technology developed to have safest communication links with the least chances hacking of data and to reduce the channel effects on the signal by generating a random key for the signal and good signaling rate with better computational abilities.

1.1 Objectives of the Mission

- Using this for long-distance communication by quantum key distribution (QKD) between satellite and the ground station in a secure way by quantum cryptography.
- Verifying the quantum entanglement distribution and quantum teleportation in space (Fig. 1).

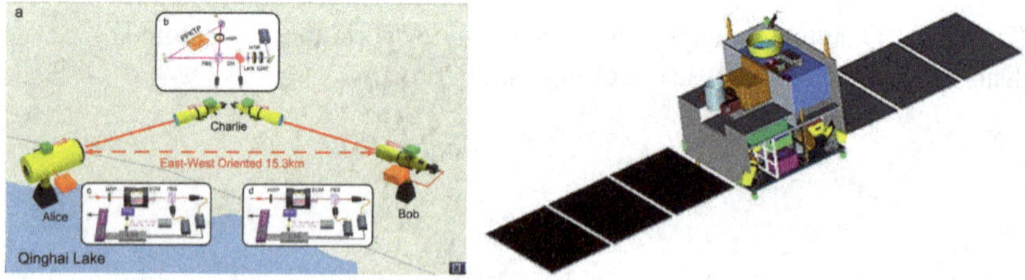

Fig. 1 QUESS satellite [2, 3]

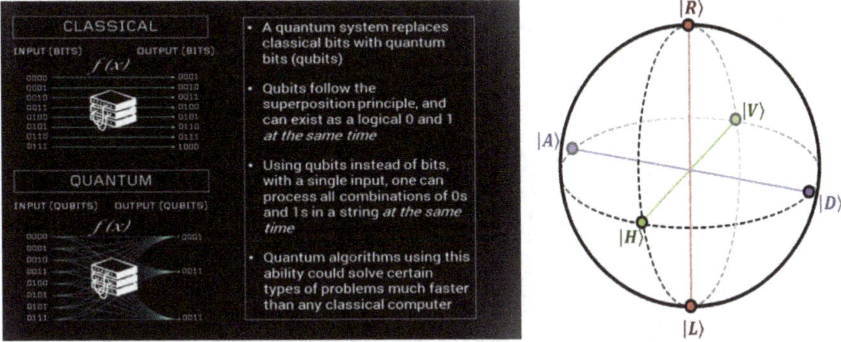

Fig. 2 Quantum mechanics and Bloch sphere

2 QUESS Satellite: A Brief Review

2.1 Quantum Mechanics

It is a branch of physics that is used to explain the nature and behavior of matter and energy on the subatomic level, which is called as quantum physics or quantum mechanics based on the properties like

- Quantized properties: Position, speed, and color occur only in specific amounts.
- Particles of light: Light can behave as a particle called as photon and even as a wave.
- Waves of matter: Matter can behave as a wave with certain frequency, e.g., electrons [4] (Fig. 2).

2.2 Quantum Entanglement

Quantum entanglement is a physical phenomenon in which, the opposite spin/energy particle pairs interact with each other considering the pair as a set, where the particles are at their own random quantum states, predicting each other's properties, that could be understood by the quantum systems only. It works on the principle of superposition. Properties like spin, energy, and position are considered for evaluation. The pairs formed are at varying values of energy such that they can be represented by Bloch sphere, unlike digital values of 1 and 0 [4] (Fig. 3).

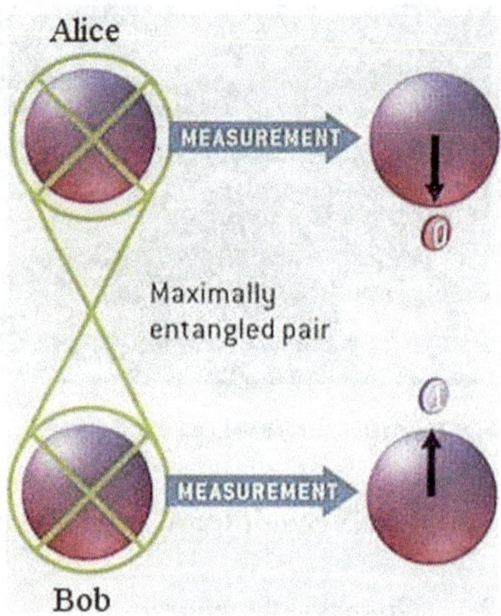

Fig. 3 Quantum entanglement

2.3 Quantum Communications

It is a field of quantum physics that deals with the processing the data which deals with the qubits/variable polarized photons of a laser, processed with the help of a quantum control processor. This is used under the principle of Sagnac effect which plays a major role in transmission and reception of the quantum signal. It uses a quantum key distribution (QKD) technique along with the quantum cryptography wherein a random key is generated for every entangled pair of photons [4] (Figs. 4 and 5).

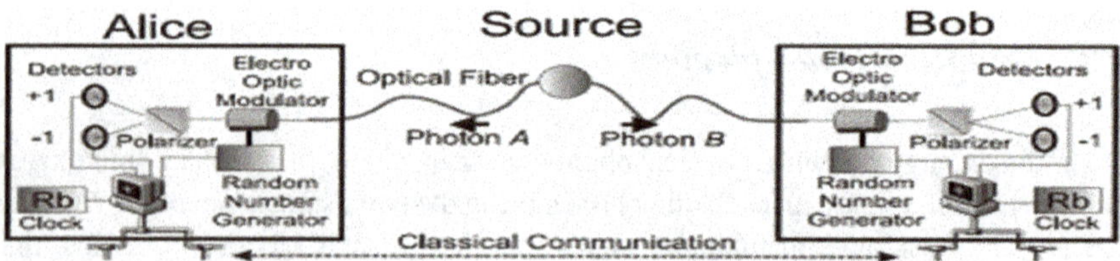

Fig. 4 Quantum communications [5]

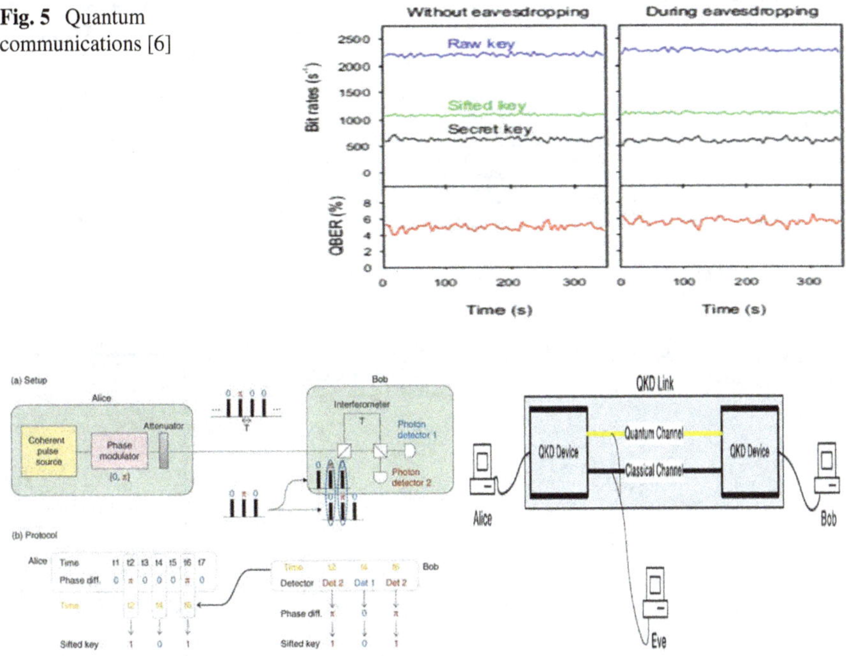

Fig. 5 Quantum communications [6]

Fig. 6 Quantum cryptography and QKD system [5]

2.4 Quantum Key Distribution (QKD) and Quantum Cryptography

Quantum cryptography is a technique used to have a security protocol for the signal that is being transmitted or received for the signal and to prevent the effect of the adversaries. The effect of the adversaries can be reduced by using the quantum key distribution technique (QKD) for quantum cryptography. In QKD, a random key is generated for the quantum entangled pair such that only the lock on the signal can be removed with the help of the second key at the receiver end only. Any adversaries if present in the quantum channel, if trying to tamper the signal, then anyone of the particle's properties is altered, which could be found by the opposite particle in the entangled pair. Hence, extra security precautions are taken [4] (Figs. 6 and 7).

2.5 Block Diagram and Principle

- It helps to send data over a quantum channel to prevent the effects of noise and for secrecy purposes so as to prevent the decryption of the key produced which cannot be opened for every pair but only opened at the receiver end.

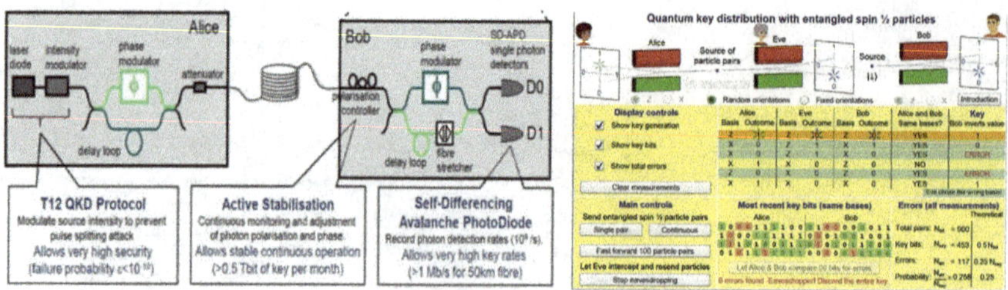

Fig. 7 QKD system

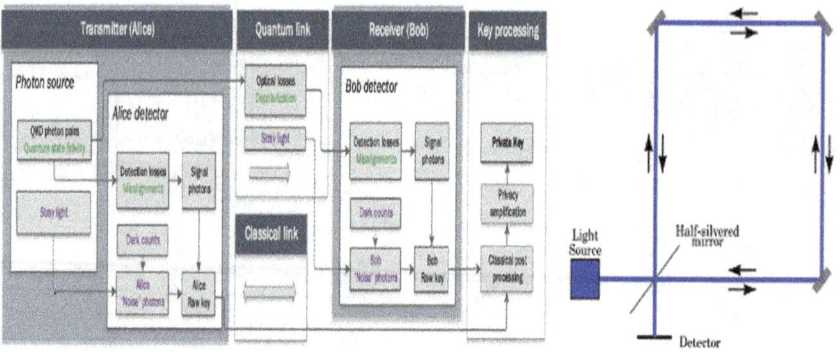

Fig. 8 Block diagram and principle [7]

- It also helps the data sent with the help of it to understand phenomena like quantum entanglement and quantum teleportation [1].
- The principle used in the experiment is Sagnac effect (Fig. 8).

3 Construction/Component Description

3.1 *Quantum Entanglement Source*

The quantum entangled pairs are advantageous than the conventional digital systems. The entangle pairs are generated from the lasers of certain wavelengths like 632.8, 1064 nm which are weak coherent pulses that are detected by photodiodes. A random key by QKD will be generated at the entanglement transmitter as a result they can be used to transmit data wirelessly. The quantum states can be easily correlated and generated at the receiver end, and they are timestamped easily without the need of timing instruments like GPS, atomic clocks (Figs. 9, 10, and 11).

Small Photon Entangling Quantum Systems are used to generate the random number generators (RNG) which act as a key at the transmitter end. Polarization of the

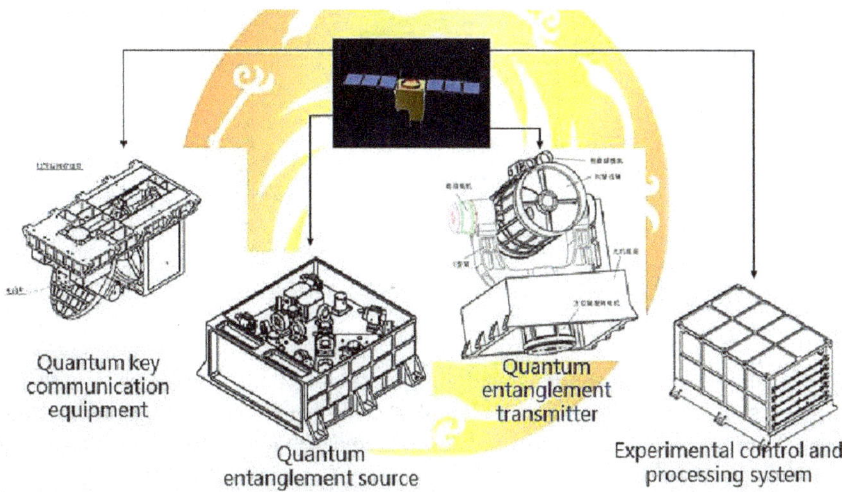

Fig. 9 Components

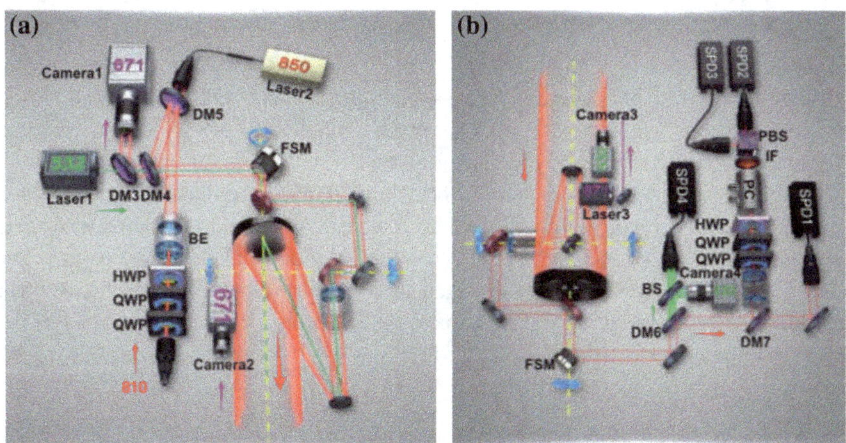

Fig. 10 Quantum entanglement source-1 [7]

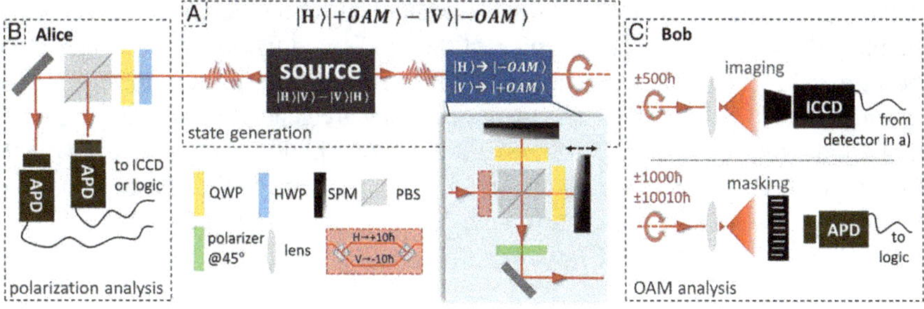

Fig. 11 Quantum entanglement source-2 [7]

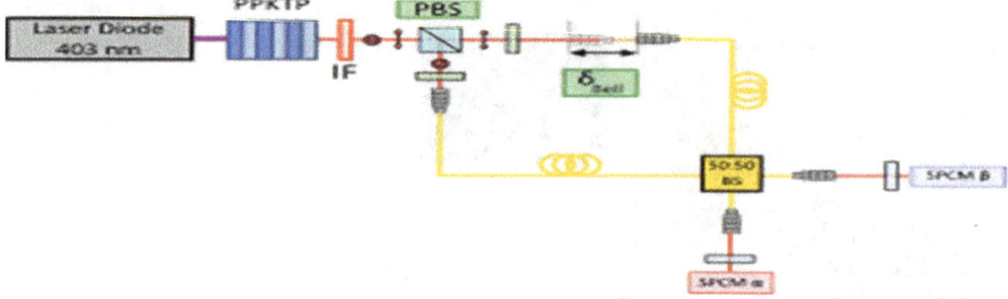

Fig. 12 PPTKP laser [8]

entangled photon pairs are generated via spontaneous parametric down-conversion (SPDC) from a bulk bariumborate (BBO) crystals as they are temperature tolerant. There are various reflecting mirror plates to increase the intensity of the plates—the single photons are currently detected Avalanche photodiodes (Si-APDs).

3.1.1 LASERS

Laser is a light beam of high directivity and gain that works on the principle of stimulated emission of a laser beam of wavelength of 1064 or 632.8 nm which is used either continuously or in a pulsed manner producing photons of varying polarization. They are detected by a homodyne detector which has photodiodes in them. The light is produced from periodically poled potassium titanyl phosphate (PPKTP) which is a 8.9-mm-long biconvex crystal where dielectrically coated in order to form a monolithic cavity. It has ring resonator to amplify the signal to a large extent. The laser signal can be transmitted or received such that the quantum bits can have error due to noise or any adversary interference, but it is considered as an attempt of the adversary to tamper the signal [8] (Fig. 12).

3.1.2 Photodiodes

A photodiode is a PIN semiconductor diode device that converts light into electrical current when it is incident. The current is generated when photons are absorbed in the photodiode at the depletion region of the diode, and the light sent comes from the mirrors, incident at a greater focus on it. Photodiodes usually have a slower response time as their surface area increases. They are generally InGaAs Avalanche photodiodes that are used to detect the lasers. They are highly sensitive to the variable photons that are incident on them. The photons arriving with a phase change is demodulated with the help of electro-optic demodulators such that any phase changes that are present controlled [1, 7, 9] (Fig. 13).

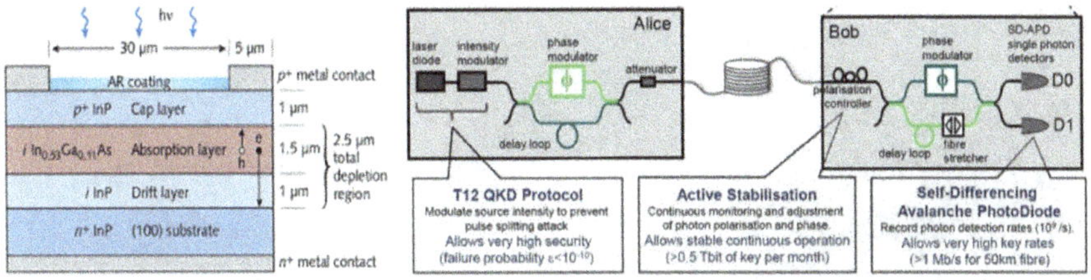

Fig. 13 Avalanche photodiodes

4 Quantum Entanglement Transmitter

It is a device that is used based on the effect of Sagnac effect which is implemented with the help of a Sagnac interferometer. Interferometer is a device that is used to analyze the light sources for a particular property of a physical quantity such that it can help to understand the characteristics of the particular physical quantity. Sagnac interferometer is used for checking the quantum nature of the pairs such that they are used to understand the quantum entanglement and quantum teleportation. Quantum key distribution (QKD) is done to the signal such that the data is given a random key, for it to be transmitter or unlocked while receiving it. It can act as Alice or Bob based on the signal that is being either transmitted or received.

The laser signal can be modulated for large distance communication that can be done by electro-optic phase modulator which is after the beam splitters of the quantum source generator. The interferometer signal has noise that can be creating the errors in the quantum signal that are to be rectified. The efficiency of the homodyne detectors beam splitters is of 99.7%. The local oscillator power is of 20 mW, and the green laser pump power was 80 mW. The laser light was introduced through a polarizing beam splitter and a Faraday rotator. It has a resolution bandwidth of 300 kHz and a video bandwidth of 300 Hz. A 10-km-long ring resonator can also be used to increase the light storage time and increases the sensitivity of the experimental setup in the frequency band between 1 Hz and about 40 Hz. A 40-kg test mass mirror and a laser power at the central beam splitter are of 90 W. The light is intensified up to 10 kW in the ring resonators. The laser signal when received at the BOB is detected by the Avalanche photodiodes [1, 7, 9].

4.1 Quantum Data Processing Unit

By silicon CMOS technology, we can have spin qubit error-correction in the near future. Quantum logic calculations are carried out with the help of one or two quantum gates in the processor due to which huge data is handled. The processor is having features like

- UART,

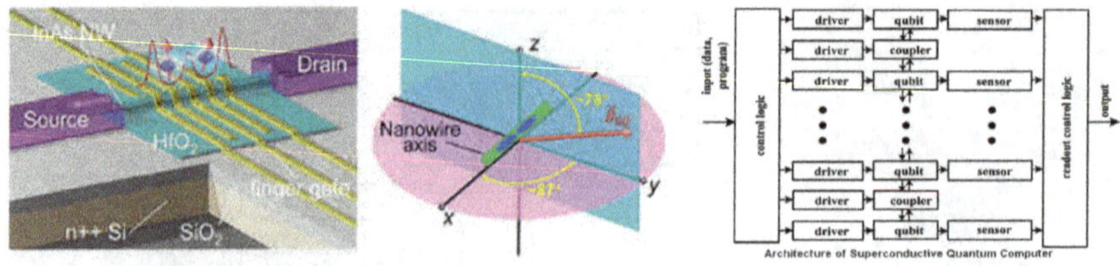

Fig. 14 Quantum processor-1

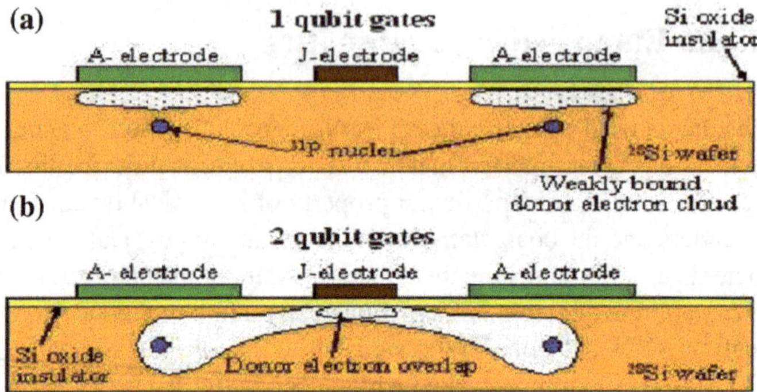

Fig. 15 2-bit qubit gate

- JTAG,
- Large storage due to entanglement (Figs. 14 and 15).

The qubits generated at high frequencies like that of microwave frequencies are used for communication which has led to solid-state qubit development. A simple qubit gates can be used to do the operations on the quantum data. It uses Pauli spin blockade principle. It is sensed and combined with ancilla qubit for parity readout and check for the errors. A silicon oxide semiconductor with microwave transmission line is used for the 2-qubit logic gates having a single electron transistor sensor. The photons from the photon multiplier tubes (PMT) received or transmitted are either decrypted or encrypted. The signal is added with noise or it has eavesdropped. The Avalanche photodiode (APD) detectors have noise which has an exponential dependence on temperature and 0.5 dB of signal loss for every degree change in temperature, hence there must cares taken for cooling actions [3] (Fig. 16).

5 Working

It works on the principle of the Sagnac effect. The signal can either be transmitted from the satellites or from the ground stations with the help of laser beam having changeable polarization states as a result they are detected with the help of homodyne

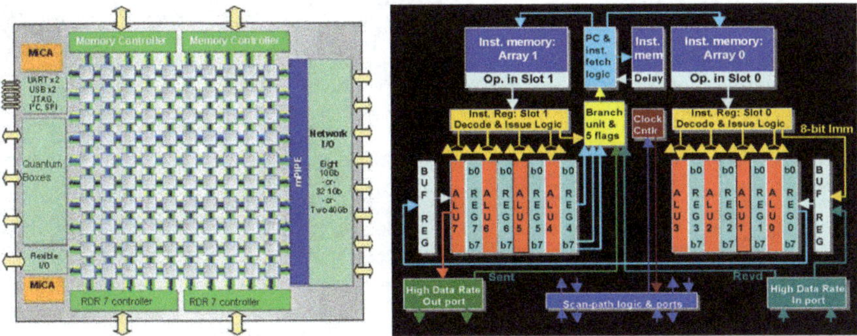

Fig. 16 Quantum control processor

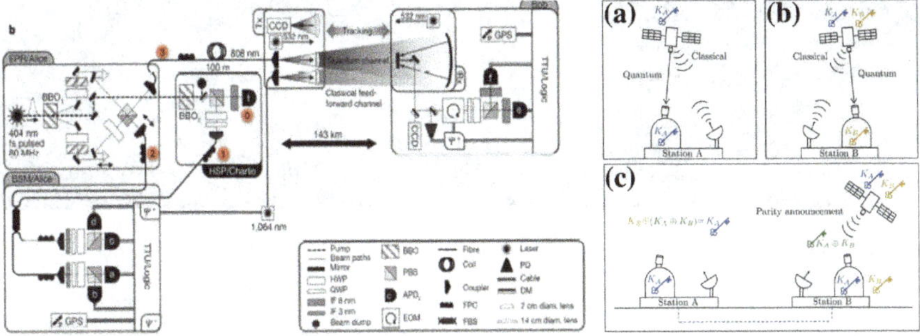

Fig. 17 Working setup [5]

detection. The lasers are of a single frequency such that they are detected when they attain a phase changes in them. The data is processed by the processor having quantum gates. The data can be uplinked or downlinked through a quantum channel between Alice and Bob via Eve (Fig. 17).

A random key will be allocated to each pair at the source or at the transmitter as a result no one can tamper the data and if anyone tries to tamper it, the detection of tampering is made possible due to change in the opposite particle's properties of the pair, as its properties are altered. It is even being used to study the quantum teleportation of particles, as they can reduce the effects of noise in the channel, as the particles can appear at two places at once simultaneously by changing the particle properties [1, 7, 9] (Fig. 18).

6 Conclusion

It is very useful for the studying the quantum entanglement and quantum teleportation which are essential for the revolutionary technology to develop which can help us

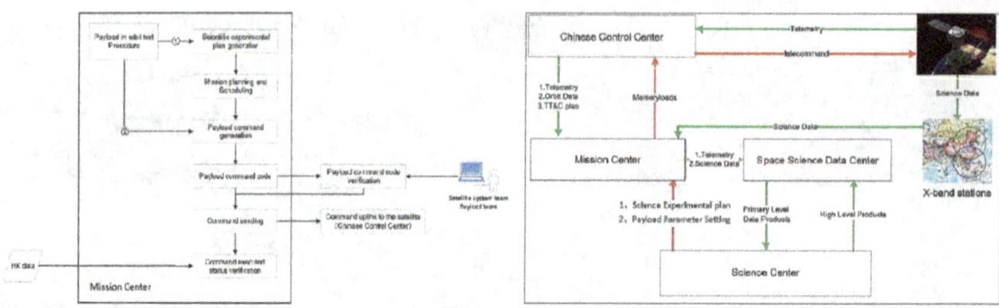

Fig. 18 Working-1

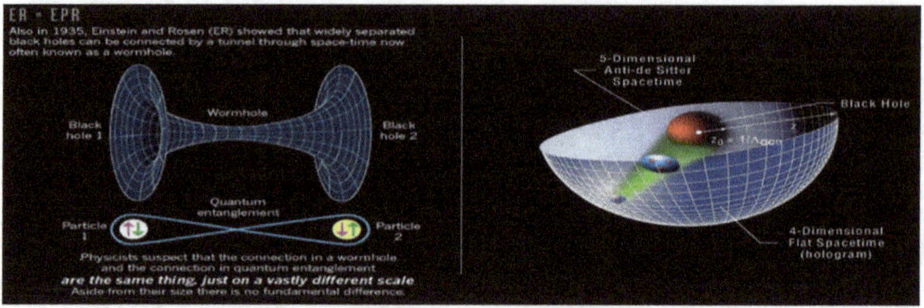

Fig. 19 Future cope/purpose

to have safest communication links and less chances of hacking of data and reduce the effects of the channel on the signal by generating a random key for the signal like a parity bit to it. The adversaries can be detected easily due to the effect of them on the signal. It can be used at present for military applications, but later it can be used for day-to-day communications and even give worldwide free access to Internet. It can be used for space applications like studying of the celestial objects and interstellar transportation reducing the cost of the expenditure for the interstellar travel. It can also be used to provide Internet access worldwide, which can be used for safe communications (Fig. 19).

References

1. M. Krenn, M. Malik, T. Scheidl, R. Ursin, A. Zeilinger, Quantum communication with photons, 5 Jan 2017. arXiv:1701.00989v1 [quant-ph]
2. S. Gupta, J. Pramanick, R. Ahamed, K. Sau, S. Pyne, R. Biswas, Quantum computation of perfect time- eavesdropping in position-based quantum cryptography quantum computing and eavesdropping over perfect key distribution, 978-1-5386-2215-5/17/$31.00 ©2017 IEEE
3. M. Calixto, Quantum computation and cryptography: an overview. Nat. Comput. **8**, 663–679 (2009). https://doi.org/10.1007/s11047-008-9094-8
4. A. Banerjee, M.R. Mathias, A. Prabhakar, Quantum key distribution—technology review. INDIA J. Defence Inf. Commun. Technol. **3**(1) (2017)
5. A. Zeilinger, Light for the quantum. Entangled photons and their applications: a very personal perspective, Royal Swedish Academy of Sciences. Phys. Scr. **92**(072501), 33 (2017)

6. J. Pan, 604 progress of the quantum experiment science satellite (QUESS) Micius project national report 2016–2018. Chin. J. Space Sci. **38**(5), 604–609 (2018). https://doi.org/10.11728/cjss2018.05
7. R. Bedington, J.M. Arrazola1, A. Ling, Progress in satellite quantum key distribution. Npj Quant. Inf. **3**, 30 (2017). https://doi.org/10.1038/s41534-017-0031-5
8. A. Bachmann, The KTP-(greenlight-) laser—principles and experiences. Minimal. Invasive Therapy **16**(1), 5–10 (2007). ISSN 1364-5706 print/ISSN 1365-2931 online # 2007 Taylor & Francis. https://doi.org/10.1080/13645700601157885
9. G. Vallone, D. Bacco, D. Dequal, S. Gaiarin, V. Luceri, G. Bianco, P. Villoresi, Experimental satellite quantum communications (Received 13 Apr 2015; revised manuscript received 26 May 2015; published 20 July 2015)
10. B.P. Williams, K.A. Britt, S. Travis, A tamper-indicating quantum seal, 21 Aug 2015. arXiv: 1508.05334v1 [quant-ph]
11. R. Alléaume, C. Branciard, J. Bouda, T. Debuisschert, M. Dianati, N. Gisinc, M. Godfrey, P. Grangier, T. Länger, N. Lütkenhaus, C. Monyki, P. Painchault, M. Peevi, A. Poppei, T. Porninl, J. Rarityg, R. Renner, G. Ribordy, M. Riguidela, L. Salvail, A. Shields, H. Weinfurter, A. Zeilinger, Using quantum key distribution for cryptographic purposes, R. Alléaumeetal./Theoret. Comput. Sci. **560**, 62–81 (2014) (©2014 Elsevier B.V. All rights reserved, 1 Aug 2013, available on 17th Sept 2014)

Simulink Model of Wireless Sensor Network in Biomedical Application

Md. Fazlul Haque Jesan, Md. Monwar Jahan Chowdhury and Saifur Rahman Sabuj

Abstract Wireless sensor networks (WSN) comprises of many locally distributed sensor nodes in human body which monitor physical data such as body temperature, pressure, and heart rate. The use of sensor nodes for medical health has greatly attracted the attention of researchers recently because sensor nodes can be used to monitor patient's health condition or assist physicians in critical situations. This can greatly improve the existing patient care in health service. In this paper, Simulink model of health monitor using WSN has been designed. Parameters such as systolic blood pressure, heart rate, and body temperature have used to build the model and show the performance characteristics of various channels, i.e., AWGN, Rayleigh fading. The result shows that body temperature has the best performance in AWGN channel and heart rate has the best performance in Rayleigh fading channel.

Keywords Wireless sensor networks · Internet of things · Health monitor · Code-division multiple access

1 Introduction

With the advanced growth of wireless communication and portable electronic devices, wireless sensor networks (WSN) has been regarded as a cutting edge technology that will change the future. WSN can be defined as a network consisting of much small number of nodes with limited monitoring, processing, and communicational capabilities that interact with each other to carry out a designed task successfully [1]. In any typical WSN systems, first sensor nodes collect the data from the surrounding environment such as temperature and humidity. After that, the data is processed,

Md. Fazlul Haque Jesan (✉) · Md. Monwar Jahan Chowdhury · S. R. Sabuj
Department of Electrical and Electronic Engineering, BRAC University, Dhaka, Bangladesh, India
e-mail: fazlul.haque.jesan@gmail.com

Md. Monwar Jahan Chowdhury
e-mail: manwarjahan247@gmail.com

S. R. Sabuj
e-mail: s.r.sabuj@ieee.org

H. S. Saini et al. (eds.), *Innovations in Electronics and Communication Engineering*, Lecture Notes in Networks and Systems 107,
https://doi.org/10.1007/978-981-15-3172-9_3

and finally, communication blocks transmit the data to other nodes for further communication. The distinct characteristics of WSN that makes it unique from other conventional wireless networks are limited computation power, small memory, high noise, and low battery-powered devices [2]. Today, WSN technology has become an important tool in many domains ranging from large industry application, healthcare monitoring, military use to smart grid, transportation, environment monitoring, etc.

WSN, Internet of Things (IoT), artificial intelligence, and machine learning together have created a new interdisciplinary technology called wireless body area network (WBAN). WBAN is a specialized WSN developed by researcher for medical application or healthcare services [3]. In WBAN, sensors are kept on a patient's body to monitor various health parameters such as temperature, blood pressure, heart rate, respiration rate, and pulse rate. After collection of data, WSN sends them to central database system without any human intervention.

Doctors can check the data from any place anywhere and take immediate steps to eliminate any emergency. As the numbers of elderly people are increasing all over the world, the health concern of them has become a crucial issue recently. WBAN can be great technology to help those senior citizens as wireless sensors can very easily install in their homes. Nodes connected with Internet and smart phones can provide real-time monitoring and send a warning signal to the near ones if any vital signs drop below the threshold limits. Remote medical help can also be provided to the patient if needed.

The main contributions of this study are summarized as follows:

- Develop a Simulink model of WSN for circadian rhythmicity of human body such as systolic blood pressure (SBP), heart rate (HR), and body temperature (BT).
- Evaluate the performance of different channels such as AWGN and Rayleigh fading.

The rest of the paper is organized as follows. Section 2 gives a brief summary of all the previous related work that has been done in the field of IoT-based eHealth. Section 3 elaborates on the system models to be developed using Simulink and discusses Gold code and Walsh code generator. Results of the model are shown in Sect. 4. Finally, Sect. 5 contains the conclusion to this paper.

2 Related Works

There has been continuous research going on in smart eHealth model based on IoT. Riazul Islam [4] emphasizes application, security, privacy, scalability, and advancement of eHealth solutions. A smart security model is also proposed here which can reduce the threats related to IoT-based wearable health products. In [5], the authors have introduced a new overlay radio scheme which sends both telemedicine and in-hospital data transfer using the same frequency band to increase spectral efficiency. They have used dirty paper coding to reduce the interference effects between the above-mentioned data transmission method. Mamoon et al. have presented spectrum aware cluster-based network model and electromagnetic interference's aware communication standards for medical services [6]. The network model improves communication accessibility between different healthcare systems. Researchers have used Internet of Things modeling language (ThingsML) to introduce code generation framework and systematic approach to the software developers as they can easily change the compilers according to their customization [7]. ThingsML approach has been a platform where people from various backgrounds, i.e., software engineer, biomedical engineers, work together seamlessly to create IoT devices which can be used for various purposes including eHealth applications. Furthermore, cloud-based health control framework for IoT devices has developed by authors in [8]. They have used ciphertext encryption, dual encryption, and Merkle hash tree to support the framework.

Saguna et al. have investigated network requirements of IoT-based eHealth monitoring used computer-aided design and optimization technique to solve the existing problems of eHealth systems [9]. Majid et al. have considered a new improved eHealth model which comprises total diabetic solution with the help of IoT structure. The model incorporates medical sensors, IoT, diabetic patients, and disease management hub into a complete eHealth solution [10]. From their experiment, it has been observed that the suggested model improves bandwidth up to 56% for a single scenario comparing to existing products. Similarly, Farshad et al. have investigated various opportunities, obstacles, and its realistic solutions to IoT-based eHealth patient monitoring system [11]. Sawand et al. have proposed architecture framework of eHealth system focusing from patient perspective [12]. Existing security threats and its solution have also extensively discussed here. Similarly, Bahar et al. have presented a patient-based eHealth system where stakeholders are connected seamlessly to each other. They have used fog computing to analyze and sense data from patient [1]. Bisio et al. have used two functions, i.e., activity recognition and movement recognition to monitor the movement of old people using the help of IoT [13]. They have compared support vector machines (SVM), decision trees, and dynamic time warping methods for the above-mentioned functions. From their research, it has been concluded that SVM shows better accuracy above 90% for both the specified function.

In [14], the authors have presented cognitive radio-based solution to address the various challenges in eHealth platform in wireless communication. Study on the performance of 5G network has evaluated in terms of eHealth and IoT in [15]. In [16], authors have investigated the issue of interference in eHealth care service network. Different prototype model and algorithm using WBAN have studied in [17–19] for better efficiency and performance optimization in wireless communication.

3 System Model

A code-division multiple access (CDMA) wireless communication system is shown in Fig. 1 [20]. It consists of three different signal generators: transmitter, channel, and receiver. At first, a random source generator generates a data according to the human body (systolic blood pressure (SBP): 85–135 mmHg; heart rate (HR): 45–95 bps; body temperature (BT): 96:40–99:90 °F) and acts like a source [21]. Data passes through spreading block. In spreading block, decimal data is converted into binary format, and then, two types of conversion are used in this paper: (i) Walsh code shown in Fig. 2 and (ii) Gold sequence shown in Fig. 3. In Walsh code, binary format of human body data and Walsh code generator data is converted into unipolar-to-bipolar. Then, both data are multiplied. In Gold sequence, binary format of human body data and Gold sequence generator data are applied into exclusive OR (XOR). Then, it is converted into unipolar-to-bipolar. Before transmitting, three signals are combined and passed through the AWGN channel or Rayleigh fading channel. In the receiver part, the signal subsequently passes through the dispreading block where

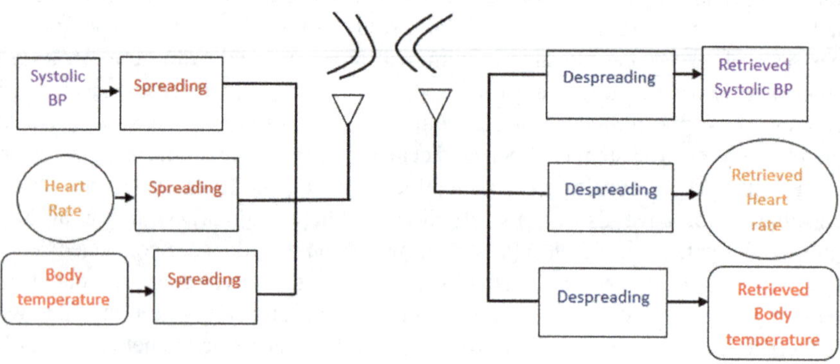

Fig. 1 Block diagram of system model

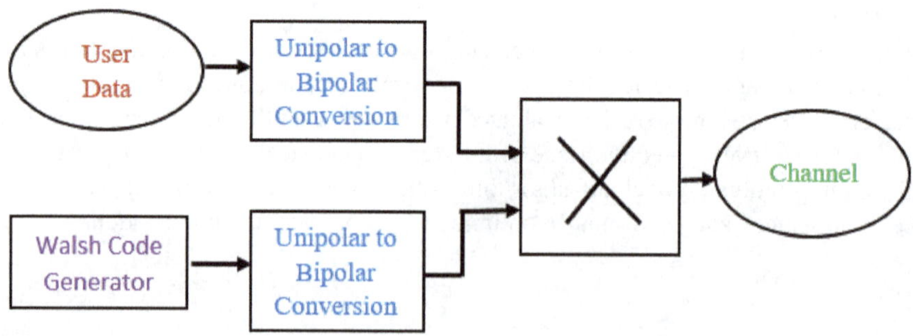

Fig. 2 Block diagram of Walsh code generator [27]

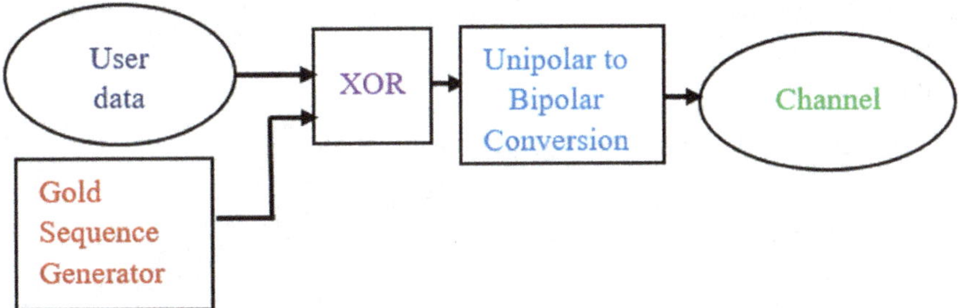

Fig. 3 Block diagram of Gold sequence generator [28]

signal follows the same process of spreading block, and then, the original signal will be retrieved.

3.1 *Walsh Code Generator*

Walsh code follows a Hadamard matrix of order N. Walsh code is expressed as W_k, where $k = 0, 1, 2, \ldots, N - 1$. The output of Walsh code generator is a row of Hadamard matrix named by Walsh code index which is integer in the range of [0, 1, …, $N - 1$] [22].

3.2 *Gold Sequence Generator*

Gold sequence generator makes a unique class of periodic m-sequences which return the minimum possible cross-correlation values [23]. Gold sequences are expressed as special pair of sequences x and y of period $N = 2^n - 1$, denoted as a preferred pair. Gold sequences can be expressed as, G(x, y) = [x, y, x ⊕ y, x ⊕ Ty, x ⊕ T^2y; _ _ _ _ _; x ⊕ T^{N-1} y], where T represents the vectors cyclically to the left by one place, and _ represents addition modulo-2. In order to generate output sequence, XOR operation of preferred polynomial (1) and preferred polynomial (2) is required [24].

4 Result and Discussion

In this study, MATLAB Simulink has been used for simulation purpose. Results show the performance analysis of packet loss with respect to signal-to-noise ratio (SNR), average path gain vector, and maximum Doppler shift. For simulation purpose, the following parameters of human body have been chosen:

- SBP: 85–135 mmHg.
- HR: 45–95 bps.
- BT: 96.4–99.90 °F.

Parameters used in Walsh code generator are as follows:

- Code length: 64 for SBP, 32 for HR, and 16 for BT.
- Code index: 60 for SBP, 28 h, and 12 BT.
- Sample time: 1/100.
- Sample per frame: 8.

Parameters used in Gold sequence generator are as follows:

- Preferred polynomial (1): [1 0 0 0 0 1 1] for SBP, [1 0 0 0 0 0 1] for HR, [1 0 1 0 0 0 1] for BT.
- Initial states (1): [0 0 0 0 0 1] for SBP, [0 0 0 0 1 1] for HR, [0 1 0 0 0 1] for BT.
- Preferred polynomial (2): [1 1 0 0 1 1 1] for SBP, [1 1 0 1 1 1 1] for HR, [1 0 0 1 0 1 1] for BT.
- Initial states (2): [0 0 0 0 0 1] for SBP, [0 0 0 0 1 1] for HR, [1 0 0 0 1 1] for BT.
- Sequence index: 0.
- Sample time: 1/100
- Shift: 0.

Figure 4 is the graphical analysis of packet loss versus SNR of human body data for Walsh code. When the SNR is 5 dB, the estimated packet loss of SBP, BT and

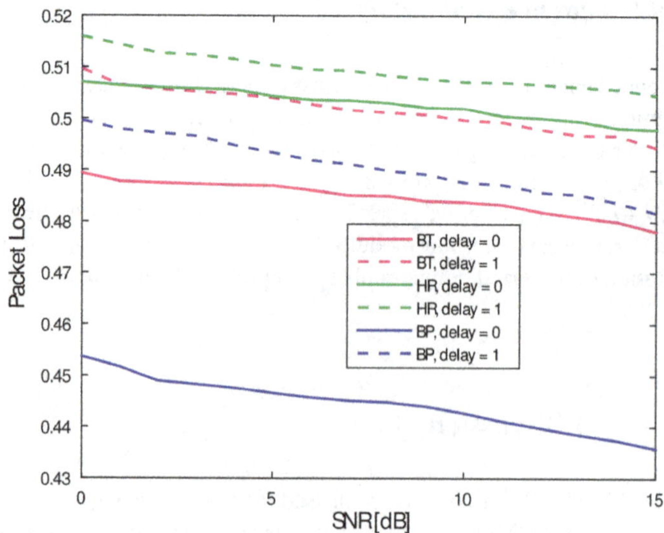

Fig. 4 Packet loss versus SNR for Walsh code

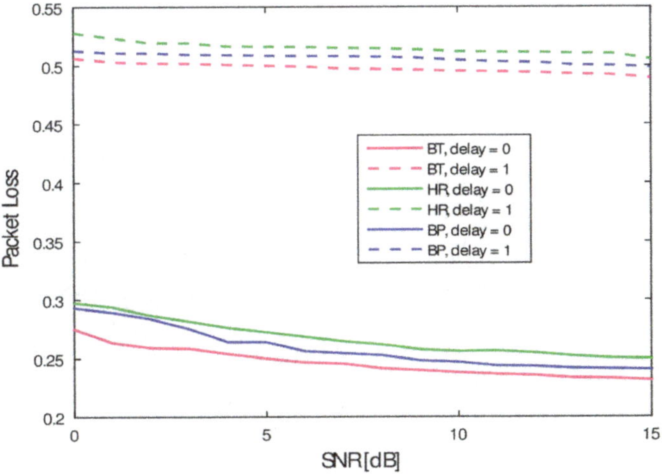

Fig. 5 Packet loss versus SNR for Gold sequence

HR are 0.4466, 0.487 and 0.5044, respectively. For SNR value of 5 dB, the system performance has 8.29% improvement in between SBP and BT. Moreover, the best performance for the delay "0" and delay "1" is SBP.

Figure 5 demonstrates the system performance comparison of packet loss versus SNR for gold sequence. From the figure, it is observed that when the SNR value is 5 dB, the packet loss of BT, SBP and HR are 0.2499, 0.2699 and 0.2724, respectively. For the value of BT and SBP, the system improvement is 7.41% at SNR of 5 dB. Furthermore, for the delay "0" and delay "1" BT has the lowest packet loss.

Figure 6 is the representation of the performance of packet loss versus average path gain for Walsh code in Rayleigh fading channel. It is observed that for the average path gain is at 3 dB, the packet loss value of SBP, HR, and BT are 0.4811, 0.4821, and 0.4848, respectively. For 3 dB average path gain the system has improvement by 0.207% for SBP and HR. Therefore, the highest packet loss is for BT and the lowest packet loss is SBP for delay "0" and delay "1." From Fig. 6, it is seen that data of packet loss with respect to average path gain is random for SBP, HR, and BT. Using curve fitting technique in MATLAB software, the mathematical expression of random data is figured out. The mathematical expressions for SBP, HR, and BT can be expressed as

$$Y_{\mathrm{SBP}} = -0.00023 * X_{\mathrm{SBP}} + 0:48 \tag{1}$$

$$Y_{\mathrm{HR}} = -0.00017 * X_{\mathrm{HR}} + 0:48 \tag{2}$$

$$Y_{\mathrm{BT}} = -0.00029 * X_{\mathrm{BT}} + 0:49 \tag{3}$$

where X is the average path gain, and Y is the packet loss.

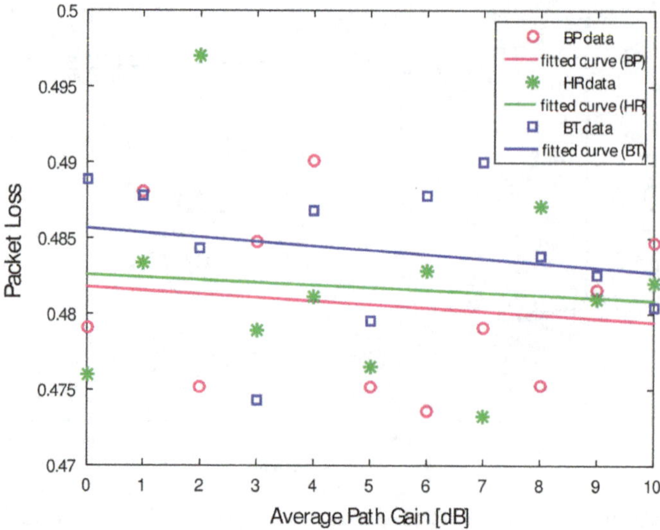

Fig. 6 Packet loss versus average path gain for Walsh code

Figure 7 is the illustration of the performance analysis of packet loss versus average path gain for Gold sequence in Rayleigh fading channel. For 4 dB average path gain value, the several packet loss values of HR, SBP, and BT are 0.4717, 0.4729, and 0.4731, respectively. Considering 4 dB average path gain, 0.253% system has improved for HR and SBP. From the above explanation, BT has the lowest packet

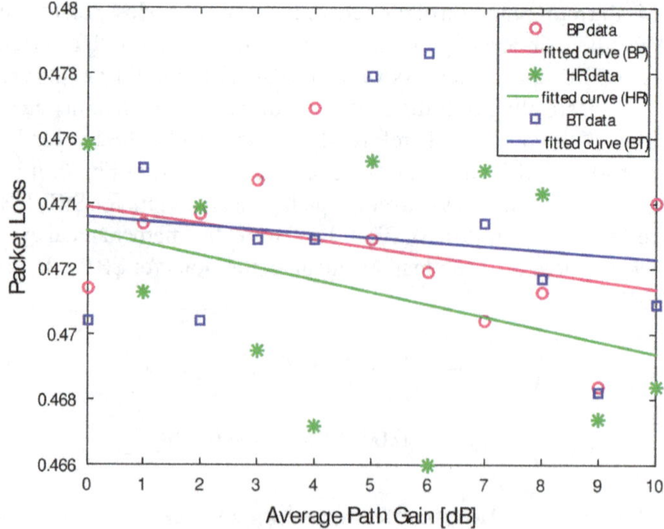

Fig. 7 Packet loss versus average path gain for Gold sequence

loss performance for delay “0” and delay “1.” The mathematical expressions for SBP, HR, and BT can be expressed as

$$Y_{\mathrm{SBP}} = -0.00025 * X_{\mathrm{SBP}} + 0.47 \tag{4}$$

$$Y_{\mathrm{HR}} = -0.00038 * X_{\mathrm{HR}} + 0.47 \tag{5}$$

$$Y_{\mathrm{BT}} = -0.00013 * X_{\mathrm{BT}} + 0.47 \tag{6}$$

Figure 8 is an observation of the performance analysis of packet loss versus maximum Doppler shift for Walsh code in Rayleigh fading channel. When the maximum Doppler shift is at 6 Hz, the respective packet loss of HR, BT, and SBP are 0.4758, 0.4760, and 0.4911, respectively. Under the consideration of delay “0” and delay “1,” for 6 Hz maximum Doppler shift, the system improvement is 0.042% for HR and BT. In short, HR demonstrates best packet loss performance in terms of BT and SBP. The mathematical expressions for SBP, HR, and BT can be expressed as

$$Y_{\mathrm{SBP}} = 0.0022 * X_{\mathrm{SBP}} + 0 : 48 \tag{7}$$

$$Y_{\mathrm{HR}} = 0.0014 * X_{\mathrm{HR}} + 0 : 47 \tag{8}$$

$$Y_{\mathrm{BT}} = 0.003 * X_{\mathrm{BT}} + 0 : 46 \tag{9}$$

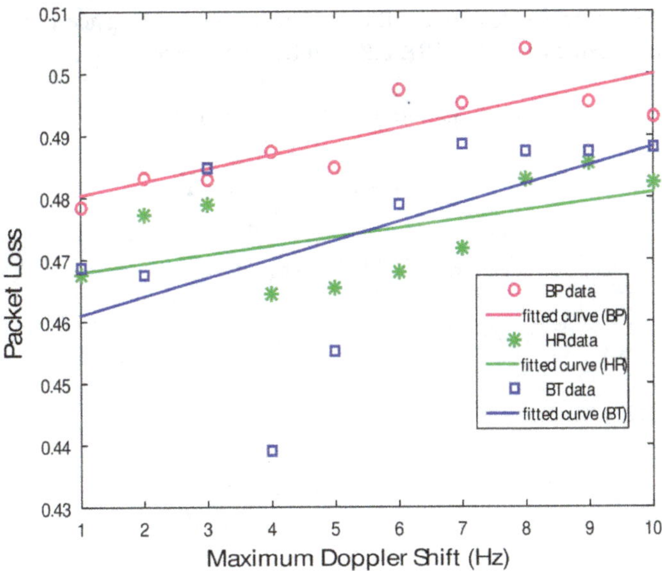

Fig. 8 Packet loss versus maximum Doppler shift for Walsh code

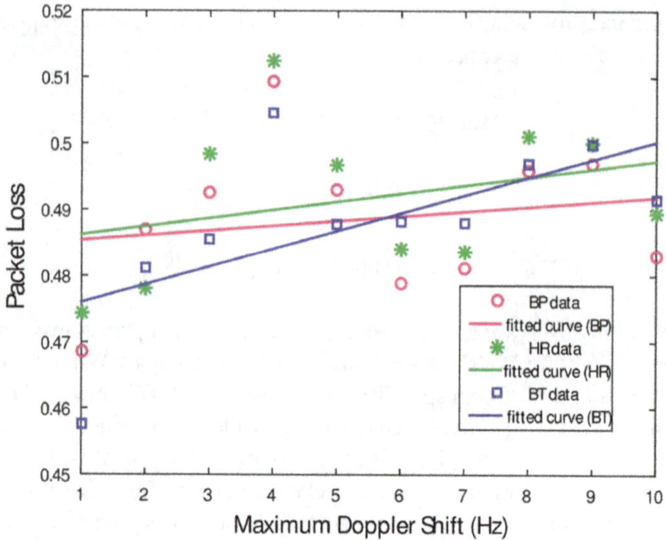

Fig. 9 Packet loss versus maximum Doppler shift for Gold sequence

Figure 9 shows the packet loss versus maximum Doppler shift performance for Gold sequence in Rayleigh fading channel. From this figure, it is shown that when maximum Doppler shift is at 3 Hz, the packet loss value of BT, SBP, and HR are 0.4810, 0.4867, and 0.4885, respectively. For the value of BT and SBP, the system improvement is up to 1.17% when the delay is "0" and "1." Therefore, the best performance of packet loss is for BT and the worst performance is for HR. The mathematical expressions for SBP, HR, and BT can be expressed as

$$Y_{\mathrm{SBP}} = 0.00072 * X_{\mathrm{SBP}} + 0.48 \tag{10}$$

$$Y_{\mathrm{HR}} = 0.0012 * X_{\mathrm{HR}} + 0 : 48 \tag{11}$$

$$Y_{\mathrm{BT}} = 0.0027 * X_{\mathrm{BT}} + 0 : 47 \tag{12}$$

5 Conclusion

In this paper, MATLAB-based Simulink model of health monitor using WSN is proposed. Systolic blood pressure (SBP), heart rate (HR), and body temperature (BT) have been taken to build the model and show the performance characteristics of various channels such as AWGN and Rayleigh fading model in this simulation. The result shows that for AWGN channel, SBP has the best performance of 8.29% for

Walsh code and BT has the best performance of 7.41% for Gold sequence. Along with that, for Rayleigh fading channel, SBP has the best performance of 0.207% (packet loss vs. average path gain) for Walsh code, and HR has the highest 0.253% (packet loss vs. average path gain) for gold sequence. Moreover, HR has the performance value of 0.042% and 1.17% (packet loss vs. maximum Doppler shift) for Walsh code and gold sequence. In future work, we plan to develop this technique under cognitive radio-based smart health solutions [25, 26].

References

1. M. Afsaneh, S.M. Ali, S.M. Paymon, S.M. Reza, Application of wireless sensor networks in healthcare system, in *Proceedings of 120th ASEE Annual Conference and Exposition* (2013), pp. 1–12
2. A. Carlos, M. Paulo, Wireless sensor networks for biomedical application, in *Proceedings 3rd Portuguese Meeting in Bioengineering* (2013), pp. 1–4
3. C. Subhajit, D. Srijan, C. Soumyadeep, B. Sayan, S. Suparna, S.G. Kali, D. Niket, Internet of things and body area network-an integrated future, in *Proceedings 8th IEEE Annual Ubiquitous Computing, Electronics and Mobile Communication Conference* (2012), pp. 396–400
4. S.M. Riazul Islam, D. Kwak, M. Hossain, K. Kyung, Md. Humaun Kabir, The internet of things for health care: a comprehensive survey. IEEE Access J. **3**, 678–708 (2015)
5. B. Bahareh, S. Shabnam, Spectrally efficient telemedicine and in-hospital patient data transfer, in *Proceedings of the International Symposium on Medical Measurements and Applications* (2017)
6. I. Al Mamoon, A.K.M. Muzahidul Islam, S. Baharun, K. Shozo, A. Ahmed, Architecture and communication protocols for cognitive radio network enabled hospital, in *Proceedings of International Symposium on Medical Information and Communication Technology* (2015), pp. 170–174
7. M. Brice, H. Nicolas, F. Franck, Model-based software engineering to tame the IoT jungle. IEEE Softw. **34**(1), 30–36 (2017)
8. Y.Y. Lo, Y.C. Pei, L.T. Yi, L.H. Jiun, Cloud-based fine gained health information access control framework for lightweight IoT devices with dynamic auditing and attribute revocation. Trans. Cloud Comput. **6**(2), 532–544 (2018)
9. K. Manh, S. Saguna, M. Karan, A. Christer, IReHMo: an efficient IoT-based remote health monitoring system for smart regions, in *Proceedings of the International Conference on E-health Networking, Application & Services* (2015), pp. 563–568
10. A.T. Majid, A.N. Waleed, J.M. Zahra, A.A. Ali, Robot assistant in management of diabetes in children based on the internet of things. IEEE Internet Things J. **4**(2), 437–445 (2017)
11. F. Farshad, F. Bahar, I. Mohamed, C. Krishnendu, From EDA to IoT e-health: promises, challenges, and solutions. IEEE Trans. Comput. Des. Integr. Circ. Syst. (2018)
12. S. Ajmal, D. Soufience, Z. Zonghua, N.A. Farid, Toward energy efficient and trustworthy ehealth monitoring system. China Commun. **12**(1), 46–65 (2015)
13. B. Igor, D. Alessandro, L. Fabio, S. Andrea, Enabling IoT for in-home rehabilitation: accelerometer signals classification methods for activity and movement recognition. IEEE Internet Things J. **4**(1), 135–145 (2017)
14. P. Phond, H. Ekram, N. Dusit, C. Serio, A cognitive radio system for e-health applications in a hospital environment. IEEE Wirel. Commun. **17**(1), 20–28 (2010)
15. M. Jose, B. Camara, Trends in wireless communication towards 5 g networks—the influence of e-health and IoT applications, in *Proceedings of the International Multidisciplinary Conference on Computer and Energy Science* (2016), pp. 1–7

16. O. Dramane, T.Q. Minh, K. Francine, A.C. Mohamed, K. Hicham, Mitigating the hospital area communication's interference using cognitive radio networks, in *Proceedings of the International Conference on e-Health Networking, Applications and Services* (2013), pp. 324–328
17. S. Qiang, L. Jing, Y. Hui, M. Zhichao, L. Ming, S. Zhichun, Adaptive cognitive enhanced platform for WBAN, in *Proceedings of the International Conference on communications in China: Wireless Networking and Applications* (2013), pp. 739–744
18. D.O. Chukwuemeka, Health monitoring using wireless sensor: a MATLAB approach. Bachelor of Engineering Thesis, Helsinki Metropolia University (2016)
19. D. Elham, W. Xianbin, A prototype body area network using IEEE 802.15.4. University of Western Ontario (2010)
20. P.I. Valery, *Spread Spectrum and CDMA: Principles and Applications* (Wiley Ltd, England, 2005)
21. Circadian Rhythm Laboratory, http://www.circadian.org/vital.html
22. MATLAB Documentation, https://www.mathworks.com/help/com/ref/walsh.html
23. F. Rodríguez Henríquez, N. Cruz Cortés, J.M. Rocha-Pérez, F. Amaro Sánchez, Generation of gold sequences with applications to spread spectrum systems (2016)
24. Gold Sequence Generator, https://www.mathworks.com/help/comm/ref/gold.html
25. R.S. Saifur, H. Masanori, Outage and energy-efficiency analysis of cognitive radio networks: a stochastic approach to transmit antenna selection. Perv. Mob. Comput. Elsevier **42**, 444–469 (2017)
26. R.S. Saifur, H. Masanori, Uplink modelling of cognitive radio network using stochastic geometry. Perform. Eval. Elsevier **117**, 1–15 (2017)
27. Asynchronous CDMA—File Exchange—MATLAB Central, https://www.mathworks.com
28. CDMA.mdl-File exchange-MATLAB Central, https://www.mathworks.com

Analysis of Power in Medium Access Control Code Division Multiple Access Protocol for Data Collection in a Wireless Sensor Network

Mohammed Salman Arafath, Shamimul Qamar, Khaleel Ur Rahman Khan and K. V. N. Sunitha

Abstract Collection of several sensor nodes in particular premises within the range of each other either directly or indirectly constitutes a sensor network. Batteries are the reason to keep these sensors alive, however, with restricted power supply. MAC protocol is considered as one of the better performance managers in the network, however, various MAC protocols in the literature were proposed for IEEE 802.11 standards. The MAC-CDMA protocol is applied to reduce the energy consumption from the battery while minimizing multiple access interference in fading channel. The S-MAC protocol is analyzed on the existing power allocation algorithms in a fading channel. A medium access protocol called S-MAC-CDMA is then analyzed to resolve the problem of contention time during carrier sense is fixed in S-MAC-CDMA. The combined S-MAC-CDMA protocol is used to adjust multiple access interference to control network load so that energy consumption can be minimized. The main parameters of performance in terms of fading effect which depends on the path chosen and power spectral efficiency for optimal frequency spectrum have been reviewed in the literature. The throughput of different MAC protocols for this chosen frequency and power spectrum has been analyzed that CDMA/TDMA MAC protocol outperforms other protocols. Throughput increases in fading environment and then saturates.

Keywords OSN · WSN · WASN · Throughput · MAC · Data collection · Fading effect · Power analysis

M. S. Arafath (✉)
College of Engineering, King Khalid Universtity, Abha, Kingdom of Saudi Arabia
e-mail: salman@kku.edu.sa

S. Qamar
College of Computer Science, King Khalid University, Abha 62529, Kingdom of Saudi Arabia
e-mail: sqamar@kku.edu.sa

K. U. R. Khan
Department of Computer Science Engineering, ACE College of Engineering, Hyderabad, India
e-mail: khaleelrkhan@gmail.com

K. V. N. Sunitha
Department of CSE, BVRIT Hyderabad College of Engineering for Women, Hyderabad, India
e-mail: k.v.n.sunitha@gmail.com

H. S. Saini et al. (eds.), *Innovations in Electronics and Communication Engineering*, Lecture Notes in Networks and Systems 107,
https://doi.org/10.1007/978-981-15-3172-9_4

1 Introduction

Wireless sensor network (WSN) uses shared medium for communication so it is prone to collisions the sensor nodes contributing to WSN in data aggregation using this shared medium may prone to collisions. There must be some mechanism of use shared medium efficiently and effectively for the requirements of quality of service. A medium access control (MAC) protocol is such mechanism to use these shared medium efficiently and effectively.

The sensor network comprises several number of nodes sensor nodes to collect data for a specified purpose. These sensor nodes are motorized by batteries with limited power [1, 2]. The main objective of all medium access control (MAC) protocols is to ensure that no node should interfere with other nodes. The retransmission in case of collision is wastage of energy. So to avoid any collision is a prerequisite of all the protocols to save the energy. The coordination to access the shared medium by the sensor nodes is done by the medium access protocol [3]. The work function of medium access protocol is to establish the link between nodes and communicate with another node efficiently, fairly, and reliably [4]. Good MAC protocol has the following performance parameters such as improved efficiency in terms of energy [5], decreased latency [6], and increased throughput with fair access to the shared medium [7] between the nodes of the WSN. The problem in S-mac protocol is overhearing idle listening duration and message collision. To solve the aforementioned problems, the literature [8–10] throws light on several existing mac protocols contrasting to other wireless networks, it is impractical to replace the fatigued battery as far as today's technology advancements is concerned, since it is one of the priority objectives of maximizing network lifetime. Moreover, the communication of sensor node within the WSN is more costly than the computation of the sensor nodes with respect to energy consumption. So, the first priority is to reduce communication and increase desire network operation (Fig. 1).

The prime contributions of this paper are as follows:

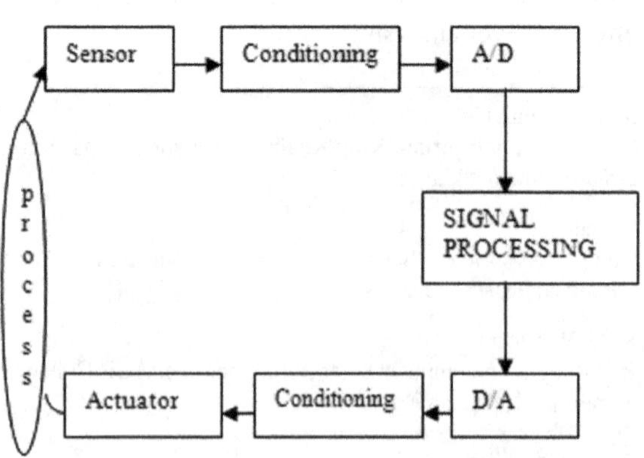

Fig. 1 Data acquisition and actuation by sensors

- The best path in WSN is selected based on node selection which in turn depends on criteria called as opportunistic selection. The opportunistic selection constitutes of both direct and relay links considering in an optimal manner. The criterion for node selection is based on lesser path loss probability (MPL) of the WSN and getting an opportunity for secondary spectrum access. In order to do this, we first characterize the signal to noise ratio over MAC-CDMA channels from end to end.
- To derive closed-form expression of MPL for both primary networks and secondary networks this paper considers two models for satellite links as well as terrestrial links. For later, this paper considers the shadowed Rician fading model and for the latter it considers the Nakagami-m fading model. Hence, this paper also throws light on the allocation policy of power to explore more opportunities for the secondary spectrum access.

2 MAC Layer Related Work

One of the prime requirements of sensor network architecture is to maximize the lifetime of the network since the nodes in the sensor network are constantly dissipating the power. In order to withstand with this situation, the planned MAC protocol must be considered seriously in a manner that can withstand battery overrun and minimizes the possible energy misuse [1, 3]. To plan a good quality MAC protocol for the WSNs, the following attributes were reported in [3]. The first attribute is the energy efficiency of all nodes in wireless sensor network. To reduce energy, we must have effective protocols in order to lengthen the lifetime of sensor network nodes. Other important attributes are the ability to expand and adaptability to adjust whenever changes in network topology and its size along with node density; it would be a successful adaptation if these parameters are handled quickly and effectively (Fig. 2).

The resource sharing method, namely dedicated, random, and demand assignment are all combined in MAC protocol for fair resource sharing. In two types of allocation, i.e., shared and dynamic, DTDMA/TDD provides both aforementioned allocation schemes in former it is fixed shared allocation of VBR and in latter it is for CBR, ABR dynamic allocation. When the base station receives the transmission requests sent by the users in predetermined dedicated slots reserved employing slotted ALOHA. Let us abbreviate BS as base station then slot allocation for mac is broadcasted by BS. Once the slot is allocated, the acknowledgments are sent denoting the successful reservations. Depending on the type of traffic allocation of slot during the call establishment process is carried out. For instance, if traffic type is CBR then the incoming CBR calls are blocked when the CBR slots are not available. Like CBR the VBR slots are also of fixed allocation type but the difference is that the unused slots can be shared with the other type of traffic classes avoiding wastage. Similarly, on the same lines, the VBR slots are not available if the arriving VBR slots are blocked. In wireless ATM networks, the ABR and UBR slots and unused CBR

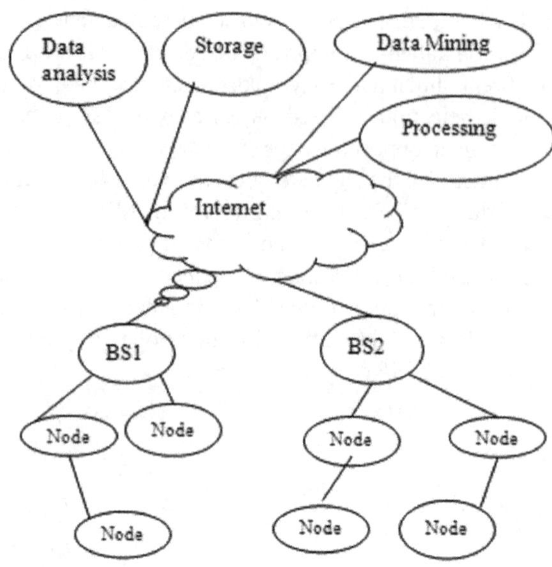

Fig. 2 Wireless sensor network structure

and VBR slots are allocated a dynamic reservation [7]; however, this protocol is not useful for the demanding user most of the time.

In a sensor network, when all the sensor nodes are interested in sending or transmitting on the channel. If all nodes execute their intension of transmitting on the channel at the same time using the orthogonal this is exploiting the whole spectrum. Among many spread spectrum techniques two techniques are used in the sensor network, namely direct-sequence spread spectrum (DSSS) and frequency-hopping spread spectrum (FHSS) since the former technique minimizes the overall interference, and the later technique is useful reducing interference as well as obstructing eavesdropping, i.e., excellent for avoiding eavesdropping. According to the algorithm FHSS, which uses the feature of switching of frequencies most often, therefore, users often send data using these varying frequencies; however, in DSSS sensor nodes transmit data using the pseudorandom number code, the signal is modulated and then transmitted. These Pseudorandom codes are orthogonal [11].

Authors of the paper [12] have projected a scheduling scheme called dynamic fair scheduling based on fair service strategy and generalized processor sharing (GPS) under the label of code division GPS (CDGPS). The CDGPS uses an approach that is different from the old archaic time-scheduling approaches because the new technique used by CDGPS uses the traffic characteristics of both link layer and adaptability of the CDMA in physical layer in order to perform fair scheduling on a time-slot basis. To enhance the capability of CDGPS in terms of efficiency, a credit-based CDGPS (C-CDGPS) scheme was proposed and analyzed to further enhance the soft link capacity for short-term fairness.

Authors in paper [13] to control interference of CDMA and congestion at same time combined the MAC layer and transport layer protocol in "one-shot" to analyze the cross-layer.

Author in paper [14] also shared the idea of minimizing the interference in CDMA signal as well as call rejection rate. In order to implement this idea, authors have set of users called as optimal set of admissible number of users with each having appropriate sending power level. However, the solution has solved to an extent by scheduling the users to satisfy their predetermined transmission.

3 Energy Constraint in MAC-CDMA

Energy constraint is the important aspect of the sensor network. While designing the sensor network and its application domain the type of sensor nodes will be decided. Every sensor node has limited power supply these nodes may drain due to power consumption. Depending upon types of sensor node used the network; the sensor network will be alive. Since the nodes in the sensor network are powered by the batteries they may be periodically changed, recharged, or replaced as these batteries get discharged with time, computation, and communication. For some special sensor nodes, none of the option is available, i.e., they get simply discarded when once energy source is discharged. Therefore, for non-rechargeable or replaceable batteries, the sensor node must have to be operated until its mission-critical time has elapsed. The duration of mission-critical time depends on the type of application. As analogy scientists observing the movements in glacial space requires their sensor to be alive for several years in contrast if used in military these sensor need to be alive for few hours or several days. As the end of the day! The essential design challenge is energy efficiency in WSN. However, the batteries that can be replaced or recharged (using solar), not significantly affect the energy consumption by nodes.

4 Codes and Power Allocation

All the sensor nodes that want to transmit data in the sensor network or the users trying to access the sensor network have to access the data channel in a manner that avoid collisions in the shared medium. CDMA is the digital technique to achieve these tasks. PSN is a pseudorandom noise which is used to code the signal data. This PSN is nothing but a particular signature sequence selected to code the signal data. We use CDMA signal as a form of direct-sequence spread spectrum (DSSS) where the original signal data is multiplied with PSN to get higher bit rate of the code and to reduce overall signal interference. It has three components or key elements.

- The PSN is used here to spread the data and the resulting wideband data is reliable and can withstand with greater resistance to intentional interference or unintentional interference, moreover, the PSN is independent of the data generated by the circuit.
- The CDMA signal takes wider bandwidth when compared with the actual bit rate required to send the information, thereby increasing the transmitted signal bandwidth.
- Synchronization between sender and receiver with PSN code is must to encode and decode the data correctly using autocorrelation function.

The actual spirit of the results behind this research project is the concept of CDMA which is based on the fact that baseband data is multiplied with the predefined signature code that results in increase in the bandwidth of the signal and giving more scope to send the signal in the increased bandwidth that is achieved than the what actual bandwidth is required to send the information with higher bit rate. At the receiver end, the same spreading code is employed to get the decoded version of the data from the received signal.

We are referring the process of extracting the data at the receiver side as correlation if and only if the PSN is exactly same as used by the sender for sending the signal. If the above condition is met, we set correlation as 1 and if correlation is +1 (plus one) we extract the data. Whenever the spreading code does not matches the above condition of correlation, we set it to −1 (minus one) we refer to this situation as spreading code does not correlate to itself. The condition is failed and the data will not be de-multiplexed and a different set of data is appeared which may have no meaning or wrong meaning. This reveals that for data to be decoded correctly at the receiver it must correlate and the spreading code or signature used at the sender must match with the spreading code used at the receiver otherwise the decoding of signal fails, i.e., encoded signal is preserved (Fig. 3).

The logic behind using PSN is numerous, to list few of them as follows: To protect our transmitted data, spreading nature is used in security. PSN seems to be random but in reality is not purely random or fully random noise. All the SPN sequences are finite and deterministic; it means that at the receiver while reconstruction the signal the receiver will have same deterministic order of the code from synchronous detection.

Using CDMA directly in WSN poses many challenges that have to be overcome in this research project. The problems and solutions where ever necessary are discussed in this and subsequent sections. Firstly, the PSN assignment problem which is a matter of debate [18–24]. It is a matter of debate to decide that the PSN code has to be assigned to the each of the terminal in the network or the each of the messages. This problem is of less importance if it is small WSN but this problem grows and becomes the anxiety for the larger WSN. To address this several protocols exist in the area of code assignment protocols, however, more or less they all proved to be less efficient or less proficient. Therefore, all the nodes in the WSN have unique code as the node itself. So, that all the neighboring nodes sense it to be unique code.

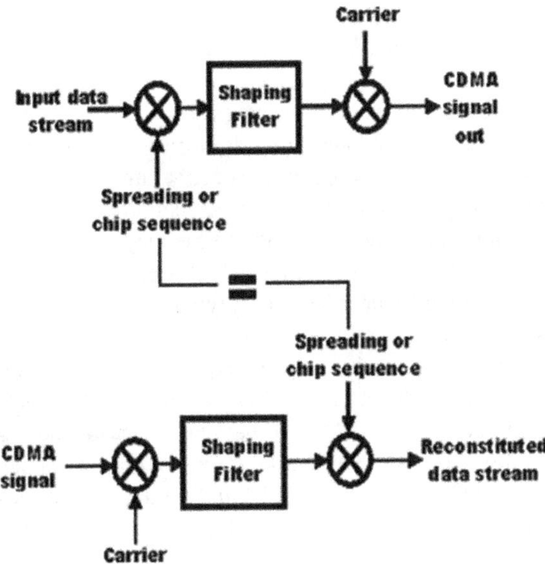

Fig. 3 CDMS system spreading codes [3]

In the proposed WSN, two types of collisions may occur, namely primary collisions and secondary collisions. A collision is said to be primary if two or more than two sensor nodes have same SPN code for encoding the data signal to transmit. This problem is purely the problem in the area of code assignment protocol. SPN is designed and signed in such a way that primary collisions are avoided. Primary collisions are problem in the communication channel to address this problem code, and assignment protocol is used not to handle primary collisions efficiently but also to reduce it as more as possible. The secondary collisions are difficult to avoid because it occurs with the fact that when two or more CDMA nodes transmitting the data using different PSN codes but within non-synchronized frames. Therefore, to conclude, it is important that code assignment protocols have to work in accordance with the topology control routing algorithms. Controlling the topology in case of new sensor nodes added or left the network is a challenge. However, still it is an open problem for many researchers. In WSN, many situations left unattended such as nodes added in the network, few nodes are dead, causing communication failures, network setup. Under these circumstances, it is compulsion that WSN should operate in these conditions a worthy MAC protocols suit well to address these situations.

To address the problem of secondary collisions, in hybrid time division multiplexing and collision detection multiple access WSN MAC protocol spread spectrum communication is used utilizing the fact that spread spectrum communication uses larger frequency range than required by the actual un-modulated original signal. It is possible to send totally different data streams by two different sensor nodes in the same slot using this protocol if and only if these sensor nodes in alike slots are not siblings. This achieves maximizing the data rate thus increasing the throughput, minimizes the end to end delay while upsurge the data security by multiplying with

PSN sequence. The beauty of this protocol is that frames are synchronized and the secondary collision problem of two nodes having different PSN codes inside non-synchronized frames can be resolved. This non-zero correlation leads to interference problem called as multiple access interference (MAI). SPN has a feature that if same versions of PSN codes are time shifted each bit they become orthogonal and can be used in virtually orthogonal codes within a CDMA system. The code is said to be truly orthogonal if these codes when multiplied bitwise (i.e., bit by bit) and when the result is added over a period of time they sum up to zero as follows

Code1: $+1-1+1-1+1+1-1-1$ and Code2: $-1-1-1-1-1+1+1-1$

Resultant code after multiplying these two codes is: $-1+1-1+1-1+1-1+1$ it sum up to 0.

In evaluating the performance of WSN, code design problem plays prominent role. Generally, in WSN, CDMA-based MAC uses the PSN codes having non-zero cross-correlation. Message passing is one of the core features of S-MAC-CDMA in this it breaks long messages into chunk of frames. These frames are distributed and sent in burst. The idea behind sending the frames in burst is to reduce overhead incur in the communication. This reduction of communication overhead is not free; it comes with the cost of sacrificing the fairness in the medium access. However, multi-hop routing algorithms may result in high latency due to periodic sleep as all the neighboring sensor nodes have their individual schedules for sleeping. The latency that happens due to periodic sleep is termed as sleep delay [15]. To improve the sleep delay, a technique called as adaptive listening was proposed. In this technique, if the neighboring nodes overhearing the transmission get wakeup for short duration at the end of the transmission to pass the data immediately if the immediate node coming after it is a destination node is in the next-hop node in the route.

ElBatt's algorithm [3] is used to achieve the minimum transmission power of the entire transmitting sensor nodes while keeping signal to noise ratio to a pre-defined threshold value in any link or transmitting path in a specific time slot. Therefore, this algorithm is analyzed by keeping maximum signal to noise ratio with minimum transmission power. It is also important to keep track changes of path losses, the algorithms [3, 16, 17] gives idea about path losses between the pair of sensor nodes. However, keeping track of path losses between pair of sensor nodes is difficult, path losses also result in degrading the throughput.

5 Throughput of MAC-CDMA in Fading Environments

In this section, this paper measures the performance in the long-distance path loss model of MAC-CDMA-based throughput and bit error rate under the practical assumptions. The analysis is done on both type of fading, i.e., shadow fading or large scale that causes due to the presence of an obstacle in the wireless medium and small scale fading that generally occurs due to variation of the signal strength over a short epoch or short distance. To deal with this fading, Rican model is used in finding the path loss. In the channel, parameters are set up in such a manner that

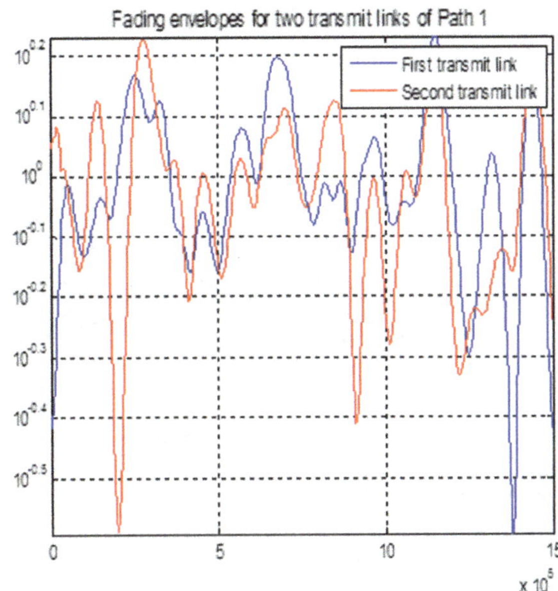

Fig. 4 Fading envelopes for different routes

it reflects the fading environment as shown in Fig. 4. In the fading channel path loss is time variant, since there are different paths from node to node.

To minimize the power consumption and to systematically and repeatedly measure the path losses between any sensor nodes, ElBatt's algorithm as mentioned in the previous section is used. Sensor node which is selected to broadcast a packet the power allocation scheme algorithm as mentioned is used to calculate the path losses and other sensor nodes measured the receive power which is also used to calculate the path losses. Using this power algorithm, the duration between two path loss is one second that transcribe to 3% of transmission is above your head (i.e., the data cannot be transmitted during this measurement) when there are 30 nodes in the change.

It is clear from Fig. 4 that in fading environments the error performance of CDMA is variable when the numbers of node are more. On the other side, when comparing the performance of CDMA using different path power allocation schemes gives real-time path loss and fading information as can be seen in Figs. 5 and 6.

In short, we conclude that if the power allocation scheme is giving us better performance then this power allocation strategy can be employed to additionally boost the performance of collision detection multiple access-based network when compared with the simple full power allocation strategies. As far as the number of sensor nodes in the chain is concerned this strategy is further reliable and extensible, moreover, the beauty of this technique is that it does not require measuring the overhead of the path losses between sensor nodes.

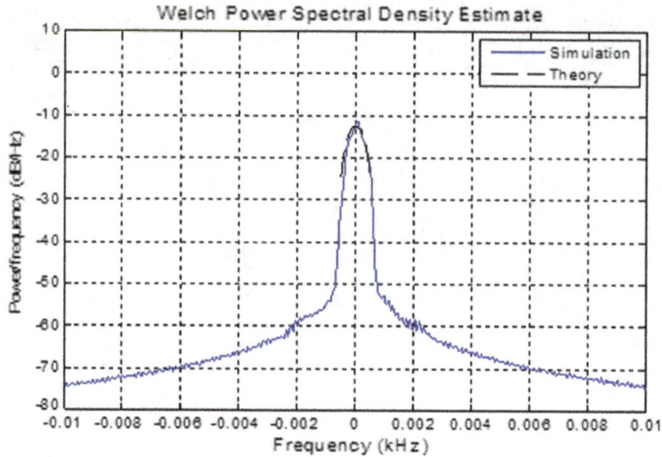

Fig. 5 Power spectrum density

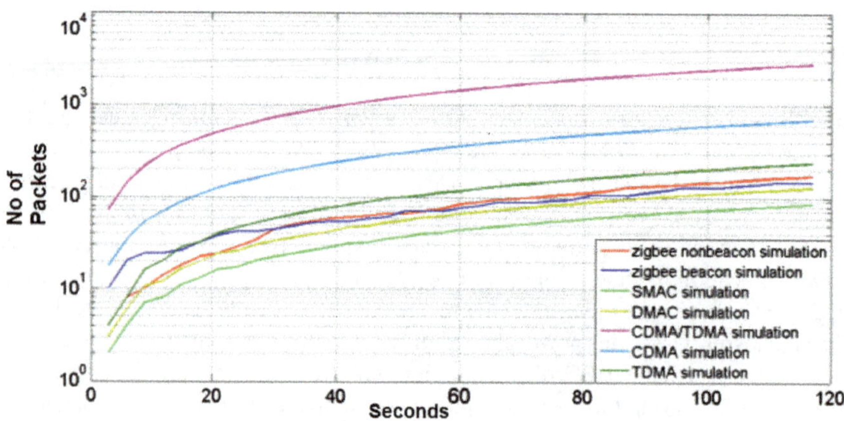

Fig. 6 Throughput of MAC-CDMA protocols in increased fading due to interference

6 Conclusion

In the near future, wireless sensor networks are expected to grow increasingly large and dense, with a necessity to support high density or hundreds of sensor nodes in a single network. Currently, multiple access techniques in IoT WSNs have been thoroughly reviewed investigated in this work and determined to have insufficient network scalability to support dense WSNs. since of this, a multiple access technique based on receiver-assigned MAC-CDMA is described, which can satisfy network scalability requirements and give the added benefits of interference mitigation and physical layer security and energy consumption based on the fading can be minimized.

Acknowledgements The authors would like to express their gratitude to Deanship of Scientific Research King Khalid University, Abha, Saudi Arabia for providing administrative, financial, and technical support under grant number G.R.P 261/2018.

References

1. M.S. Arafath, K.U.R. Khan, K.V.N. Sunitha, Incorporating privacy and security in military application based on opportunistic sensor network. Int. J. Internet Technol. Secur. Trans. (IJITST) (2017) (ISSN: (Print), ISSN: 1748-5703 (Online))
2. M.S. Arafath, K.U.R. Khan, K.V.N. Sunitha, Opportunistic sensor networks: a survey on privacy and secure routing, in *2017 2nd IEEE International Conference on Anti-Cyber Crimes (ICACC)*, 26–27 Mar 2017, pp. 41–46
3. I.F. Akyildiz, D.A. Levine, I. Joe, A slotted CDMA protocol with BER scheduling for wireless multimedia networks. IEEE/ACM Trans. Netw. **7**(2), 146–158 (1999)
4. S.S. Kulkarni, TDMA services for sensor networks, in *Proceedings of 24th International Conference on Distributed Computing Systems Workshops*, 23–24 Mar 2004, pp. 604–609
5. W. Ye, J. Heidemann, D. Estri, Medium access control with coordinated adaptive sleeping for wireless sensor networks. IEEE/ACM Trans. Netw. **12**(3), 493–506 (2004)
6. A. El-Hoiydi, Spatial TDMA and CSMA with preamble sampling for low power ad hoc wireless sensor networks, in *Proceedings of ISCC 2002, Seventh International Symposium on Computers and Communications*, 1–4 July 2002, pp. 685–692
7. V. Rajendran, K. Obraczka, J.J. Garcia-Luna-Aceves, Energy-efficient, collision-free medium access control for wireless sensor networks, in *Proceedings. ACM SenSys 03* (Los Angeles, California, 5–7 Nov 2003), pp. 181–192
8. N.A. Ali, H.M. ElSayed, M. El-Soudani, H.H. Amer, Effect of hamming coding on WSN lifetime and throughput, in *2011 IEEE International Conference on Mechatronics (ICM)* (2011)
9. L.D. Pedrosa, R.M. Rocha, WMTP: a modular WSN transport protocol: the fairness module, in *New Technologies, Mobility and Security, NTMS '08* (2008)
10. S. Bdiri, F. Derbel, O. Kanoun, Wireless sensor nodes using energy harvesting and B-Mac protocol, in *Proceedings of the 10th International Multi-conferences on Systems, Signals Devices* (Hammamet, Tunis, 18–21 Mar 2013)
11. L. Bao, J.J. Garcia-Luna-Aceves, A new approach to channel access scheduling for ad hoc networks, in *Seventh Annual International Conference on Mobile Computing and Networking*, pp. 210–221 (2001). *Performance Computing and Communications Conference* (*IPCCC*) (2011)
12. L. Xu, X. Shen, J.W. Mark, Dynamic fair scheduling with QoS constraints in multimedia wideband CDMA cellular networks. IEEE Trans. Wirel. Commun. **3**(1), 60–73 (2004)
13. A. Yener, R.D. Yates, S. Ulukus, Interference management for CDMA systems through power control, multiuser detection, and beamforming. IEEE Trans. Commun. **49**(7), 1227–1239 (2001)
14. K. Ryuji, B. Rapajic Predrag, S. Vucetic Branka, An overview of adaptive techniques for interference minimization in CDMA systems. Wirel. Pers. Commun. **1**(1), 3–21 (1994)
15. S. Pandit, G. Singh, Backoff algorithm in cognitive radio MAC protocol for throughput enhancement. IEEE Trans. Veh. Technol. **64**, 1991–2000 (2015)
16. Z. Wang, U. Mani, M. Ju, H. Che, A rate adaptive hybrid mac protocol for wireless cellular networks, in *Proceedings of IEEE International Conference on Computer Communications and Networks* (2006)
17. S.J. Isaac, G.P. Hancke, H. Madhoo, A. Khatri, A survey of wireless sensor network applications from a power utility's distribution perspective. AFRICON (2011)

18. I. Chlamtac, A. Farago, A.D. Myers, V.R. Syrotiuk, G. Zkruba, A performance comparison of hybrid and conventional mac protocols for wireless networks, in *IEEE VTC 2000* (2000), pp. 201–205
19. M.S. Arafath, K.U.R. Khan, K.V.N. Sunitha, Pithy review on routing protocols in wireless sensor networks and least routing time opportunistic technique in WSN, in *2017 10th International Conference on Computer and Electrical Engineering (ICCEE 2017)*. (University of Alberta, Edmonton, Canada, October 11–13 2017)
20. G. Dai, C. Miao, K. Ying, K. Wang, Q. Chen, An energy efficient MAC protocol for linear WSNs. Chin. J. Electron. **24**, 725–728 (2015)
21. M. Shurman, B. Al-Shua'b, M. Alsaedeen, M.F. Al-Mistarihi, K.A. Darabkh, N-BEB: new Backoff algorithm for IEEE 802.11 MAC protocol, in *Proceedings of the 37th International Convention on Information and Communication Technology*. (Opatija, Croatia, 26–30 May 2014)
22. O. Sentieys, S. Derrien, C. Huriaux, V.D. Tovinakere, Low power reconfigurable controllers for wireless sensor network nodes, in *FCCM—22nd IEEE International Symposium on Field-Programmable Custom Computing Machines* (Boston, United States, May 2014)
23. R. Rugin, G. Mazzini, A simple and efficient MAC-routing integrated algorithm for sensor network, in *IEEE International Conference on communications*, vol. 6, 20–24 June 2004, pp. 3499–3503
24. M.S. Arafath, K.U.R. Khan, K.V.N. Sunitha, Incorporating security in opportunistic routing and traffic management in opportunistic sensor network. Int. J. Adv. Intell. Paradigms [1755-0394] Intell. Comput. Syst. Inderscience publishers (in press)

Single-Feed Right-Hand Circularly Polarized Microstrip Antenna with Endfire Radiation

K. Manoj Kumar and A. Bharathi

Abstract In this paper, a novel single-feed circular polarized planar antenna with its beam parallel to the substrate plane is discussed. This antenna is composed of two structures: It consists of a vertically polarized magnetic dipole implemented using short-circuited semicircular magnetic disks and a horizontally polarized bended complementary dipole. The entire structure is implemented on a single-layer paper honeycomb substrate with relative permittivity 1.08 and 0.6 mm thickness to resonate at 2.35 GHz. It achieves wide impedance and axial ratio (AR) bandwidths with good overlap and wide beamwidth. It has shown −10 dB impedance bandwidth of 11.5% (2.22–2.49 GHz) and 3 dB AR bandwidth of 11.06% (2.20–2.46 GHz). The proposed antenna would be promising for Wi-Fi or RFID applications.

Keywords Circular polarization · Microstrip antenna · Endfire radiation · Planar antenna · Complementary dipole

1 Introduction

Circularly polarized antennas are very useful in GPS and RFID systems. Circularly polarized antennas with wide beamwidth are required to receive signals from wide range of directions. There are different ways to produce circular polarization (CP) with endfire radiation. Researchers used helical antennas [1–3] and achieved endfire radiation over wide beamwidth [4]. Crossed dipole arrays in which the elements are separated by $\lambda/4$ distance and fed with equal magnitude, and 90° out of phase inputs were also used to generate CP in endfire direction of the array. However, most of the mentioned antennas are of high profile. Another way to generate CP is using curl antennas [5, 6], where the arms of the antenna are turned around the center of the antenna like a curl. Almost all the antennas produce their beam perpendicular to the substrate plane. In order to have it parallel to substrate plane, there are few methods which either use combination of orthogonal magnetic dipole [7] or combination of aperture/loop with monopole/dipole. Orthogonal magnetic dipoles' combination

K. Manoj Kumar (✉) · A. Bharathi
ECE Department, University College of Engineering, Osmania University, Hyderabad, India
e-mail: manojkumarkomuraju@gmail.com

H. S. Saini et al. (eds.), *Innovations in Electronics and Communication Engineering*, Lecture Notes in Networks and Systems 107,
https://doi.org/10.1007/978-981-15-3172-9_5

is a complex structure with narrow impedance bandwidth. As studied in [8–12] combination of aperture and complementary dipole are employed to realize antenna with endfire beam in the substrate plane. The impedance bandwidths are less [8–12] with highest value <6.5%. This is improved with stub loading [11], but back radiation is increased heavily. This problem is resolved using a v-shaped open loop [12], but the half power beamwidth (HPBW) is 90°. The challenge faced in the design of the microstrip antenna with endfire beam is the structure of the antenna should be simple, compact, and planar. To meet these challenges, in this paper we propose a model which is combination of bended dipole and semicircular magnetic dipole. Good impedance BW and AR characteristics are observed. Radiation pattern with broad beam having HPBW of 125° is achieved.

2 Antenna Geometry

Figure 1 shows the geometry of the proposed antenna which is designed on a 82 mm × 82 mm paper honeycomb substrate of thickness 6 mm and relative permittivity $\varepsilon_r = 1.08$. It consists of semicircular patches of radius $R1$ etched on both sides of substrate, and is short-circuited at one end. It is fed with coaxial probe located at a distance $L1$ from short-circuit end. This antenna also consists a bended complementary electric dipole located at distance $L2$ and fed using stripline feeding technique from semicircular disk radiator. The complementary electric dipole is bent at angles 60° and 90°.

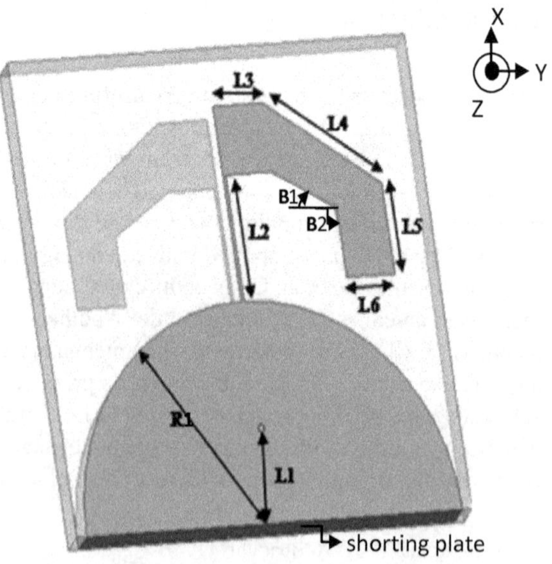

Fig. 1 Geometry of the proposed antenna $R = 38$, $L1 = 16$, $L2 = 26.5$, $L3 = 10$, $L4 = 26.5$, $L5 = 16$, $L6 = 10$ (mm), $B1 = 60°$, $B2 = 90°$

3 Principle of Operation

CP will be generated when the two orthogonal components of the antenna, i.e., electric fields in azimuthal and elevation planes, are equal. As shown in Fig. 1, two semicircular patches shorted at one end act as a disk radiator (magnetic dipole). It controls the E_ϕ component, and complementary electric dipole controls the E_θ component. For the disk radiator, component of current normal to the edge of the disk approaches to zero at the edge; this implies that tangential components of magnetic field at the edge of the disk are vanishingly small; with these assumptions, it can be modeled as a cylindrical cavity bounded by at its top and bottom by electric walls and on its edge by a magnetic wall. The solution for the wave equation in cylindrical coordinates is

$$\vec{E}_z = \vec{E}_0 J_n(k\rho)\cos n\phi \tag{1}$$

where $J_n(k\rho)$ are the Bessel functions of order n. At the edge of the disk, the surface currents must vanish, i.e.,

$$J_\rho(\rho = R1) = 0$$

where $R1$ is the radius of the disk and it is determined by the solution of the derivative of the Bessel function

$$J_1'\left(\chi_1'\right) = 0$$

$$nR_1 = \chi_1' = 1.8411 \tag{2}$$

Here n is the wave number given by $n = 2\pi/\lambda$. λ is the wavelength, J_1' is the first derivative of first-order Bessel, where χ_1' is its root. Using above equation, the radius of the disk radiator can be calculated at the desired frequency. The bended open loop is selected to generate a directive E_ϕ. To analyze radiation behavior, the structure is assumed as combination of ideal short dipoles. It will be clear for the complementary dipole symmetrical about the x-axis; all the x-component fields will cancel each other; only the y-components will be superimposed [12]. The normalized far field of the open dipole antenna is derived as

$$\vec{E}_{\text{dipole}} = \frac{\mathrm{e}^{-\mathrm{j}n\mathrm{R}}}{R}\left[\cos\left(\frac{\pi}{2}\cos B1\cos\phi\sin\theta - \delta\right)\right] \times \left(\hat{\theta}\sin\phi\cos\delta + \hat{\phi}\cos\phi\right) \tag{3}$$

δ is the leading phase in the arms of complementary dipole because of the bent angles $B1$ and $B2$ w.r.t to Y-axis.

The field of the proposed antenna is given by

$$\vec{E}_{\text{total}} = \vec{E}_{\text{disk}} + j\,\vec{E}_{\text{dipole}}$$

$$\vec{E}_{\text{total}} = \hat{\theta}\frac{\mathrm{e}^{-\mathrm{j}n\mathrm{R}}}{R}\left[\cos\phi + \mathrm{j}\cos\theta\cos\left(\frac{\pi}{2}\cos B1\cos\phi\sin\theta - \delta\right)\sin\phi\right]$$
$$+ \hat{\phi}\frac{\mathrm{e}^{-\mathrm{j}n\mathrm{R}}}{R}\left[\cos\theta\sin\phi + j\cos\left(\frac{\pi}{2}\cos B1\cos\phi\sin\theta - \delta\right)\cos\phi\right] \quad (4)$$

The radiation fields in *XZ*- and *XY*-planes are obtained by substituting $\phi = 0°$ and $\theta = 90°$, respectively, in Eq. (3). Then, the normalized radiation patterns in the *XZ*-plane can be deduced as

$$E_\theta(\theta) = 1 \quad (5)$$

$$E_\phi(\theta) = j\cos\left(\frac{\pi}{2}\cos B1\sin\theta - \delta\right) \quad (6)$$

The AR of the elevation plane is given as

$$\text{AR(dB)} = 20\left|\log_{10}\frac{|E_\theta|}{|E_\phi|}\right|(\text{dB}) \quad (7)$$

We can observe that by substituting E_ϕ and E_θ from (5) and (6) in (7), we can achieve CP radiation with endfire beam in *x*-direction. The CP direction will be decided by the probe current's loop/path as shown in Fig. (2). Current from the probe reaches to bended dipole upper arm from upper part of the magnetic disk along the feedline. Then to maintain continuity in the current loop/path, it is considered that displacement current to flow in the +*x*-direction. After that current turns toward the bottom arm of bended dipole. Then, the current now reaches bottom layer of magnetic disk through the bottom stripline. If this virtual current's path follows a right-handed helix direction with thumb pointing to +*z*-direction, the antenna

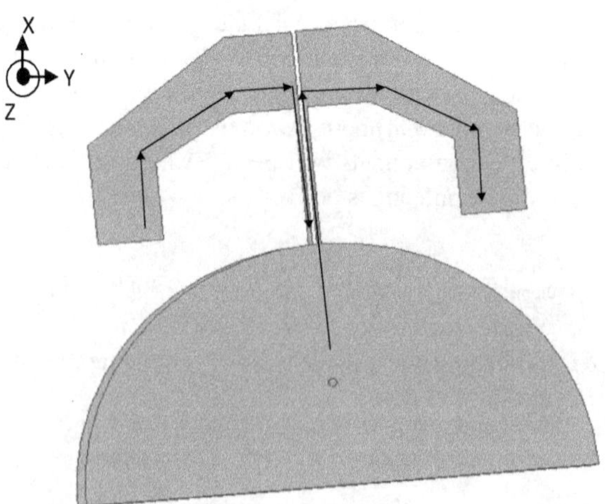

Fig. 2 Current flow path

generates right-handed circular polarization (RHCP). Else it generates left-handed circular polarization (LHCP).

4 Design Steps

1. Substrate thickness is chosen to be $h = 6$ mm to maintain planar configuration of the antenna, which is of low profile, i.e., almost 0.048 times of the guide wavelength (λ_g) at 2.35 GHz, such a way

$$\lambda_g = \frac{\lambda_0}{\sqrt{\varepsilon_{re}}}, \quad \varepsilon_{re} = \frac{\varepsilon_r + 1}{2}$$

2. The radius $R1$ is chosen with the help of Eq. (2) based on the operating frequency.
3. This antenna is fed by a coaxial probe [12]. The feed is located at a distance of *R*/3 from midpoint of shorting plate. Later, it is optimized to $L1$ for better results.
4. The length of the stripline pair ($L2$) is taken to be equal to $\lambda_g/4$, and the width of the line is chosen to be one-tenth of the height of the antenna.
5. The bended complementary dipole is connected to the stripline feed. Its width is equal to half of the wavelength and has bent angles $B1$ and $B2$. These angles are measured from *Y*-axis.
6. The design is finally optimized to achieve good impedance and axial ratio bandwidths with a broad beam in endfire direction.

5 Simulation Results

The proposed RHCP antenna is designed and simulated using HFSS 17. The impedance, AR, RHCP gain, and total gain of the antenna are observed from the simulation results. Gain of the antenna is 1.83 dB. The impedance bandwidth of the antenna for $|s_{11}| < 10$ dB is 2.22–2.49 GHz (11.49% in fraction), and the AR bandwidth of < 3 dB is 2.20–2.46 GHz (11.06% in fraction) shown in Figs. 3 and 4, respectively. There is overlap of 240 MHz between the impedance and AR bandwidths. Figures 5 and 6 show the radiation patterns of in *XY*-plane and *XZ*-planes, respectively.

6 Conclusion

In this paper, a simple, compact, and planar CP antenna with endfire beam in parallel with substrate plane is proposed. The operating principle and design procedure

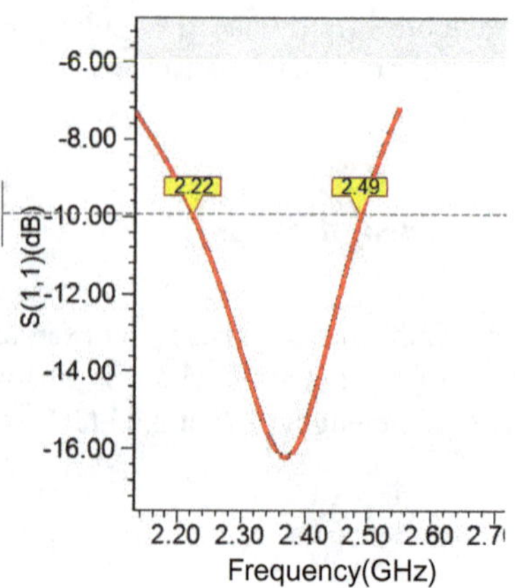

Fig. 3 Simulated S_{11} performance of the proposed antenna

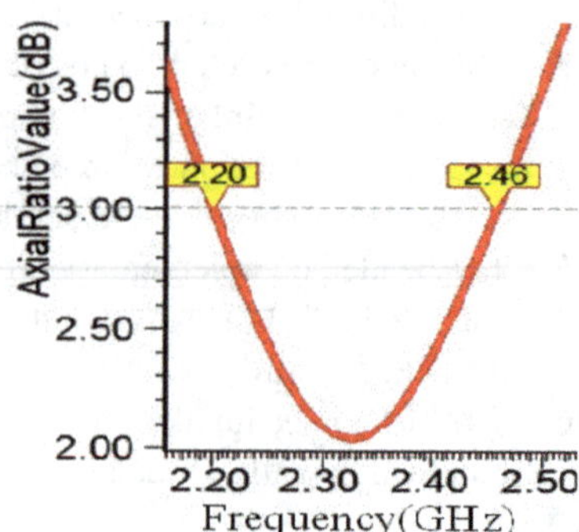

Fig. 4 Simulated AR performance of the proposed antenna

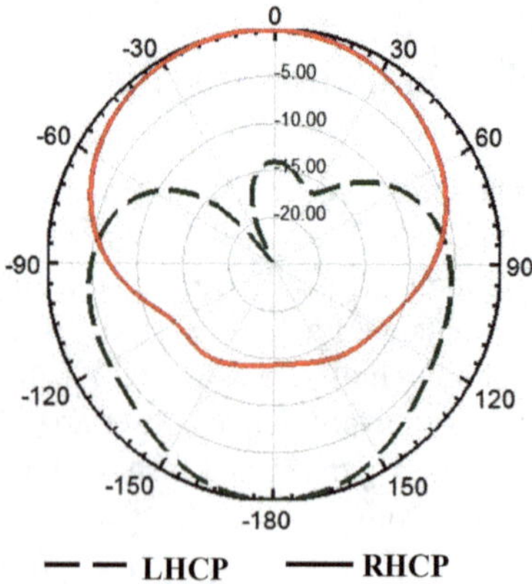

Fig. 5 Simulated radiation characteristics in *XY*-plane at $\theta = 90°$

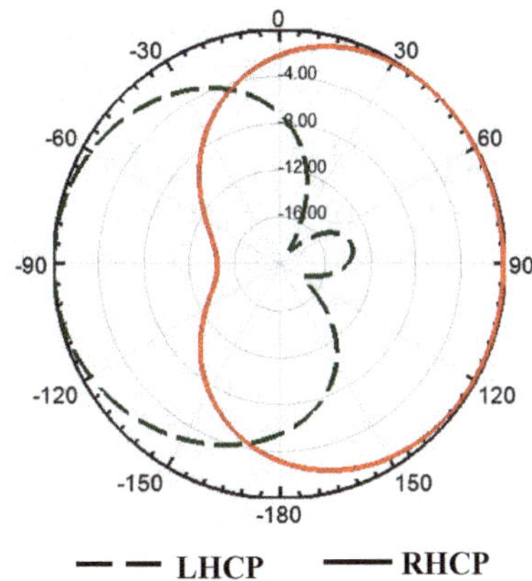

Fig. 6 Simulated radiation characteristics in *XZ*-plane at $\phi = 0°$

are explained. Further, the analysis of proposed antenna is described using source model. This antenna provides a good overlap between the impedance bandwidth and AR bandwidth up to 240 MHz and HPBW which are 202° and 125° in elevation and azimuthal planes, respectively. This antenna would be a potential candidate for RFID applications and wireless communication systems with low-elevation angle requirement.

References

1. J.D. Kraus, Helical beam antennas for wide-band applications. Proc. IRE **36**(10), 1236–1242 (1948)
2. C. Kilgus, Multi element, fractional turn helices. IEEE Trans. Antennas Propag. **16**(4), 499–500 (1968)
3. L.-S. Liu, Y. Li, Z.-J. Zhang, Z.-H. Feng, Circularly polarized patch helix hybrid antenna with small ground. IEEE Antennas Wireless Propag. Lett. **13**, 361–364 (2014)
4. W. Liu, Z. Zhang, Z. Feng, A bidirectional circularly polarized array of the same sense based on CRLH transmission line. Propag. Electromag. Res. **141**, 537–552 (2013)
5. H. Nakano, S. Okuzawa, K. Ohishi, H. Mimaki, J. Yamauchi, A curl antenna. IEEE Trans. Antennas Propag. **41**(11), 1570–1575 (1993)
6. H. Nakano, Natural and metamaterial-based spiral and helical antennas, in *Proceedings of Loughborough Antennas Propagation Conference* (Loughborough, U.K., 2014), pp. 387–390
7. W.J. Lu, J.W. Shi, K.F. Tong, H.B. Zhu, Planar endfire circularly polarized antenna using combined magnetic dipoles. IEEE Antennas Wireless Propag. Lett. **14**, 1263–1266 (2015)
8. W.H. Zhang, W.J. Lu, K.W. Tam, A planar end-fire circularly polarized complementary antenna with beam in parallel with its plane. IEEE Trans. Antennas Propag. **64**(3), 1146–1152 (2016)
9. M. Ye, X.R. Li, Q.X. Chu, Single-layer circularly polarized antenna with fan-beam endfire radiation. IEEE Antennas Wireless Propag. Lett. **16**, 20–23 (2016)
10. B. Xue, M. You, W.J. Lu, L. Zhu, Planar endfire circularly polarized antenna using concentric annular sector complementary dipoles. Int. J. RFMicrow. Comp.-Aided Eng. **26**(9), 829–838 (2016)
11. J. Zhang, W.J. Lu, L. Li, L. Zhu, H.B. Zhu, Wideband dual-mode planar endfire antenna with circular polarization. Electron. Lett. **52**(12), 1000–1001 (2016)

12. M. You, W.-J. Lu, B. Xue, L. Zhu, H.-B. Zhu, A novel planar endfire circularly polarized antenna with wide axial-ratio beamwidth and wide impedance bandwidth. IEEE Trans. Antennas Propag. **64**(10), 4554–4559 (2016)
13. R. Garg, P. Bhartia, I. Bahl, A. Ittipiboon, *Microstrip Antenna Design Handbook* (Artech House Inc., Norwood, MA, 2001)
14. H.A. Wheeler, Directive loop antenna, U.S. Patent 2 518 736, 1950
15. J.D. Kraus, *Antennas*, 2nd edn. (McGraw-Hill, New York, 1988)

Flexible RFID Tag Antenna Design

Fwen Hoon Wee, Mohamed Fareq Abdul Malek, Been Seok Yew, Yeng Seng Lee, Siti Zuraidah Ibrahim and Hasliza A. Rahim

Abstract This research is to design a radio frequency identification (RFID) high sensitivity tag antenna which operates at UHF frequency band (919–923 MHz). The major problem in designing the tag antenna is that it needs to be designed for long-range transmission with a miniaturized size. However, reducing the size of the tag antenna can cause the gain to be decreased. Another challenge in designing RFID passive tag is to ensure no huge change occurs on the resonant frequency when the tag antenna is being bent. This research had provided two methods in overcoming those problems that include the construction of a meander line structure to reduce the antenna size that can be applied to a small device. In addition, flexible substrate, polyethylene terephthalate (PET) had been chosen for tag antenna. Thus, the expected result shows high gain (1.55 dB) with a small size of flexible tag antenna.

Keywords RFID · UHF frequency band · Miniaturize tag antenna · Flexible PET

F. H. Wee (✉) · S. Z. Ibrahim · H. A. Rahim
School of Computer and Communication Engineering, Universiti Malaysia Perlis, Arau, Malaysia
e-mail: fhwee@unimap.edu.my

F. H. Wee · Y. S. Lee · H. A. Rahim
Bioelectromagnetics Research Group, Universiti Malaysia Perlis, Arau, Malaysia

Y. S. Lee
Department of Electronic Engineering Technology, Faculty of Engineering Technology, Universiti Malaysia Perlis, Arau, Malaysia

S. Z. Ibrahim
Advanced Communication Engineering Centre (ACE), Universiti Malaysia Perlis, Arau, Malaysia

M. F. Abdul Malek
Faculty of Engineering and Information Sciences, University of Wollongong in Dubai, Dubai, United Arab Emirates

B. S. Yew
Faculty of Innovative Design and Technology, Universiti Sultan Zainal Abidin (UniSZA), Kuala Terengganu, Terengganu, Malaysia

H. S. Saini et al. (eds.), *Innovations in Electronics and Communication Engineering*, Lecture Notes in Networks and Systems 107,
https://doi.org/10.1007/978-981-15-3172-9_6

1 Introduction

Radio frequency identification (RFID) is the use of radio waves to read and capture information stored on a tag attached to an object [1]. It has been widely used in various industrial sectors. For example, RFID was applied in the retail and apparel, vehicle tracking, library system and hospital. RFID system consists of two systems which are reader and tag. The reader detects the tag when tag is placed on the reader within a certain range [1]. With the rapid growth of market needs for RFID tags, flexible antennas powering up these tags have gained many interest due to their characteristics of being low cost and high quality. Compared with the RFID tags that operated in HF and LF range, UHF can perform better as it can reach longer range (0.5–5 m) and has faster data transfer rate.

A research paper on designing RFID tag antenna operates at UHF range had been studied. From the research paper, when the UHF RFID tag is bending, it cannot perform the desired frequency range and return loss [2]. Therefore, the bending performance of the designed antenna will be analyzed in this research. Other than that, some of the research paper related to UHF RFID tag formed a negative gain in the simulation when the RFID tag operated at UHF range [3]. Gain is important to be considered as the higher the gain of a tag antenna, the higher the efficiency is [4].

This research project is on the designing of flexible RFID tag antenna which works at the passive mode. The simulation will be done in CST software. Polyethylene terephthalate (PET) material is used as the substrate of the antenna design. The design is based on meander line, and it gives miniaturize characteristic and long-range reading [5]. The frequency range is maintained between 919 and 923 MHz which is at UHF range. The return loss obtained is below than 10 dB where the antenna will perform 90% radiate and have a less loss [1]. The realized gain for designed antenna should be >1 dB. Those performances are ensured to be maintained when the antenna is being bent in the simulation. The antenna designed is fabricated, and the network analyzer is used to measure the result of return loss, frequency bandwidth and gain.

2 RFID Tag Antenna Design

An antenna is a conductive structure that radiates an electromagnetic wave when a time-varying (alternating) electrical current is applied to it [6]. Any similar structure can radiate an electromagnetic wave, but it will not do so efficiently. To make an antenna radiate efficiently, it needs to be designed properly. In other words, an antenna is a particular arrangement of conductor designed to radiate an electromagnetic field in response to an applied alternating electric current. An antenna performance is affected by several parameters during design of the tag antenna to reach some of the performance characteristics, for example, resonant frequency, gain, radiation pattern, polarization, efficiency and bandwidth.

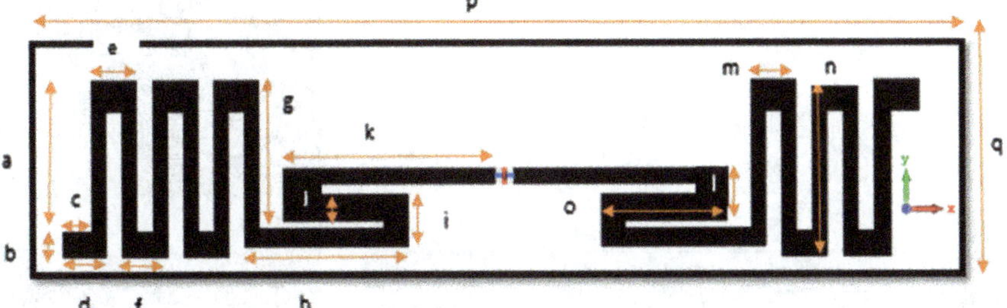

Fig. 1 Meander line shape of RFID tag antenna design

Table 1 Parameter lists of meander shape RFID tag antenna design

Length/width	mm	Length/width	mm
a	14.5	*j*	**2.5**
b	2.5	*k*	**22.0**
c	3.0	*l*	**5.0**
d	4.5	*m*	**4.7**
e	4.7	*n*	*17*
f	*4.7*	*o*	*13*
g	*16.0*	*p*	*91*
h	*17.0*	*q*	*26*
i	*5.0*		

The length/width and the measurement unit, mm are coming from the Fig 1.

2.1 Meander Dipole Antenna Software Design

A meander dipole antenna is designed to reduce the size of the antenna tag plus it is suitable to be used at UHF range. The resonant frequency can be obtained by adjusting the length and number of meander dipole antenna. Other than that the thickness of meander line also can affect the simulation result. Figure 1 and Table 1 show the meander line antenna design and its dimension, respectively, for this research project.

2.2 Meander Dipole Antenna Hardware Fabrication

Copper tape had been used as the conductor material for the radiating part of the RFID tag antenna. It was cut and pasted on the polyethylene terephthalate (PET) material based on the dimensions as the optimized simulated design. The fabricated RFID tag antenna can be seen in Fig. 2a.

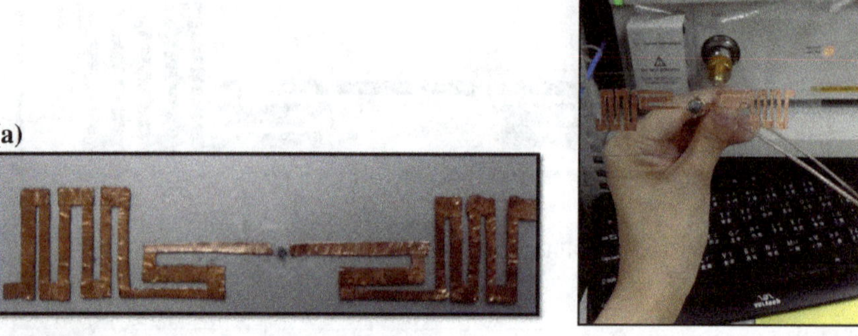

Fig. 2 Fabricated meander line shape of RFID tag antenna design. **a** prototype, **b** antenna measurement

The resonant frequency and return loss of the prototype RFID tag antenna were measured using Agilent Performance Network Analyzer (PNA) as shown in Fig. 2b. For the antenna gain result, the measurement is done using anechoic chamber which is connected with the network analyzer. The RFID tag antenna was placed in the chamber during measurement to prevent the RF signal interference and environment error from the surrounding which will affect the parametric result.

2.3 Meander Dipole RFID Tag Antenna with Bending Position

A cylinder filled with vacuum is created in order to offer a bending position to the RFID tag antenna when the antenna is attached to the cylinder. Radius R such as 30 and 50 mm had been created to provide variation results for the RFID tag antenna bending performance. Both simulated and fabricated bending RFID tag antenna can be seen in Fig. 3.

3 RFID Tag Antenna Results and Analysis

Figure 4a shows that the resonant frequency of proposed RFID tag antenna is 921 MHz which falls between UHF range (919–923 MHz). While the gain achieved is 1.55 dB, it can be claimed that it is a high gain compared to low gain (negative value) from the previous paper relating to RFID tag antenna [7].

In this research project, the flexibility of the designed antenna when $R = 30$ mm and $R = 50$ mm is analyzed. From Fig. 5, the frequency of the tag antenna for

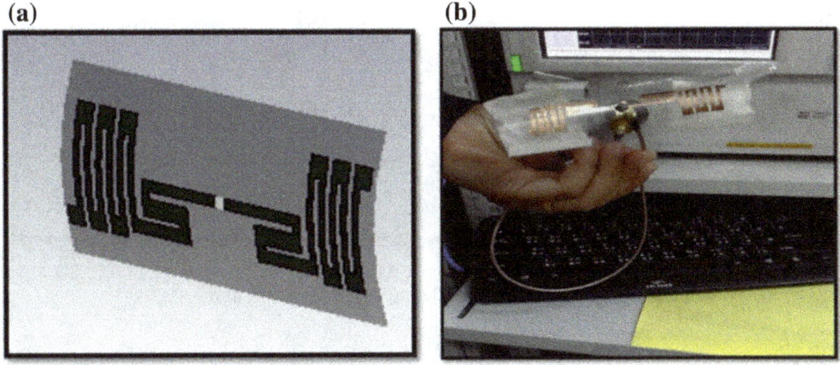

Fig. 3 Flexibility of the RFID tag antenna at $R = 30$ and 50 mm **a** simulation, **b** measurement

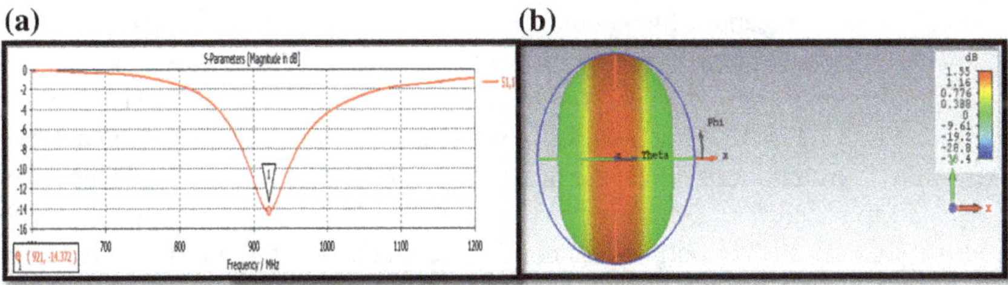

Fig. 4 RFID tag antenna parameter results **a** S11, **b** radiation pattern (gain)

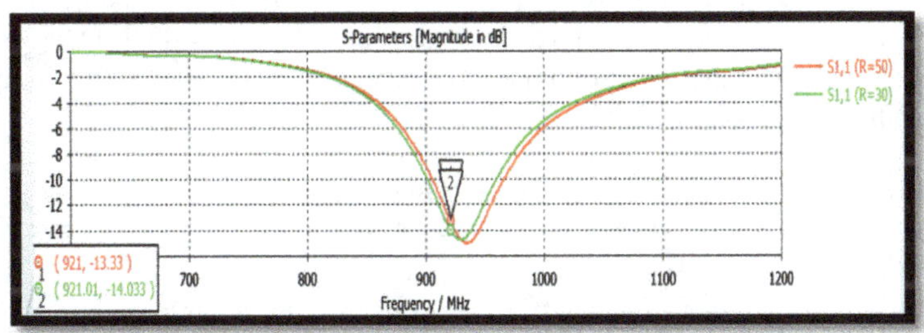

Fig. 5 S11 for 50 and 30 cm radius RFID tag antenna bending

both radii R shows a little shift in resonant frequency; however, both different radius bending RFID tag antenna is still maintain fall on the UHF range.

Based on Fig. 6, the antenna gain for both bending position of proposed RFID tag antenna shows good gain performance too with 1.49 and 1.52 dB for bending radius of 30 and 50 mm, respectively. It seems that more bend on the antenna leads to lower gain compared to less bending of the proposed RFID tag antenna.

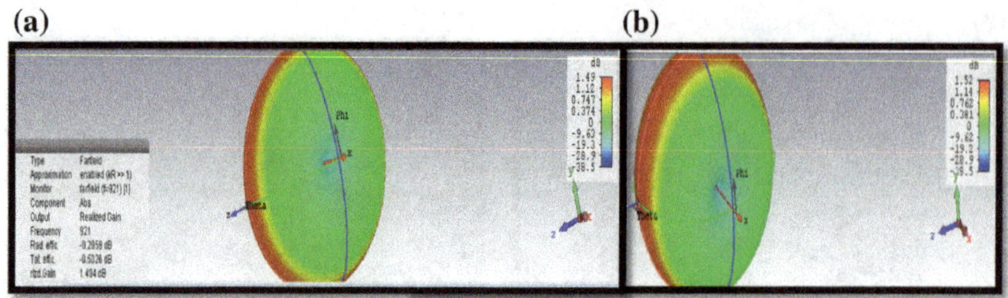

Fig. 6 Radiation pattern (gain) **a** when $R = 30$ mm, **b** when $R = 50$ mm

4 Conclusions

A passive RFID tag antenna at UHF range was designed with frequency range of 919–923 MHz. A meander line antenna was designed to miniaturize the size of the tag antenna. The antenna size for this project is 91 mm × 26 mm which is small and compact. The result for the RFID tag antenna was achieved which is 921 MHz with return loss of 14.732 dB. The flexibility of the antenna tag was also been analyzed. There were two conditions to test the flexibility of the tag antenna which is the radius of $R = 30$ mm and $R = 50$ mm. For both the simulated and fabricated result for bending position, both conditions do not give so much effect to the frequency range and return loss which is still maintaining the resonant frequency. The realized gain for RFID tag antenna was analyzed too for the UHF range. High gain is required such that the antenna tag can sense for a longer range between 0.5 m and 5 m. Some of the previous RFID tag antenna research paper did not achieve gain that is more than 1 dB. Fortunately, the value of the realized gain for this proposed RFID tag antenna was able to achieve 1.55 dB due to the low tangent loss of the substrate which had improve the gain performance.

Acknowledgements We thank our colleagues from Bioelectromagnetic (BioEM) Research Group, School of Computer and Communication Engineering, who provided insight and expertise that greatly assisted the research.

References

1. H. Lehpamer, *RFID Design Principles* (Artech House, London, 2008)
2. S. Zhang, S. Li, S. Cheng, J. Ma, H. Chang, Research on smart sensing RFID tags under flexible substrates in printed electronics, in *2015 16th International Conference on Electronic Packaging Technology (ICEPT)*, Changsha (2015), pp. 1006–1009
3. A. Atojoko, R.A. Abd-Alhameed, H.S. Rajamani, N.J. McEwan, C. H. See, P.S. Excell, Design and analysis of a simple UHF passive RFID tag for liquid level monitoring applications. in *2015 Internet Technologies and Applications (ITA)*, Wrexham (2015), pp. 484–488

4. S. Zuffanelli, in *A High-gain passive UHF-RFID Tag with Increased Read Range*. Antenna design solutions for RFID tags based on Metamaterial-Inspired Resonators and Other Resonant Structures. Springer Theses (Recognizing Outstanding Ph.D. research) (Springer, Cham, 2018)
5. A. Lewis, G. Weis, M. Randall, A. Galehdar, D. Thiel, Optimising efficiency and gain of small meander line RFID antennas using ant colony system, in 2009 IEEE Congress on Evolutionary Computation, Trondheim (2009), pp. 1486–1492
6. Colleen J. Fox, Paul M. Meaney, Fridon Shubitidze, Lincoln Potwin, Keith D. Paulsen, Characterization of an implicitly resistively-loaded monopole antenna in lossy liquid media. Int. J. Antennas Propag. (Article ID 580782), 9 (2008)
7. N.M. Faudzi, M.T. Ali, I. Ismail, N. Ya'acob, H. Jumaat, N.H.M. Sukaimi, UHF-RFID tag antenna with miniaturization techniques, in *2013 10th International Conference on Electrical Engineering/Electronics, Computer, Telecommunications and Information Technology*, Krabi (2013), pp. 1–5

Miniaturized Two-Section Branch-Line Coupler Using Open-Stub Slow-Wave Structure

Kok Yeow You, Jaw Chung Chong, Mohd Fareq Abdul Malek, Yeng Seng Lee and Sehar Mirza

Abstract This study presents the miniaturization of two-section branch microstrip lines coupler using the slow-wave technique. Regular and miniature branch-line couplers operating at 2.4 GHz are designed, fabricated, tested, and compared. Up to 38% reduced size of the miniature coupler is achieved compared to the conventional coupler. The miniaturized coupler is capable of operating from 1.85 to 3.2 GHz with a bandwidth of 1.35 GHz. Overall, the simulation and measurement results demonstrated a good agreement.

Keywords Miniature · Slow-wave · Open-stub · Isolation · Return loss · Coupling · Insertion loss · *S*-parameters · Two-section branch-line · Microstrip line · Coupler

1 Introduction

The evolution of ultra-wideband (UWB) RF component devices has led to the rapid growth of interest recently; the devices have demonstrated characteristics that are impervious to operating frequency that is over a large bandwidth ($\geq$500 MHz). To enhance the bandwidth of the couplers, two-section branch-lines couplers are applied in this study, as the bandwidth can only be increased by the cascaded branch-line [1]. However, the cascade branch structure in the coupler is large in physical size. Hence, miniaturizing the size of the coupler is one of the requests for the hybrid coupler, whereby the miniature coupler has to retain good performance. Besides, it should be

K. Y. You (✉) · J. C. Chong
Faculty Engineering, School of Electrical Engineering, Universiti Teknologi Malaysia, 81310 Skudai, Johor, Malaysia
e-mail: Kyyou@fke.utm.my

M. F. A. Malek · S. Mirza
Faculty of Engineering and Information Sciences, University of Wollongong in Dubai, Dubai, United Arab Emirates

Y. S. Lee
Department of Electronic Engineering Technology, Faculty of Engineering Technology, Universiti Malaysia Perlis, Arau, Perlis, Malaysia

H. S. Saini et al. (eds.), *Innovations in Electronics and Communication Engineering*, Lecture Notes in Networks and Systems 107,
https://doi.org/10.1007/978-981-15-3172-9_7

noted that the space for hybrid coupler installation is restricted. Thus, small size of the hybrid coupler is required to avoid proximity among RF components. Decreasing the size of microstrip line of the regular coupler involves various modification methods to be implemented, such as slow-wave microstrip line structure [1, 3]. In this work, miniaturized two-section branch-lines couplers have been designated to reduce the size of the couplers. The design concept, simulation, and experimental results are discussed in Sects. 2 and 3.

2 Circuit Design and Fabrication

The regular and miniaturized two-section branch-line couplers were printed on RT5880 substrate ($\varepsilon_r = 2.2$, tan $\delta = 0.001$, and thickness = 0.787 mm). The regular two-section branch-line coupler is designed at 2.4 GHz of center frequency, f_o according to the provided characteristic impedance, $Z_o = 50\ \Omega$, $Z_1 = 96.9\ \Omega$, $Z_2 = 37.4\ \Omega$, $Z_3 = 63.7\ \Omega$, and electrical length, $\theta_1 = \theta_2 = \theta_3 = 90°$ in [4] as shown in Fig. 1a. The corresponding width, w and length, l of every microstrip branch-line on the substrate are estimated using TX-Line program calculator and are given as $w_1 = 0.76$ mm, $w_2 = 3.87$ mm, $w_3 = 1.66$ mm, $l_1 = 23.21$ mm, $l_2 = 40.50$ mm, and $l_3 = 22.53$ mm, respectively.

The provided dimensions of the regular branch-line coupler are simulated using Microwave Office (AWR) simulator. Then, the dimensions of the conventional coupler are further adjusted to optimize its performance and size by adding slow-wave structures on the microstrip branch-lines. The slow-wave structure is created by placing various sizes (wide and narrow) of open-stubs [5] in turn on the microstrip branch-lines, as illustrated in Fig. 1b. The dimensions of the miniaturized branch-line coupler are listed in Table 1.

3 Results and Analysis

The performances of couplers are observed based on simulated S-parameters, namely return loss $|S_{11}|$, insertion loss $|S_{21}|$, coupling $|S_{31}|$, isolation $|S_{41}|$, and phase differences, $\angle|S_{21}|-\angle|S_{31}|$. On the other hand, the fabricated regular (conventional) and miniaturized couplers are experimentally tested using Keysight E5071C network analyzer from 1 to 5 GHz. The simulated results of conventional couplers (see Fig. 1a) are obtained using Microwave Office (AWR) simulator for measurement validation. The results of measurements and simulations are fairly close to each other, as shown in Fig. 2.

Performance comparisons based on measurement results between the conventional and miniaturized couplers have been shown in Fig. 3. The center frequency, f_c of the miniaturized coupler has been shifted 0.3 GHz to high frequency compared to the conventional coupler. The measured phase differences, $\angle|S_{21}|-\angle|S_{31}|$

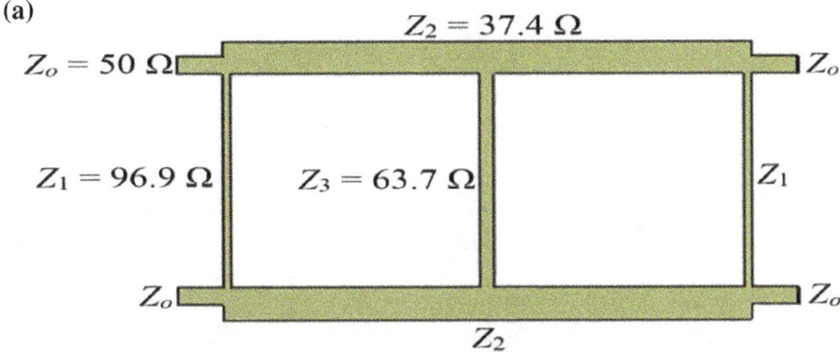

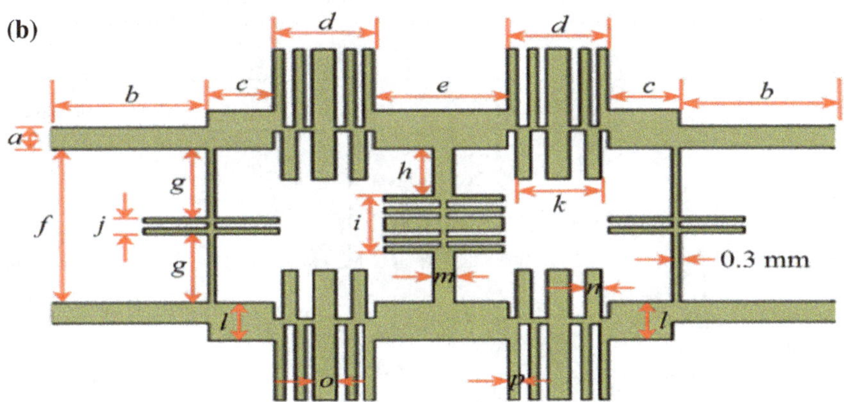

Fig. 1 **a** Regular and **b** miniaturized two-section branch-line couplers

Table 1 Dimensions of the miniature branch-line coupler in Fig. 1b

Dimensions (mm)			
Symbols	Values	Symbols	Values
a	1.9	*b*	8.2
c	3.5	*d*	5.2
e	7.0	*f*	14.1
g	6.3	*h*	4.4
i	5.2	*j*	1.5
k	4.4	*l*	3.5
m	1.2	*n*	1.8
o	1.2	*p*	0.5

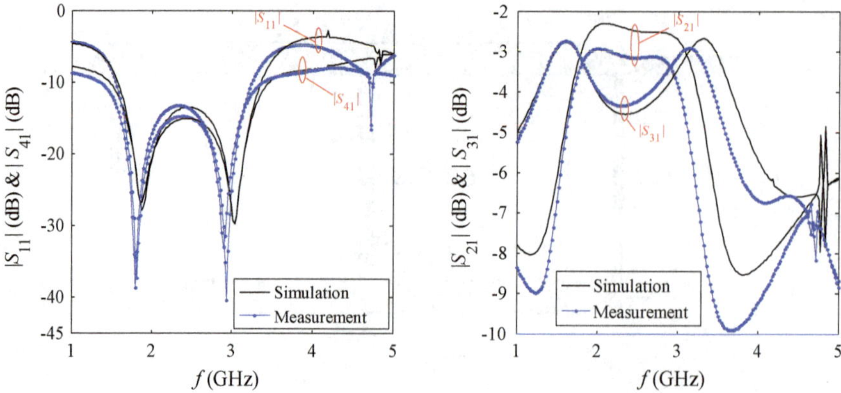

Fig. 2 Simulations and measurements results of the conventional two-section branch-line coupler

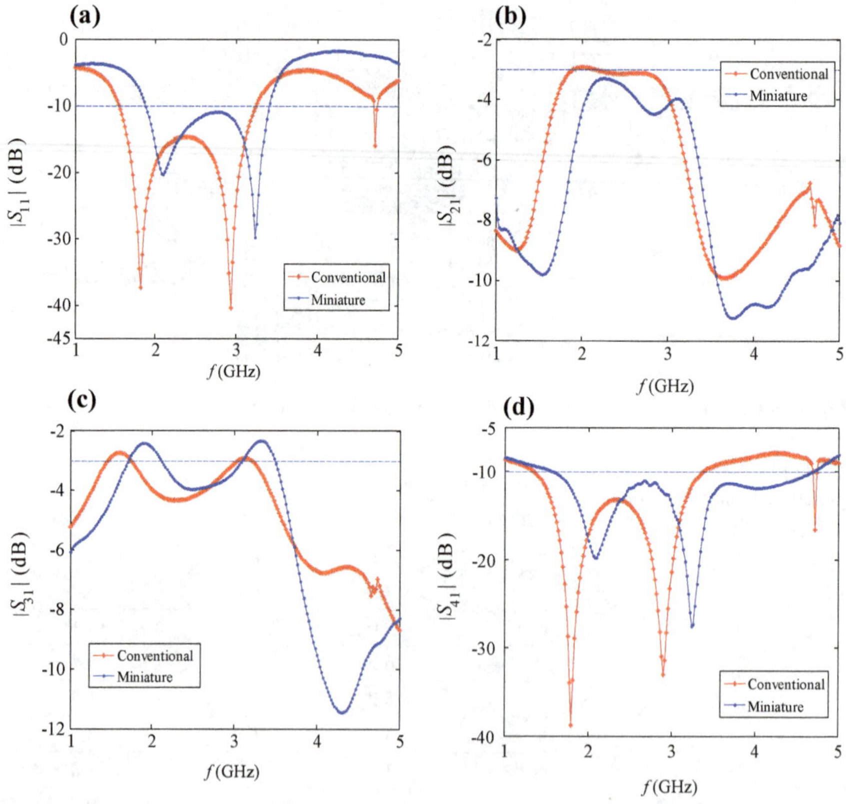

Fig. 3 Measured **a** return loss, $|S_{11}|$, **b** insertion loss, $|S_{21}|$, **c** coupling $|S_{31}|$, and **d** isolation, $|S_{41}|$ of regular (conventional) and miniaturized couplers

for the regular (conventional) and miniaturized couplers are 91.0° ± 3° and 89.0° ± 3°, respectively, over the operating bandwidth 1.6–3 GHz and 1.85–3.2 GHz, respectively, as shown in Fig. 4.

By applying capacitive slow-wave structures as illustrated in Fig. 1b, the physical length, l of the regular microstrip line in Fig. 1a can be decreased. The value of capacitance, ΔC and inductance, ΔL of the microstrip transmission line will be increased since the repeated discontinuous line structure comprises inductive and capacitive lines. In other words, this will cause a decrease in the value of the phase velocity, v for the transmission line. The smaller value of the v will lead to a smaller value of the electrical wavelength, $\lambda = v/f$. Hence, the desired quarter-wavelength at a particular center frequency can be achieved in a smaller physical length of the miniature coupler using a discontinuous microstrip line structure.

Figure 5 shows the circuit photograph of the regular and miniature two-section branch-line coupler. The dimensions, specifications, and performances (based on measurements) between the regular and miniature couplers are listed in Table 2.

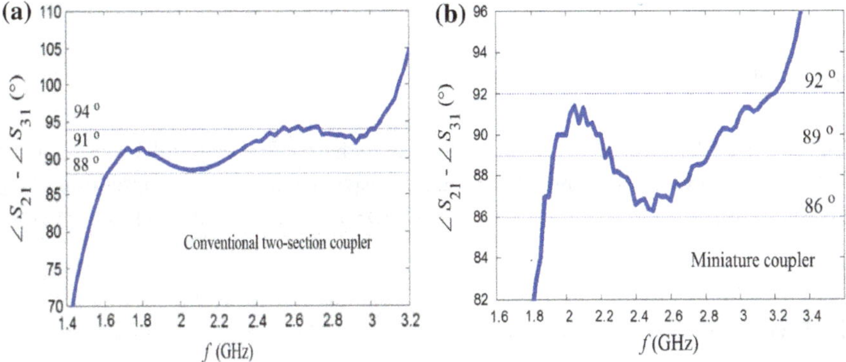

Fig. 4 Measured $\angle|S_{21}|–\angle|S_{31}|$ of **a** conventional and **b** miniaturized couplers

Fig. 5 Circuits photograph of the regular and miniature two-section branch-line couplers

Table 2 Specifications and performances of the regular (conventional) and miniature two-section branch-line couplers

Performance	Two-section branch-line coupler	
	Regular	Miniature
Area (width × length) mm^2	52.8 mm × 50 mm = 2381 mm^2	41 mm × 40 mm = 1640 mm^2
Size reduction (%)	–	38%
Bandwidth for $\lvert S_{11}\rvert < -10$ dB	1.55–3.24 GHz	1.88–3.41 GHz
Insertion loss, $\lvert S_{21}\rvert$	Min: −2.99 dB, Max: −6.64 dB Imbalanced: ±1.83 dB	Min: −3.29 dB, Max: −6.75 dB Imbalanced: ±1.73 dB
Coupling, $\lvert S_{31}\rvert$	Min: −2.67 dB, Max: −4.32 dB Imbalanced: ±1.65 dB	Min: −2.12 dB, Max: −3.92 dB Imbalanced: ±1.80 dB
$\angle\lvert S_{21}\rvert - \angle\lvert S_{31}\rvert$	Typical: 91° Imbalanced: ±3° (1.6–3.0 GHz)	Typical: 89° Imbalanced: ±3° (for 1.85–3.2 GHz)

4 Conclusion

The circuit of the two-section branch-line coupler has been successfully reduced up to 38% using an open-stub slow-wave structure. Also, the harmonic suppression of S_{21} and S_{31} at higher frequencies for the miniaturized branch-line coupler is improved.

References

1. Tang, C. W., Tseng, C. T., & Hsu, K. C. (2014). Design of wide Passband Microstrip Branch-Line Couplers with multiple sections. *IEEE Transactions on Components, Packaging and Manufacturing Technology, 4*(7), 1222–1227.
2. Sun, K. O., Ho, S. J., Yen, C. C., & van der Weide, D. (2005). A compact branch-line coupler using discontinuous microstrip lines. *IEEE Microwave and Wireless Components Letters, 15*(8), 519–520.
3. You, K. Y., Nadera, A. A., Chong, J. C., Lee, K. Y., Cheng, E. M., & Lee, Y. S. (2018). Analytical modeling of conventional and miniaturization three-section branch-line couplers. *Journal of Electrical Engineering and Technology, 13*(2), 858–867.
4. Muraguchi, M., Yukitake, T., & Naito, Y. (1983). Optimum design of 3-dB branch-line couplers using microstrip lines. *IEEE Transactions on Microwave Theory and Techniques, 31*(8), 674–678.
5. Krishna, V. V., Patel, B., & Sanyal, S. (2010). Harmonic suppressed compact wideband branch-line coupler using unequal length open-stub units. *International Journal of RF and Microwave Computer-Aided Engineering, 21*(1), 115–119

Implementation of Wireless Sensor Network Using Virtual Machine (VM) for Insect Monitoring

Mohammad Rashid Hussain, Arshi Naim and Mohammed Abdul Khaleel

Abstract Many works are being done and continued in the area of agriculture using wireless sensor network (WSN) to detect environmental monitoring, humidity monitoring, soil moisture, air quality (pollution) monitoring, insect monitoring using WSN. It is a valuable decision control and support tool for farmers but unfortunately in developing region where farmers are mostly using pest control legacy system, WSN technology will help them to use it on ad-hoc basis only when needed based on the decision by the sensor. WSN is a specific purpose computer containing hardware whose main components are memory, CPU, IO, registers, data and address bus, timer, sensor, computing logic and decision-making logic. After analyzing the image, WSN can decide to take appropriate action. It means WSN contains both control plane (signaling) and data plane (forwarding) coupled together. In this paper, we are proposing a virtual WSN (vWSN) which is based on virtualization and cloud computing technology, centrally managed device having flexible and configurable parameters with less maintenance and operational cost.

Keywords Virtual wireless sensor network (vWSN) · Virtual machine (VM) · Virtual appliances (VA) · Data plane · Control plane

1 Introduction

Many works are being done and continued in the area of agriculture using WSN to detect insect monitoring using WSN [1, 2]. The main objective of this paper is to provide alternate solution of WSN using virtual appliance vWSN, because both physical and virtual appliances can achieve the same result but differ in the sense that physical hardware (which resides on specially engineered dedicated hardware), virtual appliances run on servers. Instead of doing insect monitoring using WSN, the same result can be achieved by virtual WSN (vWSN). Numerous advantages are offered by virtual machines over physical counterparts due to the fact that it is

M. R. Hussain (✉) · A. Naim · M. A. Khaleel
College of Computer Science, King Khalid University, P.O Box: 9960, 61421 Asir, Abha, Kingdom of Saudi Arabia
e-mail: humohammad@kku.edu.sa

H. S. Saini et al. (eds.), *Innovations in Electronics and Communication Engineering*, Lecture Notes in Networks and Systems 107,
https://doi.org/10.1007/978-981-15-3172-9_8

dynamically configured. Memory, processor, virtual disk can be added or removed; Virtual machine (VM) can be shut down when not in use. Snapshot feature will help restore by saving the state of machine (server).

vWSN will be a boon for any developing country whose livelihood is mainly depending on agriculture which produces food for us without which life cannot be imagined; hence, no one can think of any development without life. It is necessary to apply technology in order to produce good quality food, grain, vegetables, etc. within limited resources (water, land, etc.). WSN or vWSN is an initiative in this direction to use the sensor instead of human intervention to decide the quantity of water, pesticide (based on the color change behavior of leaves), movement of insects, humidity, water level, pollution level, weather condition, etc. WSN is an approach to improve farming standards using technology which directly or indirectly improves the condition of farmer as well as every livelihood.

In this paper, we are trying to go one level up and introduce the use and benefit of virtual device for wireless sensor network which collects minimal information from fields and pass the information to central server which will be the main control and decision-making system. Using virtual device has many benefits like software control rather than hardware control system, because hardware control system mostly has fixed and hardcoded values with limited decision-making capability which cannot be extended. In order to add more features to a hardware-based device like WSN, device needs to be replaced whenever a new feature comes in or a new parameter needs to be added. However using virtual devices, the capability and feature can be added on runtime. These all happen using command-line configuration tool. In this paper, we are going to introduce or propose how virtual machine can be a good replacement for WSN which can be named as vWSN.

Virtual machine (VM) contains lightweight (only necessary component) operating system (OS) image file running single application. It is available from different vendors like Microsoft, Citrix and VMware but for this paper, we are concentrating on VMware solution.

vWSN is hardware virtualization of WSN. WSN is a hardware which contains components like memory, CPU, IO, controller and decision-making logic (routing table), sensor, timer, security and more. This means that WSN contains both data plane and control plane.

Using virtualization technique, the same can be shifted to virtual machine so that the physical devices (in the field) can be replaced by virtual devices which run as a normal application inside host machine operating system as a single process. Only picture or image is needed from the field which can be collected using high definition or high precision camera or any other similar devices. High precision camera can be placed somewhere in the field which can cover a broad area and image can be sent to the server where VMs are running in order to take decision, thus reducing the need of physical WSN. vWSN makes economical and manageable computing solution. Virtualization helps migrating from hardware architecture to software.

2 Materials and Methods

In this model, all the decision-making functionality will be done in virtual appliances (VA) configured using VM and deployed on VMware vSphere (Fig. 1).

As far as data collection is concerned, will be done using high precision camera and information will be sent to VA. Therefore, whatever WSN can do as a physical device sitting in the field can be done by vWSN deployed on VMware at a distant location. Any number of virtual WSN can be created as described earlier. Each vWSN can run image processing algorithm to take various decisions; based on that, the corresponding action will take place, for example, to sprinkle pesticide, urea, water, etc. in the field at required place.

2.1 Algorithm to Configure Physical WSN into VWSN

Step-1 VA with modified feature can be programmatically configured in VMware studio either using command-line interface (CLI) or web-based graphical interface.

Step-2 Select the OS and application package to include in VA, configure welcome screen, provide vendor information, etc.

Step-3 Configure virtual disk, boot script, virtual network card for the VA.

Step-4 In order to build and provision SSH connection, virtual appliance, and daemon is required to communicate. VA in VMware has .ZIP or .OVA format.

Step-5 Once VA is created, tested and verified, then may be distributed and ready for deployment on VMware hosted platform.

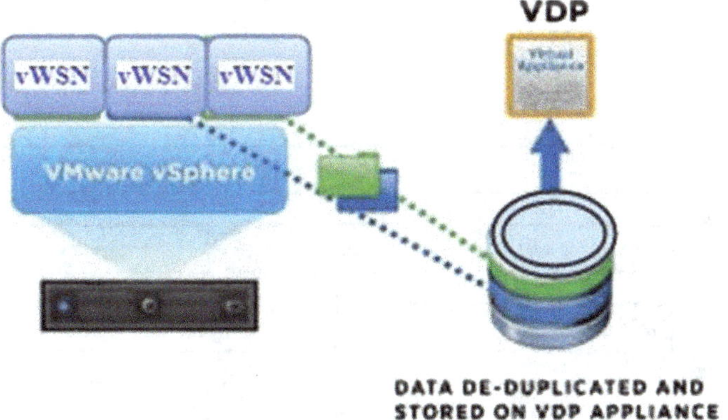

Fig. 1 VA configured using VM and deployed on VMware vSphere

Step-6 VA management infrastructure (VAMI) by VMware studio includes different components that help in managing VA to update service, to configure network and proxy setting.

2.2 Advantages of VWSN

Since vWSN is software device, administrator can add or reduce the size of memory, add or remove processor, add virtual disk or increase disk size, shut down the vWSN when not in use.

Also, the snapshot feature will help in saving the state of machine which can be restored later. The list is long but few of them including patching or upgrading application when new software version is available. vWSN can be easily managed remotely as well as locally by connecting to the server console. vWSN template can also create, modify or delete the device when not needed. This model is less expensive and more control over the device.

- Preconfigured virtual appliance (VA) will be deployed as VM running different application but similar configuration (example, some for insect monitoring, other for moisture level monitoring, humidity monitoring). VA is easy to use, can be created, deleted and modified with minimal cost using VMware technology.
- Flexible network configuration: integrating VA deployment in available network (IP address management) is the major task.
- Using VMware player VA can run and create its own VM.
- WSN mainly contains measuring (sensing) unit, communication unit and computing unit. The same can be created as a virtual appliance (VA) (Fig. 2).

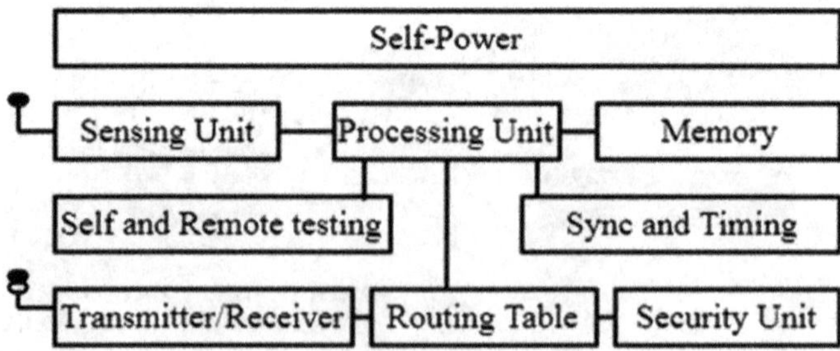

Fig. 2 Wireless sensor node location using GPS

3 Results and Discussion

Monitoring through image processing in VA.

Image processing operation can be done by any available image processing software like MATLAB. There are various proposed systems to detect pest densities by comparing images (ideal vs. infected) by continuous automatic monitoring without human intervention, hence minimizing human effort and error by simple, efficient and fast solution [3, 4].

4 Conclusions

In this paper, physical WSN replacement by virtual WSN using VMware approach is introduced. WSN is also a small computer having memory, CPU, IO, address and data bus, sensor, timer, etc. which may be interchanged by easily manageable virtual WSN which is simple, cost-effective, efficient and configurable. However, several improvements are required to actually reach the requirement. The main objective to present this paper is to separate or decouple control plane (decision making) and data plane where data collection and decision making are done by the same device. There are many scopes to elaborate vWSN in the field of load balancing, optimization, security as well as clustering techniques.

Acknowledgements The authors extend their appreciation to the Deanship of Scientific Research at King Khalid University for funding this work through Research Group Project under grant number R.G.P.1/166/40.

References

1. T. Jha, R. Hussain, wsn controlled insects monitoring: identification of onion thrips. IJCAR **6**, 5257–5260 (2014)
2. Dr. R. Hussain, G.C. D'Mello, in *WSN Controlled Insect Monitoring*
3. O. Lopez, M.M. Rach, H. Migallon, M.P. Malaumbres, A. Bonastre, J.J. Serrano, Monitoring pest insect traps by means of low-power image sensor technologies. Sensors **12**, 15801–15819 (2012)
4. N. Srivastav, G. Chopra, P. Jain, B. Khatter, Pest monitor and control system using WSN with special reference to acoustic device. ICEEE, 2013, Università di Genova, Facoltà di Economia via Francesco Vivaldi, 5 Genova Italy, January 16–18 2013 Giuseppe Cavaliere, University of Bologna (2013)
5. R. Hussain, J.L. Sahgal, P. Mishra, B. Sharma, Application of WSN in rural development, agriculture water management. IJSCE, **2**, 68–72 (2012)
6. R.P. Goldberg, Architecture of virtual machines, in *Proceedings of the Workshop on Virtual Computer Systems* (ACM Press, New York, NY, USA, 1973), Article, 74–112
7. M. Brown, VMware vCenter Server™ 6.0 Deployment Guide VMware, Inc. 3401 Hillview Avenue Palo Alto CA 94304 USA Tel 877-486-9273 Fax 650-427-5001 www.vmware.com (2015)

8. M. Durairaj, P. Kannan, A study on virtualization techniques and challenges in cloud computing. IJSTR **3**, 147–151 (2014)
9. G.H.E.L. de Lima, L.C. e Silva, P.F.R. Neto, WSN as a tool for supporting agriculture in the precision irrigation, in *ICNS, 2010, 6th International conference of Networking and Services*, Cancun, Mexico, 7–13 Mar 2010
10. Y. Liu, J. Zhang, in *Towards Continuous Surveillance of Fruit Flies Using Sensor Networks and Machine Vision*. Published by the department of networking and mobile computing Microsoft Qut Eresearch Centre in IEEE (2009)
11. B. Alka, G.P. Munkvold, Relationships of environmental and cultural factors with severity of gray leaf spot in maize. Plant Dis. **86**(10), 1127–1133 (2002)
12. X. Wu, S. Bao, Buildup and application of multi-factor spatial interpolation model in the monitoring and warning system for crop diseases and insect pests. Trans. Chin. Soc. Agric. Eng. **23**(10), 162–166 (2007)
13. R.D. Liu, N. Zhang, The monitoring and forecasting warning system of the meteorological disaster based on real-time data-driven in Yiyang. Anhui Agric. Sci. **38**(2), 788–789 (2010)
14. T. Zhong, W. Liu, C. Huang, Accelerate the construction of digital monitoring and early warning to provide support for the construction of a modern plant protection. China Plant Protect. Herald (12), 05–03 (2012)
15. X. Hu, Y. Li, Research of the agricultural disaster monitoring, early warning and prevention. Mod. Agric. Sci. Technol. **4**, 25–37 (2012)

Impact of Pointing Error on the Performance of 2-D WH/TS OCDMA in FSO

Bithi Mitra, Md. Jahedul Islam and Mir Mehedi Al Hammadi

Abstract Free space optical (FSO) communication systems are getting popularity because of high capacity and addressing high bandwidth but face some typical challenges during data transmission through free space. This paper represents an analysis of terrestrial FSO network under the impact of adverse weather conditions for a wavelength of 1.55 μm. The evaluation of pointing error for practical bit error rate (BER) is carried out through two-dimensional (2-D) wavelength-hopping/time-spreading (WH/TS) optical code division multiple access (OCDMA) technique. This approach enhances the capability of carrying a large number of simultaneous users. The BER performance is investigated in terms of proposed system parameters in the presence of an unavoidable effect of pointing error. Additionally, the optical link budget and the loss margins of the proposed system are evaluated with different pointing errors assuming it symmetrical for receivers and transmitters. It has been observed that the BER performance is highly dependent on the radial pointing errors of transmitter and receiver.

Keywords FSO · 2-D WH/TS OCDMA · Pointing errors · BER performance

1 Introduction

The development of several optical wireless technologies has emerged from the ever-increasing demand of bandwidth on communication systems. In this regard, FSO has become a very mighty and potential branch of unguided optical communication for modern age as well as for next-generation optical communication with expected data rate. It allows the transmission of signals representing data (binary data, i.e., ones or zeros) or voice via collimated beam of laser at higher data rates. The laser communication has become a worthy candidate with a communication range of the order of several thousand kilometers. Diode lasers with wavelength of 1.55 μm are available to operate at 2.5 Gbps and allow devices capable of 10 Gbps operation and

B. Mitra · Md. J. Islam (✉) · M. M. A. Hammadi
Department of Electrical and Electronic Engineering, Khulna University of Engineering & Technology, Khulna, Bangladesh
e-mail: jahed_eee@yahoo.com

H. S. Saini et al. (eds.), *Innovations in Electronics and Communication Engineering*, Lecture Notes in Networks and Systems 107,
https://doi.org/10.1007/978-981-15-3172-9_9

they permit transmitted power about 50 times more than 0.8 μm [1]. Thus, FSOs empower high data rates with compact and low-mass terminals, the use of license-free band, low immunity to radio-frequency (RF) spectrum, secured transmission, etc.

In wireless optics, the transceivers use a very narrow laser beam divergence to reduce the transmitter power and a very narrow field-of-view (FOV) telescope in receiver to minimize the noises arising from the background radiation [2]. Therefore, complexity arises during pointing the source and the destination and establishing communication between transceivers in order to provide a fixed alignment of line of sight (LOS) optics in entire communication period.

The optical code division multiple access (OCDMA) technique is brought forward by many researchers as it allows multiple access data transmission supporting multiple users to use same time slot and frequency domain. It minimizes the probability of error in system performance and possesses long haul access with high data rate. Additionally, the OCDMA with 2-D coding approach increases the number of simultaneous users with an easier network management and provides an inherent security and scalability [3]. Henceforth, the 2-D WH/TS scheme asserts an assigned code word for each user which is basically a set of binary $\{0, 1\}$ $m \times n$ matrices.

This paper represents an analysis of system BER performance in the presence of pointing error and adverse weather condition. In this analysis, 2-D WH/TS OCDMA coding approach is employed as the user address. It is found that the BER performance is affected by the change in pointing error angle for a given input power.

2 System Description

The block diagram of proposed 2-D WH/TS OCDMA technique over free space medium is illustrated in Fig. 1. It includes N number of transmitter and receiver to support N number of simultaneous users. The optical pulses of the broadband optical

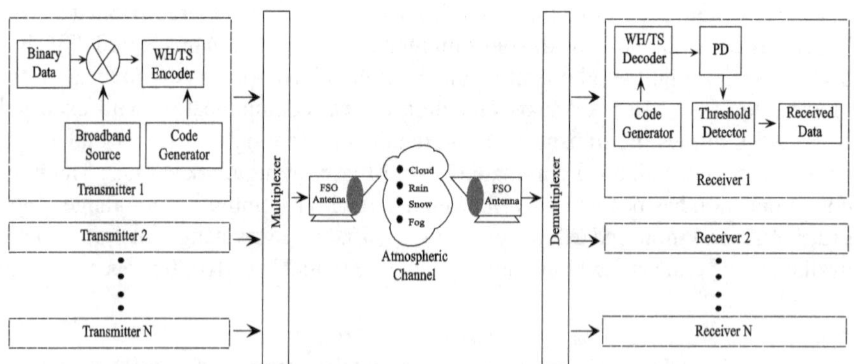

Fig. 1 Block diagram of 2-D OCDMA technique over free space medium

source are on–off keying (OOK) modulated by the binary data using the modulator. It is then encoded with the code word and formatted as 2-D WH/TS signal within an encoder. The output of each encoder is put together into the N:1 multiplexer and transmitted through transmit optics. On the contrary, the 1:N de-multiplexer receives a single input profile which is routed to one of the numerous output lines and sent them to the receivers.

In the receiver section, the signal from the de-multiplexer including multiple access interference (MAI) pulses is fed to the decoder in order to match its wavelength with the wavelength of the intended user. The chip pulses having time delays are canceled out in order to reconstruct the intended signal. Thus, the decoded signal carries both the desired signal and the MAI pulses. The photodetector then converts it into the electrical form. But at the PD, the undesirable optical beat interference (OBI) signals are appeared due to the beating among chip pulses with the same wavelengths. However, the intended data is quantified by the threshold detector and data is received in its original profile.

3 System Analysis

3.1 FSO Link Design

Considering the effect of pointing loss, the link budget equation is generally modeled as [1]:

$$P_r = P_t G_t G_r (\lambda / 4\pi L)^2 L_{geo} L_t L_r L_A \eta_t \eta_r \tag{1}$$

where P_t is the transmitted power (W), L is the distance between the transmitter and the receiver (m), λ is the actual wavelength of the beam (m), η_t and η_r are the optical efficiency of transmitter and receiver, respectively.

The geometric loss, L_{geo}, can be written as [4]:

$$L_{geo} = \{D_r / (D_t + \varphi L)\}^2 \tag{2}$$

where D_t and D_r are the aperture diameter of transmitter and receiver, respectively (m), ϕ is the divergence angle of laser beam (rad).

The transmitter antenna gain G_t is given by [1]:

$$G_t = 16/\varphi^2 \tag{3}$$

The receiver antenna gain G_r is given by [1]:

$$G_r = (\pi D_r / \lambda)^2 \tag{4}$$

The transmitter and receiver pointing loss L_t and L_r, respectively, are given by [4]:

$$L_t = e^{-G_t \gamma^2} \tag{5}$$

$$L_r = e^{-G_r \tau^2} \tag{6}$$

where γ and τ denote the transmitters' and receivers' radial pointing error angle (rad), respectively.

The atmospheric loss L_A is expressed as [4]:

$$L_A = e^{-\alpha L} \tag{7}$$

where α is loss occurred by atmospheric absorption (km^{-1}).

In this analysis, the absorption coefficient due to fog based on empirically measured data following Kruse model can be expressed as [4]:

$$\alpha = \alpha_f = \frac{3.912}{V}(\lambda/\lambda_0)^{-\delta} \tag{8}$$

where V is visibility (km) (at $\lambda = \lambda_0$), λ is the actual beam wavelength (nm), λ_0 is a reference wavelength (nm) while estimating V. δ can be found from the assumption below [5]:

$$\delta = \begin{cases} 0.585V^{1/3}, & \text{for } V < 6\text{ km} \\ 1.3, & \text{for } 6\text{ km} < V > 50\text{ km} \\ 1.6, & \text{for } V > 50\text{ km} \end{cases} \tag{9}$$

3.2 BER Performance in OCDMA System

The proposed system is structured through 2-D OCDMA technique having 2-D signature codes consisting of a time-spreading pattern and a wavelength-hopping pattern. Let at a frequency $f_{i,}$ k_i be the number of pulses due to MAI, thus the total signal current (desired and MAI currents), $I_b(b \in (0, 1))$ is given by [6]:

$$I_b^{2\text{-D}} = \Re\left(bWP_r + \sum_{i=1}^{W}\sum_{j=1}^{k_i} P_r\right) \tag{10}$$

The shot noise $i_{sb}^2 = (b \in (0, 1))$ from PD is given by [6]:

$$\left\langle i_{\text{sb}}^{2}\right\rangle = 2\text{e}B_{\text{e}}I_{\text{b}} \tag{11}$$

The thermal noise i_{th}^{2} is expressed as:

$$\left\langle i_{\text{th}}^{2}\right\rangle = 8\pi C T_{\text{n}} B_{\text{e}}^{2} k_{\text{B}} \tag{12}$$

The total variance of OBI $i_{\text{OBIb}}^{2}(b \in (0, 1))$ can be determined as [6]:

$$\left\langle i_{\text{OBIb}}^{2(2-D)}\right\rangle = 2B_{\text{e}}\tau_{\text{c}}\Re^{2}\left\{P_{\text{r}}^{2}\sum_{i}^{W}\left(bk_{i} + \binom{k_{i}}{2}\right)\right\} \tag{13}$$

where $k_i = \left(k_0, k_1, \ldots, k_{p_{\text{s}}-1}\right)$ is the interfering pulses, e is electron charge (C), B_{e} is electrical bandwidth of PD (Hz), k_{B} is Boltzmann constant (J K^{-1}), T_{n} is noise temperature of receiver (K), C is receiver capacitor (F), $\Re$ is responsively (A/W), τ_{c} is time constant (sec) = $1/B_{\text{o}}$.

The total noise variance can be summarized by:

$$\left\langle i_{\text{nb}}^{2}\right\rangle = \left\langle i_{\text{OBIb}}^{2}\right\rangle + \left\langle i_{\text{sb}}^{2}\right\rangle + \left\langle i_{\text{th}}^{2}\right\rangle \tag{14}$$

Considering the prime sequences p_{s} is less than prime number p_{h}, the total number of wavelengths similar to a pair of two codes, μ_{λ} can be calculated by [6]:

$$P_{\text{MAI}} = \mu_{\lambda}/p_{\text{s}}^{2} \tag{15}$$

$$\mu_{\lambda} = \frac{1}{\left\{\begin{matrix} p_{\text{h}} \\ p_{\text{s}} \end{matrix}\right\}} \times \left\{\binom{p_{\text{h}}-1}{p_{\text{s}}} \times \frac{p_{\text{s}}(p_{\text{s}}-1)}{p_{\text{h}}-2} + \binom{p_{\text{h}}}{p_{\text{s}}} \times \frac{(p_{\text{s}}-1)(p_{\text{s}}-2)+(p_{\text{s}}-2)}{p_{\text{h}}-2}\right\} \tag{16}$$

In our analysis, we assume N as the number of simultaneous users, i.e., interfering users are $(N-1)$. k users out of probable $(N-1)$ interfering users transmit the bit "1." So, the probability of errors can be determined by:

$$\text{BER} = \left\{\sum_{k=1}^{N-1}\binom{N-1}{k}2^{-(N-1)}\sum_{j=1}^{k}\binom{k}{j}P_{\text{MAI}}^{j}(1-P_{\text{MAI}})^{k-j}\right\} \times Q\left(\frac{I_1 - I_0}{i_{nI} + i_{n0}}\right) \tag{17}$$

where

$$Q(x) = \frac{1}{2}\text{erfc}\left(x/\sqrt{2}\right) \tag{18}$$

4 Results and Discussion

The mathematical interpretations of the previous section are used for the evaluation of the BER performance in terms of proposed system parameters. In numerical calculations, the maximum number of simultaneous users = 50, the transmission wavelength of 1.55 μm, the beam divergence angle = 1 mrad, the data rate = 1 Gigabit/s are assumed.

Fig. 2a represents the plot of BER versus visibility for different pointing error angle considering the number of user as 10 and a constant transmitted power of 23 dBm. Due to the increase in fog density, the atmospheric contents absorb some of the optical signal energy and convert it into the heat energy. The visibility increases as the contents of water molecules of fog reduce resulting in a decreasing the probability of error. But with the increase in pointing error angle, the performance degrades. For different pointing error angle say, 0.22, 0.18, and 0.11 mrad, the BER values found to be 1.44×10^{-6}, 1.15×10^{-8}, and 1.73×10^{-8}, respectively, pointing a visibility range 2 km. Fig. 2b refers to the plot of BER versus transmission length at different pointing error angle considering the number of simultaneous user as 10 and the transmitted power per bit $P_t = 23$ dBm. To maintain a BER value of 10^{-10}, the transmission length is found to be 119.35 m for a pointing error 0.22 mrad and it is 212.90 and 335.48 m for the pointing error 0.18 mrad and 0.11 mrad, respectively. That means pointing of transmitter and receiver at correct point is crucial in this case to support a longer transmission length. Fig. 3a shows the plot of BER versus transmitted power at various pointing error angle with 10 simultaneous users and 500 m link distance. It is found that system performance is poor with low transmitted power and high pointing error, i.e., the transmitted power requirement is more in case of misalignment of transmitter and receiver resulting in an increased probability of error. In Fig. 3b, the plot of BER versus number of simultaneous users at different pointing errors has been shown when the transmitted power is $P_t = 23$ dBm per bit. The BER performance becomes better with low value of error angle pointing

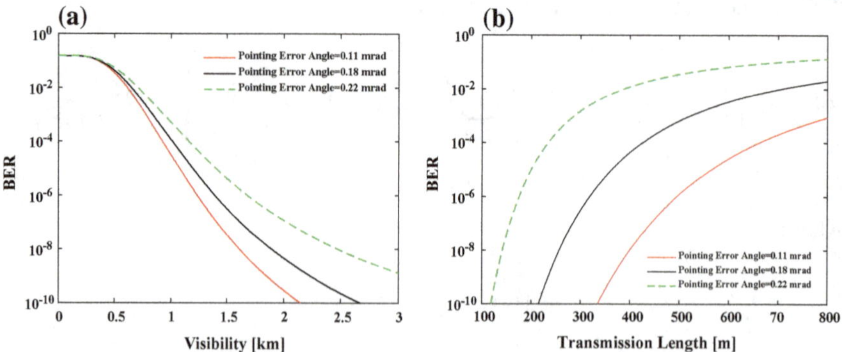

Fig. 2 **a** BER versus visibility and **b** BER versus transmission length at different pointing error angle assuming the number of simultaneous user = 10

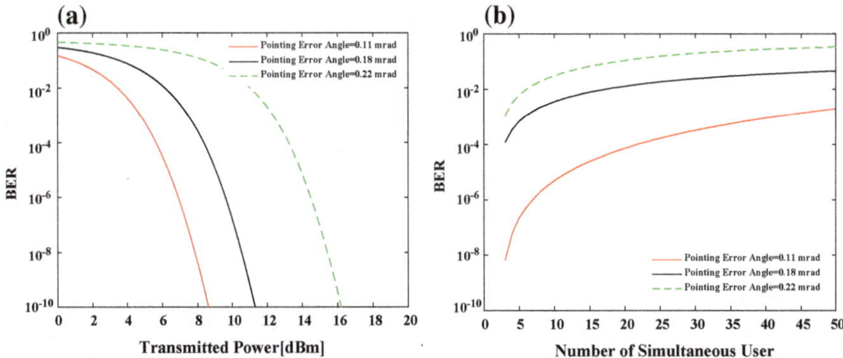

Fig. 3 **a** BER versus transmitted power, **b** BER versus number of simultaneous user at different pointing error angle assuming the number of simultaneous user = 10

the correct direction and performance deteriorates when the number of simultaneous users increases due to the adverse influence of MAI and OBI. So, the numerical explications summarize that better BER performance can be achieved by the proper alignment of LOS optics between the transceivers.

5 Conclusion

In this paper, the performance of 2-D OCDMA technique is analytically discussed in the presence of pointing errors and adverse weather conditions over FSO channel. The fog attenuates the laser beam propagating through the free space by absorbing and scattering the optical signal. As a result of that, the performance degrades, i.e., BER increases as the visibility decreases due to the increase in fog density. Also, with the increase in pointing errors of transceivers, the value of BER increases. The numerical evaluation shows the severe impact of the combined effect of pointing error and adverse weather condition on the performance of free space OCDMA communication system.

References

1. A.K. Majumdar, J.C. Ricklin, in *Free-Space Laser Communications: Principles and Advances*. (Springer Science & Business Media 2, 2010)
2. D. Kedar, S. Arnon, Urban optical wireless communication network: the main challenges and possible solutions. IEEE Opt. Commun. Suppl. IEEE Commun. Mag. S1–S7 (May 2004)
3. A.T. Pham, H. Yashima, Performance enhancement of the 2-D wavelength-hopping/time-spreading synchronous OCDM system using a heterodyne detection receiver and PPM signaling. OSA J. Opt. Networking **6**(6), 789–800 (2007)

4. M.B. EL Mashade, A.H. Toeima, M.H. Aly, Receiver optimization of FSO system with MIMO technique over log-normal channels. Optoelectron. Adv. Mat. **10** (2016)
5. F. Nadeem, T. Javornik , E. Leitgeb, V. Kvicera, G. Kandus, Continental fog attenuation empirical relationship from measured visibility data. J. Radio Eng. **19**(4) (2010)
6. N.T. Dang, A.T. Pham, Performance analysis of incoherent multi-wavelength OCDMA systems under the impact of four-wave mixing. Opt. Express **18**, 9922–9933 (2010)

Millimeter-Wave AWR1642 RADAR for Obstacle Detection: Autonomous Vehicles

Nalini C. Iyer, Preeti Pillai, K. Bhagyashree, Venkatesh Mane, Raghavendra M. Shet, P. C. Nissimagoudar, G. Krishna and V. R. Nakul

Abstract To know the nature of the surrounding is very crucial, and the profile of the terrain geometry must be known to decide the motion path in autonomous cars. The obstacles must be detected along with their coordinates with the vehicle as origin (reference) and relative velocity to determine a collision free path. Most autonomous vehicle manufacturers depend extensively on LIDAR such as Uber and Google, whereas Tesla uses RADAR as a primary sensor. RADAR is relatively less expensive and works equally well in all weather conditions such as fog, rain, snow, and dust. It can also determine relative traffic speed using Doppler frequency shift which is not possible in case of LIDAR. Hence, RADAR is a viable solution for autonomous cars considering above parameters, and it provides good results when used with secondary sensors such as ultrasonic and camera. This paper proposes a methodology for testing RADAR sensor AWR1642 in various configurations with respect to distance and angle of coverage. The testing of the above RADAR is carried out on real vehicle. The results obtained verify that RADAR provides accurate results

---

N. C. Iyer (✉) · P. Pillai · K. Bhagyashree · V. Mane · R. M. Shet · P. C. Nissimagoudar · G. Krishna · V. R. Nakul
KLE Technological University, 31 Hubli, India
e-mail: nalinic@bvb.edu

P. Pillai
e-mail: preeti_pillai@bvb.edu

K. Bhagyashree
e-mail: bhagyashree@bvb.edu

V. Mane
e-mail: mane@bvb.edu

R. M. Shet
e-mail: raghu@bvb.edu

P. C. Nissimagoudar
e-mail: sn_asundi@bvb.edu

G. Krishna
e-mail: krishna125g@gmail.com

V. R. Nakul
e-mail: nvraichur@gmail.com

H. S. Saini et al. (eds.), *Innovations in Electronics and Communication Engineering*, Lecture Notes in Networks and Systems 107,
https://doi.org/10.1007/978-981-15-3172-9_10

within 50 m range for medium-sized objects with appropriate tuning of parameters such as best range and best range resolution.

Keywords RADAR · LIDAR · Sensor · Obstacle

1 Introduction

Detecting obstacles surrounding an autonomous car is necessary to facilitate easy navigation. There are numerous technologies for obstacle detection such as camera, sonar, LIDAR, and RADAR.

RADAR is used to measure radial distance and velocity of remote objects very accurately. Additional angular information can be obtained using multiple transceiver channels. RADAR plays a critical role in detecting obstacles under various environmental conditions like rain, dust, and sunlight. Due to these reasons, RADAR is one such promising technology which is successfully used in many driver assistance systems. Frequency bands widely used for automotive applications are 24 and 77 GHz. Working with 77 GHz frequency has many advantages as follows:

- This band allows small-sized antenna for given beamwidth and gives good angular resolution with small sensor size.
- The combinations of high transmit power and high bandwidth are not allowed at 24 GHz which, on the other hand, is allowed at 77 GHz.
- Also the atmospheric absorption interference will be less.

Because of all these advantages, RADAR AWR1642 automotive radar is chosen for the study.

2 Overview of Automotive RADAR

With the transition from driver assistance to autonomous driving, the number of sensors for obstacle detection with accuracy is critical. Identifying the exact range, angle, and knowing if the obstacle is static or moving helps in taking more informed decisions [1]. These are some of the significant properties that an autonomous vehicle must possess in order to practice safe driving in various environments [2].
TI's **AWR1642BOOST** is used in the study for obstacle detection and ranging. The AWR1642 is an automotive **FMCW RADAR sensor** [3] operating in the frequency band of 76–81 GHz. It is built on low-power 45-nm process because of which 2TX, 4RX could be implemented. It is integrated with phase-locked loops (PLL), A2D converters, and also TI's high-performance C674x DSP for the signal processing [4]. An ARM R4F processor is included for calibration, control, and radio configuration.

3 Experimentation

Autonomous functions like Adaptive Cruise Control (ACC), lane departure warning, lane change assistance, watching the traffic to assist vehicle overtake, cross-traffic alert, collision warning, etc., are incorporated in majority of autonomous vehicles through RADAR sensors. The functions listed above demand for RADARs of different range of detection, and based on their detection range, they are classified as follows:

- Long-range radar (LRR),
- Medium-range radar (MRR),
- Short-range radar (SRR).

The focus of our study is obstacle detection using RADAR in autonomous vehicles which is achieved through short-range RADAR module. In this paper, we are using **single-chip frequency-modulated continuous wave (FMCW) RADAR AWR1642**. The configuration of the module and the methodology followed to extract and visualize data from AWR1642 RADAR module is discussed below.

3.1 AWR1642 RADAR Configuration

RADAR sensors need to be configured for the required range of detection for selected application [5]. The module AWR1642 used in the study supports three different configurations to cater to the above requirements. They are

1. Best range resolution,
2. Best velocity resolution,
3. Best range.

Range resolution and maximum unambiguous range for each configuration are given in Table 1.

The antenna selection option for each of the RADAR configurations is as given below:

Table 1 Configuration settings of AWR1642

Configuration	Range resolution (cm)	Maximum unambiguous range (m)
Best range resolution	3.9–4.7	3.95–10.22
Best range	6.1–97.7	5–50
Best velocity resolution	17.8–53.4	6.4–12.79

- 4Rx,2Tx (15°),
- 4Rx,1Tx (30°),
- 2Rx,1Tx (60°),
- 1RX,1Tx (none).

Configuration settings are done based on the requirement of an application. For autonomous parking assistance applications, there is a need to accurately guide the vehicle during parking as the speed of the vehicle is slow. Hence, the best velocity resolution configuration suits the application as the velocity resolution and accuracy are critical parameters. Similarly, in case of detection of closely placed objects, the best range resolution suits the application.

The study shows the usage of AWR1642 RADAR module for these different configurations. The experimentation is carried out by mounting the AWR1642 RADAR module on the real vehicle to test different configuration settings for obstacle detection. The results of each configuration are discussed in Sect. 4.

3.2 Data Extraction and Visualization

For visualization, the AWR1642 module is supported by GUI which displays the number of objects detected with their identification numbers, range, and speed dynamically on a graph. For object detection, processing this data from the AWR1642 module is necessary, and this requires CAN interface for automotive applications. One of the feasible methods is to provide an off-board CAN interface as the on-board interface of Rev-A AWR1642 module requires hardware interception [1]. The methodology adopted to facilitate this data extraction from AWR1642 is summarized below.

A controller is included as an interceptor between the RADAR module and the decision unit connected through a USB port as shown in Fig. 1.

The RADAR module used in the study has only one serial port, so configuration commands are sent on the same port by selecting 115,200 baud rate. The data from the sensor is received at 921,600 baud rate. The procedure followed to load the configuration file, and extract data is shown in Fig. 2. Python script extracts the

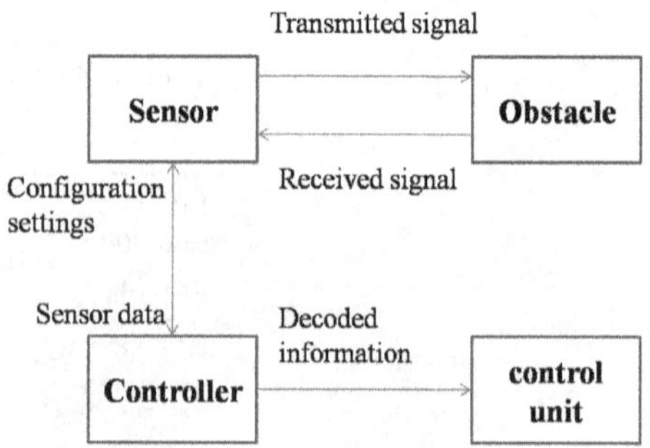

Fig. 1 Functional schematic for data extraction

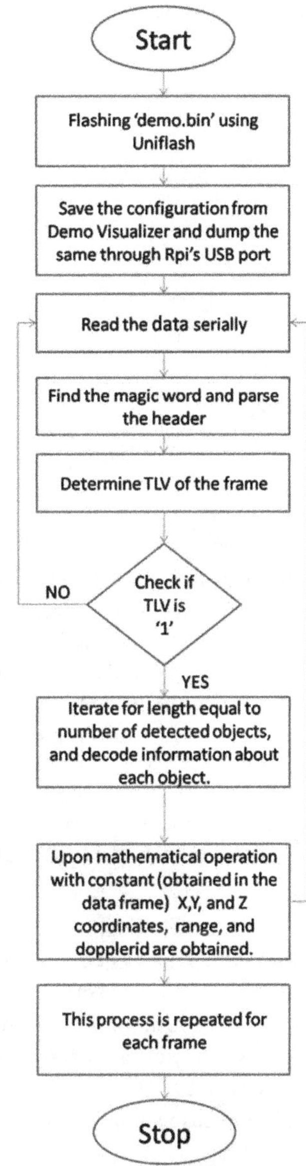

Fig. 2 Data extraction flowchart

data from AWR1642 serially. The data type obtained is **Hex String which** contains binary-coded decimal in a string format. The data type is unpacked to obtain usable information. It is decoded using the **Frame Structure** provided in the user guide [3] and is shown in Fig. 3. Using a pointer from beginning of the frame, each parameter of the frame such as Header and Detected Object is parsed. In the Detected Object field, parameters such as object coordinates are to be multiplied with specific constants to

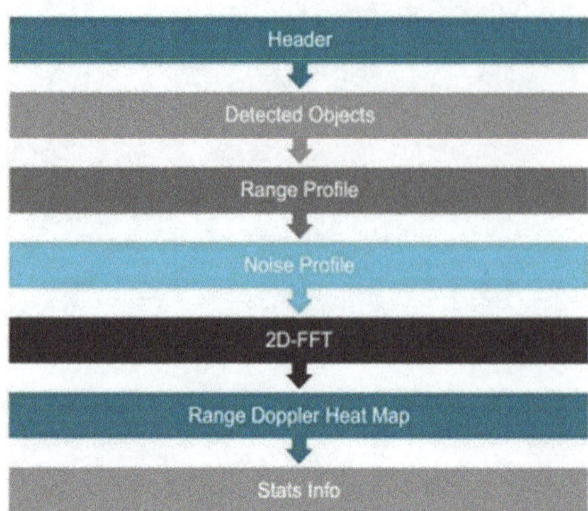

Fig. 3 Frame format

obtain values in SI units. The number of iterations depends on number of detected objects which are available in the header. Iterating through each object's description distance, angular position, relative velocity, etc., can be obtained. Averaging the values over time can help make better decisions.

The length of the frame is always a multiple of **32 bytes**. The frame format received by the RADAR sensor is illustrated in Fig. 3. In this frame format, the first two fields, i.e., Header and Detected Object, are processed for data extraction.

4 Results and Discussion

For a given application, a RADAR sensor needs to be configured for its best performance. This can be achieved based on the antenna settings along with best range settings. Best range settings can be achieved through best range resolution and range detection threshold. Based on the type of object to be detected, the range detection threshold is set, as the strength of the reflected signal is directly proportional to the size of the object and inversely proportional to its distance. This paper presents different case studies to demonstrate AWR1642 RADAR sensor performance.

Case 1: Best range resolution,
Case 2: Best range.

The results of Case 1 and Case 2 are shown in Figs. 6 and 7, respectively.

Figure 4 shows the experimental setup. Figure 5 shows the raw data received from the AWR1642 RADAR module displayed on the serial monitor and the unpacked and processed data from the controller.

The results of best range resolution are shown in Fig. 6 for range detection threshold varied from 15 to 19 dB. It can be observed form Fig. 6a that all the three objects

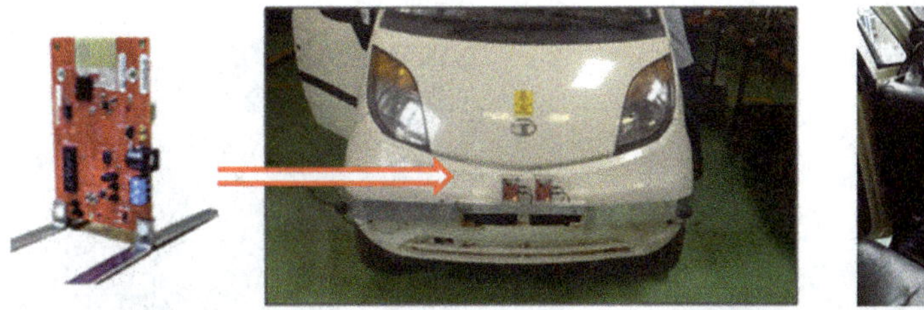

Fig. 4 Experimental setup

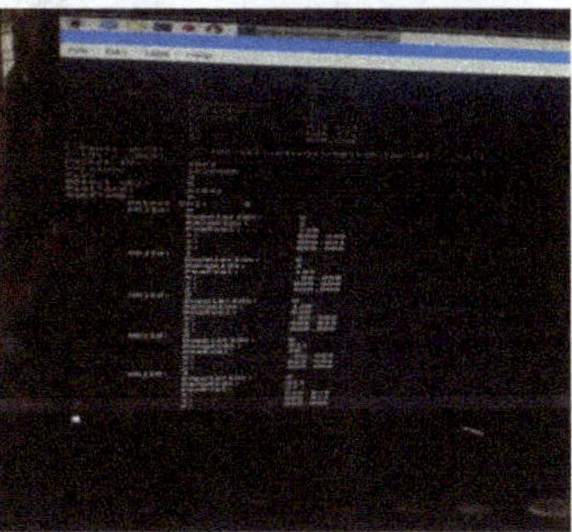

Fig. 5 Raw RADAR data (left), unpacked and processed data (right)

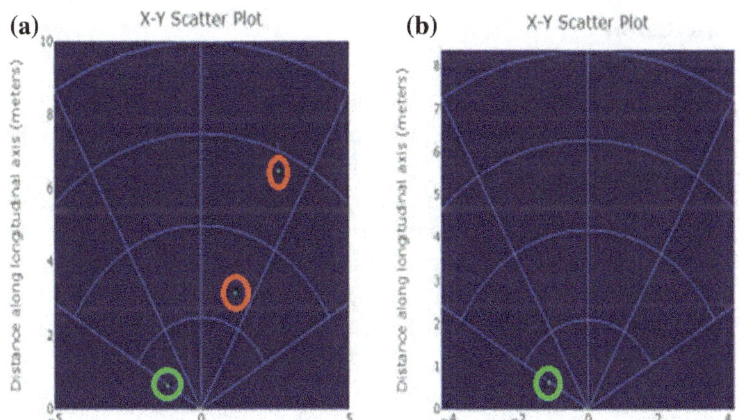

Fig. 6 Best range resolution for 15 and 19 dB threshold

are detected. From Fig. 6b it can be seen that only one object is detected among the three objects, i.e., only the closest object is detected because the reflected signal strength from that object highlighted with green is greater or equal to **19 dB**.

The results of best range configuration are shown in Fig. 7. In Fig. 7 (left), there are two objects (the bike and person) in the same bearing (same angle). The first

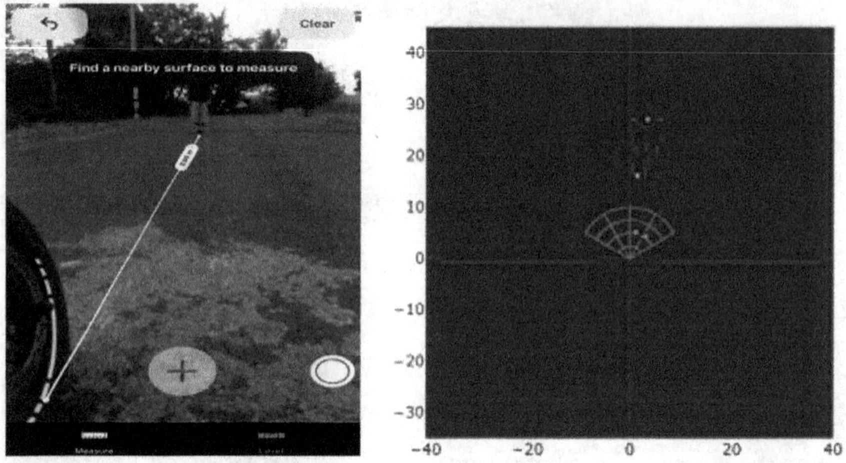

Fig. 7 Best range configuration test case

object is at 8 m from the origin (the device); another object is at 19 m which is 11 m away from the first target. Both the objects are detected **in best range configuration** which has a resolution range of 6.1–97.7 cm. Other objects in the vicinity which are big enough to reflect a signal of 15 dB or greater will also be detected.

5 Conclusion

This paper discusses the RADAR technology with 77–81 GHz that is best suited for autonomous vehicles. The raw data received from the module was unpacked and processed for decision making. The paper also discussed an alternative to extract data from the AWR1642BOOST Rev-A module without the CAN hardware interception. A vehicle is mounted with the module for experimentation to extract the real data. The results obtained are convincing and can be used further for decision making in autonomous vehicle by fusing with other sensors.

References

1. C. Blanc, R. Aufrère, L. Malaterre et al., Obstacle detection and tracking by millimeter wave RADAR. IFAC Proc. Vol. **37**, 322–327 (2004). https://doi.org/10.1016/s1474-6670(17)31996-1
2. M. Bertozzi, L. Bombini, P. Cerri, et al., Obstacle detection and classification fusing radar and vision, in *2008 IEEE Intelligent Vehicles Symposium* (2008). https://doi.org/10.1109/ivs.2008.4621304
3. Wolff CIn: Radar Basics. http://www.radartutorial.eu/02.basics/FrequencyModulatedContinuousWaveRadar.en.html
4. AWR1642 Single-Chip 77- and 79-GHz FMCW Radar Sensor datasheet (Rev. A) (2019)
5. AWR1642 Device Errata (Rev. A) (2019)

Performance Evaluation of Various Modulation Techniques for Underwater Wireless Optical Communication System

Mahin Akter, Md. Jahedul Islam and Mir Mehedi Al Hammadi

Abstract In this paper, the bit error rate (BER) performance of a single-channel underwater wireless optical communication (UWOC) system for various modulation techniques is evaluated with analytical approach. Operating wavelength of 532 nm is selected, and Si APD is used in the receiver section for the optoelectronic conversion purpose. The BER performance of different modulation techniques for UWOC system is analyzed considering different water types, e.g., pure sea water, clear ocean water, and coastal ocean water. This analysis indicates the best performance for 4-QAM in pure sea water.

Keywords UWOC · SNR · LOS · BER · Modulation techniques

1 Introduction

Underwater communication has become very important topic day by day. It has great significance for both commercial and research purposes. The scientific community-, industry-, and military-related applications demand it badly because of playing a vital role in monitoring of pollution, oil production control, climate change, offshore explorations, oceanographic research, etc. [1]. There are several ways of employing such communication system. The most commonly used technology is acoustic communication. High transmission loss, very low bandwidth, high latency, time-varying multipath propagation, and Doppler spread impede the desired application of acoustic communication [2]. The RF communication is not applied in aquatic medium due to strong attenuation. To overcome these hurdles, researchers introduced optical communication. In case of stationary and large devices, fiber optics may be used

The original version of this chapter was revised: The author affiliation "Department of Electrical and Electronic Engineering, Khulna University of Engineering & Technology, Khulna, Bangladesh" has been updated. The correction to this chapter is available at https://doi.org/10.1007/978-981-15-3172-9_75

M. Akter · Md. J. Islam (✉) · M. M. Al Hammadi
Department of Electrical and Electronic Engineering, Khulna University of Engineering & Technology, Khulna, Bangladesh
e-mail: jahed_eee@yahoo.com

H. S. Saini et al. (eds.), *Innovations in Electronics and Communication Engineering*, Lecture Notes in Networks and Systems 107,
https://doi.org/10.1007/978-981-15-3172-9_11

to obtain high data rates, but it requires superior technology and maintenance. For mobile platforms, a good initiative is the application of wireless optical link. Thus, the concept of UWOC has been emerged. Optical wave has mainly two types of properties. They are inherent and apparent properties of optic wave through the medium [3, 4]. Among them, inherent optical properties, namely absorption, scattering, and attenuation, impede the underwater optical communication significantly. The total effects of absorption and scattering are combined and indicated by attenuation coefficient. The attenuation coefficient has dependency on wavelength. The wavelength around 450–550 nm, that is, blue–green portion of the visible spectrum, is the minimum attenuation window for UWOC [1, 3]. Light source of 532 nm wavelength has the lowest attenuation in this window which is also industrially available [1]. Water type also has impact on efficient propagation of light in underwater channel as each water type has different amount of impurity in water [1, 4].

Among the various communication links for the UWOC system, the line-of-sight (LOS) link configuration is a direct and unobstructed point-to-point communication [5]. Moreover, there are different modulation techniques available which are generally used for efficient transmission of data in UWOC systems. Each technique provides different system performance in different medium as well as medium qualities. In this paper, several modulation techniques (especially, DPIM, DH-PIM, and M-PAM) are considered for the proposed UWOC system of which the performance characteristics have not been investigated before. To the best of our knowledge, the performance of these modulation techniques for UWOC in different water types is a novel analysis. In this paper, the system BER performances of different modulation techniques for the UWOC system are analyzed, where three water types [4], i.e., pure sea, clear ocean, and coastal ocean water, are considered.

2 System Description

Figure 1 shows the block diagram of the UWOC system consisting of two main sections, i.e., the transmission section and the reception section. In the transmitter, a user's binary data is modulated by an optical modulator which controls the output of the optical source. According to the control signal from the optical modulator, the optical source converts the electrical signal into optical signal and directs it toward the water channel. In this system, the wavelength of the optical beam is 532 nm.

Fig. 1 Block diagram of the underwater wireless optical communication system

During the propagation, the optical signal gets attenuated along the channel length. This optical signal is incident on the APD receiver which is then converted into electrical signal multiplied by the avalanche gain. By demodulating this electrical signal in receiver, the desired output signal is obtained.

3 System Analysis

3.1 Link Design

Power of the optical signal at the receiver end can be determined by path loss models. The received optical power P_R for LOS link can be represented as [1],

$$P_R = P_t \eta_t \eta_r \frac{A_R \cos(\theta)}{2\pi L^2[1 - \cos(\theta_d)]} \exp\left[-c(\lambda)\frac{L}{\cos(\theta)}\right] \quad (1)$$

where P_t is the optical power at the transmitter, $c(\lambda)$ is the total loss coefficient summing the optical absorption coefficient $a(\lambda)$ and the optical scattering coefficient $b(\lambda)$ in underwater channel [1, 3, 4], η_t and η_r are the transmitter and receiver optical efficiencies, L is the transmission distance of the optical beam, θ_d is the divergence angle of the optical beam, θ is the transmitter inclination angle from the axis connecting the transmitter-receiver pair, and A_R is the aperture area of the receiver. When a particular wavelength is considered, the variation of the water types produces different values of the total loss coefficient [4]. Pure sea water has the minimum impurity, while coastal ocean water has the maximum impurity dissolved in water.

3.2 The BER Calculation

At the APD, the average photocurrent is given by [5],

$$I_P = \Re P_R = M \Re_0 P_R \quad (2)$$

The responsivity $\Re$ for Si APD is 75 A/W. The gain is denoted by M which is considered to be 150. The total noise current is I_N, and its variance can be expressed as [5],

$$\langle I_N^2 \rangle = \langle I_{sh}^2 \rangle + \langle I_D^2 \rangle + \langle I_{th}^2 \rangle \quad (3)$$

I_{sh} and I_D are the shot noise current and dark current of APD whose variances are [5]

$$\langle I_{\mathrm{sh}}^2 \rangle = 2\mathrm{e}I_{\mathrm{p}}M^2F\Delta f \tag{4}$$

$$\langle I_{\mathrm{D}}^2 \rangle = 2\mathrm{e}I_{\mathrm{D}}M^2F\Delta f \tag{5}$$

Here, *e* indicates an electron charge, Δf is the bandwidth, and F is the noise figure of the APD. Here, $F = 0.5$ and $\Delta f = 5$ GHz. The signal-to-noise ratio is given by [5],

$$\mathrm{SNR} = \frac{\langle I_{\mathrm{P}}^2 \rangle}{\langle I_{\mathrm{N}}^2 \rangle} = \frac{I_{\mathrm{P}}^2}{2\mathrm{e}(I_{\mathrm{P}} + I_{\mathrm{D}})M^2F\Delta f + 4\kappa T\Delta f/R} \tag{6}$$

For optical communications, different types of modulation techniques can be implemented. The BER for various types of modulation techniques that are considered in this study are following:

BER when Binary Phase Shift Keying (BPSK) is considered [6]:

$$\mathrm{BER}_{\mathrm{BPSK}} = \frac{1}{2}\mathrm{erfc}\left(\sqrt{\mathrm{SNR}}\right) \tag{7}$$

BER when DPIM is considered [7]:

$$\mathrm{BER}_{\mathrm{DPIM}} = Q\left(\sqrt{\mathrm{SNR}/2}\right) \tag{8}$$

BER when DH-PIM is considered [7]:

$$\mathrm{BER}_{\mathrm{DH\text{-}PIM}} = \frac{4L_{\mathrm{DH\text{-}PIM}} - 3\delta}{4L_{\mathrm{DH\text{-}PIM}}}Q\left(\sqrt{2k^2\mathrm{SNR}}\right) + 3\delta Q\sqrt{2(1-k)^2\mathrm{SNR}} \tag{9}$$

BER when M-PAM is considered [8]:

$$\mathrm{BER}_{\mathrm{M\text{-}PAM}} = \frac{1}{2}\mathrm{erfc}\left(\frac{\sqrt{\mathrm{SNR}\log_2 M}}{2\sqrt{2}(M-1)}\right) \tag{10}$$

BER when H-QAM is considered [9]:

$$\mathrm{BER}_{\mathrm{H\text{-}QAM}} = \frac{2\left(1 - 1/\sqrt{H}\right)}{\log_2 H}\mathrm{erfc}\left(\frac{3\log_2 H(\mathrm{SNR})}{2(H-1)}\right) \tag{11}$$

where H and M are quadrate amplitude order and the pulse amplitude order, respectively, and the average symbol length is denoted by $L_{\mathrm{DH\text{-}PIM}}$. The range of the threshold factor is $0 < k < 1$, and $\delta > 0$ is an integer.

4 Results and Discussion

In this paper, the BER performance of a single-channel UWOC system for various modulation techniques is evaluated with analytical approach where different water types are considered. For the analysis, wavelength of the optical signal (λ) = 532 nm, optical efficiency of the transmitter and the receiver (η_t) = (η_r) = 0.8, source beam-divergence angle (θ_d) = 60°, inclination angle of the transmitter (θ) = 5°, receiver capture area (A_R) = 0.01 m^2, electron charge (e) =1.6 × 10^{-19} C, Boltzmann constant (k) = 1.38× 10^{-23} J/K, and temperature (T) = 300 K are considered; receiver load resistance (R) = 1 kΩ.

Figure 2 shows the plots of BER versus the distance of transmission for different modulation techniques in (a) pure sea water, (b) clear ocean water, and (c) coastal ocean water, respectively, for the transmitted power of 10 dBm. The BER performance of this system decreases when the link distance is increased. This situation happens due to the fact that light has to suffer underwater attenuation for longer time periods for greater link distance. As shown in Figure 2, it is also evident that for each modulation technique, the maximum channel length for optimum communication decreases with increasing impurity of water since higher impurity introduces higher optical attenuation. In each plot of Fig. 2, the fourth-order quadrature amplitude modulation technique (4-QAM) is the best-performing modulation scheme for this system. Then, 8-QAM, BPSK, DPIM, DH-PIM, and 4-PAM are performed in descending order.

Comparing the plots in the best transmission capability is achieved in pure sea water followed by clear ocean water and coastal ocean water. For example, when 4-QAM is considered, BER of 10^{-9} is maintained at the link distance of 20.38 m, 34.46 m, and 68.07 m, respectively, in pure sea, clear ocean, and coastal ocean water.

Figure 3 depicts the BER variation with respect to transmitted power for different modulation techniques in (a) pure sea, (b) clear ocean, and (c) coastal ocean water, respectively. For each analysis, the link distance is 25 m. In each figure, the best system performance achieved for 4-QAM technique requires the least amount of

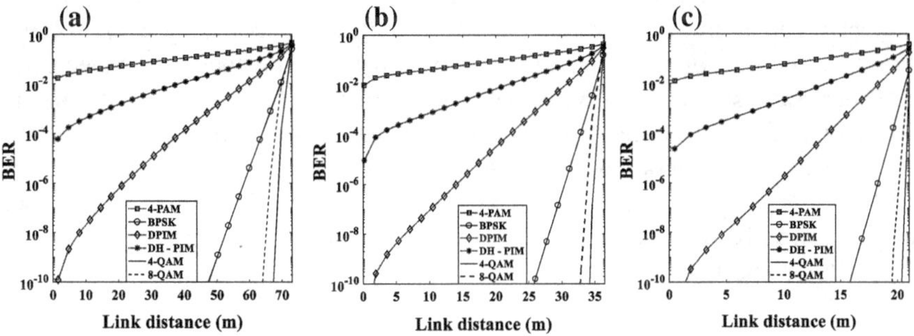

Fig. 2 Graphs showing the variation of BER as a function of link distance for different modulation schemes in **a** pure sea, **b** clear ocean, and **c** coastal ocean water when the transmitted power is 10 dBm

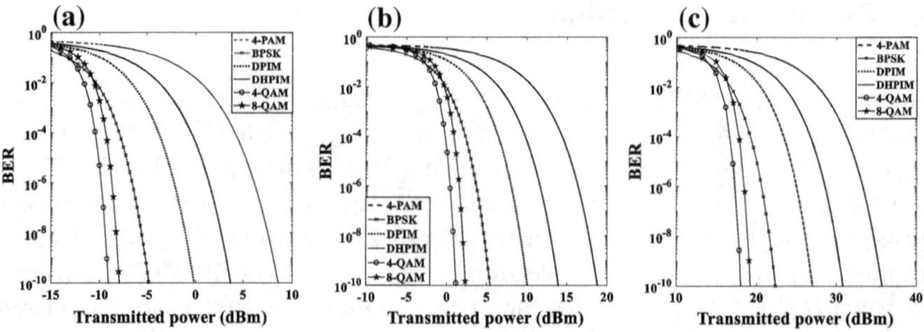

Fig. 3 Graphs showing the variation of BER as a function of transmitted power for different modulation techniques in **a** pure sea, **b** clear ocean, and **c** coastal ocean water where the link distance is 25 m

power for data transmission. From Fig. 3a, to maintain the BER of 10^{-9} in pure sea water (distance = 25 m), the required transmitted powers are −9.2 dBm for 4-QAM, −8.2 dBm for 8-QAM, −5.2 dBm for BPSK, −1 dBm for DPIM, 3.5 dBm for DH-PIM, and 8 dBm for 4-PAM modulation technique. Figure 3b, c show the performance figures in similar order, but the required transmitted power is increased with increasing water impurity. The 4-QAM modulation technique provided the best performance in each scenario. Although higher-order modulation techniques provide greater data rates, the 8-QAM technique provided lower system performance than 4-QAM. In data communication, higher-order modulations are more susceptible to noise leading to intersymbol interference which is the main performance-limiting factor. For optimum performance, the upper limit of the order of QAM technique for this system is 4.

5 Conclusion

This study is carried out for the wavelength of 532 nm which suffers minimum optical attenuation in underwater channel. Si APD is used as the optical receiver. For the LOS link configuration, the simulation results show that 4-QAM technique provides better system BER performance than other modulation techniques in each water type. Pure sea water provided the best performance among the considered water types. For example, when 4-QAM technique is considered for a 25 m link, BER of 10^{-9} is maintained at −9.2 dBm transmitted power in pure sea water, where clear ocean and coastal ocean water provided the same performance at 1.01 dBm and 17.9 dBm, respectively.

References

1. H. Kaushal, G. Kaddoum, Underwater optical wireless communication. IEEE Access 4, 1518–1547 (2016)
2. I. F. Akyildiz, Pompili D., T. Melodia, Underwater acoustic sensor networks: research challenges. Ad. Hoc. Netw. 3(3), 257–279 (2005)
3. R. M. Pope, E. S. Fry, Absorption spectrum 380–700nm of pure water. II. Integrating cavity measurements. Applied optics 36(33), 8710–8723 (1997)
4. C. Gabriel, M. A. Khalighi, S. Bourennane et al., Channel modeling for underwater optical communication. In: Proc. IEEE GLOBECOM, pp. 833–837, IEEE, Houston, TX (2011)
5. G. Keiser, Optical fiber communications. 4th edn. McGraw-Hill Co., NY (2010)
6. Y. Li et al., Performance analysis of OOK, BPSK, QPSK modulation schemes in uplink of ground-to-satellite laser communication system under atmospheric fluctuation. Opt. Commun. 317, 57–61 (2014)
7. Z. Ghassemlooy, W. Popoola, S. Rajbhandari, Optical wireless communications – system and channel modelling with MATLAB. CRC Press, Boca Raton, Florida, (2013)
8. T. Y. Elganimi, Performance comparison between OOK, PPM and PAM modulation schemes for free space optical (FSO) communication systems: analytical study. Int. J. Comp. Appl. 79(11), 22–27 (2013)
9. A. N. Z. Rashed, H.A Sharshar, Deep analytical study of optical wireless communication systems performance efficiency using different modulation techniques in turbulence communication channels. Int. J. Adv. Res. Electr. Comm. Eng. 3(3), 246–260 (2014)

MCMC Particle Filter Approach for Efficient Multipath Error Mitigation in Static GNSS Positioning Applications

N. Swathi, V. B. S. Srilatha Indira Dutt and G. Sasibhushana Rao

Abstract Global Positioning System (GPS) is the modern Global Navigation Satellite System (GNSS) all over the world. The GPS system accuracy can be corrupted by several error sources like ionosphere, troposphere, satellite clock errors, ephemeris errors, multipath and instrumental bias. GPS signals undergo reflections by objects along the path and may arrive at the receiver antenna via different paths. This leads to wrong position estimate as the signals travel feet to miles more to reach the receiver antenna than a direct line of sight signal. This problem is called as multipath. The GPS receiver tracks both the direct and reflected signal components. Multipath provokes error in both pseudorange and carrier phase measurements. In our work, the estimated multipath error using code range minus carrier phase range (CRMCPR) technique is applied as the input to the standard and Markov Chain Monte Carlo (MCMC) particle filters. For the analysis, data corresponding to the receiver placed at Department of ECE, Andhra University, is considered.

Keywords GPS · GNSS · Pseudorange · Carrier phase measurements

1 Introduction

GPS is the modern GNSS all over the world which provides user navigation solution anywhere on earth surface [1]. Accuracy of the GPS can be degraded by many error sources [2]. GPS signals undergo reflections by objects along the path and may arrive at the receiver antenna via different paths. This problem is called as multipath. Multipath depends mostly on the scenario involving the reflecting objects, the antenna, and the satellites. Signal degradation due to multipath may originate

---

N. Swathi (✉)
Department of EIE, VRSEC, Vijayawada, India
e-mail: nadipineniswathi@gmail.com

V. B. S. Srilatha Indira Dutt
Department of ECE, GIT, GITAM Deemed to be University, Visakhapatnam, India

G. Sasibhushana Rao
Department of ECE, AUCE (A), Visakhapatnam, India

H. S. Saini et al. (eds.), *Innovations in Electronics and Communication Engineering*, Lecture Notes in Networks and Systems 107,
https://doi.org/10.1007/978-981-15-3172-9_12

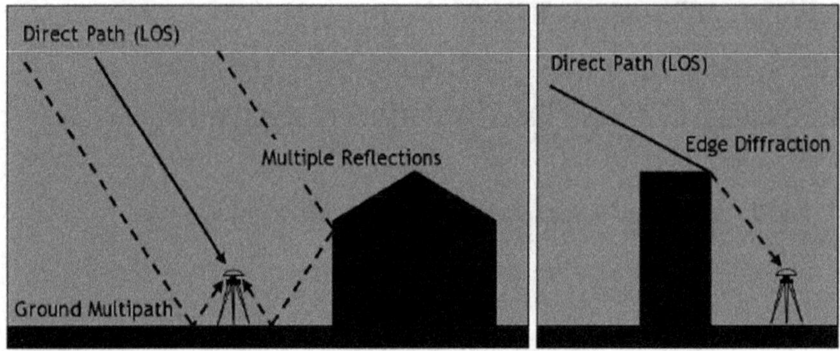

Fig. 1 Different scenarios for multipath signal propagation

from different types of reflections. The signal may be reflected by obstacles such as buildings or by the ground. Based on the geometrical conditions in the vicinity of the antenna, one satellite signal can be reflected several times. A signal may also undergo edge diffraction, i.e. the signal deviates from its original path, and the diffracted signal is finally received by the antenna. All these propagation scenarios are illustrated in Fig. 1. Several methods are available in the literature to alleviate this multipath error by filtering using digital filters [3] and adaptive filters [4]. In this paper, we are making use of Bayesian particle filtering approach for alleviating the multipath error.

2 Particle Filter (PF) Approach

This approach uses the nonlinear filtering model [5]. It includes estimating the states $X_k \in R^{n_x}$, recursively with the known measurements $Y_k \in R^{n_y}$, based on all existing measurements, $Y_{1:k} = \{Y_1, \ldots, Y_k\}$ at time index k [6]. So, our aim is to find the filtering distribution, $p(X_k|Y_{1:k})$, which can be stated recursively like

$$p(X_k|Y_{1:k}) = \frac{p(Y_k|X_k)p(X_k|X_{k-1})}{p(Y_k|Y_{1:k-1})} p(X_{k-1}|Y_{1:k-1}) \tag{1}$$

PFs use a set of N_p weighted random samples to estimate the filtering distribution by forming the set of particles $\{X_k^i, W_k^i\}_{i=1}^{N_p}$. The set N_p is obtained from the importance density distribution, $\pi\,(\cdot)$, $X_k^i \sim \pi\left(X_k|X_{0:k-1}^i, Y_{1:k}\right)$ and weighted according to

$$\tilde{W}_k^i \propto W_{k-1}^i \frac{p\left(Y_k|X_{0:k}^i, Y_{1:k-1}\right)p\left(X_k^i|X_{k-1}^i\right)}{\pi\left(X_k^i|X_{0:k-1}^i, Y_{1:k}\right)} \tag{2}$$

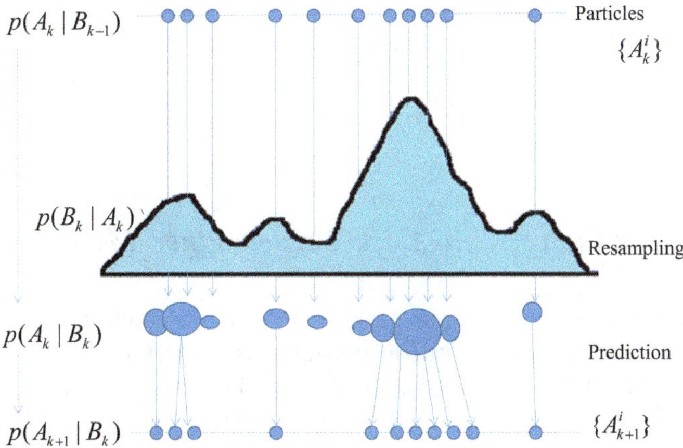

Fig. 2 Particle filter concept

This algorithm begins its operation with every new measurement Y_k [7]. First particles are generated, next their weights are calculated and normalized then minimum mean square error (MMSE) can be estimated. Particle degeneracy is a significant problem with PFs, where all but one weight tends to zero [8]. This condition makes the particles to fall to a single state point. A resampling step is applied to avoid this problem, every time the actual number of particles, defined as:

$$N_{\text{eff}} = \frac{1}{\sum_{i=1}^{N_p} \left(W_k^i\right)^2} \tag{3}$$

is below a predefined threshold. Figure 2 depicts the concept of particle filter algorithm.

2.1 *Standard Particle Filter (SPF) Algorithm*

Require: $\left\{X_{k-1}^i, W_{k-1}^i\right\}_{i=1}^{N_p}$ and Y_k Ensure: $\left\{X_k^i, W_k^i\right\}_{i=1}^{N_p}$ and $\hat{X}_k$

(1) for i = 1 to N do
(2) Produce $X_k^i \sim \pi\left(X_k | X_{0:k-1}^i, Y_{1:k}\right)$
(3) Compute $\tilde{W}_k^i \propto W_{k-1}^i \frac{p\left(Y_k | X_{0:k}^i, Y_{1:k-1}\right) p\left(X_k^i | X_{k-1}^i\right)}{\pi\left(X_k^i | X_{0:k-1}^i, Y_{1:k}\right)}$
(4) end for
(5) for i = 1 to N do
(6) Normalize weights: $W_k^i \propto \frac{\tilde{W}_k^i}{\sum_{j=1}^{N} \tilde{W}_k^j}$
(7) end for

(8) MMSE estimation: $\hat{X}_k = \sum_{i=1}^{N} w_k^i x_k^i$

(9) $\{x_k^i, 1/N\}_{i=1}^N = \text{Resample}\left(\{x_k^i, w_k^i\}_{i=1}^N\right).$

2.2 *Markov Chain Monte Carlo Particle Filter (MCMC PF)*

In engineering, nonlinearity is a frequent problem in the application of filters. The dynamic stochastic process is estimated using noisy observations. These problems are described in dynamic state space (DSS) model of an unobserved state variable which provides time-varying dynamics X_k. The probability distribution is $p(X_{k,} X_{k-1})$ where k is any physical quantity (time) [9]. The observations Y_k in the application are combined version of noise and X_k. The distribution $p(Y_k/X_k)$ represents the conditional probability of observation equation on the unknown state variable X_k. The model is represented as:

$$X_k = f(X_{k-1}) + U_k \quad \text{(State equation)}$$
$$Y_k = h(X_k) + V_k \quad \text{(Measurement equation)}$$

where U_k, V_k are state and measurement noise. State estimation problems are resolved using PFs in various applications of navigation and fault detection [10]. PFs can represent random probability densities which make them possible to overcome the problems of nonlinear and non-Gaussian estimation. Even though PFs are better for implementation, there are some limitations as it provides less diversity and computational complexity for more number of samples resulting in the divergence of PF. The less diversity is overcome by MCMC sampling process.

MCMC Algorithm

Here, MCMC PF is used to swap the standard sampling to increase the diversity and allows handling better dimensional state spaces, which SPF cannot provide.

The steps for MCMC algorithm [11]:

(1) The probability mass function of the primary state, $p(X_0)$, is known.
(2) Generate N primary particles $\left(X_{0,i}^+\right)$ based on the pdf $p(X_0)$. (i = 0, 1, ...N)
(3) For k = 1, 2,... do
(4) Do time propagation step to get prior particles (X_{ki}).
$X_{k,i^-} = f\left(X_{k-1,i}^+\right) + u_k \quad (\text{i} = 1, 2, \ldots \text{N})$
(5) Compute the weights (w_i) of each particle $\left(X_{k,i^-}\right)$.
(6) Normalize the weights as
$w_i = w_i/sum(w_j)(\text{j} = 1, 2 \ldots .\text{N})$

//sampling algorithm\\

(7) Sample u ~ U [0, 1].

(8) Get the new particles $X^*_{k,i}$ from $p(X_k/X_{k-1,i})$.
(9) If u < $\min\{1, p(Y_k/X^*_{k,i})/p(Y_k/X_{k,i^-})\}$
$X_{k,i} = X^*_{k,i}$
else $X_{k,i} = X_{k,i^-}$
(10) Return X_k.

3 Results

Results are based on the data corresponding to a dual-frequency receiver placed at Department of ECE, Andhra University (Latitude/Longitude: 17.730231° N/83.31956° E, Height: 91.6 m). Multipath error is estimated using CRMCPR algorithm [12] and is applied as input to the standard and MCMC PFs. For the analysis, data corresponding to satellite vehicles 4, 17 and 28 is considered. Figure 3 shows the multipath error inputs (MP1 and MP2) to the particle filters. Maximum multipath observed is 9.89 m at 23:10 hrs on $L2$ frequency and 6.78 m on $L1$ frequency. The multipath error mitigated using the SPF for SVPRN4 on $L1$ frequency is shown in Fig. 4. Mean value of reduced error is 0.5397 m. The estimated multipath error using CRMCPR algorithm is applied to MCMC PF, and the output of the filter is shown in Fig. 5. Figure 5a shows the MCMC particle filter output with respect to local time in hours for SVPRN 4 on $L1$. From this figure, it can be observed that the error is reduced to millimetre level. Mean value of reduced error is 5.9322e−04 m.

Figure 6 shows the SPF output with respect to GPS time in hours for SVPRN 4 on $L2$. Mean value of reduced error is 0.9605 m. Figure 7a shows the MCMC PF output with respect to local time in hours for SVPRN 4 on $L2$. From this figure, it

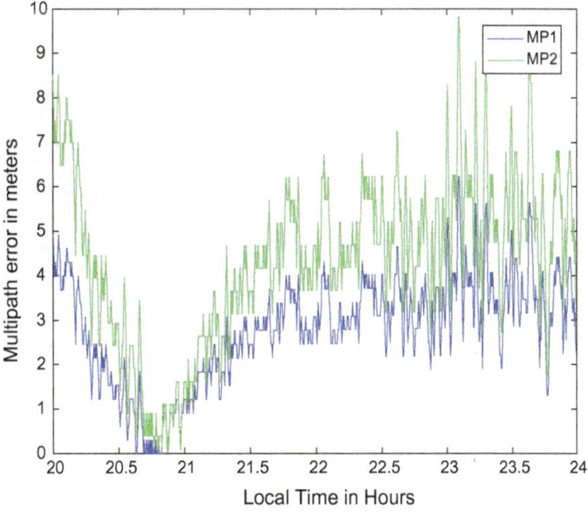

Fig. 3 Estimated multipath error on $L1$ and $L2$ frequencies for SVPRN4

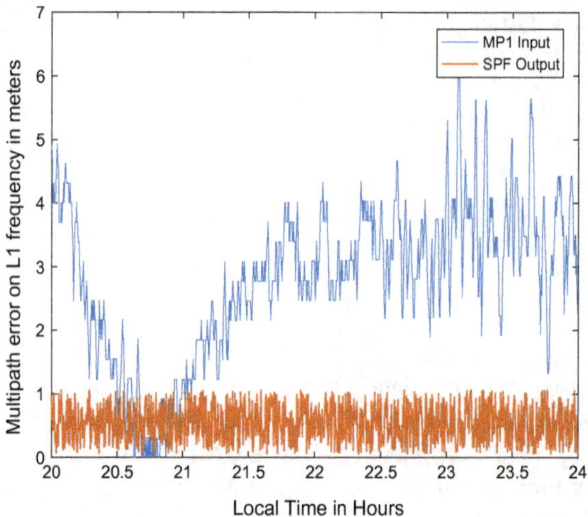

Fig. 4 Mitigated multipath error due to SPF on $L1$ frequency for SVPRN 4

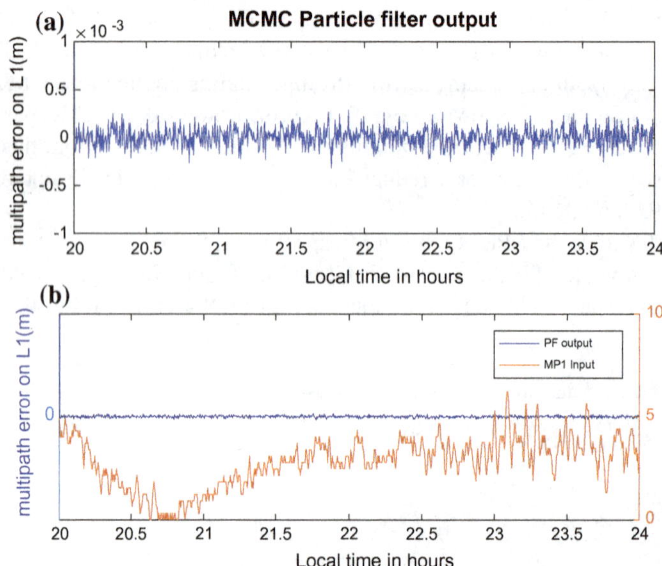

Fig. 5 **a** MCMC particle filter output (m) with respect to local time in hours and **b** MCMC particle filter output compared with the input multipath error on $L1$ for SVPRN 4

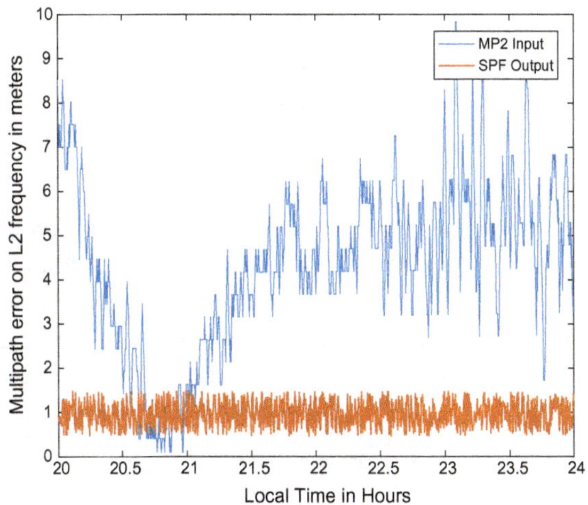

Fig. 6 SPF output compared with the input multipath error on $L2$ for SVPRN 4

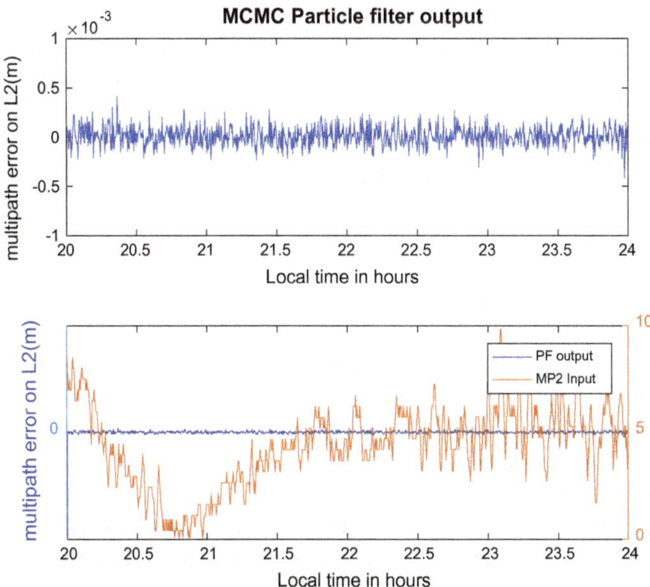

Fig. 7 **a** MCMC particle filter output (m) with respect to local time in hours and **b** MCMC particle filter output compared with the input multipath error on $L2$ for SVPRN 4

Table 1 Comparison of standard and MCMC PFs performance for different satellites

SI. No.	SV PRN	Signal type	Mean before filtering (m)	SPF		MCMC PF	
				Mean after filtering (m)	% of error reduced	Mean after filtering (m)	% of error reduced
1	4	MP1	2.7798	0.5397	80.58	**0.00059322**	**99.97**
		MP2	4.3309	0.9605	77.82	**−0.0017847**	**99.95**
2	17	MP1	3.1555	0.7702	75.59	**−0.0030014**	**99.90**
		MP2	4.9042	0.6822	86.08	**0.0042625**	**99.91**
3	28	MP1	3.8993	0.8375	78.52	**0.0061974**	**99.84**
		MP2	6.1487	0.8423	86.30	**−0.0081477**	**99.86**

The MCMC particle filter output values are kept in bold to highlight its improved performance in mitigating the multipath error compared to the standard particle filter

can be observed that the error is reduced to millimetre level. Mean value of reduced error is −1.7847e−03 m.

The multipath error mitigation efficiency of standard and MCMC PFs for SVPRNs 4, 17 and 28 is compared and the corresponding results are shown in Table 1. It is clear that, using SPF, multipath error is reduced by 80.58% and 77.82% on $L1$ and $L2$ frequencies for SVPRN 4, whereas using MCMC PF, mitigation up to 99.9% is achieved on both $L1$ and $L2$ frequencies for all the satellites.

4 Conclusions

In this research work, CRMCPR algorithm is used to estimate multipath error with the data collected from the receiver located at Andhra University College of Engineering and the results of multipath error mitigation using standard and MCMC PFs for SVPRNs 4, 17 and 28 are presented in Table 1. Using SPF, multipath error is reduced by 80.58% and 77.82% on $L1$ and $L2$ frequencies, whereas using MCMC PF, it is possible to reduce the multipath error up to 99.9% on both $L1$ and $L2$ frequencies for all the satellites. Therefore, MCMC PF is more suitable for reduction of multipath error on GPS signals.

References

1. M.S. Braasch, in *Multipath Effects, Global Positioning System: Theory and Applications*, vol. 1, ed. by B.W. Parkinson, J.J. Spilker (AIAA, 1996)
2. G. Sasi Bhushana Rao, *Global Navigation Satellite Systems* (Tata McGraw Hill publications, 2010)

3. Y. Kumandham, A.D. Sarma, A. Kumar, K. Satyanarayana, Spectral analysis and mitigation of multipath error using digital filtering for static applications. IETE J. Res. **59**(2), 156–166 (2013)
4. C.-L. Chang, J.-C. Juang, An adaptive multipath mitigation filter for GNSS applications. EURASIP J. Adv. Signal Process. **2008**, 1–10 (2008). https://doi.org/10.1155/2008/214815
5. X. Liu, P. Closas, J. Liu, X. Hu, Particle filtering and its application for multipath mitigation with GNSS receivers, in *IEEE/ION Position, Location and Navigation Symposium*, Indian Wells, CA, 2010, pp. 1168–1173. https://doi.org/10.1109/plans.2010.5507236
6. A. Giremus, J.-Y. Tourneret, V. Calmettes, A particle filtering approach for joint detection/estimation of multipath effects on GPS measurements. IEEE Trans. Signal Process. **55**(4) (2007)
7. P. Closas, C. Fernández-Prades, Particle filtering with adaptive number of particles. in *IEEE Aerospace Conference Proceedings*. https://doi.org/10.1109/aero.2011.5747439
8. M.S. Arulampalam, S. Maskell, N. Gordon, T. Clapp, A tutorial on particle filters for online nonlinear/non-Gaussian Bayesian tracking. IEEE Trans. Signal Process. **50**(2), 174–188 (2002)
9. B. Sachintha Karunaratne, M.R. Morelande, B. Moran, MCMC particle filter for tracking in a partially known multipath environment. in *IEEE International Conference on Acoustics, Speech and Signal Processing*, (2013), pp. 6332–6336. 978-1-4799-0356-6
10. M. Bocquel, H. Driessen, A. Bagchi, Multitarget tracking with Interacting Population-based MCMC-PF. in *15th International Conference on Information Fusion*, Singapore (2012), pp. 74–81
11. P. Sudhakar, D. Elizabeth Rani, Target tracking using MCMC and KLD particle filters. Int. J. Eng. Technol. (IJET) **9**(2) (2017). 10.21817
12. N. Swathi, V.B.S.S.I. Dutt, G. Sasibhushana Rao, An adaptive filter approach for GPS multipath error estimation and mitigation. Springer Lecture Notes in Electrical Engineering, vol. 372, pp. 539–546 (2016). https://doi.org/10.1007/978-81-322-2728-1_50

Investigation of Multiband Microstrip Antenna by AWR Electromagnetic Simulator

Yaqeen Sabah Mezaal

Abstract The development of wireless and computerized systems has technologically advanced the antenna design where the end-device relay signals with electromagnetic access are associated with a foremost network server in the backend. The communication amid end devices and gateway is characteristically wireless communication within RF or microwave band frequencies. In this study, an antenna size (31 mm × 31 mm) using FR-4 substrate to work under microwave wireless application is presented. Microstrip patch antenna with smallness features and quasi-fractal shape is used based on dual via ports excitation. The simulations are performed by means of Applied Wave Research (AWR) simulator that is adapted for numerous fields of electromagnetic applications. The planned antenna exhibits multiband response at 2.45, 4.71, 7.14, 8.12, and 9.62 GHz with highly compact size ($0.376\ \lambda_g \times 0.376\ \lambda_g$) at its principal frequency band that can be incorporated within any system or devices with sufficient performances. All band frequencies are within applicable strategic uses based on WLAN, C and X bands.

Keywords Microstrip antenna · Quasi-fractal · FR-4 substrate · Via ports · Multiband response · Electronic computer-aided design (ECAD)

1 Introduction

Antennas stand for devices for conveying and reception of signals through electromagnetic radiation. They exhibit copious advantages like the straightforwardness of manufacture, lightweight, and cost-effectiveness. These features make them extraordinarily predominant and nice-looking for the microwave circuit scholars since the early days they come into view. If the antenna dimension is a notable limitation for an application, their huge dimensions in various cases can be unfitting to be incorporated within numerous wireless systems and devices [1, 2]. Moreover, these microstrip antennas may undergo from narrow bandwidth. To overwhelm this

Y. S. Mezaal (✉)
Medical Instrumentation Engineering Department, Al-Esraa University College, Baghdad, Iraq
e-mail: yakeen_sbah@yahoo.com

H. S. Saini et al. (eds.), *Innovations in Electronics and Communication Engineering*, Lecture Notes in Networks and Systems 107,
https://doi.org/10.1007/978-981-15-3172-9_13

limitation, slot structures with dissimilar outlines were engaged to design broadband/wideband printed antennas [3, 4]. As a result, it looks plausible to combine fractal geometry and slot structure techniques to design multiband antennas with boosted resonant bandwidths [5–7].

An assortment of fractal structures has been used to structure miniature and multiband antennas based on unique fractal characteristics in terms of self-similarity and space filling [8, 9].

A miniaturized microstrip patch antenna based on unbalanced slits has been used as a receiver for Global Positioning Systems (GPS) [10]. It can be incorporated into countless convenient wireless systems. This antenna uses, to some extent, square microstrip patch resonator and one probe fed with four orthogonal edge slits. It has a very compact size and high-grade radiation characteristics to be in service with the GPS system. The consequential circular polarization bandwidth with axial ratio $\leq$3 dB has been within the requirements of this application along with an average gain of 4.5 dB essential for the GPS L1 wireless relevance. Single band microstrip patch antenna for 5G wireless communication has been specified in [11]. This projected antenna has been apposite for the millimeter wave frequency. The single band antenna has innovative H and E slots printed on the radiating patch with 50 ohms microstrip feeder. This antenna is simulated using Rogers RT5880 dielectric substrate. It exhibits an input reflection of −40.99 dB at 60 GHz for 5G wireless relevance. Microstrip antennas are reported in [12], to operate as single- or dual-band device using FR 4 substrate and coaxial feed. Square slot loading has been created to induce dual-band frequency response. The simulated input reflection for all antennas has practical performance to be in use for GPS, Bluetooth, and GSM applications. Co-planar waveguide (CPW) slot antenna printed on FR4 substrate has been designed and simulated by CST simulator using single feed to be operated as dual-band circuit with enhanced bandwidths as stated in [13]. The designed antenna is based on structured slot of 2nd order of peano geometry. The resulting antenna has 36 mm × 45 mm size that is applicable for convenient terminal uses. Another dual-band microstrip antenna is investigated in [14] for use in wireless applications. This microstrip antenna has a printed slot in form of 2nd iteration of Cantor fractal on the ground plane of an FR4 substrate. Parametrical examinations have been done to observe the consequences of geometrical limitations on the antenna S11 responses. The simulations and measurements reveal that the antenna has a tolerable dual-band response with widespread resonant frequency ratios. In [15], various printed slot antennas have been simulated by CST electromagnetic simulator. The designed antennas are based on square fractal slots printed on the ground plane of FR4 substrate with single microstrip feed. These antennas in this study exhibit either single band or dual-band response. However, multiband antenna has been realized by 90° rotation for fractal slot structure.

In this research work, an investigation of simulation of multiband antenna has been detailed in terms of input reflection, phase response, group delay, and radiation patterns. This antenna has been designed based on quasi-fractal microstrip resonator, FR4 substrate, complete ground plane and dual via ports. The simulated antenna has a small-scale size in mm units and strategic bands with tolerable performances that can be adopted for many cellular and satellite communication systems.

2 Antenna Toplogy

Generally, microstrip antenna has three layers as explained by Fig. 1. The upper layer represents microstrip or transmission line that typically has a thickness (t) of 35 μm with specific dimensions and topology. The transitional layer stands for a substrate layer identified by dielectric material with dielectric constant ($\in_r$) and thickness signified by h. The most frequently consumed substrate material is FR4 substrate. The third layer in the bottom is characteristically complete conductor ground plane [16].

Figure 2 depicts the top view of the proposed patch antenna with distributed regular slots. It has modeled by Applied Wave Research (AWR) simulator by accumulating uniform symmetrical resonator as quasi-fractal resonator [2]. AWR simulator stands for electronic design automation (EDA) or as well called electronic computer-aided design (ECAD) that provides a computer-based environment for simulating electromagnetic system design, digital and analog circuits. General applications take account of all sorts of wireless communications systems and defense electronics involving electronic warfare, radar, and guidance systems [17]. The general procedural steps for designing any microstrip antenna with via ports are depicted in Fig. 3.

The AWR simulator has meshing structurer with feasibilities to select substrate type and dimensions in addition to operating frequencies and microstrip antenna design. As grid size and step frequency are selected smaller, the outcomes of simulation accuracy in relation to measurements will be bigger. Accordingly, the simulation has been run under high accuracy with a grid size of 0.5 mm and a step frequency of

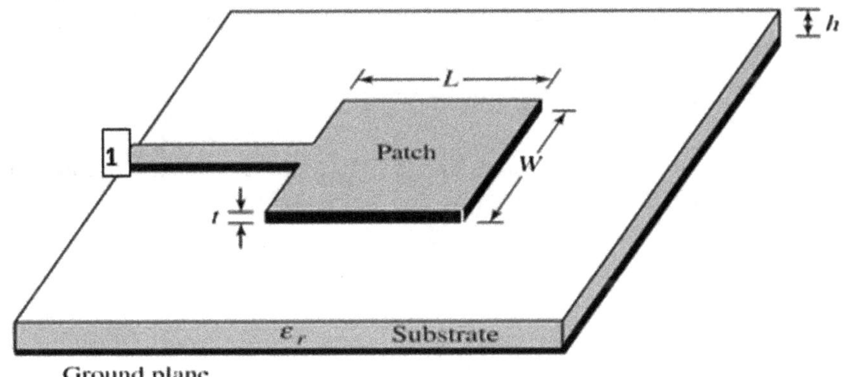

Fig. 1 Typical microstrip antenna

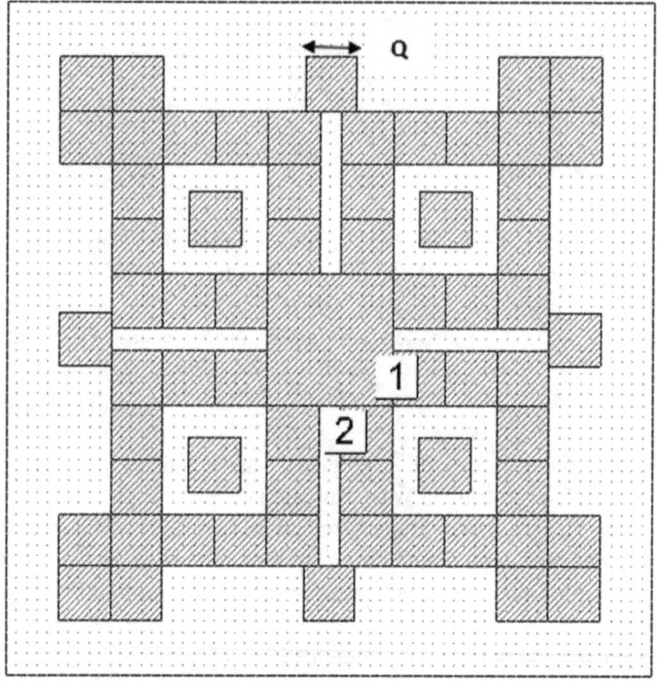

Fig. 2 Suggested microstrip antenna

0.01 GHz. The lowest square patch generator viably initiates the antenna structure as in Fig. 2. It possesses a length (Q) of 2.5 mm. The full length of this microstrip quasi-fractal resonator has been 21 mm. Via Ports (1) and (2) as situated in Fig. 2 are used for circular polarization. FR4 substrate has been used in the simulation with $\epsilon_r = 4.4$ and $h = 1.6$ mm.

A radio wave or microwave device must have dimensions compliant with the guided wavelength at fundamental resonant frequency (f), calculated by [16]:

$$\lambda_g = \frac{c}{\epsilon_{\text{re}}\ f} \tag{1}$$

The constant of effective relative dielectric ϵ_{re}, for the simulated antenna, has been determined based on Eq. (2) [16]:

$$\epsilon_{\text{re}} = \frac{\epsilon_{r+1}}{2} \tag{2}$$

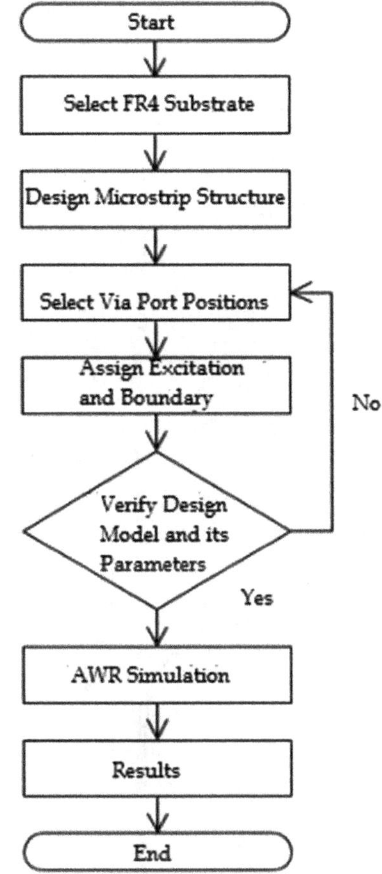

Fig. 3 Procedural steps for antenna design with via ports

3 Simulation Results and Discussion

Figure 4 describes the S11 (input reflection) response of the designed quasi-fractal antenna. It gives details about multiband frequency performance for the antenna at 2.45, 4.71, 7.14, 8.12, and 9.62 GHz in 1–10 GHz frequency full range with input reflection of deeper than −10 dB for all resonances. The multiband response is due to the self-similarity of quasi-fractal antenna with regular slots that electromagnetically motivate the multiple resonances as similar as fractal antennas [5, 6].

Figure 5 explains the angle response of the simulated antenna. Its results possess various active and linear trails within 1 to 10 GHz sweeping frequencies.

Figure 6 shows the group delay of the simulated antenna. Understandably, there are negative and positive delay magnitudes attributable to the transmission line of the quasi-fractal resonator and potential resonance modes with akin in microstrip quasi-fractal structure.

The resultant electrical specifications at each band frequency have been given in Table 1 regarding input reflection (S11), frequency ratio, bandwidth, phase, and

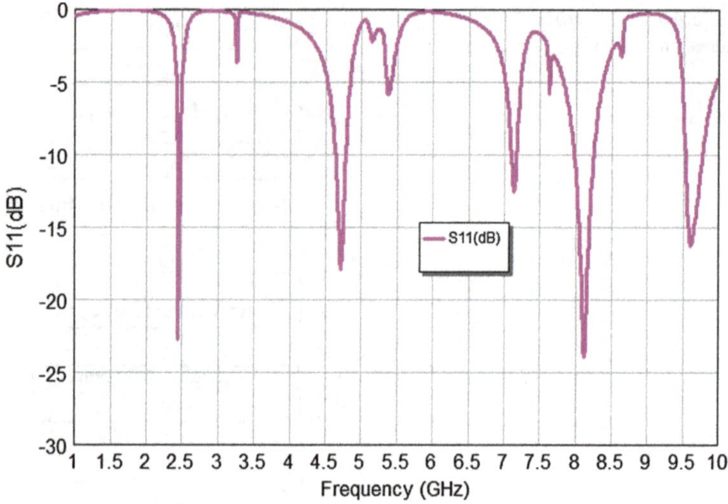

Fig. 4 Input reflection response of microstrip quasi-fractal antenna

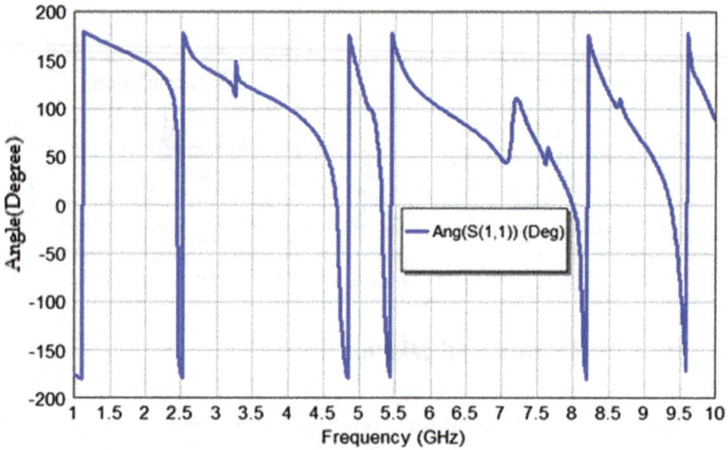

Fig. 5 Phase response of the suggested antenna

group delay. Frequency ratios from 2nd to 5th operating frequency with respect to the 1st operating frequency are not integers that give evidence for absent harmonics. It is noticeably that the last band at 9.61 GHz has the maximum bandwidth of about 0.24 GHz with maximum phase magnitude of 179° and reasonable group delay of 2.329 ns. The highest input reflection values are −23.873 dB at fourth operating resonance followed by −22.7 dB at first operating resonance.

In the case of antenna, side length is less than 0.25 λ_g, and the antenna is technically impracticable since bandwidth and radiation immunity besides gain are going to be contracted based on guided wavelength and antenna electrical specification [1, 2].

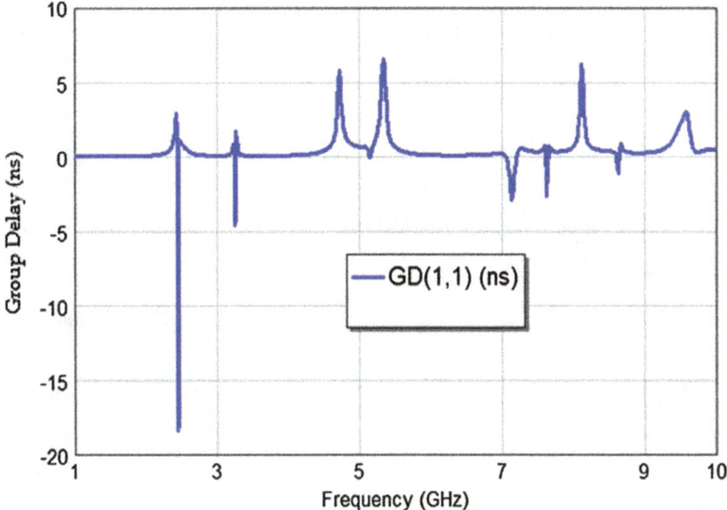

Fig. 6 Group delay result of suggested microstrip antenna

Table 1 Simulated consequences for modeled multiband microstrip antenna

Electrical parameter	Frequency (GHz)				
	$F1 = 2.45$	$F2 = 4.71$	$F3 = 7.14$	$F4 = 8.12$	$F5 = 9.61$
Input reflection (dB)	−22.7	−17.852	−12.513	−23.873	−16.214
−10 dB bandwidth range (GHz)	2.4377–2.4731	4.6373–4.7784	7.1–7.1724	7.9813–8.2464	9.529–9.7704
Frequency ratio	…	1.92	2.914	3.314	3.922
Phase magnitude (°)	71.36	−53.22	80.27	176	179
Group delay (ns)	−18.35	5.83	−2.849	6.261	2.329

Parenthetically, the simulated antenna size in terms of guided wavelength has been (0.376 × 0.376) λ_g at its fundamental frequency of 2.45 GHz, which is feasibly integrated within copious handy and subjective wireless and computerized systems.

Figures 7 and 8 explain the consequences of principle plane cut left-hand circular polarization (PPC-LHCP) along with principle plane cut right-hand circular polarization (PPC-RHCP) at $\varphi = 0°$ that can be determined in theory based on the following equations [2, 17]:

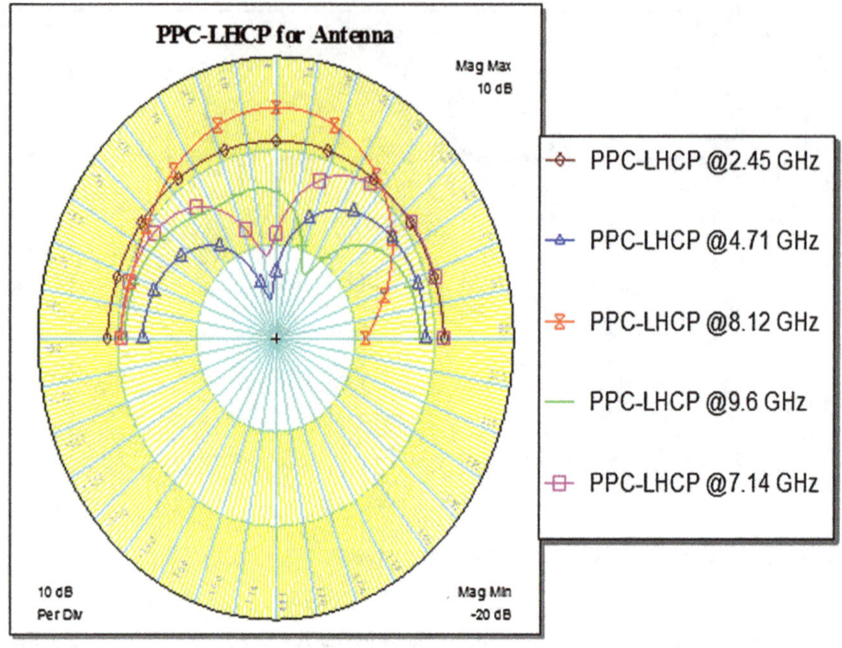

Fig. 7 PPC-LHCP radiation consequences for multiband antenna

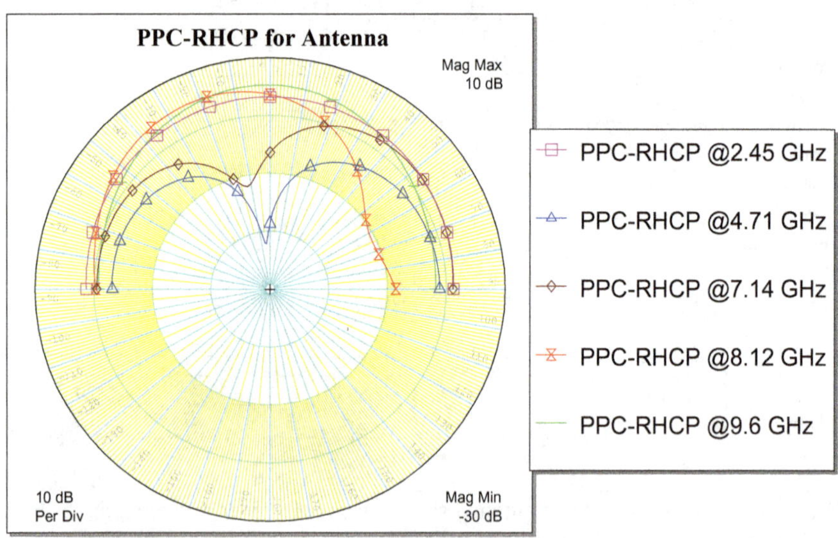

Fig. 8 PPC-RHCP radiation consequences for multiband antenna

$$\text{RHCP} = \frac{E\theta + jE\varphi}{\sqrt{2}} \tag{3}$$

and

$$\text{LHCP} = \frac{E\theta - jE\varphi}{\sqrt{2}} \tag{4}$$

Figures 7 and 8 for PPC-LHCP and PPC-RHCP have consequences from −90 to 90° at simulated φ for each band frequency with specific radiation pattern. All of them are relevant in antenna design tentatively.

Figure 9 clarifies PPC total power (PPC-TPWR) from $E\varphi$ and $E\theta$ fields irrespective of polarization under $\varphi = 0°$ based on each band frequency correspondingly.

4 Conclusion

By using FR4 substrate material, a new antenna is simulated and investigated utilizing a quasi-fractal microstrip structure for modern wireless applications. This antenna has been modeled using AWR simulator with study of various electromagnetic issues of this device. The planned antenna exhibits multiband response at 2.45, 4.71, 7.14, 8.12, and 9.62 GHz with highly compact size (0.376 λ_g × 0.376 λ_g) at

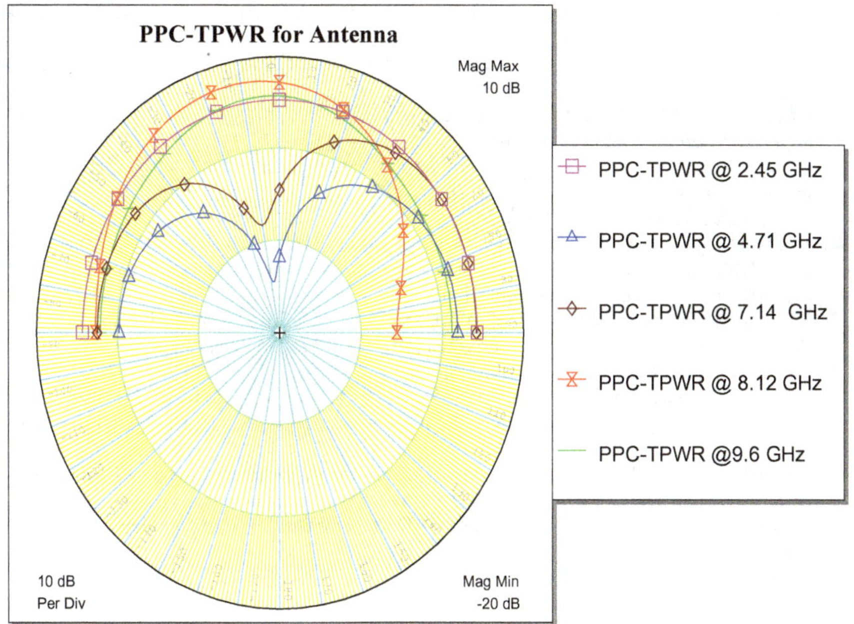

Fig. 9 PPC-TPWR radiation consequences for multiband antenna

its principal frequency that can be incorporated within any system or devices with sufficient performances. All band frequencies are within applicable strategic uses based on WLAN, C and X bands. As future work, this design can be manufactured and demonstrated with feasible high agreement since the simulation has been run under high accuracy with a grid size of 0.5 mm and a step frequency of 0.01 GHz. As well, AWR simulator has numerous interferences with various vector network analyzers (VNAs) and microstrip etching machines.

References

1. D.G. Fang, *Antenna Theory and Microstrip Antennas* (CRC Press, 2017)
2. Y.S. Mezaal, New compact microstrip patch antennas: design and simulation results. Indian J. Sci. Technol. **9**(12) (2016)
3. J.K. Ali, S.F. Abdulkareem, A circle-based fractal slot antenna for dual-band wireless applications, in *IEEE Mediterranean Microwave Symposium (MMS'2013)*, Saida, Lebanon, 2–5 Sept 2013
4. J.K. Ali, A new dual band e-shaped slot antenna design for wireless applications, in *PIERS Proceedings*, Suzhou, China, 12–16 Sept 2011
5. W.-L. Chen, G.-M. Wang, C.-X. Zhang, Bandwidth enhancement of a microstrip-line-fed printed wide-slot antenna with a fractal-shaped slot. IEEE Trans. Antennas Propag. **57**(7) (2009)
6. Y.J. Sung, Bandwidth enhancement of a wide slot using fractal-shaped Sierpinski. IEEE Trans. Antennas Propag. **59**(8) (2011)
7. B. Mirzapour, H.R. Hassani, Size reduction and bandwidth enhancement of snowflake fractal antenna. IET Microwaves Antennas Propag. **2**(2), 180–187 (2008)
8. D.H. Werner, S. Gangul, An overview' of fractal antenna engineering research. IEEE Antennas Propag. Mag. **45**(1) (2003)
9. N. Popržen, M. Gaćanović, *Fractal Antennas: Design, Characteristics And Application.* Regular paper (2011)
10. J.K. Ali, A new compact size microstrip patch antenna with irregular slots for handheld GPS application. Eng. Technol. J. **26**(10) (2008)
11. J. Saini, S.K. Agarwal, Design a single band microstrip patch antenna at 60 GHz millimeter wave for 5G application, in *Conference on Computer, Communications and Electronics (Comptelix)*, Jaipur, India, 1–2 July 2017
12. V. Dongre, A. Kumar Mishra, Design of single band and dual-band microstrip antennas for GSM, GPS and Bluetooth applications, in *International Conference & Workshop on Electronics and Telecommunication Engineering (ICWET 2016)*, Mumbai, India, 26–27 Feb 2016
13. S.F. Abdulkarim, A.J. Salim, J.K. Ali, A.I. Hammoodi, M.T. Yassen, M.R. Hassan, A compact Peano-type fractal based printed slot antenna for dual-band wireless applications, in *IEEE International on RF and Microwave Conference (RFM)*, 9–11 Dec 2013
14. J. Ali, S. Abdulkareem, A. Hammoodi, A. Salim, M. Yassen, M. Hussan, H. Al-Rizzo, Cantor fractal-based printed slot antenna for dual-band wireless applications. Int. J. Microwave Wirel. Technol. 263–270 (2016)
15. Y.S. Mezaal, S.A. Hashim, Design and simulation of square based fractal slot antennas for wireless applications. J. Eng. Appl. Sci. **13**(17), 7266–7270 (2018)
16. R. Waterhouse, *Microstrip Patch Antennas: A Designer's Guide* (Springer Science & Business Media, 2013)
17. https://www.awr.com/. Accessed: 01.06.2019

A Voltage Dependent Meander Line Dipole Antenna with Improve Read Range as a Passive RFID Tag

Md. Mustafizur Rahman, Ajay Krishno Sarkar and Liton Chandra Paul

Abstract This paper presents a general outline for passive RFID backscatter transponder antenna in practical use. The read range of tag antenna is additionally investigated. Meander line dipole antenna (MLDA) has been proposed as a tag antenna due to its tiny size and greater radiation efficiency. The radiation efficiency, gain, input resistance, and loss power were investigated in accordance with the variation of material conductivity. In this letter, it was overlooked that the power loss is inversely proportional to the square root of the conductivity of the material. It is observed that the RF input power and tag received voltage are the deciding factors for making the read range maximum. The passive tag could achieve a range of 6 m by a voltage multiplier circuit with it. For example, the read range could be achieved 3 and 5 m when the voltage doubler and quadrupler circuit were used. The design and analysis of the proposed antenna have been performed utilizing simulation software FEKO.

Keywords Meander line dipole antenna · Radiation efficiency · Read range · RF input power · Tag voltage

1 Introduction

An RFID system consists of a reader, a transponder and a host computer is shown in Fig. 1. A reader sends a continuous signal that is reflected back by the tag. A passive tag is a huge cheaper than active tag due to the absence of battery [1]. Meander line antenna is a newly introduced antenna which is made from a dipole

Md. Mustafizur Rahman (✉)
Bangladesh Army University of Engineering and Technology, Qadirabad, Natore, Bangladesh
e-mail: mustafizur.170710@gmail.com

A. Krishno Sarkar
Rajshahi University of Engineering and Technology, Kazla, Bangladesh
e-mail: sarkarajay139@gmail.com

L. Chandra Paul
Pabna University of Science and Technology, Pabna, Bangladesh
e-mail: litonpaulete@gmail.com

H. S. Saini et al. (eds.), *Innovations in Electronics and Communication Engineering*, Lecture Notes in Networks and Systems 107,
https://doi.org/10.1007/978-981-15-3172-9_14

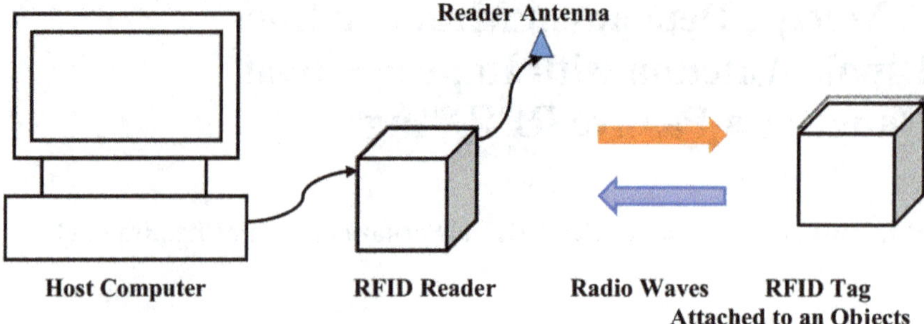

Fig. 1 Basic components of a RFID system

antenna by twisting the straight dipole. In RFID system for different applications, MLD antenna makes its remarkable consideration due to its efficiency and size. Some papers familiarized with the efficiency and gain of the antenna [2–5]. Lately, Koch fractal dipole antenna was a good choice for tag antenna, but low efficiency and complex arrangement has degraded its use [6]. Even though, design in [7–9] has depicted only the frequency and gain of MLD antenna. Therefore, a low profile MLD antenna with a small read range has been studied, but these reported letters have not given the proper idea about the tag voltage elsewhere in the open literature. Some factors are additionally important making an RFID tag antenna such as frequency band, size, objects, orientation, cost, and power for the tag. Notably, in this article, the material selection, read range, cost, and tag power have been discussed. Besides these criteria, the antenna matching technique was also discussed.

2 Analysis and Design of Meander Line Dipole Antenna

Meander line antenna is a repetitive structure of a dipole antenna as shown in Fig. 2. The length of the twisted dipole antenna is a bit lower than that of the straight dipole antenna.

In RFID system, the cost of tag must be low, thus depends both on antenna structure and materials for its construction. It is desirable to know how materials change the antenna performance. The chart is shown in Fig. 3 for better understanding the design process of a tag antenna.

2.1 Calculation of Efficiency

An RFID system, the antenna performance parameters are very important to choose the material for making the meander line dipole antenna. For MLD antenna, the current in parallel lines will always be equal and opposite in direction. From Fig. 4,

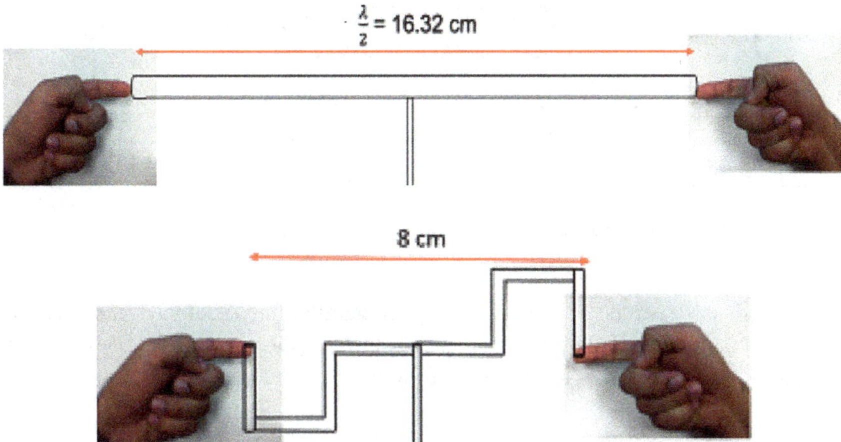

Fig. 2 Half wave dipole antenna before and after twisting

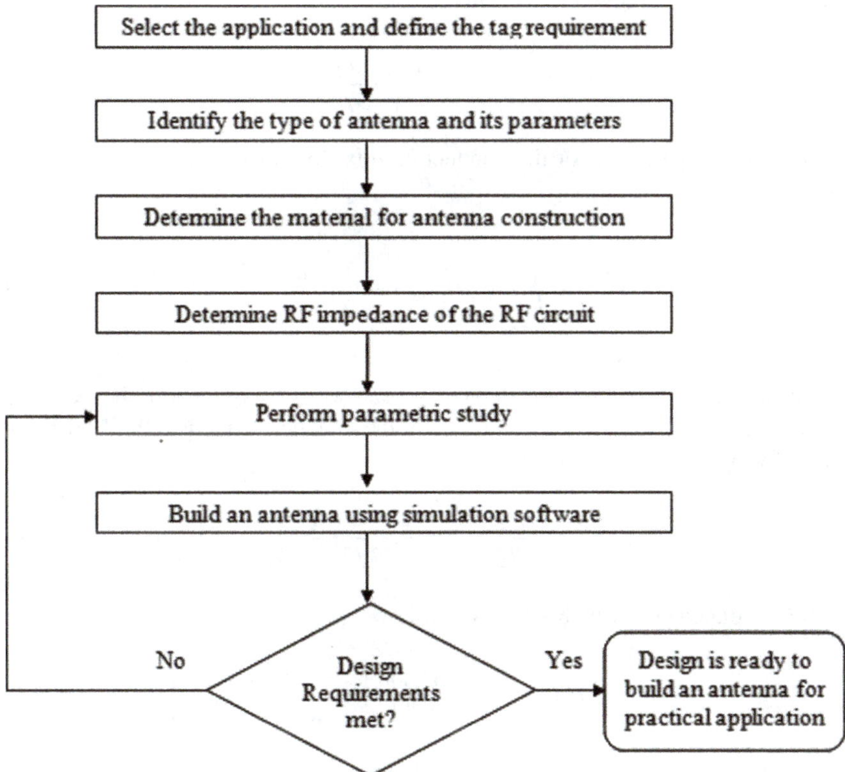

Fig. 3 Flowchart summarizing for RFID tag antenna design process

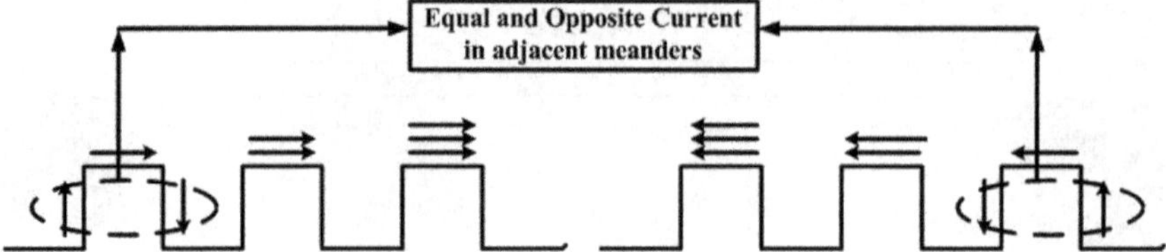

Fig. 4 Meander line dipole antenna indicating current distribution

it can be visually perceived that the current in parallel lines is always canceled out, and resulting in the radiation efficiency is only through the non-parallel lines [10–12]. The current distribution of MLDA is as follows:

$$I(z) = I_0 \sin k\left(\frac{L}{2} - |z|\right) \tag{1}$$

The resistance per length of the conductor is as follows

$$R_{\text{loss}} = \frac{1}{C}\sqrt{\frac{\omega\mu_0}{2\sigma}} \tag{2}$$

where C is the perimeter of the conductor. Total loss of power can be written as follows:

$$P_{\text{loss}} = \int_{\frac{L}{2}}^{\frac{-L}{2}} I^2(z) R_{\text{loss}} \mathrm{d}z = \frac{1}{2} I_0^2 \frac{L}{C}\sqrt{\frac{\omega\mu_0}{2\sigma}} \tag{3}$$

According to the T. Endo the radiation resistance is the product of the resistance of half-wave dipole antenna and the ratio of the total horizontal length of MLDA to the half wavelength of a dipole antenna.

$$\frac{R_{\text{radiation}}}{R_{\text{dipole}}} = \frac{L - 2wm}{\lambda/2} \tag{4}$$

Total radiated power by MLDA is as follows

$$P_{\text{rad}} = \frac{1}{2} I_0^2 R_{\text{radiation}} = \frac{1}{2} I_0^2 \left\{\frac{2(L - 2wm)}{\lambda} R_{\text{dipole}}\right\} \tag{5}$$

Radiation efficiency is expressed as follows

$$\eta = \frac{P_{\text{rad}}}{P_{\text{rad}} + P_{\text{loss}}} = \frac{2\frac{(L-2wm)}{\lambda} R_{\text{dipole}}}{2\frac{(L-2wm)}{\lambda} R_{\text{dipole}} + \frac{L}{C}\sqrt{\frac{\omega\mu_0}{2\sigma}}} \tag{6}$$

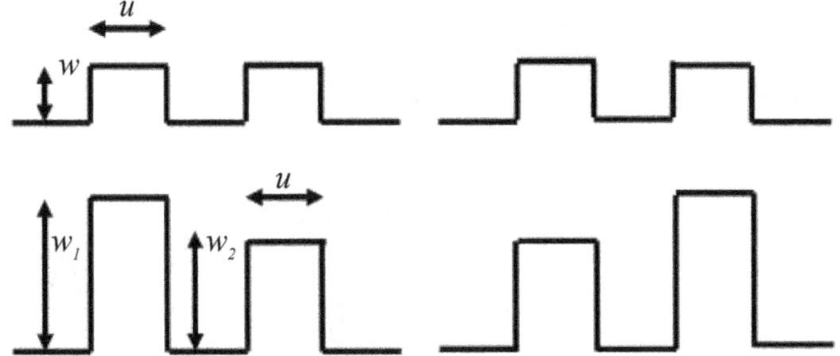

Fig. 5 Meander line dipole antenna with equal (above) and unequal (below) vertical length

where L is the entire length of the wire, w is the vertical length of the MLDA, m is the no of meander section, σ is the conductivity of the materials, μ is the permittivity of free space, and ω is the wavelength of the antenna [13–16].

From (6), it was observed that the radiation resistance does not depend on the conductivity of the materials, only the loss resistance depends on the conductivity. If the vertical length w of the MLDA antenna is not uniform, then the radiated power of the MLDA can be expressed as follows:

$$P_{\text{rad}} = \frac{1}{2} I_0^2 \left\{ \frac{2L}{\lambda} R_{\text{dipole}} - \frac{4w_1}{\lambda} R_{\text{dipole}} - \frac{4w_2}{\lambda} R_{\text{dipole}} \cdots \right\} \tag{7}$$

where the w_1, w_2 … are the vertical length of MLDA (see Fig. 5).

Two pictures are shown in Fig. 5, wherein the above figure, horizontal segments have equal length u and vertical segments have equal length w. Total length can be represented in terms of m, i.e., $L = (10u + 8w)$ when $m = 4$.

In other figures, all the horizontal segments are of equal length u and all vertical segments are of unequal length w_1, w_2, respectively. Total length L of the wire can be expressed in terms of number of meander section m, i.e., $L = (10u + 4w_1 + 4w_2)$ when $m = 4$.

2.2 Determination of RF Impedance

The major consideration of an antenna is the antenna's impedance. The antenna impedance must be in shape to the tag chip. An RFID tag will have a tag antenna and a chip, both with complex impedance (see Fig. 6). For example, if the calculated impedance of the IC chip is $Z_C = R_C - jX_C$, the antenna impedance should be $Z_A = R_A + jX_A$ for conjugate matching. Therefore, for the maximum amount of power delivered to the antenna, the load impedance must satisfy the conjugate matching condition [17–21].

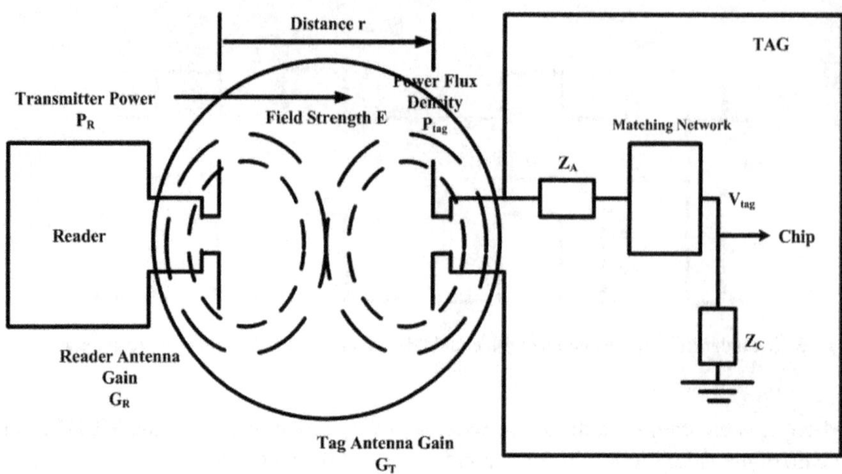

Fig. 6 Forward power transfer for RFID systems

$$Z_\text{A} = Z_\text{C}^* \tag{8}$$

2.3 *Operating Read Range and Tag Voltage Evaluation for RFID System*

Let P_R, G_R, A_R be the power, gain and effective area of the reader antenna and P_T, G_T, A_T be the same quantities for the transponder or tag antenna. The tag antenna has an effective isotropic radiated power and a power density at distance r is appearing in Fig. 6.

The isotropic radiated power and power density are expressed as follows

$$P_\text{EIRP} = P_\text{R} G_\text{R} \tag{9}$$

$$S_\tau = \frac{P_\text{EIRP}}{4\pi r^2} \tag{10}$$

The tag antenna extract power P_T given in terms of the A_T is as follows

$$P_\text{T} = S_\tau . A_\text{T} = \frac{P_\text{R} G_\text{R} A_\text{T}}{4\pi r^2} \tag{11}$$

The gain of the tag antenna is as follows

$$G_T = \frac{4\pi A_T}{\lambda^2} \quad (12)$$

where, λ is the wavelength. Therefore, the available tag power at the tag antenna can be expressed as follows

$$P_{tag} = P_T = \frac{P_R G_R G_T \lambda^2}{(4\pi r)^2} \quad (13)$$

This is called Friis transmission equation [16]. It is considered that the practical value of the chip resistance is R_C. Therefore, the tag voltage is predicated as follows

$$V_{tag} = \sqrt{P_{tag}.R_C} \quad (14)$$

The read range of the tag antenna has been initiated by the following Eq. (15)

$$r = \sqrt{\frac{P_R G_R G_T \lambda^2}{(4\pi)^2 P_{tag}}} \quad (15)$$

If Γ_t and Γ_r are the reflection coefficients of the transponder antenna and the reader antenna, respectively, and e_p denotes the polarization efficiency. Equation (15) can be rewritten as follows

$$r = \sqrt{\frac{P_R G_R G_T \lambda^2 (1 - |\Gamma_T|^2)(1 - |\Gamma_R|^2) e_p}{(4\pi)^2 P_{tag}}} \quad (16)$$

The maximum read range will be achieved if the P_{tag} is the minimum threshold power for the tag antenna.

$$r\text{max} = \sqrt{\frac{P_R G_R G_T \lambda^2 (1 - |\Gamma_T|^2)(1 - |\Gamma_R|^2) e_p}{(4\pi)^2 P_{th}}} \quad (17)$$

If antenna's impedance is almost matched to their source and load, an ideal form of (17) is defined as follows

$$r\text{max} = \sqrt{\frac{P_R G_R G_T \lambda^2}{(4\pi)^2 P_{th}}} \quad (18)$$

where $P_{tag} = P_{th}$.

Normally, the path gain factor is much less than 1. However, this path loss occurs in free space [22, 23].

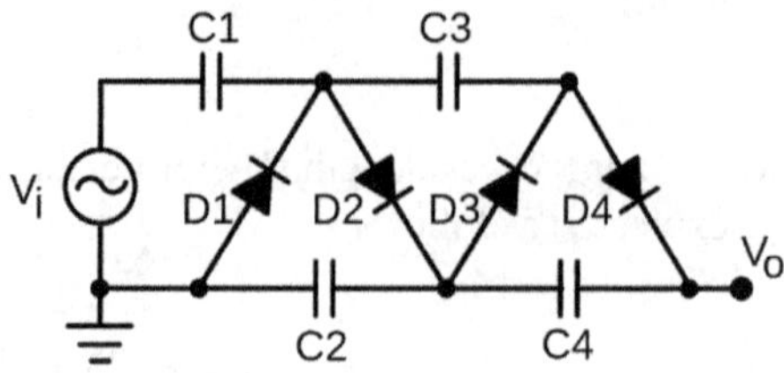

Fig. 7 Voltage multiplier circuit to power the tag chip

2.4 *Reader Received Power and DC Supply Voltage for Tag Chip*

In RFID systems, the received power by the reader is expressed as follows

$$P_{\text{Rreceived}} = \frac{P_{\text{R}} G_{\text{R}}^2 \lambda^2 \Delta s}{(4\pi)^3 r^4} \tag{19}$$

where Δs is the radar cross-section.

In the passive RFID system, there is no battery to power up the tag chip. Therefore, it is the desired necessary RF power and received voltage to power the tag. Tag voltage may be increased by way of the usage of voltage multiplier circuit (see Fig. 7).

2.5 *Proposed Meander Line Dipole Antenna*

In fact, RFID system consists of two parts, one is named the reader and another is named transponder or tag. The tag act consists of an antenna and a low power CMOS IC [16]. The IC has one very important part is RF to dc rectifier circuit. In Fig. 6, the antenna impedance $Z_{\text{A}} = R_{\text{A}} + jX_{\text{A}}$ must be equal to the chip impedance $Z_{\text{C}} = R_{\text{C}} - jX_{\text{C}}$ and the chip impedance varies with the frequency.

It is considered that the Hagg4 Alien series IC has a resistance of 1800 Ω and the capacitance is 0.95 pF. The IC has an operating frequency is varying from 840 to 960 MHz. In this composition, the designed antenna must have a frequency of which range is varied from 840 to 960 MHz. To attain this, the proposed antenna shape is viewed in Fig. 8. Dimension of MLDA is displayed in Table 1.

3 Simulation Result and Discussion

In this article, the design of the MLD antenna was done by the simulation software FEKO. After designing, the antenna performance parameters have been calculated by using the MOM method. The designed antenna and its dimension are revealed in Fig. 9 in which the gap between the arms of the dipole is 2 mm and the radius of

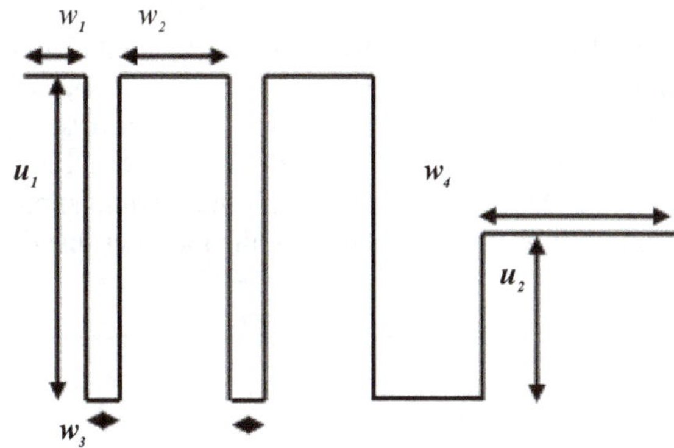

Fig. 8 Structure of meander line dipole antenna with half arm

Table 1 Antenna dimension with half arm

Meanders m	Horizontal length (mm)				Total physical length (mm)	Total wire length (mm)	Vertical length (mm)	
	w_1	w_2	w_3	w_4			u_1	u_2
4	2.7	5.4	1.4	12	34.7	144.7	20	10

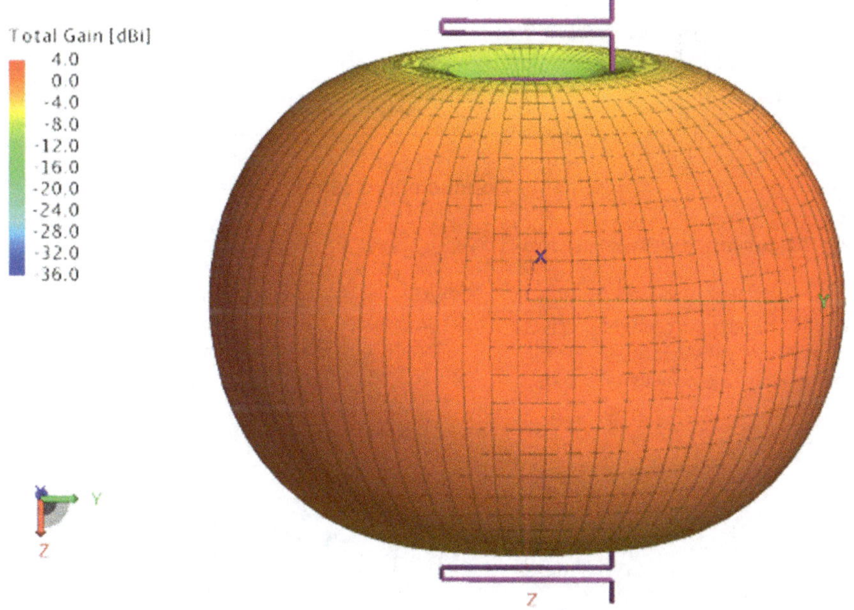

Fig. 9 3D view of a proposed meander line dipole antenna

the wire is 0.3 mm. First of all, the conductivity of the material was varied, ranging from 10^3 to 10^8 S/m. It was found that the efficiency of the antenna was increasing with the increase in conductivity due to the change of loss power (see Fig. 10). From (3), it was observed that the power loss is inversely proportional to the square root of the conductivity. From the simulation result, it was visible that the gain of the antenna withal incrementing with the incrimination in conductivity is exhibited in Fig. 11. The radiation pattern of the meander line dipole antenna is identical to the

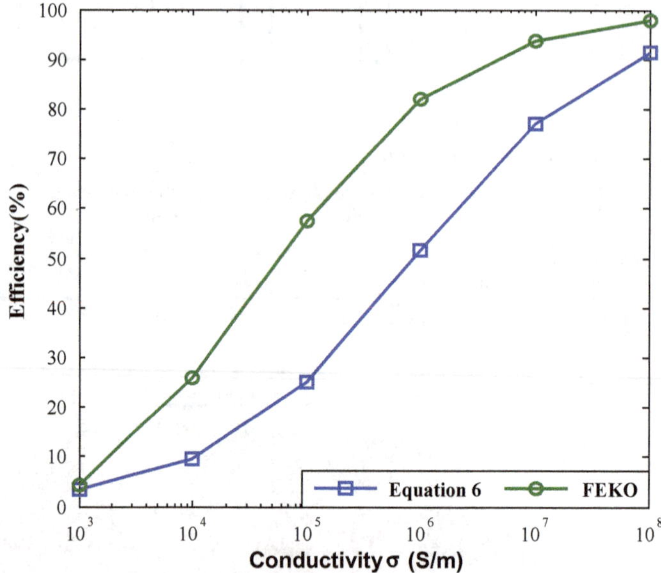

Fig. 10 Efficiency versus conductivity curve

Fig. 11 Simulation result of gain (dBi) of MLDA with the variation of conductivity

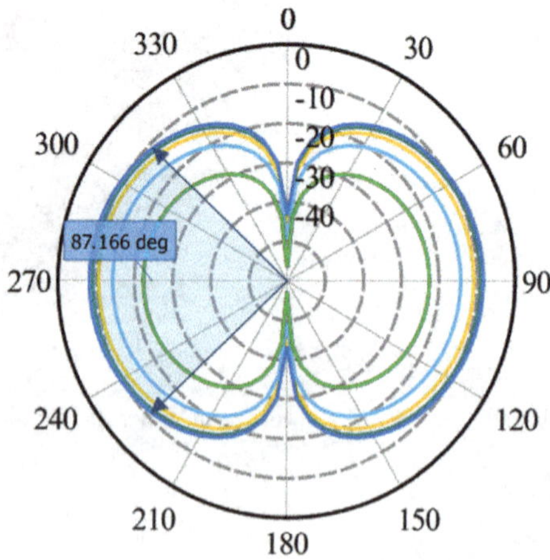

radiation pattern of the conventional dipole antenna. There are some materials which have a conductivity ranging from 3.5×10^7 to 6.30×10^7. Most of the materials such as silver, copper, gold, and aluminum are lying in this conductive range. In this article, it is important to identify what materials need to be used. Table 2 shows the antenna performance parameters. Those materials gave the better performance but their results are nearly the same. In Fig. 12, the curves of reflection co-efficient are superimposed with each other. Analogous results were shown in Fig. 13 for antenna impedance. Consequently, among these materials copper was preferred for making the antenna due to its lower cost.

From Table 2, it is noticed that the impedance of the antenna is $Z_A = 21.95 + 1.43j$ Ω and the impedance of the IC is $Z_C = 21 - 189j$ Ω if the frequency of the antenna is 920 MHz, where the value of the capacitor is 0.95 pF and the resistor is 1800 Ω. Hence, the matching network may be required. For matching load, 32.45 nH inductor is connected in series with the antenna impedance. Higher the reader output power,

Table 2 Antenna performance parameters

Material's name with conductivity σ (S/m)	Efficiency (η%)	Reflection co-efficient Γ (dB)	Impedance Z (Ω)
Carbon-Graphene (10^8)	97.91	−10.59	21.79 + 1.25j
Silver (6.30×10^7)	97.38	−10.67	21.92 + 1.40j
Copper (5.96×10^7)	97.31	−10.68	21.95 + 1.43j
Annealed Copper (5.96×10^7)	97.28	−10.68	21.95 + 1.43j
Gold (4.10×10^7)	96.77	−10.76	22.07 + 1.54j
Aluminum (3.50×10^7)	96.52	−10.79	22.14 + 1.61j

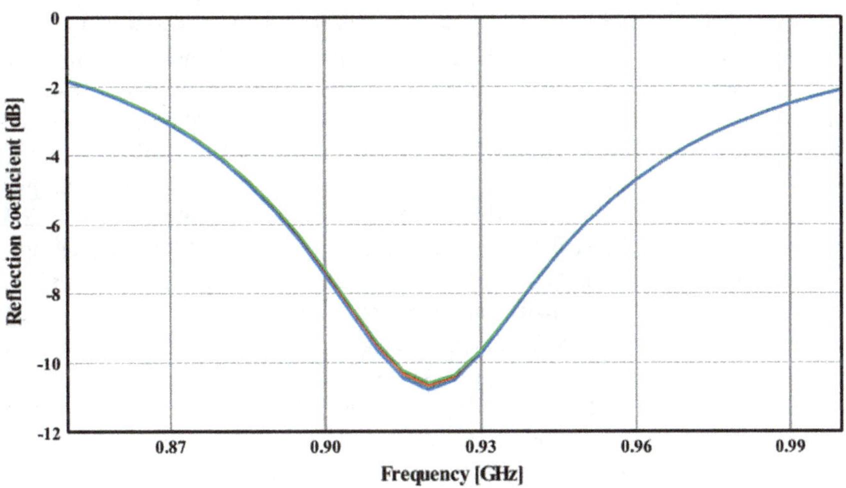

Fig. 12 Simulation result of reflection co-efficient for different materials

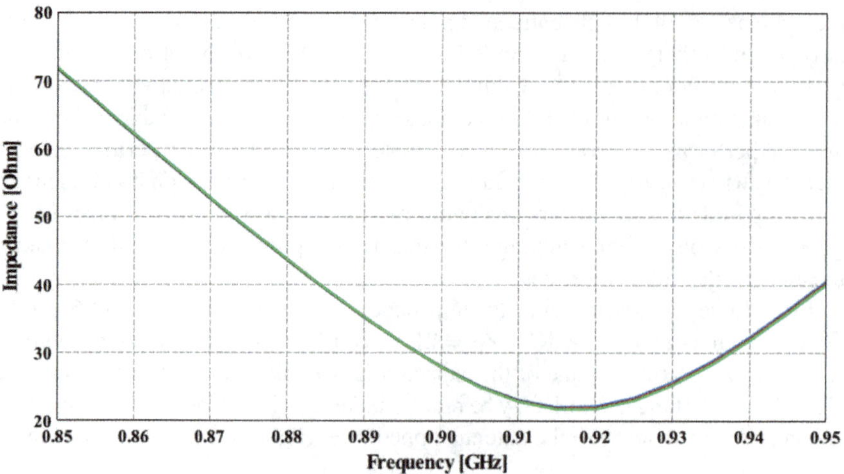

Fig. 13 Simulation result of impedance for different materials

the higher the read range. It is considered that the reader antenna is dipole antenna and then the isotropic radiated power of the reader antenna was found 29.15 dBm or 822.24 mW.

Alike, if the reader antenna is Yagi–Uda, the isotropic radiated power of the reader antenna was found 35.14 dBm or 3.26 W due to its higher gain. From (13), the tag available power versus distance for both two reader antennas is demonstrated in Fig. 14.

Besides this, tag received power is required to calculate the tag voltage. From (14), the tag voltage versus reader range is shown in Fig. 15 for both the reader antennas.

The RF input power within the range of 10 μW (−20 dBm) to 50 μW(−13 dBm) is obligatory to power on the tag. The minimum requirement of voltage V_{tag} is 1.2 V_{rms} for rectifying the signal. As mentioned earlier, for the maximum read range, two requirements have to be met. The read range must be 1.5 m for dipole reader when RF power is 1.23 dBm and 3 m when RF power is −8.30 dBm. Nevertheless, at this time, the tag voltages are 1.54 and 0.61 V, respectively (see Table 3). Additionally, 0.61 V is not sufficient to power on the tag, a modification is needed. A voltage multiplier (doubler) circuit can be used and the tag voltage could be maintained as 1.22 V. Therefore, the read range can be extended by using voltage doubler and quadrupler circuit. The read range will be 3 or 5 m if the voltage doubler and quadrupler circuit are used (see Table 3). To make the chip less weight and cost effective, voltage doubler circuit is appropriate for making the ac to dc converter circuit. In this case, the reader maximum read range will be 3 m (see Figs. 14 and 15).

Similar case is notified for Yagi–Uda reader (see Table 4) and the reader maximum range could be 5.5 m when voltage doubler circuit was used (see Figs. 14 and 15).

The received power by the tag has two parts including the reflected power and the available power. Therefore, a small fraction of energy returned to the reader antenna.

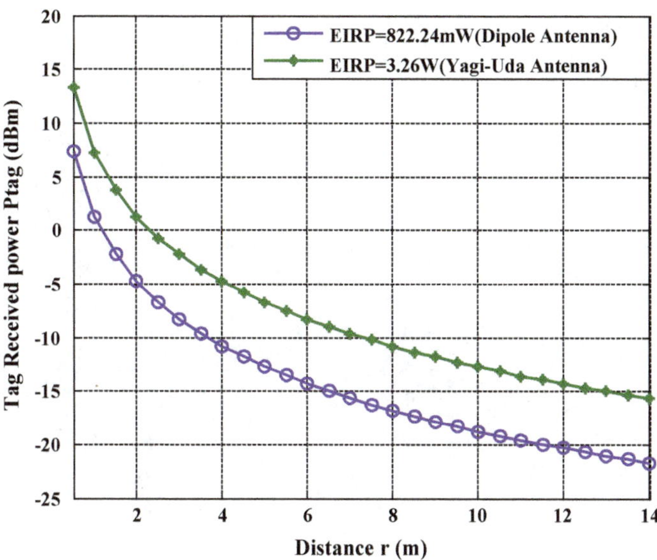

Fig. 14 Tag received power versus distance

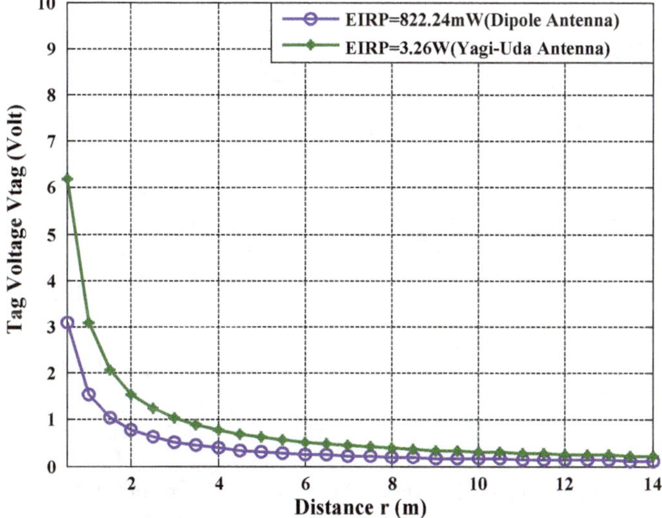

Fig. 15 Tag voltage versus distance

Table 3 Read range, tag received power, and tag voltage for dipole antenna

Read range r(m)	Tag received power P_{tag} (dBm)	Tag voltage V_{tag} (V)
0.5	7.25	3.09
1.5	1.23	1.54
2	−4.78	1.03
2.5	−6.72	0.77
3	−8.30	0.61
3.5	−9.64	0.51
4	−10.80	0.44
4.5	−11.82	0.38
5	−12.74	0.30
5.5	−13.56	0.28
6	−14.32	0.25
6.5	−15.01	0.23

Table 4 Read range, tag received power, and tag voltage for Yagi–Uda antenna

Read range r(m)	Tag received power P_{tag} (dBm)	Tag voltage V_{tag} (V)
0.5	13.26	6.17
1.5	7.24	3.08
2	3.72	2.05
2.5	1.22	1.54
3	−0.71	1.23
3.5	−2.29	1.02
4	−3.63	0.88
4.5	−4.79	0.77
5	−5.82	0.68
5.5	−6.73	0.61
6	−7.56	0.56
6.5	−8.32	0.51

From (19), the reader received power versus distance for both two reader antennas as seen in Fig. 16. From Fig. 16, it is seen that the reader received power for both two antennas are overlay with each other.

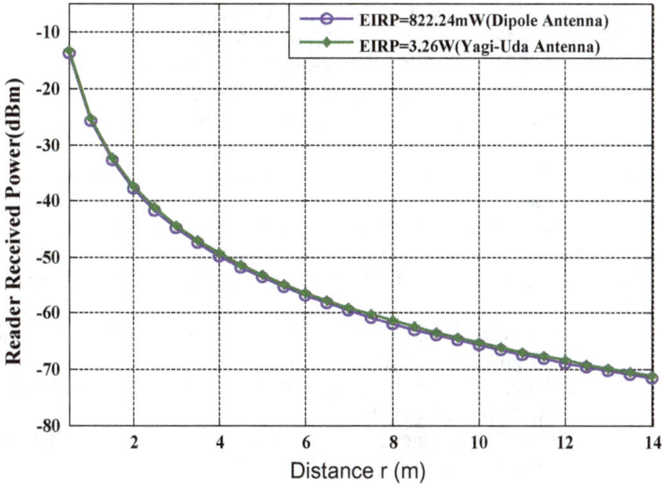

Fig. 16 Reader received power versus read range

4 Conclusion

This letter has demonstrated that the design procedure of RFID tag antenna with improving read range and higher radiation efficiency. A technique was proposed to increase the read range of tag antenna by using a voltage multiplier circuit. Read range of 3 m at the operating frequency of 920 MHz has been achieved by connecting the voltage doubler with the transponder circuit. An equation for calculating the radiation efficiency of MLD antenna that depends on the antenna's geometry and materials conductivity has been derived and presented. Analytical results are compared with the result obtained from simulation software FEKO. With a noble compromise between voltage multiplier circuit, RF input power and size of the tag antenna, the proposed tag antenna can be used in the different applications with a higher read range.

References

1. K. Finkenzeller, *RFID Handbook*, 2nd edn. (John Wiley and Sons, 2003)
2. C.-C. Lin, S.-W. Kuo, H.-R. Chuang, A 2.4-GHz printed meander-line antenna for USB WLAN with notebook-PC housing. IEEE Microw. Wirel. Compon. **15**(9) (September 2005)
3. A. Galehdar, D.V. Thiel, S.G. O'Keefe, S.P. Kingsley, Efficiency variations in electrically small, meander line RFID antennas. IEEE Antennas Propag. Soc. Int. Symp. **3**, 201–205 (2007)
4. G. Marrocco, Gain-optimized self-resonant meander line antennas for RFID applications. IEEE Antennas Wirel. Propag. Lett. **2**, 302–305 (2003)
5. M. Takiguchi, O. Amada, Improvement of radiation efficiencies by applying folded configurations to very small meander line antennas, in: IEEE Topical Conference on Wireless Communication Technology, Honolulu, HI, USA, pp. 342–334 (2003)

6. Best, On the resonant properties of the Koch fractal and other wire monopole antenna. IEEE Antennas Propag. Soc. Int. Symp. June 22–27, pp. 856–859 (2003)
7. K.V.S. Rao, P.V. Nikitin, S.F. Lam, Antenna design for UHF RFID tags: a review and a practical application. IEEE Trans. Antennas Propag. **2**, 3870–3876 (2005)
8. K.T. Mustafa, Small size multiband meander line antenna for wireless application. IEEE Antenna & Propagation Conference, Loughborough, UK, pp. 321–325 (17–18 March 2008)
9. H. Zongjian, Z. Guohua, L. Zhihua, Research on passive meander-lined radio frequency indentificationtag antenna on ultrahigh frequency, in: IEEE 5th International Conference on Biomedical Engineering and Informatics, pp. 234–238. China (2012)
10. M. Martinez-Moreno, C.A. Sanchez-Diaz, J.E. Gonzalez-Villarruel, G. Perez-Lopez, B. Tovar-Corona, Dipole antenna design for UHF RFID tags. Int. Conf. Electron. Commun. Comput. 220–224 (2009)
11. Y. Tikhov, Y. Kim, Y. Min, Compact low cost antenna for passive RFID transponder. IEEE Antennas Propag. Soc. Int. Symp. 1015–1018 (July 2006)
12. Y. Yamada, W.H. Jung, N. Michishita, Extremely small antennas for RFID tags. IEEE Int. Conf. Commun. Syst. 15 (October 2006)
13. Kam C., MHz UHF RFID tag antenna design, fabrication and test, M.S. Thesis, California Polytechnic State University, San Luis Obispo, pp. 902–928 (August 2011)
14. O.O. Olaode, W.D. Palmer, W.T. Joines, Effect of meandering on dipole antenna resonant frequency. IEEE Antenna Wirel. Propag. Lett. **11**, 122–125 (2012)
15. T. Endo, Y. Sunahara, S. Satoh, T. Katagi, Resonant frequency and radiation efficiency of meander line antennas. Electron. Commun. Jpn. Part 2 (Electron.) **83**, 52–58 (2000)
16. M. Shapari, D.V. Thiel, The impact of reduced conductivity on the performance of wire antennas. IEEE Trans. Antennas Propag. **63**, 4686–4692 (November 2015)
17. X. Qing, C.K. Goh, Z.N. Chen, Measurement of UHF RFID tag antenna impedance. IEEE Int. Work. Antenna Technol. 1–4 (March 2009)
18. H. Zhu, Y.C.A. Ko, T.T. Ye, Impedance measurement for balanced UHF RFID tag antennas. IEEE Radio Wirel. Symp. 128–131 (January 2010)
19. P.V. Nikitin, K.V.S. Rao, R. Martinez, S.F. Lam, Sensitivity and impedance measurements of UHF RFID chips. IEEE Trans. Microw. Theory Technol. **57**, 1297–1302 (May 2009)
20. S. Shao et al., Design approach for UHF RFID tag antennas mounted on a plurality of dielectric surfaces. IEEE Antenna Propag. Mag. 121–128 (October 2014)
21. S. Shao et al., Broadband textile-based passive UHF RFID tag antenna for elastic material. IEEE Antenna Propag. Lett. 67–74 (October 2014)
22. R. Chakraborty, S. Roy, V. Jandhyala, Revisiting RFID link budgets for technology scaling: range maximization of RFID tags. IEEE Trans. Microw. Theory Technol. **59**, 496–503 (February 2011)
23. C.A. Balanis, *Antenna Theory Analysis and Design*, 3rd edn. (Copyright 2005 by John Wiley & Sons, Inc., 2007)

Evaluation of Latency in IEEE 802.11ad

Garima Shukla, M. T. Beg and Brejesh Lall

Abstract It has increasingly become apparent that bandwidth of transmissions will no longer be a constraint in new generations of wireless communications. This is expected to drastically lower the latency in the transmissions and bring about improvement in the QoS of the transmissions. However, the new radio access technologies (e.g. mmWave) will have issues such as beamforming, beam tracking, and resultant overheads which can cause bottlenecks in reduction of latency. New techniques will then be required to overcome these bottlenecks. Medium access is no longer being considered only in terms of either contention-based or contention-free modes; hybrid approaches such as in IEEE 802.11ad are now being implemented. Moreover, the traditional concepts of transmission control, reliability, etc., require new thinking in view of the change brought about by the new generations of wireless transmissions. This paper evaluates the impacts on latency in IEEE 802.11ad which is the settled protocol for short-distance LAN in 60 GHz mmWave domain.

Keywords Latency · 5G · Millimetre wave · Beamforming · Beam tracking · Contention free · Contention based

1 Introduction

Low latency is one of the key requirements in new generation wireless networks especially 5G [1]. Even in non-QoS applications, latencies are expected to be low, as far as feasible. It has been remarked that a round-trip latency of 1 ms on account of speed of propagation of light (300 km/ms) alone restricts the maximum distance of the receiver to approximately 150 kms. Wireless networks involve much lesser distances, i.e. of the order of 100s of metres and even of the order of metres in mmWave personal area networks.

G. Shukla (✉) · M. T. Beg
Faculty of Engineering and Technology, Jamia Millia Islamia, New Delhi 110025, India
e-mail: garima.upadhyaya@gmail.com

B. Lall
Indian Institute of Technology, New Delhi, New Delhi 110016, India

H. S. Saini et al. (eds.), *Innovations in Electronics and Communication Engineering*, Lecture Notes in Networks and Systems 107,
https://doi.org/10.1007/978-981-15-3172-9_15

The latency in a Wi-fi network depends on the equipment specification as well as the load on that equipment. The propagation latency for Wi-fi ranges is under a millisecond, and for personal area network it is much lower. However, the data is packed in successive physical data units (PDUs), associated with various layers of the network, processed by the application, encoded/decoded, etc., which means that the latency of a Wi-fi system is several orders of magnitude larger than the propagation latency. Round-trip time (RTT) is an accepted measure of latency. In typical Wi-fi's in common use, there is not more than 2–3 ms of latency. In congested networks, it can increase. Higher latency (of higher order of magnitude) may be due to poor connection or interference.

2 Latency Considerations

The observed latency in wireless networks is due to channel characteristics, network constraints, processing in various network layers, transmission control, acknowledgements scheme used, etc. The aspects on which allowable latency decisions are to be taken concern the following:

1. Fundamental limits of performance of the particular wireless network.
2. Development of new waveforms and medium access control techniques.
3. Extent of use of spatial diversity through MIMO and other techniques.
4. The allowable impact of interference on signal reception, the quality of radio resource, and mobility considerations of users and base stations in infrastructure and non-infrastructure wireless networks.
5. Design of network architecture, layer protocols.
6. Status of network infrastructure including the limitations on packet gateways, connectivity between base stations through backhaul and functioning of core network.
7. Requirement of low overheads for control information vis-à-vis transmission of data.
8. Mission-critical applications, e.g. smart grid, industry automation and control, robotics, tactile Internet, and vehicular communication.
9. Virtualization of network functions.
10. Ensuring that a given service is available to the assortment of users.

3 Measuring Latency

There are several metrics for latency in wireless networks. Some of these are:

(a) End-to-end (E2E) latency [2]: This delay figure includes the channel transmission delay, queuing delay, and application delay. This paper will analyse the E2E latency.

(b) User plane latency (3GPP) [3]: This is defined as one-way time of successful delivery of an application layer packet/message from the entrance in radio protocol layer to the entrance in radio protocol of the radio interface, in both uplink or downlink in the unloaded network (assuming the user equipment (UE) is in active state). Typical minimum figures for user plane latency are 4 ms for 5G radios for a single user.
(c) Control plane latency (3GPP) [4]: This is the transition time from a "battery efficient" state (e.g. idle state) to the state wherein there is data transfer (e.g. active state) on a continuous basis. The minimum requirement for control plane latency is 20 ms.

4 Tradeoffs in Latency Considerations

(a) **Trade-off with reliability**: Ensuring low latencies requires that usually there is a trade-off with reliable receipt of communications. Thus, this could involve choosing a data encoding rate much lower than Shannon capacity for efficient resource allocation. Further, hybrid automatic repeat request (HARQ), one of the popular techniques of enhancing reliability, can lead to low spectral efficiency. Often there is a requirement of mixing bursty traffic with low latency traffic which leads to congestion and contention-based access control. Reliability also depends on maximizing energy resources.
(b) **Energy considerations**: The practice of choosing multiplicity of low-power transmissions instead of one high-reliable and high-power transmission to conserve energy resources also contributes to attaining less than optimum reliability.
(c) **Channel feedback**: For closed loop-based channel feedback more bandwidth and system resources are required for channel training and estimation. In the process, more accurate CSI is obtained; however, the data transmission latency is increased. In case of open loop-based feedback systems, broadcasting is efficient but requires more downlink resources. The downlink can be supplemented by means of relay nodes or distributed antennas.
(d) **User density and considerations of antennas, bandwidth and blocklength**: User density in most of the derivations in information theory is assumed to be low. However, in dense deployments of users, the assumption of infinite bandwidth availability for large blocklengths no longer holds as the available bandwidth has to be divided amongst a large number of users. Full buffer users in fixed numbers may not be available. A many-user information theory has been proposed [4] where the number of users increases without bound with the blocklength.

5 Latency in IEEE 802.11ad

IEEE 802.11ad is expected to have very low latency due to possible data rates of up to 7 Gbps and high spatial reuse in transmissions. This allows reduced control and data transmission times simultaneously with ease of communication between multiple pairs of users. Beamforming and beam tracking have an important role to play in ensuring that this is possible. Beamforming ensures high gains to offset distance loss due to use of mmWaves. It also allows for high spatial diversity as adjacent beams are isolated and communication between proximate pairs of transmitters and users can continue unhindered. However, full beamforming training is to be ensured before initial channel access and during the event of significant change in channel conditions. Beam tracking is required in the event of channel conditions changing enough to cause a degradation in performance.

The IEEE 802.11ad protocol incorporates both beamforming and beam tracking. Medium access is in multiples of beacon intervals, as shown in Fig. 1. Beamforming is achieved through the process of sector level sweep (SLS) followed by an optional beam refinement process (BRP). The SLS phase utilizes low rate modulation and coding scheme (MCS) at the PHY layer. The best sector is selected, and thereafter, within this "coarse" direction, in the BRP phase, antenna settings found during SLS are fine-tuned. Antenna training incorporates achieving highest directional gain. Data transfer is conducted during the data transfer time in both contention-free and contention-based modes.

Beamforming and beam tracking add random latency which increases with the number of nodes in the system. Each node may experience the conditions necessitating beamforming afresh and continuous beam tracking. This may also be occasioned by nobility of nodes. Beyond a certain number of nodes, the transmissions may have to change from contention-free to contention-based access. In contention-based access, latency is enhanced due to the use of backoff timer and congestion control algorithms. Even in contention-free access, the system will have high scheduling overhead. In either case, there will be a high overhead for beamforming because the channel will change by the next time a user can access the channel [5, 6].

Also, the previous discussion is valid for establishing point-to-point communication. In case point-to-multipoint communication is required, then a new round of search and beamforming procedure takes place. The lengthy preamble, sent at the lowest data rate, for every transmission is a limiting factor for improving latencies.

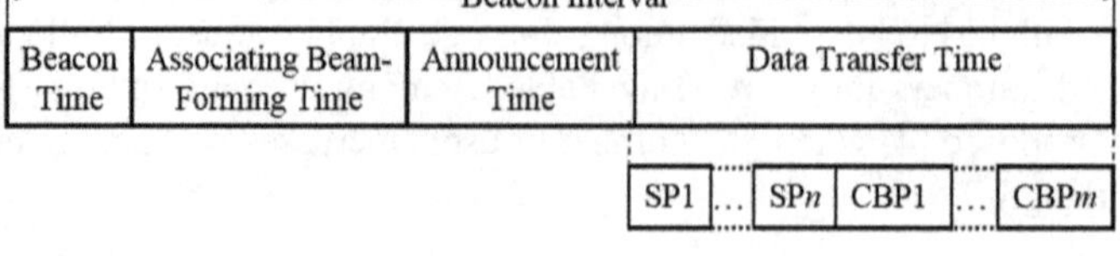

Fig. 1 IEEE 802.11ad beacon interval for medium access

The guard symbols in each SC-FDMA symbol and cyclic prefixes before every symbol in OFDM systems are sent at the full-speed data rate, and their overhead is lower than that of the preamble [7–11].

6 Evaluating Latency in IEEE802.11ad

An evaluation of the latency for a single link in a 5G network operating at 60 GHz was undertaken. The user node was first considered stationary and then mobile. Further, the number of nodes was also increased to assess the impact on latency. The ns-3 discrete event simulator was selected for the study [12]. The following parameters were considered during the evaluation (Table 1).

The link distance was varied keeping all other conditions unchanged. The delay has a close coupling with the propagation conditions as seen in Fig. 2.

As can be seen from Fig. 2, there is a gradual decline in throughput while the link distance is increased up to nearly 3 m. This gradual decline is on account of distance loss, enhanced possibility of interference with other directional beams. Beyond 3 m, there is a sharp fall in throughput on account of insufficient SINR due to high attenuation of mmWaves.

The number of nodes was also varied to study the impact on latency, after dividing the service period allocations between the stations. The topography comprised distances of up to 2 m between any two sending receding nodes along with a suitable choice of the polygon containing the stations at the vertices. Thus, the Access Point

Table 1 Simulation parameters for the study

S. no.	Simulation parameter	Simulation value
1	Antenna	Directional 60 GHz antenna with 12 sectors Tx 10 dBm = 10 mW Tx gain, Rx gain = 0 Energy detection threshold = −79 + 3 dBm
2	Socket type and transmission rate	Universal Datagram Protocol (UDP), on-off application, 100 Mbps data rate. The UDP packet delivery does not involve receipt of SYN and ACK packages and is more of a "broadcast". The UDP gives an optimistic consideration of the latency figures
3	Channel	Constant speed propagation delay model, Friis propagation loss model
4	MAC	802.11ad, service period set to 3200 ms for participating nodes up to four station nodes evaluated in the paper
5	Modulation and coding scheme	OFDM 6756.75 Mbps
6	Mobility models	Constant speed mobility model stationary model for the user

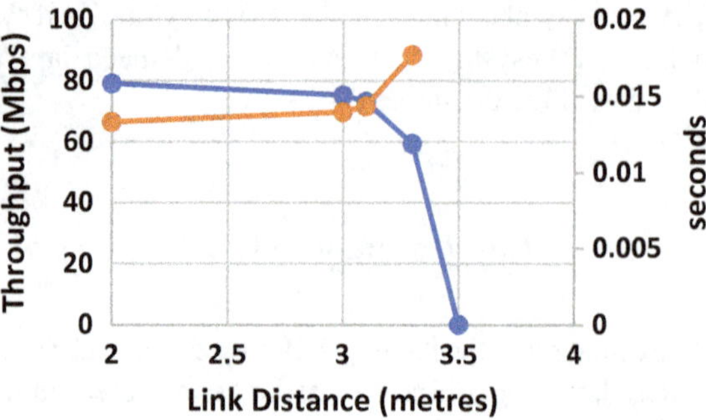

Fig. 2 Variation of throughput delay with link distance

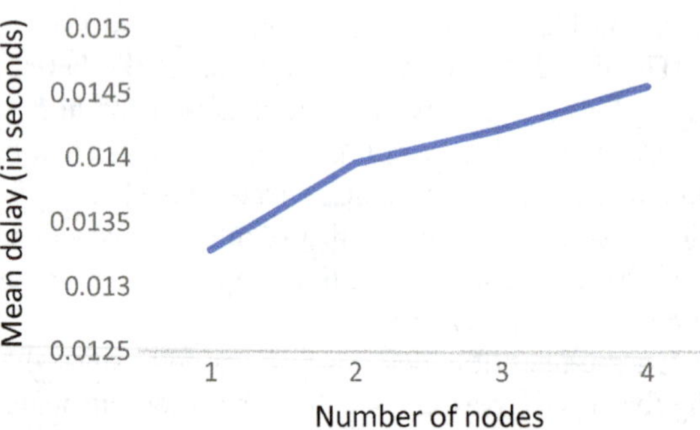

Fig. 3 Variation of delay with number of station nodes

and two nodes were kept at the vertices of a triangle, the AP and three nodes were kept at vertices of a square, and the AP and four nodes were kept at the vertices of a pentagon, and the mean delay was recorded as shown in Fig. 3.

Figure 3 shows that the mean delay rises with the increase in the number of nodes. This is in account of increased requirements of control transmissions, contention-free and contention-based channel access, interference due to the beam from other nodes. It is important that the mean delay be kept to a minimum, given the constraints. Thus, more appropriate network parameters have to be found out to support network densification.

7 Conclusion

The elaborate architecture of IEEE 802.11ad protocol makes the system quite susceptible for unnecessary enhancement in delays due to node densification and increase in link distance. However, the impact can be minimized through a judicious use of contention-free and contention-based access periods, provided the link does not degrade completely due to the attenuation inherent in mmWaves.

It is also important that the various congestion control mechanisms, channel monitoring and spatial diversity, be also utilized simultaneously.

The latency observed in the given configuration in the ns-3 module is more than 10 ms and will be studied to reduce it to below 5 ms by tuning the default parameters of the simulator.

Further work will be undertaken to understand the impact of congestion control algorithms, queuing disciplines, dynamic service period allocation, spatial diversity on delay in IEEE 802.11ad transmissions.

References

1. G. Yang, M. Xiao, H. Vincent, Poor low-latency millimeter-wave communications: traffic dispersion or network densification? arXiv:1709.08410v2 [cs.NI] 14 Mar. 2018
2. R. Ford, M. Zhang, M. Mezzavilla, S. Dutta, S. Rangan, M. Zorzi, Achieving ultra-low latency in 5G millimeter wave cellular networks arXiv:1602.06925v1 [cs.NI] 22 Feb. 2016
3. S. Rangan, T.S. Rappaport, E. Erkip, Millimeter-wave cellular wireless networks: potentials and challenges. Proc. IEEE **102**(3), 366–385 (2014)
4. M. Akdeniz, Y. Liu, M. Samimi, S. Sun, S. Rangan, T. Rappaport, E. Erkip, Millimeter wave channel modeling and cellular capacity evaluation. IEEE J. Sel. Areas Commun. **32**(6), 1164–1179 (2014)
5. J. Gozalvez, 5G tests and demonstrations [mobile radio]. IEEE Veh. Technol. Mag. **10**(2), 16–25 (2015)
6. G. Fettweis, The tactile internet: Applications and challenges. IEEE Veh. Technol. Mag. **9**(1) (2014)
7. P. Popovsk, V. Brau, H.-P. Mayer, P. Fertl, Z. Ren, D. Gonzales-Serrano, E.G. Str¨om, T. Svensson, H. Taoka, P. Agyapong et al., EU FP7 INFSO-ICT-317669 METIS, D1. 1 scenarios, requirements and KPIs for 5G mobile and wireless system (2013)
8. P.K. Agyapong, M. Iwamura, D. Staehle, W. Kiess, A. Benjebbour, Design considerations for a 5g network architecture. IEEE Commun. Mag. **52**(11), 65–75 (2014)
9. S. Dutta, M. Mezzavilla, R. Ford, M. Zhang, S. Rangan, M. Zorzi, Frame structure design and analysis for millimeter wave cellular systems, in arXiv:1512.05691 [cs.NI] (2015)
10. P. Kela, M. Costa, J. Salmi, K. Leppanen, J. Turkka, T. Hiltunen, M. Hronec, A novel radio frame structure for 5G dense outdoor radio access networks, in *Proceedings of IEEE 81st Vehicular Technology Conference (VTC Spring)* (2015), pp. 1–6
11. E. Lahetkangas, K. Pajukoski, J. Vihriala, G. Berardinelli, M. Lauridsen, E. Tiirola, P. Mogensen, Achieving low latency and energy consumption by 5G TDD mode optimization, in *Proceedings of IEEE International Conference on Communications Workshops (ICC)* (2014), pp. 1–6
12. M. Mezzavilla, S. Dutta, M. Zhang, M. R. Akdeniz, S. Rangan, 5g mmwave module for the ns-3 network simulator, in *Proceedings of the 18th ACM International Conference on Modeling, Analysis and Simulation of Wireless and Mobile Systems (MSWiM '15)*

An Approach for Real-Time Indoor Localization Based on Visible Light Communication System

Dharmendra Dhote and Manju K. Chatopadhyay

Abstract Indoor localization is an emerging research area in recent times. During the last few years, it is observed that the visible light communication (VLC) has become a strong candidate for wireless communication. It has been gaining popularity in indoor localization systems. In this work, we investigate the VLC-based system for real-time positioning. We developed a small prototype of a VLC-based indoor localization system in order to determine the real-time position of the receiver. For prototyping, LEDs are used as transmitter and photodiode as receiver, controlled by Raspberry Pi (RPi) microcontroller. These LEDs are modulated at different frequencies for unique identification. Received signal strength (RSS) method is used for distance calculation from the transmitter to the receiver. The calculated distance is sent to the controller where the individual frequency of the transmitter will be extracted. Trilateration algorithms are used for estimating the receiver position. The proposed system performs well in real time.

Keywords VLC · RSS · DRMS · Indoor localization

1 Introduction

Many technologies (viz. Wi-Fi, Bluetooth, RFID, Zigbee, and GPS) have been in use to estimate the position of indoor environment. Each technology has its own advantages and disadvantages. In recent years, visible light communication has been gaining attention for indoor localization [1, 2].

VLC-based indoor localization involves two modules—a transmitter and a receiver. Generally, LED is used as a transmitter unit because of its high bandwidth and fast switching capability [3]. Photodiode (PD) is used as a receiver unit due to its high reception bandwidth.

Transmitter module requires multiple LEDs which can be modulated by different techniques. On/Off keying is a simple technique in which LED is modulated by turning LED on and off in order to transmit the bit 1 and 0. In Ref. [4], author used the

D. Dhote · M. K. Chatopadhyay (✉)
School of Electronics, Devi Ahilya Vishwavidyalaya, Indore, Madhya Pradesh 452017, India
e-mail: mkorwal@yahoo.com

H. S. Saini et al. (eds.), *Innovations in Electronics and Communication Engineering*, Lecture Notes in Networks and Systems 107,
https://doi.org/10.1007/978-981-15-3172-9_16

NRZ-OOK modulation technique and achieved a speed of 550 Mbit/s in real time. A pulse position modulation (PPM) uses a position of pulse and identifies the transmitter symbol [4], time-division multiplexing (TDM) [5], and frequency-division multiplexing (FDM) [6]. A code-division multiplexing (CDMA) is a multiple access technology, wherein each LED is assigned its unique code and the receiver can decrypt the data.

Receiver module consists of photodiode that receives the LED signals. After extracting the signal, position techniques are used to locate the target device. Different position techniques have been studied to estimate the location. (i) Time of arrival (TOA)—in Ref. [7] author used white LED as transmitter and assumed that the transmitter and receiver are perfectly synchronized. (ii) Time difference of arrival (TDOA)—the positional accuracy of 3.2 cm has been proposed in Ref. [8]. However, the author used simulation for the experiment. In the practical environment, this accuracy could not be obtained. (iii) Angle of arrival (AOA) measures angle between transmitters and receivers, and using geometric property, it estimates the distance. (iv) Received signal strength (RSS) is an asynchronous technique; using the attenuation property of the transmitter signal, the distance can be obtained [6].

In general, it is noticed that the TOA is a simple method but it requires synchronization between transmitter and receiver unit. It is tedious to obtain in practice and, hence, is error-prone. In TDOA, synchronization is required between transmitters. This is less difficult than TOA but clock resolution and PD response can cause errors here. RSS has the advantage of asynchronous operation but is subject to path loss, especially, when receivers are not in the line of sight to the transmitter. However, RSS is easy to use as compared to TOA and TDOA [7–10].

Triangulation is a common positioning algorithm that requires angle or distance between transmitters and receivers. Two derivations are there—lateration and angulation. Lateration involves TOA, TDOA, and RSS for position estimation, and angulation uses AOA to estimate the target device's position.

2 Proposed Methodology

2.1 RSS

RSS received method is used to calculate the transmitter to receiver distance. When receiver (PD) is in the line of sight, then using Lambertian characteristics, the relation between transmitted and received power is given as:

$$P_{\mathrm{r}} = \begin{cases} P_{\mathrm{t}} \frac{(m+1)A}{2\pi d^2} \cos^m(\emptyset) T_{\mathrm{s}}(\psi) g(\psi) \cos(\psi) \\ \qquad 0 \leq \psi \leq \boldsymbol{\psi}_c \\ \qquad 0, \quad \psi \leq \boldsymbol{\psi}_c \end{cases} \tag{1}$$

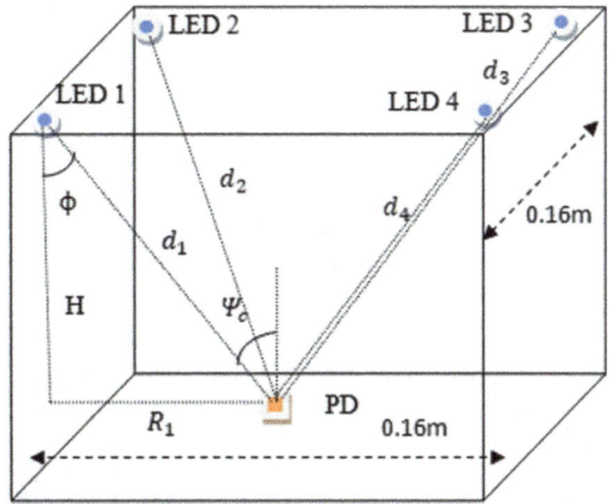

Fig. 1 Transmitter and receiver model with four LEDs and a photodiode (PD)

where P_t is the transmitted power, A is the physical area of the detector in the photodiode (PD), m is the order of Lambertian emission, φ is the irradiance angle, ψ is the incidence angle, $T_s(\psi)$ is the gain of the optical filter, $g(\psi)$ is the gain of the optical concentrator, Ψ_c is the field of view at the PD, and d is the transmitter to receiver distance.

In our experiments, FOV is taken 60°, so m can be calculated as (Fig. 1):

$$m = \frac{-\log 2}{\log(\cos(\Phi_{0.5}))} = \frac{-\log 2}{\log(\cos(60^\circ))} = 1 \tag{2}$$

We assume no optical filter and concentrator is used, by substituting the value of m and $(T_s(\psi) = g(\psi) = 1)$ in Eq. (1), we can obtain equation as

$$P_r = P_t \frac{A}{\pi d^2} \cos(\Phi) \cos(\Psi) \tag{3}$$

Using Eq. (3), the linear distance from transmitter to receiver is calculated as follows:

$$R_n = \sqrt{d_n^2 - H^2} \tag{4}$$

In order to calculate the physical distance between the transmitter and the receiver, the signal strength at the receiver must be known. Equation (2) is used to find the signal strength. Rearranging Eq. (3) gives an equation in terms of diagonal distance d between the LED and the photodiode.

$$d = \sqrt[2]{\frac{AH^2 P_t}{\pi P_r}} \tag{5}$$

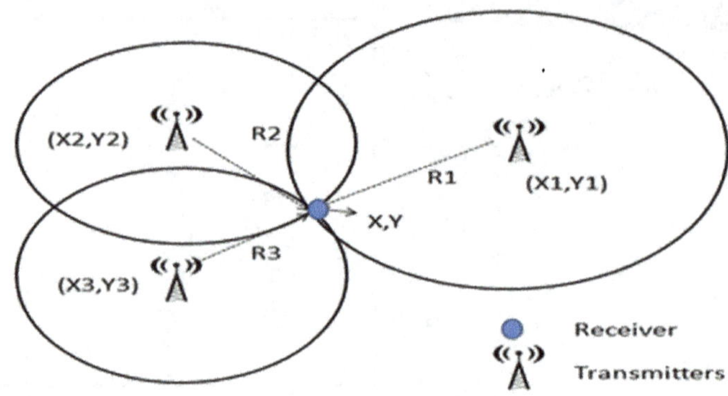

Fig. 2 Trilateration model

2.2 *Trilateration*

In our work, circular trilateration is utilized to estimate the location of the receiver. Transmitter locations are known and the distance from each transmitter to the receivers is obtained using RSS technique.

As shown in Fig. 2, distance between the ith transmitter and the receiver is given as R_i. Each circle as shown in Fig. 2 is a set of the receiver's possible locations from a single range measurement, which can be expressed by following equation [11]:

$$(X_i - x)^2 + (Y_i - y)^2 = R_i^2$$

where $i = 1, 2, \ldots, n$ and n is the number of transmitter points.

$$\begin{aligned} R_i^2 - R_1^2 &= (x - X_i)^2 + (y - Y_i)^2 - (x - X_1)^2 + (y - Y_1)^2 \\ &= X_i^2 + Y_i^2 - X_1^2 - Y_1^2 - 2x(X_i - X_1) - 2y(Y_i - Y_1) \end{aligned}$$

Then, the equations can be described into matrix form as:

$$\mathbf{AX} = \mathbf{B}$$

where $X = \begin{bmatrix} x \; y \end{bmatrix}^{\mathrm{T}}$

$$\mathbf{A} = \begin{bmatrix} X_2 - X_1 & Y_2 - Y_1 \\ \vdots & \vdots \\ X_n - X_1 & Y_n - Y_1 \end{bmatrix}$$

and

$$\mathbf{B} = \frac{1}{2}\begin{bmatrix} \left(R_1^2 - R_2^2\right) + \left(X_2^2 + Y_2^2\right) - \left(X_1^2 - Y_1^2\right) \\ \vdots \\ \left(R_1^2 - R_n^2\right) + \left(X_n^2 + Y_n^2\right) - \left(X_1^2 - Y_1^2\right) \end{bmatrix}$$

The least squares solution of the system is then given by:

$$X = (A^T A)^{-1} A^T B \tag{6}$$

The flow diagram has two sections shown in Fig. 3: first section is the hardware section which consists of LEDs, PD, ADC converter, and controller. The second is the computation section which performs the calculation in order to estimate PD location.

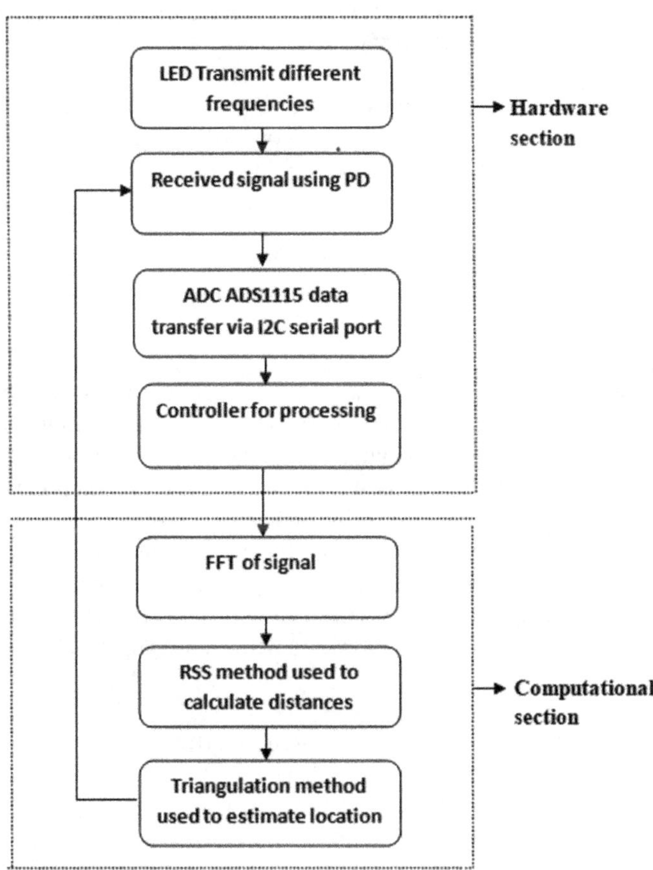

Fig. 3 Flow diagram of proposed approach for indoor localization

3 System Description

This section describes the components used to perform the experiments. Majorly, there are four units, viz. transmitter, receiver, A/D converter, and controller. Each is discussed below:

Transmitter: In our work, LEDs are used as transmitter unit which are modulated by a controller. Frequency-division multiplexing generates a unique frequency by each LED. The LED has 20 mw maximum power and can operate up to 5 V.

Receiver: Photodiode (PD) OPT101 from Texas Instrument is used a receiver unit. It has a responsivity of 0.45 V/W at visible light of 650 nm. It generates an output voltage.

A/D converter: The photodiode output is analog voltage. Before processing the data in Raspberry Pi, it needs to be converted to digital form. Texas Instrument ADC (ADS1115) is used as the A/D converter. It has 16 bits resolution output and programmable data rate of 8 SPS to 860 SPS. Data is sent via I2C—compatible serial interface.

Controller: Raspberry Pi 3 model is used as a controller to process the digital data. RPi CPU has 1.2 GHz 64/32-bit quad-core ARM Cortex-A53 processor and 1 GB RAM. It has capability to compute FFT and to do other computations on a single processor.

4 Experiments and Results

Initially, these experiments were performed inside the closed box to avoid the external light effect. Dimensions of closed box are 0.16 m $\times$ 0.16 m $\times$ 0.20 m with area of 0.0256 m^2. LEDs are used as transmitter units and have been adjusted over different frequencies. Four LEDs have been used and positioned at four corners to perform the experiment. Table 1 represents the position and frequency of different LEDs. During the experiments, photodiode is placed at the center of the area. Samples were taken in real time during with time gap of 1 s. FFT is performed over the collected sample and different frequency components were extracted. Figure 4 shows the collected input samples, and Fig. 5 shows the extracted frequency components.

Table 1 LED position and frequency

LED ID	Frequency (HZ)	Position in cm
LED 1	50	[0, 0]
LED 2	145	[0, 16]
LED 3	125	[16, 16]
LED 4	70	[16, 0]

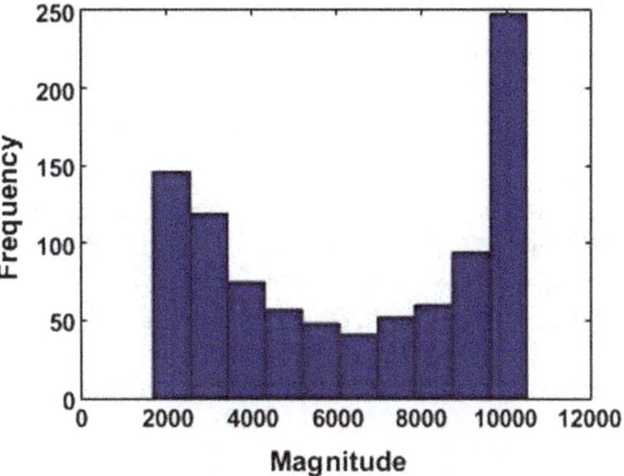

Fig. 4 Histogram of input data

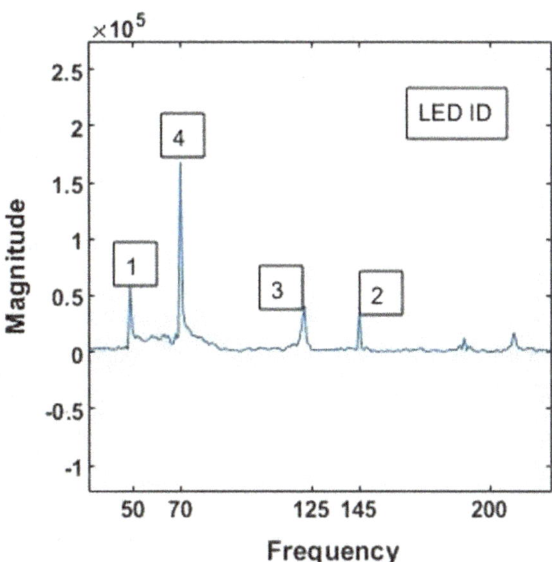

Fig. 5 Frequency plot output of FFT

After frequency extraction, RSS method calculates the distance of receiver from each transmitter. Subsequently, trilateration method estimates the location in 2D space.

The proposed approach works in real time and the location is updated after every 4 s. The distance root mean square error (DRMS) is measured to define the two-dimensional accuracy. The horizontal position error () can be computed from the direction of the coordinate axis's known position. The DRMS measurement shows that the probability of being within the circle is 63–68% [12].

$$\mathrm{DRMS} = \sqrt{\sigma_x^2 + \sigma_y^2} \tag{7}$$

To calculate the 95% confidence region for an object to be located in space, we can modify Eq. (7) as follows:

$$2\text{DRMS} = 2\sqrt{\sigma_x^2 + \sigma_y^2} \tag{8}$$

The distance root mean square error distribution is shown in Fig. 6, and the DRMS distribution in the different location is shown by boxplot in Fig. 7. Figure 8 shows the estimated location of the object in the space where the large blue circle represents the position accuracy computed from DRMS.

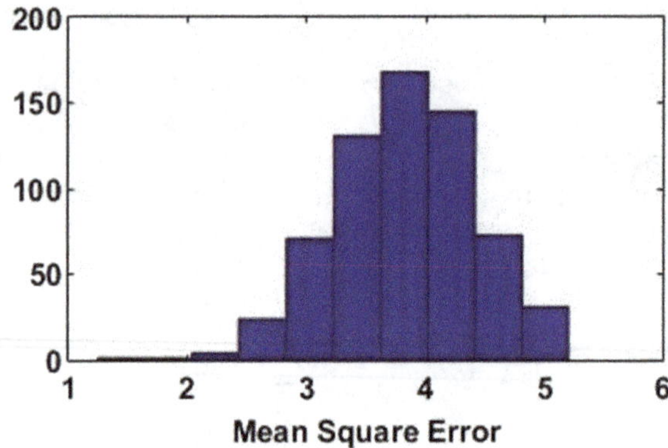

Fig. 6 Distance mean square distribution of location at $x = 16$ and $y = 8$

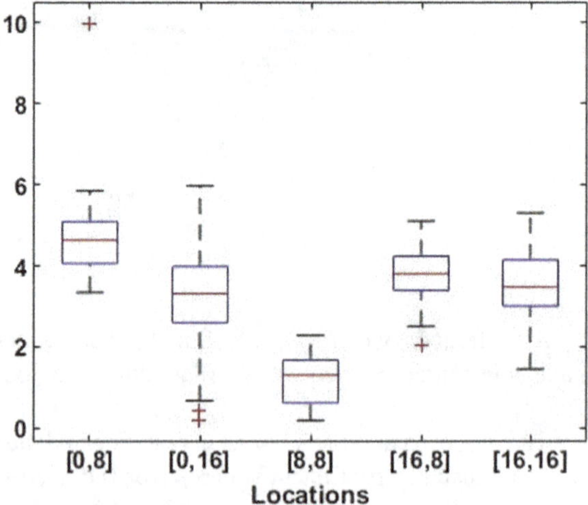

Fig. 7 Boxplot represents mean square distribution of different locations

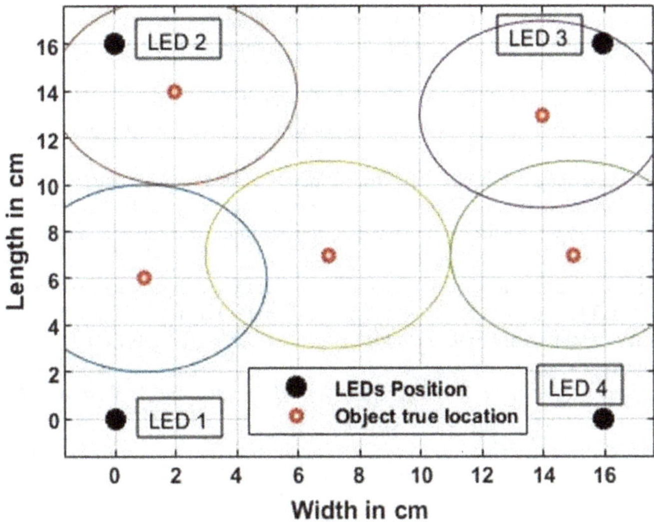

Fig. 8 Position accuracy based on DRMS

5 Conclusion

The proposed approach is tested on a small scale and it works well in real time. The obtained results successfully show that our approach can locate the object within confidence region. Obtained distance root mean square error is 0.04 m in the experiments. To extend it, one may perform this experiment over large scale to check the versatility and robustness. Furthermore, other algorithms may be tested to increase the accuracy.

References

1. P.H. Pathak, X. Feng, P. Hu, P. Mohapatra, Visible light communication, networking, and sensing: a survey potential and challenges. IEEE Commun. Surv. Tutorials **17**(4), 2047–2077 (2015)
2. A. Jovicic, J. Li, T. Richardson, Visible light communication: opportunities, challenges and the path to market. IEEE Commun. Mag. **51**(12), 26–32 (2013)
3. T. Komine, M. Nakagawa, Fundamental analysis for visible-light communication system using LED lights. IEEE Trans. Consum. Electron. **50**(1), 100–107 (2004)
4. H. Li, X. Chen, J. Guo, H. Chen, A 550 Mbit/s real-time visible light communication system based on phosphorescent. White light LED for practical high-speed low-complexity application. Opt. Express **22**(22), 27 (2014)
5. S. Yang, D. Kim, H. Kim, Y. Son, S. Han, S, Indoor positioning system based on visible light using location code, in *Fourth International Conference of Communications Electronics*, 360–363 (2012)

6. Y.C. See, N.M. Noor, Y.M. Calvin Tan, Investigation of indoor positioning system using visible light communication, in *IEEE Region 10 Annual International Conference, Proceedings/TENCON* (2017)
7. T.Q. Wang, Y.A. Sekercioglu, A. Neild, J. Armstrong, Position accuracy of time-of-arrival based ranging using visible light with application in indoor localization systems. J. Light. Technol. **31**(20), 3302–3308 (2013)
8. T.-H. Do, J. Hwang, M. Yoo, TDoA based indoor visible light positioning systems, in *Fifth International Conference on Ubiquitous and Future Networks (ICUFN)*, 456–458 (2013)
9. T.-H. Do, M. Yoo, An in-depth survey of visible light communication based positioning systems. Sensors **16**(5), 678 (2016)
10. C.Y. Wang, L. Wang, X.F. Chi, S.X. Liu, W.X. Shi, J. Deng, The research of indoor positioning based on visible light communication. China Commun. **12**(8), 85–92 (2015)
11. Chichester indoor positioning methods using VLC LEDs, in *Short-Range Optical Wireless* (Wiley, UK, 2015), pp. 225–262
12. E.D. Kaplan, C.J. Hegarty, *Understanding GPS - Principles and Applications* (Artech House, 2006)

Performance Analysis of 3 × 8 Multiband Antenna Arrays with Uniform and Non-uniform Inputs for RADAR Applications

P. Sai Vinay Kumar and M. N. Giri Prasad

Abstract Implementation of uniform and non-uniform inputs on antenna arrays has been performed and considered two 3 × 8 multiband antenna arrays. First antenna is operating in the frequencies of 1.61 and 2.492 GHz covering the L-band and S-band frequencies which are used for RADAR applications. The second antenna is resonating at four frequencies of 1.176, 1.575, 1.6 and 2.492 GHz covering the L-band and S-band frequencies which are used for navigational applications. Stacked patch configuration is used for both antennas. Performance of the two antenna arrays in terms of gain and side lobe level is compared. Both the antennas are designed on FR4 epoxy substrate which is having a thickness of 1.6 mm. Coaxial feed has been used for antenna elements, and each element of the array is fed by an independent coaxial feed. Beam shaping mechanism is obtained with the help of uniform and non-uniform inputs. Commercial available 3D model simulator tool ANSYS HFSS has been used to simulate the antenna array.

Keywords Antenna array · Non-uniform inputs · Coax feed · Multiband

1 Introduction

With the advent of the microstrip antennas and low profile technologies in the electronics, the size of the communication systems has reduced enabling the use of RADAR application in many new areas for identifying the object. Though microstrip antennas which are known for their small size are used to get small size for the communication systems, further miniaturization is always needed along with high gain for long coverage area. One of the best techniques to increase the gain of the antenna is to develop antenna array but care has to be taken such that the array size will not affect the overall size of the communication system, so to meet the needs of the current systems we need an antenna array with considerable gain in a compact structure.

P. Sai Vinay Kumar (✉) · M. N. Giri Prasad
Department of Electronics and Communication Engineering, JNTU Ananthapuramu, Anantapur, Andhra Pradesh, India
e-mail: vinnie.polepalli@gmail.com

H. S. Saini et al. (eds.), *Innovations in Electronics and Communication Engineering*, Lecture Notes in Networks and Systems 107,
https://doi.org/10.1007/978-981-15-3172-9_17

A triple-band antenna array for X, Ku and K bands has been proposed in [1] where slots have been used to generate multiple resonating frequencies. An antenna array with series and parallel feed elements has been proposed in [2] which work for the frequency of 9.3–9.4 GHz. A double-sided antenna array radiating at two frequency bands is proposed in [3], one side of the antenna radiates for single band and the other side of the antenna array radiates for two bands. A cavity-backed antenna array is shown in [4] where to increase the gain of the antenna an air cavity is introduced in between the radiating element and ground plane.

In this paper, two antenna arrays were proposed with different number of operating frequencies. FR4 substrate has been used to design the antennas with a length and width of 80 mm. The board's thickness is 4 mm for each layer, and relative dielectric constant is $\varepsilon_r = 4.4$. First antenna is operating in the frequencies of 1.61 and 2.492 GHz covering the L-band and S-band frequencies which are used for RADAR applications. The dimensions of the lower patch are 39.4 mm × 40 mm, and the dimensions of the upper patch are 27 mm × 29.1 mm. The second antenna is resonating at four frequencies of 1.176, 1.575, 1.6 and 2.492 GHz covering the L-band and S-band frequencies which are used for navigational applications [5]. The dimension of the lower patch is 55.2 mm × 55.2 mm, the dimension of middle patch is 39.4 mm × 40 mm and the dimensions of the upper patch are 26 mm × 28.1 mm. To maintain the compact nature, interelement spacing in the antenna array is chosen to be 0.5 times the free space wavelength [6, 7]. Simulation studies were performed on the two antenna arrays by implementing uniform and non-uniform inputs at different levels, and the two arrays have been analysed in terms of gain and side lobe level at different frequencies of operation.

2 Antenna Design and Configuration

Proposed antenna array is twenty-four element antenna array aligned in 3 × 8 matrix format. We considered two different antennas with different number of operating frequencies. One array is having dual frequency of operation which is achieved by using two rectangular radiating elements using stacked patch technique, and the other array is having quad frequency of operation which is achieved by using three rectangular radiating elements using stacked patch technique. Coaxial feed has been given to antenna elements and each element is fed by an independent coaxial feed. Commercial available 3D model simulator tool ANSYS HFSS has been used to simulate the antenna array. To obtain the best performance, the interelement spacing in between the antenna elements is taken as 0.5λ [8].

The dimensions of the radiating elements were calculated based upon the following equations. Equations (1)–(5) are used to calculate the dimensions of the rectangular patch [9, 10]

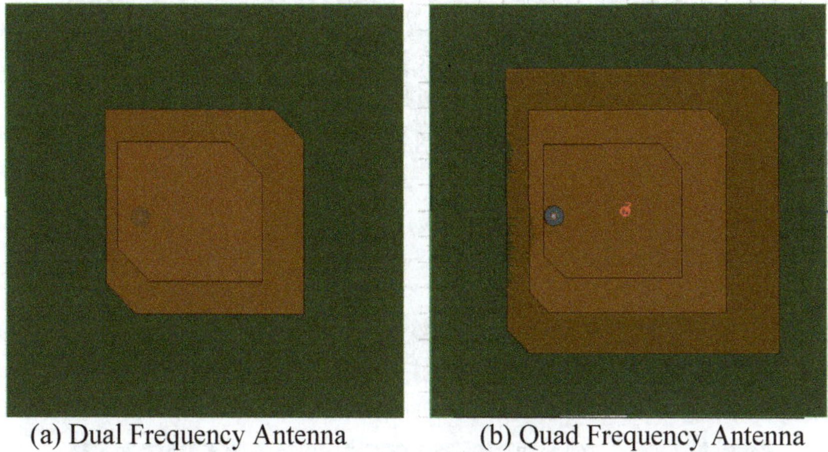

Fig. 1 Antenna element

$$W = \frac{c}{2f_o\sqrt{\frac{(\varepsilon_r+1)}{2}}} \tag{1}$$

$$\varepsilon_{\text{eff}} = \frac{\varepsilon_r + 1}{2} + \frac{\varepsilon_r - 1}{2}\left[1 + 12\frac{h}{W}\right]^{\frac{1}{2}} \tag{2}$$

$$L_{\text{eff}} = \frac{c}{2f_o\sqrt{\varepsilon_{\text{eff}}}} \tag{3}$$

$$\Delta L = 0.412h\frac{(\varepsilon_{\text{eff}} + 0.3)\left(\frac{W}{h} + 0.264\right)}{(\varepsilon_{\text{eff}} - 0.258)\left(\frac{W}{h} + 0.8\right)} \tag{4}$$

$$L = L_{\text{eff}} - 2\Delta L \tag{5}$$

The simulated antenna structures of the antenna elements and antenna arrays are shown in Figs. 1 and 2, respectively.

Fabricated models of two antennas are given in Fig. 3.

3 Results

Antenna arrays are analysed using HFSS software, uniform and non-uniform inputs have been implemented on the antenna arrays and the effect on gain at different frequencies is studied. For non-uniform array antenna considered pascal triangle weights as the non-uniform inputs for the array, pascal triangle weights are calculated for eight iterations to get a series of eight weights as the antenna array considered having eight-element row. The calculated weights are 1, 7, 21, 35, 35, 21, 7, 1. These

(a) Dual Frequency Antenna Array

(b) Quad Frequency Antenna Array

Fig. 2 Antenna array

weights are given as input to the array and the effect on gain and side lobe levels is analysed.

Figure 4 depicts the return loss of arrays. The working frequency of the dual-frequency antenna is 1.60 and 2.492 GHz, and the working frequency of the quad antenna is 1.176, 1.575, 1.60 and 2.492 GHz.

Figure 5 depicts the gain of the dual-frequency antenna array at 1.6 GHz for uniform and non-uniform inputs. It is observed that the side lobes reduced by 13 dB and gain of the antenna array reduced by 1.1 dB for array with non-uniform inputs.

Figure 6 depicts the gain of the dual-frequency antenna array at 2.492 GHz for uniform and non-uniform inputs. It is observed that the side lobes reduced by 4 dB and gain of the antenna array reduced by 1.2 dB for array with non-uniform inputs.

Figure 7 depicts the gain of the quad-frequency antenna array at 1.176 GHz for uniform and non-uniform inputs. It is observed that there is a slight reduction in side

(a) Dual Frequency Fabricated Antenna

(b) Quad Frequency Fabricated antenna

Fig. 3 Fabricated antenna models

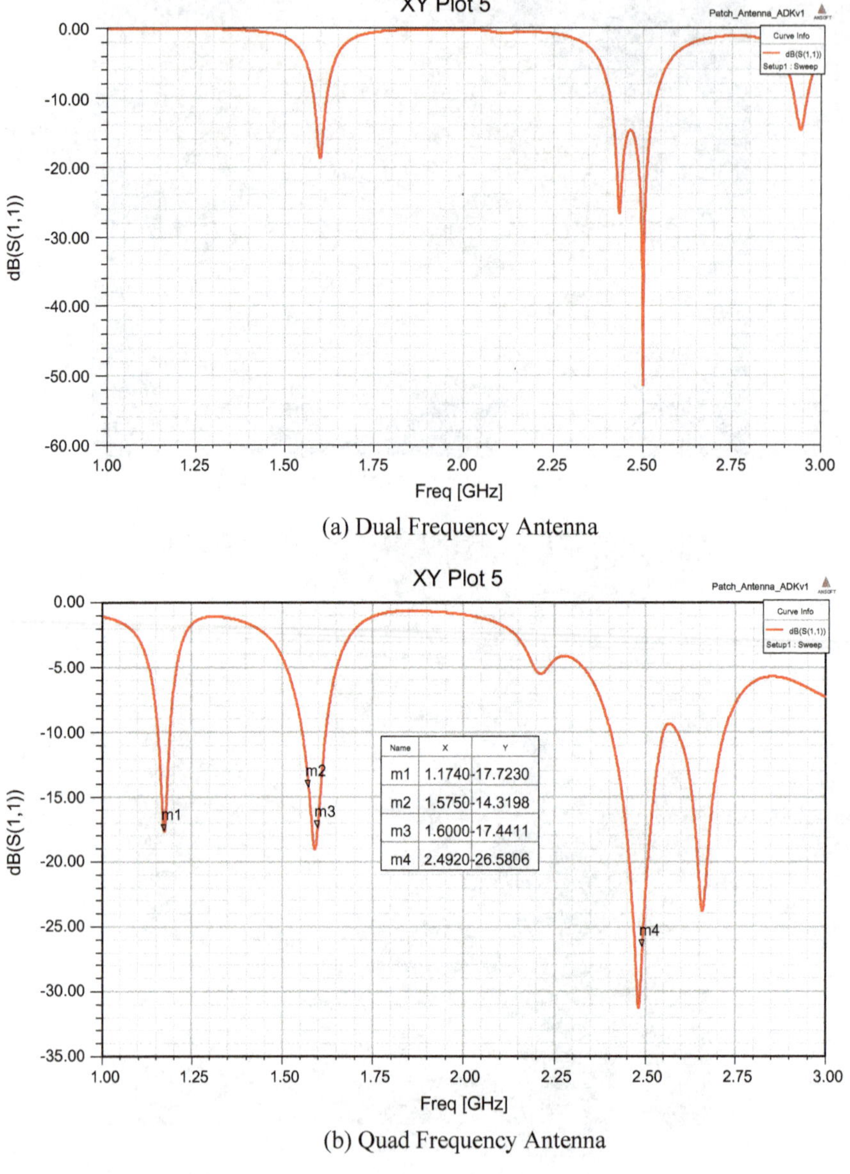

(a) Dual Frequency Antenna

(b) Quad Frequency Antenna

Fig. 4 Return loss

lobes as the side lobe level is already very high and gain of the antenna array reduced by 1.1 dB for array with non-uniform inputs.

Figure 8 depicts the gain of the quad-frequency antenna array at 1.575 GHz for uniform and non-uniform inputs. It is observed that the side lobes reduced by 1 dB and gain of the antenna array reduced by 1.1 dB for array with non-uniform inputs.

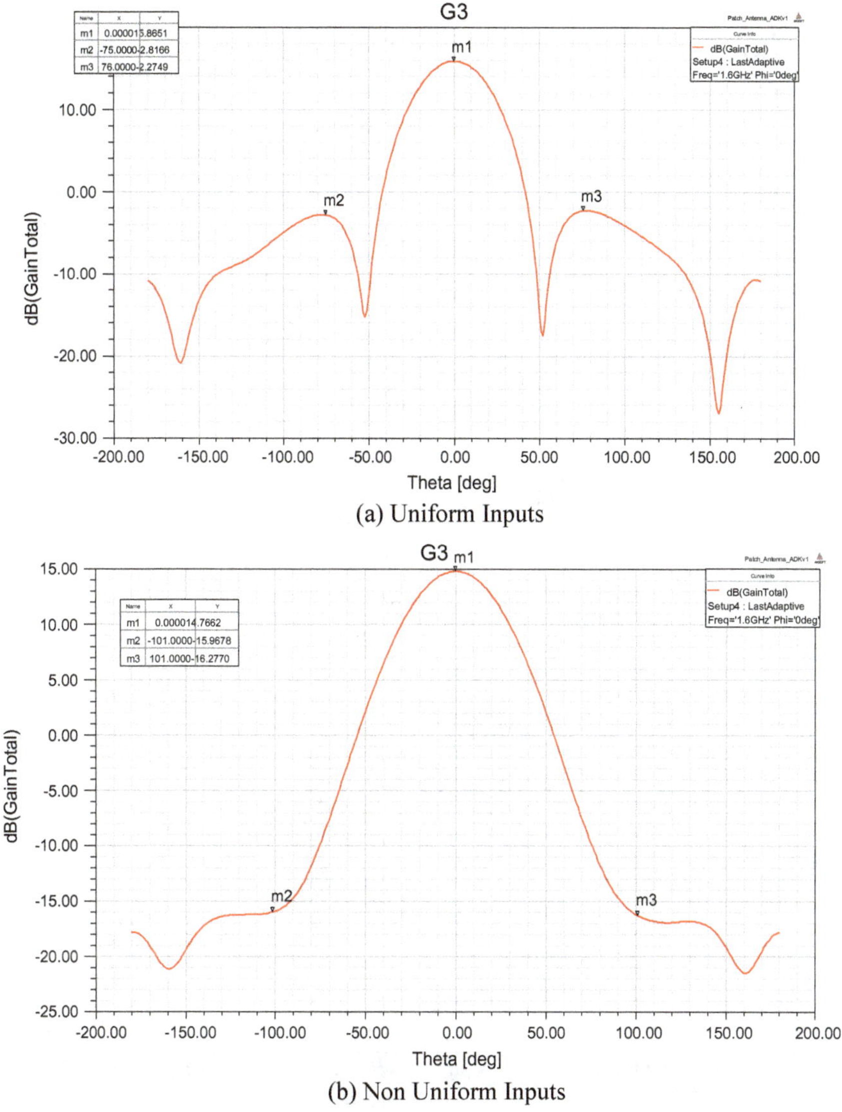

(a) Uniform Inputs

(b) Non Uniform Inputs

Fig. 5 Gain at 1.6 GHz

Figure 9 depicts the gain of the quad-frequency antenna array at 1.6 GHz for uniform and non-uniform inputs. It is observed that there is a slight reduction in side lobes as the side lobe level is already very high and gain of the antenna array reduced by 1.1 dB for array with non-uniform inputs.

Figure 10 depicts the gain of the quad-frequency antenna array at 2.492 GHz for uniform and non-uniform inputs. It is observed that the side lobes increased by 1 dB and gain of the antenna array reduced by 1.1 dB for array with non-uniform inputs.

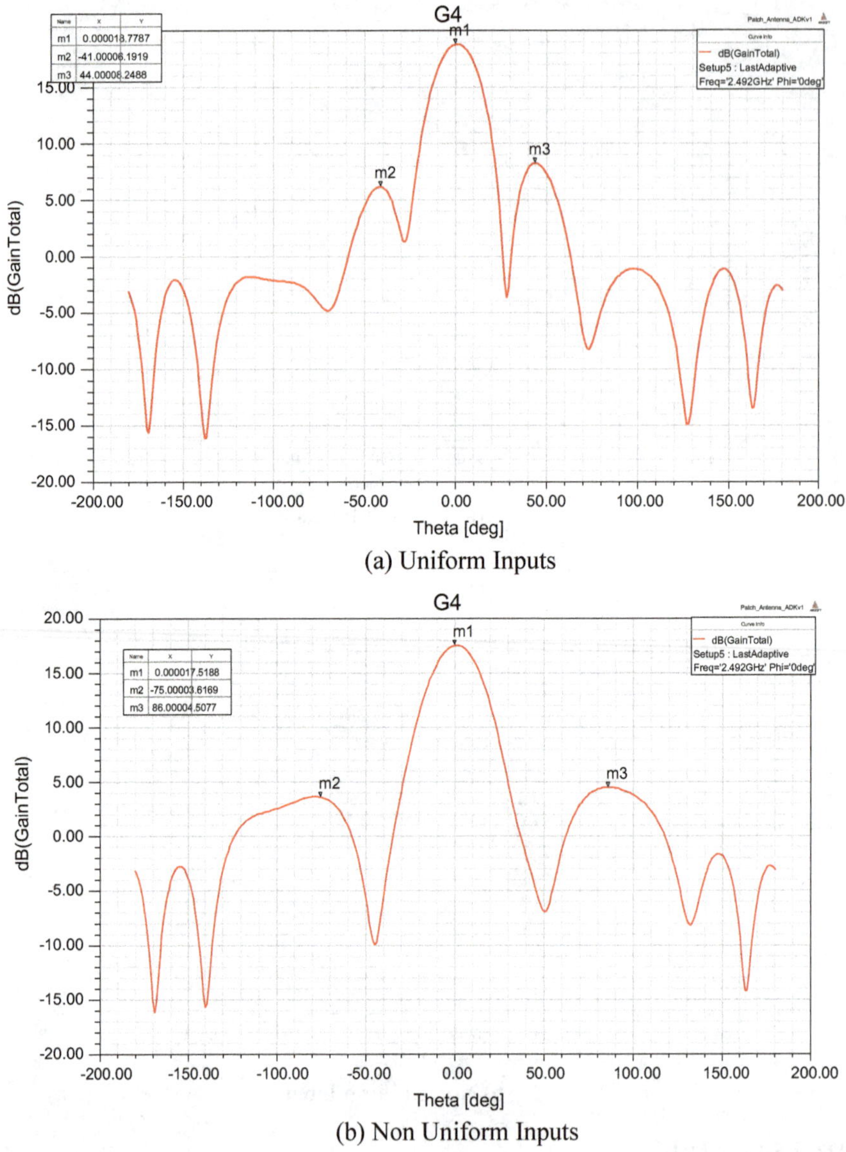

(a) Uniform Inputs

(b) Non Uniform Inputs

Fig. 6 Gain at 2.492 GHz

4 Conclusion

In this paper, two 3 × 8 antenna arrays are proposed of which one is resonating for two frequencies and the second one is resonating at four different frequencies. Performance of the two antenna arrays using uniform and non-uniform inputs is

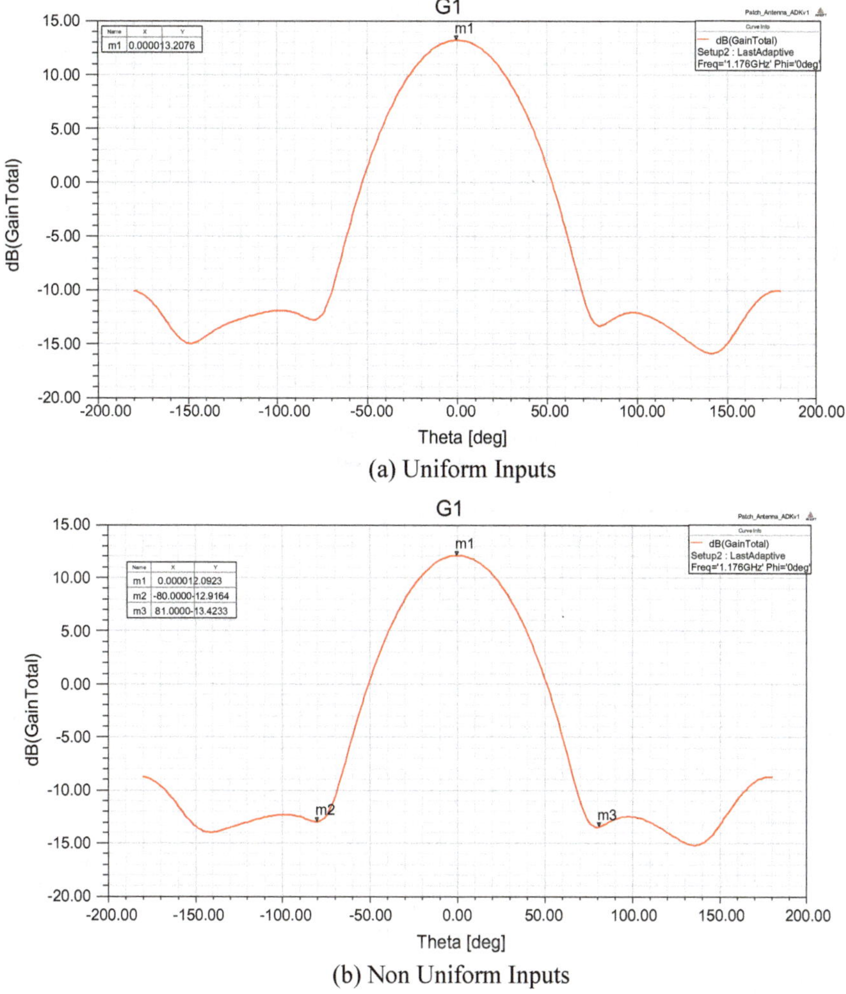

(a) Uniform Inputs

(b) Non Uniform Inputs

Fig. 7 Gain at 1.176 GHz

observed. For the studies, we have considered pascal triangle weights as the non-uniform inputs for the array. The calculated weights are 1, 7, 21, 35, 35, 21, 7, 1. These weights are given as input to the array, and the effect on gain and side lobe levels is analysed. These performance studies are necessary for antenna array for enhancement of the performance and beam shaping of the antenna arrays.

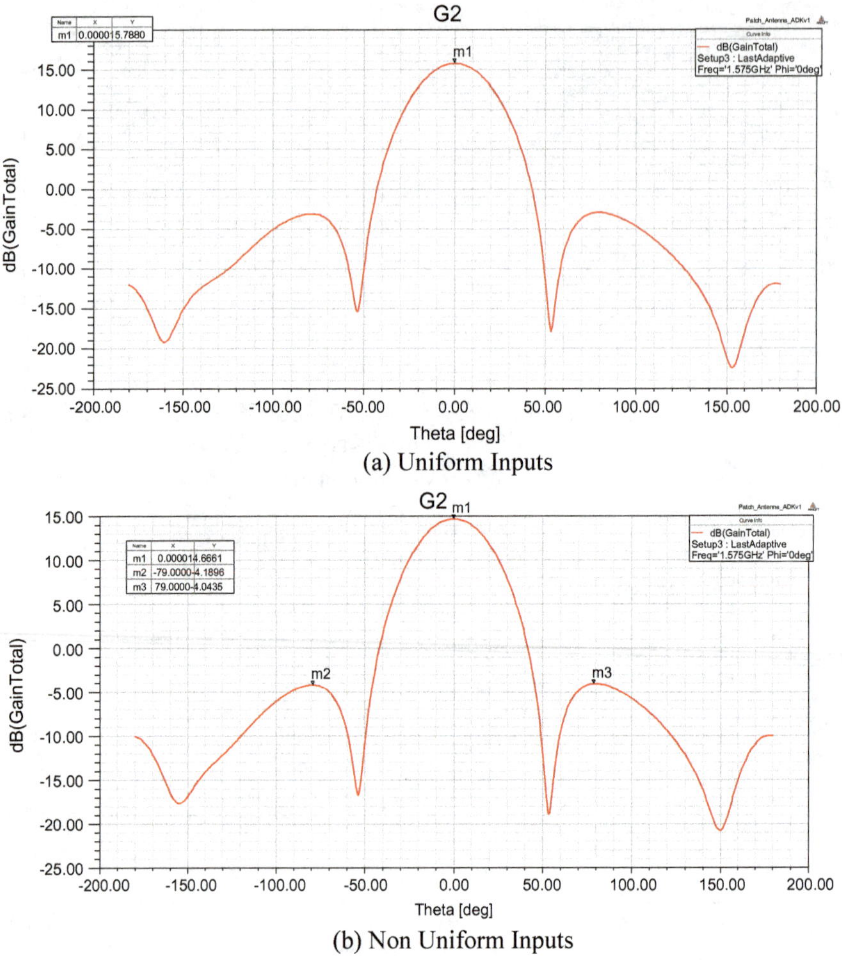

(a) Uniform Inputs

(b) Non Uniform Inputs

Fig. 8 Gain at 1.575 GHz

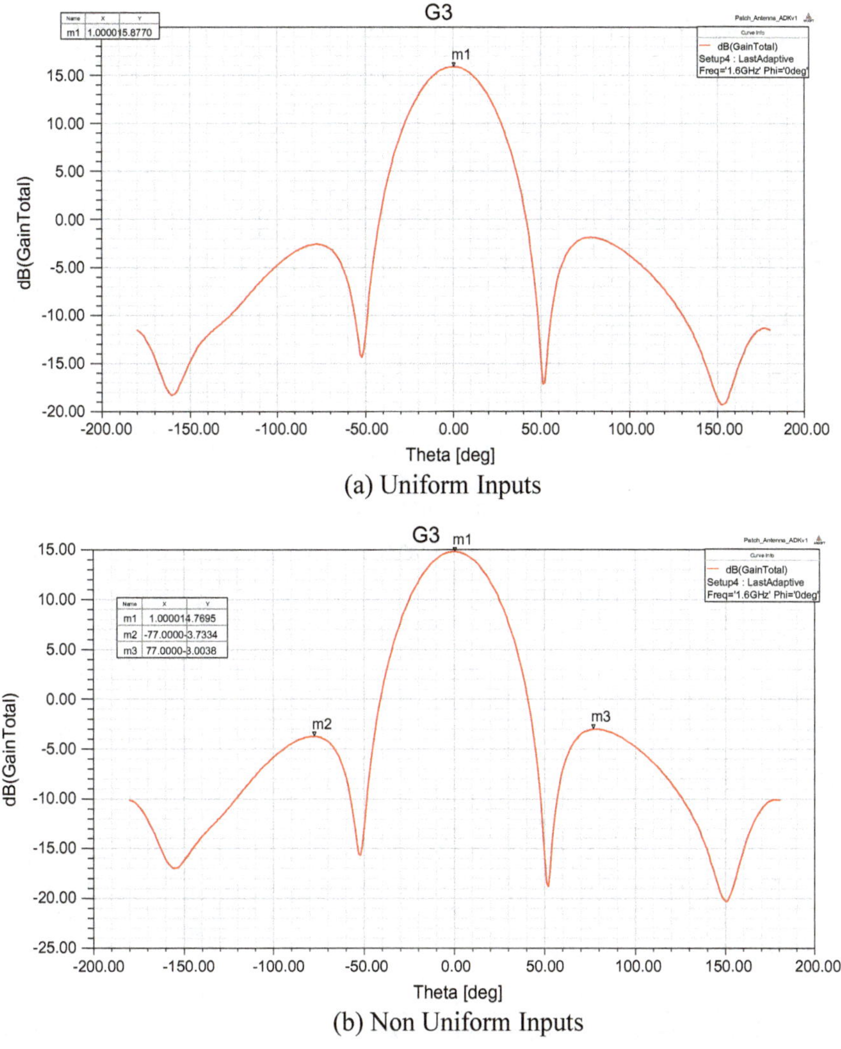

Fig. 9 Gain at 1.6 GHz

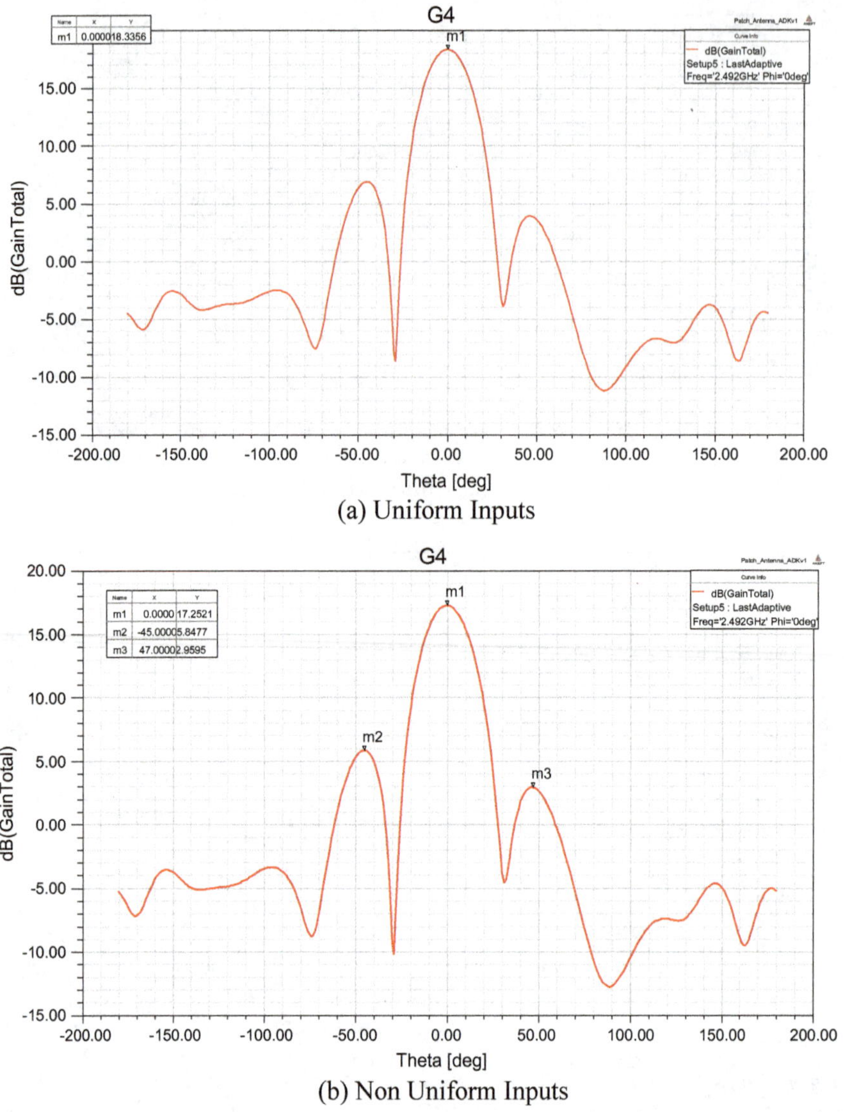

Fig. 10 Gain at 2.492 GHz

References

1. M.A. Motin et al., Design and simulation of a low cost three band microstrip patch antenna for the X-band, Ku-band and K- band applications, in *2012 7th International Conference on Electrical and Computer Engineering*, 20–22 Dec., Dhaka, Bangladesh (2012)
2. Y. Harrou, Y. Sun, Statistical monitoring of linear antenna arrays. Eng. Sci. Technol. Int. J. **19**(4), 1781–1787 (2016)

3. A.H. Hussein, M.A. Metawe'e, H.H. Abdullah, Hardware implementation of antenna array system for maximum SLL reduction. Eng. Sci. Technol. Int. J. **20**(3), 965–972 (2017)
4. M. Pehlivan, Y. Asci, K. Yegin, C. Ozdemir, X band patch array antenna design for marine radar application, in *22nd International Microwave and Radar Conference*, 14–17 May 2018, Poznan, Poland
5. M. Angeline, A. Flashy, V. Shanthi, Microstrip circular antenna array design for radar applications, in *International Conference on Information Communication and Embedded Systems*, 27–28 Feb. 2014, Chennai, India
6. S. Feng, M.P. Jin, Broad-band cavity-backed and probe-fed microstrip phased array antenna in X-band, in *IEEE International Symposium on Antennas and Propagation & USNC/URSI National Radio Science Meeting*, 9–14 July 2017, San Diego, CA, USA
7. E. Kusuma Kumari, A.N.V. Ravi Kumar, Wideband high-gain circularly polarized planar antenna array for L band radar, in *2017 IEEE International Conference on Computational Intelligence And Computing Research*, Tamil Nadu College of Engineering (2017) ISBN: 978-1-5090-6620-9
8. E. Kusuma Kumari, A.N.V. Ravi Kumar, Development of an L band beam steering cuboid antenna array, in *2017 IEEE International Conference on Computational Intelligence and Computing Research*, Tamil Nadu College of Engineering (2017) ISBN: 978-1-5090-6620-9
9. N.A. Rao, S. Kanapala, Wideband circular polarized binomial antenna array for L-band radar, in *Microelectronics, Electromagnetics and Telecommunications*, ed. by G. Panda, S. Satapathy, B. Biswal, R. Bansal. Lecture Notes in Electrical Engineering, vol. 521 (Springer, Singapore)
10. S. Kanapala, N.A. Rao, Beam steering cuboid antenna array for L band RADAR, in *Microelectronics, Electromagnetics and Telecommunications*, ed. by G. Panda, S. Satapathy, B. Biswal, R. Bansal. Lecture Notes in Electrical Engineering, vol. 521 (Springer, Singapore)

Performance of BLDC Motor for Enhancing the Response of Antenna's Positioner Using PI Controller

Bhaskaruni Suresh Kumar, D. Varun Raj and Segu Praveena

Abstract Brushless DC motor (BLDC) has applications in field of telemetry for tracking Antennas. Antenna's positioners are typically used for dynamic tracking and telemetry data purpose, such as data from missiles or aircraft. To these antenna's positioners, BLDC motor is connected through a gear reducer. Based on error generated from potentiometer, these motors need to rotate in required direction. BLDC motor can work continuously or intermittently. For antenna's positioner, both continuous and intermittent types are required. Control system design for BLDC motor using PI controller helps in reducing position error and improves response towards the antennas. PI controller should help in improving control systems parameters such as stability, disturbance attenuation and accurate set point. Therefore, designing PI controller is challenging for an engineer for improving response towards tracking position of the antenna. BLDC motor which is current controlled and chopper fed is simulated using MATLAB/Simulink for antenna's positioner.

Keywords BLDC motor · Antenna's positioner · PI controller

1 Introduction

Antenna's positioner is used for tracking missiles, satellites and warships [1, 2]. The antenna's positioner is used to convert input command to output position. A controller is coupled part of a positioning system which comprises the tracking controller, dual axis (elevation-over azimuth), antenna's positioner (pedestal), two port control units, one or two microwave receivers as shown in Fig. 1.

B. Suresh Kumar (✉)
Associate Professor, CBIT, Gandipet, Hyderabad, India
e-mail: bskbus@gmail.com

D. Varun Raj
PG Scholar, CBIT, Gandipet, Hyderabad, India
e-mail: duddyalavarun1269@gmail.com

S. Praveena
Asst. Professor, MGIT, Gandipet, Hyderabad, India
e-mail: veenasureshb@gmail.com

H. S. Saini et al. (eds.), *Innovations in Electronics and Communication Engineering*,
Lecture Notes in Networks and Systems 107,
https://doi.org/10.1007/978-981-15-3172-9_18

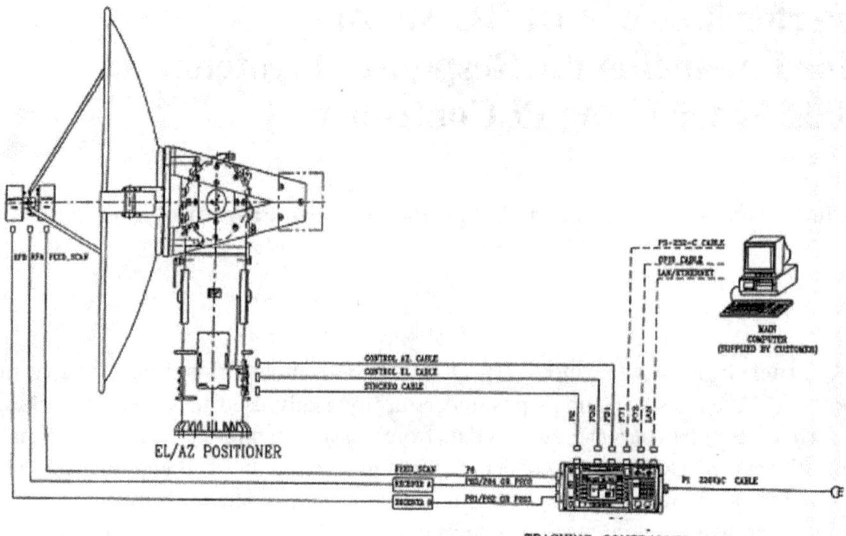

Fig. 1 Antenna's positioner

A GPS receiver is attached to the tracking antenna, the GPS receiver output is the missile location in XYZ or longitude, latitude, altitude (LLA) coordinates, is connected to a advanced VHF transmitter which in turn downloads the data via a wide-beam antenna to the tracking station. Tracking station is equipped with a advanced VHF receiver passing the data further to the controller. The controller, knowing its own location, calculates the azimuth and elevation angles to the target missile and points the main dish antenna towards it (Fig. 2).

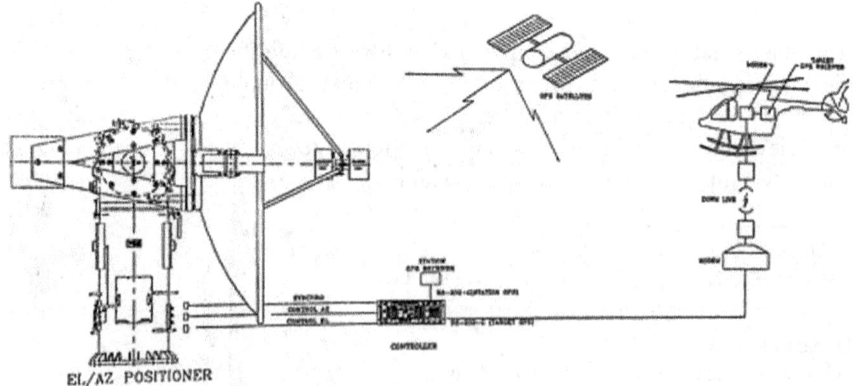

Fig. 2 Antenna's positioner tracking an object

The major tracking antennas/receivers' configurations is single-channel monopulse (SCM) antenna.

1.1 Single-Channel Monopulse (SCM) Antenna

The monopulse configuration antenna provides a sum (sigma) signal which is maximum when the antenna is pointed on boresight, and two difference signals are (Delta AZ, Delta El) which are minimum at that point. The system components are activated synchronously to produce azimuth and elevation angular error readouts between the antenna electrical boresight and RF source line of sight.

The SCM signal is fed to tracking receiver which preforms "delta over sigma" normalization functions. The receiver output is "AM Tracking". This AM signal is filtered by the controller's input filter. The controller continuously samples the filtered AM tracking signal and produces DC error signal to the following equations: (Figs. 3 and 4)

$$\text{DC\_err\_AZ} = \text{AM(T1)} - \text{AM(T3)} \tag{1}$$

$$\text{DC\_err\_EL} = \text{AM(T2)} - \text{AM(T4)} \tag{2}$$

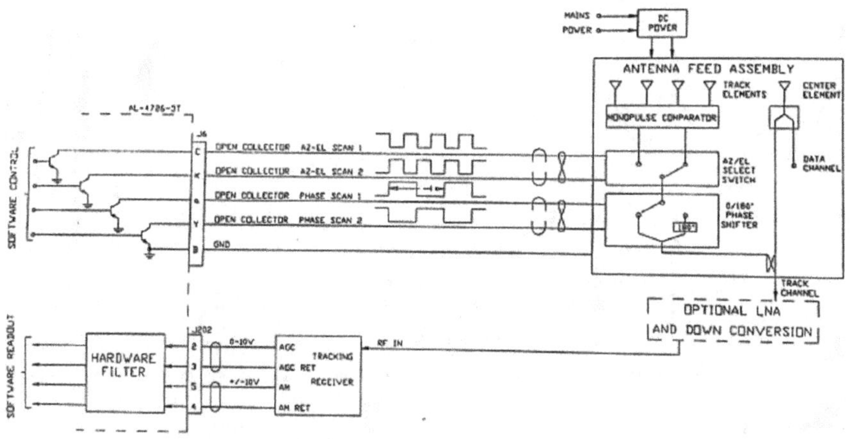

Fig. 3 Single channel monopulse antenna

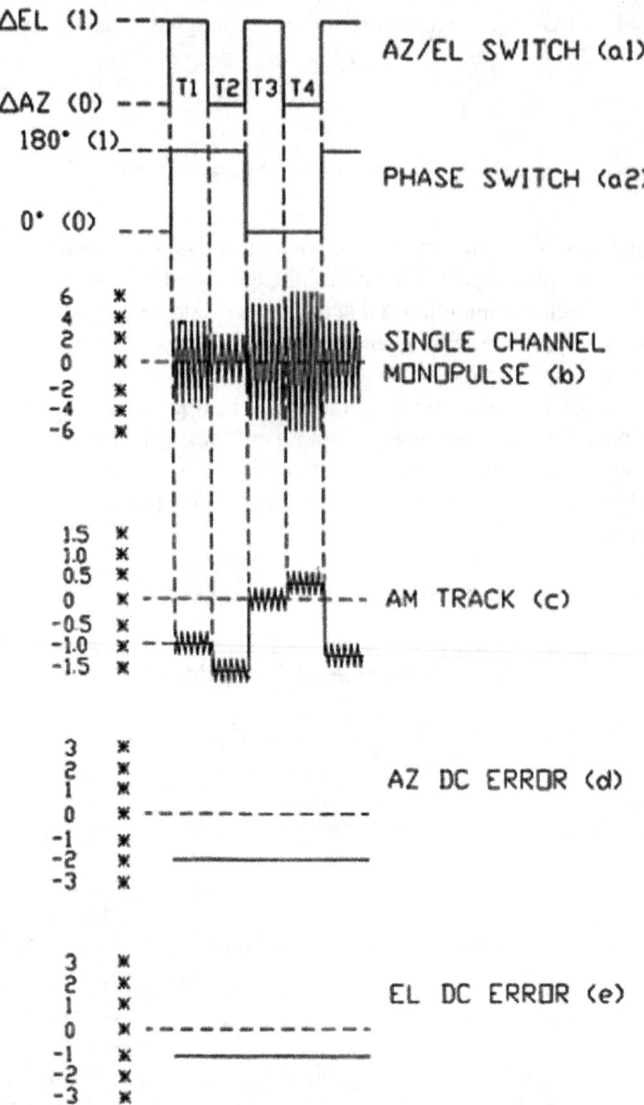

Fig. 4 AM signal generation of SCM

1.2 BLDC Motor

Generally, antenna's positioner is connected to electrical motors through gear reducer. There are many electrical motors which support antenna, and among those, BLDC motor has more advantages over stepper and DC motor [3–6]. In DC motor, brushes make commutation complicated, whereas in BLDC motor, commutation is done by electronic switches.

Figure 5 represents a position control scheme for BLDC motor. There are three loops: the outermost loop called the position loop; the second loop is the speed loop; and the innermost loop called the current control loop. The position of the rotor is compared with the reference value, and the rotor position error is amplified through a PI controller.

To obtain a better dynamic response, speed has to be controlled and this is done by controlling current or torque of BLDC motor as shown in Fig. 3 the output of PI controller becomes reference current input to the motor [7–11]. This reference current is compared with stator current which is measured using ammeters and produces required speed to motor (Fig. 6; Table 1).

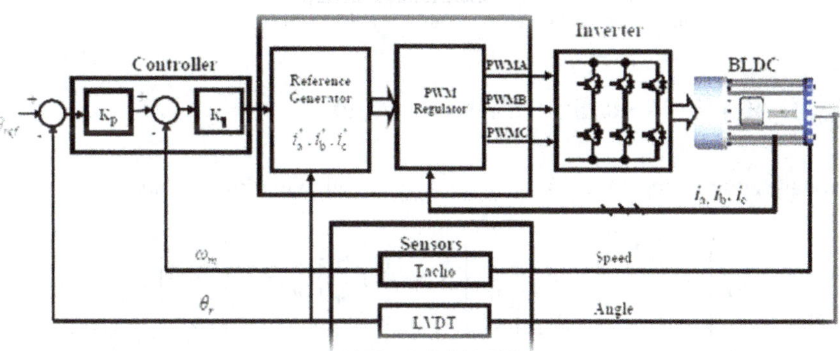

Fig. 5 BLDC motor control loops

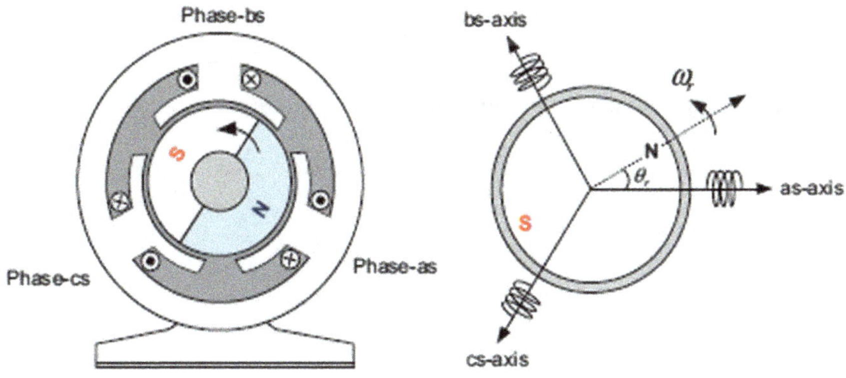

Fig. 6 Modelling of BLDC motor

Table 1 BLDC motor parameters

S. No.	Naming	Values	Units
1	Supply voltage	120	V
2	Stator resistance (R_a)	3.5	Ω
3	Stator inductance (L_a)	8	mh
4	Back emf	120	V
5	Rotor inertia (J)	0.8	kg m^2
6	Azimuthal range	N* 360 continuous	°
7	Elevation up range	183–186	°
8	Elevation down range	−3 to −6	°
9	k_p1(proportional controller) k_I1(integral controller) of speed control loop		
10	k_P2(proportional controller) k_I2(integral controller) of position control loop		

2 Dynamic Modelling of BLDC Motor

Voltage equation of stator winding is given by following Eqs. (3) and (4)

$$\begin{bmatrix} v_a \\ v_b \\ v_c \end{bmatrix} = \begin{bmatrix} R_s & 0 & 0 \\ 0 & R_s & 0 \\ 0 & 0 & R_s \end{bmatrix} \begin{bmatrix} i_a \\ i_b \\ i_c \end{bmatrix} + \frac{d}{dt} \begin{bmatrix} L_{aa} & L_{ab} & L_{ac} \\ L_{ba} & L_{bb} & L_{bc} \\ L_{ca} & L_{cb} & L_{cc} \end{bmatrix} \begin{bmatrix} i_a \\ i_b \\ i_c \end{bmatrix} + \begin{bmatrix} e_a \\ e_b \\ e_c \end{bmatrix} \tag{3}$$

$$v_{as} = v_{ao} - v_{no}, v_{bs} = v_{bo} - v_{no}, \quad \text{and, } v_{cs} = v_{co} - v_{no}$$

Torque equation of BLDC motor is given by Eq. (4)

$$T_e = [e_a i_a + e_b i_b + e_c i_c]/\omega_m \tag{4}$$

Speed and angular position relation are given by Eq. (5)

$$\frac{\mathrm{d}\theta_r}{\mathrm{d}t} = \frac{p}{2}\omega_m \tag{5}$$

3 PI Controller

The output of receiver, AM, is compared with rotor position using sensors. Error generated from this is passed through PI controller which helps in improving the performance of the system. Transfer function of PI controller is given in Eq. (6). Figure 7 shows the block diagram of the given transfer function.

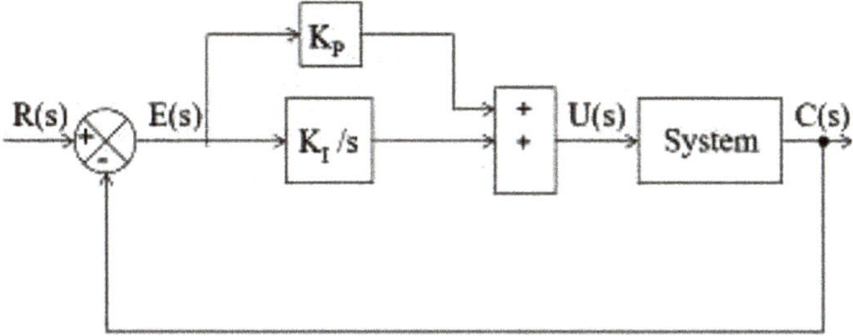

Fig. 7 Block diagram of PI controller

$$\frac{U(S)}{R(s)} = k_p + \frac{k_i}{S} \tag{6}$$

4 Simulation Results

The Simulink model of DC-link current-controlled BLDC motor is shown in Fig. 8. Simulink model consists of BLDC motor, inverter block, PI controller and PWM. It consists of two loops, inner control loop and outer control loop (Fig. 9).

Case 1: BLDC motor operated with $k_P1 = 0.001, k_I1 = 0.3$
Figure 10 shows stator emf of all the phases in which speed is set after 2 s. Because of transients, stator emf is not trapezoidal up to 1.5 s.

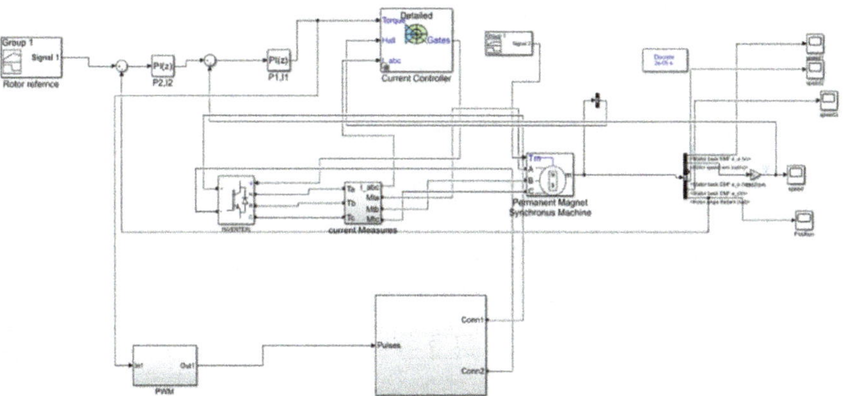

Fig. 8 Simulink model of BLDC motor

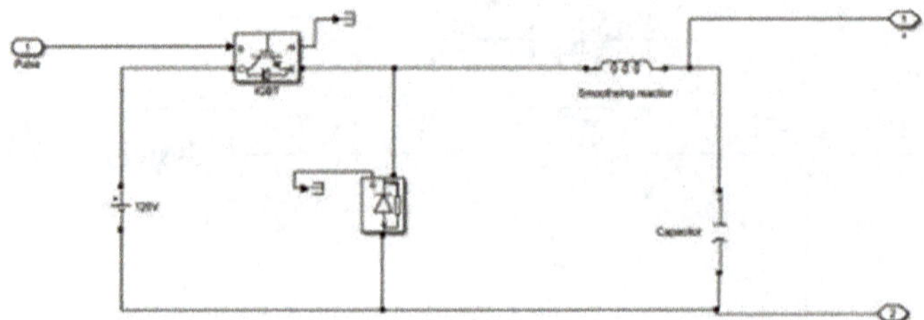

Fig. 9 DC chopper simulink model

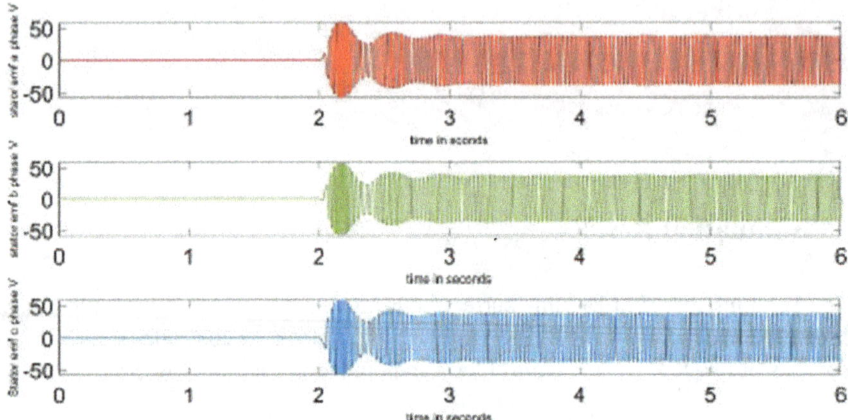

Fig. 10 Stator emf of all phases

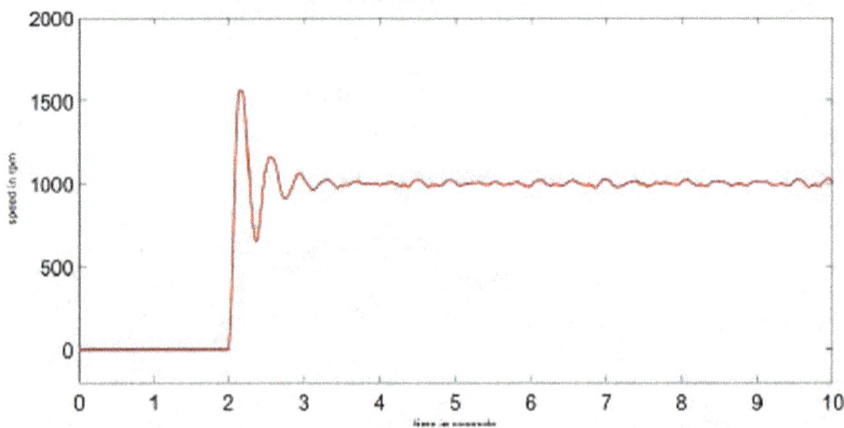

Fig. 11 Speed of rotor

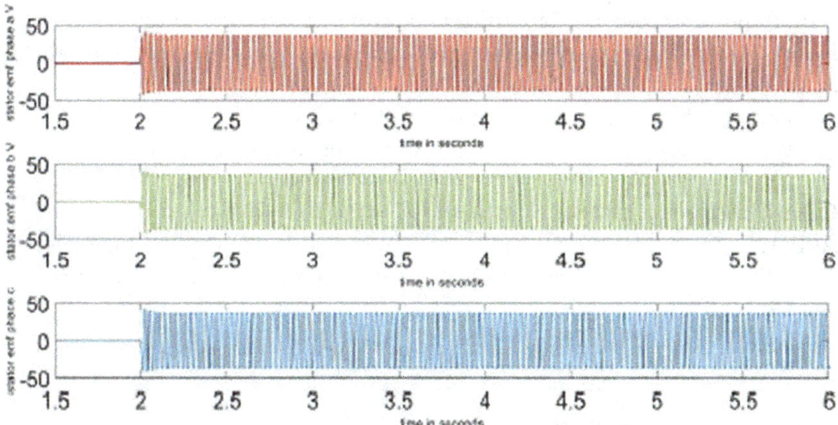

Fig. 12 Stator emf

Table 2 k_P1, k_I1 values of speed control loop

S. No.	k_P1, k_I1 values	Percentage of peak over shoot (%)	Settling time (s)
1	$k_P1 = 0.01, k_I1 = 0.3$	54.837	4
2	$k_P2 = 0.3, k_I2 = 8$	15.2	0.2

Case 2: BLDC motor operated with $k_p1 = 0.3, k_I1 = 0.3$

Figure 12 shows stator emf attains its shape in very time due to increase in PI values (Table 2).

From Figs. 11 and 13, it shows as the PI value increases, oscillations have been reduced with improved speed response (Fig. 12).

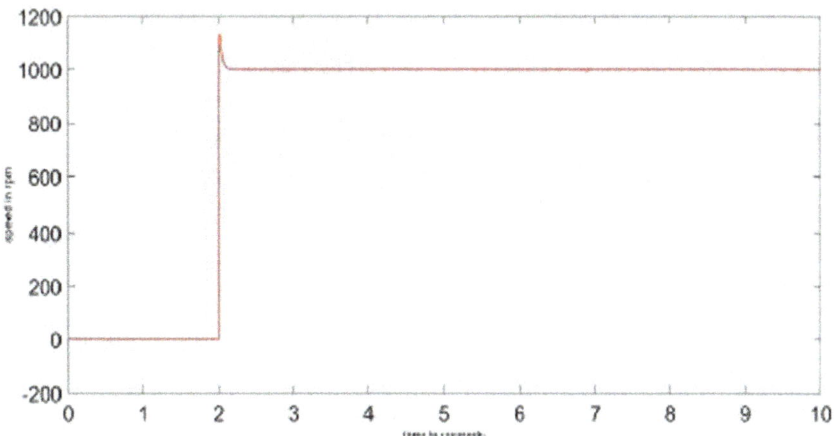

Fig. 13 Speed of rotor

Case 3: position control of BLDC motor with $k_P2 = 12, k_I2 = 0.3$

Figure 14 shows stator emf for short period. Once rotor attains its position, all other quantities (speed, stator emf) become zero.

Figure 15 show rotor attains its position after 2.2 s

Figure 16 shows speed becomes zero after rotor attains its position.

Case 4: position control of rotor with $k_P2 = 32, k_I2 = 0.6$

As PI values are increased, emf of stator Emf comes to zero in very less time (Fig. 17).

As PI values increase, rotor attains its position in very less time (Table 3; Figs. 18 and 19).

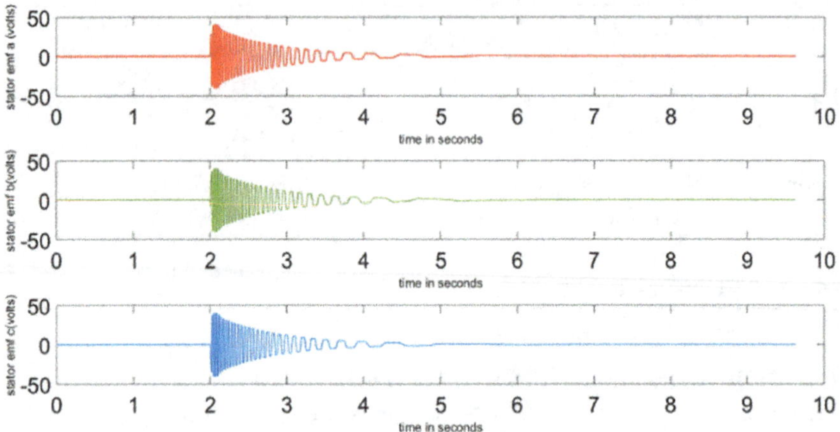

Fig. 14 Stator emf of all phases

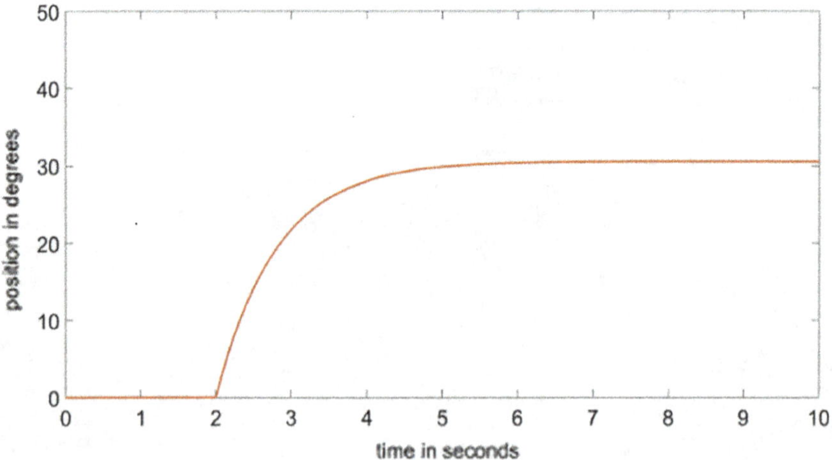

Fig. 15 Rotor position at 30° position

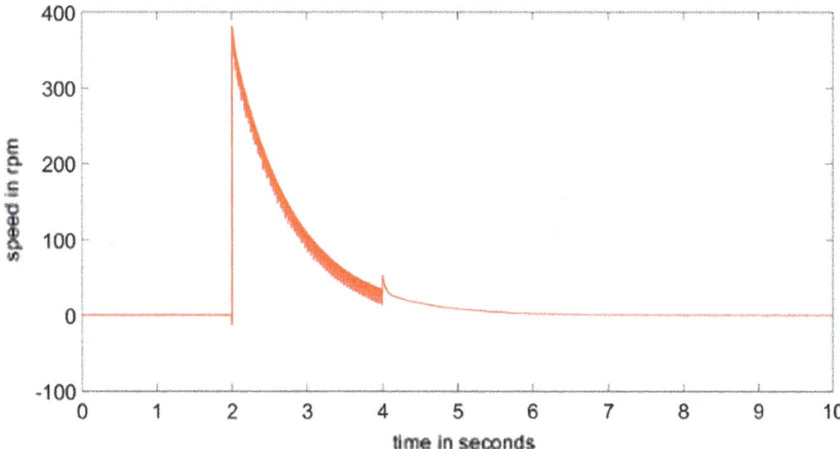

Fig. 16 Speed of rotor

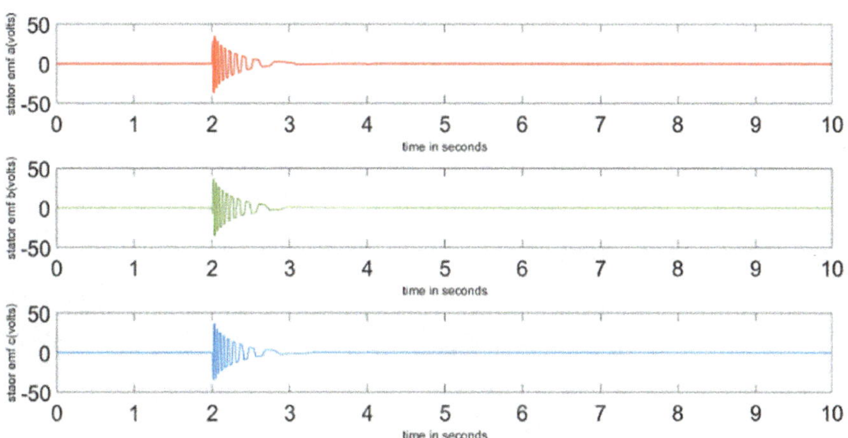

Fig. 17 Stator emf

Table 3 k_P2, k_I2 values of position control loop

S. No.	PI values	Settling time (s)
1	$k_P2 = 12$, $k_I2 = 0.3$	2.2
2	$k_P2 = 32$, $k_I2 = 0.6$	0.6

5 Conclusions

In this paper, position control of antenna is demonstrated with the help of current-controlled BLDC motor using MATLAB/Simulink. There are two loops, position

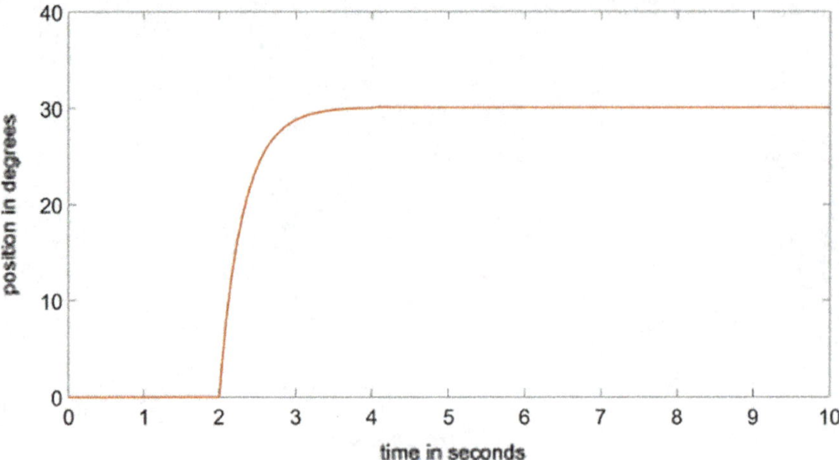

Fig. 18 Rotor position

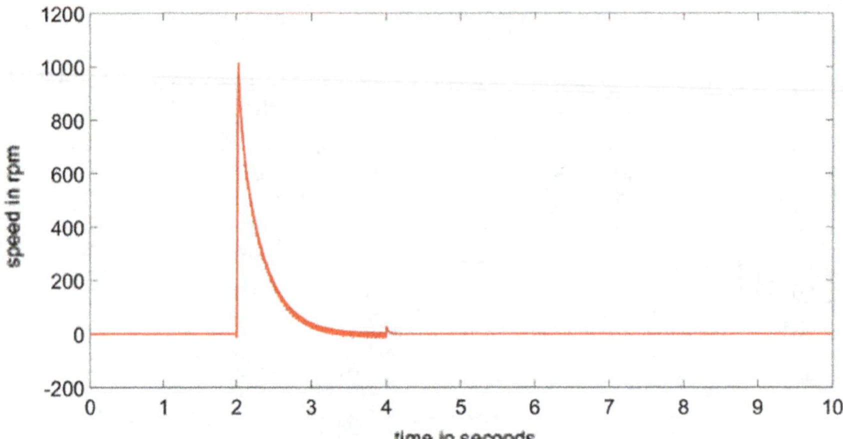

Fig. 19 Speed of rotor

control loop and speed control loop. As PI values of these loops increase, there is a reduction in peak overshoot, settling time and speed of response has Improved.

References

1. S.A. Fandakli, H.I. Okumuş, Antenna azimuth position control with PID, fuzzy logic and sliding mode controllers, in *INISTA* (2016)

2. O. Sharon Shobitha, K.L. Ratnakar, G. Sivasankaran, Precision control of antenna positioner using P and Pi controllers, International
3. P. Sarala, S.F Kodad, B. Sarvesh, Analysis of closed loop current controlled BLDC motor drive, (ICEEOT) (2016)
4. T.J.E. Miller, *Brushless Permanent Magnet and Reluctance Motor Drive* (Oxford, 1989)
5. M. Mubeen, Brushless DC motor primer. Motion Tech Trends (2008)
6. P. Yedamale, Hands-on workshop: motor control part 4 -brushless DC (BLDC) motor fundamentals, in *Microchip AN885* (2003)
7. S.-H. Kim, Brushless direct current motors. Electric Motor Control (2017)
8. IIT Kharagpur, Industrial Automation and Control, Chapter- 7, electrical actuators: BLDC motor drives, NPTEL, 2009, https://nptel.ac.in/courses/108105063/35#. J. Eng. Sci. Innov. Technol. (IJESIT) **4**(3) (2015)
9. P. Agarwal, A. Bose, Brushless DC motor speed control using proportional-integral and fuzzy controller, **5**(5) (2013)
10. B.C. Kuo, *Automatic Control Systems*, 3rd edn. (Prentice-Hall Inc., 1975)
11. Y. Chen, J. Tang, D.-s. Cai, X. Liu, School of Energy Science and Engineering, University of Electronic Science and Technology of China (1) School of Electrical and Information Engineering, Xihua University (2)

LoRa Transmission Over Rayleigh Fading Channels in Presence of Interference

Bharat Chaudhari and Marco Zennaro

Abstract Although LoRa (long range) technology is designed to tackle interference and power consumption issues, interference from other LoRa and non-LoRa networks limits its performance. In practice, LoRa performance also gets constrained by multipath fading. This paper presents the study of LoRa transmission over Rayleigh fading. We considered different scenarios of LoRa networks along with simultaneous parallel transmissions, change in transmission bandwidths, and Doppler shifts. Results demonstrate deviation in the BER because of multipath and interference. It is also evident that even for a change in transmission bandwidth or Doppler shift for Rayleigh fading channel, LoRa performance gets impacted.

Keywords IoT · LPWAN · LoRa · Rayleigh fading · Interference

1 Introduction

With the emergence of the Internet of Things (IoT) and machine to machine (M2M) communications, exponential growth in the sensor node deployment is expected. Different industry projections show that around 50 billion nodes would be connected to the Internet in the next few years [1]. Such progression allows any things such as sensors, vehicles, robots, machines, or any such objects to connect to the Internet. These technologies enable sensing nodes to send the sensed data and parameters to the remote centralized server, which brings out intelligence for taking an appropriate decision or actuating action.

In general, IoT applications require the energy efficient and low complex nodes for long-range applications that are to be deployed on the scalable networks. Currently, traditional wireless technologies such as IEEE 802.11 (WLAN), Bluetooth, IEEE 802.15.1 (ZigBee), and LR-WPAN are being used for sensing applications in

B. Chaudhari (✉)
School of Electronics and Communication Engineering, MIT World Peace University, Pune 411038, India
e-mail: bsc@ieee.org

M. Zennaro
The Abdus Salam International Centre for Theoretical Physics, Strada Costiera, Trieste, Italy

H. S. Saini et al. (eds.), *Innovations in Electronics and Communication Engineering*, Lecture Notes in Networks and Systems 107,
https://doi.org/10.1007/978-981-15-3172-9_19

short-range environments, whereas wireless cellular technologies such as 3G and 4G are used for long-range applications. WLAN and Bluetooth technologies were designed for data communication, whereas ZigBee and LR-WPAN were designed for wireless sensing applications in the local environments. These technologies can be used for the communication distance ranging from a few meters to a few hundred meters maximum, depending on the line of sight/non-line of sight signal paths, interference, transmit power, etc. Wireless cellular networks are designed for voice and data communication, not primarily for wireless sensing applications. Though they are used for sensing in some cases, their performance in terms of performance metrics used in the wireless sensor networks may not be acceptable.

The most critical requirements of wireless IoT/M2M devices are minimal electrical power consumption, longer transmission distance, scalability, acceptable RF interference, low cost, and easy deployment. To support these requirements, recently, a new paradigm of low power wide area networks (LPWAN) has been evolved. LPWAN technologies are promising and can be deployed for applications such as environmental monitoring including water, humidity, temperature, pollution, waste management; in smart utilities such as utility meters, power transformer monitoring; in smart city application such as smart lighting, parking management, intelligent transport system, and many other. Some of the prominent technology candidates for LPWAN are LoRa [2], SIGFOX [3], Weightless [4], narrowband IoT (NB-IoT) [5], long-term evolution for machines (LTE-M) [6], and extended coverage GSM (EC-GSM) [7]. Some of these solutions are 3GPP, and others are non-3GPP. Some of them are proprietary. They work in unlicensed bands or licensed bands or both. LoRa, SIGFOX, and Weightless are the non-3GPP standards, whereas NB-IoT, LTE-M, and EC-GSM are from 3GPP. LPWAN non-3GPP technologies work in unlicensed bands.

LoRa is a proprietary modulation technology [2]. It is based on the use of chirped spread spectrum (CSS) modulation by sweeping the carrier frequency tone linearly according to the variation in the input data. LoRa is a physical layer technology with some link layer functionalities, specially designed for low power and long-range transmissions. To support the LoRa on Internet, LoRa alliance has developed the LoRaWAN [8] which includes the network and upper layer functionalities. LoRa can be used for long distance and low power transmission. However, due to design compromises and in spite of corrective measures such as low data rates, channel hopping, and adaptive data rates, LoRa has some limitations. With higher spreading factors (SF), long air time makes the transmission prone to packets collisions and also interferes with other LoRa transmissions [9]. LoRa is based on wideband linear frequency modulation in which carrier frequency varies for the defined extent of time.

Radio frequency (RF) signal propagation in the wireless channels is susceptible to multipath fading and shadows, making the signal detection a very complex process and also deteriorates the performance of the network. When the transmitted signal fades during propagation, its amplitude and phase fluctuate over the time. If the sending and receiving stations both or one of them is mobile, the frequency changes due to the Doppler shift making the scenario worse because of the frequency offset. In urban areas, most of the wireless transmission suffers from such fading as there can be many obstacles in the transmission path as well as mobility of nodes. The receiving station gets the time-delayed replicas of the same signal. When the received signal fades away during the transmission, its phase and amplitude fluctuate over time. Though the features of LoRa make it comparatively more complacent to the resistant against multipath and Doppler effect, it still limits the performance.

Recently, several researches have been reported on the LoRa and its performance analysis. LoRa transmissions with different spreading factors are quasi-orthogonal [10] and allow multiple transmissions with different spreading factors (SFs), simultaneously. These transmissions with the same value of spreading factor exhibit cross-correlation properties and deteriorate the link performance [11]. CSS simulation for the long-range transmissions [12, 13] studied the interference sensitivity and estimation of bit error rate. The analytical study presented along with the verified expression of BER [14] shows the dependence of BER on SNR at the receiving node. The LoRa performance is evaluated with a gateway and dense deployment of end nodes and quantified the minimum SNR thresholds for obtaining the required performance [15]. Theoretical analysis of the achievable LoRa throughput on the uplink along with co- and inter-SF interferences is presented [16] and derived the expressions for perfect and imperfect SF orthogonality. LoRa transmission reliability is evaluated by simulations for the different fading channels [17] and demonstrated that the LoRa configuration impacts the energy/reliability trade-off and the best one strongly depends on the type of channel.

From the literature study, it is learnt that although LoRa is designed with robust characteristics to interference and multipath fading, however, its performance is degraded due to these reasons. During the design and deployment of LoRa-based LPWAN, it is necessary to consider the impact of interference and multipath fading to achieve maximal performance. Furthermore, such measure will also increase battery efficiency with the reduced number of retransmissions. The contribution of this paper is to analyze the performance of LoRa for interference from other LoRa transmissions for Rayleigh fading channels in an urban environment. To measure the level of interference, we carried out extensive simulations considering multiple static and dynamic LoRa nodes deployed in urban settings. It is shown that performance of LoRa varies with change in spreading factors, bandwidth, and mobility of the node. Rest of the paper is structured as follows: Sect. 2 covers LoRa and its system-level modeling for Rayleigh fading. Simulations and results are presented and discussed in Sect. 3, followed by conclusions in Sect. 4.

2 LoRa and Rayleigh Fading

LoRa can transmit with various data rates on a fixed channel bandwidth by using quasi-orthogonal spreading factors. As LoRa is a PHY layer implementation, it can be integrated with existing network architectures. The data rates in LoRa can be written as

$$R_b = \frac{\mathrm{SF} \cdot B}{2^{\mathrm{SF}}} \mathrm{bits/s} \quad (1)$$

where SF is spreading factor varying from 6, 7, … 0.12, and B is channel bandwidth (125, 250, 500 kHz) which is equal to the chip rate. The spreading factor for a LoRa link may be varied as per the requirement of on-air-time (OTA) and the distance between the sending and receiving nodes. The simultaneous transmissions in the same frequency channel using different SFs are possible; however, it suffers interference because of their imperfect orthogonally. The symbol period $T_s = 2^{\mathrm{SF}}/B$ s. The sensitivity of the receiver is defined as [2]

$$S = -174 + 10\log_{10} B + \mathrm{Noise\,Figure} + \mathrm{SNR} \quad (2)$$

where −174 represent the thermal noise at the receiving node and SNR is signal to noise ratio requirement at the transmitting node. LoRa uses Hamming code-based forward error correction technique for error control. For the block of four data bits, the redundant bits for error control vary and the corresponding code rates can be 4/5, 4/6, 4/7, or 4/8.

In the real-time deployments, there can be several static or dynamic obstacles such as building, hills, trees, moving objects like vehicles, human beings, animals, etc., between the sending and receiving nodes. On the other hand, the LoRa modes may be static or mobile or nomadic. Such scenarios form multipath fading channels during communication. In urban deployments, though the LoRa has some capabilities to tackle such interference, it increases the signal to interference plus noise ratio (SINR) threshold requirement on the higher side, upsetting the energy budget. The mobile object also produces Doppler effect because of the shift in the frequency. In this work, the Rayleigh fading is considered as our interest is to study and analyze LoRa transmissions in the urban environment. By assuming the separate delay power profile and the Doppler spectrum, for band limited discrete multipath Rayleigh fading channels, the output y_i can be expressed as

$$y_i = \sum_{n=-N_1}^{N_2} s_{i-n} g_n \quad (3)$$

where $\{s_i\}$ denotes the set of input samples to the channel, and $\{g_n\}$ is the set of weights as given by

$$g_n = \sum_{k=1}^{K} a_k \mathrm{sinc}\left[\frac{\tau_k}{T_{\mathrm{sa}}} - n\right] \quad -N_1 \le n \le N_2 \quad (4)$$

where T_{sa} is the set of the input sample period, $\{\tau_k\}$, where $1 \le k \le K$ is the set of path delays. K is the total number of paths in the multipath fading channel. $\{a_k\}$ is the set of complex path uncorrelated gains of the multipath fading channel. N_1 and N_2 are chosen for $\{g_n\}$ is small when n is less than $-N_1$ or greater than N_2. Based on Jakes model [17], the Rayleigh fading channel gain is given by

$$h(t) = \frac{1}{\sqrt{L}} \sum_{l=1}^{L} \mathrm{e}^{j(2\pi f_{\mathrm{D}} \cos(\phi_{\mathrm{l}})t + \theta_{\mathrm{l}})} \tag{5}$$

where L represents the number of NLOS paths, f_{D} is the Doppler shift, estimated as $f_{\mathrm{D}} = \upsilon_{\mathrm{r}} c / f_{\mathrm{c}}$, where υ_{r} is a relative velocity between node and base station, c is the velocity of light, and f_{c} is the carrier frequency. ϕ_{l} is the initial phase, and θ_{l} is an incident angle of the lth path. The coherence time of the channel is inversely proportional to Doppler shift ($T_{\mathrm{c}} = 1/f_{\mathrm{D}}$). For the symbol duration, T_{s} greater than T_{c} or υ_{r} greater than $f_{\mathrm{c}}^{\mathrm{BW}}/(f_0 s^{\mathrm{SF}})$, fast fading can occur, resulting in higher attenuation of the signal.

3 Simulations, Results, and Discussion

To evaluate the performance of LoRa on Rayleigh fading channels, we simulated a LoRa transmission in the presence of interference from other LoRa transmissions. We considered static as well as dynamic nodes for different Doppler shifts and transmission bandwidths to evaluate the network performance. It is assumed that there is no interference from non-LoRa transmissions. The parameters and metrics used for the simulations are as given in Table 1.

The simulation creates the LoRa interference scenario with Rayleigh fading. We undertake simulations for one SF as LoRa signal transmission and other SF transmissions as interfering transmissions. A node simulates LoRa traffic with a specific spreading factor (In our case, for the first scenario, SF = 7 and second scenario SF = 8), and this transmission is interfered by LoRa transmissions of SF*s* ranging from 7 to 12, simultaneously.

Figure 1 shows the change in BER for Doppler shift of 5°. Initial Doppler shift is assumed to be 0° and BER of 0.01. This scenario can be considered as static to moving nodes case. When the nodes are mobiles, the Doppler shift deteriorates the performance of the networks. The results show that for different values of SINR,

Table 1 Parameters and metrics

Parameter	Symbol	Value
Signal bandwidth	BW	125, 250, 500 kHz
Spreading factor	SF	6–12
Code rate	CR	4/7
Desired bit error rate	BER	0.01
Signal to interference plus noise ratio	SINR	−30 to 0 dB
Doppler shift		0°, 5°, and 10°

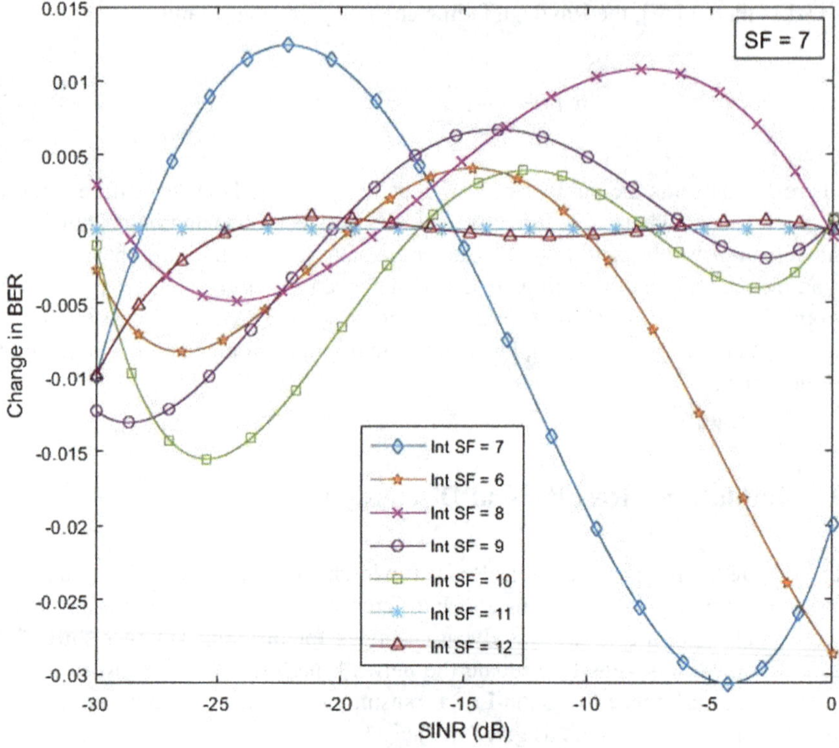

Fig. 1 Change in BER due to Doppler shift of 5°

there is variation in BER values for 0° shift to 5° shift. The *Y*-axis of the figure shows the difference in BER (not BER values).

Although LoRa uses chirp spread spectrum, which employs wideband linear frequency chirps and transmits on the entire allocated bandwidth so that it can tackle noise and multipath, however, the results illustrate the variation BER and hence the degradation in the performance of the network. The results also show the dependence of SF and Doppler shift on BER. With the increase in interfering SF, BER variation is at the lower side.

LoRa transmission has three transmission bandwidth options as 125, 250, and 500 kHz. To study the impact of bandwidth variation of BER, in the second scenario, we simulated for variation in the bandwidths of 125 and 500 kHz in the presence of 10° Doppler shift for Rayleigh fading. The impact of Doppler shift on BER for different bandwidth is shown in Fig. 2. From the results, the variation in BER values between no Doppler shift, i.e., 0° and 10° shift can be seen for different values of SINR. The results also show that for higher values of interfering spreading factors, the variation in BER value is small because of longer air time of the symbols and hence reducing the impact of Doppler in such cases.

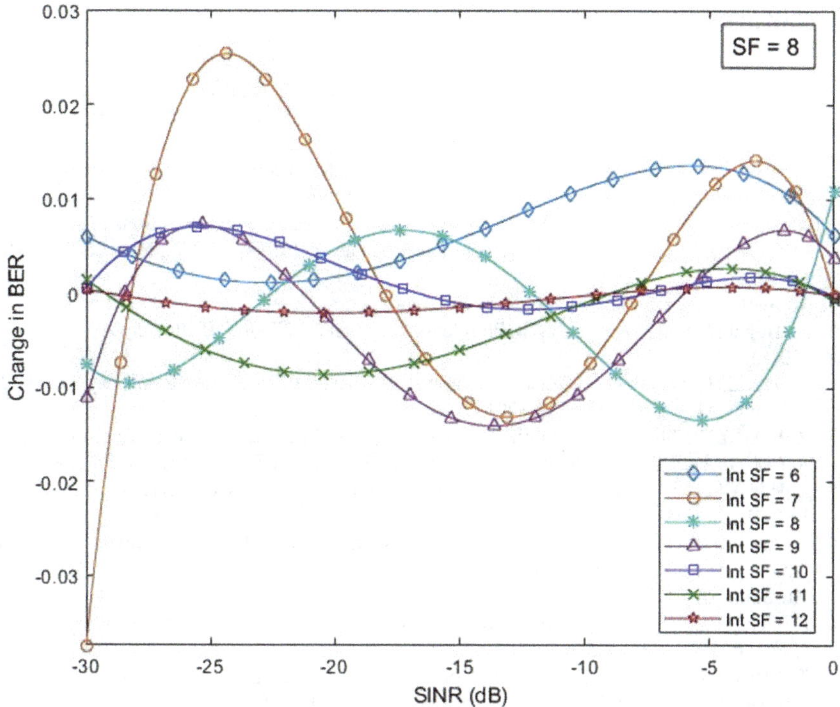

Fig. 2 Change in BER for the bandwidth of 500 and 125 kHz for Doppler shift of 10°

4 Conclusions

LPWAN technologies are promising candidates for IoT and M2M communications. These technologies are widely getting deployed in the smart city and other urban area applications. By design, LoRa has capabilities to mitigate the impact of radio frequency interference, but due to Rayleigh fading nature of channels and also mobility of nodes in the urban environment, the deployment may become a challenge. The study was carried out to measure such impact on LoRa transmission. The simulation results show that the Doppler shift degrades the BER, and hence, it increases the minimum SINR threshold to obtain the same performance. The results also demonstrated the effect of quasi-orthogonal spreading factors, transmission bandwidth, and Doppler shift on network performance. The study reveals that while designing LoRa networks or during scalability, one must consider the impact of these parameters to maintain the required performance of the networks.

References

1. J. Hrishikesh, K. Lee, W.S. Lee, A. Raha, Y. Kim, V. Raghunathan, Powering the internet of things, in *International Symposium on Low Power Electronics and Design* (ACM, 2014), pp. 375–380
2. Semtech. LoRa Modulation Basics. AN1200.22 Revision 2, Camarillo, CA, USA: https://www.semtech.com/uploads/documents/an1200.22.pdf
3. M. Centenaro, L. Vangelista, A. Zanella, M. Zorzi, Long-range communications in unlicensed bands: the rising stars in the IoT and smart city scenarios. IEEE Wirel. Commun. **23**(5), 60–67 (2016)
4. Weightless Open Standards. http://www.weightless.org
5. A. Hoglund et al., Overview of 3GPP release 14 enhanced NB-IoT. IEEE Netw. **31**(6), 16–22 (2017)
6. A. David, E. Dahlman, G. Fodor, S. Parkvall, J. Sachs, LTE release 12 and beyond. IEEE Commun. Mag. **51**(7), 154–160 (2013)
7. Z.A. Díaz, P. Merino: The 3GPP NB-IoT system architecture for the Internet of Things, in *International Conference on Communications Workshops* (IEEE, 2017), pp. 277–282
8. LoRA Alliance Technical Marketing Workgroup 1.0, "A technical overview of LoRa and LoRaWAN," White Paper (2015)
9. E. Aras, N. Small, G. Sankar Ramachandran, S. Delbruel, W. Joosen, D. Hughes, Selective jamming of LoRaWAN using commodity hardware, in *MobiQuitous 2017* (IEEE, Melbourne, Australia, 2017), pp. 363–372
10. O. Georgiou, U. Raza, Low power wide area network analysis: Can LoRa scale? IEEE Wirel. Commun. Lett. **6**(2), 162–165 (2017)
11. T. Elshabrawy, J. Robert, The impact of ISM interference on LoRa BER performance, in *Internet of Things (GCIoT) 2018 Global Conference* (IEEE, 2018), pp. 1–5
12. B. Reynders, S. Pollin: Chirp spread spectrum as a modulation technique for long range communication, in *Symposium on Communications and Vehicular Technologies (SCVT)* (2016), pp. 1–5
13. B. Reynders, W. Meert, S. Pollin, Range and coexistence analysis of long range unlicensed communication, in *23rd International Conference on Telecommunications (ICT)* (IEEE, 2016), pp. 1–6
14. H. Mroue, A. Nasser, B. Parrein, S. Hamrioui, E. Mona-Cruz, G. Rouyer, Analytical and simulation study for LoRa modulation, in *25th International Conference on Telecommunications (ICT)* (2018), pp. 655–659
15. D. Croce, M. Gucciardo, S. Mangione, G. Santaromita, I. Tinnirello, Impact of LoRa imperfect orthogonality: analysis of link-level performance. IEEE Commun. Lett. **22**(4), 796–799 (2018)
16. A. Waret, M. Kaneko, A. Guitton, N. El Rachkidy, LoRa throughput analysis with imperfect spreading factor orthogonality. IEEE Wirel. Commun. Lett. **8**(2), 408–411 (2018)
17. X.-C. Le, B. Vrigneau, M. Gautier, M. Mabon, O. Berder, Energy/reliability trade-off of LoRa communications over fading channels, in *25th International Conference on Telecommunications (ICT)* (IEEE, 2018), pp. 544–548

Orchestrator Controlled Navigation of Mobile Robots in a Static Environment

Rameez Raja Chowdhary, Manju K. Chattopadhyay and Raj Kamal

Abstract The paper presents an orchestrated controlled model for automated guided vehicle (AGV) like mobile robots (RBs) to perform a transportation task in environment such as a warehouse or a factory. The AGV uses graph-based approach for path planning. It uses a proposed navigation algorithm for following the path trajectory. It also avoids a collision during transportation between start and end points. The proposed model also prevents the AGV from a deadlock situation. The performance of model has been validated experimentally.

Keywords AGV · Orchestrated control · Navigation of mobile robots · Networked robot · Warehousing

1 Introduction

AGV transportation system plays an important role in logistics centres [1], cargo terminals [2], modern warehouses [3] and industrial environment [4]. An AGV transportation system is an autonomous system. It transports the load between the given source and destination point [5–8]. An AGV control system uses AGV according to the load handling capacity. The system determines the navigation path between the source and destination points. AGV navigation through the path requires avoiding of the collisions and deadlock situations [9–16]. AGV systems improve the load handling efficiency and throughput of manufacturing plant. The appropriate design

R. R. Chowdhary (✉)
Department of Electronics and Telecommunication Engineering, Institute of Engineering and Technology, Devi Ahilya University, Indore 452001, India
e-mail: rameez.chowdhary@gmail.com

R. R. Chowdhary · M. K. Chattopadhyay
School of Electronics, Devi Ahilya University, Indore 452001, India
e-mail: manju.elex@gmail.com

R. Kamal
Department of Electronics and Telecommunication Engineering, Prestige Institute of Engineering and Science, Indore 452001, India
e-mail: dr_rajkamal@hotmail.com

H. S. Saini et al. (eds.), *Innovations in Electronics and Communication Engineering*, Lecture Notes in Networks and Systems 107,
https://doi.org/10.1007/978-981-15-3172-9_20

increases cargo-carrying capabilities. The advantages of systems over the operator-controlled load carrying system are as follows [17–19]: High efficiency, navigation accuracy and reduced operating cost.

The AGV system divided into centralised and decentralised controlled system. The centralised system uses a master. It performs all necessary tasks related to a mission. The tasks such as mission plan, service assignment, path planning and motion coordination. The decentralised approach does not use a master for vehicle management. Here, vehicles independently plan their path and navigation sequence during navigation. Various researchers have developed different methods for centralised [20, 21] and decentralised [22, 23] controlling of the AGVs in a warehouse environment. The centralised and decentralised approaches have their own advantages and disadvantages. Hence, the present study uses the orchestrated approach for controlling the navigation of AGVs within a warehouse-like environment.

1.1 Orchestrated AGV Systems

Chowdhary et al. [24–27] have used orchestrated approach in different areas of networked RBs, for example, task assignment model, vehicle status monitoring system and intelligent robotic system. The approach has many advantages over the centralised and decentralised approaches [24, 25]. Our proposed model uses orchestrated approach to control the motions of mobile RBs. The orchestrator allocates the tasks to the RBs and then monitors the execution status of the tasks. Similarly, each RB timely informs the execution status of the allocated task to the orchestrator (ORCH).

We present a model for path planning and navigation of multiple AGVs like mobile RBs in a warehouse-like environment. Our model uses two-layer architecture for path planning and controlling the navigation of the RBs. First, the model uses a D* Lite algorithm for path determination between the start (S) and end (E) points. Second, the model uses a navigation algorithm to follow the path while avoiding the collision.

The three important benchmarks, namely mean navigation time (MT), mean navigation speed (MS) and inter message communication load of RBs have been used to analyse the effectuation of introduced three experiments. The results of our model are compared with the results of other researcher's works, keeping in view, the same three benchmarks.

2 Navigation Algorithm

Our main contribution in this paper is to design and implement a navigation algorithm for AGV like mobile RBs. This section explains the architecture and working of our proposed algorithm. The design of the algorithm is based on the orchestration principle. The ORCH allocates the navigation tasks to the RBs at the start. The ORCH only provides the S and E point for navigation. It does not provide a navigation path for each RBs. D* Lite algorithm computes the path between the S and E points.

Similarly, the RBs use the navigation algorithm to follow the path. The algorithm includes a collision avoidance method.

The algorithm uses a right-hand turning method for avoiding the collision. The method avoids collision between the two moving RBs in opposite direction onto the same path. Figure 1 is a pictorial demonstration of a right-hand collision avoidance method for the two blue RBs, BR1 and BR2. When the alert zone of the two moving RBs onto the same path touches each other, both the RBs apply the right-hand turning method to avoid collision between them. The BR1 and BR2 in Fig. 1 show the initial position of the RBs before applying the right-hand method. The blue-shaded oval shape represents the alert zone for the BR1 and BR2. The blue solid line arrow shows the path (Primary path of RBs between the *S* and *E* points) of the BR1 and BR2. A_d denotes the radius length of alert zone in the navigation algorithm. The dashed line yellow RBs show the new position of the BR1 and BR2 after applying the right-hand method. The yellow dashed line arrow shows the path (A temporary path for avoids the collision between the RBs) for both BR1 and BR2 to avoid the collision from each other. When the BR1 and BR2 find himself in a safe zone, they use path correction method for correcting their path coordinates. In other words, the RBs use path correction rule to return to their allocated paths.

Moving Object Method: The algorithm uses an innovative technique for detecting a moving object (or obstacle or RB). Each RB moves with the same speed in the experimental arena. Therefore, the RBs travel a constant distance in a period (constant time). The algorithm exploits this property of the RBs to detect the moving obstacles. The algorithm uses the following equation to identify the moving obstacles.

$$C_d \leq (O_d - N_d) \tag{1}$$

where, C_d represents a constant distance that can be determined using a formula *constant distance* = *constant speed* × *periodic time*. N_d denotes a new distance

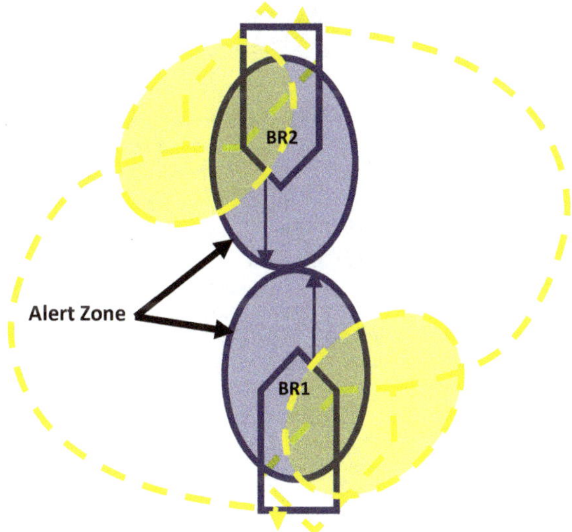

Fig. 1 Right-hand method for avoiding the collision between two RBs

measured in current periodic cycle and O_d denotes an old distance measured in previous periodic cycle.

Stationary Object Method: When the RB detects a stationary obstacle on the path and it finds a space either on the right side or on the left side of the obstacle. The RB turns right or left for avoiding the collision with the obstacle.

MTH Method: When the RB encounters a sudden obstacle and the distance of the obstacle is closer than the minimum collision avoidance distance (M_{TH}), the RB stops itself and waits for the fixed time. The RB again checks all sensor readings after the waiting duration. If RB still does not find any obstacle on previous position (Demonstrates a dynamic obstacle situation), then it follows the previous path. However, the RB detects the obstacle at the same position. The RB analyses the sensor reading to detect any empty space for the navigation. However, if the RB still is unable to detect the path, then the RB uses the stationary object method.

Algorithm 1: Navigation Algorithm.

```
function SensorReadings( )
F_d = front sensor pair distance;
R_d = rig□t sensor pair distance;
L_d = left sensor pair distance;

function PathCorrection( )
if( x ≠ current sensor value of x )
  correct x position;
else if(y ≠ current sensor value of y)
  correct y position;
else
  follow path;

function IdentifyMovingObject( )
if( C_d ≤ ( O_d − N_d) )
  MovingObject = 1;
else
  StationaryObject = 1;

function FindPath( )
SensorReadind( );
if( F_d ≥ A_d )
  go straight;
  PathCorrection( );
if( R_d ≥ A_d )
  Right-hand collision avoidance method;
  PathCorrection( );
if( L_d ≥ A_d )
  Left-hand collision avoidance method;
  PathCorrection( );
else
  if( flag == 1 )
    Wait( PVR );
  SensorReadind( );
  if(( F_d ≤ A_d )&&( R_d ≤ A_d )&&( L_d ≤ A_d ))
    SendOC( x,y );
    goto D* Lite Algorithm for a new path;

function MTH( )
flag = 0; // To stop multiple execution of wait
function.
Wait( PVR );
SensorReadind( );
if( ( F_d ≥ A_d )&&( R_d ≥ A_d )&&( L_d ≥ A_d ) )
  follow path;
else
  FindPath( );

function MovingObject( )
Right-hand collision avoidance method;
PathCorrection( );

function StationaryObject( )
flag = 1;
FindPath( );

function Main( )
flag = 0, MovingObject = 0, StationaryObject
        = 0:
PathCoordinates( );
Start
IdentifyMovingObject( );
SensorReadind( );
if( ( ( F_d ≥ A_d )&&( R_d ≥ A_d )&&( L_d ≥
A_d ))&&(N_d > A_d) )
  follow path;
  PathCorrection( );
else if( N_d ≤ M_TH )
  MTH( );
else if((MovingObject == 1)&& (N_d ≤ A_d) )
  SensorReadind( );
  PathCoordinates( );
  Check coordinates of next two cells;
  if( AnyS□arpTurn == 1 )
    PriorityCA( );
    PathCorrection( );
  else
    MovingObstical( );
else if((StationaryObject == 1)&& (N_d
        ≤ A_d) )
  StationaryObject( );
else if((x == x_e )||( y == y_e ))
  PathCoordinates( );
  if((x == x_e )&&( y == y_e ))
    Stop Moving:
else
  goto Start:
```

2.1 D* Lite Algorithm

Koenig et al. [28] proposed the D* Lite algorithm in 2002. We have adopted the D* Lite algorithm for the path determination between the *S* and *E* point. The algorithm is a novel path re-planning algorithm, which is similar to the D* algorithm in navigation strategy. The algorithm is algorithmically different from the A* and D* algorithms. The algorithm is a lightweight algorithm, because it requires less number of line codes to implement, compared to the A* and D* algorithms [28]. It is easier to implement, as it uses only single tiebreaker condition to compare the priority. This also helps to update the priority queue and eliminate the requirement of nested conditional statements in the code. It makes the code simpler increased ingenuity.

3 Experiments and Result

Figure 2 shows the pictorial presentation of actual experimental arena with black stationary objects acting as obstacle. The RBs cannot access the cell that contain obstacle. The length and width of experimental arena are 270 cm and 225 cm, respectively. The area of each cell is 45 cm × 45 cm. The ORCH keeps the master copy of this graph map into its database. Similarly, each RB keeps the copy of the same graph map in its memory for determining the path independently. Additionally, when the RB finds any new stationary obstacle during navigation, then the RBs sends the coordinates of obstacle to the orchestrator. The orchestrator updates the position of the new obstacle in its database and broadcast this information in robotic network. Thus, each RB present in the network updates the new obstacle location into their local map.

The ORP robotic platform has been used to conduct the experiments [26]. The ORCH and four RNs *R*1, *R*2, *R*3 and *R*4 have been used in experiments. All RBs move with same speed and the Max. speed of each RB is 0.39 m/s. Each experiment is repeated five times to check the consistency of the navigation algorithm.

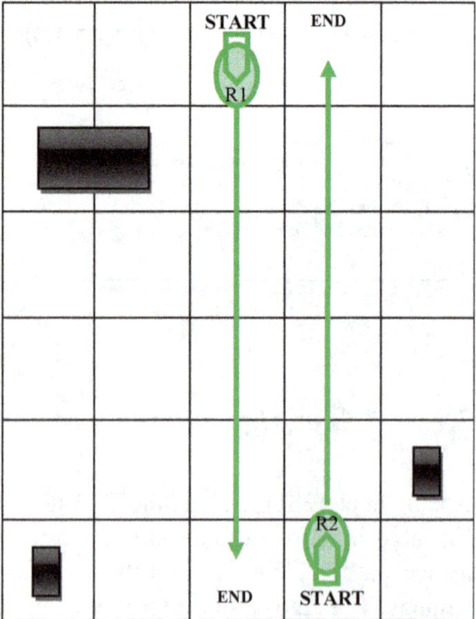

Fig. 2 Navigation of RB *R*1 and *R*2 in environment of Experiment 1

Table 1 Navigation time of the RBs in Experiment 1

RB *R*1			RB *R*2		
MT	SD	MS	MT	SD	MS
7.5	0.10	0.36	7.6	0.10	0.35

3.1 Experiment 1

Experimental arena for experiment one is shown in Fig. 2. The orchestrator allocates the navigation task to two different RBs. The RB *R*1 navigates from top to down and the RB *R*2 navigates from down to top in the experimental environment. Both RB uses the D* Lite to generate their respective paths from the *S* to *E* point. The top to down arrow shows the path for the RB *R*1, similarly the down to top arrow shows the path for the RB *R*2. Any moving obstacle does not come in the paths of both the RBs, because orchestrator uses only two different RBs. Moreover, their paths are different. The MT with standard deviation (SD) and MS of both RB are given in Table 1. The path length for the RBs in experiment one is 2.7 m.

3.2 Experiment 2

Experiment two is similar to experiment one, because the ORCH uses same two RBs viz. the RBs *R*1 and *R*2. Moreover, it allocates almost the same task compared to experiment two for both RBs. Unlike experiment one, both RBs use same path to execute the allocated task. A bidirectional green arrow line shows the path for both RBs in Fig. 3. Both the RBs start the execution of the allocated task at the same time. The blue circle shows the possible intersection area for both RBs, because both RBs uses same path in opposite direction. Whenever the alert zones of RBs touch each other in the intersection area, the RBs use collision avoidance rule for moving obstacle. The pictorial presentation of collision avoidance between two moving RBs is shown in Fig. 1. The path length for the RBs in our experiment two is same as our experiment one. Table 2 gives the MT and MS values for Experiment 2.

3.3 Experiment 3

The ORCH uses four RBs in experiment three. Figure 4 shows the diagram for experimental environment for experiment three. The *S* and *E* points for each RB are shown in Fig. 4. The ORCH allocates a navigation task to each RB. The RBs *R*1 and *R*2 run on same path, but in opposite direction, similarly the RBs *R*3 and *R*4 runs on the same path, but the RB *R*3 run from left to right and the RB *R*4 run from right to left as shown in Fig. 4. All RBs use the navigation algorithm to navigate from their *S*

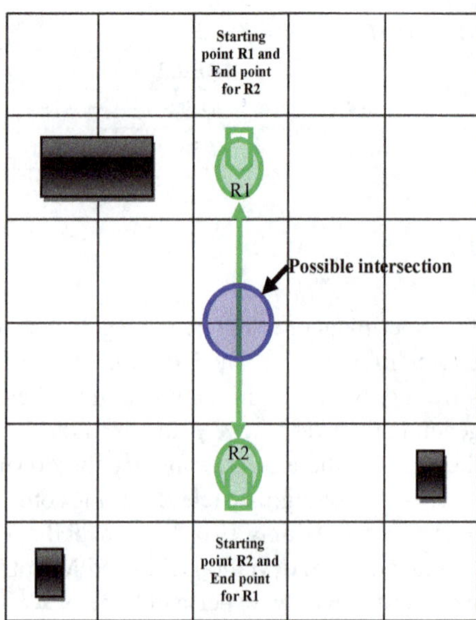

Fig. 3 Navigation of RB $R1$ and $R2$ in environment of Experiment 2

Table 2 Navigation time of the RBs in Experiment 2

RB $R1$			RB $R2$		
MT	SD	MS	MT	SD	MS
8.4	0.17	0.32	8.5	0.16	0.318

to E point. The blue circle in Fig. 4 shows the possible intersection area for all four RBs. All RBs reach at this point on the same time, because all RBs are moving with same speed. When their alert zones touch each other, all RBs use collision avoidance rule for moving obstacles. Table 3 gives the MT for all four RBs, namely $R1$, $R2$, $R3$ and $R4$ used in experiment three. The path length for the RBs in experiment three is 2.25 m. The MS of all four RBs is same that is 0.33 m/s.

Figure 5a shows the RBs $R1$, $R2$, $R3$ and $R4$ and their respective starting positions. Figure 5b shows the initial motion of the RBs $R1$, $R2$, $R3$ and $R4$ from their respective starting points. Figure 5c–e demonstrates the moving collision avoidance of the BRs $R1$, $R2$, $R3$ and $R4$ from each other using navigation algorithm in experimental arena. Figure 5f shows the RBs $R1$, $R2$, $R3$ and $R4$ at their respective end points.

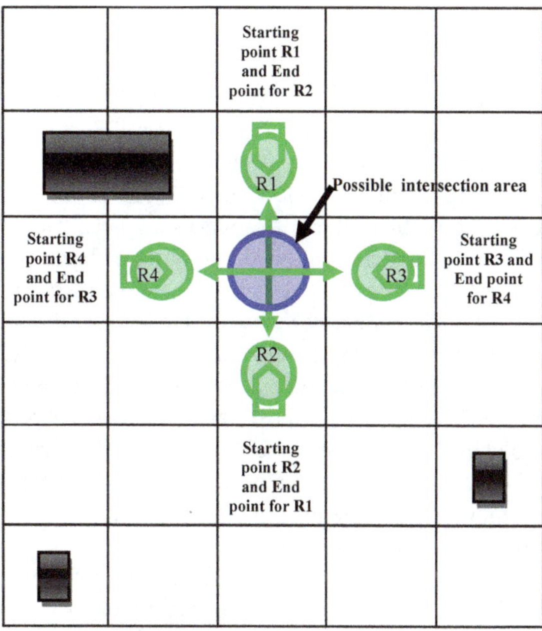

Fig. 4 Experiment 3 using four RBs

Table 3 Navigation time of the RBs in Experiment 3

RB *R*1		RB *R*2		RB *R*3		RB *R*4	
MT	SD	MT	SD	MT	SD	MT	SD
6.8	0.15	6.8	0.15	6.8	0.15	6.8	0.15

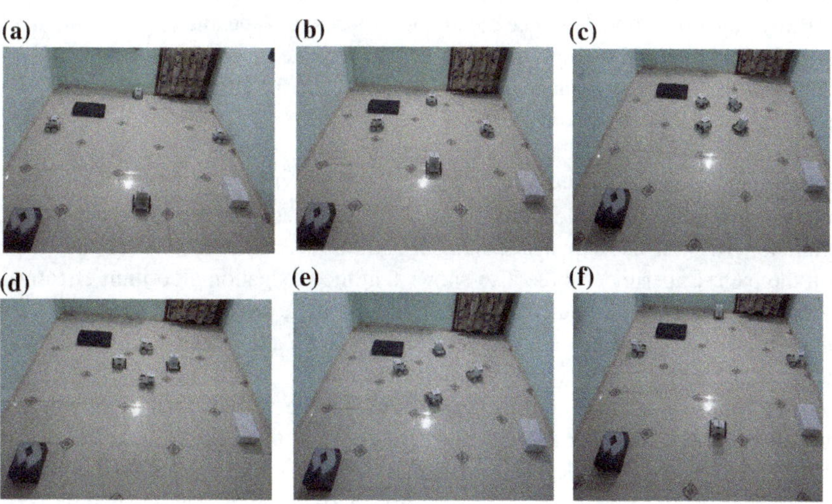

Fig. 5 Navigation RBs in Experiment 3

4 Discussion

The results of all three experiments are discussed here. The D* Lite algorithm efficiently determines the path in all the three experiments. The algorithm always gives the optimum path between the *S* and *E* points. This can be verified from Figs. 2, 3 and 4. Moreover, the algorithm needed less number of code lines in implementation compared to the A* and D* algorithms. This makes it good choice for small embedded devices. All the RBs uses navigation algorithm to navigate between the *S* and *E* points. The algorithm also avoids the collision between RBs during navigation between the *S* and *E* points. Furthermore, it stops the RBs at minimum safe distance from the obstacle, in any unavoidable situation.

Table 1 give the results for experiment one. The MT for both RBs is almost identical, because the RBs navigate on the different but identical path. The lower SD value (0.10 s for both RBs) again shows the consistency of the navigation algorithm. The MS of the RBs *R*1 and *R*2 is 0.36 m/s and 0.35 m/s, respectively. The MS of the both RBs is almost equal to the maximum speed of the RBs. This is due to the straight navigation path of experiment one.

The results of experiment two are given in Table 2. Experiment two and one use almost same experimental scenario; however, in experiment two, both RBs navigate on the same path, but in opposite direction. Hence, the RBs use the collision avoidance technique in experiment two for effectively completes the allocated task. The MT and SD value of experiment two is greater than the values obtained in experiment one. The large MT values are due to the collision avoidance technique used by the RBs.

Experiment one and two use same path length, i.e. 2.7 m. The MS of both RBs of experiment two is less as compared to the speed of both RBs of experiment one. The decline in the speed of the RBs is due to the collision avoidance technique. When a RB uses the collision avoidance technique, it turns to a specific angle on the track. The diagrammatic representation of collision avoidance scenario is shown in Fig. 1. Hence, the collision avoidance technique slows down the RBs and it is the proposed reason for reduced speed of RBs.

Experimental results of experiment three are given in Table 3. The MT of all four RBs of experiment four is almost identical to MT of experiments two. The results of these three experiments show that the MT of RBs depends on the number of moving collision (RBs). It also depends on number of turns faced by the RBs while navigating on the path. Experiment three also shows that the navigation algorithm effectively gives the access of a particular cell to all four RBs at the same time. The scenario also demonstrates how the navigation algorithm effectively avoids any deadlock situation during access of the same cell by all four RBs.

The results of all three experiments show that the RBs can navigate with their maximum speed between the *S* and *E* points, when the navigation path does not have any turns and obstacles. Hence, the RBs of experiment one run with maximum speed because their path does not have turns and obstacles. The navigation speed of the RBs in experiments two, and three is almost same. The reason for this is, all these experiments use similar path between the S and E points that have at least one moving obstacle. In other words, the numbers of turns in the navigation paths of the RBs affect the average navigation speed of the RBs. The techniques used by the RBs to avoid the collision between the RBs affect the navigation speed of the RBs. Additionally, the navigation speed of the RBs from experiments two and three are almost same. This means, the number of moving obstacles also decrease the speed of the RBs and it depend on the number of obstacles faced by the RBs.

The results of experiments show that the navigation speeds of the RBs, which use the navigation algorithm, depend on the number of turn. The speed also depends on the number of moving obstacles between their S and E points. The claim can be verified from the experiments two and three. The number of RBs in experiment two and three are 2 and 4, respectively. But the MS of all these RBs is almost same. In other words, increase in the number of the RBs in the experimental arena does not slowdown the MS. The MS of RBs is affected by the number of intersection points in the path of the RBs. MS of the RBs is reduced, when two or more RBs want to access the same intersection point of their respective path at the same time. However, the navigation speed of the RBs does not decrease when two or more RBs access the same intersection point of their respective path at the different time.

The RBs using the navigation algorithm do not exchange message with each other during collision avoidance and resolve the conflicts of the path access. Therefore, the navigation algorithm reduces the communication overhead to zero. The researchers have been using the scenario of the experiment four to demonstrate a situation of the deadlock. Experiment four shows that the navigation algorithm successfully resolves the deadlock situation. The navigation algorithm also adds the fault-tolerant capability.

Table 4 gives comparison of present work with other researcher's work using benchmarks defined in Sect. 1.1. Their experimental and simulation approaches are conceptually similar to our present approach. However, their simulations, experimental environment and the number of RBs are different from our present experiments and environment. Therefore, the results of present experiments are qualitatively compared with the results of researchers [29]. Researchers used a method that was based on negotiation for solving the path access conflicts between the RBs at the intersection points. The RBs exchanged the information with each other for avoiding the collision in their experiment. While in our approach, we use the navigation algorithm to resolve similar situation. Our proposed approach does not exchange any information between the RBs.

The task execution time of the researchers depends on the number of RBs involved in experiments. It linearly increases with the number of RBs. On the other hand, in the present approach, the mean navigation time of the RBs in all three experiments depend on the number of turns. It also depends on the common intersection points on the path encountered by the RBs at the same time. Similar to other approaches in Table 4, the present approach is able to resolve the deadlock situations. In addition, the present method adds the fault-tolerant ability.

Table 4 Comparative summary of other researchers work

Approach and Author	Path access resolution method	Type of experiments	Task execution time
Digani et al. [29] used a decentralised approach	• Negotiation method for solving the path access conflicts at the intersection points • It uses message communication between RBs	• Simulations are conducted using 5, 10, 15 and 20 AGV	• Elapsed time (experiment completion time) of simulations depend on the number of RBs involved in the experiments • It linearly increases with the number of RBs
Present approach uses the orchestrated approach	• Navigation algorithm resolves path access conflicts without any communication between the RBs	• ORP platform is used to perform experiments	• Mission (Task) execution time depends on the number of intersection points which two or more RBs want to access at the same time • It does not depend on the number of RBs that access the intersection point at different time

5 Conclusion

The paper presented the design and implementation method for path planning and collision avoidance model. The D* Lite algorithm plans the path for the RBs and it gives an optimal path between the S and E points. This can be verified from the results of the experiments. The navigation algorithm also has been designed and implemented in this paper. The results of experiments show that the algorithm efficiently navigates the RB between the S and E points. The communication overload due to navigation algorithm for resolving the path access conflicts is zero. The navigation algorithm with ORCH approach makes the system more fault tolerant.

The navigation time (or experiment completion time) depends on path length travelled by the RBs and the number of turns in navigation path. It does not depend on the number of RBs present in the arena that are performing different tasks and their paths do not intersect. In our future work, we will improve the navigation algorithm for handling more RBs for same path.

References

1. V. Jaiganesh, D. Kumar, J. Girijadevi, Automated guided vehicle with robotic logistics system. Proc. Eng. **97**, 2011–2021 (2014)
2. L. Schulze, S. Behling, S. Buhrs, Automated guided vehicle systems: a driver for increased business performance, in *Proceedings of the International Multi Conference of Engineers and Computer Scientists*, Hong Kong, 2, (2008), pp. 1275–1280
3. K.C.T. Vivaldini, G. Tamashiro, J.M. Junior, M. Becker, Communication infrastructure in the centralized management system for intelligent warehouses, in *International Workshop on Robotics in Smart Manufacturing* (2013), pp. 127–136
4. Q.S. Kabir, Y. Suzuki, Increasing manufacturing flexibility through battery management of automated guided vehicles. Comput. Ind. Eng. **117**, 225–236 (2018)
5. W. Khalil, R. Merzouki, O.B. Belkacem, Modelling for optimal trajectory planning of an intelligent transportation system, in *7th IFAC Symposium on Intelligent Autonomous Vehicles* (2010)
6. S.A. Mnubi, Motion planning and trajectory for wheeled mobile robot. Int. J. Sci. Res. **5**(1), 2319–2706 (2016)
7. S. Carvalhosa, A.P. Aguia, A. Pascoal, Cooperative motion control of multiple autonomous marine vehicles: collision avoidance in dynamic environments, in *Proceedings of IAV 2010-7th Symposium on Intelligent and Autonomous Vehicle* (Lecce, Italy, 2010)
8. R. Olmi, C. Secchi, C. Fantuzzi, Coordination of multiple robots with assigned paths, in *Proceedings of the 7th IFAC Symposium on Intelligent Autonomous Vehicles* (2010)
9. L. Abdenebaoui, H.J. Kreowski, Decentralized routing of automated guided vehicles by means of graph-transformational swarms, in *Dynamics in Logistics*. Lecture Notes in Logistics (2017), pp. 457–467
10. H.F. Wang, C.M. Chang, Facility layout for an automated guided vehicle system. Proc. Comput. Sci. **55**, 52–61 (2015)
11. A. Cesett, C.P. Scotti, G. Di Buò, M. Babini, R. Donnini, G. Angione, L. Lattanzi, C. Cristalli S. Longhi, Field robot supporting the activities of a household appliances laboratory, in *7th IFAC Symposium on Intelligent Autonomous Vehicles* (2010), pp. 539–544
12. D. Bechtsis, N. Tsolakis, D. Vlachos, S.J. Srai, Intelligent autonomous vehicles in digital supply chains: a framework for integrating innovations towards sustainable value networks. J. Clean. Prod. **181**, 60–71 (2018)
13. M. Mattei, V. Scordamaglia, Path planning for wheeled mobile robots using core paths graphs, in *7th IFAC Symposium on Intelligent Autonomous Vehicles*, vol. 7(1) (2010), pp. 360–365
14. A. Codas, M. Devy, C. Lemaire, Robot Localization algorithm using odometry and RFID technology, in *7th IFAC Symposium on Intelligent Autonomous Vehicles*, vol. 43(16) (2010), pp. 569–574
15. D. Bechtsis, N. Tsolakis, D. Vlachos, E. Iakovou, Sustainable supply chain management in the digitalization era: the impact of automated guided vehicles. J. Clean. Prod. **142**(4), 3970–3984 (2017)
16. S. Chopra, G. Notarstefano, M. Rice, Magnus, Distributed version of the hungarian method for a multirobot assignment. IEEE Trans. Robot. **33**(4), 932–947 (2017)
17. A. Krnjak, I. Draganjac, S. Bogdan, T. Petrovi´c, Z. Kovaˇci´c, Decentralized control of free ranging AGVs in warehouse environments, in *Proceedings of IEEE International Conference Robotics Automation (ICRA)* (2015), pp. 2034–2041
18. A. Franchi, P. Stegagno, M.D. Rocco, G. Oriolo, Distributed target localization encirclement with a multi-robot system, in *7th IFAC Symposium on Intelligent Autonomous Vehicles* (2010), pp. 151–156
19. R. Yanz, S.J. Dunnett, L.M. Jackson, Novel methodology for optimising the design, operation and maintenance of a multi-AGV system. Reliab. Eng. Syst. Saf. **178**, 130–139 (2018)
20. T. Miyamoto, K. Inoue, Local and random searches for dispatch and conflict-free routing problem of capacitated AGV systems. Comput. Ind. Eng. **91**, 1–9 (2016)

21. B. Li, X. Hong, D. Xiao, G. Yu, Y. Zhang, Centralized and optimal motion planning for large-scale AGV systems: A generic approach. Adv. Eng. Softw. **106**, 33–46 (2017)
22. G. Demesure, M. Defoort, A. Bekrar, D. Trentesaux, M. Djemaï, Decentralized motion planning and scheduling of AGVs in FMS. IEEE Trans. Industr. Inf. **14**(4), 1744–1752 (2018)
23. R. Zhang, K. Cai, Supervisor localization for large-scale discrete-event systems under partial observation. Int. J. Control, 1–13 (2018)
24. R.R. Chowdhary, M.K. Chattopadhyay, Rajkamal, Study of an orchestrator for centralized and distributed networked robotic systems, in *Proceedings of the International Conference on Information Engineering, Management and Security* (2015), pp. 139–143
25. R.R. Chowdhary, M.K. Chattopadhyay, Rajkamal, Comparative study of orchestrated, centralised and decentralised approaches for orchestrator based task allocation and collision avoidance using network controlled robots. Accepted in Elsevier's J. KSU—Comput. Inf. Sci. (2019)
26. R.R. Chowdhary, M.K. Chattopadhyay Rajkamal, Orchestration of robotic platform and implementation of adaptive self-learning neuro-fuzzy controller. J. Electron, Des. Technol. **8**(3), 17–29 (2017)
27. R.R. Chowdhary, M.K. Chattopadhyay Rajkamal, IoT model based battery temperature and health monitoring system using electric vehicle like mobile robot. Accepted in J. Adv. Robot. (2019)
28. S. Koenig, M. Likhachev, D* Lite, in *Proceedings of the National Conference on Artificial Intelligence* (2002)
29. V. Digani, L. Sabattini, C. Secchi, C. Fantuzzi, Hierarchical traffic control for partially decentralized coordination of multi AGV systems in industrial environments, in *IEEE International Conference on Robotics and Automation (ICRA)*, Hong Kong (2014), pp. 6144–6149

Multi-band Hybrid Aperture-Cylindrical Dielectric Resonator Antenna for Wireless Applications

Chandravilash Rai, Amit Singh, Sanjai Singh and Ashutosh Kumar Singh

Abstract In this article, multi-band hybrid aperture-cylindrical dielectric resonator antenna (CDRA) is presented. Regular pentagon-shaped aperture is used to generate to radiating mode ($HEM_{11\delta}$ and $HEM_{12\delta}$ mode). Lower-frequency band is linearly popularized, and upper-frequency band is dual popularized (combination of linear and circular). In both of the structure (spoon-type microstrip line and rectangular-type microstrip line), spoon-type microstrip line is resonated at frequency band 6.40–7.50 GHz with fractional bandwidth 15.82% and rectangular-type microstrip line at band 6.95–7.75 GHz with fractional bandwidth 11.26%, and VSWR is calculated nearly at resonance frequency.

Keywords Cylindrical dielectric resonator antenna (CDRA) · Aperture coupled · Multi-band · Rectangular microstrip line · Spoon-type microstrip line

1 Introduction

Dielectric resonator antenna (DRA) was first proposed by Robert Richtmyer and after him in 1983, L. C. Long et al. design and testing of DRA. Simultaneously, many researchers tried their luck in the improvement of this antenna. DRA has been the most popular in the last 30 years due to some smart features like high gain, large bandwidth, high radiation efficiency and no metallic losses in high frequency [1].

DRA can be of any shape and size but cylindrical and rectangular DRA are most popular choice because of additional degree of freedom which means small and thick DRA can be used same as large and thin DRA [1]. In this paper, we have used cylindrical dielectric resonator antenna (CDRA) due to easy manufacturing and availability. Another feature of CDRA has three modes (TEmnp, TMmnp, HEmnp) which is useful to achieve the desirable radiation pattern [2].

C. Rai (✉) · A. Singh · S. Singh · A. K. Singh
Department of Electronics and Communication Engineering, Indian Institute of Information Technology Allahabad, Allahabad, India
e-mail: rse2018508@iiita.ac.in

S. Singh
e-mail: ssingh@iiita.ac.in

H. S. Saini et al. (eds.), *Innovations in Electronics and Communication Engineering*, Lecture Notes in Networks and Systems 107,
https://doi.org/10.1007/978-981-15-3172-9_21

Dual band/multi-band antenna is most popular in wireless communications due to advantage like a single antenna can be used for different applications. Systematically reduce the interference, free of orientation in transistor and receiver [2]. Mainly three techniques are used to obtain dual characteristics (liner and circular polarization) in DRA: First is the addition of parasitic element, second is hybrid DRA (combination of more than one element), and third one is higher mode generating [3, 4]. Some researchers (Leung and Fang) proposed a dual band circular polarized antenna but main problem of this antenna is feeding, and after that hybrid aperture coupled antenna are more popular [5].

In this proposed work, multi-band, hybrid aperture CDRA is used. Mainly two important feature of this proposed antenna, first is hybrid mode generated (HEM11δ and HEM12δ) and second one is spoon-type microstrip line generate additional λ/4 path delay orthogonal electric field at 6.95 GHz. The proposed antenna is useful at different frequency band 6.45–7.4 GHz and 6.90–7.60 GHz.

2 Geometry and Design of Antenna

Figure 1 shows systematic diagram of proposed DRA; first Fig. 1a shows the spoon-like microstrip line and second Fig. 1b has rectangular microstrip line; both of the antennas are simulated on the marginal cost FR4 epoxy substrate with ($\varepsilon_{r\,sub} = 4.4$, tan $\delta = 0.02$), height $h = 1.6$ mm, length $L_G = 50$ mm and width $W_G = 50$ mm substrate is etched in the shape of regular pentagon. Microstrip line (50 Ω) is deposited below the substrate. A material alumina is used to make CDRA ($\varepsilon_{r\,CDRA} = 9.8$, tan $\delta =$ 0.002) with radius $R = 13$ and height $H = 12$. Table 1 shows the optimized dimension of parameter of proposed antenna.

In CDRA, the resonance frequency of the hybrid mode $HEM_{11\delta}$ is calculated by the following formulas [6]:

$$f_r = \frac{6.321c}{2\pi R\sqrt{\varepsilon_{reff} + 2}}\left[0.27 + 0.36\left(\frac{R}{2H_{eff}}\right) + 0.02\left(\frac{R}{2H_{eff}}\right)^2\right] \tag{1}$$

where R is the radius, ε_{eff} is the effective permittivity, H_{eff} is the height of CDRA but H_{eff} effective height, and effective permittivity (ε_{eff}) can be calculated by formulas

$$\varepsilon_{r\,eff} = \frac{H_{eff}}{\frac{H}{\varepsilon_{rCDRA}} + \frac{H_s}{\varepsilon_{r\,sub}}} \tag{2}$$

There are no stabile formulas for calculating the resonance frequency of $HEM_{12\delta}$ mode so resonance frequency of mode $HEM_{12\delta}$ finding with help of mode $HEM_{11\delta}$

$$\text{fHEM}_{12\delta} \geq 1.9\text{fHEM}_{11\delta} \tag{3}$$

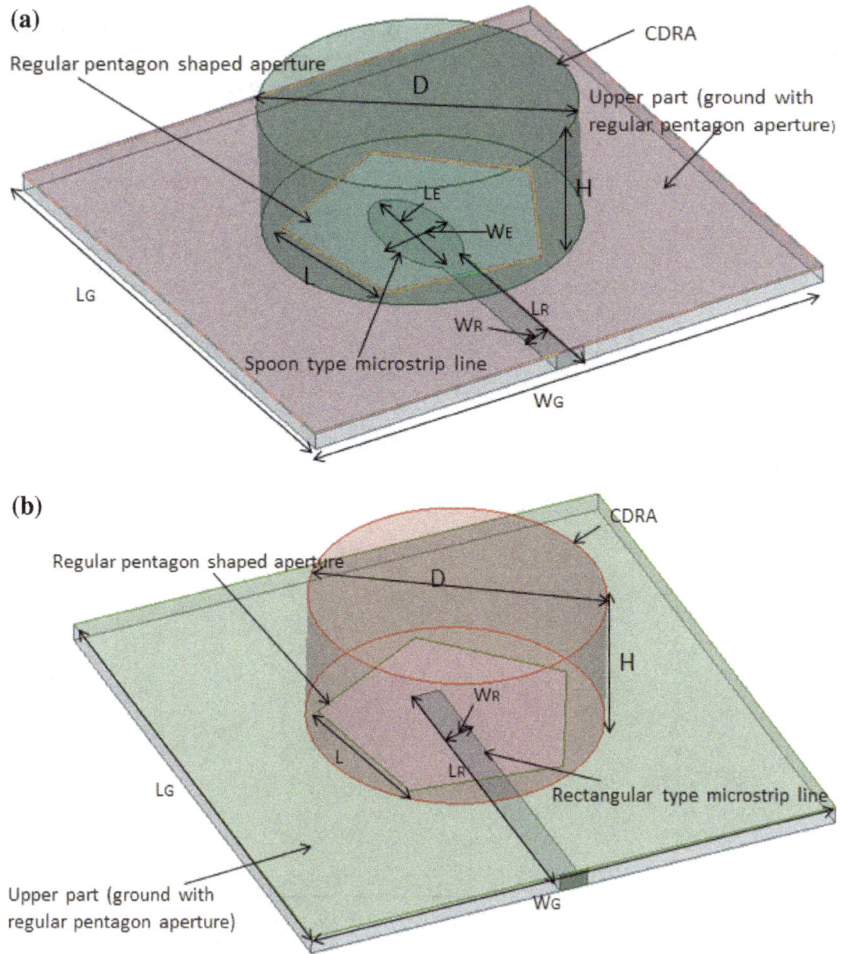

Fig. 1 **a** Systematic diagram of spoon-type microstrip line proposed radiator, **b** systematic diagram of rectangular-type microstrip line proposed radiator

but the resonance frequency of $HEM_{12\delta}$ highly depends on aspect ratio ($D/2H$) aperture (L) in regular pentagon shape derived from the radius of circle ($R = 12$).

3 Working of Antenna

The Ansoft HFSS EM simulator has been used in the investigation of proposed radiator. Figure 2 shows the return loss graph of proposed antenna with regular pentagon aperture with spoon-type microstrip line structure and rectangular microstrip

Table 1 Optimized dimension of different parameter of proposed antenna

Symbol (spoon type)	Dimension (mm)	Symbol (rectangular type)	Dimension (mm)
L_G	50	L_G	50
W_G	50	W_G	50
H	12	H	12
D	26	D	26
w_R	2.6	W_R	2.6
L_R	24	L_R	3.1
w_E	2		
L_E	3.1		

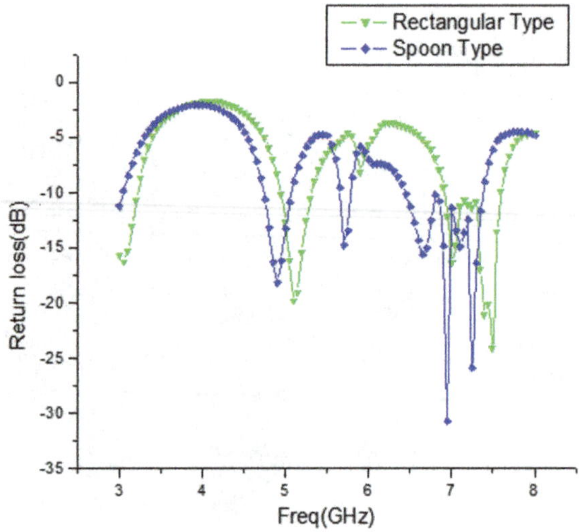

Fig. 2 Return loss versus frequency curve

line structure. Here, spoon-type microstrip line is responsible for multi-band generation. In the below figure, both antennas resonate at lower-frequency band at 4.9 and 5.10 GHz due to CDRA and upper band at 6.95 and 7.50 GHz due to shape of spoon-type and rectangular-type microstrip line. It is fact that the excitation principle of CDRA acts as a horizontally situated magnetic dipole for $HEM_{11\delta}$ mode generation [7]. Pentagon-shaped slot regularly places as a horizontally placed magnetic dipole. For generation $HEM_{12\delta}$ mode in CDRA feeding structure is playing role for horizontally placed electric dipole [7]. Proposed antenna resonate at 4.90 GHz in mode $HEM_{11\delta}$ and 5.10 GHz in mode $HEM_{12\delta}$ show in Fig. 3, electric field line *X*-polarized and *Y*-polarized are excited by regular pentagon aperture shape, so its generate $HEM_{11\delta}$ and $HEM_{12\delta}$ mode. Where, $HEM_{11\delta}$ is strong coupling and $HEM_{12\delta}$ is week coupling.

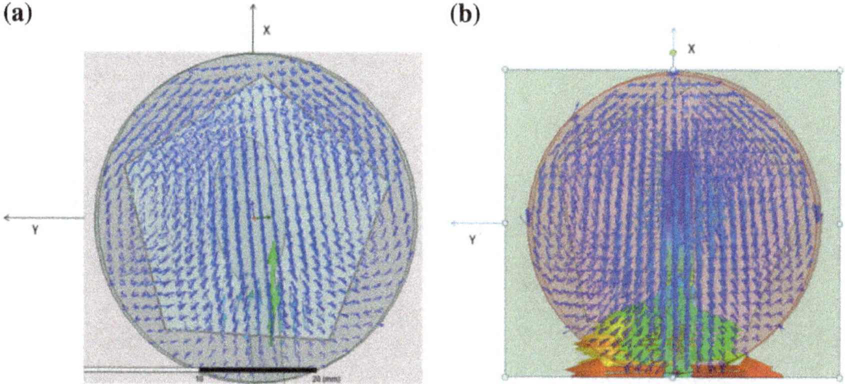

Fig. 3 E-field distribution in CDRA **a** at 6.95 GHz in spoon type, **b** at 7.50 GHz in rectangular type

4 Result and Discussion

Optimized result of return loss is shown in Fig. 2, where spoon-type microstrip is resonating in multi-band and rectangular type is resonating at dual band but bandwidth of spoon type is large. Field pattern is shown in Fig. 4 where co and cross polarization nearly maintain 3 dB difference. Figure 5 shows that both types of simulated result gain are more than 0 dB at 6.95 and 7.50 GHz, and Fig. 6 shows the VSWR closed to 1.

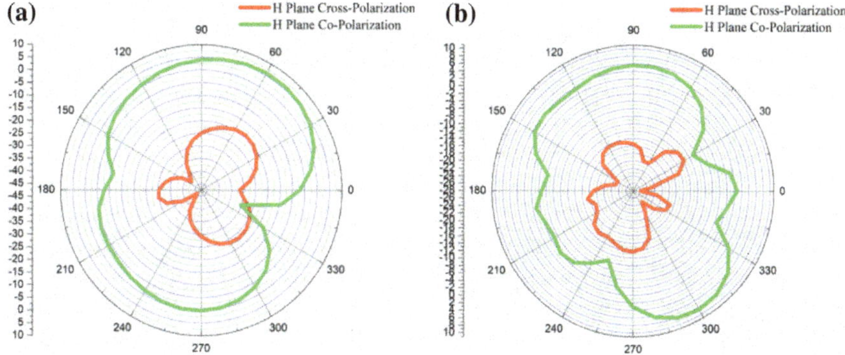

Fig. 4 Far field pattern representation **a** at 6.95 GHz in spoon type, **b** at 3.68 GHz in rectangular type

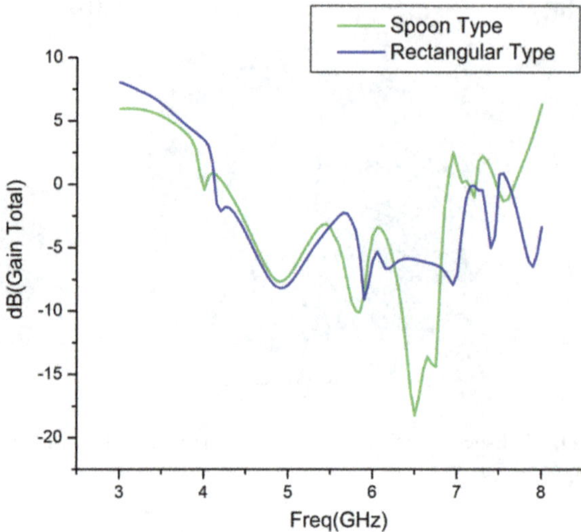

Fig. 5 Total gain versus frequency curve

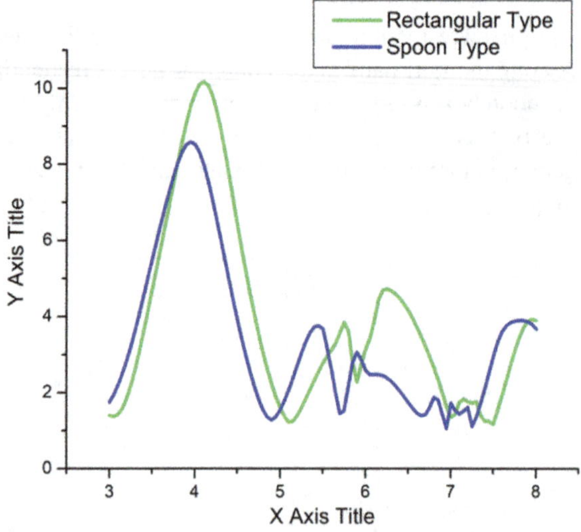

Fig. 6 VSWR versus frequency curve

5 Conclusion

This manuscript presents a multi-band hybrid aperture CDRA, dual band characteristic is achieved by rectangular-type micro strip line, and multi-band is achieved by spoon-type micro strip line. In this work, hybrid mode is achieved by regular pentagon-shaped aperture. The main purpose of this antenna is wireless communication.

References

1. A. Petosa, *Dielectric Resonator Antenna Handbook* (Artech House, Norwood, 2007)
2. K.M. Luk, K.W. Leung, *Dielectric Resonator Antenna* (Research Studies Press Ltd., Baldock, Hertfordshire, 2003)
3. H.M. Chen, Y.K. Wang, Y.F. Lin, S.C. Lin, S.C. Pan, A compact dual-band dielectric resonator antenna using a parasitic slot. IEEE Antennas Wirel. Propag. Lett. **8**, 173–176 (2009)
4. Y. Din, K.W. Leung, On the dual-band DRA-slot hybrid antenna. IEEE Trans. Antennas Propag. **57**, 624–630 (2009)
5. Y.F. Li, H.M. Chen, C.H. Lin, Compact dual-band hybrid dielectric resonator antenna with radiating slot. IEEE Antennas Wirel. Propag. Lett. **8**, 6–9 (2009)
6. H.S. Ngan, X.S. Fang, K.W. Leung, Design of dual-band circularly polarized dielectric resonator antenna using a high-order mode, in *Proceedings of IEEE-APS APWC* (2012) 424–427
7. D. Guha, P. Gupta, C. Kumar, Dual band cylindrical dielectric resonator antenna employing $HEM_{11\delta}$ and $HEM_{12\delta}$ mode excited by new composite aperture. IEE Trans. Antennas Propag. **63**, 433–438 (2015)

An Enhanced Dynamic Cluster Head Selection Approach to Reduce Energy Consumption in WSN

C. Sudha, D. Suresh and A. Nagesh

Abstract Wireless sensor network is an emerging technology, and it is employed in various fields from health care to military. WSN is basically working as sensor node and connecting sensor nodes to each other. Each node requires power to maintain data transmission to sink. Each node has a definite quantity of power. In many cases, energy usage of nodes is not equal. WSN lifetime is decreasing appropriate to their partial, a lesser amount of power sources; energy becomes essentially the majority invaluable useful resource for sensor nodes in such networks. The objective is to decrease energy consumption and increase the lifetime of sensor network and improve the communication among the wireless network sensor nodes to sink or base station. In this work, we present new clustering approach called enhanced dynamic cluster head selection approach (EDCHSA) that is ready to manage delay-sensitive applications with discrete best effort applications and meets every interruption and energy constraint. EDCH selection approach is being analyzed by specifying a set of parameters, such as the packets received, end-to-end delay, throughput; these measurements bring an analysis to decide which approach among both work better solution to increase lifetime and reduce the risk of node failures in WSN.

Keywords WSN · Clustering · EDCHSA · Energy consumption

C. Sudha (✉)
Annamalai University, Chidambaram, Tamil Nadu, India
e-mail: csudhahyd@gmail.com

D. Suresh
Department of IT, Annamalai University, Chidambaram, Tamil Nadu, India
e-mail: deiveekasuresh@gmail.com

A. Nagesh
Department of CSE, Mahatma Gandhi Institute of Technology, Hyderabad, India
e-mail: akknagesh@rediffmail.com

H. S. Saini et al. (eds.), *Innovations in Electronics and Communication Engineering*, Lecture Notes in Networks and Systems 107,
https://doi.org/10.1007/978-981-15-3172-9_22

1 Introduction

In co-operative wireless sensor network (WSN), an effective distribution of energy may be executed by the way of organizing sensor nodes into clusters. They incorporate a large number of sensors distributed indiscriminately in areas typically hostile and or inaccessible to humans [1]. For massive quantity of sensor nodes in a wireless sensor networks, clustering involves group of sensor nodes into one place which are referred to as clusters. An entire network is dividing addicted to clusters, and for every cluster, a head node is selected known as cluster head (CH). At cluster head node, facts are maintained something like the associated recognized node and its cluster [2]. CH of a cluster imparts among every one of the sensor nodes of its very own cluster utilizing the stored perceptive that it has. Cluster head collects statistics from every node of it's possess cluster along with filters the statistics, then compresses that statistics in the direction of transmit. This compacted as well as non-redundant data is transmitted to sensor nodes of unlike cluster; otherwise, base station cluster head transmits understand-a-way to nodes of one-of-a-type clusters both through manner of the gateways or by way of associated CH of that cluster organization.

Toward select for CH of a cluster, a procedure includes selection among each and every one of sensor nodes of that cluster and this method transparency on the community as for the duration of this approach additional energy is spent by way of the network sensor nodes. During the efficient CH determination process, it is extremely complex toward reloading energy of the wireless sensor network nodes. More than a few schemes are proposed with the aid of the researchers which do not forget the barriers of the sensor nodes resembling battery usage and memory boundaries, energy utilization [3].

The remainder of this paper is ordered as follows. Related work regarding cluster head selection is discussed in Part 2 and enhanced dynamic cluster head selection approach (EDCHSA) proposed work analyzed in Part 3. Part 4 consists of simulation result. Part 5 includes conclusion based on the above study discussed.

1.1 Problem Declaration

One of the main problems or issues in WSN is unreasonable CH assortment and energy use. This makes errands for ventures and instructive elements. Therefore, forward energy utilization coping is one among the method to amplify the system network lifetime. Primarily shrinking the system energy utilization will help in increasing the system network lifetime. Once the separation among sensor nodes persists, the energy use of the system is high as well maintaining single CH will increase the workload. The aim of the proposed approach is increasing the sensor network lifespan and scale back the chance of node failures. Node failure may cause the network breakdown. It can be reduced by minimizing the load on nodes and squat the battery power usage.

2 Literature Review

See Table 1.

Table 1 Summarization of available related studies undergone in cluster head selection approaches

Authors name	Year	CH selection approach	Parameter used	Energy factor	Data aggregation	Scalability	Achieved
Heinzelman et al. [2]	2000	LEACH	Random CH node selection	No	Yes	Limited	Load distribution in network
Heinzelman et al. [4]	2002	LEACH—C	Average	Yes	Yes	Good	Achieves more rounds in n/w
Fahmy and Younis [5]	2004	Extended HEED	Prob/Energy	Yes	Yes	Good	Very difficult to obtain the network lifetime
Wang et al. [6]	2005	ACW	Minimal back off value	Yes	Yes	Limited	Cluster head distribution becomes uniform
Chu et al. [7]	2006	CIPRA	ID based	Yes	Yes	Good	It reduces the amount of transmission
Manimala and Senthamil [8]	2013	LEACH F	Energy	Yes	Yes	Limited	Delay is small
Jia et al. [9]	2016	DCHSM algorithm	Distance	Yes	Yes	Good	Imbalance of energy usage improves information redundancy
Dala [1]	2018	EDDCH technique	Average residual energy	Yes	Yes	Very good	Data transmission is effective and energy efficient
Darabkh and Zomot [10]	2018	MOD-CEED	Residual energy, distance	Yes	Yes	Very good	Efficiently in extending the network lifetime
Rubel et al. [3]	2018	Priority CH selection	Minimum distance, residual energy	Yes	Yes	Very good	Better performances for critical applications

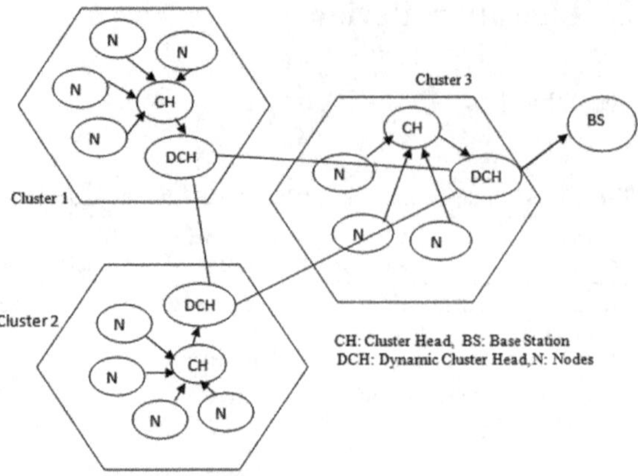

Fig. 1 CH selection and enhanced dynamic cluster head selection approach (EDCHSA)

3 The Proposed Approach

The study of concerning energy utilization, wireless communication network lifetime, node failures and network coverage is completed; however, the typical energy utilization and network system lifetime are still not in satisfactory level. We need a system or process to improve system network lifetime and decrease the node failures. The proposed enhanced dynamic cluster head selection approach for wireless sensor network will help to increase the lifetime and decrease the failures.

In this approach, the same energy level sensor nodes are gathered in an individual distinctive group. In the group, one of the nodes can be a cluster head which can be the nearest to the sink node and others as node. Dynamic cluster head is also a part of cluster and selected based on the required energy rate, average residual energy. CH and DCH can be in a single cluster. CH and DCH are preferred based on average energy levels of the connected sensor nodes. The function of cluster head is to get the accumulated data from sensor nodes and share all the data with DCH. Dynamic cluster head communicates the data information with its counterpart of the other cluster. The segregated data from the DCH will go to the sink node. Figure 2 shows the cluster head and enhanced dynamic cluster head selection approach (Fig. 1).

3.1 *EDCHS Approach*

The process of enhanced dynamic CH selection approach is given below:

Node Development: The sensor nodes area unit deployed within the observation sector and also the target sensor node will collect the locality information and energy data of all the sensor nodes.

Cluster Formation: Energy of the entire sensor node is calculable and sensor hub that has more energy among the cluster members, and we consider CH based on smallest amount of distance from sink chosen as cluster head (CH) in cluster one.

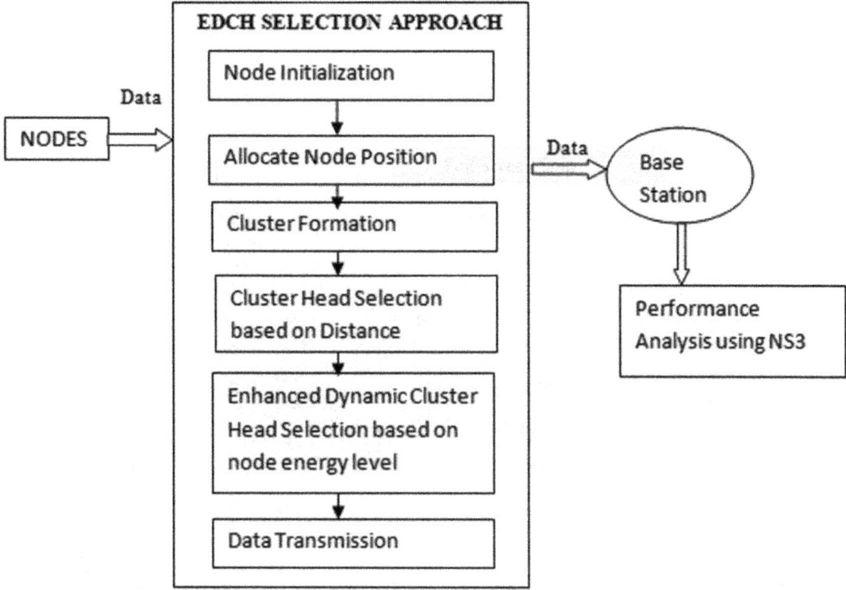

Fig. 2 The process for CH and enhanced dynamic cluster head selection approach

Enhanced Dynamic Cluster Head Selection: From the set of cluster head (CH), we need to choose dynamic energy efficient cluster head, the EDCH is based on energy rate distance rate. It will consider which cluster heads have sufficient residual energy and more neighbor nodes.

Average Residual Energy: The computation quantity of energy leftover within a node at the current occurrence of the time is called residual energy. The leftover energy of network considered through using the following Formula (1) given below:

$$R.\text{Energy} = \frac{\sum(\text{Node Initial Energy} - \text{Energy consumed by nodes})}{\text{Number of packets transmitted}} \tag{1}$$

Data Transmission: We organize data distribution based on the packet load. The packets are scheduled based on the priority mechanism, where some packets need to push in a queue to deliver toward a destination. For this process, we customize the RREQ packet format to justify the packet scheduling.

Performance Analysis: We evaluate the performance of cluster by packet delivery ratio, throughput ratio and delay. Whole number of data packets effectively delivered to the receiver is called as packet delivery ratio (PDR). Total number of data packets that are successfully passed through is called throughput ratio. Delay means the total delay occurred from sender to receiver to deliver the packet. We used NS-3.23 to simulate our efficient approach. We show the sensor nodes and communication range

Table 2 Parameters considered

Number of nodes	100
Routing protocol	AODV/light path
Agent	UDP
Application	CBR
Communication range	250 unit
Traffic CBR	8 Kbps per flow
Number of flows	50
Pause time	5 s
Maximum speed	10 unit per second
Network interface type	Physical wireless
Node placement	Random
Number of traffic flows	2–5
MAC protocol	IEEE 802.11 p
MAC rate	2 Mbps

in animated format for which we use NetAnim 3.106. Comparing the efficiency of the proposed efficient approach with the existing cluster head selection techniques shows better PDR, throughput, average residual energy and delay. Table 2 shows the parameters that we have implicit for simulation of our proposed CH selection approach.

4 Simulation Result

In this section, we will evaluate the performance of our EDCH selection approach and compare it with EDDCH and LEACH in terms of the wireless sensor network lifespan. We analyze our selection approach by NS-3.23 (Figs. 3, 4, 5, 6, 7 and 8).

5 Conclusion

The proposed enhanced dynamic cluster head selection approach for wireless communication network will help to minimize the energy level by analyzing the level of energy consumption in each node. The proposed work occupied with four important features, together with average residual energy and PDR, throughput and end-to-end delay. The proposed method performance is compared with efficient dynamic deputy cluster head selection [EDDCH] technique and LEACH. It is proven that the proposed method overcomes the inequality usage of node energy level and improves data redundancy during data transmission between nodes as well as improves the wireless sensor network lifetime.

```
Waf: Leaving directory `/home/roopa/Desktop/ns-allinone-3.23/ns-3.23/build'
'build' finished successfully (8.294s)
The nodes in the network sector C1 are 2, 3, 4, 5, 6,
The nodes in the network sector C2 are 7, 8, 9, 10, 11,
The nodes in the network sector C3 are 12, 13, 14, 15, 16,
The nodes in the network sector C4 are 17, 18, 19, 20, 21,
The nodes in the network sector C5 are 22, 23, 24, 25, 26,
The nodes in the network sector C6 are 27, 28, 29, 30, 31,
The nodes in the network sector C7 are 32, 33, 34, 35, 36,
The nodes in the network sector C8 are 37, 38, 39, 40, 41,
The nodes in the network sector C9 are 42, 43, 44, 45, 46,
The nodes in the network sector C10 are 47, 48, 49, 50, 51,

The random Cluster head choosen is 4th node of network sector C1
The Cluster head for network sector C1 is node number 5

The random Cluster head choosen is 1th node of network sector C2
The Cluster head for network sector C2 is node number 7

The random Cluster head choosen is 5th node of network sector C3
The Cluster head for network sector C3 is node number 16

The random Cluster head choosen is 2th node of network sector C4
The Cluster head for network sector C4 is node number 18

The random Cluster head choosen is 3th node of network sector C5
The Cluster head for network sector C5 is node number 24

The random Cluster head choosen is 1th node of network sector C6
The Cluster head for network sector C6 is node number 27

The random Cluster head choosen is 5th node of network sector C7
The Cluster head for network sector C7 is node number 36

The random Cluster head choosen is 4th node of network sector C8
The Cluster head for network sector C8 is node number 40

The random Cluster head choosen is 5th node of network sector C9
The Cluster head for network sector C9 is node number 46

The random Cluster head choosen is 2th node of network sector C10
The Cluster head for network sector C10 is node number 48
```

Fig. 3 Cluster formation and cluster head selection in NS-3.23 simulation tool

```
The nodes in the network sector C7 are 32, 33, 34, 35, 36,
The nodes in the network sector C8 are 37, 38, 39, 40, 41,
The nodes in the network sector C9 are 42, 43, 44, 45, 46,
The nodes in the network sector C10 are 47, 48, 49, 50, 51,

The random Cluster head choosen is 4th node of network sector C1
The Cluster head for network sector C1 is node number 5

The random Cluster head choosen is 1th node of network sector C2
The Cluster head for network sector C2 is node number 7

The random Cluster head choosen is 5th node of network sector C3
The Cluster head for network sector C3 is node number 16

The random Cluster head choosen is 2th node of network sector C4
The Cluster head for network sector C4 is node number 18

The random Cluster head choosen is 3th node of network sector C5
The Cluster head for network sector C5 is node number 24

The random Cluster head choosen is 1th node of network sector C6
The Cluster head for network sector C6 is node number 27

The random Cluster head choosen is 5th node of network sector C7
The Cluster head for network sector C7 is node number 36

The random Cluster head choosen is 4th node of network sector C8
The Cluster head for network sector C8 is node number 40

The random Cluster head choosen is 5th node of network sector C9
The Cluster head for network sector C9 is node number 46

The random Cluster head choosen is 2th node of network sector C10
The Cluster head for network sector C10 is node number 48

g = 7
The choosen Hybridclusterhead in 1st sector is of network sector C2 with node no. 7
g = 18
The choosen Hybridclusterhead in 2nd sector is of network sector C4 with node no. 18
g = 40
The choosen Hybridclusterhead in 3rd sector is of network sector C8 with node no. 40

**9**RecvSink = 0.0.0.0:9
```

Fig. 4 The enhanced dynamic cluster head selection approach is based on the energy rate and distance rate. EDCHSA will reduce the network overhead issues

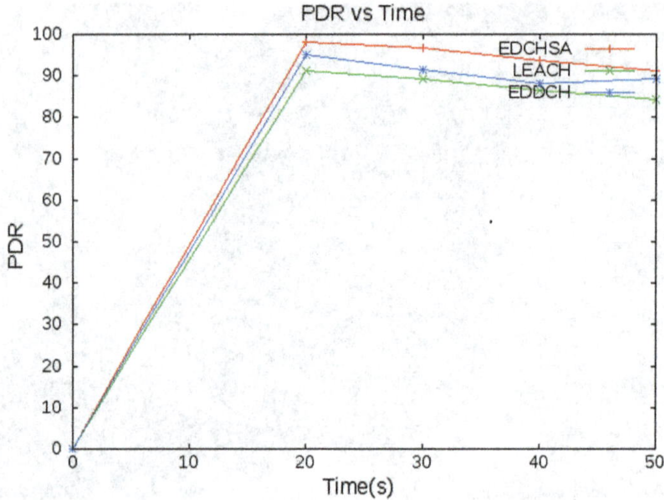

Fig. 5 Comparison of PDR for proposed EDCHSA than the existing cluster head selection approaches

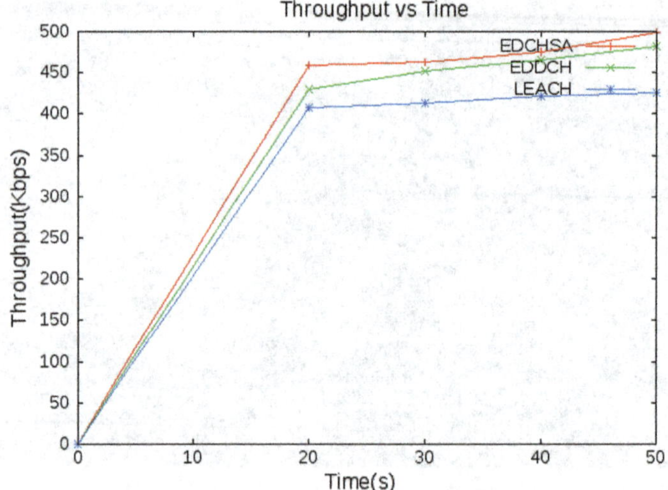

Fig. 6 Performance evaluation between delay versus time

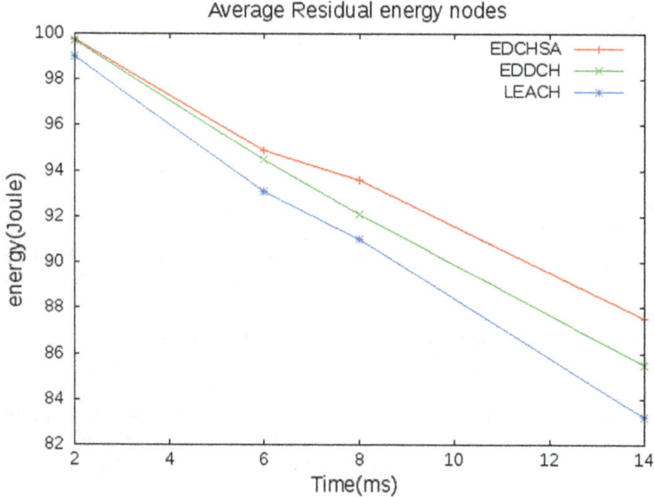

Fig. 7 Performance of throughput is made high estimate to existing ones EDDCH and LEACH

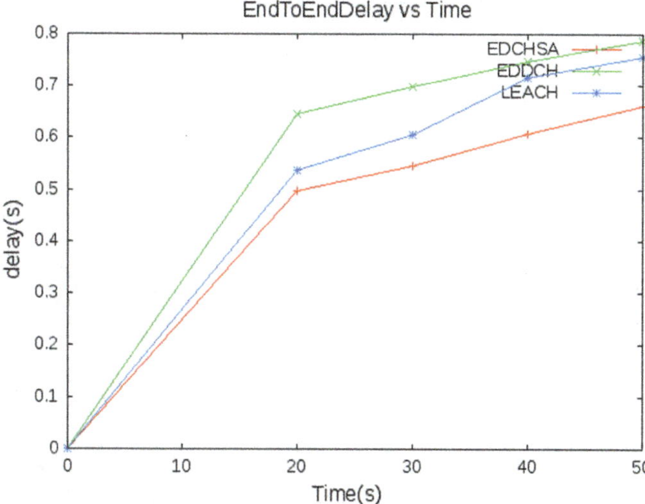

Fig. 8 Comparison of average residual energy for proposed EDCHSA than the existing cluster head selection approaches

References

1. L.B. Dalal An efficient dynamic deputy cluster head selection method for wireless sensor networks. IRJET **05**(12), 187–192 (2018)
2. W.R. Heinzelman, A. Chandrakasan, H. Balakrishnan, Energy-efficient communication protocol for wireless microsensor networks, in *Proceedings of 33rd Annual Hawaii International*

Conference on System Sciences (2000), pp. 1–10
3. M.S.I. Rubel, N. Kandil, N. Hakem, *Priority Management with Clustering Approach in Wireless Sensor Network* (IEEE, 2018), pp. 7–11
4. W.R. Heinzelman, A. Chandrakasan, H. Balakrishnan, An application specific protocol architecture for wireless mincrosensor networks. IEEE Trans. Wirel. Commun. **1**(4), 660–670 (2002)
5. Fahmy, O. Younis, Distributed clustering in ad-hoc sensor networks: a hybrid, energy-efficient approach, in *Proceedings of the IEEE Conference on Computer Communications (INFOCOM)* (Hong Kong, 2004)
6. L.-C. Wang, C.-W. Wang, C.-M. Liu, Adaptive contention window-based cluster head selection algorithm for wireless sensor networks, in *2005 IEEE 62nd VTC-2005-Fall*, vol. 3 (2005), pp. 1819–1823
7. E. Chu, T. Mine, M. Amamiya, A data gathering mechanism based on clustering and in-network processing routing algorithm: CIPRA, in *The Third International Conference on Mobile Computing and Ubiquitous Networking, ICMU* (2006)
8. P. Manimala, R. Senthamil, A survey on leach-energy based routing protocol. Int. J. Emerg. Technol. Adv. Eng. (IJETAE) **3**(12), 657–660 (2013)
9. D. Jia, H. Zhu, S. Zou, Dynamic cluster head selection method for wireless sensor network **16**(8), 2746–2754 (2016)
10. K.A. Darabkh, J.N. Zomot, *An Improved Cluster Head Selection Algorithm for Wireless Sensor Networks* (IEEE, 2018), pp. 65–70

Security Enhancement by Preventing Wormhole Attack in MANET

Anjali B. Aswale and Radhika D. Joshi

Abstract The Mobile Ad hoc Network (MANET) is a special type of wireless infrastructure-less network, without any central administration, formed temporarily by group of mobile nodes. With well-known advantages, handful of MANET characteristics like no central administration, dynamic topology and limited resources is responsible for its vulnerable behavior. Lack of defense mechanism causes various attacks (like Wormhole) on network stack and hence requires higher degree of security during data exchange. To enhance security, Rivest, Shamir, Adleman (RSA) and advanced encryption standard (AES) algorithms are combined. The existing security algorithm achieves secure communication at the cost of energy. The proposed technique uses Modified Rivest, Shamir, Adleman (MRSA) and AES for secure and energy-efficient data transmission over public channels. It is implemented and tested in MATLAB and NetSim. Results show improvement in network lifetime.

Keywords MANET · Wormhole attack · Security · Cryptography · RSA · AES · Energy consumption

1 Introduction

With technological advancements and increasing portable computing devices, networking and communication domains are gaining importance and are expecting to hold key role in numerous applications. In this pervasive computing age, every single user will use number of computing platforms to access required information. Characteristics and nature of these devices make wireless communication as an easy solution for interlinking. Cooperative communication in wireless environment provides centralized network functionality. Such system network is mentioned as MANET.

A. B. Aswale (✉) · R. D. Joshi
Electronics and Telecommunication, COEP, Pune, India
e-mail: aswaleab17.extc@coep.ac.in

R. D. Joshi
e-mail: rdj.extc@coep.ac.in

H. S. Saini et al. (eds.), *Innovations in Electronics and Communication Engineering*, Lecture Notes in Networks and Systems 107,
https://doi.org/10.1007/978-981-15-3172-9_23

Infrastructure-less MANET is a dynamic creation of mobile nodes with random and short-lived topology. This network is also referred as 'On-Demand', 'Self-Configured', 'Self-Administrative Network' [1, 2].

Node mobility forces undesirable challenges which need to be conquered in order to experience the benefits of MANET [1]. Mobile nodes are battery operated, so efficient utilization of energy is the prime concern. This energy conservation could be the design optimization criteria for development in MANET. Battery life of nodes also affects the transmission range of individual node in the network. Limited transmission range requires the cooperation of intermediate nodes to support out-of-range data transmission, which invites possible security attacks [2]. MANET uses Ad hoc On-demand Distance Vector (AODV) routing protocol with IEEE802.11b wireless standard for communication. AODV forwards data to the destination, and hence, it should be robust and adaptive to network specifications like topology, traffic and network size.

Mobility and bandwidth capacity impose more constraints on MANET. Dynamic topology causes frequent disconnections which increases routing overhead to somehow propagate updated route information to networking devices, while limited bandwidth implies sharing of transmission medium by devices which affects the successful delivery of information. Both challenges force retransmission of data. Multi-hop communication requires the cooperation of intermediate nodes. If any of these is the misbehaving node, it can hamper the network performance. Such situation is described as attack on the network. Denial of service (DoS), sinkhole and wormhole are possible types of attack [1].

The paper focuses on prevention of wormhole attack in order to achieve secure and energy-efficient data transmission, using hybrid cryptography algorithm. Mathematical complexity of RSA along with faster and energy-efficient AES algorithm can serve MANET requirements of security and effective energy utilization. So to enhance MANET characteristics, RSA and AES are combined in suggested algorithm.

The content of the paper is organized as follows: Sect. 2 addresses wormhole attack. Section 3 presents the existing work. Section 4 gives the details of the proposed hybrid algorithm. Section 5 provides an exposure to experimentation and results of the proposed technique. The final conclusion and future scope are given in Sect. 6.

2 Routing Protocol and Security Issues

Ad hoc network provides connectivity among the wireless nodes. In MANET, nodes can roam freely and can join or leave the network at any time. Being self-configurable network, individual node act as both host and router, and due to limited transmission range, network necessitates cooperation of honest intermediate nodes. If intermediate node is misbehaving, network performance affects. Hence, routing and security are the main concern in MANET. MANET uses AODV routing protocol [3].

2.1 Wormhole Attack

Wormhole is consequential and major security attack which can disrupt network performance partially or make the whole system unavailable. In wormhole attack, assaulter node receives the data at one end and tunnels it to the second location. AODV protocol in MANET uses broadcasting mechanism to find out the route to the destination and to forward the data. Malicious nodes in wormhole sends back Route Reply (RREP) with higher residual power and lesser hop count to the destination [4, 5]. Wormhole attack can be launched in two ways, one is out-of-band mode (Fig. 1) and the other is in-band mode (Fig. 2). In out-of-band channel, assaulter node simply catches the data packet with the help of specialized hardware with higher power level to directly transmit data to other end with additional delay, whereas in other mode, assaulter node hijacks good-behaving nodes and use them to forward data to the destination.

Fig. 1 Out-of-band channel

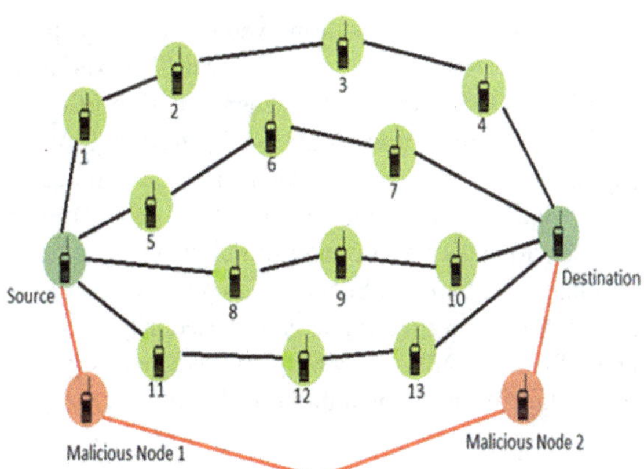

Fig. 2 In-band channel

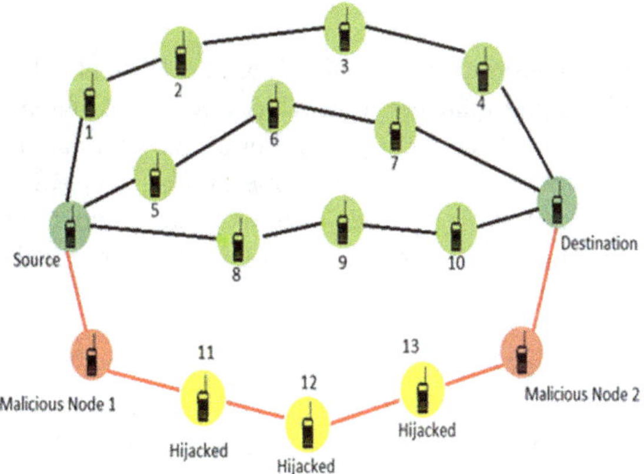

Wormhole prevents detection of any other intermediate node to the destination so as to avoid selection of other path for data transmission. Wormhole attack works at the network layer. Either specialized hardware or encapsulation method is used to launch the attack. Assaulter node encapsulates the data packet at transmitting end and sends off in tunnel; at the receiving side, second assaulter node will decapsulate the data. Due to encapsulation, intermediate nodes cannot change hop count field of data packet and it seems to be directly coming from the first malicious node with one hop count.

In wormhole, two malicious nodes are not directly connected, just an illusion is created of being directly connected and makes the network believe that it is connected [5, 6].

3 Overview of the Existing Cryptography Algorithm

Security is the prerequisite nowadays, and cryptography [7, 8] is one way, where exchange of data is done in secure form. The existing algorithms like Data Encryption Standard (DES), RSA, AES, hash function-based algorithms provide best secure communication at individual level with some limitations. RSA is one of the popular asymmetric cipher model, which provides higher computational complexity so as to provide higher degree of security. At the same time, AES is symmetric cipher model with faster and energy-efficient processing. AES is currently used in wide applications. National Security Agency (NSA) allows AES for classified data of top secret. So to improve the degree of security and eventually, network lifetime, AES and RSA are combined. Comparative study is given in papers [9, 10]. Overview of these algorithms is given in the following section.

3.1 Asymmetric Key Cryptography Algorithm: RSA

Asymmetric cryptographic algorithm RSA [11–13] works on number theory and uses prime numbers to provide security. Plaintext and ciphertext in RSA are the integers between 0 to $(n - 1)$ for particular value of n. n is the value obtained from the multiplication of two prime numbers. In RSA, both transmitter and receiver individual must be aware of the n value. e and d are the encryption and decryption key values which are public and private, respectively.

3.1.1 RSA Key Generation

Unlike symmetric algorithms, public key algorithm requires the computation of pair keys e, d. In asymmetric key technique, all keys must be random and need to be selected from certain set of values. Inverse of chosen key is nothing but the private key.

1. Choose large primes p and q;
2. Multiply two prime numbers to calculate n, $n = p \times q$;
3. Calculate $\varphi(n)$, $\varphi(n) = (p - 1)(q - 1)$;
4. Choose random e from a set $\{1, 2, \ldots, \varphi(n - 1)\}$ such that $gcd\ (e, \varphi(n)) = 1$;
5. Compute d such that $d \cdot e \equiv 1 \bmod \varphi(n)$. d is calculated using Extended Euclidean Algorithm (EEA).

Remarks:

- Choose large prime numbers; $p, q \geq 2^{512}$;
- Hence, $n \geq 2^{1024}$.

3.1.2 RSA Encryption and Decryption

- Encryption: Given $\{e, n\}$, Plaintext $M \in \{0, 1, \ldots, n - 1\}$;

$$C = M^e \bmod n \tag{1}$$

- Decryption: Given $\{d\}$, Ciphertext $C \in \{0, 1, \ldots, n - 1\}$.

$$M = C^d \bmod n = (M^e)^d \bmod n = M^{ed} \bmod n \tag{2}$$

3.1.3 Computational Complexity of RSA

Attacker in the network has public key e and n. Now, it wants to know d. Decryption key d can be calculated using EEA. Still, the attacker cannot compute d because of the computational complexity of $\varphi(n)$. n with 2^{1024} is almost 300 digit number, and no one can factorize that number hence cannot calculate $\varphi(n)$ in turn fails to decrypt the plaintext.

3.2 Symmetric Key Cryptography Algorithm: AES

National Institute of Standards and Technology (NIST) published Advanced Encryption Standard (AES) in 2001, given by Rijindael, which is considered to replace block cipher named DES for numerous applications. Compared to other cryptographic algorithms, AES is a bit complex to understand, but it is most effective for secure transmission [14]. It is probably the most important crypto-algorithm.

AES is a block cipher [12], that encrypts all 128 bits of data path in one round. AES accepts a plaintext input of 128 bits, and the key length used can be of 128, 192 or 256 bits. 128-bit input is represented as 4×4 square matrix of 16 bytes called as state array. At each stage of encryption or decryption process, this state array is modified, and the final stage output will be cipher matrix [15].

Key of AES is represented in similar 4×4 matrix forms in columnar order. This 128 bit key is expanded using key schedule algorithm for number of operations. Each column of 4 byte forms a word. At the end of key schedule, there are total of 44 words each of 32 bits. Depending on key length of AES, there are N rounds of operation. For the key length of 128 bits, there are total 10 rounds, whereas there are 12 and 14 rounds for 192 and 256 bit key length, respectively. Each round in AES consists of the following four-stage transformation:

- Substitute Byte operation;
- Shift Rows operation;
- Mix Column operation;
- Add Round Key operation.

The final round of AES operation consists of only three operations, and at the beginning of the first round of AES, sub-key is added, called as key whitening.

AES uses key expansion algorithm, to generate 44 words from 128 bit key. Each new generated word depends on immediate preceding word. The initial key forms the first four words, namely $w0$, $w1$, $w2$ and $w3$. To get $w4$, complex function g is used, which performs the following three sub-functions:

- Perform one byte circular left shift of 16 byte $w3$, in the first RotWord sub-function;
- Each byte of this modified word is mapped to a new value using SubWord function, which is similar to the S-Box transformation;
- The final sub-function performs XORing of recent word with Round Constants Rcon[j], defined for each round of operation.

4 Proposed Hybrid Algorithm

In order to deal with energy-efficient security issue, combination of RSA and AES algorithm can give an edge in terms of security and network lifetime. In this process, strengths of both algorithms are combined (RSA is difficult to decrypt due to computational complexity and AES is faster and energy efficient), while disadvantages are eliminated (RSA is slower and energy-consuming algorithm, and AES key maintenance is a challenging task) [10, 15]. The proposed algorithm technique is shown in Fig. 3.

Hybrid cryptographic approach is given in papers [16, 17]. In the presented hybrid algorithm, plaintext is encrypted using AES encryption process, while the key of the AES algorithm is encrypted using RSA algorithm. AES algorithm is more efficient for large amount of plaintext with less power requirements; hence, AES is used to

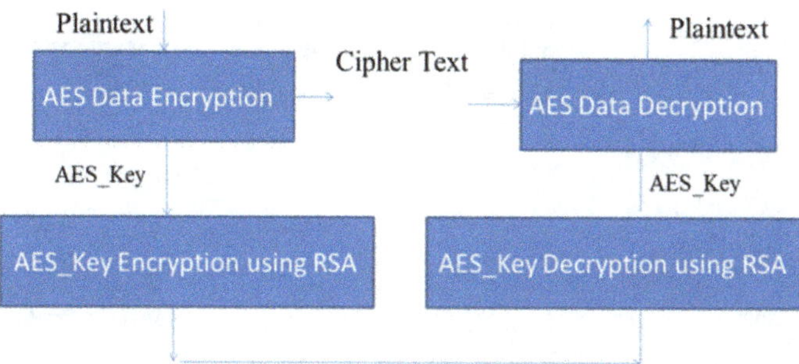

Fig. 3 Hybrid algorithm

encrypt plaintext of source node. AES has fixed size keys of 128 or 192 or 256 bits. Though RSA is power hungry algorithm, it can be used to encrypt fixed amount of data which results in increased security.

As described in Sect. 3.1.3, RSA is a mathematically complex algorithm if chosen prime numbers are large. Larger prime numbers result in greater *n*, and hence, range for selection of *e* and *d* will increase. As the size of *e* and *d* enlarges, processing time at encryption and decryption level improves. This increased processing will reflect in expanded encryption time. So to reduce encryption time of RSA, a methodology is suggested in the following section.

4.1 Methodology for Fast Exponent Calculation

Problem in practice is to calculate exponential values of very large numbers during encryption and decryption process of RSA. This calculation increases the time complexity of the algorithm and hence affects the parameters named energy consumption. To solve this problem, a method of fast mathematical calculation can be implemented to improve the performance of the system. Say to calculate x^6, normal and easy method for exponential calculation is shown in Table 1.

Table 1 Exponent calculation methods

Method I	Method II
$x \cdot x = x^2$	$x \cdot x = x^2$
$x \cdot x^2 = x^3$	$x \cdot x^2 = x^3$
$x \cdot x^3 = x^4$	$x^3 \cdot x^3 = x^6$
$x \cdot x^4 = x^5$	–
$x \cdot x^5 = x^6$	–
No. of operations = 5	No. of operations = 3

Table 2 Fast exponent calculation

Solution	Operation	Binary method solution
$x \cdot x = x^2$	Square	$(x^1)^2 = x^{10_2}$
$x \cdot x^2 = x^3$	Multiply	$x^1 \cdot x^{10_2} = x^{11_2}$
$x^3 \cdot x^3 = x^6$	Square	$(x^{11})^2 = x^{110_2}$
$x^6 \cdot x^6 = x^{12}$	Square	$(x^{110})^2 = x^{1100_2}$
$x \cdot x^{12} = x^{13}$	Multiply	$x^1 \cdot x^{1100_2} = x^{1101_2}$
$x^{13} \cdot x^{13} = x^{26}$	Square	$(x^{1101})^2 = x^{11010_2}$

In order to get the final outcome, both methods are compared in terms of number of operations required to perform. When the second method is used for x^8 calculation, almost 50% of operation is reduced. If $x^{2^{1024}}$ is to be calculated, it is not possible using the first method. Also, the second method requires 1024 multiplications to get the result. The first method imposes linear complexity, while the logarithmic complexity is imposed by the second one. Hence, the algorithm used for RSA exponent calculation is square and multiply algorithm.

4.2 Square and Multiply Algorithm for Exponent Calculation

This method of exponent calculation is also called as 'Binary method' or 'Montgomery Ladder' technique. Working of this algorithm is described with the help of example. Example: Calculate x^{26}.

From the given example in Table 2, it is clear that square and multiply sequence can be used to calculate x^{26} with reduced calculations and minimum amount of time. But to do so, one needs to know the sequence of square and multiply operations. A simple idea of square and multiply technique with the knowledge of correct sequence of operation helps to improve system performance.

These rules when to multiply and to square depend on binary representation of the exponent. Solution for same problem with the proposed method would be—binary method of exponent calculation. Binary representation of 26 is $(11010)_2$. Rule depends on the binary bits, and the goal is to generate the binary pattern of particular exponent. Binary method solution for x^{26} is presented in column 3 of Table 2.

Squaring of binary number is nothing but left shift of given binary number by 1 bit, and multiplication is the addition of exponent in binary form. The algorithm illustrated in Fig. 4 will reduce the time complexity of mathematical calculations and will improve the performance of network in terms of energy efficiency.

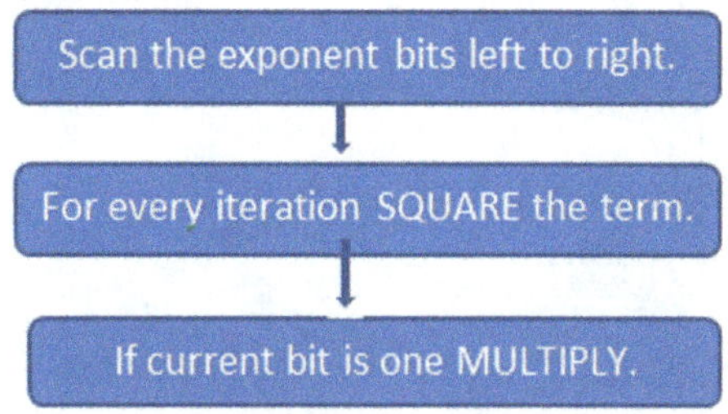

Fig. 4 Fast exponent calculation algorithm

5 Experimentation and Results

To consider mathematical complexity of cryptographic algorithms as well as real-time scenarios, two platforms are used, namely MATLAB and NetSim. RSA and AES algorithms are implemented on MATLAB, which is then interfaced with NetSim for analysis.

To consider actual scenario, malicious nodes are added in the network scenario of 20 nodes; number of malicious nodes (2, 4, 6) are varied, and the effect of malicious node on throughput is analyzed as shown in Fig. 5.

Individual RSA and AES are implemented in MATLAB to check normal functionality of both. After successful implementation of individual algorithms, hybrid algorithm is implemented and result of the encrypted data is shown in Fig. 6, provided encrypted data and key are of size 128 bit.

Results are taken with different data sizes (512 bits, 1024 bits), and the comparison of encryption time for RSA and AES is done. This effect of different data size on encryption time can be observed in Fig. 7. It can be seen that with AES, required encryption time is much less when compared to RSA. RSA selects random value of e from the set of values as mentioned in Sect. 3.1.1 (step 4). This process hence takes random amount of time. Therefore, result of encryption time for different data size is random and is considered to observe the nature of algorithms.

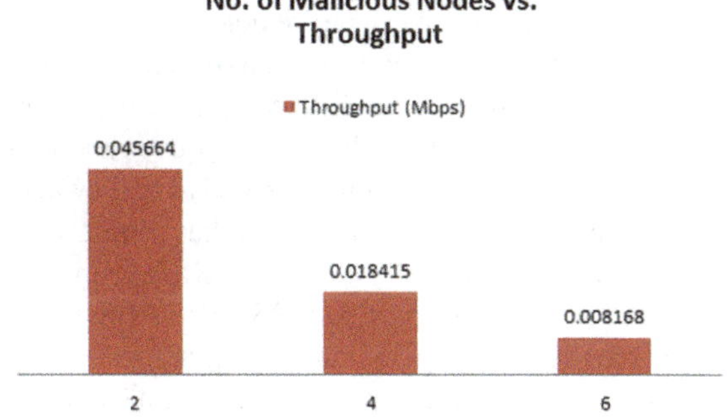

Fig. 5 Malicious nodes versus throughput

```
>> AES_RSA_KEY('ABCDEFGHIJKLMNOP')

rows =

     1


PLAINTEXT:
ABCDEFGHIJKLMNOP

CIPHERTEXT:
0(¶□?,   2@ýÅìA□)A

ENCRYPTED_KEY:

    30595048     46215706     7212112 32826142     23985224     48570937     60700215     32656585

ans =

    [16x2 char]

fx >> |
```

Fig. 6 MATLAB result of hybrid algorithm for 128 bit input data

Fig. 7 Encryption time of RSA and AES

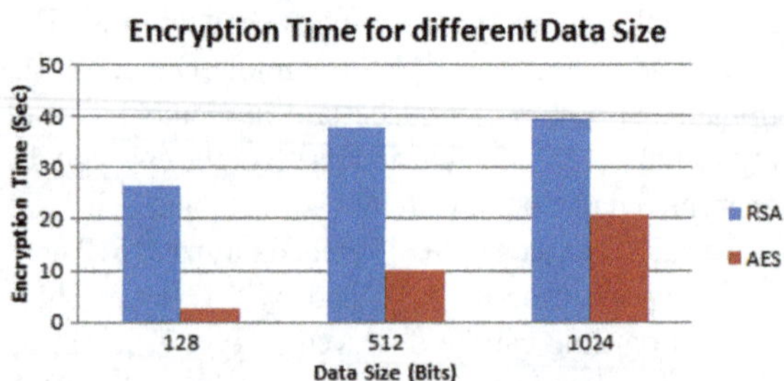

Well-known relation between energy, power and time is given in Eq. 3.

$$Energy = Power * Time \tag{3}$$

From this relation, it can be predicted that with increase in encryption time, energy consumption increases with constant power. For the node power of 100 mW, results of energy consumption for different data size are shown in Fig. 8.

Figures 7 and 8 show that RSA requires more encryption time and additional consumption of energy when compared to AES, whereas Sect. 3.1.3 discussed strong computational ability of RSA and is thus robust. Also, AES proved to be faster and energy-efficient solution compared to RSA.

Considering the pros of both RSA and AES, hybrid algorithm is implemented to boost security. Result of hybrid cryptography in terms of encryption time is presented in Fig. 9. RSA being computationally complex needs to be enhanced. It is modified using the proposed Binary method which is described in Sect. 4.2 and is named as modified RSA (MRSA).

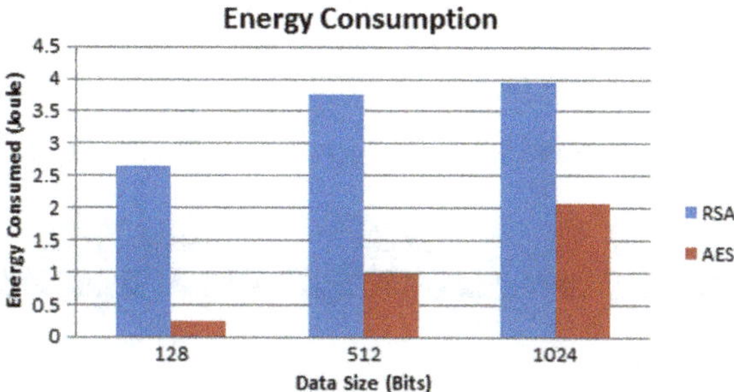

Fig. 8 Energy consumption of RSA and AES

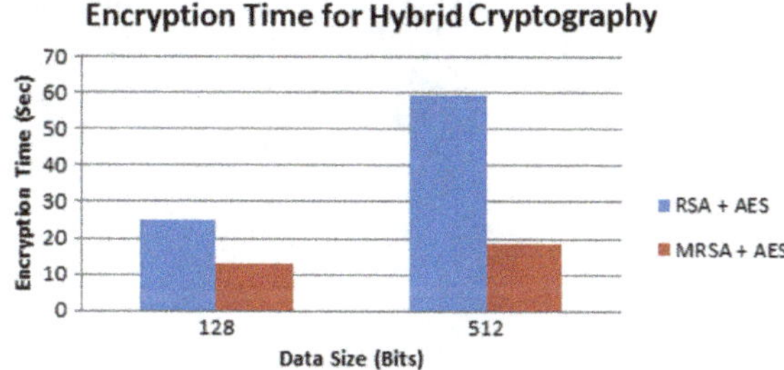

Fig. 9 Encryption time of hybrid algorithm

Figure 9 shows comparison of conventional hybrid cryptography and modified hybrid cryptography. It can be observed that there is reduction in encryption time and consecutive reduction in energy consumption. Figure 10 represents precision of the average energy comparison of both the methods (for 128 bits). This modified algorithm in MATLAB is interfaced with NetSim. Figure 11 shows encrypted result after getting input payload of source node from NetSim. Shown ciphertext is fed back to NetSim for transmission of data over insecure channel.

In order to observe effect of modification on other network parameters like throughput and delay, experimentation is carried out for two cases. A scenario (case 1) without malicious nodes and with no security incorporation is considered to analyze network behavior. Another scenario (case 2) with consideration of malicious nodes and security (with modified hybrid cryptography) is also simulated to observe effects on performance of the network.

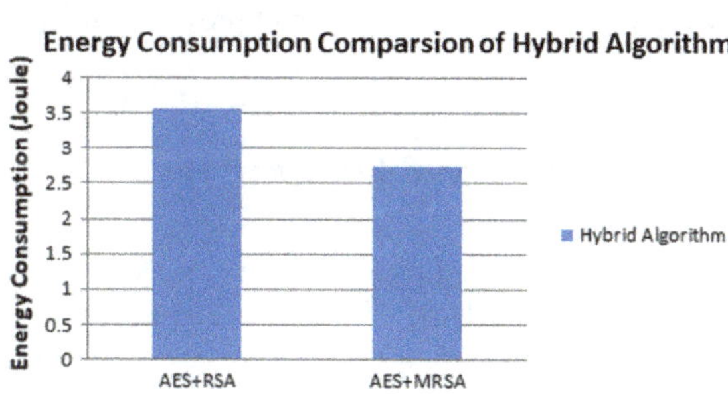

Fig. 10 Energy consumption of hybrid algorithm

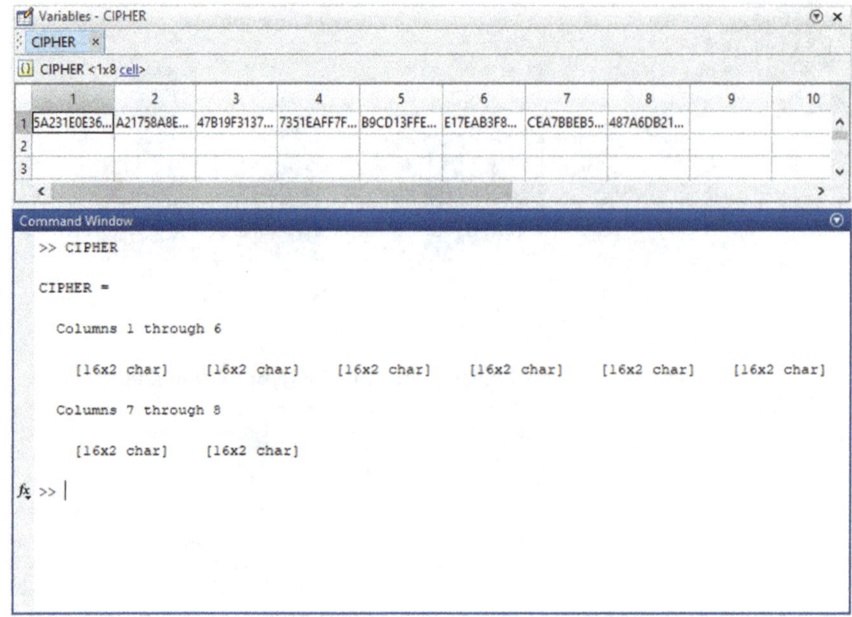

Fig. 11 Result of MATLAB–NetSim interfacing

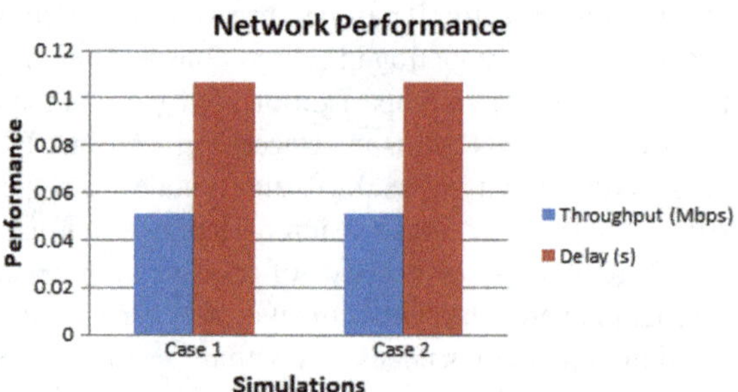

Fig. 12 Throughput and delay comparison

From Fig. 12, it is clear that with modified hybrid technique, the performance in terms of throughput and delay is maintained, and also, energy consumption is reduced significantly along with improved security.

6 Conclusion and Future Scope

Novel hybrid cryptography is implemented to address challenges (security and energy consumption) in MANET. Hybrid algorithm used MRSA and AES, to secure data over the network and to increase the energy efficiency, which ultimately improves network lifetime.

Modified algorithm showed 14% reduced energy consumption without deteriorating the network behavior. This work can be further extended to improve throughput and to minimize end-to-end delay.

References

1. M. Ilyas, *The Handbook of Ad Hoc Wireless Networks* (CRC press, 2002)
2. G. Aggelou, *Mobile Ad Hoc Networks: From Wireless LANs to 4G Networks* (McGraw-Hill Professional, 2004)
3. C. Perkins, E. Belding-Royer, S. Das, Ad hoc on-demand distance vector (AODV) routing. Technical Reporting (2003)
4. A. Patel, N. Patel, R. Patel, Defending against wormhole attack in MANET, in *2015 Fifth International Conference on Communication Systems and Network Technologies* (IEEE, 2015), pp. 674–678
5. P. Sahu, S.K. Bisoy, S. Sahoo, Detecting and isolating malicious node in aodv routing algorithm. Int. J. Comput. Appl. **66**(16), 8–12 (2013)
6. S.A. Sharifi, S.M. Babamir, A new approach to detecting and preventing the worm hole attacks for secure routing in mobile ad-hoc networks based on the SPR protocol, in *2016 IEEE 10th International Conference on Application of Information and Communication Technologies (AICT)* (IEEE, 2016), pp. 1–5
7. M. Imran, F.A. Khan, T. Jamal, M.H. Durad, Analysis of detection features for wormhole attacks in manets. Procedia Comput. Sci. **56**, 384–390 (2015)
8. A. Bhardwaj, S. Som, Study of different cryptographic technique and challenges in future, in *2016 International Conference on Innovation and Challenges in Cyber Security (ICICCS-INBUSH)* (IEEE, 2016), pp. 208–212
9. P. Mahajan, A. Sachdeva, A study of encryption algorithms AES, DES and RSA for security. Glob. J. Comput. Sci. Technol. (2013)
10. P. Patil, P. Narayankar, D. Narayan, S.M. Meena, A comprehensive evaluation of cryptographic algorithms: Des, 3DES, AES, RSA and Blowfish. Procedia Comput. Sci. **78**, 617–624 (2016)
11. S. Sankaranarayanan, G. Murugaboopathi, Secure intrusion detection system in mobile ad hoc networks using RSA algorithm, in *2017 Second International Conference on Recent Trends and Challenges in Computational Models (ICRTCCM)* (IEEE, 2017), pp. 354–357
12. W. Stallings, *Cryptography and Network Security, 4/E* (Pearson Education India, 2006)
13. C. Thirumalai, Review on the memory efficient RSA variants. Int. J. Pharm. Technol. **8**(4), 4907–4916 (2016)
14. F.J. D'souza, D. Panchal, Advanced encryption standard (AES) security enhancement using hybrid approach, in *2017 International Conference on Computing, Communication and Automation (ICCCA)* (IEEE, 2017), pp. 647–652
15. G. Singh, Supriya, A study of encryption algorithms (RSA, DES, 3DES and AES) for information security. Int. J. Comput. Appl. **67**, 33–38 (2013)
16. A. Sharma, D. Bhuriya, U. Singh, Secure data transmission on MANET by hybrid cryptography technique, in *2015 International Conference on Computer, Communication and Control (IC4)* (IEEE, 2015), pp. 1–6
17. S.S. Jathe, V. Dhamdhere, Hybrid cryptography for malicious behavior detection and prevention system for manets, in *2015 International Conference on Computational Intelligence and Communication Networks (CICN)* (IEEE, 2015), pp. 1108–1114

UWB Antenna with Artificial Magnetic Conductor (AMC) for 5G Applications

S. Kassim, Hasliza A. Rahim, Mohamedfareq Abdulmalek, R. B. Ahmad, M. H. Jamaluddin, M. Jusoh, D. A. Mohsin, N. Z. Yahya, F. H. Wee, I. Adam and K. N. A. Rani

Abstract This paper presents the design of an ultra-wideband (UWB) antenna for Internet of Things (IoT) applications that operate within 5G operating frequencies. One of the IoT-based devices' architecture is wireless body area networks (WBANs). WBAN allows computer device to communicate with human body signal by trading digital information like electrical conductivity. Fifth generation (5G) is the state-of-the-art generation mobile communication. A higher data speed it offers will improve data communication efficiency in WBAN system. One of the biggest challenges foreseen for the wearable UWB antenna is the antenna bandwidth. The challenge is to warrant a wideband performance throughout the operating frequency, and a trade-off with a high dielectric in proposed substrate is essential. This paper presents design and parametric analysis of an antenna using a typical industry-preferred Rogers

---

S. Kassim · H. A. Rahim (✉) · R. B. Ahmad · M. Jusoh · D. A. Mohsin · F. H. Wee · I. Adam
Bioelectromagnetics Research Group (BioEM), School of Computer and Communication Engineering, Universiti Malaysia Perlis (UniMAP), Kampus Pauh Putra, 02600 Arau, Perlis, Malaysia
e-mail: haslizarahim@unimap.edu.my

M. Abdulmalek
Department of Engineering and Information Science, University of Wollongong in Dubai, Block 15, Dubai Knowledge Village, Dubai, UAE

M. H. Jamaluddin
Faculty of Electrical Engineering, Universiti Teknologi Malaysia (UTM), Utm Skudai, 81310 Johor, Malaysia

D. A. Mohsin
Department of Computer Technical Engineering, Electrical Engineering Technical College, Middle Technical University (MTU), Baghdad, Iraq

N. Z. Yahya
Physics Section, School of Distance Education, Universiti Sains Malaysia (USM), 11800 George Town, Penang, Malaysia
e-mail: norzakiah@usm.my

K. N. A. Rani
Bioelectromagnetics Research Group (BioEM), Faculty of Engineering Technology, Universiti Malaysia Perlis, 02100 Padang Besar, Perlis, Malaysia

H. S. Saini et al. (eds.), *Innovations in Electronics and Communication Engineering*, Lecture Notes in Networks and Systems 107,
https://doi.org/10.1007/978-981-15-3172-9_24

material (RO4350B) substrate with wider bandwidth as compared to 5G frequencies, 10.125–10.225 GHz. This paper also exhibits bandwidth improvement with the presence of artificial magnetic conductor (AMC) as a metasurface. A typical UWB patch antenna was initially designed before being integrated with AMC through a parametric analysis. This paper analyzes the frequency, gain, directivity and antenna efficiency before and after optimization. This paper successfully demonstrates a slotted Y-shaped antenna design with coplanar waveguide (CPW) using a Rogers material (RO4350B) as a substrate and the bandwidth improvement by 15.6% with the AMC as a metasurface.

Keywords UWB antenna · Wide bandwidth · Rogers material (RO4350B) · Artificial magnetic conductor (AMC) · Wearable antenna

1 Introduction

Future 5G architecture is envisaged as highly dense, diversified, versatile as well as a unified technology with an extraordinary bandwidth availability for almost unlimited upgradation [1, 2]. Such future technology integrated with IoT-based applications is WBAN that offers endless benefits in the medical monitoring system. It is anticipated that the future wireless access networks also will be combined with radio-over-fiber (RoF), making it possible to realize high-speed data transmission [3]. Through such application, smart wearable devices are created, having capability of sensing data, controlling actuators, communicating with external devices and recharging wirelessly. Integration of electronics into smart textiles has been recognized as key enabler for this future revolutionized technology where one of the key features is wearable textile antenna [4]. The smart textile will be embedded into day-to-day garments, and the trend has been increasing rapidly for other applications such as localization [5] and energy harvester [6]. The off-body communication antennas [7] play pivotal role in realizing the future wearable 5G IoT network where it provides seamless established communication between on-body sensors and other external devices, such as base station and wireless fidelity (WiFi) router.

Antenna design becomes one of the key considerations to deploy 5G front-end systems. Several popular topologies for wearable planar antenna have been proposed in the literature, like textile monopole antennas [8–11], textile planar inverted-F antennas (PIFAs) [12, 13] and patch antennas [7, 14, 15]. The microstrip patch antennas (MPAs) have been regarded as the best choice in smart wearable devices for off-body communication in enhancing the front-to-back ratio (FBR) and specific absorption rate (SAR) values. The MPA offers low profile, planar, lightweight, easy to integrate with clothing, robustness against human body and obtrusive body communications, making it one of the prominent designs for WBAN applications. In addition, due to its full rear ground plane, the MPA enables shielding against the effects of the body, reducing power absorption and dielectric coupling, besides influencing its SAR level [7, 11, 16]. The effect of antenna toward human body and tissue is such of importance since public worries about the negative health effect caused by the radio frequency radiation exposure [17]. Despite its various advantages, such antenna suffers from a very narrow bandwidth. Several techniques were introduced to overcome this limitation by using different substrate and conductive materials with varied thickness, partial ground or non-ground structure in UWB antennas [18] and CPW structures [19, 20]. Coplanar waveguide (CPW) structures are among the commonly used technique to excite additional resonances and providing a wideband feature within the antenna structure. For optimum performances of wearable antenna, UWB is used, whereas it requires 3.1–10.6 GHz according to the Federal Communication Commission (FCC) in 2002 with the minimum bandwidth of 500 MHz. UWB is preferred in WBAN or to be specific as medical sensor, i.e., blood pressure, glucose level, or electrocardiogram (ECG), that requires a continuous signal in short range with high data rate transmission where in 5G, high data rate is crucial in combating the interference caused by multipath fading [21].

In this paper, the antenna topology is designed based on CPW integrated with AMC. The appropriate CPW geometry dimension resulted in wider bandwidth coverage when utilizing RO4350B as the substrate material.

2 Antenna Topology and Material Specifications

The workflow of the antenna design can be summarized as the following subsections. The scope includes designing an antenna and optimizing the design before proceeding to the fabrication. The details of material used and stage involved in this research work are explained in the subsections below. The technique used in this design including the suitable design and unit cell was also presented. The relative permittivity for Rogers material (RO4350B) substrate is $\varepsilon_r = 3.66$ with thickness of material $t = 5$ mm. The selection of substrate material heavily depends on the deployment location of the antenna. There are some wearable antennas which are inserted in wireless sensors for sport applications. The whole module is boxed in a container made of plastic, and the antenna is not in direct contact with the body. In such applications, every sort of substrate is possible because antenna would be printed in the same substrate with sensor and transceiver.

The patch antenna structure was designed with Rogers material (RO4350B) as a substrate with a relative permittivity of $\varepsilon_r = 3.66$ and ShieldIt conductive textile as the radiator. ShieldIt Super electrotextile, manufactured by LessEMF Inc., was used to form the conducting parts of the antenna with a thickness, H_t, of 0.17 mm. The estimated conductivity of ShieldIt Super is $\sigma = 1.18 \times 105$ S/m [14]. The thickness of RO4350B is $t = 5$ mm. Since the initial design of antenna is using conventional microstrip patch antenna, the improvement of patch will be crucial in order to be used in 5G frequency band. Moreover, slot will be added in order to broaden the bandwidth. The specification of proposed antenna can be observed in Table 1. The addition of AMC widens the bandwidth as it tunes to the desired resonant frequency. The implementation of full ground plane also reduces the radiation towards human.

Table 1 Design specification of the proposed UWB antenna with AMC

Specification	Details
Operating frequency (GHz)	8–11
Gain (dB)	>6 dB
Reflection coefficient (*S11*)	<−10 dB
Substrate	RO4350B
Feeding technique	Thin microstrip line
Efficiency	>80%

To design and simulate antenna, CST microwave simulation software was selected due to the comparable results as compared to the measurement as well as the user-friendly features. The critical parameters considered when designing the antenna are the operating frequency, gain and input return loss or input reflection coefficient, *S11*, as given in Table 2. The most important aspect of this design is the enhancement of bandwidth. The 5G technology uses spectrum above 6–100 GHz, and the UWB is declared by at least 500 MHz of bandwidth or at least 50% fractional bandwidth [12]. Thus, this antenna will be operating from 8 to 11 GHz with a bandwidth of 3 GHz. This structure of antenna is five layers of antenna, consisting of two layers of substrates and three layers of Shieldt Super textile conductor as the conductive part. Figures 1 and 2 show the conventional patch antenna design and simple improvement of patch layer of antenna. Table 2 gives the dimension of this improved patch antenna with CPW. Meanwhile, the design of proposed UWB slotted Y-shaped antenna with integrated CPW is shown in Fig. 3. A square patch with concentric ring-shaped slot AMC was selected with the design parameters given in Table 3 while Fig. 4 shows the front and side view of 2 × 2 AMC unit cell. 2 × 2 AMC is chosen due to the substrate's total dimension. In order to prevent a mismatch in antenna, total width of AMC will follow the width of substrate in patch. Figure 5 shows the front and side view of the proposed antenna.

The thickness of antenna is high due to the thickness of substrates. The location of AMC is in the middle between two substrates.

3 Results and Discussion

3.1 Analysis of Conventional Patch Antenna with Improvement

The conventional antenna was simulated in order to obtain the desired frequency range. Figure 6 and Table 4 show the S11 and bandwidth improvement, respectively. The simulated S11 indicated the bandwidth enhancement of 1.7 GHz for the improved patch antenna with CPW. The bandwidth of improved patch antenna is observed to be enhanced, doubled than the bandwidth of the conventional one (Fig. 6).

Table 2 Dimension of improved patch antenna with CPW

W	L	w_1	w_2	w_3	l_1	l_2	l_3	H_s	H_t
36 mm	40 mm	3 mm	16.5 mm	16.5 mm	13.5 mm	14.5 mm	1 mm	5 mm	0.17 mm

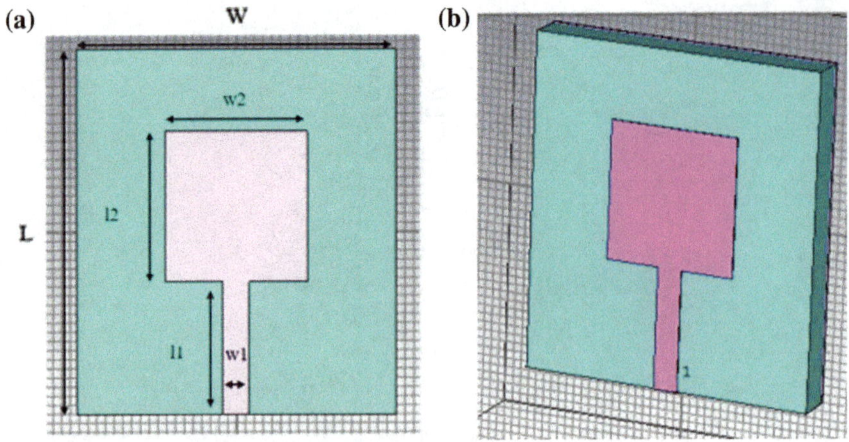

Fig. 1 Geometry of the patch antenna structure of **a** without slot and **b** with slot

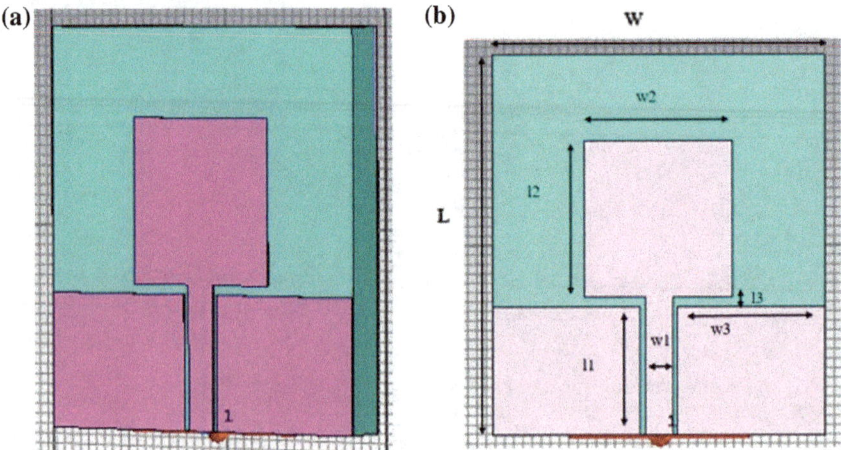

Fig. 2 Details of improved conventional patch antenna with CPW **a** side view and **b** dimensions

3.2 UWB Antenna Topology with AMC

The use of AMC is to ensure the enhancement of bandwidth, gain and radiation pattern. Since the antenna will be a part of WBAN, the important aspect to be considered is the radiation pattern. Therefore, with AMC, the performance of antenna is expected to be good, especially in terms of radiation pattern. In AMC, −90° till 90° shows the bandwidth where the wave is being reflected back to the top radiator. The wider the bandwidth, more electromagnetic (EM) wave is reflected to the front. Thus, this characteristic will affect the operating frequencies, gain directive and efficiency of an antenna significantly.

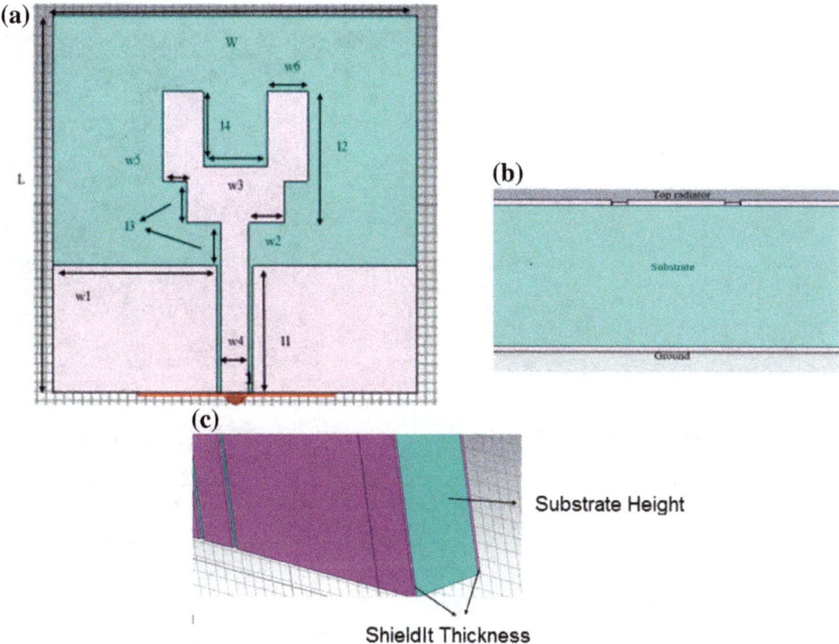

Fig. 3 Details of proposed UWB slotted Y-shaped antenna with integrated CPW **a** topology, **b** bottom view and **c** cross section

Table 3 Dimension of unit cell

W	r_1	r_2	r_3	r_4	Height
20	8	5	0.5	1	5

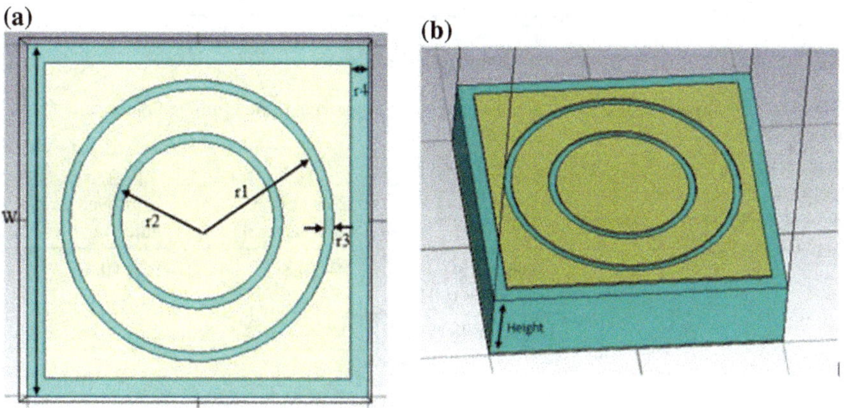

Fig. 4 Square patch with concentric ring-shaped slot AMC unit cell **a** topology and **b** perspective view

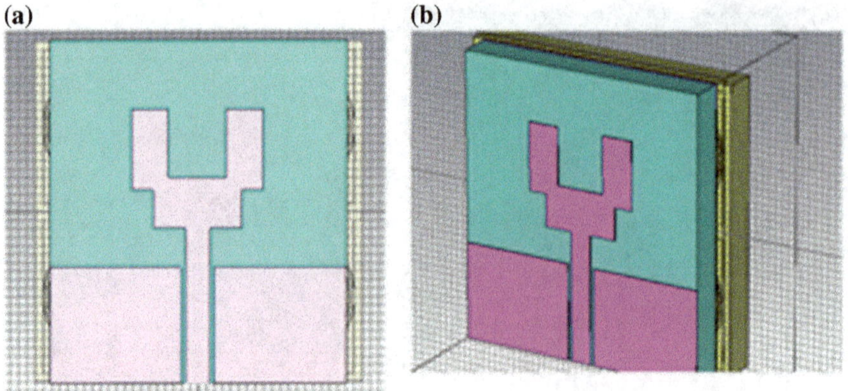

Fig. 5 Proposed UWB slotted Y-shaped antenna with AMC **a** topology and **b** perspective view

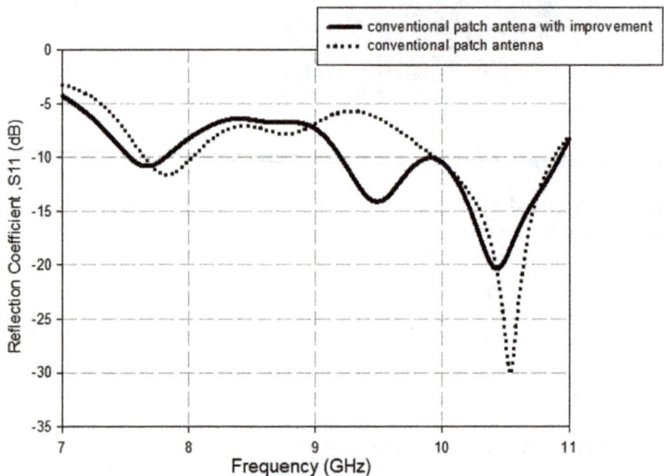

Fig. 6 *S11* comparison between conventional patch antenna and the improved patch antenna with CPW

Table 4 Enhancement of bandwidth for improved patch antenna with CPW

	Conventional patch antenna	Improved patch antenna
Frequency <−10 dB (GHz)	10–10.8	9.2–10.9
Bandwidth (GHz)	0.8	1.7

Table 5 Key parameters with presence of AMC

	Without AMC		With AMC	
Frequency (GHz)	8.5	10.2	8.5	10.2
Realized gain (dB)	6.5	8.65	6.57	8.14
Directivity (dBi)	5.59	9.12	5.50	7.93
Efficiency (%)	83.3	86.2	85.5	86.8

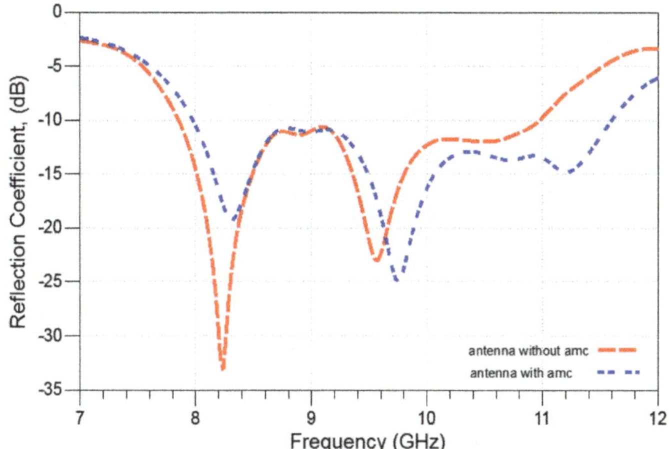

Fig. 7 *S11* of antenna with AMC and without AMC

The simulation antenna patch with AMC as a metasurface and the performance of gain, directivity, reflection coefficient, efficiency and radiation pattern of the antenna is presented in Fig. 7 and Table 5. The result shows that the bandwidth is increased by a maximum of 15.6% compared to without the AMC. Figure 7 also shows that the *S11* achieves the desired range of frequencies which is from 8 to 10 GHz. This is due to the fact that 5G technology indeed needs higher frequencies band in order to achieve the high-speed data rate. Thus, the future potential frequencies are in the frequency range of 6–100 GHz. The result exhibits that the proposed antenna with AMC features an enhanced bandwidth by 15.6% more than the design without AMC. Table 5 presents the improvement of efficiency when AMC is integrated into the proposed design. At 8.5 GHz, there is an increase of 3% radiation efficiency from 83% to nearly 86% for the proposed antenna with AMC. Gain at 8.5 GHz is comparable, with AMC and without AMC. At 10.2 GHz, there is a slight drop in gain but still met the 6 dB baseline. The efficiency of proposed antenna with AMC is greater than 80% and further proved that such design met the specification that was set initially. For radiation pattern, the patch with AMC exhibits good direction compared to without AMC, as shown in Figs. 8 and 9.

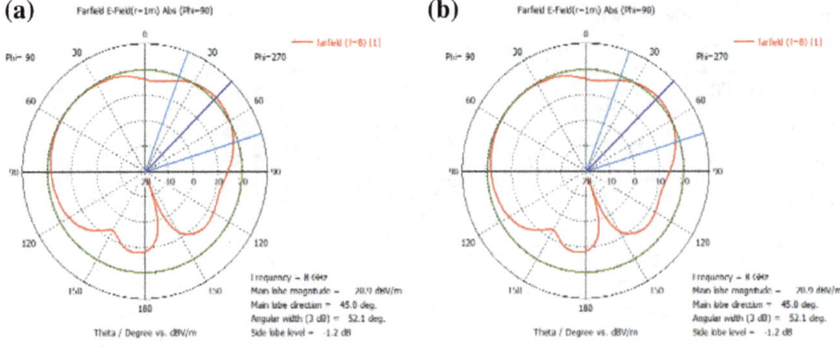

Fig. 8 Far field of antenna without AMC at 8.5 GHz **a** E-field and **b** H-field

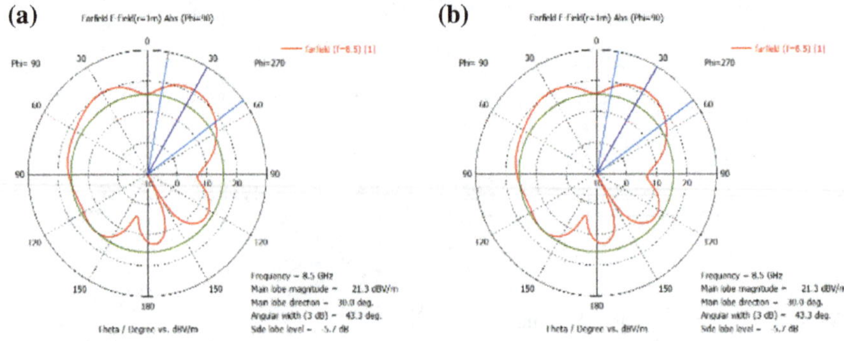

Fig. 9 Far field of antenna with AMC at 8.5 GHz **a** E-field and **b** H-field

In [11], the AMC backing is proved to be effective in improving the front-to-back (FBR) of the antenna. These patterns are evolved in two fields which are E-field and H-field. The main lobe direction indicates the reflection wave to the front while the side lobe indicates the wave that radiated to the back.

4 Conclusion

The paper presents the design of UWB antenna with AMC for future 5G application. This paper highlights the analysis of the key parameters of antenna such as range of frequency, gain, directivity and antenna efficiency before and after the optimization of the proposed design. The paper has successfully demonstrated the UWB slotted Y-shaped patch antenna design with CPW using Rogers material (RO4350B) as a substrate, and the bandwidth improvement is achieved by 15.6% with the AMC as a metasurface.

References

1. T.S. Rappaport, et al., Millimeter-wave mobile communications for 5G cellular: it will work! IEEE Access (2013), pp. 335–349
2. F. Jilani, Q.H. Abbasi, A. Alomainy, Inkjet-printed millimetre-wave PET-based flexible antenna for 5G wireless applications, in *2018 IEEE MTT-S International Microwave Workshop Series on 5G Hardware and System Technologies (IMWS-5G)*, Dublin, Ireland, Aug 2018, pp. 1–3
3. N.A. Al-Sharee et al., Development of a new approach for high-quality quadrupling frequency optical millimeter-wave signal generation without optical filter. Prog. Electromagn. Res. **134**, 189–208 (2013)
4. R. Del-Rio-Ruiz, J. Lopez-Garde, J. Legarda, Planar textile off-body communication antennas: a survey. Electronics **8**(6), 714 (2019)
5. M.I. Jais, et al., 1.575 GHz dual-polarization textile antenna (DPTA) for GPS application, in *2013 IEEE Symposium on Wireless Technology & Applications (ISWTA)*, Kuching, Malaysia, Sept 2013, pp. 376–379
6. I. Adam, M. Abdulmalek, N. Mohd Yasin, H.A. Rahim, Double band microwave rectifier for energy harvesting. Microw. Opt. Technol. Lett. **58**(4), 922–927 (2016)
7. E.A. Mohammad, et al., Dual-band circularly polarized textile antenna with split-ring slot for off-body 4G LTE and WLAN applications. Appl. Phys. A **124**(2) (2018)
8. H.A. Rahim, F. Malek, I. Adam, S. Ahmad, N.B. Hashim, P.S. Hall, Design and simulation of a wearable textile monopole antenna for body centric wireless communications, in *PIERS 2012*, Moscow, Russia, Aug 2012, pp. 1381–1384
9. H.A. Rahim, F. Malek, I. Adam, S. Ahmad, N.B. Hashim, P.S. Hall, On-body textile monopole antenna characterisation, in *Proceedings of Progress in Electromagnetics Research Symposium (PIERS)*, Moscow, Russia, Aug 2012, pp. 1377–1380
10. H.A. Rahim, F. Malek, I. Adam, S. Ahmad, N.B. Hashim, P.S. Hall, Effect of different substrates on a textile monopole antenna for body-centric wireless communications, *IEEE 2012 Symposium on Wireless Technology and Applications (ISWTA)*, Bandung, Indonesia, Sept 2012, pp. 245–247
11. H.A. Rahim, M. Abdulmalek, P.J. Soh, G.A.E. Vandenbosch, Evaluation of a broadband textile monopole antenna performance for subject-specific on-body applications. Appl. Phys. A Mater. Sci. Process. **123**(1), 1–6 (2017)
12. G. Gao, C. Yang, B. Hu, R. Zhang, S. Wang, A wide-bandwidth wearable all-textile PIFA with dual resonance modes for 5 GHz WLAN applications. IEEE Trans. Antennas Propag. **67**(6), 4206–4211 (2019)
13. S. Yan, V. Volskiy, G.A.E. Vandenbosch, Compact dual-band textile PIFA for 433-MHz/2.4-GHz ISM bands. IEEE Antennas Wirel. Propag. Lett. **16**, 2436–2439 (2017)
14. L.A.Y. Poffelie, P.J. Soh, S. Yan, G.A.E. Vandenbosch, A high-fidelity all-textile UWB antenna with low back radiation for off-body wban applications. IEEE Trans. Antennas Propag. **64**(2), 757–760 (2016)
15. K.N. Paracha, et al., A low profile, dual-band, dual polarized antenna for indoor/outdoor wearable application. IEEE Access, Feb 2019
16. P.J. Soh, G.A.E. Vandenbosch, F.H. Wee, A. Van Den Bosch, M. Martínez-Vázquez, D. Schreurs, Specific absorption rate (SAR) evaluation of textile antennas. IEEE Antennas Propag. Mag. **57**(2), 229–240 (2015)
17. F. Malek, K.A. Rani, H.A. Rahim, M.H. Omar, Effect of short-term mobile phone base station exposure on cognitive performance, body temperature, heart rate and blood pressure of Malaysians. Sci. Rep. **5**, Aug 2015
18. F. Guichi, M. Challal, Ultra-wideband microstrip patch antenna design using a modified partial ground plane, in *2017 Seminar on Detection Systems Architectures and Technologies (DAT)*, Algiers, Algeria, pp. 1–6, Feb 2017
19. I.B. Vendik, A. Rusakov, K. Kanjanasit, J. Hong, D. Filonov, Ultrawideband (UWB) planar antenna with single-, dual-, and triple-band notched characteristic based on electric ring resonator. IEEE Antennas Wirel. Propag. Lett. **16**, 1597–1600 (2017)

20. B. Mukherjee, et al., Coplanar waveguide fed ultra-wide band printed slot antenna with dual band-notch characteristics, in *2017 8th Annual Industrial Automation and Electromechanical Engineering Conference (IEMECON), Bangkok*, pp. 314–317, Aug 2017
21. Z. Sembiring, M. F.A. Malek, H. Rahim, Low complexity OFDM modulator and demodulator based on discrete Hartley transform, in *2011 Fifth Asia Modelling Symposium*, pp. 252–256, May 2011

Flexible UWB Compact Circular Split-Ring Slotted Wearable Textile Antenna for Off-Body Millimetre-Wave 5G Mobile Communication

H. W. Lee, Hasliza A. Rahim, Mohamedfareq Abdulmalek, R. B. Ahmad, M. H. Jamaluddin, M. Jusoh, D. A. Mohsin, F. H. Wee, I. Adam, N. Z. Yahya and K. N. A. Rani

Abstract A flexible ultra-wideband (UWB) compact circular split-ring slotted wearable textile antenna for off-body 28 GHz fifth-generation (5G) mobile communication is proposed. The proposed antenna is implemented using low-cost felt textile substrates and copper. The proposed 5G wearable antenna of compact circular split-ring slotted with enhanced bandwidth of 0.5% with the resonance frequency of 28 GHz is presented. The S_{11} for patch antenna with slot exhibited 43.4% more than the patch antenna without slot. The results also exhibited that the bending angle of 10° and 20° perform better return loss than in flat condition, up to 14% for patch antenna with slot against without the slot.

H. W. Lee (✉) · H. A. Rahim · R. B. Ahmad · M. Jusoh · D. A. Mohsin · F. H. Wee · I. Adam
Bioelectromagnetics Research Group (BioEM), School of Computer and Communication Engineering, Universiti Malaysia Perlis (UniMAP), Kampus Pauh Putra, 02600 Arau, Perlis, Malaysia
e-mail: leehuoywen@yahoo.com

H. A. Rahim
e-mail: haslizarahim@unimap.edu.my

M. Abdulmalek
Department of Engineering and Information Science, University of Wollongong in Dubai, Block 15, Dubai Knowledge Village, Dubai, UAE

M. H. Jamaluddin
Faculty of Electrical Engineering, Universiti Teknologi Malaysia (UTM), UTM Skudai, 81310 Johor, Malaysia

D. A. Mohsin
Department of Computer Technical Engineering, Electrical Engineering Technical College, Middle Technical University (MTU), Baghdad, Iraq

N. Z. Yahya
Physics Section, School of Distance Education, Universiti Sains Malaysia (USM), 11800 George Town, Penang, Malaysia
e-mail: norzakiah@usm.my

K. N. A. Rani
Bioelectromagnetics Research Group (BioEM), Faculty of Engineering Technology, Universiti Malaysia Perlis, 02100 Padang Besar, Perlis, Malaysia

H. S. Saini et al. (eds.), *Innovations in Electronics and Communication Engineering*, Lecture Notes in Networks and Systems 107,
https://doi.org/10.1007/978-981-15-3172-9_25

Keywords UWB antenna · Millimetre-wave textile antenna · 5G mobile communication · Wearable textile antenna · Split-ring slotted

1 Introduction

It is envisioned that the future 5G deployment will be highly dense, diversified, versatile and a unified technology with massive bandwidth spectrum [1]. This revolutionized technology is foreseen to be merged with a variety of wireless body area networks (WBANs) applications [2]. Millimetre wave has a high potential to be utilized for high-speed wireless broadband communications, and the spectrum band is between 30 and 300 GHz which is now tested on 5G technology. One of the key enablers of such technology is wearable antenna [3–12], embedded into the smart textile system. Wearable antenna in 5G can provide information that requires very fast data transmission in order to combat the interference caused by the multipath fading [13]. However, one of the challenging issues of the design of the antenna is the performance of the antenna degrades when the antenna is in bending condition in terms of bandwidth, resonant frequency and reflection coefficient. Thus, it is paramount to ensure that the main criteria of body-worn antenna are met which include small size, flexible, lightweight and conformal. Thus, such kind of antenna that offers better performance for 5G system is urgently needed.

The aim of this work is to design the flexible UWB compact circular split-ring slotted wearable textile millimetre-wave antenna for off-body 28 GHz 5G mobile communication with enhanced bandwidth of about 0.5% comparable to the textile millimetre-wave antenna without the slot. The targeted single band frequency range that used for this project is 28 GHz. The bandwidth can be enhanced further by adding slot to the antenna.

2 Antenna and Material Specifications

The design of the proposed wearable textile antenna is shown in Fig. 1. A commercial electromagnetic solver, CST Microwave Studio software, was used in the design and optimization process of the antenna. The proposed antenna is aimed to be used in the 28 GHz of the 5G frequency. The antenna design is deployed on a felt textile as a substrate while copper as conducting material. There are three elements in this antenna design which are ground plane, substrate and patch design. The substrate of the antenna is flexible felt textile with height, *h*, of 1.6 mm, the dielectric constant, ε_r, is 2.2 and loss tangent is 0.044. The substrate is sandwiched between patch and ground plane, with thickness of a flexible copper tape is 0.035 mm [12]. For the top radiator structure, the design of split-ring slot containing two split-rings, which are outer ring and inner ring, is embedded. The radius of the outer ring, st = 0.48 mm, while the radius of the inner ring, stt = 0.2 mm.

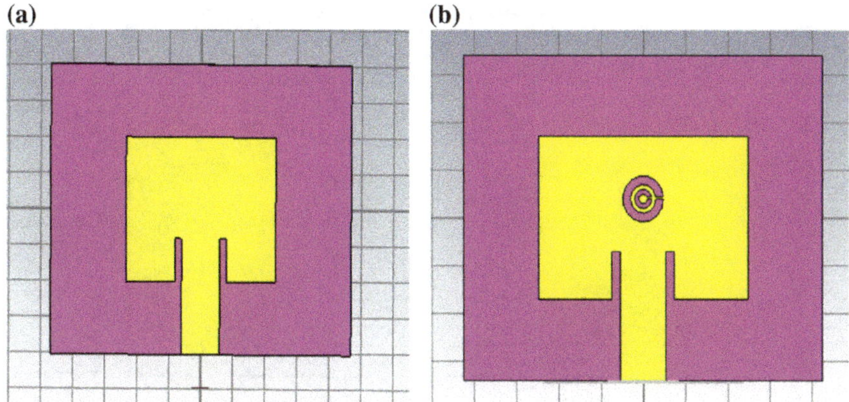

Fig. 1 Geometry of the patch antenna structure of **a** without slot and **b** with slot

3 Results and Discussion

3.1 Reflection Coefficient, S_{11}, for Antenna

S-parameter is the relationship of ports and input-output of electrical system. S_{11} represents the amount of power that is reflected from antenna, and it is also known as reflection coefficient or return loss. The antenna will be radiated best at the frequency that the value of S_{11} below −10 dB which means at least 90% of input power is delivered to the device and less than 10% of reflected power [3]. The proposed antenna was patch antenna with circular split-ring slot. The comparison between the patch antenna with and without slotted split-ring is shown in Fig. 2. The result shows that both antennas achieved the resonant frequency at 28 GHz. The S_{11} for patch antenna without slot is −22.01 dB. Meanwhile, the return loss for patch antenna with slot is −31.56 dB, achieving 43.4% more than the patch without the slot. Likewise, the 10-dB impedance bandwidth of slotted patch antenna slightly outperforms the unslotted patch antenna by 0.5% (Table 1).

3.2 Critical Parametric Assessments

Four critical parameters have been analysed including length of patch, L_p, width of patch, W_p, radius of the outer ring of slot, st and radius of the inner ring of slot, stt by evaluating the changes in the impedance bandwidth. When each of the parameters is varied, the others remain constant. The length of the patch antenna is simulated from 3.148 to 3.648 mm with variation of every 0.5 mm every length as shown in Fig. 3a. The results show that the S_{11} is shifted downwards when the length of patch increases from 3.148 to 3.448 mm. However, when L_p increases from 3.498

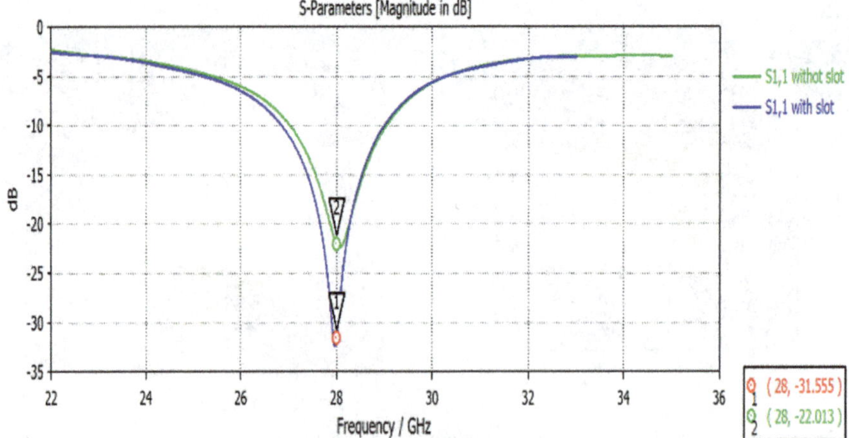

Fig. 2 S_{11} of patch antenna with and without split-ring slot

Table 1 Design specification of the proposed millimetre-wave textile antenna

Design specification	Value
Operating frequency (GHz)	28 GHz
Percentage of bandwidth	≥0.5%
Realized gain (dB)	>4 dB
Resonant frequency	28 GHz
Type of antenna	Single patch with and without slot antenna
Material and substrate	Felt textile and copper

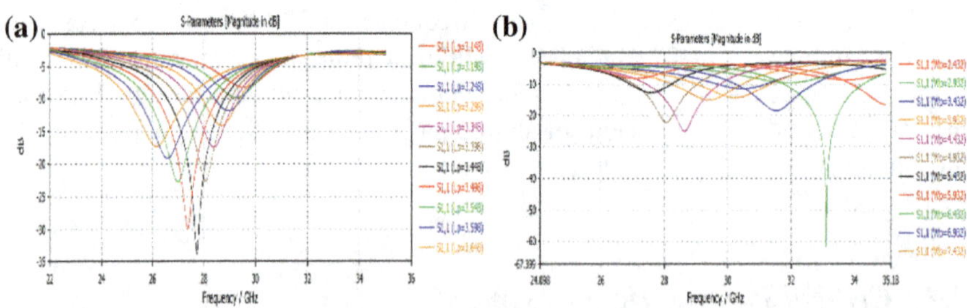

Fig. 3 Parametric study on **a** length of patch, L_p and **b** width of patch, W_p

to 3.648 mm, the S_{11} is shifted upwards, up to 4.1% from 27.5 to 26 GHz. Hence, the length of 3.398 mm illustrates the optimized result among others for frequency at 28 GHz.

Meanwhile, the width of the patch antenna is varied from 2.432 to 7.432 mm for 11 different widths with the increase of 0.5 mm each as shown in Fig. 3b. The results

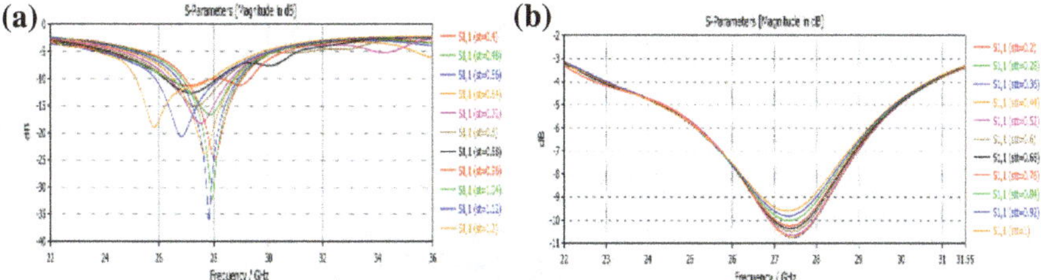

Fig. 4 Parametric study on radius of **a** outer ring of slot, st and **b** inner ring of slot, stt

show that the S_{11} shifted downwards to 18% at 27 GHz with reduced frequency when the length of patch increases up to 7.432 mm. Hence, the width of 4.932 mm illustrates the best result among others for resonant frequency of 28 GHz.

Since adding the slot on the antenna is one of the effective techniques to broaden the bandwidth of the antenna, a slot of split-ring on patch antenna was proposed and studied. The other parameters remained constant. The radius of the outer ring of slot, st, from 0.4 to 1.2 mm was varied, while radius of the inner ring of slot, stt, was varied from 0.2 to 1 mm. Figure 4a shows the parametric study on radius of the outer ring of slot, st, while Fig. 4b shows the parametric study on radius of the inner ring of slot, stt. Each slot width is increased by 0.08 mm, while the inner split-ring remained constant when simulating outer split-ring and vice versa. The resonant frequency varies from 25 to 28 GHz for patch antenna with slot as the width of slot for outer ring varies from 0.4 to 1.2 mm. The results exhibit that as the slot width increases, S_{11} degrades, and this obviously shows a poorer impedance matching and thus worsens the return loss values. For the radius of the inner ring of slot, there are no significant changes for resonance frequency from radius 0.2 to 1 mm, where it depicts minimum variation up to 0.1%. The return loss is also poorer with larger width of the ring radius by 13%.

3.3 Antenna Bending Variation

The curvature effect of wearable antenna was simulated according to the possibility of antenna location on body. The examples of location on body are arm, wrist, legs, neck and shoulder. The bending angle is defined by angle of arc that formed by bending the ground, substrate and patch over a cylinder model with radius, r. The bending angles from 0° to 50° for different antenna designs were simulated. The 0° of bending represents flat condition. The rest bending angles are equivalent to 24.38, 30.48, 40.6, 69.8 and 121.9 mm of curvature radius values. These values were chosen as representative of the bending angle to cover the arm curvature in a regular body [12].

Based on the result in Fig. 5, the S_{11} shifted downwards when the bending angle increases for patch antenna without slot. The bending angles of 40° and 50° for patch

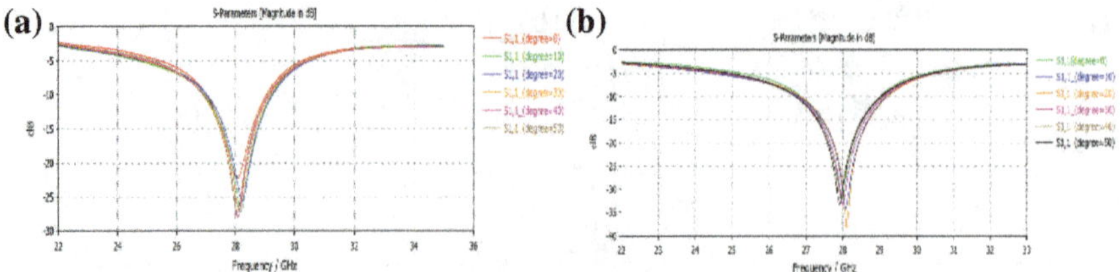

Fig. 5 Simulated reflection coefficient for the different values of the curvature angles, 0°–50° **a** without slot and **b** with slot

antenna without slot outperform return loss values of flat condition. It is suggested that the antenna in bending condition for patch antenna without slot can perform better under bending and on arm conditions. For patch antenna with slot, the bending angles of 10° and 20° perform better return loss than in flat condition, up to 14%. This can conclude that the performance of the antenna with slot is better in the small curvature conditions of 24.38 and 30.48 mm.

3.4 Substrate Thickness Variation

The substrate thickness was assessed for both patch antenna and patch antenna with slot at bending of 10° in order to analyse its effect towards the S_{11}. The substrate thickness varied from Hs = 0.1 to Hs = 1 mm was analysed, see Fig. 6. The results show that the resonant frequency varies from 24 to 28 GHz for both patch antennas when substrate thickness increases. The greater the substrate thickness, the lower the S_{11}. The return loss of patch antenna with slot is better than patch antenna without slot for the various substrate thicknesses in bending condition. When the substrate thicknesses increases, the resonant frequency shifts to the left. Hence, the value of Hs = 0.35 mm was selected for the proposed antenna.

The radiation pattern measures the capabilities of antenna to receive or transmit on a certain direction. The present radiation pattern in Fig. 7a shows angular width (3 dB) = 81.0°. The angular width is depending on presence of the left and right parasitic that is relying on 0 dBi. Meanwhile, the radiation pattern in Fig. 7b shows

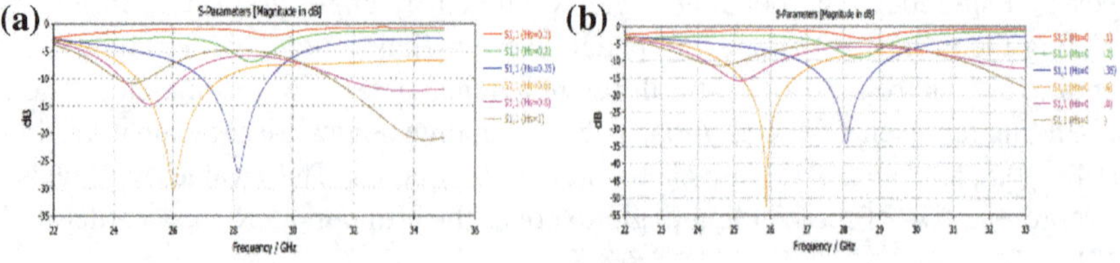

Fig. 6 Simulated reflection coefficient for the different substrate thickness values of **a** patch antenna without slot and **b** patch antenna with slot

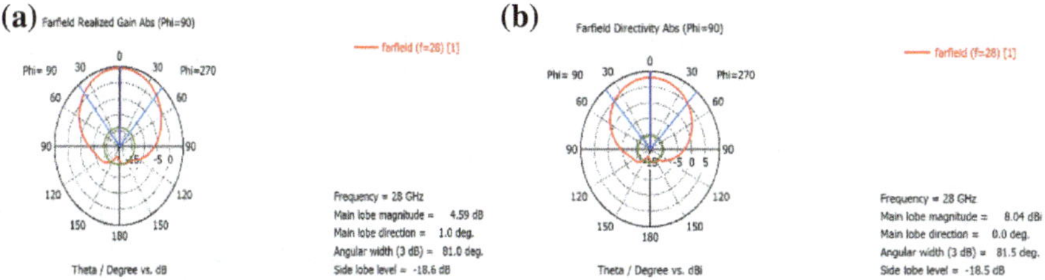

Fig. 7 Radiation pattern for patch antenna **a** without slot and **b** with slot

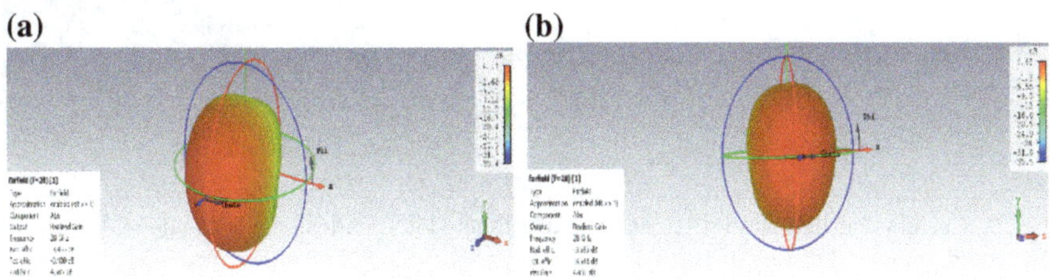

Fig. 8 Realized gain for patch antenna **a** without slot and **b** with slot

angular width (3 dB) = 81.5° for patch with slot, 0.5° more than the patch without the slot.

Figure 8 shows the realized gain of antenna. The gain is 4.59 dB for patch antenna without slot. Meanwhile, realized gain for patch antenna with slot is 4.45 dB. The gain slightly decreases, suggesting that the presence of the slot may deteriorate with minimum variation up to 3%.

3.5 *Current Distribution*

The surface current distributions on both patch antennas are simulated for further examining the whole proposed antenna. The current distribution was simulated at 28 GHz for both antennas. Based on the simulation result in Fig. 9a–d, the orientation of the current is towards *Y* direction and on the surface of the patch antenna. The current intensity is mainly focused on the feed line, centre of patch and around the split-ring slots. The current intensity for patch antennas with bending of 10° is relatively lesser than the patch antennas without bending.

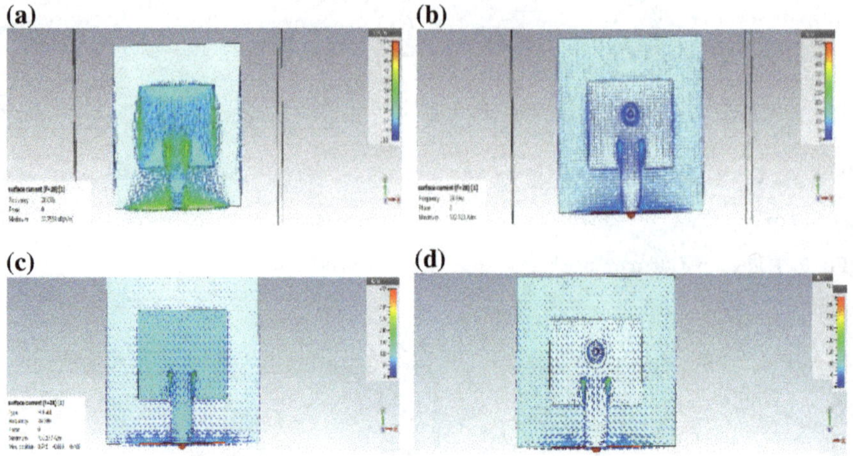

Fig. 9 Current distribution in patch antenna **a** without slot, **b** with slot, **c** without slot bending of 10° and **d** with slot bending of 10°

4 Conclusion

A flexible UWB compact circular split-ring slotted wearable textile antenna for off-body 28 GHz 5G mobile communication is presented. The proposed antenna exhibits enhanced bandwidth of 0.5% against similar structure without the slot. The S_{11} for patch antenna with slot is 43.4% more than the patch antenna without slot. The S_{11} shifted downwards when the bending angle increases for patch antenna without slot. For patch antenna with slot, the bending angles of 10° and 20° perform better return loss than in flat condition, up to 14%.

References

1. S.F. Jilani, Q.H. Abbasi, A. Alomainy, Inkjet-printed millimetre-wave PET-based flexible antenna for 5G wireless applications, in *2018 IEEE MTT-S International Microwave Workshop Series on 5G Hardware and System Technologies (IMWS-5G)*, Dublin, Ireland, Aug 2018, pp. 1–3
2. R. Del-Rio-Ruiz, J. Lopez-Garde, J. Legarda, Planar textile off-body communication antennas: a survey. Electronics **8**(6), 714 (2019)
3. M.I. Jais, et al., 1.575 GHz dual-polarization textile antenna (DPTA) for GPS application, in *IEEE Symposium on Wireless Technology & Applications (ISWTA)*, Kuching, Malaysia, Sept 2013, pp. 376–379
4. E.A. Mohammad, et al., Dual-band circularly polarized textile antenna with split-ring slot for off-body 4G LTE and WLAN applications. Appl. Phys. A **124**(2) (2018)
5. H.A. Rahim, F. Malek, I. Adam, S. Ahmad, N.B. Hashim, P.S. Hall, Design and simulation of a wearable textile monopole antenna for body centric wireless communications, in *PIERS 2012*, Moscow, Russia, Aug 2012, pp. 1381–1384

6. H.A. Rahim, F. Malek, I. Adam, S. Ahmad, N.B. Hashim, P.S. Hall, On-body textile monopole antenna characterisation, in *Proceedings of Progress in Electromagnetics Research Symposium (PIERS)*, Moscow, Russia, Aug 2012, pp. 1377–1380
7. H.A. Rahim, F. Malek, I. Adam, S. Ahmad, N.B. Hashim, P.S. Hall, Effect of different substrates on a textile monopole antenna for body-centric wireless communications, in *IEEE 2012 Symposium on Wireless Technology and Applications (ISWTA)*, Bandung, Indonesia, Sept 2012, pp. 245–247
8. H.A. Rahim, M. Abdulmalek, P.J. Soh, G.A.E. Vandenbosch, Evaluation of a broadband textile monopole antenna performance for subject-specific on-body applications. Appl. Phys. Mater. Sci. Process. **123**(1), 1–6 (2017)
9. G. Gao, C. Yang, B. Hu, R. Zhang, S. Wang, A wide-bandwidth wearable all-textile PIFA with dual resonance modes for 5 GHz WLAN applications. IEEE Trans. Antennas Propag. **67**(6), 4206–4211 (2019)
10. S. Yan, V. Volskiy, G.A.E. Vandenbosch, Compact dual-band textile PIFA for 433-MHz/2.4-GHz ISM bands. IEEE Antennas Wirel. Propag. Lett. **16**, 2436–2439 (2017)
11. H.A. Rahim, M. Abdulmalek, P.J. Soh, K.A. Rani, N. Hisham, G.A.E. Vandenbosch, Subject-specific effect of metallic body accessories on path loss of dynamic on-body propagation channels. Sci. Rep. **6** (2016)
12. R. Sanchez-Montero, P.-L. Lopez-Espi, C. Alen-Cordero, J.-A. Martinez-Rojas, Bend and moisture effects on the performance of a U-Shaped slotted wearable antenna for off-body communications in an industrial scientific medical (ISM) 2.4 GHz band. Sensors **19**, 1804 (2019)
13. Z. Sembiring, M.F.A. Malek, H. Rahim, Low complexity OFDM modulator and demodulator based on discrete Hartley transform, in *2011 Fifth Asia Modelling Symposium*, May 2011 (pp. 252–256)

Achievable Throughput of Energy Detection Spectrum Sensing Cognitive Radio Networks

Anitha Bujunuru and Srinivasulu Tadisetty

Abstract Upgrading usages of wireless communication applications have many constraints on the utilization of accessible wireless spectrum. Cognitive radio (CR) technology is an emanating and auspicious solution to the issue of insufficient licensed spectrum. The spectrum sensing is the majority demanding issue in cognitive radio applications to find out the accessible spectrum bands which can be utilized by secondary user without providing any unfavorable intervention to the primary user. SU will sense the existence of PU and utilizes the spectrum for data transmission if the spectrum is free without providing any harmful interference to the PU. To achieve this, SU must require an adequate amount of sensing time, which in turn reduces the transmission slot. Thus, the total average throughput of the SU must reduce with increase in sensing time. The performance of simple energy detection (ED) spectrum sensing is analyzed in terms of total error probability and receiver operator characteristics. Simulation results of variation of throughput with SNR and effect of increasing sensing time on throughput are presented using MATLAB under AWGN channel.

Keywords Cognitive radio · Spectrum sensing · SNR · Sensing time · Throughput

1 Introduction

The radio spectrum available is limited, and wide increase in usage of wireless communication rises to the issue of spectrum scarcity. Most of the pre-allocated radio spectrum is underutilized by the primary user which creates holes are also called spectrum holes. Spectrum holes are the unutilized chunk of spectrum by the licensed user at the given specific time. Therefore, most of the pre-allocated radio spectrum is underutilized due to the uninspired approach of spectrum management schemes (fixed spectrum assignment schemes) and can be solved using cognitive

---

A. Bujunuru (✉)
ECE Department, Guru Nanak Institutions Technical Campus, Hyderabad, India
e-mail: anitha.shanala@gmail.com

S. Tadisetty
ECE Department, Kakatiya University, Warangal, India

H. S. Saini et al. (eds.), *Innovations in Electronics and Communication Engineering*, Lecture Notes in Networks and Systems 107,
https://doi.org/10.1007/978-981-15-3172-9_26

radio [1]. CR is a smart dynamic spectrum management system that adjusts the environment conditions [1, 2]. Cognitive radio is defined as "It could be a radio for wireless communication that automatically detects the available channels and depends on the interaction with the environment to communicate more effectively to prevent interfacing to authorized users."

CRN nodes can be classified as primary (licensed) users (PUs) and secondary (cognitive or unlicensed) users (SUs). PU has absolute liberty to access the particular licensed spectrum band, whereas SU detects unutilized chunks of spectrum momentarily through its PU and opportunistically utilizes them. CRN enables unlicensed users for exploiting the spatially and/or temporally underutilized spectrum by communicating over the licensed bands. CRN is an overlay network with dynamic spectrum access, where SU should have spectrum sensing capability for sensing whether there is presence of PU before transmission, thereby provides spectrum efficiency and improves network performance.

2 Cognitive Radio Modules

The important tasks of cognitive radio are spectrum sensing, spectrum sharing, spectrum management and spectrum mobility which are shown in Fig. 1. Spectrum sensing is a procedure of identifying the unused spectrum portions by secondary user. Spectrum management is the procedure of assigning available portion of spectrum to the user. Spectrum mobility is a task of exchanging of frequency of operation of cognitive users. Spectrum sharing is a method of sharing the available primary user spectrum with the secondary user [3]. Depending on the sensing results, SUs can get status of the channels to access.

Spectrum sensing is a procedure of identifying the unused spectrum portions by continuous monitoring of primary user and makes use of the free spectrum by unlicensed users. The process of making use of the spectrum whenever the user required is known as dynamic spectrum access (DSA) which improves the spectrum

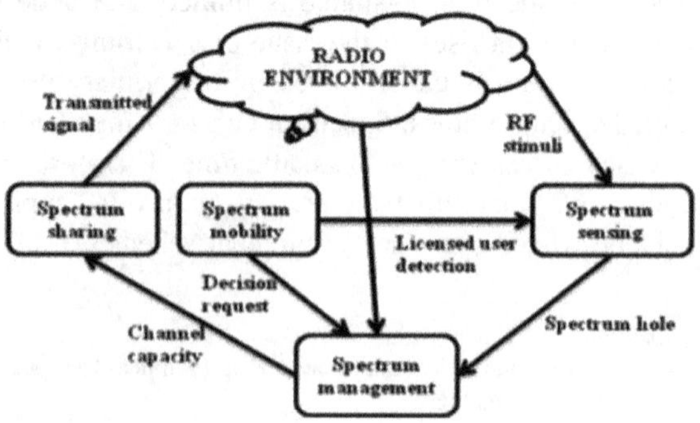

Fig. 1 Cognitive cycle modules

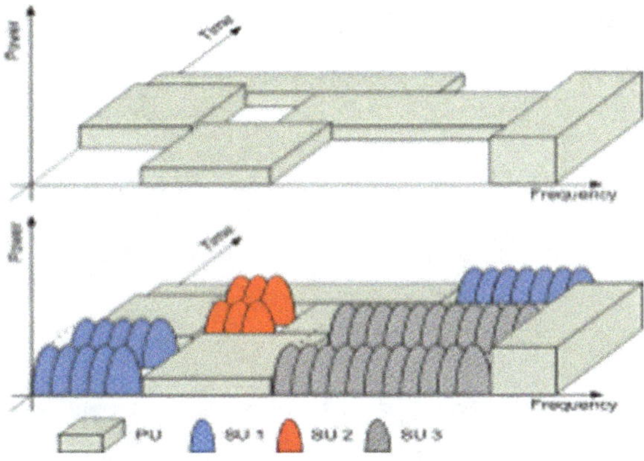

Fig. 2 Dynamic spectrum access of wireless spectrum

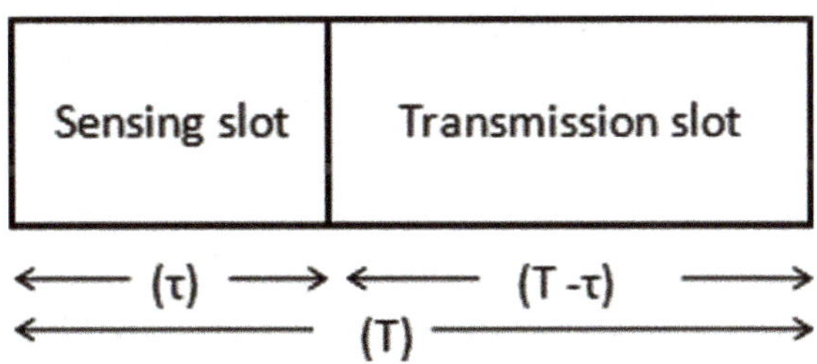

Fig. 3 Frame structure of CR user

efficiency. The dynamic spectrum access of wireless spectrum by the secondary users is shown in Fig. 2.

Dynamic spectrum access cognitive radio frame structure consists of two slots, sensing slot and transmission slot. Frame structure of CR user of T duration is shown in Fig. 3. During the sensing slot, CR user senses the presence of PU for the duration of τ. If the PU is not utilizing spectrum, then SU can transmit its data during transmission slot for the period of $(T - \tau)$.

3 Spectrum Sensing

The main intention of spectrum sensing is to know the appearance of primary user, so that SU can use the channel without providing intervention to the primary user. Spectrum sensing means finding spectrum holes of primary user continuously using the channel parameters such as transmit power, noise and interference levels.

SU acquires the signal from the primary user and then applies different sensing methods to find out the residence of primary user by comparing the obtained signal with the threshold. If the received signal is less than threshold, then SU takes the decision that PU is available; otherwise, PU is not utilizing the channel. The flow of spectrum sensing is shown in Fig. 4.

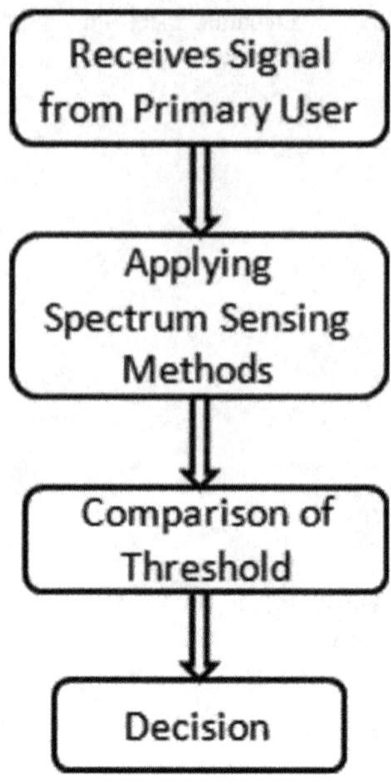

Fig. 4 Flow of spectrum sensing

3.1 *Energy Detection Spectrum Sensing*

The most common and simple sensing method is ED spectrum sensing. It is a non-coherent detection process that can sense the appearance of primary user based on the energy of the sensed primary signal [2]. The SU can calculate the average energy (E) of PU signal and compare with the threshold value (λ). If calculated energy $E \geq \lambda$, SU takes the decision that PU is available (H_1); otherwise, SU makes a decision that PU is absent (H_0).

Figure 5 shows the basic model of energy detector [4]. PU received signal energy can be calculated by using Eq. (1)

$$E = \frac{1}{T}\int_{-\infty}^{\infty} s^2(t)\mathrm{d}t \quad (1)$$

PU detection is done with hypothesis test given in Eq. (2).

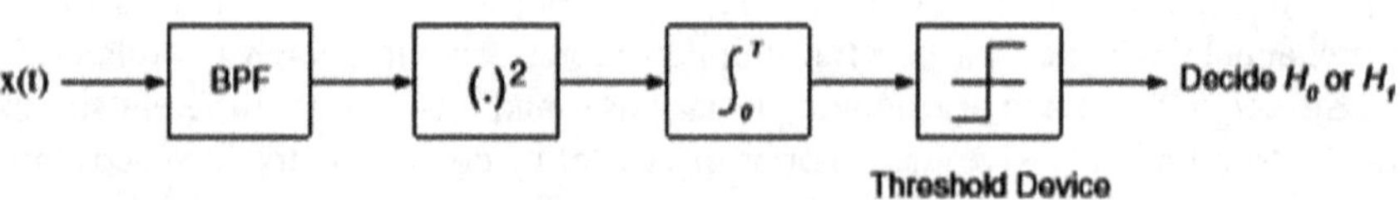

Fig. 5 Block diagram of energy detector

This technique does not necessitate any of the prior knowledge

$$\left.\begin{array}{ll} y(t) = n(t), & H0 \\ y(t) = h * s(t) + n(t), & H1 \end{array}\right\} \tag{2}$$

The energy detector output is the collected signal energy which is calculated by

$$E = \sum_{n=0}^{N} y(n)^2 \tag{3}$$

where number of sample is n and $y(n)$ is the signal collected by the SU. Decision of the energy detector is obtained by calculated energy (E) of the energy detector and the threshold value (λ) [5, 6], i.e.,

If $E \geq \lambda$, PU signal is available
If $E \leq \lambda$, PU signal is not available

The performance measures of probability of detection (P_d) and probability of false alarm (P_f) are calculated by utilizing the below equations.

$$P_\mathrm{d} = Q\left(\frac{\lambda - N(\delta_n^2 + \delta_s^2)}{\sqrt{2N(\delta_n^2 + \delta_s^2)^2}}\right) \tag{4}$$

$$P_\mathrm{f} = Q\left(\frac{\lambda - N\delta_n^2}{\sqrt{2N\delta_n^4}}\right) \tag{5}$$

where $Q(.)$ represents the Q-function, δ_n^2 is the variance of noise and δ_s^2 is the variance of the PU signal. Sensing threshold depends on noise power and is calculated by

$$\lambda = \left(Q^{-1}(P_f)\sqrt{2N} + N\right)\delta_n^2 \tag{6}$$

Two calculations of probability of detection (P_d) and probability of false alarm (P_f) are also represented by using SNR (γ) and are given in following equations.

$$P_\mathrm{d} = Q\left(\frac{\overline{\lambda} - N(1 + \gamma)}{\sqrt{2N(1 + \gamma)^2}}\right) \tag{7}$$

$$P_\mathrm{f} = Q\left(\frac{\lambda - N\delta_n^2}{\sqrt{2N\delta_n^4}}\right) \tag{8}$$

where $\overline{\lambda}$ is the average threshold and is given by $\overline{\lambda} = \frac{\lambda}{\delta_n^2}$.

Advantage

(i) This technique does not necessitate any of the previous information of PU.
(ii) Implementation is also easier.

Drawback

(i) It cannot perform well at low SNR conditions [7].
(ii) Selection of threshold is difficult.
(iii) It cannot discriminate PU signal and noise [8, 9].

4 Analysis of Throughput

This section will discuss the throughput analysis of CR user. To maintain the QoS of licensed user, a high probability of detection is considered [10]. As stated earlier, the conventional frame consists of sensing slot and transmission slot, as shown in Fig. 3. SU can transmit its data in two cases, one is in the absence of PU, which does not create any interference to the PU. The second is in the presence of PU, which will interfere PU transmission [11, 12].

The throughput of SU under the absence of PU is denoted as $C_{1,}$ and in the presence of PU, it is mentioned as C_2.

$$C_1 = \log_2(1 + \text{SNR}_\text{s}) \tag{9}$$

$$C_2 = \log_2\left(1 + \frac{\text{SNR}_\text{s}}{1 + \text{SNR}_\text{p}}\right) \tag{10}$$

where SNR_s is the SNR of SU, and SNR_p is the SNR of SU received from PU.

Case 1:

If the licensed user is absent and SU does not provide any false alarm, then the average throughput of SU is measured using Eq. (11).

$$R_1 = \frac{T - \tau}{T} C_1\big(1 - P_f\big)P_0 \tag{11}$$

where P_0 is the probability in the absence of PU.

Case 2:

When SU wrongly detects the presence of PU and tries to access the spectrum in presence of PU, the average throughput of SU is given by R_2 in Eq. (12).

$$R_2 = \frac{T - \tau}{T} C_2(1 - P_\text{d})P_1 \tag{12}$$

where P_1 is the probability in the presence of PU.

The total average achievable throughput of the SU is R given by Eq. (13).

$$R = R_1 + R_2$$

$$R = \frac{T-\tau}{T} C_1(1 - P_f)P_0 + \frac{T-\tau}{T} C_2(1 - P_d)P_1 \tag{13}$$

5 Results and Discussions

In this section, performance detection is analyzed using receiver operating characteristics curves of energy detection. Simulation result of SNR versus total probability of error is shown in Fig. 6, and the probability of error decreases with increase in SNR. Simulation outcome of Pd against Pf is shown in Fig. 7. Simulation results have proved that probability of false alarm and probability detection are inversely proportional to each other.

Simulation results of average throughput of SU under the absence of PU transmission C_1 and in the presence of PU transmission C_2 are shown in Fig. 8. By utilizing the available spectrum of PU, throughput of the SU C_1 will increase with increase in SNR. Throughput C_2 shows the utilization of SU in the presence of PU transmission which introduces interference to the PU transmission.

To achieve a better sensing performance, CR user must require an adequate amount of sensing time. As sensing time increases, the total average throughput of CR decreases, and simulation results are shown in Fig. 9.

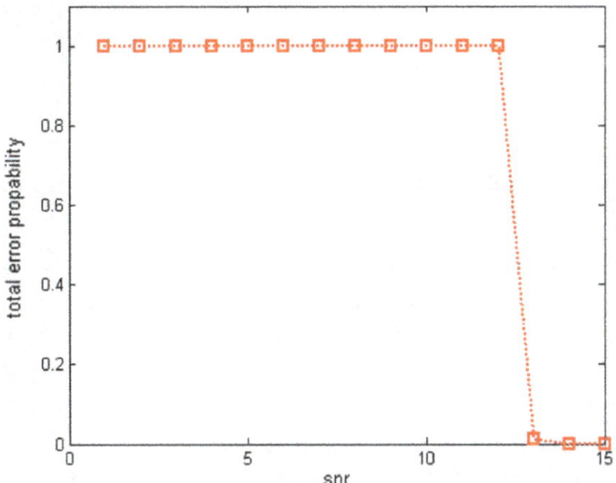

Fig. 6 Simulation results of total error probability with SNR

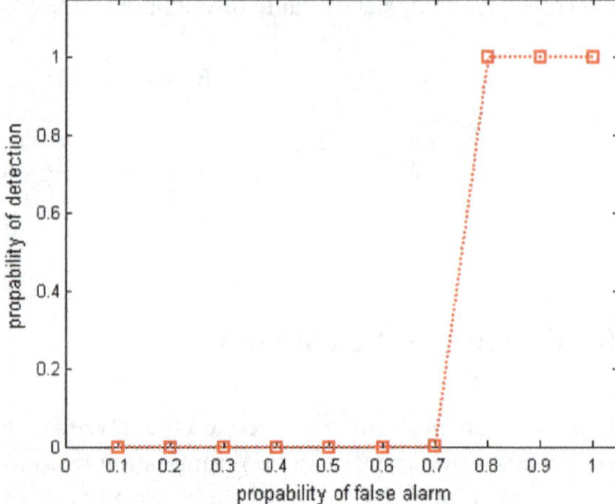

Fig. 7 Receiver operating characteristics curves of sensing techniques

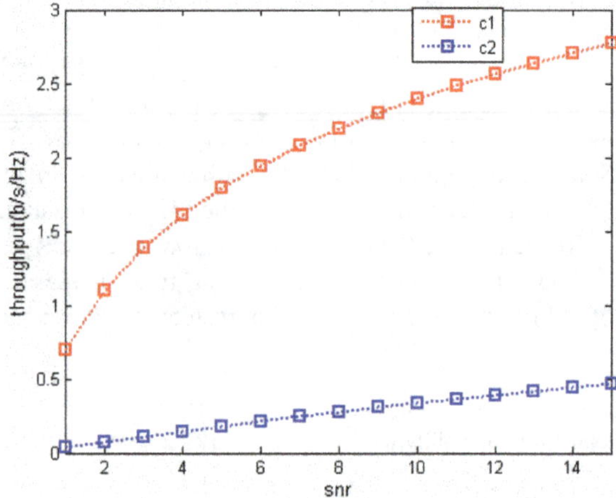

Fig. 8 Average throughput of SU with SNR

6 Conclusion

In this paper, we have analyzed the performance of energy detection spectrum sensing in terms of total error probability and receiver operator characteristic curves. Total error probability is decreased with increasing snr and remains constants after certain value. The performance metrics of probability of detection are analyzed with probability of false alarm (P_{fa}).

Throughput of SU is analyzed under the presence and absence of PU. To avoid the interference on the PU transmission, sensing performance of SU must be precisely high which takes longer sensing time. However, an increase in sensing time will result in an adequate reduction in transmission time, which leads to reduced throughput. Therefore, always there exist a trade-off between perfect sensing and throughput.

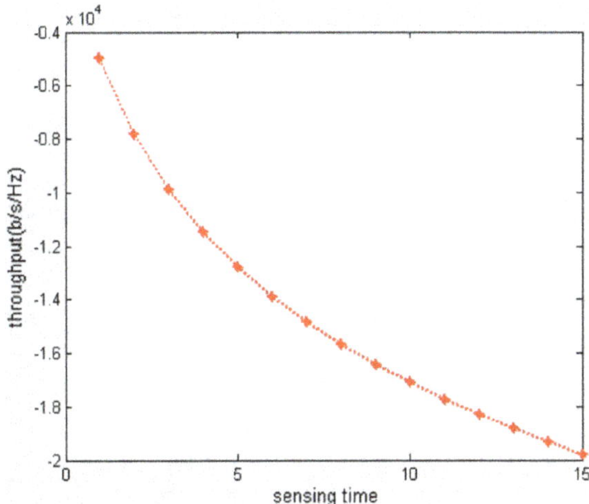

Fig. 9 Total throughput of SU with sensing time

References

1. M. Riahi Manesh, M.S. Apu, N. Kaabouch, W.-C. Hu, Performance evaluation of spectrum sensing techniques for cognitive radio systems, in *2016 IEEE 7th Annual Ubiquitous Computing, Electronics & Mobile Communication Conference* (UEMCON)
2. M. Gupta, G. Verma, R.K. Dubey, Cooperative spectrum sensing for cognitive radio based on adaptive threshold, in *2nd International conference on Computational Intelligence &Communication Technology* (2016)
3. A. Ahmad, S. Ahmad, M.H. Rehmani, N.U. Hassan, A survey on radio resource allocation in cognitive radio sensor networks (IEEE, 2015)
4. G. Ghosh, P. Das, S. Chatterjee, Simulation and analysis of cognitive radio system using MATLAB. Int. J. Next-Gener. Netw. (IJNGN) **6**(2) (2014)
5. R. Umar, A.U.H. Sheikh, A comparative study of spectrum awareness techniques for cognitive radio oriented wireless networks. Phys. Commun. 148–170 (2013)
6. S. Ziafat, W. Ejaz, H.U. Jamal, Spectrum sensing techniques for cognitive radio networks: performance analysis, in *2011 IEEE MTT-S International Microwave Workshop Series on Intelligent Radio for Future Personal Terminals* (2011)
7. L.M.G. Díaz, L.M. Marrero, J. Torres, performance comparison of spectrum sensing techniques for cognitive radio networks, in *VII Simposio de Telecomunicaciones*, Mar 2016
8. B. Sridhar, T. Srinivasulu, A novel high resolution spectrum sensing algorithm for cognitive radio applications. IOSR J. Electron. Commun. Eng. (IOSR-JECE), **8**(4), 30–38 (2013). e-ISSN: 2278-2834, p-ISSN: 2278-8735
9. A.M. Fanan, N.G. Riley, M. Mehdawi, M. Ammar, M. Zolfaghari, Survey: a comparison of spectrum sensing techniques in cognitive radio, in *Int'l Conference Image Processing, Computers and Industrial Engineering (ICICIE)* (2014), pp. 15–16
10. W. Wang, Spectrum sensing for cognitive radio, in *3rd International Symposium on Intelligent Information Technology Application Workshops* (2009), pp. 410–412
11. Y.-C. Liang, Y. Zeng, E.C. Peh, A.T. Hoang, Sensing-throughput tradeoff for cognitive radio networks. IEEE Trans. Wirel. Commun. **7**(4), 1326–1337 (2008)
12. B. Wang, K.J.R. Liu, Advances in cognitive radio networks: a survey. IEEE J. Sel. Top. Sign. Process. **5**(1), 5–23 (2011)

A Review on UWB Metamaterial Antenna

Ambavaram Pratap Reddy and Pachiyaannan Muthusamy

Abstract This paper deals with a brief review of advancements in the metamaterial field for the generation of the UWB frequency response during the past years. Extraordinary progress can be observed in the metamaterial field research which led for this technique to implement in many antenna applications. These are nonnatural engineered materials which have some special artificial electromagnetic properties. Ultra-wideband radio technology generally used for short-range communications covers a wide portion of the frequency spectrum and also consumes a very small energy level, the UWB range of frequencies 3.1–10.6 GHz. Metamaterials have many advantages like excellent beam performance, small size, low cost, radiated power enhancement and consume low power. The comparison of the proposed antennas is given in table; it is observed that the proposed antennas provide compact and enhanced bandwidth and show better improvement.

Keywords Antenna · Metamaterial (MTM) · Ultra-wideband (UWB) · DNG · MNG

1 Introduction

The word metamaterial (MTM) is derived from meta and material words of which the Greek word meta means beyond. So, metamaterials can be simply said as not just a material but as beyond a material. These are artificially engineered materials such that they have negative values for electromagnetic properties, whereas the natural materials available have only +ve values for electromagnetic properties. A visionary speculation on metamaterials has been first given by Vector Veselago in 1967 as the materials which are having −ve value for the electromagnetic properties [1]. Conventional materials will allow wave propagation in a forward-directed way only

A. P. Reddy (✉) · P. Muthusamy
Advanced RF Microwave & Wireless Communication Laboratory, Vignan's Foundation for Science, Technology & Research, Guntur, Andhra Pradesh, India
e-mail: pratap.phd5001@gmail.com

P. Muthusamy
e-mail: pachiphd@gmail.com

H. S. Saini et al. (eds.), *Innovations in Electronics and Communication Engineering*, Lecture Notes in Networks and Systems 107,
https://doi.org/10.1007/978-981-15-3172-9_27

which is also called as right-hand propagation and so the materials are called as right-handed materials, but the metamaterials have this peculiar property of supporting the wave propagation in backward direction which is also called as left-hand propagation, i.e., left-handed materials [2]. Based on Snell's laws and Doppler effect which are used to represent transition of wave propagation from one medium to another also have new approach to metamaterials properties like zero negative refractive index. In natural materials, rays get refracted along the normal of the interface, but in the LHM, ray will get refracted far away from the position where the normal is present. This will result in the production of a focus point inside the metamaterials. In general, we will be using four types of structures to engineer the metamaterials: S-structure symmetrical-ring structure, complimentary split-ring structure and omega structure, etc. It took larger than 30 years to realize the concepts proposed by Veselago; until then, no one was able to release and demonstrate a LH material experimentally by Smith et al. [3]. The UWB range of frequencies 3.1–10.6 GHz and the comparison of proposed antennas are given in table.

2 Classification of Metamaterials

The investigation of artificial materials or metamaterials started in the late 90s. In the year 1967, Russian physicist Victor Veselago described negative-index materials and demonstrated that these MTMs have the capability to transmit the light through it. Author illustrated that in MTMs, the directions of the wave propagation and phase velocity are in opposite directions which is not a general case and is in opposition with the natural materials [4].

In the year 1999, John Pendry demonstrated practical way to design metamaterials. MTM is no natural but is engineered materials with special properties in terms of electromagnetic properties which are not existing in the nature. In Fig. 1, according to Victor Veselago, all these materials are grouped into four different groups depending upon their parameters like permittivity and permeability.

2.1 DPS Materials

From Fig. 1, double-positive materials are the ones for which the ε and the μ are positive. The examples for these types of materials are generally available dielectric materials. The positive refractive index for these types of materials is available, and the waves follow right-hand rule to propagate in the forward-directed way.

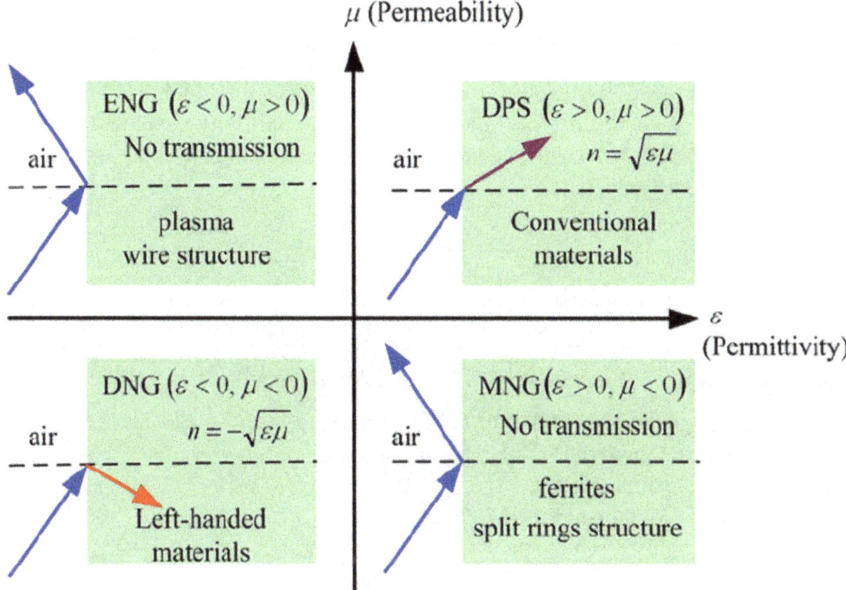

Fig. 1 Metamaterials classification

2.2 *ENG Materials*

From Fig. 1, epsilon-negative materials are the ones for which the ε is negative and the μ is positive for these materials, and the examples are vertical metallic vias inserted between radiating surface and the ground plane. These materials produce shunt inductance, and the radiation pattern due to current at these materials is similar to that of dipole antenna. In Fig. 1, showing ENG, TL unit cell model is constructed using the combination of shunt capacitance and series inductance.

2.3 *DNG Materials*

The double-negative materials (DNG) have negative permeability and negative permittivity and both ε and μ less than zero. These materials can be released only but are not naturally found.

2.4 MNG Materials

From Fig. 1, permeability-negative materials are the ones for which the ε is positive and the μ is negative, i.e., for this kind of materials, permittivity >zero and the value of permeability <zero ($\varepsilon > 0$, $\mu < 0$). Few gyrotropic materials exhibit these properties at some particular frequencies.

3 Types of Metamaterials

Three major classes of metamaterials are as follows.

3.1 Electromagnetic Metamaterials (EM)
3.2 Photonic Metamaterials (PM)
3.3 Acoustic Metamaterials (AM)
3.4 Mechanical Metamaterials (MM)

3.1 Electromagnetic Metamaterials

Electromagnetic metamaterials (EM) are having the composition of traces and particles in a dielectric matrix; these EM metamaterials have a zero or negative refractive index. Applications like beam stereos, antenna radomes, modulators, lenses, microwave couplers and band-pass filters are widely using these materials. The single-negative metamaterials define electromagnetic MTMs which have either negative permeability or negative permittivity, but not both. The double-negative metamaterial defines backward media define as negative refractive index and both negative permittivity (ε) and negative permeability (μ) [5].

3.2 Photonic Metamaterial

It deals with design of optical frequencies using photonic metamaterials. Photonic metamaterial has zero indexes of refraction, and this metamaterial is the current research area in the field of optics [6].

3.3 Acoustic Metamaterials

Acoustic metamaterials are making up of two or more different materials with different mass densities and bulk modulus; these types of metamaterials have negative effective mass densities and bulk modulus. These are the artificially fabricated metamaterial which is designed to direct, manipulate and control sound waves in liquid, solids, and gases. Any type of sound waves can be directly controlled by controlling the bulk modulus and mass density [7].

3.4 Mechanical Metamaterials

Mechanical metamaterial is the artificial composite metamaterial which is consisting of different types of mechanical properties; this metamaterial is having negative Poisson's ratio, negative elastic modulus, zero shear modulus and frictional properties. This is made up of material with inclusion of secondary materials or a controlled pored structure. The researchers of Northwestern University and Harvard University have interest in new research application of the mechanical MTMs in the field of Aerospace and Defense [8].

4 Metamaterial Structures for UWB Applications

Ultra-wideband (UWB) which covers the range of frequencies 3.1–10.6 GHz has been assigned for the radio application by the FCC in February 2004. Nowadays, handheld devices and mobile communication are in a blooming path of advancement with the growing users and demands. Most of the mobile communication systems are depending on the WLAN technology for satisfying the needs of high data rate. MTM can be a new solution for the growing need of high bandwidth with a compact antenna. Different MTM approaches were already been conceptualized and are in use like high impedance surfaces (HIS) and resistive impedance surfaces (RIS) which are similar to MTM and can be called as metasurfaces. Both these RIS and HIS surfaces use the basic concept of placing homogeneous periodic structure on a natural material such that it exhibits special electromagnetic behavior. By introducing these periodic structures and along with them, some conducting pins or vias in between the homogeneous periodic structure and the plane of ground can develop additional impedances and capacitances in the antenna. The operational bandwidth of the antenna will depend upon the inductance and capacitance [9]. The detailed different UWB metamaterial antenna configurations are reported in Table 1.

5 Conclusion

The metamaterial fields have a great future with its wide range of potential applications which consists of civilian and defense applications. This field does not only constrain to RF engineers but also to the material engineers and basic physics researchers.

Table 1 Comparison of different UWB metamaterial antennas

Ref	Type of EBG	Frequency in GHz	Size (mm)	BW	Number of units cells	Via process
[10]	ELV EBG	3.74–8.8	32 × 32 × 1.0	5.06	n/a	n/a
[11]	Pin diode based	2.85–11.85	20 × 28 × 0.22	9.0	n/a	n/a
[12]	n/a	0.69–2.84	70 × 70 × 1.58	2.15	n/a	n/a
[13]	n/a	3.1–10.9	24 × 16 × 0.8	7.8	n/a	n/a
[14]	TELI EBG	3.0–12	23 × 29 × 1.52	9.0	n/a	n/a
[15]	n/a	3.4–11	25 × 28 × 0.7	7.6	n/a	n/a
[16]	Fractal EBG	0.5–11.3	23 × 6.0 × 0.8	10.8	4	Yes
[17]	Minkowsi EBG	0.5–5.5	14 × 6.2 × 0.8	5.0	4	Yes
[18]	Long-Li EBG	0.4–4.7	13.4 × 5.2 × 1.6	4.3	4	Yes
[19]	Bowtie EBG	2.25–4.7	10 × 6.9 × 0.8	2.45	2	Yes
[20]	n/a	5.75–5.85	29.3 × 26 × 1.6	0.09	2	Yes
[21]	W-chen EBG	2.78–5.16	50 × 30 × 1.6	2.38	1	No
	n/a	3.9–12.3	25 × 15 × 1.6	8.4	2	No

Many researchers have developed many applications and concepts for utilizing these MTMs concepts. There is a good scope for further research and can be used as a building area of research. So, there is a wide area of research development for creation of new technologies and future enhancement with the development of structured fabrication that offers exciting possibilities for the new component design of devices and salient features improvement.

References

1. C. Caloz, T. Ioth, *Electromagnetic Metamaterials: Transmission Line Theory and Microwave Applications* (Wiley-IEEE, Piscataway, 2005)
2. V. Veselago, The electrodynamics of substances with simultaneously negative values of ε and μ. Sov. Phys. Uspekhi **10**(4), 509–514 (1968)
3. D.R. Smith, W.J. Padilla, D.C. Vier, S.C. Nemat-Nasser, S. Schultz, Composite medium with simultaneously negative permeability and permittivity. Phys. Rev. Lett. **84**(18), 4184–4187 (2000)
4. R.W. Ziolkowski, A. Kipple, Causality and double-negative metamaterials. Phys. Rev. E **68**, 026615 (2003)
5. E. Nader, R.W. Ziolkowski, A positive future for double-negative metamaterials. IEEE Trans. Microw. Theory Tech. **53**, 1535–1556 (2005)
6. R. Paschotta (2008–18). *Photonic Metamaterials Encyclopedia of Laser Physics and Technology*, I & II (Wiley-VCH Verlag, 2009), p. 1
7. R.V. Craster, S. Guenneau (eds.), *Acoustic Metamaterials: Negative Refraction, Imaging, Lensing and Cloaking*, vol. 166 (Springer Science & Business Media, 2012)
8. M. Ashby, *Material Selection in Mechanical Design*, 4th edn. (Butterworth-Heinemann, Oxford, UK, 2010)
9. M.M. Islam, M.T. Islam, M. Samsuzzaman, M.R.I. Faruque, Compact Metamaterial Antenna for UWB Applications. Electron. Lett. **51**, 1222–1224 (2015)

10. D.S. Chandu, S.S. Karthikeyan, A novel broadband dual circularly polarized microstrip-fed monopole antenna. IEEE Tran. Antennas Propag. **65**(3), 1410–1415 (2017)
11. S. Nikolaou, M.A.B. Abbasi, Design and development of a compact UWB monopole antenna with easily-controllable return loss. IEEE Trans. Antennas Propag. **65**(4), 2063–2067 (2017)
12. Mingjian Li, Nader Behdad, A compact, capacitively fed UWB antenna with monopole-like radiation characteristics. IEEE Trans. Antennas Propag. **65**(3), 1026–1037 (2017)
13. M. Gulam Nabi Alsath, M. Kanagasabai, Compact UWB monopole antenna for automotive communications. IEEE Trans. Antennas Propag. **63**(9), 4204–4208 (2015)
14. M.S. Khan, et al., A compact CSRR-enabled UWB diversity antenna. IEEE Antennas Wireless Propag. Lett. **16**, 808–812 (2017)
15. M.R. Singha, D. Vakula, Directive beam of the monopole antenna using broadband gradient refractive index metamaterial for ultra-wideband application, in *IEEE Access*, vol. 5, (2017), pp. 9757–9763
16. M. Alibakhshi-Kenari, M. Naser-Moghadasi, Novel UWB miniaturized integrated antenna based on CRLH metamaterial transmission lines. Int. J. Electron. Commun. (AEÜ) **69**, 1143–1149 (2015)
17. R.A. Sadeghzadeh, Low profile antenna based on CRLH-TL with broad bandwidth. Microw. Opt. Technol. Lett. (MOTL) **58**(1), 27–31 (2016)
18. M. Alibakhshi-Kenari et al., Traveling-wave antenna based on metamaterial transmission line structure for use in multiple wireless communication applications. Int. J. Electron. Commun. (AEÜ) **70**, 1645–1650 (2016)
19. M. Alibakhshi-Kenari, M. Naser-Moghadasi, UWB miniature antenna based on the CRLH-TL with increasing the gain for advanced electromagnetic requirements. Adv. Electromagn. **3**(1), 61–65 (2014)
20. A.A. Ibrahim, M.A. Abdalla, CRLH MIMO antenna with reversal configuration. Int. J. Electron. Commun. (AEÜ) **70**, 1134–1141 (2016)
21. S.K. Sharma, R.K. Chaudhary, A compact zeroth-order resonating wideband antenna with dual-band characteristics. IEEE Antennas Wirel. Propag. Lett. **14**, 1670–1672 (2015)

Investigating Combinational Dispersion Compensation Schemes Using DCF and FBG at Data Rate of 10 and 20 Gbps

Md. Asraful Sekh, Mijanur Rahim and Abdul Touhid Bar

Abstract We have investigated combinational dispersion compensation techniques using dispersion compensation fiber (DCF) and fiber Bragg grating (FBG) for the data rates of 10 and 20 Gbps. The performance characteristics in terms of eye diagrams, Q factors, gains of received signals are obtained and analyzed for different system configurations. Pre-, post-, and symmetrical-compensation techniques using DCF are well-known techniques of fiber dispersion compensation. FBG is also used for this purpose. The main problem faced in these schemes is either low output power and/or high signal distortion at moderate or higher data rate. Though erbium-doped fiber amplifier (EDFA) to reduce the signal attenuation is generally utilized, in the symmetrical dispersion compensation using DCF, output signal is, however, distorted enough to retrieve the signal. In this paper, we have proposed combinational dispersion compensation schemes using DCF and FBG together. A combinational dispersion compensation scheme over 25–100 km optical link at 10 Gbps and over 25–65 km link at 20 Gbps are investigated. Results show improved performance parameters compared to other schemes.

Keywords Dispersion compensation technique · Dispersion compensation fiber (DCF) · Fiber Bragg grating (FBG) · Erbium-doped fiber amplifier (EDFA)

1 Introduction

In optical fiber communication system, information is transmitted by sending light through an optical fiber. An EM carrier wave in the form of light is modulated to carry the information. Among several advantages, the huge bandwidth and the low cost are essential in higher data rates communication [1]. The chromatic dispersion or group velocity dispersion (GVD) adversely affect high data rate fiber optic communication

Md. A. Sekh (✉) · M. Rahim · A. T. Bar
Optical Fiber Communication and Optical Networks Lab, Department of Electronics and Communication Engineering, Aliah University, IIA/27, New Town, Kolkata 700160, India
e-mail: asekh@yahoo.com

M. Rahim
e-mail: mijanur.ece@gmail.com

H. S. Saini et al. (eds.), *Innovations in Electronics and Communication Engineering*, Lecture Notes in Networks and Systems 107,
https://doi.org/10.1007/978-981-15-3172-9_28

system using single-mode optical fibers [2]. The dispersed signal creates problem when the broadening pulses begin to overlap, and as a result of this, there is ambiguity of data retrieval which is known as intersymbol interference (ISI) [1–4].

In this paper, an investigation of combinational dispersion compensation technique using dispersion compensation fiber (DCF) and fiber Bragg grating (FBG) for the data rates of 10 and 20 Gbps has been carried out. The performance characteristics in terms of eye diagrams, *Q* factors, and gains of received signals are obtained and analyzed. Results are compared with the other two techniques of dispersion compensation using DCF and FBG individually. Schemes using DCF for three different configurations as pre-, post-, and symmetrical-compensation techniques are usually employed for dispersion compensation [5, 6]. FBG is also used for to achieve high gain. The main problem faced in these techniques is either low quality factor or low gain at moderate or higher data rate. Though erbium-doped fiber amplifier (EDFA) to reduce the signal attenuation is generally used, in the symmetrical dispersion compensation using DCF, output signal is, however, distorted enough to retrieve the signal. An enhanced performance has been achieved by the proposed combinational dispersion compensation technique using DCF and FBG together. Systems using DCF only and FBG only with several grating lengths were analyzed earlier. A combinational dispersion compensation scheme over 25–100 km and 25–65 km optical link is investigated for data rates of 10 Gbps and 20 Gbps, respectively. A comparison between the proposed scheme with conventional schemes is made in terms of few tables and figures.

2 Mathematical Formulation

The FBG operation follows Fresnel reflection, where light travels between media having different refractive indices and may both reflect and refract at the interface. The reflected wavelength is λ_b, which is known as the Bragg wavelength [1, 2].

$$\lambda_b = 2n_e\Lambda. \tag{1}$$

where n_e is effective refractive index of the grating in the fiber core and Λ is grating period and

$$\Delta\lambda = \left[\frac{2\delta n_0 n}{\pi}\right]\lambda_b. \tag{2}$$

where $\Delta\lambda$ is bandwidth, δn_0 is difference between the two refractive indexes ($n_2 - n_1$), and n is the fractional power in the fiber core.

$$P_B(\lambda_b) \approx \tanh^2\left[\frac{Nn(V)\delta n_0}{n}\right]. \tag{3}$$

where $P_B(\lambda_B)$ is the peak reflection power and N is the number of periodic variations and V is the V number of the fiber.

DCF is mostly used for chromatic dispersion compensation. The idea behind using DCF is sequenced with the SMF having opposite signs of dispersion coefficients. The equivalent wavelength-dependent dispersion can be clarified through [7]

$$D_{eq}(\lambda) = [RD_{SMF}(\lambda) + D_{DCF}(\lambda)]/(1 + R). \tag{4}$$

where R is the length ratio between SMF and DCF, D_{SMF} and D_{DCF} are the dispersion coefficients of the SMF and the DCF, respectively. From Eq. (4), when D_{eq} becomes zero, one can get the ideal compensation condition and for $R = L_{SMF}/L_{DCF}$, Eq. (4) would become [8–10].

$$D_{SMF}L_{SMF} + D_{DCF}L_{DCF} = 0. \tag{5}$$

For externally modulated sources, transmission distance limited by chromatic dispersion is

$$L < \frac{2\pi C}{16|D|\lambda^2 B_T^2}. \tag{6}$$

where L is length of fiber in km, c is speed of light (m/s), λ is wavelength in meter, and B_T is Bit rate in Gbps.

3 Simulation Model and Parameter Settings

Figure 1 shows the block diagram used in the simulation of combinational dispersion

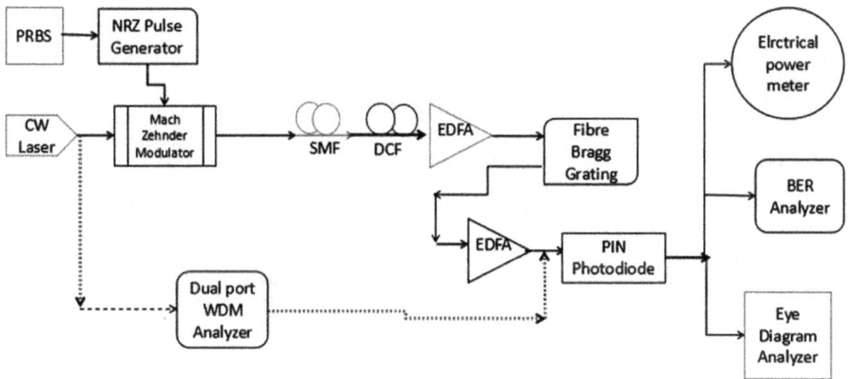

Fig. 1 Dispersion compensation using DCF and FBG

compensation schemes using DCF and FBG over 100 km optical link at data rate of 10 and 20 Gbps. The parameters used for simulation are—CW laser power of 5 dBm, extinction ratio of Mach–Zehnder modulator of 30 dB, transmission distance up to 100 km, attenuation of SMF 0.2 dB/km, DCF dispersion 16.75 ps/nm/km, and dispersion slope 0.075 ps/nm/km. The length of DCF is 5 km, and the effective index of 1–5 mm length FBG is 1.45. Length of EDFA is 5 m with a numerical aperture of 0.24.

4 Results and Discussion

Three different schemes of dispersion compensation using DCF only, FBG only, and DCF and FBG together has been carried out for comparative performance analysis. Figures 2 and 3 show comparative illustration of Q factor variation with transmission distance at data rate of 10 Gbps and 20 Gbps, respectively. It is observed that the combinational scheme results higher Q factor at both data rates. Figures 4 and 5 show a comparative illustration of gain variation with transmission distance at data rate of 10 Gbps and 20 Gbps, respectively. It is observed that combinational scheme improves gain compared to scheme using DCF only but degrades compared to scheme using FBG only at both data rates. Similar observation can be drawn when output power is considered at both data rates. Figure 6 shows the eye diagrams for DCF-FBG combinational scheme at 10 Gbps and 20 Gbps respectively showing better

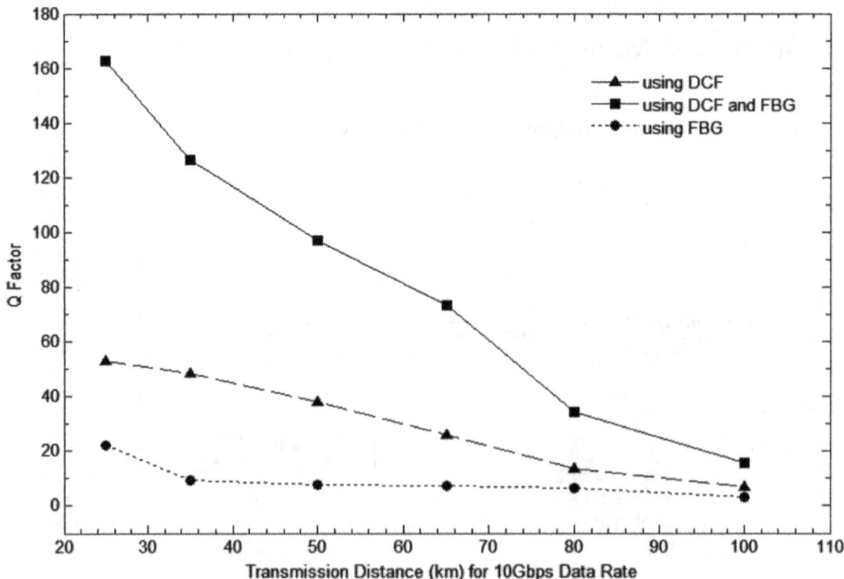

Fig. 2 Q factor versus transmission distance (km) for 10 Gbps data rate

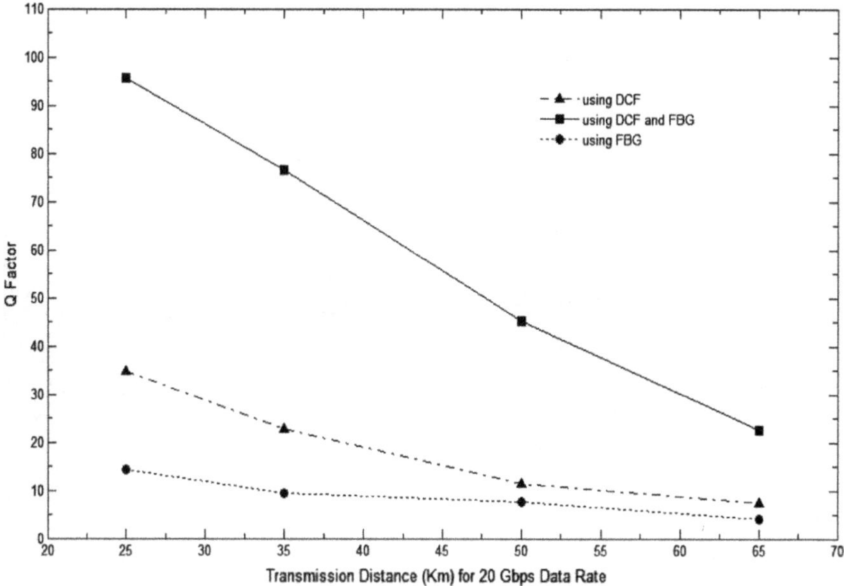

Fig. 3 *Q* factor versus transmission distance (km) for 20 Gbps data rate system

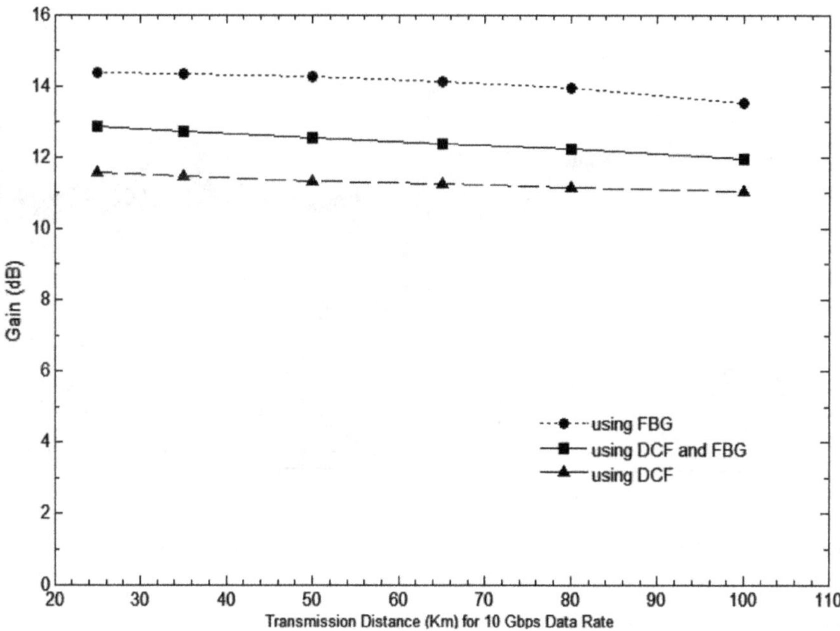

Fig. 4 Gain (dB) versus transmission distance (km) for 10 Gbps data rate

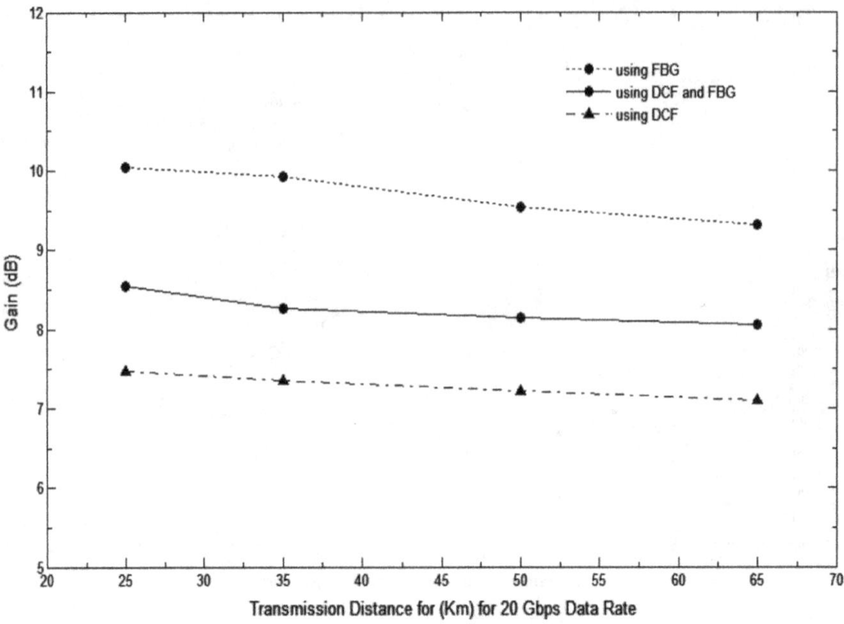

Fig. 5 Gain versus transmission distance (km) for 20 Gbps data rate

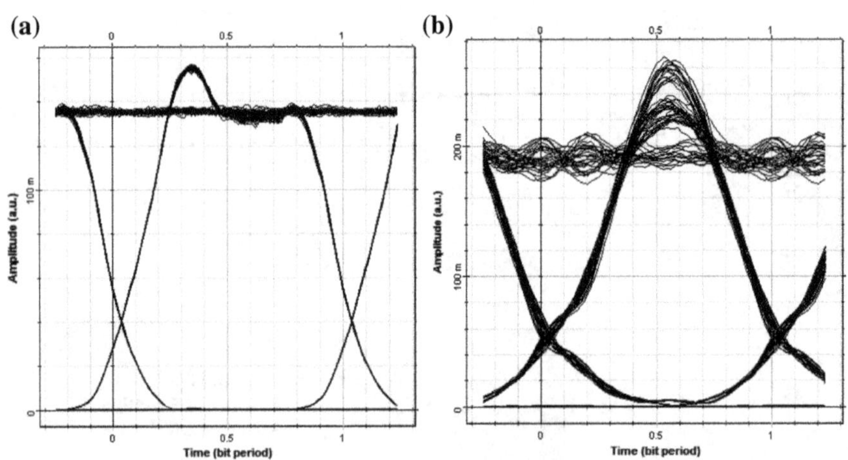

Fig. 6 Eye diagrams of DCF-FBG combinational schemes at (a) 10 and (b) 20 Gbps

results at 10 Gbps in terms of eye opening, eye height and noise performance. Better performance parameters in terms of Q factor and link distance enhancement are achieved both at 10 and 20 Gbps system when DCF-FBG combinational scheme is used. Tables 1 and 2 summarize the detailed results of all three schemes.

Table 1 Performance parameters for three schemes at 10 Gbps

Performance characteristics	Quality factor			Output gain (dB)			Output power (dBm)		
Transmission distance (km)	Using DCF and FBG	Using DCF	Using FBG	Using DCF and FBG	Using DCF	Using FBG	Using DCF and FBG	Using DCF	Using FBG
20	162.62	53.17	22.32	12.87	11.56	14.38	17.86	16.95	19.37
35	126.69	48.37	9.53	12.73	11.48	14.35	17.69	16.67	19.35
50	96.97	38.22	7.75	12.54	11.32	14.27	17.52	16.35	19.26
65	73.46	25.96	7.41	12.39	11.25	14.14	17.38	16.28	19.14
80	34.45	13.59	6.55	12.24	11.16	13.95	17.24	16.16	18.95
100	15.38	6.88	3.276	11.97	11.03	13.54	16.97	16.05	18.54

Table 2 Performance parameters for three schemes at 20 Gbps

Performance characteristics	Quality factor			Output gain (dB)			Output power (dBm)		
Transmission distance (km)	Using DCF and FBG	Using DCF	Using FBG	Using DCF and FBG	Using DCF	Using FBG	Using DCF and FBG	Using DCF	Using FBG
25	95.6	34.87	14.41	8.54	7.47	10.04	16.72	20.03	20.03
35	76.5	22.83	9.34	8.26	7.36	9.92	16.34	19.91	19.91
50	45.23	11.34	7.57	8.14	7.22	9.54	16.12	18.54	18.54
65	22.47	7.46	4.02	8.05	7.11	9.32	15.89	17.32	17.32

5 Conclusion

We have investigated three dispersion compensation schemes using DCF only, using FBG only, and combinational scheme using DCF-FBG together at different transmission distance and different data rates. The simulated transmission system has been analyzed on the basis of eye diagram, quality factor, gain, and output power. It is found that the DCF-FBG combinational scheme shows superior results compared to other two dispersion compensation schemes when high data rate and longer link distance are considered.

References

1. G. Keiser, *Optical Fiber Communications* (Wiley Online Library, Hoboken, 2003)
2. G. Agrawal, *Applications of Nonlinear Fiber Optics* (Academic Press, Cambridge, 2001)
3. Hu, B., Jing, W., Wei, W., Zhao, R.: Analysis on dispersion compensation with DCF based on Optisystem, in *2010 2nd International Conference on Industrial and Information Systems (IIS)*, vol. 2 (IEEE, 2010), pp. 40–43
4. R. Kashyap, *Fiber Bragg Gratings*, 2nd edn. (Academic Press, Cambridge, 2010)
5. M. Rahim, A.T. Bar, A. Begam, M.A. Sekh, Investigation of dispersion compensation methods for the data rates 2.5 and 10 Gbps using standard and dispersion compensated fibers. Int. J. Emerg. Technol. Innov. Res. (IJETIR) **6**(5), 731–734 (2019)
6. M. Rahim, A.T. Bar, A. Begam, M.A. Sekh, Fiber optic link design for 10 Gbps system and its performance characteristics, in *National Conference on Atomic, Molecular and Nano Sciences (NCAMNS-2019)* at Aliah University, 3–4 Apr 2019. https://doi.org/10.13140/rg.2.2.11431.39842/1
7. C.-C. Chang, A.M. Vengsarkar, D.W. Peckham, A.M. Weiner, Broadband fiber dispersion compensation for sub-100-fs pulses with a compression ratio of 300. Opt. Lett. **21**(15), 1141–1143 (1996)
8. T. Xu, G. Jacobsen, S. Popov, J. Li, S. Sergeyev, A.T. Friberg et al., Analysis of chromatic dispersion compensation and carrier phase recovery in long-haul optical transmission system influenced by equalization enhanced phase noise. Optik-Int. J. Light Electron Opt. **138**, 494–508 (2017)
9. T.F. Hussein, M.R.M. Rizk, M.H. Aly, Opt. Quant. Electron. **51**, 103 (2019). https://doi.org/10.1007/s11082-019-1823-y
10. Liu, Z., Xu, T., Saavedra, G., Bayvel, P.: 448-Gb/s PAM4 transmission over 300-km SMF-28 without dispersion compensation fiber, in *Optical Fiber Communication Conference*, p. W1 J.6. Optical Society of America, San Diego, California (2018)

Embedded Systems

A Low-Power FinFET-Based Miller Op-Amp Design with g_m Enhancement and Phase Compensation

Mohammed Kursheed, C. H. Kiran Kumar and Ravindrakumar Selvaraj

Abstract This paper presents the investigation, design, and implementation of a highly efficient and low-power op-amp design using multi-gate device. The op-amp is based on class-AB–AB Miller structure. The multi-gate-based opamp design utilizes the gm enhancement through local common-mode feedback. The combination of both Miller and phase-lead compensation is achieved. The problem with the CMOS leakage is addressed. From the analysis, it has been found that the FinFET-based design is better in performance than CMOS.

Keywords CMOS · FinFET · Op-amp · Miller op-amp · Leakage · Low power · Class AB–AB

1 Introduction

The operational amplifier (op-amp) becomes the important block in analog acquisition unit. From arithmetic unit to filtering unit, the op-amp is required. The use of op-amp is to provide higher gain, low output impedance, and higher output swing. For higher output voltage swing, 2-stage Miller op-amps can be used. Through cascading the transistors, the resistance will have output swing and is suitable for open-loop operation. Varying the supply voltage will affect the CL and the phase margin. The slew rate will have symmetrical positive and negative values. To stabilize the CL values, Miller compensated op-amps are used. The internal dominant pole decreases their PM as CL grows. The two-stage class-A conventional Miller op-amp circuit is shown in Fig. 1. The higher static power consumption will increase the parameters

M. Kursheed (✉) · C. H. Kiran Kumar
Malla Reddy College of Engineering and Technology, Hyderabad, Telangana, India
e-mail: kursheed012@gmail.com

C. H. Kiran Kumar
e-mail: ckkmtech11@gmail.com

R. Selvaraj
Sri Shakthi Institute of Engineering and Technology, Coimbatore, Tamil Nadu, India
e-mail: gsravindrakumar7@gmail.com

H. S. Saini et al. (eds.), *Innovations in Electronics and Communication Engineering*, Lecture Notes in Networks and Systems 107,
https://doi.org/10.1007/978-981-15-3172-9_29

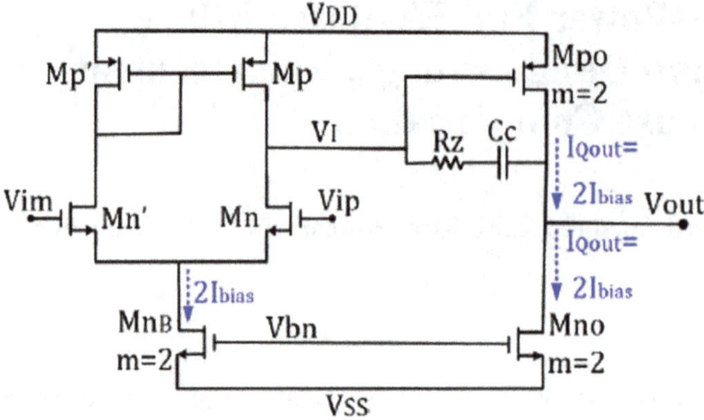

Fig. 1 One kernel conventional two-stage class-A Miller op-amp

of the slew rate. Qu et al. [1] reported a Design- Oriented Analysis (DOA) method. The method is based on the Miller compensation scheme.

The CL values only have stable relative narrow growth with the Miller compensated op-amps with internal dominant pole that will decrease their PM as CL grows. In Fig. 1, the two stages of class-A conventional Miller op-amp with some highly nonsymmetrical SR are presented. The nMOS (pMOS) input stage is with two-stage Miller and has negative SR − (SR+) that is in general limited to an approximate value $SR - = IQ_{out}/(C_L + C_c) = 2I_{bias}/(C_L + C_c)$. Where C_c is the Miller compensation capacitor, and I_{bias} and IQ_{out} are the bias current of the unit transistors and the output branch. The higher static power consumption of expense will increase IQ_{out} SR.

2 Literature Survey

Surkanti et al. [2] presented the class-AB amplifier of low-voltage low-transistor-count wide swing multistage pseudo-configuration. Mita et al. presented a true-class-AB amplifier. Sutula et al. [3] presented a circuit based on class-AB operational trans-conductance amplifier (OTA) with nonlinear current amplifiers. Lopez-Martin et al. [4] presented OTA to drive larger capacitive loads to uplift the single-stage folded cascade operational trans-conductance amplifier. The circuits are adaptively biased with the input differential pair. To increase the gain–bandwidth (GBW) product, the current folding stage in class-AB operation with dynamic current boosting is implemented. Pourashraf et al. [5] presented super class-AB op-amps for extremely high slew-rate improvement. It avoids open-loop gain degradation. The OTA with improved output current and slew rate is presented in the literature [5]. It operates in class-AB and has higher gain–bandwidth product as reported. The input stage is dynamically biased with cascade transistors with adaptive biasing. A buffer amplifier for LCDs should have faster driving capability and rail-to-rail common-mode input voltage range [6]. This high-performance device [7] of ultra-low-power amplifier based on three-stage CMOS OTAs was effective. By increasing the number of stages, the Miller capacitance can be reduced or removed [8]. But the power and area should be optimized and the Miller compensation was employed to drive huge capacitive load. CMOS-based operational amplifier with rail-to-rail voltage of 0.9-V, 0.5-μA was presented by Stockstad and Yoshizawa [9]. The low-voltage input stage of the CMOS amplifier uses the body effect for threshold voltage modulation. Grasso et al. [10] presented a biasing scheme for CMOS differential amplifier with low supply voltage and strong inversion mode transistors. Monsurro et al. [11] describe the biasing technique for minimum supply CMOS amplifiers for bulk terminals. Aguado-Ruiz et al. [12] presented the measurement results and post-layout simulation of buffers. Figueiredo et al. [13] presented the differential CMOS amplifier with self-biasing techniques. Cabrera-Bernal et al. [14] presented the operational trans-conductance amplifiers (OTAs) for bulk-driven high-performance architecture. The designs are mainly based on CMOS, and less research are carried out for FinFET devices [15]. The operational amplifiers are implemented in multigate devices play a vital role in several applications [16, 17]. The work on FinFET devices for DWT architecture and memory have been increased in recent years [18, 19].

3 Existing Methodology

3.1 Circuit Operation

In the past, several architectures were implemented for op-amp using differential pair and shell based blocks. One such architecture is shown in Fig. 2a. The equivalent transistor level implementation is presented in Fig. 2b. It can be observed that the first stage constitutes the differential pair which makes it a composite unit along with a resistive block. The LCMFB is a resistive composite stage. The stage is connected with a shell circuit formed by transistors Mp1, Mp1', Mn1, and Mn1'.

It can be known that $(r_o m_p \| r_o M_n) \gg R$. So the $(V_X - V_{X'})$ is given by

$$V_X - V_X' = A_X\left(V_{ip} - V_{im}\right) = A_x V_i = g_{mD} R V_i. \tag{1}$$

The above is the small-signal differential voltage with V_i the differential input voltage. The current at the output of first stage is

$$I_{outI} = A_x g_{mp1} V_i = g_{meff} V_i. \tag{2}$$

The composite first stage has the effective trans-conductance proportional to the gain A_I of the input stage $g_{meff} = A_x g_{mp1}$

$$A_i = g_{meff}\ R_{oI=} g_{meff} r_o M_p 1 \| r_o\ M_{nI}. \tag{3}$$

(a)

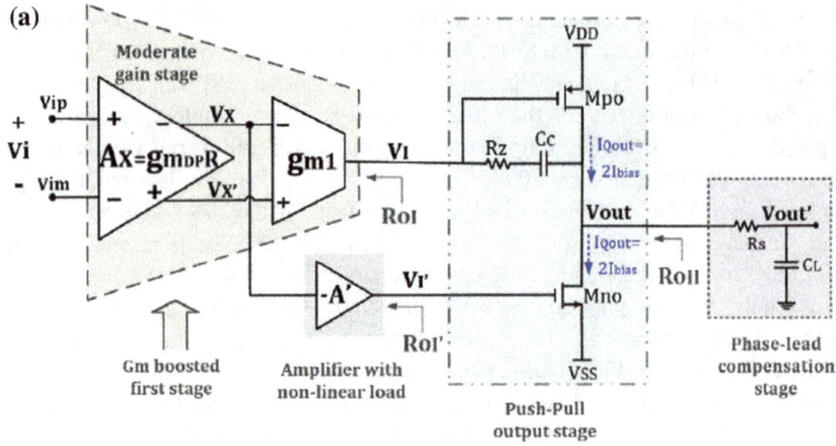

(b)

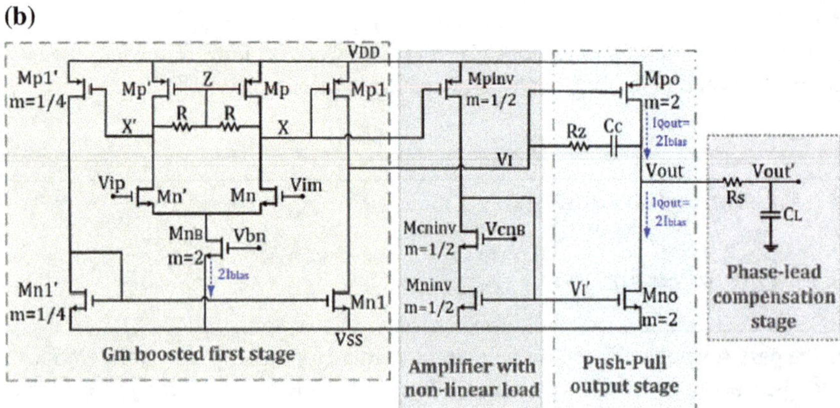

Fig. 2 **a** Architecture of the proposed class-AB–AB op-amp. **b** Transistor level implementation

The resistance at output node of the composite stage is $R_{\text{oI}} = r_{\text{o}}M_{\text{p1}}||r_{\text{o}}M_{n1}$. So the effective trans-conductance is improved by the gain factor A_{X}. The gain factor A_{x} is due to linear load R. The resistance value boosts the g_{m} value. The peak currents in the shell transistors are maximized by the voltage variations $(V_{\text{X}} - \text{V}_{\text{X}}^{'})$ which is directly proportional to the I_{bias} and R. The peak currents are greater than I_{bias}. So at node $v1$ of class-AB circuit shows a maximum current $I_{\text{out}}\text{I}_{\text{MAX}} \gg 2I_{\text{bias}}$. A push–pull in output will limit the current. The pMOS transistors at the output are driven by the V_I voltage. The quiescent voltage of amplifier A generates a quiescent current $I_{\text{Qout}} = 2I_{\text{bias.}}$ For small signal, the gain and output impedance are low, and for large input signal, it provides large output impedance.

It is assumed that the pMOS and nMOS transistors have the same trans-conductance gain $g_{\text{m.}}$ The c open-loop gain is given by

$$A_{\text{OLDC}} = g_{\text{meff}} R_{\text{oI}} g_{\text{mII}} R_{\text{oII}} = g_{\text{mDR}} R(g_{\text{m}} r_{\text{O}}/2) \tag{4}$$

At high-frequency pole as nodes X and X' which is introduced in the class-AB-AB amplifier, the g_m is boosted at the first stage. Thus, to improve the PM a phase-lead compensation resistor R_s in addition to Miller compensation is connected between the output V_{out} and C_L that is required in order to add a left half-plane (LHP) zero.

The parasitic capacitance at node X (C_X) is calculated as,

$$C_X = C_{gsMp1} + C_{gsMpinv} + C_{gdMp} + C_{gdMn} + C_{dbMn} + C_{dbMp}. \quad (5)$$

3.2 Operation of the Nonlinear Load

The boundary between triode and saturation is in closed saturation that operates in transistor M_{ninv}. Under quiescent conditions, the cascade voltage $V_{cn}B$ is selected in it. Thus, M_{ninv} and Mno operate as a current mirror at DC. The quiescent output current in Mno is the same as the quiescent current in Mpo ($I_{Qout} = 2I_{bias}$). To minimize the power, the size of transistors is scaled down. For small-signal current, the amplifier $A^{'}$ has unity gain magnitude, and M_{ninv} has an impedance of 1/gm. The value of 1/gm is low and leads to negative output current for large signals. For nonlinear load, M_{ninv} enters triode operation and develops large voltage variations at node $V_I^{'}$.

4 Proposed Methodology

4.1 FinFET Technology

The channel length below 90 nm will face problems in a Complementary Metal Oxide Semiconductor (CMOS) technology such as gate leakage current, under the threshold leakage current and drain to induce the lowering barrier (DIBL) current. These problems in the CMOS technology forced the scientist to find new devices. The alternate devices require higher mobility, greater stability, and the scalability against the process. Hence, the short-channel effects happen when scaling of deep submicron transistors for memory are done. The short-channel effects, power supply variations, and threshold changes will affect the functionality of the logics.

For the design of high-performance circuits in signal processing and memory, CMOS devices have many challenges below 45 nm technology. The FinFET device is the alternate solution to solve the problems faced in the CMOS technology. The parameter values used in this work are shown in Table 1. 32 nm CMOS and FinFET technology using predictive technology models were used for simulation (Fig. 3).

Table 1 Parameter values of the proposed circuit Fig. 2b

Parameter	Value	Parameter	Value
$(W/L)_{NMOS}$ (μm/μm)	20/0.7	C_c (pF)	10
$(W/L)_{NMOS}$ (μm/μm)	100/0.7	C_X (pF)	0.4
g_{mDP} (μA/V)	109	R_z (KΩ)	6
g_{mP1} (μA/V)	137	R_{OI} (KΩ)	435
g_{mP0} (μA/V)	428	g_{mnII} (μA/V)	866
g_{mn0} (μA/V)	438	R_{oII} (KΩ)	125
g_{meff} (μA/V)	1724	g_{mpinv} (μA/V)	73.8
R (KΩ)	150	G_{mninv} (μA/V)	74.3

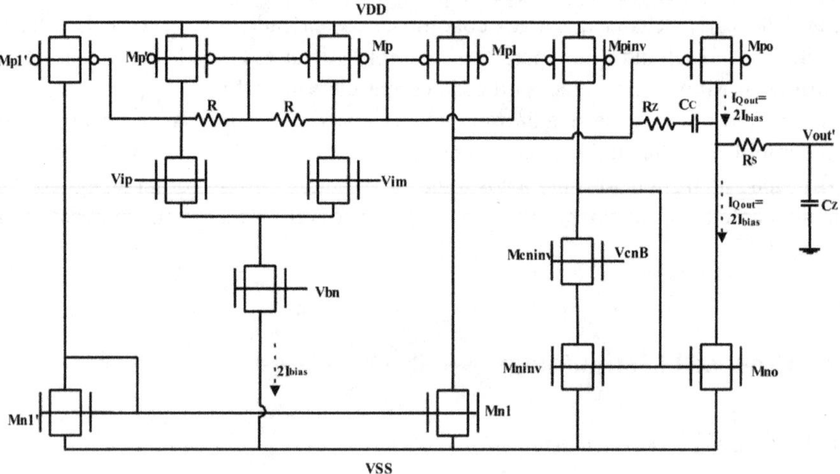

Fig. 3 Implementation of class-AB–AB op-amp using FinFET

5 Result and Discussion

The circuits are designed in 32 nm CMOS and FinFET technology. SPICE simulation was used. The circuit is designed with dual supplies $V_{DD} = +0.9$ V and $V_{SS} = -0.9$ V, threshold voltages $V_{th}P \approx V_{th}N \approx 0.45$ V, and $I_{bias} = 5$ μA. Figure 2b shows the present circuit designed in 180-nm CMOS technology. SPICE simulation is used. The circuit is designed with dual supplies $V_{DD} = +0.9$ V and $V_{SS} = -0.9$ V, threshold voltages $V_{th}P \approx V_{th}N \approx 0.45$ V, and $I_{bias} = 5$ μA. In Table 2 shows the transistors' dimensions, output resistances, and small-signal trans-conductance gains (R_{oI}, R_{oII}, and g_m). Table 2 gives the power analysis of the design. Proposed amplifier gives

Table 2 Comparison table of CMOS- and FinFET-based circuits

	Avg power	Peak power	Avg current	Peak current
Conventional	1.3851E−04	6.4035E−03	7.5373E−05	2.2801E−01
Existing	2.1254E−02	8.3194E−02	−1.5181E−02	−1.3871E−02
Proposed	1.0667E−04	8.8211E−04	−7.5643E−05	−1.2260E−05

the minimum power consumption. The implementation of class-AB–AB op-amp using FinFET utilizes less average power when compared to the CMOS technology (Table 2).

6 Conclusion

Implementation of a highly efficient and low-power op-amp design using multi-gate device based on class-AB–AB Miller structure is presented. The gm enhancement through local common-mode feedback is implemented. The CMOS–FinFET-based design is compared. From the results, it has been found that the proposed method consumes less power compared to the existing method.

References

1. W. Qu, S. Singh, Y. Lee, Y.S. Son, G.H. Cho, Design oriented analysis for miller compensation and its application to multistage amplifier design. IEEE J. Solid-State Circ. **52**(2), 517–527 (2017)
2. P.R. Surkanti, P.M. Furth, Converting a three-stage pseudoclass AB amplifier to a true-class-AB amplifier. IEEE Trans. Circ. Syst. II Exp. Briefs **59**(4), 229–233 (2012)
3. S. Sutula, M. Dei, L. Terés, F. Serra-Graells, Variable-mirror amplifier: a new family of process-independent class-AB single-stage OTAs for low-power SC circuits. IEEE Trans. Circ. Syst. I Reg. Papers **63**(8), 1101–1110 (2016)
4. A. Lopez-Martin, M.P. Garde, J.M. Algueta, A. Carlos, R.G. Carvajal, J. Ramirez-Angulo, Enhanced single-stage folded cascade OTA suitable for large capacitive loads. IEEE Trans. Circ. Syst. II, Exp. Briefs **65**(4), 441–445 (2018)
5. S. Pourashraf, High current efficiency class-AB OTA with high open loop gain and enhanced bandwidth. IEICE Electron. Exp. Lett. **14**(17), 2017–0719 (2017)
6. A.D. Grasso, D. Marano, F. Esparza-Alfaro, A.J. Lopez-Martin, G. Palumbo, S. Pennisi, Self-biased dual-path push-pull output buffer amplifier for LCD column drivers. IEEE Trans. Circ. Syst. I Reg. Papers **61**(3), 663–670 (2014)
7. A.D. Grasso, D. Marano, G. Palumbo, S. Pennisi, Design methodology of subthreshold three-stage CMOS OTAs suitable for ultra-low-power low-area and high driving capability. IEEE Trans. Circ. Syst. I Reg. Papers **62**(6), 1453–1462 (2015)
8. M. Tan, W.-H. Ki, A cascade Miller-compensated three-stage amplifier with local impedance attenuation for optimized complex-pole control. IEEE J. Solid-State Circ. **50**(2), 440–449 (2015)
9. T. Stockstad H. Yoshizawa, A 0.9-V 0.5 μm rail-to-rail CMOS operational amplifier. IEEE J. Solid-State Circ. **37**(3), 286–292 (2002)

10. A.D. Grasso, P. Monsurró, S. Pennisi, G. Scotti, A. Trifilietti, Analysis and implementation of a minimum-supply body-biased CMOS differential amplifier cell, IEEE Trans. Very Large Scale Integr. (VLSI) Syst. **17**(2), 172–180 (2009)
11. P. Monsurró, G. Scotti, A. Trifiletti, S. Pennisi, Biasing technique via bulk terminal for minimum supply CMOS amplifiers. Electron. Lett. **41**(14), 779–780 (2005)
12. J. Aguado-Ruiz, A. Lopez-Martin, J. Lopez-Lemus, J. Ramirez-Angulo, Power efficient class AB op-amps with high and symmetricals low rate. IEEE Trans. Very Large Scale Integr. (VLSI) Syst. **22**(4), 943–947 (2014)
13. M. Figueiredo, R. Santos-Tavares, E. Santin, J. Ferreira, G. Evans, J. Goes, A two-stage fully differential inverter-based self-biased CMOS amplifier with high efficiency. IEEE Trans. Circ. Syst. I Reg. Papers **58**(7), 1591–1603 (2011)
14. E. Cabrera-Bernal, S. Pennisi, A.D. Grasso, A. Torralba, R.G. Carvajal, 0.7-V three-stage class-AB CMOS operational transconductance amplifier. IEEE Trans. Circ. Syst. I Reg. Papers **63**(11), 1807–1815 (2016)
15. V.M. Senthilkumar, S. Ravindrakumar, A low power and area efficient FinFET based approximate multiplier in 32 nm technology, in *Springer-International Conference on Soft Computing and Signal Processing* (2018)
16. S. Pourashraf, J. Ramirez-Angulo, A.J. Lopez-Martin, R.G. Carvajal, Super class AB OTA without open-loop gain degradation based on dynamic cascade biasing. Int. J. Circ. Theory Appl. **45**(12), 2111—2118 (2017)
17. V.M. Senthilkumar, A. Muruganandham, S. Ravindrakumar, N. S. Gowri Ganesh, FINFET operational amplifier with low offset noise and high immunity to electromagnetic interference. Microprocess. Microsyst. **71**, 102887 (2019)
18. V.M. Senthilkumar, S. Ravindrakumar, D. Nithya, N.V. Kousik, A vedic mathematics based processor core for discrete wavelet transform using FinFET and CNTFET technology for biomedical signal processing. Microprocess. Microsyst. **71**, 102875 (2019)
19. K. Prasanth, M. Ramireddy, T. Keerthi priya, S. Ravindrakumar, High speed, low matchline voltage swing and search line activity TCAM cell array design in 14 nm FinFET technology. In: *Emerging Trends in Electrical, Communications, and Information Technologies*, vol 569, ed. by T. Hitendra Sarma, V. Sankar, R. Shaik (Springer, Singapore, 2020). Lecture Notes in Electrical Engineering

A Solar Tracking and Remote Monitoring System Using IoT

Fariha Khatoon and Sandeep Kumar

Abstract In this paper, for smart management and control of solar tracking system, a prototype is built to test or check the management and control of the system. The newly build prototype is developed for many applications with the aim of being a powerful tool for the learning of the smart solar energy system. The Internet of Things incorporates everyday objects using the Internet to extend into the real world. Here, we facilitate IoT technology for supervising solar PV (photovoltaic) power generation which can enhance the maintenance, monitoring, and performance of the plant. This will provide tracking of the solar panel and turning it in the direction of the sunlight. This is all possible using LDR sensors. The IoT automatically keeps track of the amount of voltage supply received by the solar panel in direction of sunlight. The proposed system displays the usage of the power of solar PV online. Finally, its application is discussed further.

Keywords Arduino · LDR · Solar PV · IoT

1 Introduction

Several technologies recently enabled trends of Internet of Things (IoT) which includes the availability of ubiquitous wireless communication, sensing capabilities, communication, onboard computing with low power devices, low cost, and small form factor. IoT has many applications which diverse such as smart grids, smart cities, transportation, habitat monitoring, environmental, industrial automation, office, home, health care, and medical [1].

77.9% of electricity is generated using fossil fuel and nuclear by the traditional method. These methods are the causes of global warming and climate change and are the heavily polluting environment. We may have heavy natural disasters due to

F. Khatoon (✉) · S. Kumar
Department of Electronics and Communications, Sreyas Institute of Engineering and Technology, Hyderabad, Telangana, India
e-mail: farihakhatoon@gmail.com

S. Kumar
e-mail: er.sandeepsahratia@gmail.com

H. S. Saini et al. (eds.), *Innovations in Electronics and Communication Engineering*, Lecture Notes in Networks and Systems 107,
https://doi.org/10.1007/978-981-15-3172-9_30

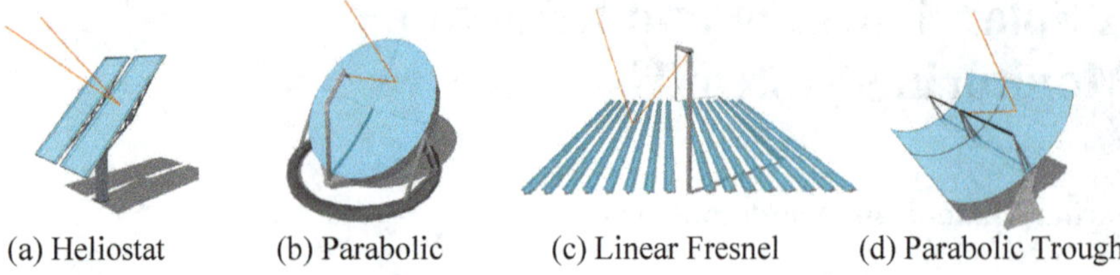
(a) Heliostat (b) Parabolic (c) Linear Fresnel (d) Parabolic Trough

Fig. 1 Solar concentrator systems

the usage of fossil energy sources. Therefore, we need to go with eco-friendly energy (solar, wind, hydropower, and geothermal) sources even more than that we are using now. The solar energy provides more efficiency than any other renewable energy and is harmless to the ecosystem. To get more efficiency from the solar panel, different technologies have been researched and one of them is by solar tracking system which controls the system by aligning it with the sun which makes solar tracking key for getting more efficiency [2]. Our aim is to provide a powerful system for solar tracking fields with less cost and easy implementation.

2 Literature Survey

2.1 Solar Concentration Technology

There are different types of renewable energies such as biomass, waste energy, wind energy, hydropower, ocean energy, solar energy, and geothermal energy [3]. The technologies which solar energy employs reach in three different ways: photophysically, photochemically, and thermally which are photovoltaic, photosynthesis, and heat engine or process heating [4].

To obtain the above task, we use lenses or mirror to concentrate solar flux on the receiver [5]. The mirrors are of four type's parabolic trough, heliostat, dish, or linear Fresnel as shown in Fig. 1 [6]. There are two main technologies for solar concentration: concentrated photovoltaic (CPV) and concentrated solar thermal power (CSP). In the last few decades, the scientists and researchers have produced the combination of both CSP and CPV which is said as concentrated photovoltaic thermal (CPVT) [7]. The CPVT is more potential due to its unique features [8].

2.2 Solar Tracking

With enough precision and by means of solar position algorithm, we can define the change in direction of the sun because the sky is moving and earth is in constant rotation [9]. The issue for system efficiency is the system demand for the mechanism for the system alignment with the sun [10]. Regarding the rotation and position, solar trackers are classified. Linear concentration system is the system which has like parabolic troughs or linear Fresnel [11]. It has subcategories such as tilted-axis systems, horizontal and vertical.

Whereas point-focus system which has two rotation axes is employed with dish modules and heliostats [12]. Depending on the rotation axis position, the subcategories are polar trackers, target aligned, and Azimuth–elevation [13]. Regarding control type, solar trackers are classified into two types: passive and active solar trackers [14]. Based on thermal expansion, a couple of actuators are composed which work against each other are said to be passive solar tracker [15]. Where on the other hand active trackers can be classified in electro-optical sensor data and time-controlled PC based, auxiliary bi-facial solar cell and microprocessor based, and fourth is the combination of all three [16].

The closed-loop and open-loop controllers both can be found with respect to control of active solar trackers [17]. In the literature survey, we studied about solar equations or a hybrid, bi-facial cells, electro-optical sensors, maximum beam, control based on continuous movements, closed-loop, open-loop, and one- and two-axis tracking systems [18]. Even with these many variations in the solar tracking system, there is no indication of a system which can adapt to any solar technology with location and time-independent [19]. Related to this a new sun tracker is produced whose main aim is to avoid solar tracking problems with time and location [20].

3 Methodology

The proposed system uses the concept of solar tracking using LDRs which use light intensity to measure the direction of sunlight and rotate the solar panel into that direction. The hardware of the system consists of main blocks such as IoT module, motor driver, LDR sensors, and LCD. The microcontroller used in the proposed system is Arduino Nano to which all the other elements are attached. The main need of a microcontroller is to understand the light intensity from the LDR sensor and rotate the solar PV panel in that direction. The other use of Arduino is to send the data using an IoT module onto the web page [21] (Fig. 2).

The solar PV panel provides data about how much amount of voltage is produced to the microcontroller. An extra battery is used in case of grid failure the system does not stop working. The higher amount of voltage cannot be given directly to Arduino. Because Arduino Nano uses the only 5 V to work. If higher voltage is provided to the microcontroller, the Arduino can get damage. The temperature is also measured and can be analyzed on a web page. If the temperature increases more than the threshold level, the buzzer starts beeping which is an alert about high heat that can damage the system [22].

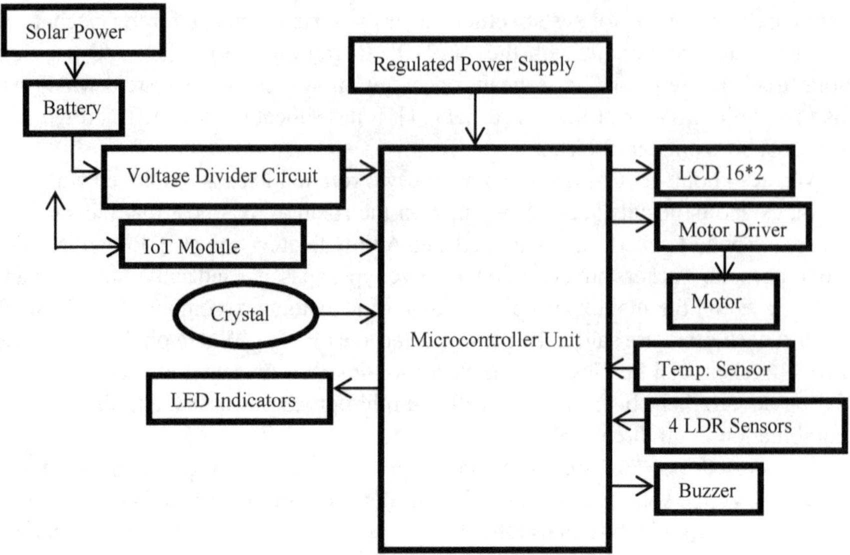

Fig. 2 Block diagram of proposed system

The proposed system uses four LDR sensors in which two LDR sensors are used to rotate motor no. 1 and two LDR sensors are used to rotate motor no. 2. The above-mentioned devices are connected to the analog pins of the Arduino. The analog pins are the inputs to the microcontroller. Let us further study digital pins which are the outputs from the microcontroller. The IoT module used in the system is ESP8266. ESP8266 offers a complete and self-contained Wi-Fi networking solution, allowing it to either host the application or to offload all Wi-Fi networking functions from another application processor. When ESP8266 hosts the application, and when it is the only application processor in the device, it is able to boot up directly from an external flash. It has integrated cache to improve the performance of the system in such applications and to minimize the memory requirements. The two pins of ESP8266 are used which are transmitter and receiver pins. The IoT module is used to transmit the data such as temperature, solar PV panel voltage, and LDR sensor values [23] (Fig. 3).

To provide the value of LDR on the system at that instant LCD is used. The LCD is connected to the digital pins of the Arduino. In the given system, the LCD acts as output, and the other output device is the motor driver which is used to rotate the solar PV panel. The dual axis solar tracker uses two motors in which motor no. 1 operates according to two LDRs and the motor no. 2 operates according to rest two LDRs. The motor no. 1 is connected in the vertical direction to the solar PV panel and rotates in clockwise and anticlockwise directions according to the LDR values. The motor no. 2 is connected in horizontal direction to the solar panel and is used to flip the solar panel front and backward direction [24].

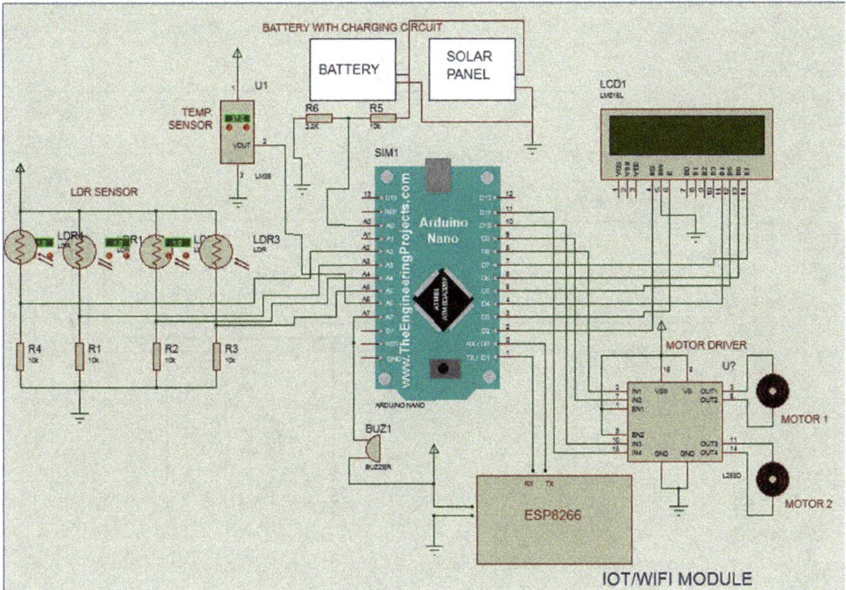

Fig. 3 Circuit diagram of solar remote monitoring and tracking system

The system consists of four LDRs in which the two LDRs which rotate motor no. 1 are opposite to each other. The rest two LDRs which rotate motor no. 2 are opposite to each other. The motor driver used in the system is the L293D IC [25].

4 Control and Management of the System

The connection between the Arduino Nano and IoT (Internet of Things) server is established by initializing power supply to both of them. Initially, the power supply is provided to all the devices of the system. After establishing the communication, the inputs from all the sensors are updated over IoT using IoT module. The system consists of four LDRs whose values are provided to the Arduino Nano which is later on updates the values over IoT using IoT module. If any of the LDR senses the light intensity less then provided values, the motor starts rotating the solar panel in clockwise direction which can be said as it rotates the solar PV panel in the direction of sunlight. If there is a condition that if the voltage passes the threshold level, this triggers the buzzer and the system alarms. The values such as resistance of LDR and temperature, and battery voltage are updated on the web page automatically [26] (Fig. 4).

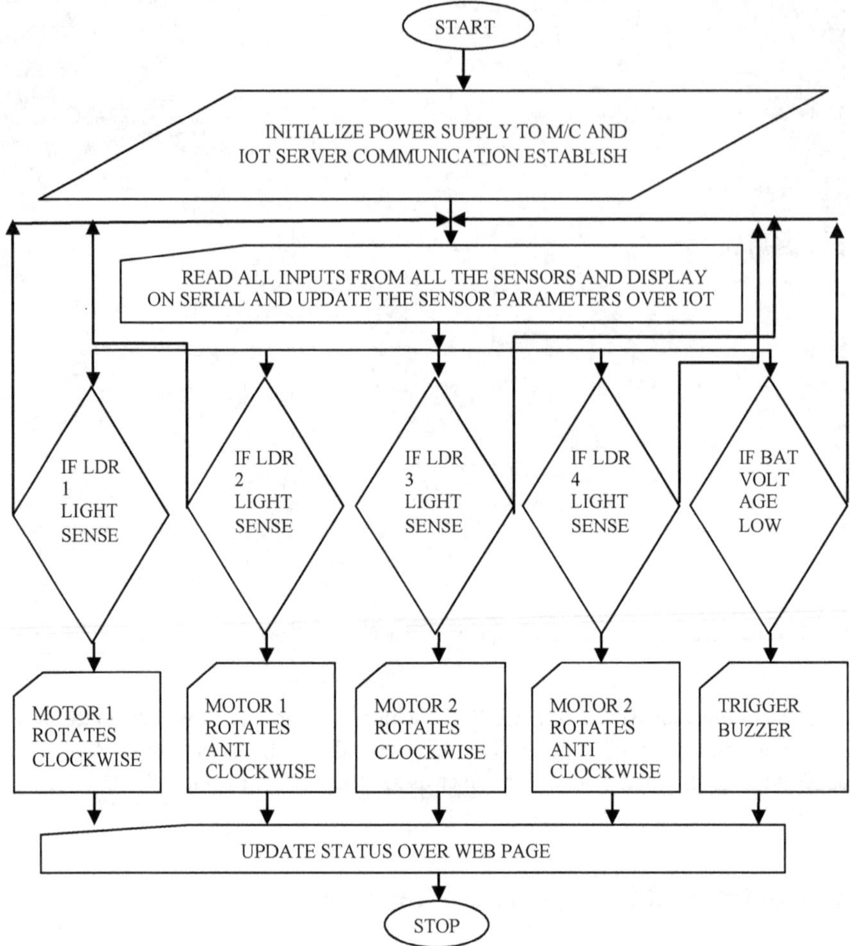

Fig. 4 Operation of proposed system

5 Result

The proposed system consists of the main block of IoT module/Wi-Fi module which is used to update the values over the Internet on a provided web page. The values which are provided on the web page consist of a data logger which provides not only the data of that day but also the data from previous days. This helps in analyzing the data and voltage of the solar PV panel every day. The given webpage consists of six fields in which data of six devices are provided [27]. The devices are

a. Solar PV panel
b. Temperature
c. LDR sensors (Fig. 5).

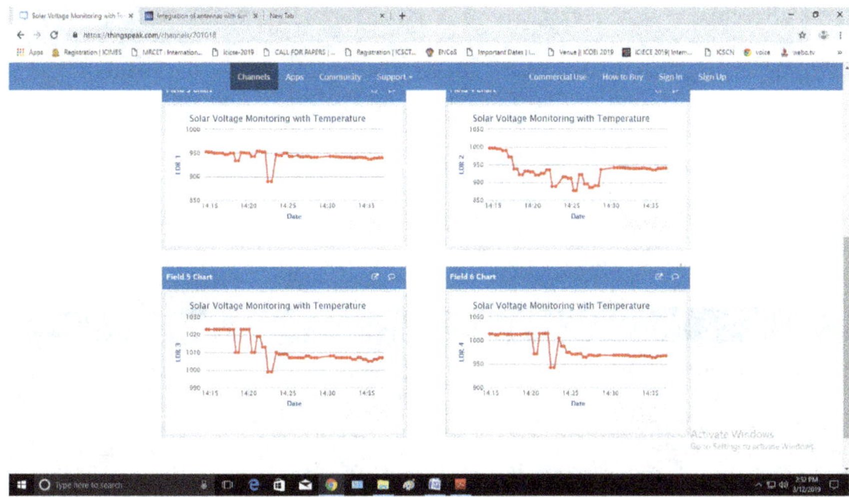

Fig. 5 Design of web page

The values of the system are monitored on the web page and using the Internet can be checked all around the world. The main advantages of the proposed work is to monitor the system with a less human effort. The values of the LDR determine the direction of sunlight and the direction of the solar PV panel. The voltage of solar panel defines the efficiency and the time at which we are getting a higher amount of current in a day, and the temperature determines the heat of the sunlight and if any damage can be caused by heat to the system [28].

The LDR sensor used in the system is generally in four directions. Consider them as North, South, East, and West in which the LDRs are placed. The direction is determined by the values of LDRs the lesser the LDR value the solar panel will be rotated in the opposite direction (Fig. 6).

The result of the proposed system shows that as the values of LDR change with change in sunlight, the system shifts the direction of solar PV panel. According to Fig. 7, its very clear the power generation is more in case of sun tracking solar system as compared to fixed solar system.

Let us take an example, at 14:25 the field 3 which represents LDR 1 is just below 950 Ω, the field 4 which represents LDR 2 is below 900 Ω, the field 5 which represents LDR 3 is below 1010 Ω, and the field 6 which represents LDR 4 is below 1000 Ω. The above readings show that the LDR 2 has lesser value, so the solar panel will be rotated in the opposite direction to the LDR 2. The voltage produced in the solar panel depends on the heat generated by sunlight. The proposed system is a prototype which uses a smaller solar panel which produces approximately 0.1 v. The higher the temperature the more it produces voltage. There is a change in voltage in point form because the voltage is produced at a higher efficiency due to the rotation of solar panel in sunlight direction (Fig. 8).

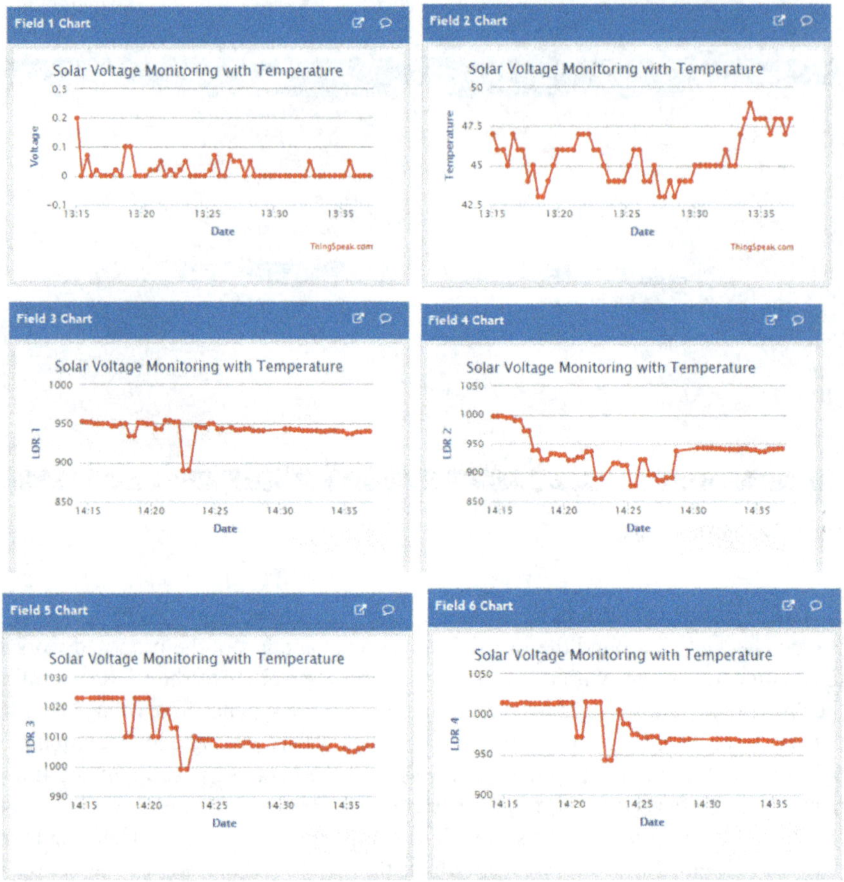

Fig. 6 Fields in the web page consisting of data

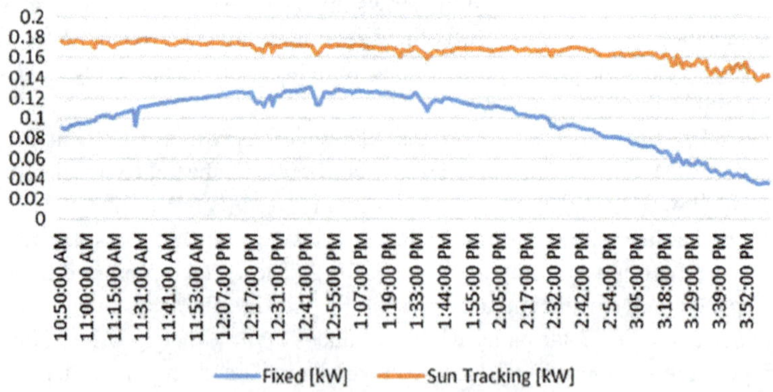

Fig. 7 Fixed versus sun tracking power generation

Fig. 8 Proposed system in real time

6 Conclusion

The proposed system is a prototype built to control and manages the solar tracking system. The system was initially built to test if the system can work in a given environment. The aim of the project was to make a powerful tool which can have a wide range of application. The Internet of Things is used to supervise the solar PV power generation which can enhance the performance, management, and maintenance of the system. The system will help in tracking and turning the solar panel in the direction of sunlight using LDRs. The IoT helps in tracking the voltage supply of the solar PV panel.

References

1. K. Abas, K. Obraczka, L. Miller, Solar-powered, wireless smart camera network: An IoT solution for outdoor video monitoring. Comput. Commun. **118**, 217–233 (2018)
2. J.A. Carballo, J. Bonilla, L. Roca, M. Berenguel, New low-cost solar tracking system based on open source hardware for educational purposes. Solar Energy **174**, 826–836 (2018)
3. I. Bisaga, N. Puźniak-Holford, A. Grealish, C. Baker-Brian, P. Parikh, Scalable off-grid energy services enabled by IoT: a case study of BBOXX SMART Solar. Energy Policy **109**, 199–207 (2017)
4. N. Sahraei, S. Watson, S. Sofia, A. Pennes, T. Buonassisi, I. Marius Peters, Persistent and adaptive power system for solar-powered sensors of the Internet of Thing. Energy Procedia **143**, 739–741 (2017)
5. N. Sahraei, E.E. Looney, S.M. Watson, I. Marius Peters, T. Buonassisi, Adaptive power consumption improves the reliability of solar-powered devices for the internet of things. Appl. Energy **224**, 322–329 (2018)

6. J. Bito, R. Bahr, J.G. Hester, S. Abdullah Nauroze, A. Georgiadis, M.M. Tentzeris, A novel solar and electromagnetic energy harvesting system with a 3-D printed package for energy efficient internet-of-things wireless sensors. IEEE Trans. Microw. Theor. Tech., 1–12 (2017)
7. W. Li, T. Yang, F.C. Delicato, P.F. Pires, Z. Tari, S.U. Khan, A.Y. Zomaya, On enabling sustainable edge computing with renewable energy resources. IEEE Commun. Mag. 94–101 (2018)
8. T. Jang, G. Kim, B. Kempke, M.B. Henry, N. Chiotellis, C. Pfeiffer, D. Kim, Y. Kim, Z. Foo, Hyeongseok Kim, A. Grbic, D. Sylvester, H.-S. Kim, D.D. Wentzloff, D. Blaauw, Circuit and system designs of ultra-low power sensor nodes with illustration in a miniaturized GNSS logger for position tracking: Part II—data communication, energy harvesting, power management, and digital circuits. IEEE Trans. Circ. Syst.–I **64**, 2250–2262 (2017)
9. A. D´ıaz, R. Garrido, J.J. Soto-Bernal, A filtered sun sensor for solar tracking in HCPV and CSP systems. IEEE Sens. J. **19**, 917–925 (2019)
10. B.K. Hammad, R.H. Fouad, M. Sami Ashhab, S.D. Nijmeh, M. Mohsen, A. Tamimi Adaptive control of solar tracking system. IET Sci. Measur. Technol. **8**, 426–431 (2014)
11. J. Wu, X. Chen, L. Wang, Design and dynamics of a novel solar tracker with parallel mechanism. IEEE/ASME Trans. Mechatron. **21**, 88–97 (2016)
12. A. Narbudowicz, O. O'Conchubhair, M.J. Ammann, D. Heberling, Integration of antennas with sun-tracking solar panels. Electron. Lett. **52**, 1325–1327 (2016)
13. X. Zhang, J. Du, C. Fan, D. Liu, J. Fang, L. Wang, A wireless sensor monitoring node based on automatic tracking solar-powered panel for paddy field environment. IEEE Internet Things J. **4**, 1304–1311 (2017)
14. Z. Zhen, Z. Zengwei, S. Li, W. Jun, P. Wuchun, L. Zhikang, W. Lei, C. Wei, S. Yunhua, The effects of inclined angle modification and diffuse radiation on the sun-tracking photovoltaic system. IEEE J. Photovolt. **7**, 1410–1415 (2017)
15. B. Asiabanpour, Z. Almusaied, S. Aslan, M. Mitchell, E. Leake, H. Lee, J. Fuentes, K. Rainosek, N. Hawkes, A. Bland, Fixed versus sun-tracking solar panels: an economic analysis. Clean Technol. Environ. Policy **19**, 1195–1203 (2017)
16. E. Kiyak, G. Gol, A comparison of fuzzy logic and PID controller for a single-axis solar tracking system. Renew. Wind Water Solar (2016)
17. H. Geun Lee, S.-S. Kim, S.-J. Kim, S.-J. Park, C.-w. Yun, G.-p. Im, Development of a hybrid solar tracking device using a gps and a photo-sensor capable of operating at low solar radiation intensity. J. Korean Phys. Soc. **67**, 980–985 (2015)
18. M. Natarajan, T. Srinivas, Study on solar geometry with tracking of collector. Appl. Solar Energy **51**(4), 274–282 (2015)
19. S.A. Orlov, ShI Klychev, Compensation of axis errors of Azimuth and Zenith Moving concentrators in programmable solar-tracking systems. Appl. Solar Energy **54**(1), 61–64 (2018)
20. S. Kumar, S. Alam M. Nelanti, A study on smart home automation based on IOT. Int. J. Adv. Innov. Res. **5**(1), 37–43 (2018). ISSN: 2394-7780
21. S. Kumar, P. Raja G. Bhargavi, A comparative study on modern smart irrigation system and monitoring the field by using IOT, in *IEEE International Conference on Computing, Power and Communication Technology, (GUCON)*, 28th–29th Sept. 2018, pp. 637–641
22. Soumya, S. Kumar, Health care monitoring based on internet of things, in *The (Springer) International Conference on Artificial Intelligence & Cognitive Computing (AICC)*, Hyderabad, 2nd–3rd Feb. 2018
23. M. Hassnuddin, S. Kumar, Advance green energy scheduling in smart grid using IOT, in *International IEEE Conference on Recent Advances in Energy-Efficient Computing and Communication (ICRAECC-2019)*, 7th–8th Mar. 2019
24. F. Khatoon, S. Kumar, M. Niyaz Ali Khan, A study on solar remote monitoring using the internet of thing, in *2nd International Springer/Elsevier Conference on Nano Science & Engineering Applications*, 4th–6th Oct. 2018
25. S. Kumar, S. Anirudh, IOT and RF-ID Based E-Passport System, in *7th (Springer) International Conference on Innovation in Electronics and Communication Engineering (ICIECE-2018)*

26. S. Kumar, V. Taj Kiran, S. Swetha, P. Johri, IoT based smart home surveillance and automation, in *IEEE International Conference on Computing, Power and Communication Technology (GUCON)*, 28th-29th Sept. 2018, pp. 795–799
27. F. Khatoon, S. Kumar, A study on E-nose and air purifier system, in *IEEE International Conference on Innovative Technologies in Engineering (ICITE-2018)*, (2018)
28. S. Kumar, H. Dalmia, A study on internet of things applications and related issues. Int. J. Appl. Adv. Sci. Res. **2**(2), 273–277 (2017). ISSN: 2456-3080

Secured Electronic Voting Machine Using Biometric Technique with Unique Identity Number and IOT

Kone Srikrishnaswetha, Sandeep Kumar and Deepika Ghai

Abstract Elections play an important role in our democratic country as people can select a person as a leader for the government. This paper was about a new proposed methodology having a highly secured process. This consists of mainly Aadhar, biometric and IOT. Aadhar ID is a unique card for everyone, biometric face recognition for security and IOT for the safe and immediate results. This proposed system has automatic counting of votes: highly data secured system, sending of data immediately and safe voting.

Keywords EVM · Secured components · Face recognition · Unique ID card · Cascade classifier technique · IOT

1 Introduction

India is a country where elections play a major part in the government. It is a method of selecting a candidate for the government to rule [1–6]. This paper explains about secured voting and the process of using biometric for better elections. Selecting a candidate for the government was important, so people have trust in the electronic voting machine as it gives the correct voting [5–12]. So the EVM should be designed so has to be highly secured and safe. They are many technologies like polling booths, ballots and punch cards, e-voting, block chains, using a microcontroller, network security, i-voting, online voting, and GSM module. But still, they are many challenges in EVM at present days [13–22]. So this paper explains a proposed methodology of secured EVM using face recognition biometric, a unique card with IOT [15, 24–28]. Biometric face recognition is useful for the identification of people, and voter cannot

K. Srikrishnaswetha (✉) · S. Kumar
Department of Electronics and Communications, Sreyas Institute of Engineering and Technology, Hyderabad, Telangana, India
e-mail: Srikrishnaswetha24@gmail.com

S. Kumar
e-mail: er.sandeepsahratia@gmail.com

D. Ghai
School of Electrical and Electronics Engineering, Lovely Professional University, Jalandhar, India
e-mail: deepika.21507@lpu.co.in

H. S. Saini et al. (eds.), *Innovations in Electronics and Communication Engineering*, Lecture Notes in Networks and Systems 107,
https://doi.org/10.1007/978-981-15-3172-9_31

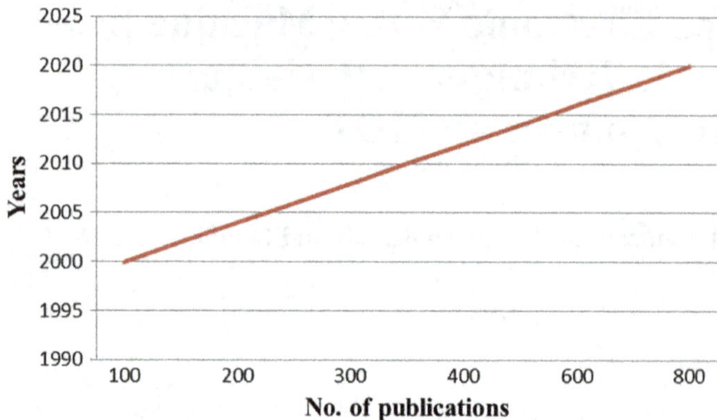

Fig. 1 Data for using EVM past 20 years

repeat their voting. Aadhar was the unique ID number for everyone. IOT is useful for sending of results immediately [22, 23, 29–31].

Data for using the technology in EVM from the last 20 years in the research community were increasing as shown in Fig. 1, and data are taken from Science Direct.

2 Literature Analysis

See Table 1.

3 Problem Statement

According to the literature analysis, there are a few major technical problems in the voting process. First, manually counting of vote's process is not accurate and secure as shown in Fig. 2 [4, 6, 21].

Second, fingerprint biometric is not safe and secured at present days due to fake fingerprint as shown in Fig. 3. A person who does not have hands is not allowed for voting due to this process [12, 15].

Third, missing of votes was a great loss for the voting process. In India, 2019 elections, nearly 21 billion people lost their vote and few got a chance of double voting in few places. This happened due to the improper registration of voter's details and no proper ID proof of a person as shown in Fig. 4a, b.

Another problem is recounting and declaration of delay in voting. In the gap of the voting process and results, there is a chance of hacking [7–11]. Existing machines are not connected to the online. So the transportation also takes time after the polling of all the phases is completed.

Table 1 Comparison of literature work of voting machine

S. No.	Author name	Year	Methodology	Remarks
1	Shahzad et al. [1]	2019	EVM, blockchain, i-voting, e-voting	This is about blockchain which is a solution at the polling process and security purpose. It is useful at the authentication of votes
2	Salah et al. [2]	2018	IOT, Network and data security	This paper explains about three levels of high-, low- and intermediate-level layers
3	Han et al. [3]	2018	Online voting, privacy preservation, end-to-end verification.	Here, it was about the process of the e-voting system
4	Karim et al. [4]	2017	Automatic calculation of results, integrated database, and EVM.	In this, the system is used for smart voting. They used fingerprint by Arduino. Registration was good, and it generates SMS to the voter. It gives high data security
5	Rahman et al. [5]	2017	EVM, Arduino and fingerprint scanner	This is for mainly securing purpose and to make the process faster. The fingerprint is used for detecting the voter is authorized or not
6	Kalaiselvi et al. [6]	2016	EVM, identity card and biometric	Fingerprint and Aadhar used for the transparent voting process
7	Anish et al. [7]	2015	GSM module, Arm 7 Cortex	This is used for sending and receiving results
8	Ashok et al. [8]	2014	LCD, microcontroller, sensors	This explains the accuracy of votes and display of results
9	Pandey et al. [9]	2013	EVM and online process	Aadhar was for identifying people. The easy process online

4 Proposed Methodology

The proposed methodology consists of two steps, namely registration of voter details and voting process, as shown in Figs. 5 and 6. The hardware components used are Raspberry Pi3 works by using virtual network computing (VNC) viewer which is visual desktop-sharing system of PI, webcam, buttons for voting, monitor and connecting cable and software used was Python version 3.5 and (SSH) Secure Shell

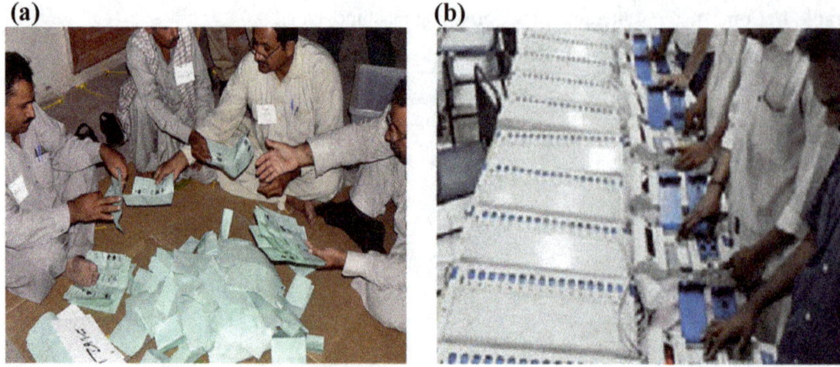

Fig. 2 **a** Hand counting, **b** digital ballots

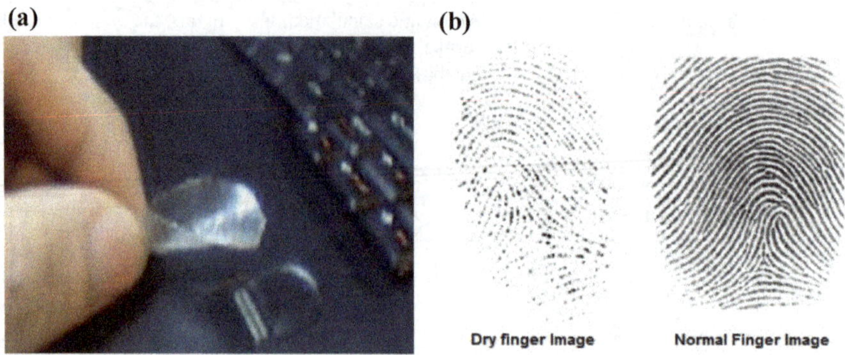

Fig. 3 **a** Fake fingerprint, **b** improper fingerprint

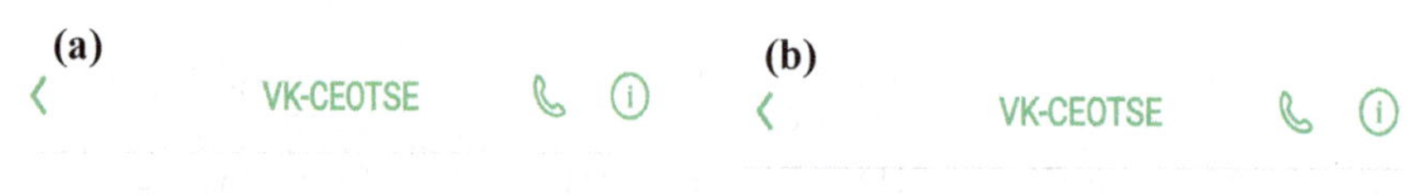

Fig. 4 **a** Voter ID, **b** voter ID for the same person

for access to the PI terminal. From the above literature analysis, we proposed a new methodology for counting problems, biometric, registration of voter and declaration of results.

In the proposed methodology, the webcam with good quality is used to build a real-time face detector and achieve better accuracy. Monitor/screen is used for entering Aadhar number, face recognition and to identifying the authorized & unauthorized voters with automatic counting of votes. Hardware consists of monitor, voting buttons, connecting cable and wires and webcam as shown in Fig. 7. VNC viewer works by connecting to the server with IP address using username and password as shown in Fig. 8.

Algorithm (Registration)

Step1 Enter ID number
Step2 Capturing of images of face
Step3 Captured successfully

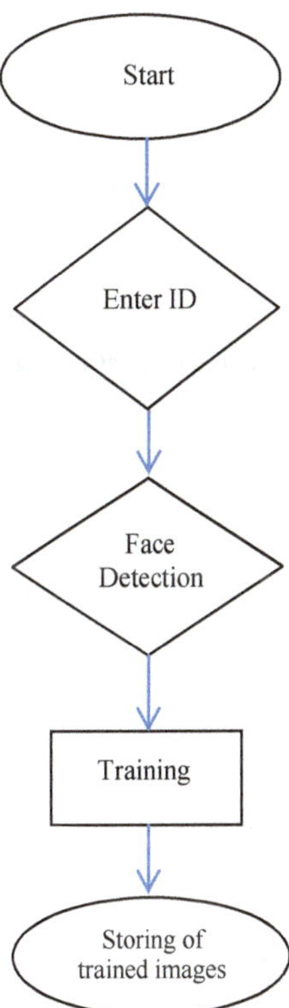

Fig. 5 Registration method

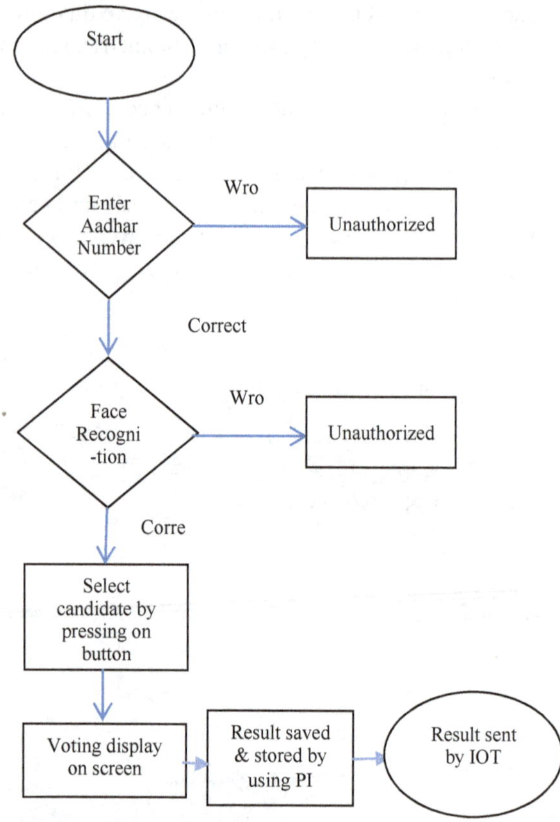

Fig. 6 Voting process of the proposed methodology

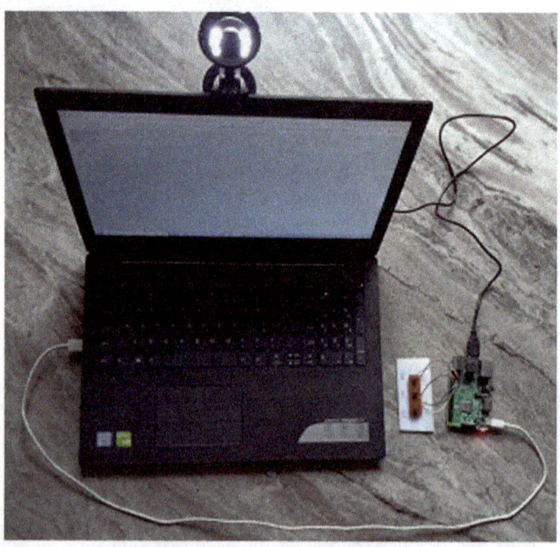

Fig. 7 Hardware setup

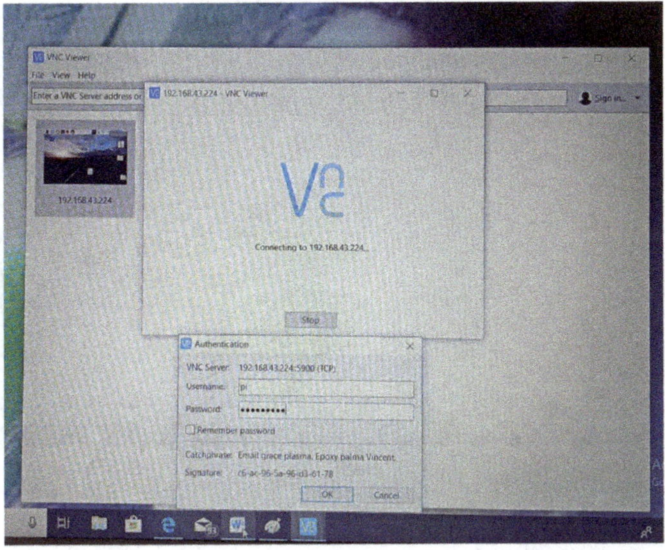

Fig. 8 Connection with IP address

Step4 Images are stored with IDs in haarcascade path.
Step5 Successfully trained
Steps6 Registration completed

Registration of voter details: Registration was important step and is used to register details of the voter. Figure 9 shows the entering of ID number in register process, and Fig. 10 shows the capturing of images while registering of details, and after that, the captured images are trained and stored in the dataset. These are used for the voting process for face recognition. Voting process: According to the proposed methodology, voting process consists of three steps ,i.e., entering of Aadhar, face recognition, storing and sending of data using IOT.

(1) Entering of Aadhar: The process starts by clicking on the start button. Immediately it displays to enter Aadhar number; if it is correct, we can go with the next step or else the person is unauthorized.
(2) Face recognition: After entering the Aadhar number, if the person is authorized immediately the webcam will on and face recognition starts. If the face was matched with the registered images, then it displays authorized and the person can go with voting by clicking on one of the voting buttons. After this, the result will display on the screen. Face recognition works by using Haarcascade frontal face algorithm (Fig. 11).

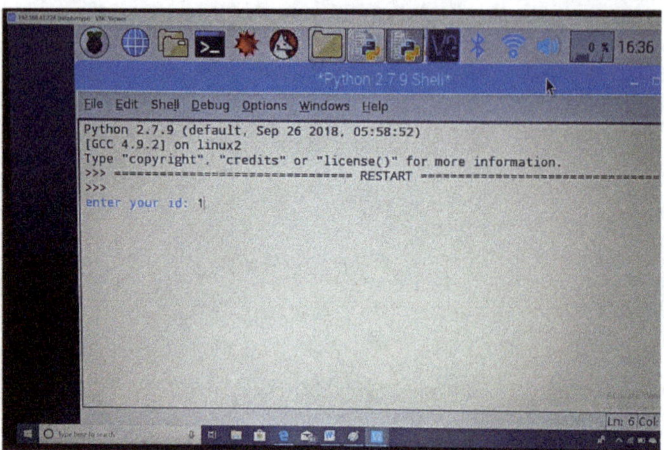

Fig. 9 Entering of ID number

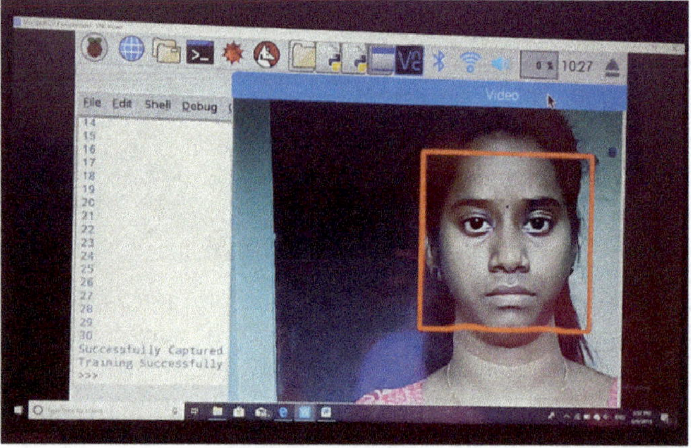

Fig. 10 Results of successfully capturing of images

(3) Haarcascade frontal face algorithm: It is an object or human face detecting algorithm in which cascade can train images from many of positive images and negative images. Every image has 24 × 24 base windows to calculate features. Haarcascade identifies features with rectangular boxes; white indicates the lighter position and black indicates darker positions. Integral images are that which can calculate the images pixel values at origin (x, y) and can calculate by adding of all image pixel values of the present pixel. How the pixel calculation is carried out is explained below:

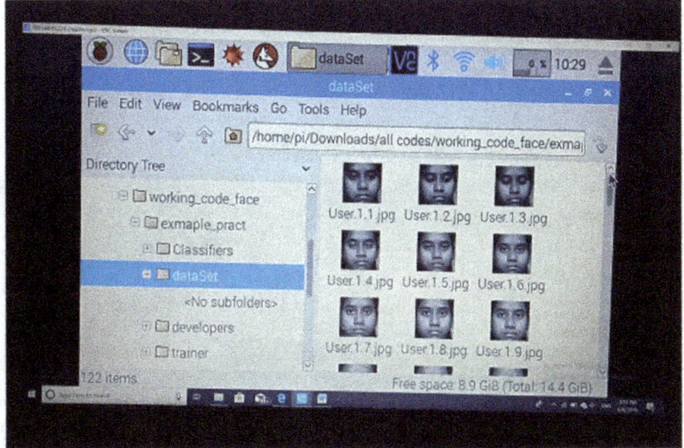

Fig. 11 Storing and training of images

6	3	2	8	5
3	**9**	**1**	**2**	1
1	**6**	**0**	**5**	8
4	**3**	**6**	**5**	7
5	3	1	7	3

The first table is the numerical values from the image of the left side.

$$9 + 1 + 2 + 6 + 0 + 5 + 3 + 6 + 5 = 37; \quad 37/9 = 4.11$$

It requires nine operations. 100 * 9 = 900.

1	7	13	21	22
7	**20**	**25**	**35**	41
15	**36**	**40**	**55**	72
23	**46**	**56**	**76**	111
24	48	60	79	107

The second table is the integral image values of the left side.

$$(76 - 20) - (24 - 5) = 37; \quad 37/9 = 4.11$$

It requires four operations. 56 + 100 * 4 = 456 (operations)

The image can be reduced by using this integral image. By converting a large image into small, accuracy increases. The rectangular frame on the image is calculated by (x, y) coordinates, width and height of the frame.

$$[Y : y + h, X : x + w]$$

Algorithm (IOT)

Step1 After voting process, the data is stored in Raspberry Pi in Excel format
Step2 Data are sent to the main location through e-mail.
Step3 If else, data can be checked in Raspberry Pi by the authorized person.
Step4 It can be send manually also.
Step5 Results are released without delay.

Algorithm (Voting Process)

Step1 Enter Aadhar ID number
Step2 Immediately opening of webcam.
Step3 Capturing of live face and recognizing
Step4 Display of authorized or an unauthorized by matching the trained images and live data.
Step5 Selecting of voting button
Step6 Voting done successfully if authorized.
Step7 Display of results.

Storing and sending of data using IOT: By using IOT, data of result are stored and immediately sent to the office through mail. By this, late declaration of results can be avoided.

5 Result

Experimental results is performed in three categories, i.e., authorized, unauthorized and face spoofing as shown in Figs. 12, 13, 14, 15, 16 and 17.

1. Authorized

The process starts with entering Aadhar number as shown in Fig. 12. If the number entered was correct, then immediately face recognition starts. If the person was authorized, he/she can go with voting process as shown in Fig. 13. Finally, votes are stored in excel format as in Fig. 14, and the result was sent through a mail immediately after voting as shown in Fig. 15.

2. Unauthorized person as shown in Fig. 16.
3. Figure 17 proved that images are not detectable; only live detection was possible in this methodology. In the process, if a person was unauthorized, then he/she was not eligible for voting, and even though they click on voting button, it does not take the vote as shown in Fig. 16.

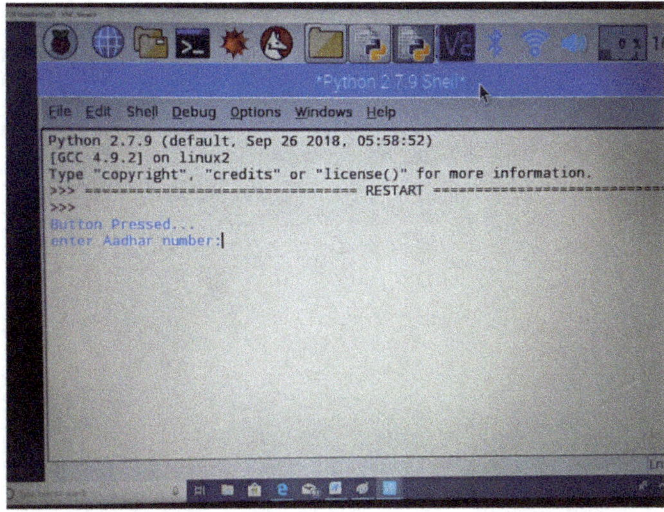

Fig. 12 Enter Aadhar ID

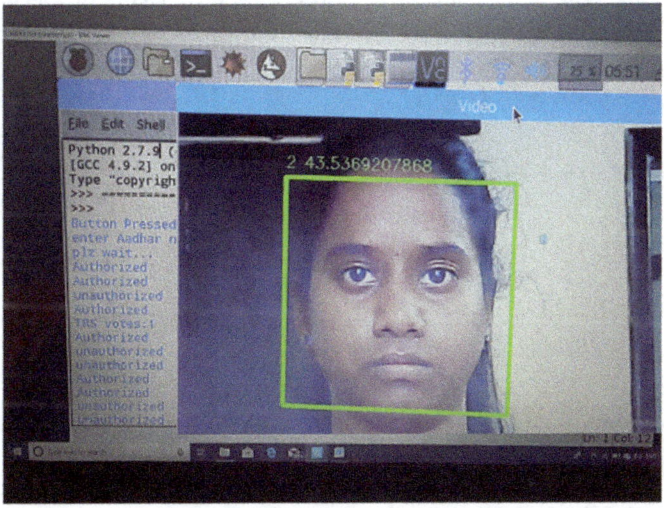

Fig. 13 Display of authorized and voting

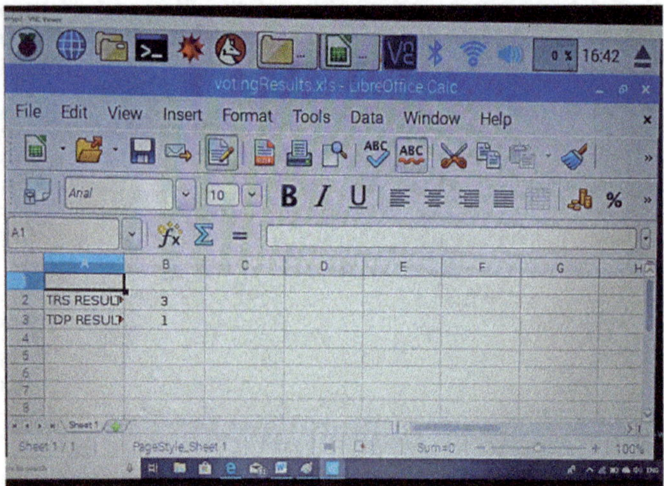

Fig. 14 Result stored in Excel format

Finally, the proposed methodology works better than the existing technology for the identification of a human face. The results are shown in Table 2.

$$\text{Accuracy} = 100 * ((\text{TP} + \text{TN})/\text{N})$$

$$(\text{FAR}) = \text{TP}/(\text{TP} + \text{FN})$$

$$(\text{FRR}) = \text{TP}/(\text{TP} + \text{FP})$$

From the above comparison, we can say that our proposed methodology was better in accuracy, and it reduces error percentage.

6 Conclusions

In our country still, the process of elections is not secured. This paper proposed a highly secured process for fair voting. This proposed method can go with automatic counting of votes, face recognition, better registration process, storing and sending of results and declaration of results. So, by this, we can avoid duplicate voting, wrong registration, fake biometric and late declaration of results for the better elections for the better country.

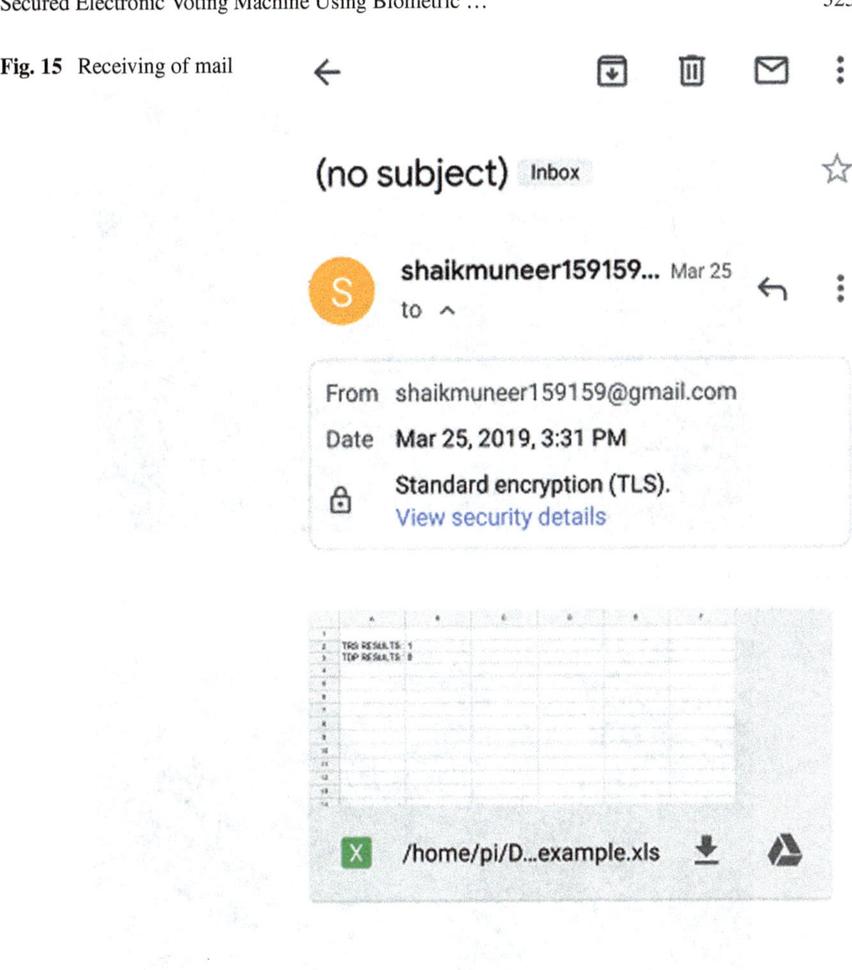

Fig. 15 Receiving of mail

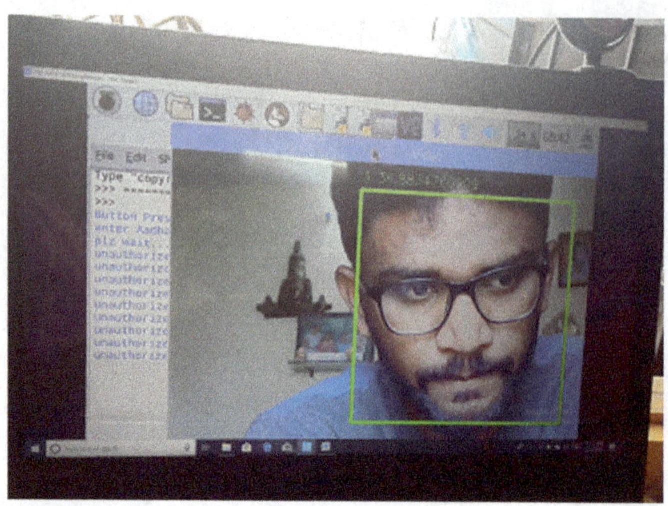

Fig. 16 Unauthorized person

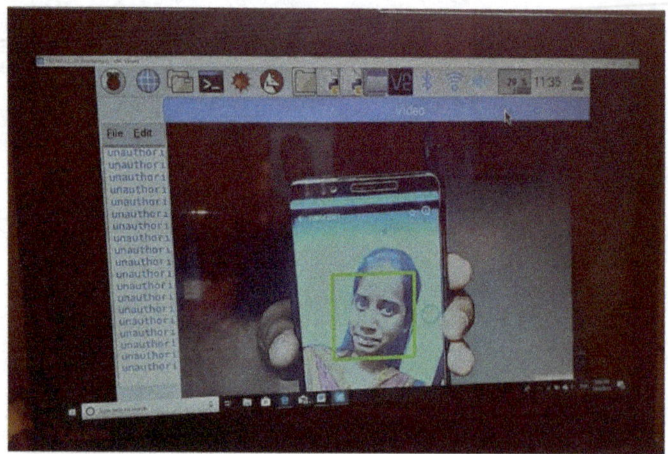

Fig. 17 Images cannot be detectable

Table 2 Comparison of different techniques parameters

S. no.	Method	Accuracy %	Error %	FAR (precision) %	FRR (recall) %
1	Skin + edge	66.2	33.78	19.8	–
2	E + skin + edge	82.7	17.2	22.8	–
3	RGB	69	–	43.05	–
4	RGB + CBCR	77	–	36.14	–
5	RGB + H	83.5	–	33.82	–
6	RGB + H + CBCR	90.83	–	28.29	–
7	Proposed	96.2	3.8	0.04	0.02

Acknowledgement The authors acknowledge that the photographs/images used in this paper are their own and approved for publishing online.

References

1. B. Shahzad, J. Crowcroft, Trustworthy electronic voting using adjusted blockchain technology. IEEE Access **7**, 24477–24488 (2019)
2. M. Ahmad Khan, K. Salah, IoT security: review, blockchain solutions, and open challenges. Future Gener. Comput. Syst. **82**, 395–411 (2018)
3. X. Yang, X. Yi, S. Nepal, A. Kelarev, F. Han, A secure verifiable ranked choice online voting system based on homomorphic encryption. IEEE Access **6**, 20506–20519 (2018)
4. M. Karim, S. Khan, A proposed framework for the biometric electronic voting system, *IEEE International Conference on Telecommunications and Photonics*, vol. 1(1) (2017), pp. 12–19
5. R. Rezwan, A. Rahman, Biometrically secured electronic voting machine, in *IEEE Region 10 Humanitarian Technology Conference* (2017), pp. 127–131
6. V.K. Kalaiselvi, F. Trindade, J. Porfirio A. de Carvalho, R. Cantanhede, Smart voting, in *Second International Conference On Computing and Communications Technologies(ICCCT'17)* (2016), pp. 1–10
7. S. Anandaraj, R. Anish, P.V. Devakumar, Secured electronic voting machine using biometric, in *IEEE International Conference on Innovations in Information, Embedded and Communication Systems (ICIIECS)* (2015), pp. 15
8. Kumar, D. Ashok, T. Ummal Sariba Begum, A novel design of an electronic voting system using a fingerprint. Int. J. Innov. Technol. Creative Eng. **1**(1), 12–19 (2011)
9. A. Himanshu, G.N. Pandey, Online voting system for India based on AADHAAR ID, in *11th International Conference on ICT and Knowledge Engineering (ICT&KE)* (2013), pp. 1–4
10. Kumar, D. Ashok, T. Ummal Sariba Begum, Electronic voting machine—a review, in *IEEE International Conference on Pattern Recognition, Informatics and Medical Engineering (PRIME)* (2012), pp. 41–48
11. K. Srikrishna Swetha, S. Kumar, A study on smart electronics voting machine using face recognition and Aadhar verification with IOT, in *7th International Conference on Innovations in Electronics & Communication Engineering (ICIECE—2018)*, vol. 65, (Springer, 2018), p. 8795
12. K. Srikrishna Swetha, S. Kumar, Comparison study on various face detection techniques, in *4th International Conference on Computing, Communication, and Automation (ICCCA-2018)* (IEEE, 2018), pp. 95–99

13. R.P. Jacobi, F. Trindade, J.P.A. de Carvalho, R. Cantanhede, JPEG decoding in an electronic voting machine, in *IEEE 13th Symposium on Integrated Circuits and Systems Design* (2000), pp. 177–182
14. M.M. Islam, M.S.U. Azad, M.A. Alam, N. Hassan. Raspberry Pi and image processing based electronic voting machine (EVM). Int. J. Sci. Eng. Res. **5**(1), 1506–1510 (2014)
15. D. Karima, Pr. T. Victor, Dr. R. Faycal, An improved electronic voting machine using a microcontroller and a smart card, in *9th International Design and Test Symposium* (IEEE, 2014), pp. 219–224
16. S.M. Hussain, C. Ramaiah, R. Asuncion, S. Azeemuddin Nizamuddin, R. Veerabhadrappa, An RFID based smart EVM system for reducing electoral frauds, in *5th International Conference on Reliability, Infocom Technologies and Optimization (ICRITO) (Trends and Future Directions)* (2016), pp. 371–374
17. A. Das, M. Pratim Dutta, S. Banerjee, C.T Bhunia, Cutting edge multi stratum secured electronic voting machine design with inclusion of biometrics RFID and GSM module, in *IEEE 6th International Conference on Advanced Computing* (2016), 667–673
18. S. Lavanya, Trusted secure electronic voting machine, in *International Conference on Nanoscience, Engineering, and Technology* (2011), pp. 507–209
19. S. Gawhale, V. Mulik, P. Patil, N. Raut, IOT based E-voting system. Int. J. Res. Appl. Sci. Eng. Technol. (IJRASET) **5**(5) (2017)
20. S. Kumar, S. Singh, J. Kumar, Live detection of face using machine learning with multi-feature method, in *Wireless Personal Communication Springer Journal (SCI)* Published https://doi.org/10.1007/s11277-018-5913-0
21. S. Kumar, S. Singh, J. Kumar Automatic live facial expression detection using genetic algorithm with Haar wavelet features and SVM, in *Wireless Personal Communication Springer Journal (SCI)* Published https://doi.org/10.1007/s11277-018-5923-y
22. Soumya, S. Kumar, Health care monitoring based on internet of things, in *The (Springer) International Conference on Artificial Intelligence & Cognitive Computing (AICC)*, 2nd–3rd Feb. 2018 Hyderabad (Scopus Indexed-Published)
23. S. Kumar, S. Singh, J. Kumar, A study on face recognition techniques with age and gender classification, in *IEEE International Conference on Computing, Communication and Automation (ICCCA)* (2017), pp. 1001–1006
24. S. Kumar, S. Singh J. Kumar, A comparative study on face spoofing attacks, in *IEEE International Conference on Computing, Communication and Automation (ICCCA)* (2017), pp. 1104–1108
25. S. Kumar, S. Singh J. Kumar, Gender classification using machine learning with multi-feature method, in *IEEE 9th Annual Computing and Communication Workshop and Conference (CCWC)*, Las Vegas, USA, 7th–9th Jan. 2019
26. S. Kumar, A. Sony, R. Hooda, Y. Singh, Multimodel biometric authentication system for automatic certificate generation, in J. Adv. Sch. Res. Allied Educ. **16**(3) (2019)
27. S. Kumar, S. Singh J. Kumar, Automatic face detection using genetic algorithm for various challenges. Int. J. Sci. Res. Mod. Educ. **2**(1) (2017), 197–203. ISSN: 2538-4155
28. S. Kumar, Deepika, M. Kumar, An improved face detection technique for a long distance and near-infrared images. Int. J. Eng. Res. Mod. Educ. **2**(1), (2017), pp. 176–181. ISSN: 2455–4200
29. S. Kumar, S. Singh J. Kumar, A multiple face detection using hybrid features with SVM classifier, in *The Springer Conference on Data Communication and Networks (Co-located with GUCON 2018)*. ISBN: 978-981-13-2254-9 (Springer Journal- Published)
30. S. Kumar, S. Sharma, *Image Compression Based on Improved Spiht and Region of Interest*. M. Tech diss. (2011)
31. S. Kumar, H. Dalmia, A study on internet of things applications and related issues. Int. J. Appl. Adv. Sci. Res. **2**(2), 273–277 (2017). ISSN: 2456-3080

Electricity Management in Smart Grid Using IoT

Mohammed Hassnuddin, Sandeep Kumar and Hemlata Dalmia

Abstract Internet of things will be expected to grow 50 billion devices till 2020 according to the stats of Cisco. As we know that today's grid is electromechanical and usage of electricity is getting hiked day on a daily basis, then there will be a hike in the electricity cost. In order to overcome the problem in our paper, a new methodology has been described which aims to cut down the prices during high peak hours. We developed a prototype to describe both the concepts of demand-side response and the Internet of things. PC will act as a broadcaster, and it will broadcast the values over the Internet. Coordination node is fully equipped with the logic that will control the home appliances during peak and off-peak hours according to the threshold.

Keywords IoT · Energy coordination node · Broadcaster · PC · Arduino · Web page

1 Introduction

Electricity plays a crucial role in our daily life, and its usage has been getting hiked day-by-day basis [1]. We were aware that energy is generated from natural resources like coal, water, and wind,. which are limited and we cannot produce the excess energy beyond the availability of resources [2]. With an increase in demand, there will be a hike in price. We have different electricity boards, but it won't display how much electricity price has been released for a particular time. This can be overcome by designing smart appliances which are connected to the Internet all the time, and they will track the value from the broadcaster side. Tracking of value can be done

M. Hassnuddin (✉) · S. Kumar · H. Dalmia
Department of Electronics and Communications, Sreyas Institute of Engineering and Technology, Hyderabad, Telangana, India
e-mail: hassnuddin1407@gmail.com

S. Kumar
e-mail: er.sandeepsahratia@gmail.com

H. Dalmia
e-mail: dalmiahemlata@gmail.com

H. S. Saini et al. (eds.), *Innovations in Electronics and Communication Engineering*, Lecture Notes in Networks and Systems 107,
https://doi.org/10.1007/978-981-15-3172-9_32

by the use of the Internet [3–5]. The smart grid is nothing but electricity grid which is embedded with smart sensors and Internet, so that exchange of data takes place between the customer and broadcaster [6, 7]. Smart grid has different characteristics which are self-resolver, consumer-friendly, robustness, and better storage options [8–14]. Self-resolver is nothing but it consists of centralized communication grid, i.e., whenever there is a problem at supplying unit, there will be a communication which takes place between the main unit and subunit and that problem gets resolved without the interference of user [15–19]. It is customer-friendly because different pricing techniques have been introduced which are time of use (TOU), a day-ahead pricing, and real-time pricing. In a time of use, the price will vary for hours, but in real time, the prices will change for every 15 min of duration. In the day-ahead pricing, the prices are shared with the customer a day before the broadcasting [20–24]. We have better storage options to store electricity [25, 26]. The research work on this technology has been increasing rapidly with the use of IoT [27–35]. Initially, we don't have information about the price, but after embedding grid with the Internet, we can see the exchange of data [14–17]. Figure 1 shows the variation between the published papers with IoT and without IoT. Data has been extracted from a well-known site ScienceDirect.

As we know that world's population is raising at a rate of 215,120 persons per day, our world's population will reach 11 billion people till 2019. Many families, particularly in India and China, are crossing their poverty line and moving in the category of the middle class, and now they can afford home appliances like the fridge and air conditioners. There is a hike in demand for these products which in turn related to power which means there is a hike in power or electricity requirement [33]. In order to overcome the drawback of the present grid, a new infrastructure of

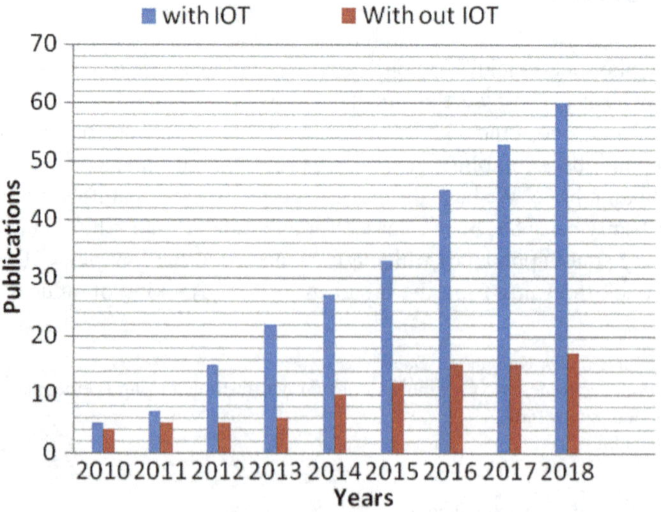

Fig. 1 Papers published with and without IoT during years 2010–2018

electricity has been built to keep up the production and management of electricity done. As a consequence, new and larger demands are being made for electric power, and new electric infrastructure is being built at a rapid pace in most underdeveloped countries to keep up.

2 Literature Analysis

See Table 1.

3 Problem Statement

As per the above literature analysis, we have found many problems which are given below: Today's electric grid is electromechanical, and we don't have any Web sites to show the current pricing on time-to-time basis. Another problem is one-time pricing, and we have constant pricings it won't change time to time. Tracking of price cannot be possible by the smart devices as we not embedded our devices with the Internet. Energy monitoring portals not available at the user end to see the current price.

We have different electricity board Web site, but we won't get information about the electricity price how much it is. User only finds the total usage and total production, but user cannot see live data.

We can clearly see that we cannot see the current pricing of electricity on the Web site. Bill is calculated based on prices set with respect to the unit usage which are tabulated back side of bill which is shown in Fig. 2a. When we check with the charges, it shows constant price for first 50 units of electricity, but if we use beyond that, the charge value get changes.

The bill won't get generate for every month or in every 36 days, after these days there will be a hike in electricity bill because overall units are generated more than 257. It will generate more bill due to consumption of more units as shown in Fig. 2b.

4 Proposed Methodology

In our proposed method, we will describe that the devices will get turn off and turn on based on set condition. All devices cannot be turned on and off; we need to consider devices like we can install our devices in printers into power strips and vary the usage of printers during peak and non-peak hours as they are most rarely used device. If we program our AC/heaters in such a way, then they can run during non-peak hours. We run washer dryers during non-peak times. If we really need these devices in peak hours, we can alternatively switch to solar during peak demand so that we skip from high cost. A basic block diagram of demand response has been shown in Fig. 3.

Table 1 Comparison proposed techniques for demand response of smart grid

S. no.	Author name	Year	Methodology	Remarks
1	Abid and Hasan [1]	2018	Fuzzy controller and integer linear programming	Non-flexible appliances can make decisions based on the provided energy levels. The proposed mechanism is introduced to control and optimize the energy-side management
2	Etxegarai et al. [2]	2018	Dynamic pricing	Dynamic pricing like time of use, real-time pricing, and day-ahead pricing have been introduced
3	Wang [3]	2018	Game theory and the genetic algorithm	In this paper, they aim about the generation of electricity and its storage options, where consumer can shift to different load options according to the load schedule from the broadcaster
4	Mortaji et al. [4]	2017	Load control using IoT	It overviews about the direct load control using the Internet. Forecasted load model has been developed to broadcast the data over the Internet
5	Elma and Selamoğullari [5]	2017	Demand response of smart grid	It overviews about the load shifting and pricing mechanism. IoT has been introduced to control the home appliances with respect to prices
6	Wang et al. [6]	2017	IoT	Security measures and the causes involved in the smart grid have been explained in this paper
7	Kim et al. [13]	2013	Data collection for smart grids	It discussed smart meters and interfacing of wireless sensors to it to collate the data and transfer that data to the server end using the Internet
8	Ashraf Mahmud [8]	2017	Price suggestion unit	The provided new technique for the outdated current grid

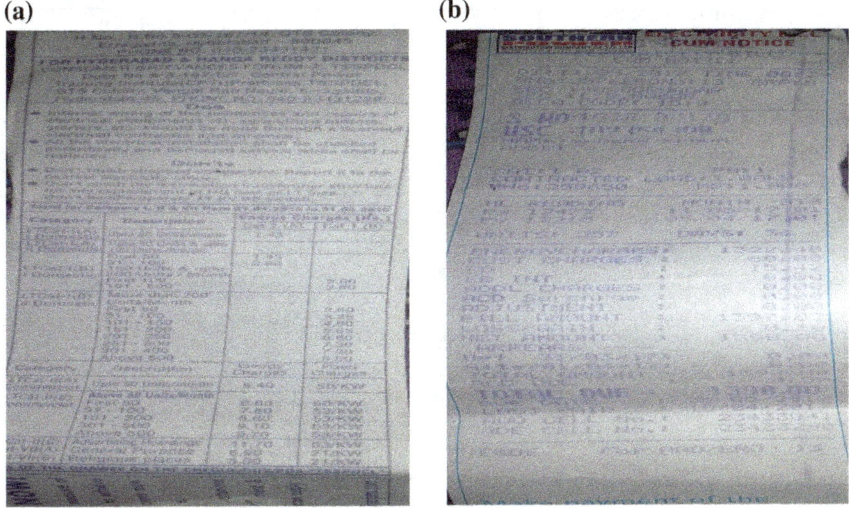

Fig. 2 **a** Electricity bill with different unit price, **b** electricity bill

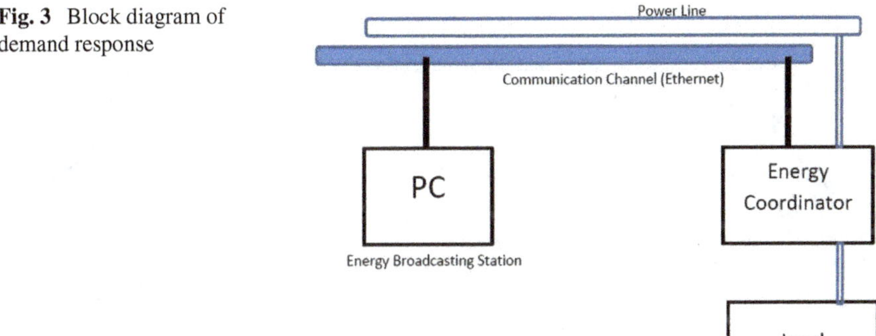

Fig. 3 Block diagram of demand response

It consists of a personal computer; it acts as a broadcaster unit which will broadcast different prices which will be called a server. Energy coordinator node is appended with logic which will take the decision to turn on or turn off the device.

A basic block diagram of energy coordinator node has been shown in Fig. 4. Coordinator consists of Ethernet module to establish connections through the Internet, and it consists of Arduino and relay board to dump the proposed algorithm so that the necessary actions can be done with the incoming information from the server side. Energy coordinator node consists of a router which will be acting as a bridge between two networks. Router and Ethernet have been connecting by wired cable called as CAT 5 cable through RJ 45 jack.

The basic flow chart has been shown in Fig. 5 which describes the working of demand response.

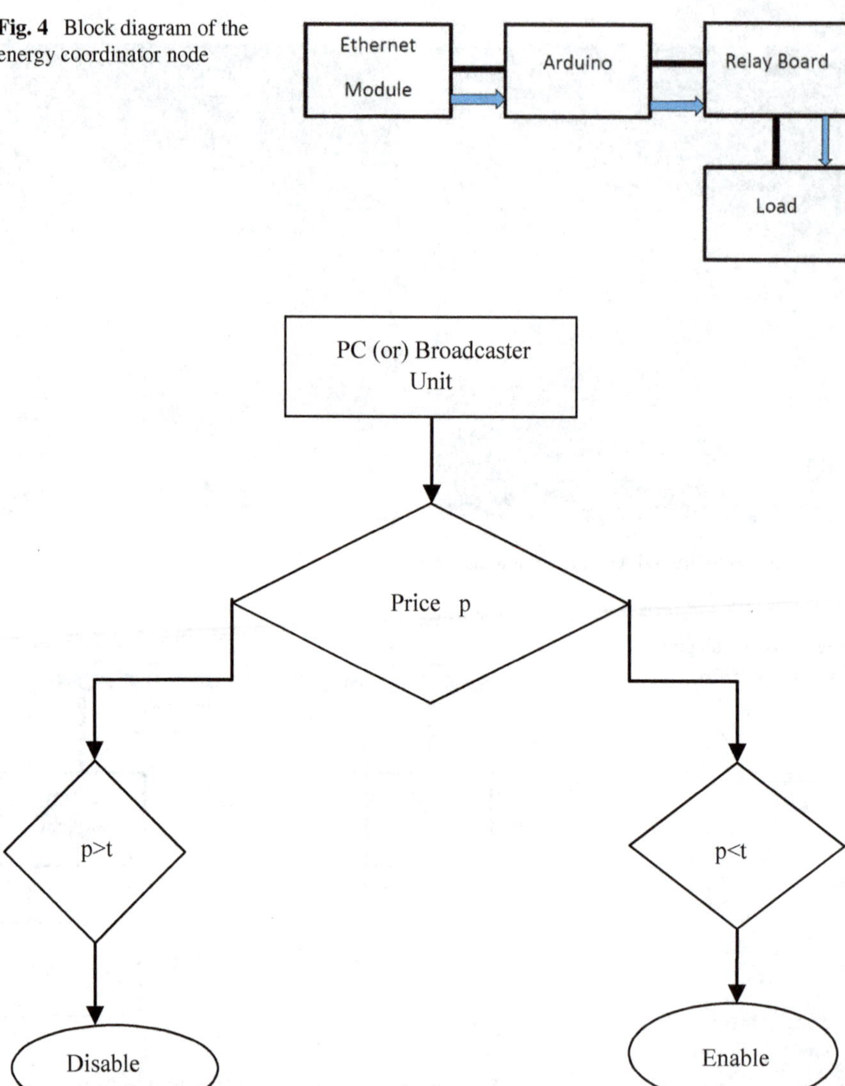

Fig. 4 Block diagram of the energy coordinator node

Fig. 5 Flowchart of demand response of smart grid

5 Results

When user switched to time of use plans, it will charge different pricing for different hours, considering three conditions, peak time, moderate, and off-peak time, and if we are using our home appliances in peak hours, it means the cost is high during this period and will lead to getting high electricity prices. We designed devices in such a way that it will track the pricing value. Whenever device encounters with the high cost, it will turn off that device, and whenever we get less cost, the device will get turn on. The results with different values have been shown.

Assuming 7R/S as high, 3R/S as moderate and 2R/S as less cost, if we are not using smart devices in our home, then it won't track the pricing value if broadcaster released 7R/S for entire day then as per average electricity usage per month of a single household will be 90 units then bill generated will be 630R/S. If we embedded home appliances with the Internet, they will track the prices from the broadcaster side and devices will get off during high peak; it will work according to condition.

If we considered 2R/S to 6R/S as less cost, then as per average utilization of the energy, household bill will be generated between 180R/S and 540R/S.

The pricing value is broadcasted from the server side which is opened by using the IP address which can be obtained by login into the router web page which has been shown in Figs. 6 and 7.

The IP address is obtained from the router home page need to consider it as a domain name and need to open it from any browser to broadcast values the web page after login with IP has been shown in Fig. 8. The output results for different pricing values have been shown in Fig. 9. The results are shown in Table 2.

Four parameters need to be identified during electricity management in smart grid power used, TOU, RTP, and day-ahead pricing.

- The first parameter defines how much electricity used during the billing period. We can reduce this by improving efficiency and lowering power consumption.
- The second parameter defines the variation of cost from time to time. It is high peak during morning and low peak during night times.
- Penalty is depends on power factor, if power factor is less penalty will be more (Table 3).

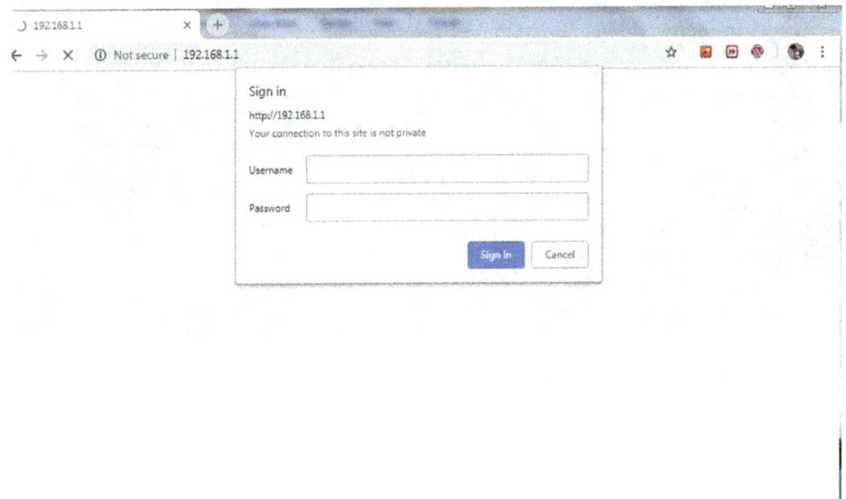

Fig. 6 Basic IP web page of router

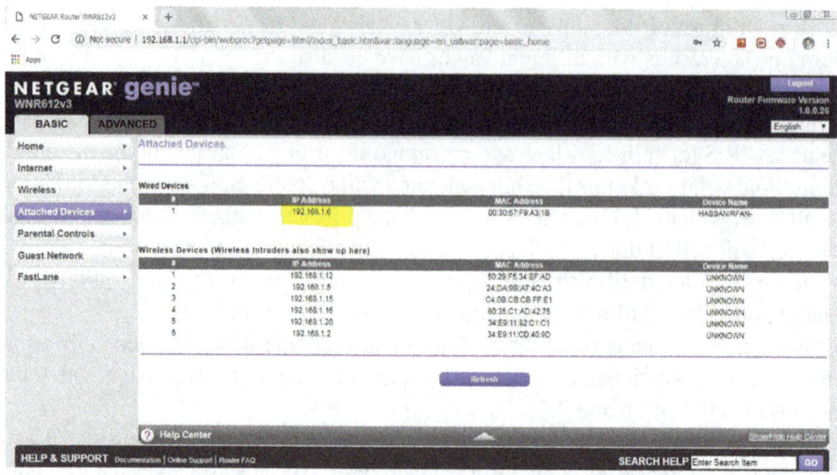

Fig. 7 Basic home page of router

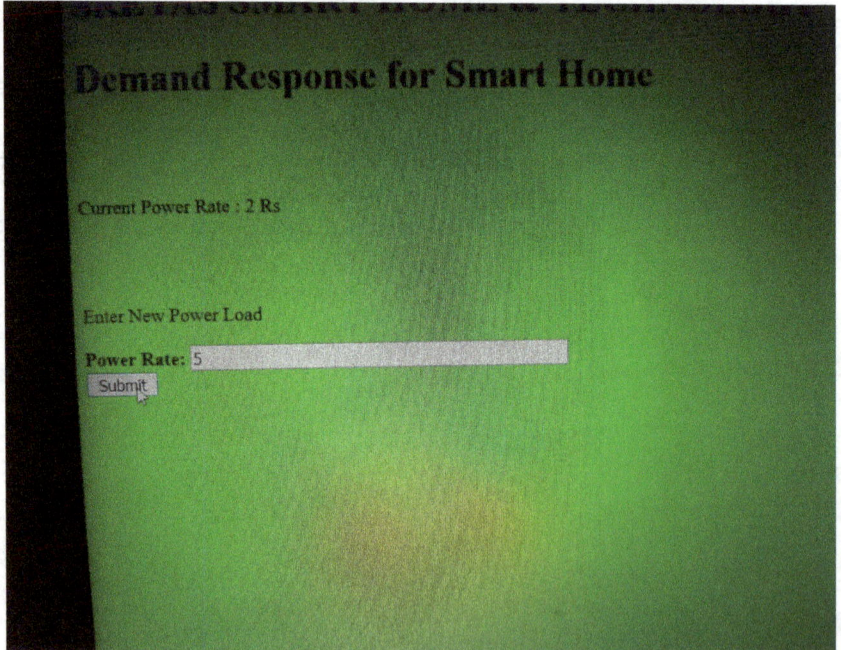

Fig. 8 Energy coordinator node Web site

(a) Loads are on for price value < 3

(b) Loads are on and off for price value < 7

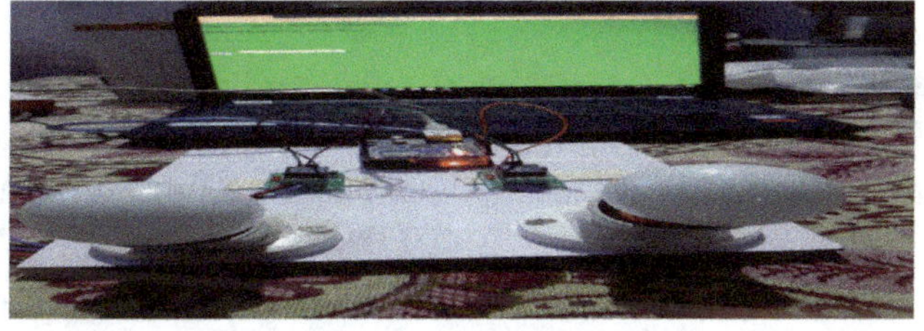

(c) Loads are off for price value >7

Fig. 9 Hardware implementation results

Table 2 Results for different pricing values

S. no.	Pricing value	Loads	The average bill generated without TOU	The average bill generated with TOU
1	6	ON and OFF	630R/S	<540R/S
2	7	OFF and OFF	630R/S	NO BILL
3	2	ON and ON	630R/S	Between 180 and 540

Table 3 Parameter comparison

TOU	RTP	Day-ahead pricing
Bill is less compared with today's method (180–540)	Not applicable We cannot estimate the savings	Constant bill Good efficiency (depends on fixed cost)

6 Conclusions

Our proposed methodology will reduce the cost usage of electricity as loads will toggle between peak and off-peak hours. We proposed a new methodology of basic client–server infrastructure model in a smart grid that totally depends on a dynamic pricing strategy to provide benefits for the consumer.

Smart devices with the Internet of things (IoT) can be able to detect price from broadcaster side and makes an intelligent decision on their own for demand-side management.

As per statistical records of the USA, the estimated cost savings will be more than 8 billion dollars after implementing these devices. The consumer can manage electricity usage by accessing energy monitoring portals. The data privacy is good as portals are password protected. In the future, many IoT-based smart grids' working models can be integrated for demand-side management using cloud computing, data analytics, and Big Data.

References

1. A.H. Abid, A. Hasan, An approach for demand-side management of non-flexible load in academic buildings, in *2018 15th International Conference on Smart Cities: Improving Quality of Life Using ICT & IoT (HONET-ICT)*, Islamabad (2018), pp. 26–28
2. A. Etxegarai, A. Bereziartua, J.A. Dañobeitia, O. Abarrategi, G. Saldaña, Impact of price-based demand response programs for residential customers, in *First IEEE International Conference on Smart Grid Communications* (2018), p. 1
3. K. Wang, Green energy scheduling for demand side management in the smart grid. IEEE Commun., 2473-2400 (2018)
4. H. Mortaji, S.H. Ow, M. Moghavvemi, H.A.F. Almurib, Load shedding and smart-direct load control using the internet of things in smart grid demand response management. IEEE Trans. Ind. Appl. **53**(6), 5155–5163 (2017)
5. O. Elma, U.S. Selamoğullari, An overview of demand response applications under smart grid concept, in *2017 4th International Conference on Electrical and Electronic Engineering (ICEEE)*, Ankara (2017), pp. 104–107
6. K. Wang, H. Li, Y. Feng, G. Tian, Big data analytics for system stability evaluation strategy in the energy internet. IEEE Trans. Ind. Inform. **13**(4), 1969–1978 (2017)
7. K. Wang, Y. Wang, X. Hu, Y. Sun, D.J. Deng, A. Vinel, Y. Zhang, Wireless big data computing in smart grid. IEEE Wirel. Commun. **24**(2), 58–64 (2017)
8. A.S.M. Ashraf Mahmud, Real-time price savings through price suggestions for the smart grid demand response model, in *IEEE International Conference on Smart Grid Communications* (2017), 5090-5938
9. L. Raju, S. Gokulakrishnan, P.R. Muthukumar, IoT based autonomous demand side management of a micro-grid using Arduino and multi-agent system. IEEE Trans. Ind. Inform. **99**, 1 (2017)
10. A. Rondoni, Simulation of a decentralized optimal demand response algorithm, in *IEEE PES Innovative Smart Grid Technologies Latin America (ISGT LATAM)* (2015)
11. M.R. Alam, M. St-Hilaire, T. Kunz, A modular framework for cost optimization in smart grid, in *2014 IEEE World Forum on the Internet of Things (WF-IoT)*, Seoul (2014), pp. 337–340
12. S. Mahajan, Q. Zhu, Y. Zhang, S. Gjessing, T. Basar, Dependable demand response management in the smart grid: a Stackelberg game approach. IEEE Trans. Smart Grid **4**(1), 120–132 (2014)
13. K. Kim, H. Bang, S. Jin, Efficient data collection for the smart grid using wireless sensor networks, in *2013 IEEE 2nd Global Conference on Consumer Electronics (GCCE)*, Tokyo (2013), pp. 231–232
14. X. Fang, S. Misra, G. Xue, D. Yang, Smart grid—the new and improved power grid: a survey. IEEE Commun. Surv. Tutor. **14**(4), 944–980 (2012)
15. Y. Zhang, R. Yu, M. Nekovee, Y. Liu, S. Xie, S. Gjessing, Cognitive machine-to-machine communications: visions and potentials for the smart grid. IEEE Netw. Mag. **26**(3), 6–13 (2012)

16. K. Wang, X. Hu, H. Li, P. Li, D. Zeng, S. Guo, A survey on energy internet communications for sustainability. IEEE Trans. Sustain. Comput. **2**(3), 231–254 (2017)
17. Y. Jiaxi, M. Anjia, G. Zhizhong, *Cyber Security Vulnerability Assessment of the Power Industry* (IEEE, 2006)
18. Y. Yan, Y. Qian, H. Sharif, D. Tipper, A survey on smart grid communication infrastructures: motivations, requirements, and challenges. IEEE Commun. Surv. Tutor. **15**(1), 5–20 (2013)
19. A. Valdes, S. Cheung, Intrusion monitoring in process control systems, in *Proceedings of the 42nd Annual Hawaii International Conference on System Sciences HICSS* (2009), pp. 1–7
20. D. Watts, Security and vulnerability in electric power systems, in *35th North American Power Symposium* (2003), pp. 559–566
21. R. Berthier, W.H. Sanders, H. Khurana, Intrusion detection for advanced metering infrastructures: requirements and architectural directions, in *2010 First IEEE International Conference on Smart Grid Communications* (2010), pp. 350–355
22. C. Efthymiou, G. Kalogridis, Smart grid privacy via anonymization of smart metering data, in *First IEEE International Conference on Smart Grid Communications* (2010), pp. 238–243
23. G. Kalogridis, C. Efthymiou, S.Z. Denic, T.A. Lewis, R. Cepeda, Privacy for smart meters: towards undetectable appliance load signatures, in *First IEEE International Conference on Smart Grid Communications* (2010), pp. 232–237
24. S. Maharjan, Y. Zhang, S. Gjessing, D.H. Tsang, User-centric demand response management in the smart grid with multiple providers. IEEE Trans. Emerg. Top. Comput. **5**(4), 494–505 (2017)
25. I. Atzeni, L.G. Ordonez, G. Scutari, D.P. Palomar, J.R. Fonollosa, Demand-side management via distributed energy generation and storage optimization. IEEE Trans. Smart Grid **4**(2), 866–876 (2013)
26. I. Atzeni, L.G. Ordonez, G. Scutari, D.P. Palomar, J.R. Fonollosa, Noncooperative and cooperative optimization of distributed energy generation and storage in the demand-side of the smart grid. IEEE Trans. Signal Process. **61**(10), 2454–2472 (2013)
27. S. Kumar, H. Dalmia, A study on internet of things applications and related issues. Int. J. Appl. Adv. Sci. Res. **2**(2), 273–277 (2017). ISSN: 2456-3080
28. S. Kumar, V.T. Kiran, S. Swetha, P. Johri, IoT based smart home surveillance and automation, in *IEEE International Conference on Computing, Power and Communication Technology (GUCON)*, 28–29 Sept 2018, pp. 795–799
29. S. Soumya, S. Kumar, Health care monitoring based on internet of things, in *The (Springer) International Conference on Artificial Intelligence & Cognitive Computing (AICC)*, Hyderabad, 2–3 Feb 2018
30. M. Hassnuddin, S. Kumar, Advance green energy scheduling in smart grid using IoT, in *International IEEE Conference on Recent Advances in Energy-Efficient Computing and Communication (ICRAECC-2019)*, 7–8 Mar 2019
31. F. Khatoon, S. Kumar, M.N.A. Khan, A study on solar remote monitoring using the internet of thing, in *2nd International Springer/Elsevier Conference on Nano Science & Engineering Applications*, 4–6 Oct 2018
32. S. Kumar, S. Anirudh, IoT and RF-ID based E-passport system, in *7th (Springer) International Conference on Innovation in Electronics and Communication Engineering (ICIECE-2018)*
33. F. Khatoon, S. Kumar, A study on E-nose and air purifier system, in *IEEE International Conference on Innovative Technologies in Engineering (ICITE-2018)*, Apr 2018
34. S. Kumar, S. Alam, M. Nelanti, A study on smart home automation based on IoT. Int. J. Adv. Innov. Res. **5**(1), 37–43 (2018). ISSN: 2394-7780
35. S. Kumar, P. Raja, G. Bhargavi, A comparative study on modern smart irrigation system and monitoring the field by using IoT, in *IEEE International Conference on Computing, Power and Communication Technology, (GUCON)*, 28–29 Sept 2018, pp. 637–641

GaN-Based High-Electron Mobility Transistors for High-Power and High-Frequency Application: A Review

P. Murugapandiyan, V. Rajya Lakshmi, N. Ramkumar, P. Eswaran and Mohd Wasim

Abstract The advancement of high-speed RF power electronics requires wide band gap semiconductor materials due to its high electron velocity, high breakdown voltage which makes the device to achieve high speed and high power simultaneously. GaN-based high-electron mobility transistors have demonstrated outstanding performance in the last two decades. This review article presents the recent development in GaN HEMT technologies for high-power microwave applications.

Keywords GaN · HEMT · Millimeter wave · High power · Breakdown voltage · Cut-off frequency

1 Introduction

GaN-based high-electron mobility transistors are optimistic applicant for future high-power and high-frequency applications. GaN-based microwave transistors are used in military applications, mobile communications, satellite communication, TV broadcasting and high-speed data transmission. Table 1 lists the major parameters of the semiconductor materials. For high-power and high-frequency operation of the transistors, large breakdown voltage and high electron velocity are required. Even though, SiC electrical properties are similar to GaN, due to critical issues in heterojunction formation and extreme process condition bounds the applications of SiC. On the other hand, GaN material can easily form the heterojunctions with other nitride alloys and thus enabling it to design high-electron mobility transistors (HEMTs) which have

P. Murugapandiyan (✉) · V. R. Lakshmi · N. Ramkumar
Department of Electronics and Communication Engineering, Anil Neerukonda Institute of Technology and Sciences, Visakhapatnam, India
e-mail: murugavlsi@gmail.com

P. Eswaran
Department of Electronics and Communication Engineering, SRM Institute of Science and Technology, Chennai, India

M. Wasim
Department of Electronics and Communication Engineering, Lovely Professional University, Jalandhar, India

H. S. Saini et al. (eds.), *Innovations in Electronics and Communication Engineering*, Lecture Notes in Networks and Systems 107,
https://doi.org/10.1007/978-981-15-3172-9_33

Table 1 Semiconductor materials properties

Material properties	Si	GaAs	SiC	GaN	Diamond
Band gap E_g (eV)	1.1	1.4	3.3	3.4	5.5
Permittivity ε_r	11.8	13.1	10	9.0	5.5
Electron mobility $\mu_n\left(\mathrm{cm^2V^{-1}s^{-1}}\right)$	1350	8500	700	1200	1900
Saturation velocity $v_{sat}\left(10^7\ \mathrm{cm/s}\right)$	1.0	1.0	2.0	2.5	2.7
Critical electric field E_{cr} (MV/cm)	0.3	0.4	3.0	3.3	5.6
Thermal conductivity $\theta\left(\mathrm{WK^{-1}m^{-1}}\right)$	150	43	330	130	2000
JFoM/JFoM$_{Si}$	1.0	1.3	20.0	27.5	50.4
BFOM/BFOM$_{Si}$	1.0	14.4	11.9	20.0	82.0

been used in various fields. Moreover, the availability of polarization-induced charge carriers (2DEG) at the heterojunction interface enables the device fabrication simple, because without the need of external doping high sheet charge density available in the 2DEG. Due to the aforementioned key features, GaN material is widely used in electronics in the past three decades. Since the first GaN-based MESFET demonstrated in 1993 [1], several substantial researches have been carried out for verifying the existence of 2DEG with higher mobility by using AlGaN/GaN HEMT structure [2–4]. And hence, the HEMT structure becomes the main structure of the GaN-based transistors [5]. For high-power switching applications, the operating frequency range is varying from kHz to MHz, where the breakdown voltage is in the range of 10–1000 V. Another important application of GaN-based transistors is high-power RF applications where the device gives maximum output power while operating at high frequency. The high power handling capability and high-frequency operation of the device are related to the breakdown voltage of the device and saturation velocity of the electron in 2DEG. Therefore, GaN-based semiconductor devices are the future forerunner of high-power microwave applications.

2 Basic Properties of III-Nitride Semiconductors

The III-N compound semiconductors are majorly classified as follows:. 1. Binary (AlN, GaN and InN) compound materials, 2. ternary compound materials ($Al_xGa_{(1-x)}N$, $In_xAl_{(1-x)}N$ and $In_xGa_{(1-x)}N$) and 3. quaternary alloys ($In_xAl_{(1-x)}Ga_{(1-x-y)}N$). The GaN is the interesting material among all in the nitride family as it can be grown with high quality on sapphire, SiC and Si substrates. This is the major reason why GaN material is used as the building blocks of a wide range of nitride-based devices such as RF power transistor, high-frequency MMICs, laser diode light-emitting diodes, MEMS and power conversion. However, GaN alone is not creating the devices. It includes the AlN, AlGaN, InGaN or InAlGaN layers for creating heterostructures for functional electronics and optoelectronics. In this section, we revisit the III-N semiconductors, in particular, more relevant to electronic devices.

Table 2 Band gap E_g and lattice constant a and c of binary nitride and together with bowing parameter b of ternary nitride alloys [6, 7]

Material	E_g (eV)	a (Å)	c (Å)	Alloy	b (eV)
AlN	6.14	3.112	4.982	AlGaN	0.7
GaN	3.42	3.189	5.185	InGaN	1.4
InN	0.64	3.545	5.703	InAlN	5.36

2.1 Band Structure and Lattice Constant of III-N Semiconductors

As a consequence of the strength of the metal-nitrogen bonds, the III-nitrides are characterized, except for InAlN, by larger band gaps and shorter lattice constants compared to other III-V compound semiconductors families, like arsenides, phosphides, antimonides and their alloys. The band gap and lattice constant of binary nitride materials are listed in Table 2. Band gaps and lattice constant of nitride semiconductor alloys are shown in Fig. 1 along with other III-V semiconductor materials. The energy band gaps of $Al_xGa_{(1-x)}N$, $In_xAl_{(1-x)}N$ and $In_xGa_{(1-x)}N$ are shown in the following quadratic equation:

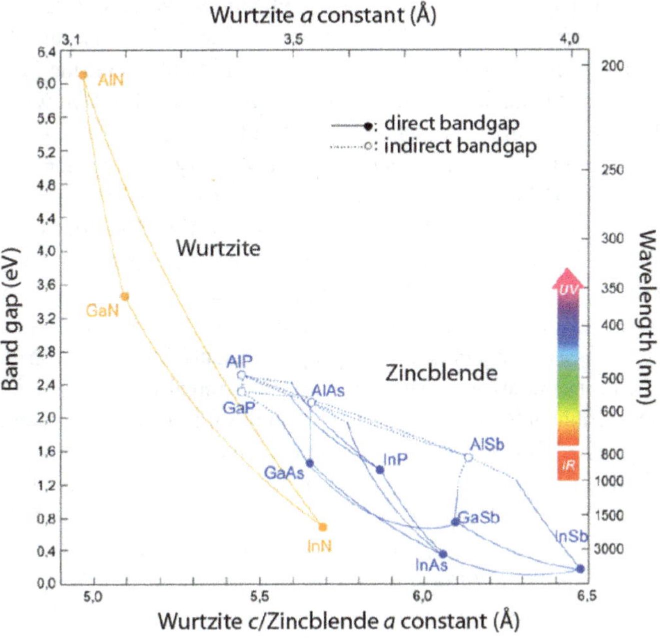

Fig. 1 Band gap of wurtzite III-nitrides and classical zincblende III-V semiconductors as a function of lattice constants

$$E_{g,A_xB_{1-x}N} = xE_{g,AN} + (1 - x)E_{g,BN} - b_{ABN}x(1 - x) \quad (1)$$

where A and B stand for Al, Ga or In depending on the compound desired and *b* is the bowing parameter. The III-N family bowing parameter is listed in Table 2.

A second distinctive feature of III-nitrides with respect to classical III-V semiconductors is their crystal structure. III-nitride adopts the wurtzite structure, which is the stable phase, whose main characteristics are the hexagonal symmetry and the lack of an inversion center. The basic wurtzite cell is illustrated in Fig. 1. Classical III-Vs share instead of the more symmetric, cubic zincblende structure. The '*a*' and '*c*' constants describing the unit cell of binary III-nitrides are reported in Table 2. The lattice constants of the alloys can be easily obtained by linear interpolation of the corresponding binary compounds, without the need of introducing any bowing parameters. As III-nitride devices and heterostructures are generally grown along the *x*-axis, the '*a*' lattice constant is the most meaningful one when considering heterostructures.

3 Advantages of III-N Semiconductor

The nitrides have several features among the large semiconductor materials, which make the interest for various electronics applications where compact devices are required to efficiently handle high power levels. The wide band gap of GaN- and III-nitride-based alloys makes the breakdown voltage of III-nitride-based devices generally higher than analogous devices of the same size fabricated with Si or classical III-Vs. At the same time, with respect to other wide band gap materials like SiC, GaN is characterized by a higher saturation velocity, which means that parasitic resistances can be reduced, making the devices more energetically efficient. A final strength of III-nitride semiconductors is that they easily allow for heterostructure-based devices. Therefore, it is possible to play with band engineering for improving device performance, as for classical III-Vs, for example through the introduction of a quantum well (QW) or 2DEGs, while this is not possible for SiC- or diamond-based devices. As a consequence of their properties, GaN-based electron devices have found applications mainly in two fields: power electronics and microwave transistors. Table 1 shows Johnson's figure of merit and Baliga's figure of merit of semiconductor materials [8]. The figure of merits is normalized to Si.

In the field of power electronics, GaN has allowed the realization of devices with high breakdown voltage (V_{BR}) and very low specific on-resistance (R_{on}). While a high breakdown voltage (V_{BR}) is required to the device handling high power densities, a low R_{on} is necessary to minimize the power dissipation in the ON-state. The figure of merit that best measures the potentials of a semiconductor for power electronics applications is Baliga's figure of merit (BFOM) [9]:

$$\text{BFOM} = \epsilon\mu E_{BR}^3 \quad (2)$$

where ε is the dielectric constant of semiconductor, μ is the electron mobility in the drift region and E_{BR} the electric field for the breakdown. As BFOM describes the best possible trade-off between V_{BR} and R_{on}, an equivalent and suitable expression, more suitable for comparing real devices, is the following one:

$$\text{BFOM} = \frac{V_{BR}^2}{R_{on}} \quad (3)$$

The theoretical BFOM for the principal semiconductors is listed in Table 1, and the summary of V_{BR} and R_{on} data for published devices is presented in Fig. 2a. From this, it can be seen that both theoretically and practically, GaN-based devices offer reduced R_{on} and thus lower dissipation. At present, high-performance GaN-based P-N diode has been demonstrated with 3.7 kV breakdown voltages [10], and GaN-based vertical transistors with low specific on-resistance and breakdown voltage as high 1.5 kV have been reported [11]. This indicates that GaN holds great potential for low/medium voltage power circuits and is being considered seriously for improving the efficiency of DC-DC converters [12], LED [13] and motor drivers [14], etc.

For microwave transistors, which are the objectives of this research work, GaN is an interesting material as it combines a high breakdown voltage and is associated with high saturation electron velocity. Analogously to power devices, a high breakdown voltage is fundamental for increasing the power density a transistor can handle. A high saturation velocity is instead advantageous to reduce the electron transit time under the gate electrode or across the base region, and therefore the switching time. Combining these two features, GaN-based microwave transistors and amplifiers hold the promise of delivering very high output power levels and at the same time high cut-off frequencies. The suitability of a material for power microwave applications can be measured by means of Johnson's figure of merit (JFoM) [25]:

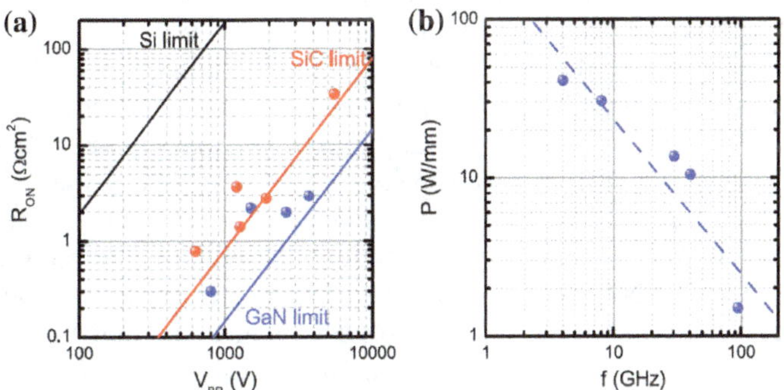

Fig. 2 **a** Dots: specific ON-resistance versus breakdown voltage for SiC [15–18] and GaN devices [10, 11, 13, 19], lines: theoretical limits for Si, SiC and GaN. **b** Best reported output power density at various microwave frequencies for AlGaN/GaN HEMTs [20–24]

$$\text{JFoM} = \frac{E_{\text{BR}} \upsilon_{\text{sat}}}{2\pi} \quad (4)$$

where υ_{sat} is the saturated velocity. JFoM quantities the trade-off between V_{BR} and device speed, which is quantified by the cut-off frequency f_{T}. Thus, an equivalent definition suitable for comparing device is the following one:

$$\text{JFoM} = V_{\text{BR}} f_{\text{T}} \quad (5)$$

The JFoM for the semiconductors is listed in Table 1. It can be seen that GaN JFoM is ten times that of GaAs and is surpassed only by diamond. However, if we consider that diamond is not easy to dope and does not easily allow for heterostructures, GaN emerges as the ultimate semiconductor for RF applications. GaN-based heterostructure has been used for the realization of 2DEG-based devices, in particular, HEMTs, which have demonstrated in the last 15 years breakdown voltages much higher than III-V-based HEMTs or bipolar transistors [26]. As a direct consequence, nitride-based HEMTs have reached record power outputs and efficiencies and are currently available on the market for RF systems requiring high power levels, like cellular base stations. Figure 2b summarizes the highest power levels reached by AlGaN/GaN HEMTs, which have been the workhorse for the RF community in the past years.

4 Advanced Fabrication Technologies for Enhancing High-Frequency and High-Power Operation

Eblabla et al. [27] fabricated 300 nm AlGaN/GaN HEMT on Si for X-band application. The device features T-gate technology for achieving low gate resistance and enhanced mobility. Zhao [28] experimentally verified the impact of AlGaN back barrier on AlGaN/GaN HEMT characteristics. The inclusion of AlGaN back barrier layer resulted in enhanced 2DEG density, reduction in the buffer leakage current and high breakdown voltage. Jung et al. [29] designed Al_2O_3 passivated recessed T-gate InAlN/GaN HEMT for improving the transconductance and sheet resistance. And also, the passivated device surfaces improved the mobility of the electrons in the channel. Wang et al. [30] demonstrated SiN passivated device surface with recessed T-gate InAlN/GaN HEMT DC and RF characteristics, A significant reduction in the subthreshold slope (SS) as the result of good aspect ratio by recessed gate structure. Moreover, the improved drain current density and transconductance were observed with good $I_{\text{on}}/I_{\text{off}}$ ratio. Yue et al. [31] proposed InAlN/GaN MOSHEMT with regrown n + GaN ohmic contact. The proposed HEMT achieved a good $I_{\text{on}}/I_{\text{off}}$ ratio, drastic reduction in contact resistances, lower the DIBL (drain-induced barrier lowering) and significant improvements in current gain cut-off frequency. Chunjiang et al. [32] fabricated AlGaN/GaN HEMT structure for millimeter wave applications.

Fe-doped GaN channel with asymmetric gate position improved the breakdown voltage of the device by reducing the buffer leakage currents. Jie et al. [33] designed T-gate AlGaN/GaN with heavily doped source/drain region. The designed HEMT structure had shown very low contact resistance and significant improvement in cut-off frequency. Higashiwaki and Matsui [34] proposed the AlGaN/GaN HEMT with high Al content in barrier layer with SiN passivation T-gate (Ti/Pt/Au) structure. Improved RF characteristics with high-sheet charge carrier density are achieved by the proposed device structure. Tsou et al. [35] presented SiN passivated device surface T-gate InAlN/GaN HEMT. The T-shaped gate reduced the parasitic resistance and enabled high cut-off frequency. Lee et al. [36] designed InAlN/GaN MOSHEMT with AlGaN back barrier for improving the breakdown voltage of the device by reducing the gate leakage current and buffer leakage current. The device achieved higher mobility with improved breakdown voltage. More breakdown voltage is achieved by keeping the gate terminal nearer to source.

5 Conclusion

From the literature review, to optimize the GaN-based HEMT structure for high-power microwave application, the following essential key factors are identified: The carrier confinement can be improved by employing a double heterojunction structure instead of a single heterojunction. Two basic approaches to improve the carrier confinement and mobility of the GaN-based HEMTs are: AlN spacer layer and InGaN or AlGaN back barrier. The wide band gap AlN spacer layer is situated between the barrier, and channel improves the gate leakage characteristics of the HEMT as well as provides effective conduction band offset for enabling large polarization-induced high sheet carrier density in the GaN channel. This spacer layer also reduces the impurity scattering which leads to enhance the carrier transport of channel. The double heterojunction HEMT by using AlGaN/InGaN back barriers confines more number of sheet charges in the 2DEG quantum well, which also minimizes the buffer leakage current and short-channel effects (subthreshold slope), and as a consequence, the breakdown voltage of the HEMT is improved. T-shaped gate structure with optimized stem height and foot width and passivation layer reduces gate capacitance and resistance. The ultra-scaled gate length with minimum drain–source spacing enhances the carrier mobility under the gate, which improves the RF characteristics of HEMT devices. This heavily doped n-type GaN source and drain regions with ohmic contacts are helped to reduce the access resistances. It also minimizes the source–drain separation which in turn enhances the electron velocity under the gate. To minimize the short-channel effects and to improve the maximum frequency oscillation, recessed gate enhancement mode HEMT structure is preferred.

Acknowledgements The authors acknowledge the Nanoelectron Devices and Circuits Laboratory of Electronics and Communication Engineering Department at Anil Neerukonda Institute of Technology and Sciences, Andhra Pradesh, India, for providing all facilities to carry out this research work.

References

1. M. Asif Khan, J.N. Kuznia, A.R. Bhattarai, D.T. Olson, Metal semiconductor field effect transistor based on single crystal GaN. Appl. Phys. Lett. **62**(15), 1786 (1993)
2. O. Ambacher, J. Smart, J.R. Shealy, N.G. Weimann, K. Chu, M. Murphy, W.J. Schaff, L.F. Eastman, R. Dimitrov, L. Wittmer, M. Stutzmann, W. Rieger, J. Hilsenbeck, Two dimensional electron gases induced by spontaneous and piezoelectric polarization charges in N- and Ga-face AlGaN/GaN heterostructures. J. Appl. Phys. **85**(6), 3222–3233 (1999)
3. M.A. Khan, Q. Chen, C.J. Sun, M. Shur, B. Gelmont, Two-dimensional electron gas in GaN-AlGaN heterostructures deposited using trimethylamine-alane as the aluminum source in low pressure metalorganic chemical vapor deposition. Appl. Phys. Lett. **67**(10), 1429–1431 (1995)
4. M.A. Khan, J.N. Kuznia, J.M. Van Hove, N. Pan, J. Carter, Observation of a two dimensional electron gas in low pressure metalorganic chemical vapor deposited GaN-$Al_xGa_{1-x}N$ heterojunctions. Appl. Phys. Lett. **60**(24), 3027–3029 (1992)
5. M. Asif Khan, A. Bhattarai, J.N. Kuznia, D.T. Olson, High electron mobility transistor based on a GaN-AlGaN heterojunction. Appl. Phys. Lett. **63**(9), 1214 (1993)
6. E. Sakalauskas, H. Behmenburg, C. Hums, P. Schley, G. Rossbach, C. Giesen, M. Heuken, H. Kalisch, R.H. Jansen, J. Bläsing, J. Dadgar, A. Krost, R. Goldhahn, Dielectric function and optical properties of Al-rich AlInN alloys pseudomorphically grown on GaN. J. Phys. D Appl. Phys. **43**, 365102 (2010)
7. I. Vurgaftman, J.R. Meyer, Band parameters for nitrogen-containing semiconductors. J. Appl. Phys. **94**, 3675 (2003)
8. O. Ambacher, J. Majewski, C. Miskys, A. Link, M. Hermann, M. Eickhoff, M. Stutzmann, F. Bernardini, V. Fiorentini, V. Tilak, B. Schaff, L.F. Eastman, Pyroelectric properties of Al(In)GaN/GaN hetero- and quantum well structures. J. Phys.: Condens. Matter **14**, 3399 (2002)
9. B.J. Baliga, Power semiconductor device figure of merit for high-frequency applications. IEEE Electron Device Lett. **10**, 455 (1989)
10. I.C. Kizilyalli, A.P. Edwards, H. Nie, D. Bour, T. Prunty, D. Disney, 3.7 kV vertical GaN PN diodes. IEEE Electron Device Lett. **35**, 247 (2014)
11. H. Nie, Q. Diduck, B. Alvarez, A.P. Edwards, B.M. Kayes, M. Zhang, G. Ye, T. Prunty, D. Bour, I.C. Kizilyalli, 1.5-kV and 2.2-mΩ-cm^2 vertical GaN transistors on bulk-GaN substrates. IEEE Electron Device Lett. **35**, 939 (2014)
12. J. Das, J. Everts, J. Van den Keybus, M. Van Hove, D. Visalli, P. Srivastava, D. Marcon, K. Cheng, M. Leys, S. Decoutere, J. Driesen, G. Borghs, A 96% efficient high-frequency DC-DC converter using E-mode GaN DHFETs on Si. IEEE Electron Device Lett. **32**, 1370 (2011)
13. D. Disney, H. Nie, A. Edwards, D. Bour, H. Shah, I.C. Kizilyalli, Vertical power diodes in bulk GaN, in *Proceedings of the International Symposium on Power Semiconductor Devices and IC's (ISPSD)* (2013), p. 59
14. T. Morita, S. Tamura, Y. Anda, M. Ishida, Y. Uemoto, T. Ueda, T. Tanaka, D. Ueda, 99.3% efficiency of three-phase inverter for motor drive using GaN-based gate injection transistors, in *Proceeding of the IEEE Applied Power Electronics Conference and Exposition* (2011), p. 481
15. D. Sheridan, A. Ritenour, V. Bondarenko, P. Burks, J. Casady, Record 2.8 mΩ-cm^2 1.9 kV enhancement-mode SiC VJFETs, in *Proceedings of the International Symposium on Power Semiconductor Devices and IC's (ISPSD)* (2009), p. 335

16. S. Balachandran, C. Li, P.A. Losee, I.B. Bhat, T.P. Chow, 6 kV 4H-SiC BJTs with specific on-resistance below the unipolar limit using a selectively grown base contact process, in *Proceedings of the International Symposium on Power Semiconductor Devices and IC's (ISPSD)* (2007), p. 293
17. A. Furukawa, S. Kinouchi, H. Nakatake, Y. Ebiike, Y. Kagawa, N. Miura, Y. Nakao, M. Imaizumi, H. Sumitani, T. Oomori, Low on-resistance 1.2 kV 4H-SiC MOSFETs integrated with current sensor, in *Proceedings of the International Symposium on Power Semiconductor Devices and IC's (ISPSD)* (2011), p. 288
18. T. Nakamura, Y. Nakano, M. Aketa, R. Nakamura, S. Mitani, H. Sakairi, Y. Yokotsuji, High performance SiC trench devices with ultra-low RON, in *Proceedings of the International Symposium on Power Semiconductor Devices and IC's (ISPSD)* (2011), pp. 26.5.1
19. I.C. Kizilyalli, A.P. Edwards, H. Nie, D. Disney, D. Bour, High voltage vertical GaN p-n diodes with avalanche capability. IEEE Trans. Electron Devices **60**, 3067 (2013)
20. Y.F. Wu, A. Saxler, M. Moore, R.P. Smith, S.T. Sheppard, P.M. Chavarkar, T. Wisleder, U.K. Mishra, P. Parikh, 30-W/mm GaN HEMTs by field plate optimization. IEEE Electron Device Lett. **25**, 117 (2004)
21. Y.F. Wu, M. Moore, A. Saxler, T. Wisleder, P. Parikh, 40-W/mm double field-plated GaN HEMTs, in *Proceedings of IEEE Device Research Conference* (2006), p. 152
22. Y.F. Wu, M. Moore, A. Abrahamsen, M. Jacob-Mitos, P. Parikh, S. Heikman, A. Burk, High-voltage millimeter-wave GaN HEMTs with 13.7 W/mm power density, in *IEDM Technical Digest* (2007), p. 405
23. T. Palacios, A. Chakraborty, S. Rajan, C. Poblenz, S. Keller, S.P. DenBaars, J.S. Speck, U.K. Mishra, High-power AlGaN/GaN HEMTs for Ka-band applications. IEEE Electron Device Lett. **26**, 781 (2005)
24. D.F. Brown, A. Williams, K. Shinohara, A. Kurdoghlian, I. Milosavljevic, P. Hashimoto, R. Grabar, S. Burnham, C. Butler, P. Willadsen, M. Micovic, W-band power performance of AlGaN/GaN DHFETs with regrown n+ GaN ohmic contacts by MBE, in *IEDM Technical Digest* (2011), p. 461
25. E.O. Johnson, Physical limitations on frequency and power parameters of transistors. RCA Rev. **26**, 163 (1965)
26. U.K. Mishra, L. Shen, T.E. Kazior, Y.F. Wu, GaN-based RF power devices and amplifiers. Proc. IEEE **287**, 96 (2007)
27. R.C. Fitch, D.E. Walker, A.J. Green, S.E. Tetlak, J.K. Gillespie, R.D. Gilbert, K.A. Sutherlin, W.D. Gouty, J.P. Theimer, G.D. Via, K.D. Chabak, G.H. Jessen, Implementation of high-power-density X-band AlGaN/GaN high electron mobility transistors in a millimeter-wave monolithic microwave integrated circuit process. IEEE Electron Device Lett. **36**(10) (2015)
28. S.L. Zhao, Analysis of the breakdown characterization method in GaN-based HEMTs. IEEE Trans. Power Electron. **31**(2) (2016)
29. J.W. Chung, O.I. Saadat, J.M. Tirado, X. Gao, S. Guo, T. Palacios, Gate-recessed InAlN/GaN HEMTs on SiC substrate with Al_2O_3 passivation. IEEE Electron Device Lett. **30**(9) (2009)
30. R. Wang, P. Saunier, X. Xing, C. Lian, X. Gao, S. Guo, G. Snider, P. Fay, D. Jena, H. Xing, Gate-recessed enhancement-mode InAlN/AlN/GaN HEMTs with 1.9-A/mm drain current density and 800-mS/mm transconductance. IEEE Electron Device Lett. **31**(12) (2010)
31. Y. Yue, Z. Hu, InAlN/AlN/GaN HEMTs with regrown ohmic contacts and f_T of 370 GHz. IEEE Electron Device Lett. **33**(7) (2012)
32. R. Chunjiang, L. Zhonghui, Y. Xuming, W. Quanhui, W. Wen, C. Tangsheng, Z. Bin, Field plated 0.15 μm GaN HEMTs for millimeter-wave application. J. Semicond. **34**(6) (2013)
33. H. Jie, L. Ming, T. Chak-Wah, L. Kei-May, $L_g = 100$ nm T-shaped gate AlGaN/GaN HEMTs on Si substrates with non-planar source/drain regrowth of highly-doped n^+ GaN layer by MOCVD. Chin. Phys. B **23**(12) (2014)
34. M. Higashiwaki, T. Matsui, AlGaN/GaN heterostructure field-effect transistors with current gain cut-off frequency of 152 GHz on sapphire substrates. Jpn. J. Appl. Phys. **44**(16), L 475–L 478 (2005)

35. C.-W. Tsou, C.-Y. Lin, Y.-W. Lian, S.S. Hsu, 101-GHz InAlN/GaN HEMTs on silicon with high Johnson's figure-of-merit. IEEE Trans. Electron Devices **62**(8) (2015)
36. H.-S. Lee, D. Piedra, M. Sun, X. Gao, S. Guo, T. Palacios, 3000-V 4.3-mΩ cm^2 InAlN/GaN MOSHEMTs with AlGaN back barrier. IEEE Electron Device Lett. **33**(7) (2012)

Elapsed Time Counter (ETC) for Power Monitoring System

Vaibhav Sugandhi, Nalini C. Iyer, Aishwarya Pattar and Saroja V. Siddamal

Abstract This paper proposes an application of elapsed time counter (ETC) for power monitoring system as a relatively new arrival with reference to DS1682 IC embedded system. The main goal of the system is to monitor real-time power consumption of any random appliances or different rooms in building with logging of events. The core features of the DS1682 are being fundamentals for the embedded system with ATmega328P controller support to computation and analysis. The ETC IC is dedicated for event counting and elapsed time counting with no external crystal oscillator or any digital clock devices. The ragged applications in industry need power monitoring system with event counting and elapsed time logging, which is optimizing the power consumption analysis in room/building and(or) specific application. The current RTC-based system is incompetent with a vigorous change in environmental factors, which leads to improper analysis and impact. This paper focuses on design and development of ETC-based embedded board for real-time power monitoring and event logging with no real-time clock (RTC).

Keywords DS1682 · Elapsed time counter · Event logging · Elapsed time recorder · Event counter · Power state analysis · Power consumption analysis · Power monitoring

V. Sugandhi (✉) · N. C. Iyer · A. Pattar · S. V. Siddamal
School of Electronics and Communication Engineering, KLE Technological University, Hubli 580031, India
e-mail: vaibhavksugandhi5566@gmail.com

N. C. Iyer
e-mail: nalinic@bvb.edu

A. Pattar
e-mail: aishwarya.pattar363@gmail.com

S. V. Siddamal
e-mail: sarojavs@bvb.edu

H. S. Saini et al. (eds.), *Innovations in Electronics and Communication Engineering*, Lecture Notes in Networks and Systems 107,
https://doi.org/10.1007/978-981-15-3172-9_34

1 Introduction

Many counter-based devices are implemented from the past years, but the storing data capacity had to be externally interfaced. Sometimes there might be an error while keeping a track of many interrupts that occur during the count. To effectively solve the issue of counting the events along with their elapsed time, DS1682 can be the beneficial IC. Continuous as well as precise counting along with the time interval can be fetched through the event pin. To make the IC more effective, the communication is through the I2C protocol standards which can be used to interface external devices easily. Cumulatively, a single IC can do the evaluation of total counts and the duration of the event time since the start of the first event count. The total active period of the event is the main count the user intends to notice for deciding upon the particular application. The collective period of all events duration is up counted in elapsed time counter register and stored in EEPROM located inside DS1682. The detailed approach to this research is explained in methodology with pictorial representation. We explored the work that has been done in the area of elapsed time counter which is the duration for which the event was active. As per the datasheet of elapsed time counter [1], the necessary features are extracted and analysis of the DS1602 [2], DS1682 [3] and DS1683 IC [4] are done. With respect to DS1602, the main notice was that the manufacturing of the IC was stopped and there were no more applications built on the features of DS1602 [5]. Hence, the opted DS1682 as a research model includes EEPROM for event counter and elapsed time recording.

As a result of datasheet analysis with electrical parameters and programming compatibilities, the ATmega328P microcontroller is used for communication, analysis and information display. The vigorous internal and external features of ATmega328P are becoming supporting for development of this embedded board. The power unit of the system is designed with lithium-ion batteries and some DC–DC buck convertors, after finding the best features and electrical parameters of lithium-ion technology. The charge time and discharge time are competitively efficient with respect to other batteries. The long-run applications with portable or retro-fit architecture are more benefited with this technology. The split-core transformer technology is being one more advantageous to our proposed design, which helps to monitor the power flow in a conductor with non-invasive method of sampling data. The voltage variations from sensor are modulated and re-mapped to analytical scale for betterment of results. The power consumption of building or different rooms is done with wired energy meters and with pulse sense technology. This method leads to accurate and commercial way of energy monitoring, but no analytics or wireless communication is interfaced with it. The main problem identified while survey is all about no information about power failure period and power losses due to unknown errors.

2 Proposed Methodology

ETC for power monitoring system is mainly inculcated with the DS1682 IC. The IC works with respect to the event triggering and elapsed time counting [1]. The DS1682 can be used in two ways depending on which period of power cycle we intended to measure, which depends on the application. Selecting the less duration count is beneficial, since time elapse would be very less and data analysis could be faster. I2C standard protocol-based SDA and SCL lines are used to interact with internal registers. The serial bus is connected to microcontroller to trigger an event start. Battery pack with 7.4 V is provided through the DC–DC buck convertor for power regulations. The DS1682 internal event count is displayed on the I2C-based 20 × 4 LCD module through the EEPROM content. The glitch created while shifting the content to the counter is overcome through the switch de-bouncer algorithm in programming.

Initially, the ATmega328P-AU is pre-loaded with the boot loader through the ISP programming cables which are set on the testing board, so that the IC can communicate with the computer. Once the process is done, the microcontroller is ready to use and it is a one-time activity. The interaction between the IC and the controller is by I2C bus and event pin. The main goal of the proposed system is to focus on elapsed time monitoring, which is indeed monitoring the power ON or OFF period from the pre-set value of 80 W. As mentioned in introduction part, the split-core sensor is used for power value measurement; hence, the controller is having option to setting through program. The accuracy and data efficiency of the designed system are improved by utilizing the 10 bit successive approximation resolution (SAR)-type ADC in ATmega328P.

As shown in Fig. 1, the block-level architecture of system is representing the interaction of peripherals with microcontroller. The microcontroller is interconnected with other peripherals like DS1682, I2C LCD module, SCT current sensor and battery circuit. As an objective of the methodology, the ATmega328P is always kept in active mode for every single event monitoring. Furthermore, the detailed framework is used for better representation of ETC for power monitoring system.

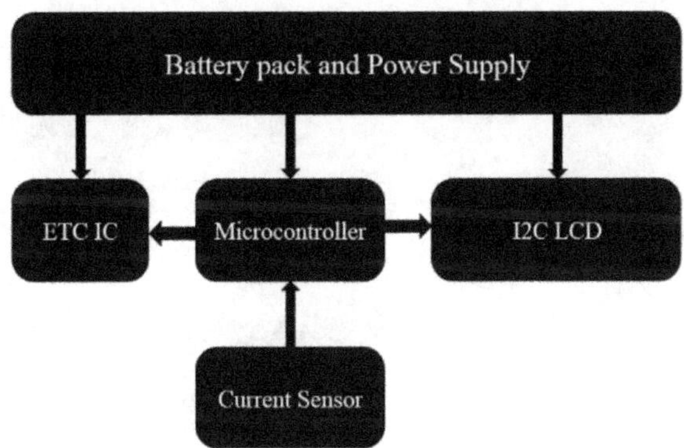

Fig. 1 Block-level architecture of system

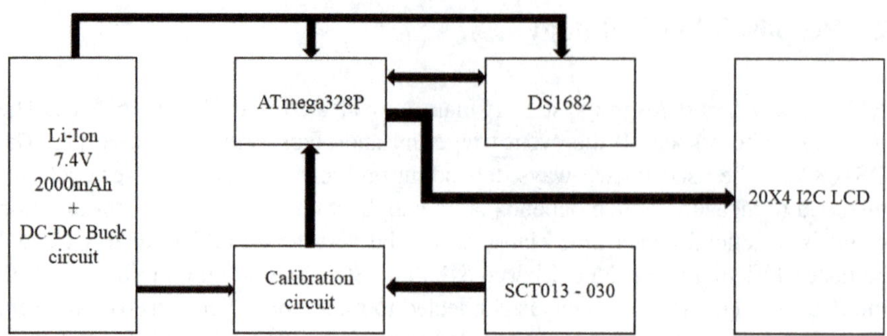

Fig. 2 Process methodology

As depicted in Fig. 2, the basic propagation of system actions is listed. Upon power-up condition, the system is initialized and I2C and other communication protocols are activated. The last pre-loaded value in EEPROM will be read from DS1682 for first screen display. The total events generated so far and total elapsed time will be stored in volatile variable for computation. Since split-core current sensor is powered up with reference voltage of 5 V, which monitors current flow in conductor subjected to monitor power consumption. Once power value reaches pre-set value, then it will trigger the event interrupt from microcontroller. The same sensor is responsible for resettling event as power flow goes less than pre-set.

3 System Architecture

Figure 3 represents proposed framework or system architecture from the block-level design. The individual peripherals of the design are listed with specifications and model number, which ease to demonstrate the working process. The ATmega328P being interconnected with each and every blocks, the communication and data analysis is done within it using pre-designed library for DS1682 ETC computation [6]. The

Fig. 3 Architecture of ETC for power monitoring system

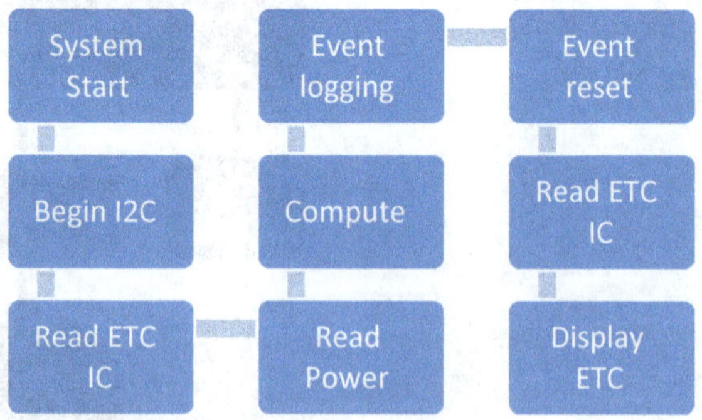

Table 1 Design specification of ETC power monitoring system

Parameters	Values
Operating voltage	4.5–5.5 V
External battery pack	7.4 V 2600 mAh
I2C LCD module	20 × 04
SCT sensor	30 A/1 A
Power consumption	2.75–3.5 Wh
Input voltage	6.2–8.2 V DC
Input current	2.6–10 Ah

SCT013-030 modeled split-core current sensor is used for monitoring the real-time power flow in a subjected conductor with non-invasive method. DS1682 IC is being the core part of the embedded system, which holds the elapsed time for all the events it gets from controller. The lithium-ion cells and DC–DC buck circuit are used for power supply design. And nonetheless, I2C-based LCD module is used for display of all information. The DS1682 is interconnected with ATmega328P through the I2C protocol interfacing. The evaluation data is stored in the internal EEPROM of the IC which can be accessed and read on the display. The triggering can happen as active high/low of event pin based on the user's input. The external rechargeable batteries help in increasing the durability of the product. The crystal oscillator facilitates the frequency synchronization to ATmega328P microcontroller and other peripherals connected. Initially, the boot loader can be burned into the controller through the SPI protocol and programmed for dedicated application. Depending on the application, the device can be modified and made user-friendly. The DS1682 can be made to work continuously for about a 3 years through the Li-ion battery with higher capacity.

Table 1 shows the design parameters of ETC for power monitoring system, which includes typical values of electronic modules used. It also helps to tune the system for upgrading of higher-level applications. The 30 A/1 A SCT sensor helps to monitor 6.9 kWh load at a given time, but if any application needs to monitor beyond this scale, SCT sensor can be upgraded to 100 A/1 A or more. Input voltage and current are depended on Li-Ion battery pack used in the system, which helps to keep system optimistically ON for all the time. While no power changes or no event triggering, the controller makes use of deep sleep mode option to consume less power from battery. To extend battery life or avoid frequent charging, battery pack can be upgraded to 10 Ah or beyond.

4 Result and Discussion

The design of ETC for power monitoring system is implemented using DS1682, and it is validated for its functionality as shown in Figs. 4 and 5. The plug-in mechanism of the ETC for power monitoring system helps the user to monitor the usage of power

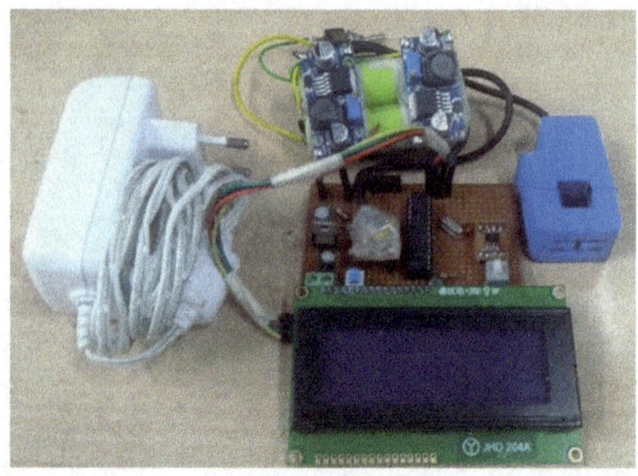

Fig. 4 Final working model without casing

Fig. 5 Final working model with casing

at any convenient location. Since the IC has 34 years of working ability for counting elapsed time, the embedded board remains durable and long-running once setup. The deep sleep technique helps to use the kit for many months without recharging batteries connected. When the power usage is more than the threshold limit or pre-set value, an event is generated by controller which is monitored indicating the user that the appliance is drawing more energy or turned on with full load. This can be used to indicate the unnecessary wastage or consumption of electricity at the installed location, so that immediate action can be taken.

The model developed and proposed in this paper is POC for validation, which can be modified, manufactured and implemented in real-time power monitoring for proposed work. Along with 12 V/2 A adaptor is provided to user, so that timely recharge of batteries can be maintained.

Table 2 illustrated first testing analysis of proposed system. It includes one cycle of event triggering and resettling. As discussed in the introduction, pre-set value of this system is 80 W, which triggers event as per the power consumption of load. The sampling time T1 and T2 are below pre-set value, hence no event triggered from

Table 2 Test results of proposed system

Sl. no.	Power readings in Watts	Sampling time Tx	Event state	Elapsed time in seconds
1	45	T1	–	0.00
2	70	T2	–	0.00
3	85	T3	Triggered	1.25
4	125	T4		2.50
5	250	T5		3.75
6	175	T6		5.00
7	120	T7		6.25
8	90	T8		7.50
9	65	T9	Resettled	8.75
10	40	T10		8.75

controller to start ETC. But sampling time T3 to T9 event triggered and started ETC to count elapsed time, which results in 8.75 s elapse for power consumption beyond pre-set value.

5 Applications

The applications of the "ETC for power monitoring system" is being clear with its title as mentioned, but still to elaborate the sectors and areas where this embedded board can be enhanced with applications [7] is listed and explained below,

1. ***Real-time power monitoring***: The power consumption of any device or machinery in the industry is measured and monitored for power calculations to enhance optimization by decreasing unexpected power losses.
2. ***Machinery warranty tracking***: The warranty period of any machines is being updated and evaluated based on usage time, rather than dates on the calendar.
3. ***Equipment's performance analysis***: The performance of any device or equipment is clearly monitored with no RTC, but time logging.
4. ***Rental equipments***: The rent for bulky generators in big events and special occasions are measured by usage period than the number of hours rented.
5. ***Product servicing and security***: The service period of any electronic devices is converted to usage-based cycle than a checklist of calendar dates.
6. ***High temperature and vibration machines***: The performance of crystal oscillator is tended to fail in such conditions where this embedded system plays a vital role.

The above-mentioned applications are more specific and focused areas, but the embedded board is designed as per the objective and developed which is not only limited to these but also expandable for any other real-time applications, where event logging is significantly used for monitoring, analytics and management. The flexibility of sensor replacement and user programming enhance this product to industrial applications with industrial standard norms taken care. From the consumer electronic devices to industrial machineries power monitoring and event logging applications, this device works without major modification due to easy customization options in it.

6 Conclusion and Future Scope

Elapsed time counter for power monitoring is developed which monitors real-time power consumption of appliances. Low power consumption is one of the salient features of the system as it uses deep sleep mode concept. Due to retro-fit mechanism, this system can be fit into any host device. The system can be made feasible with any device as per the usage with slight modifications. Furthermore, the designed system is very stable, reliable and easy to use and requires less cost.

The existing system for power monitoring uses the same sensor technology to read real power and voltage sensor along with. Due to voltage and current readings, it is possible to compute phase angle, which helps to determine real power. The system proposed in this paper is measuring current only and results in apparent power, but the objective is to develop elapsed time computing device for electronic power monitoring; hence, the efficiency of the connected load will be determined in this system.

With these results and discussion, the designed system is effective for power monitoring and event logging in wide application range. The customizable, retro-fit and portable device can be installed without any electrical equipment's or tools, which helps to user for interchanging the monitoring load as per the need, which is completely not possible in existing system.

The future scope is not limited to applications mentioned in the proposed paper, but can be extended as follows,

1. ***Heart rate monitoring system***: The heart rate of human is measured with oximetry technology using specific and dedicated sensors. But these sensors fail to perform in the specific conditions like low temperature and other environment. Hence, ETC is used for this application [5].
2. ***IoT enabled power monitoring***: Since the test case provided good results, we have planned to make next POC with IoT technology which enables end user to track the power consumption information at their fingertips.

References

1. https://www.maximintegrated.com/en/app-notes/index.mvp/id/506. Accessed on Dec 2018

2. Dallas Semiconductor Maxim, *DS1602 Elapsed Time Counter*, DS1602 datasheet. Revised July 1999
3. Maxim Integrated, *Total-Elapsed-Time Counter with Alarm*, DS1682 datasheet. Revised Oct 1999
4. Maxim Integrated, *Total-Elapsed-Time and Event Recorder with Alarm*, DS1683 datasheet. Revised June 2012
5. Github.com, *DS1682 + ESP8266 Based Pulse Counter with Deepsleep*, 8 May 2017. [Online]. Available: https://gist.github.com/whatnick/1f6ae0b5f7ea5dbd636c444dd12375aa. Accessed on Nov 2018
6. Github.com/ETC_Master, *ETC_Master_Library*, ETC DS 1882 IC library
7. V.V. de Araújo, R.A. Hernandez, E. Simas, A. Oliveira, W.L.A. de Olivera, Dedicated hardware implementation of a high precision power quality meter, in *Proceedings of IEEE International Instrumentation and Measurement Technology Conference (I2MTC)*, May 2015, pp. 393–398

Application of Smart Appliance Using Internet of Things

Md. Saiful Islam Milon, Monirul Islam Pavel, M. Samiul Ehsan, Sadman Hoque Sadi and Saifur Rahman Sabuj

Abstract Nowadays, it is a growing trend for our electrical appliances to be much more automated with the use of sensors and Internet of things (IoT)-based remote control, one particular example being the home juice maker. In this paper, we design a system for home juice maker to have smart features with the use of numerous advanced sensors and Internet connectivity to enable IoT applications. For experimental setup, we propose a Raspberry Pi-3-based smart juice maker which through the use of IoT which is capable of taking commands remotely from a phone application via MySQL servers. In order to the quality, pH and temperature sensor are used to maintain the freshness. The prediction model of ARIMA is implemented to acknowledge the further pH values in different temperature where the best case shows 1.63% MSE, and in the worst case, it gets 12.72% error.

Keywords Internet of things · Automated juice maker · Raspberry Pi-3 · Relay · Pumps · pH sensors · ARIMA

1 Introduction

As technology makes progress, the devices we use in everyday lives, are getting more sophisticated, and thus things are becoming complicated to use for many. Due to this reason, the trend of making smart electronics is raising for building the devices more capable of setting the proper configurations and changing them when it is required without any human intervention using a set of different sensors. Although smart devices are being used in many areas such as manufacturing and automobiles, here we are focusing on their application in households. Another big trend in current

Md. S. I. Milon (✉) · M. S. Ehsan · S. H. Sadi · S. R. Sabuj
Department of Electrical and Electronic Engineering, BRAC University, Dhaka, Bangladesh
e-mail: md.saiful.islam.milon@g.bracu.ac.bd

S. R. Sabuj
e-mail: s.r.sabuj@ieee.org

M. I. Pavel
Department of Computer Science and Engineering, BRAC University, Dhaka, Bangladesh
e-mail: monirul.islam.pavel@g.bracu.ac.bd

H. S. Saini et al. (eds.), *Innovations in Electronics and Communication Engineering*, Lecture Notes in Networks and Systems 107,
https://doi.org/10.1007/978-981-15-3172-9_35

technology is IoT, which electively means the connection of all electronic devices to the Internet to achieve better control, monitoring and automation. In terms of household appliances, these features can be used to bring a significant improvement to our lives, alleviating many tasks which are rather tedious and time consuming (i.e., cooking, washing clothes, etc.). The usage of IoT and smart appliances can make these tasks much easier and add greater amount of leisure time to people or simply allow them to be more productive at home. These technologies can be broadly applied to almost every home appliance, but for the purposes of this research, we have focused on one particular electric device, which is the juice maker. In this paper, automated juice maker utilizes IoT for remote control and monitoring. The juice maker we have designed utilizes a Raspberry Pi-3 along with a MySQL database and an Android application. The device itself mainly functions through the usage of four DC water pumps and three sensors which are used for measuring: water supply, temperature and pH value. The two of the sensors—pH and temperature—highlights the additional notable features of our juice maker which is the ensuring of product quality and healthiness as we have found to be directly related to the pH value of any juice based on its temperature and avor.

2 Related Works

IoT-enabled electrical devices designed for households are an area which is being widely researched with continuous development. The addition of IoT has the potential to bring a significant improvement in the general livelihoods of people. However, due to the core nature of how IoT works, we require easy device interoperability and interfacing, and for that, we require well-developed network architectures specially designed for this purpose along with the proper network protocols. Several architectures have already been carried out to this regard where each has its own pros and cons as established in survey paper [1], while the communication part, commonly referred to as machine to machine (M2M), communication has been covered in the survey paper [2]. Many practical works have also been carried out toward the introduction of IoT-enabled or interfaced devices in homes. Many of these projects focus on adding IoT functionality toward existing electrical devices to enable remote control and sometimes monitoring. The user generally accesses the IoT functions through smartphones as seen in [3] or through a Webpage user interface (UI) like in [4]. In case of [5], we observe the same functionality with Webpage UI. As monitoring is also a common application of IoT systems, the authors of [6] emphasize on power efficiency that adds to the practicality of IoT devices, through having portable wireless network nodes with long-lasting batteries. However, power efficiency can also be tackled through monitoring of the power usage of other electrical devices and switching them 'on' or 'off' in an intelligent manner as seen in [7, 8]. Furthermore, [9] has developed a relatively unique IoT node which is designed to monitor the house environment, along with some work being done on the reduction of wireless signal degradation. Some papers put more emphasis on establishing the current status of IoT systems to lay a solid foundation and then go on for farther development. References [10, 11] present the current architectures which are in use for IoT and then continue on a prototype model of adding IoT functionalities to existing devices. In case of reference [10], while the latter explores the current problems which are faced in the industry and suggests solutions to them. Several research papers on the healthiness aspect of fruit beverages have also been written, particularly in terms of the distribution. One such part is the pasteurizing of the fruit juice, where we get rid of the harmful bacteria and/or microorganisms living in the juice. The authors of [12] did a research analysis regarding the effectiveness of high hydrostatic pressure (HHP) and carbon dioxide (CO_2) for the pasteurization process for fruit juices of various pH levels; ending with the conclusion that the HHP process is very successful for high pH value juice, even at relatively low pressures, whereas for fruit juice, lower pH value HHP on its own is quite acceptable, but its effectiveness increases exponentially with the addition of CO_2. Meanwhile, the researchers in [13] did a similar analysis on pasteurization; however, they instead based their analysis on all existing published papers which were freely available to carry out a meta-analysis with the goal of formulating a linear model based on basic Bigelow equation. The process of pasteurization at focus was heat treatment, and the research concluded in having made a successful model for predicting the effectiveness of heat treatment against *Alicyclobacillus acidoterrestris* that is found in fruit beverages. Some researchers such as in [14] studied the effects of packaging and long-term storage on the fruit juice. Here, the particular juice studied was roselle–mango juice blend which had its various attributes measured (e.g., anthocyanins, vitamin C, total phenols) while in a variation of conditions (glass or plastic storage, and low or high temperature). Their conclusion was that the use of plastic or glass for storage made no difference, but cold storage served much better for keeping the various elements of the drink intact (e.g., vitamin C and anthocyanins) (Fig. 1).

3 Experimental Setup

The proposed system model of smart appliance is built based on the usage of IoT, in order to enable remote control and monitoring over the devices using various smart gadgets (e.g., smartphone, tablet, smartwatches and smart television). A UI has been developed on the Android platform, and a Raspberry Pi-3 is used to adequately process any commands received from the user. In this particular case, the Raspberry Pi-3 utilizes the built-in hardware and ingredients to make the correct avor and sweetness of juice as instructed while ensuring juice quality through temperature and pH sensors used in conjunction with an autoregressive integrated moving average (ARIMA)-based model. Table 1 shows the whole working process.

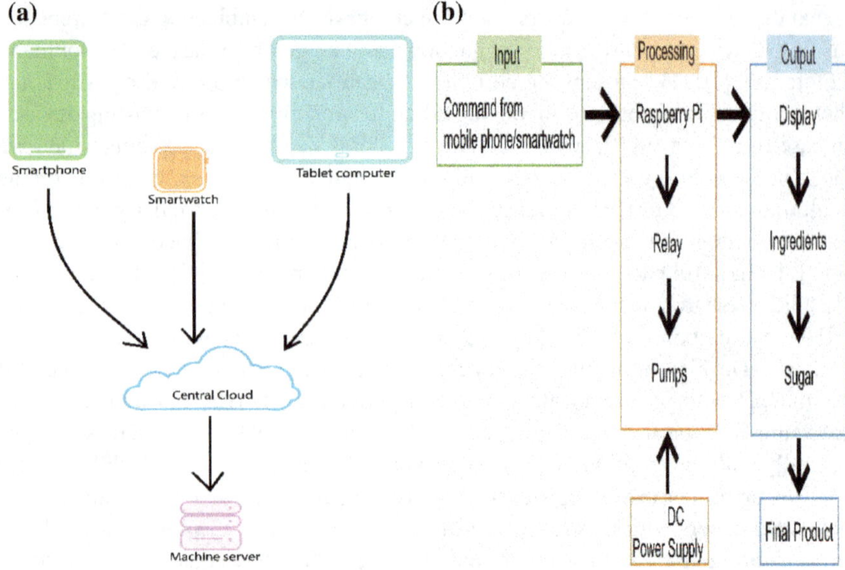

Fig. 1 **a** Proposed model [15], **b** system design for automated juice maker

Table 1 Proposed algorithm

```
Function Insert (flavor, sugar_syrup, amount, user info[ ])
    Form url of insertion;
    Send → http request;
  begin php
    Connect with database;
    Handle http request;
    Split column wise;
    Send → Database;
  end php
  begin getRequest (flavor, sugar_syrup, amount, user info[ ])
    Quantity ← waterflow_sensor();
   while ( Quantity  amount)
      if (flavor == n)
        relay switch = select (n);
        motor_n = ON;
      end if
      if (sugar_syrup == m)
        m ← to.convert(time)
        for ( time =1; time  m; time++)
          relay switch = select (m);
          motor_m = ON;
        end for
      end if
   end while
  end getRequest
    Total_Quantity = Total_Quantity-amount;
    Juice temp = temp( );
    pH = pH( );
    urllib (Total_Quantity, Juice temp, pH);
Function end
```

3.1 Hardware Implementation

The hardware setup is formed using four-channel relay, pH sensor, DS18B20 digital temperature sensor, water control module, four mini water pumps, DC motors, a display, power source and Raspberry Pi-3. Four mini water pumps consisting of DC motors with water flow control module are used for pouring the juice based on user requirements. A 5 V four-channel relay module was connected to the Raspberry Pi-3 and to the different pumps, and one wire is needed for each connection to the pumps. One battery is used which is connected with four relays and pumps. When command is given from application and the power is turned on, the switch in a relay channel would turn on, which goes on to turn the DC motor of the water pump on. Then, the water pump takes liquids in from two of the four bottles and discharges the liquids into a glass or cup. A pH sensor is used for measuring hydrogen-ion activity, determining the acidity or alkalinity of the drink, whereas the DS18B20 digital temperature sensor is utilized for monitoring the temperature. Our pH sensor shows values with respect to temperature, which is used for juice quality monitoring, as any kind of juice has a xed pH range; and if the measured pH value goes beyond the threshold, then the taste may deteriorate or indicate that the sample juice is not safe for consumption at that time. Furthermore, all modules and sensors are then configured with Raspberry Pi-3 which is shown in Fig. 2c. By default, a Raspberry Pi-3 does not support analog sensor, hence all analog sensors which generate analog value (e.g., DS18B20 and pH sensors) are connected with an ADC converter board to convert analog values into digital forms. Raspberry Pi-3 has built-in Wi-Fi sensors which having been connected to the Internet, receives commands from a MySQL server, hence powers the relay based on the selected juice. The pump pours water as per the received command via the use of a water flow sensor. While the process is going on, the LCD screen shows the selected juice. In the case of adding the sugar and flavoring, we use concentrated syrups (i.e., liquid ingredients). The syrup is pumped into the cup or glass using one of the four pumps, which is dedicated for this purpose only, in the same manner as the water.

3.2 Software and Communication System Implementation

The software implementation part is divided into three major areas—Python programming for sensing and giving output, an Android application for sending order request based on the concept of IoT and PHP backend programming for server-side processing. Python is the core programming language for Raspberry Pi-3 as it is Linux-based operating system. The coding architecture is drawn based on the GPIO pins of Raspberry Pi-3, and an Android application was developed as most smart devices (e.g., smartphones, tabs, watches and televisions) are having Android as their operating system. Each button of the application consists of a URL which

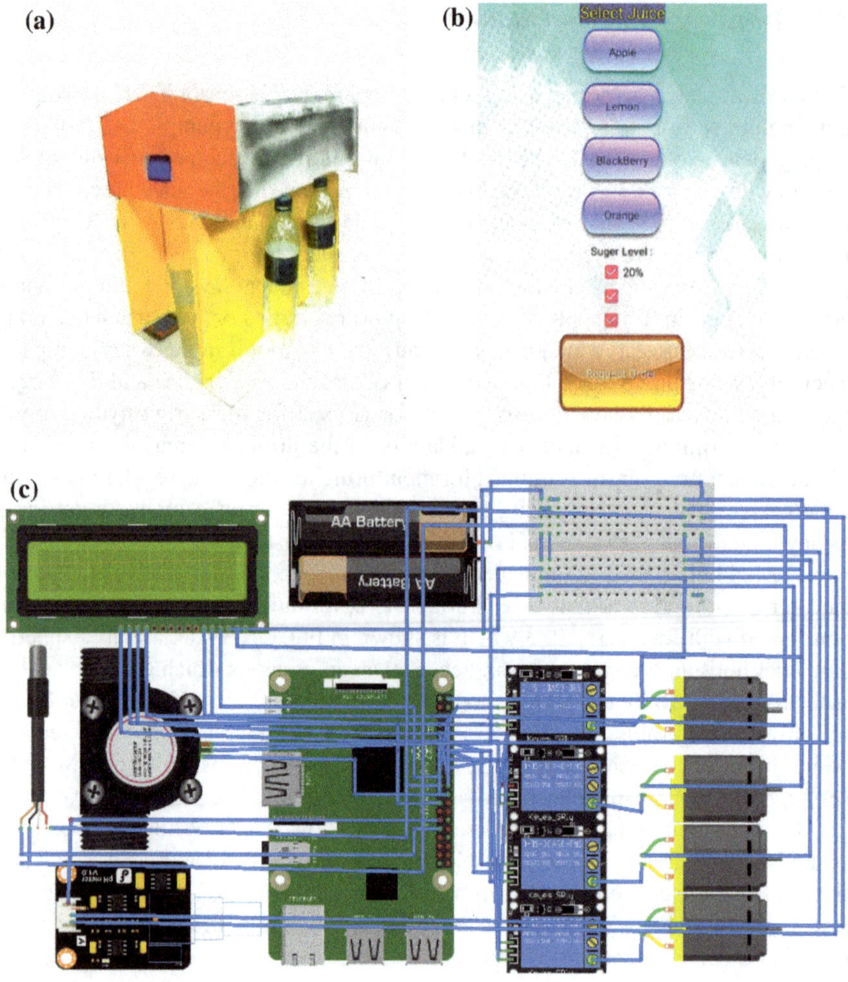

Fig. 2 **a** Proposed model, **b** Android app's user interface [15], **c** circuit diagram for juice maker [15]

contains the juice's code, amount and sugar level in percentage (e.g., IoT juicemaker.com?order.php?juice=2&amount=350&sugar=20). Then, the PHP backend is applied, which handles the URL request. It handles the request code, then splits and posts it to the MySQL database causing our PHP backend program to make a connection with this database. The Raspberry Pi-3 fetches data whenever a new row is inserted with a valid value for order, utilizing its MySQLdB library connection. Lastly, the proposed algorithm is applied as shown in Table 1, where relay switch will be turned on when the parameters that are requested from the Android application are reached in MySQL database and fetched to Raspberry Pi. The proposed algorithm uses a final confirmation via the Android application to turn on when the users approach the juice machine.

3.3 ARIMA-Based Prediction System

The pH values are obtained according to temperature at five-second intervals; and this prediction system is done with the application of stationary data ARIMA [16, 17]. The model has three combined part AR, I and MA, which are denoted by p, d, q that refer to lag order, degree of differencing and order of moving average, respectively. The system is fitted by an ARIMA (3, 1, 0) with lag value of 3 for performing autoregression, a model that is gained by trial and error process as it shows the least mean square errors. The value denotes time series stationary, where moving average is considered as 0. After the model is prepared, fitting the function, the prediction is done through a pre-built predict() function. To evaluate this function, training set is processed with ARIMA and cross validated with testing set, and accuracy is gained from these cross validation checking.

4 Result and Discussion

The proposed device with Android application and IoT can be modeled for all home appliances. The communication between Raspberry Pi-3 and MySQL database via PHP backend program and the mobile application has less than 1% data loss during processing type. The entire compact device has been designed to be used and carried easily. In order to ensure the liquid pH value, juice is tested time to time along with its temperature. The average range of pH values of our sample fresh juices from three different temperatures is apple 4.1–5.21, orange 3.98–4.8, blueberry 3.6–4.07 and lemon 3.3–3.96. These values were gained from measurements in 10, 20 and 25 °C temperature, respectively, within 3 h after the juices were freshly made. Crossing the stated pH range may effect the quality or refer to an impurity of the juice avors. Figure 3a–c shows the predicted pH values obtained from ARIMA model. The accuracy depends on how less the mean square error (MSE) is. Table 2 depicts the MSE of the prediction system based on temperature. The MSE is the error rate in the results of the ARIMA model, which is gained by checking cross validation of training and testing set. When the temperature is 10°, orange juice's pH value has the lowest MSE, and apple juice has the highest MSE. Again, when the temperature is 20°, orange juice has 2.65% which is the lowest MSE, and again, apple juice has the highest error rate with 9.08%. With the temperature 25°, lemon juice's predicted pH value has 1.21% MSE, and orange juice has the highest MSE with 4.49% error rate. Apart from these researches, the effect of fruit beverages on our dental health is also taken into account. Both [18, 19] have studied the matter using samples of 20 children who have had the pH level of their dental plaque measured at different

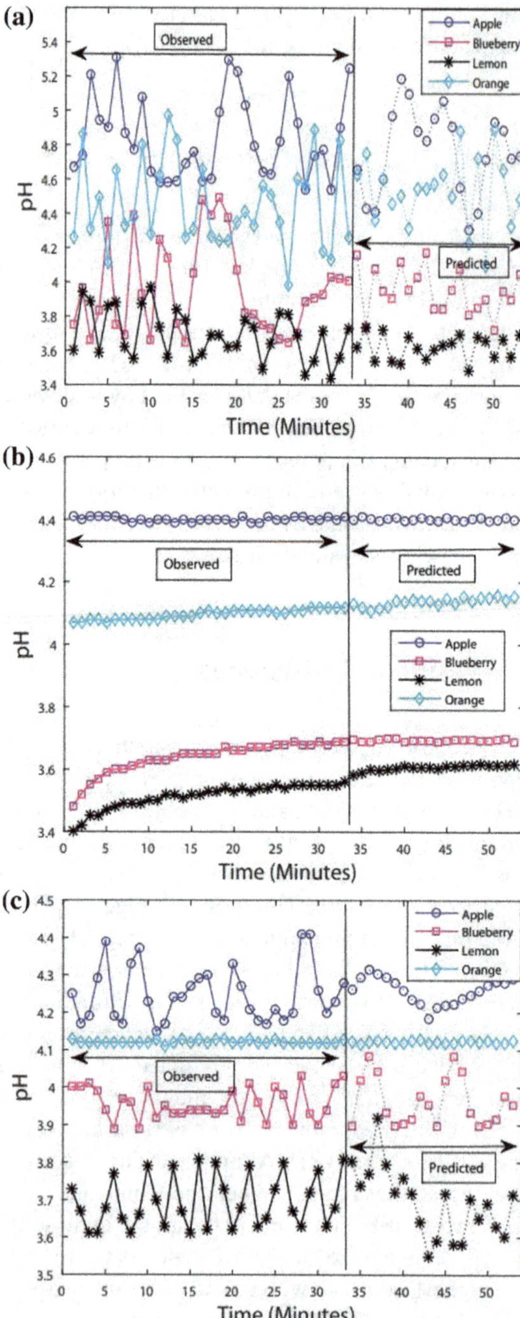

Fig. 3 pH value at **a** 10°, **b** 20°, **c** 25°

Table 2 Mean square error comparison

Sample	MSE (10°) (%)	MSE (20°) (%)	MSE (25°) (%)
Orange juice	3.02	2.65	4.49
Blueberry juice	3.84	4.54	1.63
Apple juice	12.72	9.08	2.17
Lemon juice	3.07	3.87	1.21

time intervals. The main objective was whether the fruit juices caused the pH level of plaque to fall below a certain threshold in which case the teeth get damaged. Based on the study of [18] which involved two fresh juices and two ready-made ones, no such danger was observed as the pH level of the plaque never went below the threshold. Authors of [18] present the mean plaque pH levels and pH difference after collecting. It is found that range of apple juice's pH is 6.18–6.49, whereas the paper shows apple juice's pH range is 4.1–5.21. However, [19] made an additional note from their similar research based on mango juice exclusively that commercially available mango fruit juice is more likely to harm dental health than fresh juice. In [19], authors divided the same juices in two different sets and have taken the value of pH using five-minute interval, from 0 to 60 min. The result shows the differences of pH values of different juices from time to time. Our proposed system works better as it knows the future value based on the prediction system. As a result, less time is wasted, and early decisions can be made when gaining the predicted value.

5 Conclusion

In this paper, we outline the concept and implementation of IoT-based smart appliances, with an IoT juice maker as our research implementation. The automated juice maker is designed to remotely receive user input from an Android application and process the correct ingredients to create the desired type of juice. Furthermore, juice quality is ensured through the use of a temperature and pH sensor. Overall, through the use of a simple Android application and an IoT-based system with smart sensors, we have been successful in creating an automated juice maker that anyone is able to use. In terms of future work, the model for our IoT juice maker can be integrated into other devices such as ice machines and electrical devices at home can be controlled with the use of phone, tablet or smartwatch applications. Lastly, since our model utilizes cloud storage, users with particular medical conditions (e.g., diabetes) can greatly be benefited since all the home appliances would automatically control his or her sugar intake.

References

1. P.P. Ray, A survey on internet of things architectures. J. King Saud Univ. Comput. Inf. Sci. **30**(3), 291–319 (2018)
2. A. Al-Fuqaha, M. Guizani, M. Mohammadi, M. Aledhari, M. Ayyash, Internet of things: a survey on enabling technologies, protocols, and applications. IEEE Commun. Surv. Tutor. **17**(4), 2347–2376 (2015)
3. D. Wang, D. Lo, J. Bhimani, K. Sugiura, AnyControl—IoT based home appliances monitoring and controlling, in *2015 IEEE 39th Annual Computer Software and Applications Conference*, vol. 3, July 2015 (IEEE, 2015), pp. 487–492
4. B.P. Kulkarni, A.V. Joshi, V.V. Jadhav, A.T. Dhamange, IoT based home automation using Raspberry PI. Int. J. Innov. Stud. Sci. Eng. Technol. **3**(4), 13–16 (2017)
5. T.A. Abdulrahman, O.H. Isiwekpeni, N.T. Surajudeen-Bakinde, A.O. Otuoze, Design, specification and implementation of a distributed home automation system. Procedia Comput. Sci. **94**, 473–478 (2016)
6. S. Pirbhulal, H. Zhang, M.E. Alahi, H. Ghayvat, S. Mukhopadhyay, Y.T. Zhang, W. Wu, A novel secure IoT-based smart home automation system using a wireless sensor network. Sensors **17**(1), 69 (2017)
7. P. Kumar, U.C. Pati, IoT based monitoring and control of appliances for smart home, in *2016 IEEE International Conference on Recent Trends in Electronics, Information & Communication Technology (RTEICT)*, May 2016, pp. 1145–1150
8. P. Sindhuja, M.S. Balamurugan, Smart power monitoring and control system through internet of things using cloud data storage. Indian J. Sci. Technol. **8**(19), 1 (2015)
9. H. Ghayvat, S. Mukhopadhyay, X. Gui, N. Suryadevara, WSN- and IOT-based smart homes and their extension to smart buildings. Sensors **15**(5), 10350–10379 (2015)
10. R.K. Kodali, S. Soratkal, L. Boppana, IOT based control of appliances, in *2016 International Conference on Computing, Communication and Automation (ICCCA)*, Apr 2016, pp. 1293–1297
11. P.P. Gaikwad, J.P. Gabhane, S.S. Golait, A survey based on smart homes system using internet-of-things, in *2015 International Conference on Computation of Power, Energy, Information and Communication (ICCPEIC)*, Apr 2015, pp. 0330–0335
12. L. Wang, J. Pan, H. Xie, Y. Yang, D. Zhou, Z. Zhu, Pasteurization of fruit juices of different pH values by combined high hydrostatic pressure and carbon dioxide. J. Food Prot. **75**(10), 1873–1877 (2012)
13. L.P. Silva, U. Gonzales-Barron, V. Cadavez, A.S. Sant'Ana, Modeling the effects of temperature and pH on the resistance of *Alicyclobacillus acidoterrestris* in conventional heat-treated fruit beverages through a meta-analysis approach. Food Microbiol. **46**, 541–552 (2015)
14. B. Mgaya-Kilima, S.F. Remberg, B.E. Chove, T. Wicklund, Physiochemical and antioxidant properties of Roselle-mango juice blends; effects of packaging material, storage temperature and time. Food Sci. Nutr. **3**(2), 100–109 (2015)
15. M. Chowdhury, M. Jahan, M. Jesan, F. Haque, M. Milon, S. Islam, Fundamental applications of internet of things. B.Sc. thesis, BRAC University, 2018
16. S.M. Kamruzzaman, M.I. Pavel, M.A. Hoque, S.R. Sabuj, Promoting greenness with IoT-based plant growth system, in *Computational Intelligence and Sustainable Systems* (2019), pp. 235–253
17. S.D.O. Domingos, J.F. de Oliveira, P.S. de Mattos Neto, An intelligent hybridization of ARIMA with machine learning models for time series forecasting. Knowl.-Based Syst. **175**, 72–86 (2019)
18. P.E. Chaly, M. Rajkumar, C. Reddy, N.A. Ingle, Effect of fruit juices on pH of dental plaque—a clinical study. J. Int. Oral Health **3**(6), 1–5 (2011)
19. G. Bhawna, K. Vijender, P. Anuradha, Effect of consumption of mango fruit, fresh mango juice and commercially available mango juice on dental plaque ph at different time intervals. Int. J. Curr. Adv. Res. **5**(4), 726–729 (2016)

Realization of a Continuous-Time Current-Mode Tow-Thomas-Equivalent Biquad Using Bipolar Current Mirrors

Ashish Gupta, Agha A. Husain and Amendra Bhandari

Abstract Design of analog integrated circuits is feasible using current and voltage-mode form of signal processing. The state-of-the-art analog integrated circuit design has received considerable advancement due to development in current-mode processing which has now started dominating traditional voltage-mode designs. Several design techniques and various circuits are available in literature that can help in designing different types of current and voltage-mode signal processing circuits that can be suitably implemented both in CMOS as well as BiCMOS technologies. The current-mode form of signal processing is an attractive approach because mathematical operations can be easily implemented and they can be operated at higher bandwidth as compared to voltage-mode circuits. The main intent of this paper is to present here the design of continuous-time biquadratic filter operating in current mode which are capable of operating at high frequencies and are suitably implemented in VLSI technology.

Keywords Continuous-time filters · Biquadratic filters · Complementary current mirrors · Tow-Thomas (TT) biquad · Current-mode signal processing

1 Introduction

An electric filter is a network [1–3] which has the capability to shape the spectrum of the input signal to obtain an output signal with the desired frequency content. Thus, it may have passbands and stop bands in which frequency components are, respectively, transmitted and rejected at the outputs [4–7]. The main challenges faced during the design of analog filters operating at higher frequencies (in the range of

A. Gupta (✉) · A. A. Husain
Department of ECE, I.T.S. Engineering College, 46, KP-III, Greater Noida, UP 201308, India
e-mail: ashishguptaas@its.edu.in

A. A. Husain
e-mail: aghaasimhusain@its.edu.in

A. Bhandari
Department of ECE, KCCITM, KP-III, Greater Noida, UP 201308, India
e-mail: amendrabhandari@gmail.com

H. S. Saini et al. (eds.), *Innovations in Electronics and Communication Engineering*, Lecture Notes in Networks and Systems 107,
https://doi.org/10.1007/978-981-15-3172-9_36

MHz as proposed by [8–12]) are: (i) able to operate at high frequency, (ii) automatic control of parameters against changing operating conditions, and (iii) fabrication tolerances.

So, at higher operating frequencies, the output should be obtained in the form of current given by the relation: $I_{\text{out}} = g_{\text{m}} * V_{\text{in}}$, where g_{m} is the transconductance parameter of the active device.

Thus, for application as continuous-time filters, transconductances should ensemble the following properties: (i) simpler, linear, and have wider frequency response, (ii) large output and input–output impedance to simplify circuit design, (iii) circuit should operate at low voltage so as to conserve power and to make it compatible with digital technologies, and (iv) facilitate electronic tunability such that the transconductance parameter depends either upon DC bias voltage or current.

Based upon the technology choice and operating frequency of transconductance circuits [18], the circuit can be designed for frequency >50 MHz (in CMOS technology), >500 MHz (by using bipolar technology) or for frequency >1 GHz (by the use of GaAs technology). Thus, it is feasible to design high-frequency continuous-time filters for telecommunication circuits. Further, because transconductances and capacitors are the only components required for realizing a filter, g_{m}-C filters can readily be simulated in completely integrated form, well suited for integration with the digital system [13–17].

For active simulations of filters, it is required that the transconductors are identical and grounded capacitors are used for simple IC layout and processing, and also, the implementation of an integrated analog filter based on analog gate arrays appears to be a distinct possibility.

Additionally, wide transconductance bandwidth coupled with the reduced parasitic effects of circuit and device on filter makes it possible for the circuits to operate at much higher frequency range.

2 Proposed Current-Mode Tow-Thomas-Equivalent Biquad Circuit

A current-mode Tow-Thomas-equivalent biquad [19] is obtained by cascading a non-ideal (lossy) integrator followed by ideal (lossless) integrator to obtain five desired filter transfer functions namely: (i) low-pass response, (ii) high-pass response, (iii) band-pass response, (iv) band-stop response, and (v) all-pass response.

Figure 1 gives the functional block diagram, and Fig. 2 represents the circuit realization of a current-mode Tow-Thomas-equivalent biquad. The necessary mathematical steps for deriving the current transfer functions are shown in the next section.

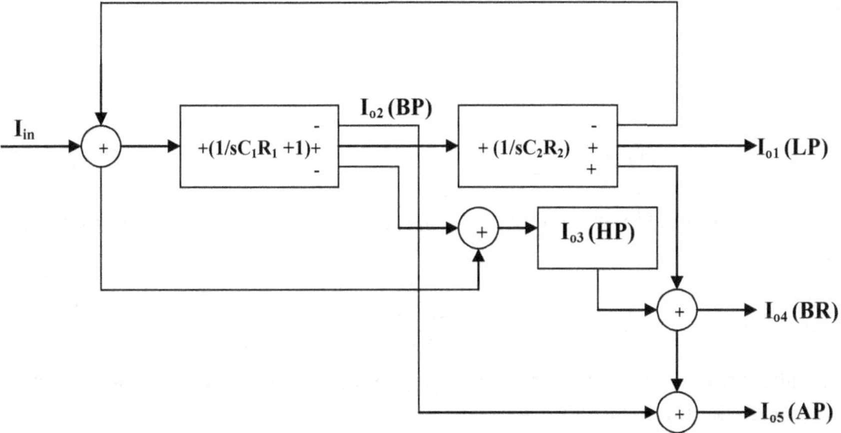

Fig. 1 Functional block diagram of a current-mode Tow-Thomas-equivalent biquad

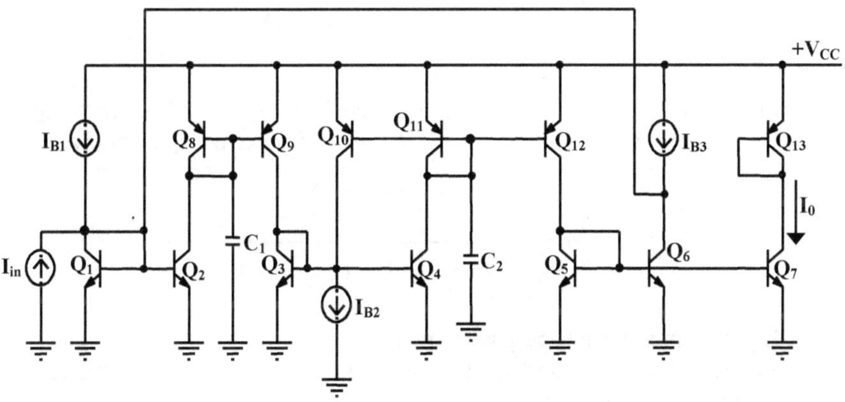

Fig. 2 Circuit realization of current-mode Tow-Thomas-equivalent biquad

3 Mathematical Analysis

From Fig. 1, we can write the following three equations:

$$I_1 = I_{\text{in}} - I_{\text{o}1} \tag{1}$$

$$I_{\text{o}1} = \frac{I_{\text{o}2}}{sC_2R_2} \tag{2}$$

$$I_{\text{o}2} = \frac{I_1}{1 + sC_1R_1} \tag{3}$$

From Eqs. (2) and (3), we have,

$$I_{o2} = sC_2R_2I_{o1} \tag{4}$$

$$I_1 = (1 + sC_1R_1)I_{o2} \tag{5}$$

From Eqs. (4) and (5), we obtain the following expression for current I_1 in terms of current I_{o1}.

$$I_1 = sC_2R_2(1 + sC_1R_1)I_{o1} \tag{6}$$

On substituting I_1 from Eq. (6) in Eq. (1), we obtain an expression for the current I_{o1} in terms of input current I_i as:

$$\frac{I_{o1}}{I_i} = \frac{\left(\frac{1}{C_1C_2R_1R_2}\right)}{s^2 + s\left(\frac{1}{C_1R_1}\right) + \left(\frac{1}{C_1C_2R_1R_2}\right)} \tag{7}$$

Thus, Eq. (7) represents a low-pass response.

Similarly, when the value of I_{o1} is substituted from Eq. (7) in Eq. (4), we obtain the expression for the ratio of the currents I_{o2} and input current I_i as follows:

$$\frac{I_{o2}}{I_i} = \frac{s\left(\frac{1}{C_1R_1}\right)}{s^2 + s\left(\frac{1}{C_1R_1}\right) + \left(\frac{1}{C_1C_2R_1R_2}\right)} \tag{8}$$

Thus, Eq. (8) represents a band-pass response.

Again, from Fig. 1, we can write the following current equations for the currents I_{o3}, I_{o4}, and I_{o5} in terms of input current I_i.

$$I_{o3} = I_1 - I_{o2} = sC_1R_1I_{o2} \tag{9}$$

Now, substituting the value of I_{o2} from Eq. (8) in Eq. (9), we obtain an expression for the current transfer I_{o3}/I_i.

$$\frac{I_{o3}}{I_i} = \frac{s^2}{s^2 + s\left(\frac{1}{C_1R_1}\right) + \left(\frac{1}{C_1C_2R_1R_2}\right)} \tag{10}$$

Thus, Eq. (10) represents a high-pass response.

Similarly, the currents I_{o4} and I_{o5} can be written as:

$$I_{o4} = I_{o3} + I_{o1} \tag{11}$$

Now, substituting the value of the currents I_{o3} and I_{o1} from Eqs. (7) and (10), respectively, in Eq. (11), we obtain the expression for the ratio of the current I_{o4} and current I_i.

$$\frac{I_{o4}}{I_i} = \frac{s^2 + \left(\frac{1}{C_1 C_2 R_1 R_2}\right)}{s^2 + s\left(\frac{1}{C_1 R_1}\right) + \left(\frac{1}{C_1 C_2 R_1 R_2}\right)} \quad (12)$$

Thus, Eq. (12) represents a notch response.

Similarly, for the current $I_{o5,}$ we have the following expression:

$$I_{o5} = I_{o4} - I_{o2} \quad (13)$$

Now, substituting the value of the currents I_{o4} and I_{o2} from Eqs. (8) and (12), respectively, in Eq. (13), we obtain an expression for the current I_{o5} in terms of input current I_i.

$$\frac{I_{o5}}{I_i} = \frac{s^2 - s\left(\frac{1}{C_1 R_1}\right) + \left(\frac{1}{C_1 C_2 R_1 R_2}\right)}{s^2 + s\left(\frac{1}{C_1 R_1}\right) + \left(\frac{1}{C_1 C_2 R_1 R_2}\right)} \quad (14)$$

Thus, Eq. (14) represents an all-pass response.

The expressions for the filter's cut-off frequency (ω_o) and the filter's quality factor (Q) can be obtained from the above current transfer function and they are, respectively, given as:

$$\omega_o = \frac{1}{\sqrt{C_1 C_2 R_1 R_2}} \quad (15)$$

$$Q = \sqrt{\frac{C_1 R_1}{C_2 R_2}} \quad (16)$$

where

$$R_1 = \frac{kT}{q}\frac{1}{I_{B1}} \text{ and } R_2 = \frac{kT}{q}\frac{1}{I_{B2}} \quad (17)$$

From Fig. 2, it can be seen that R_1 and R_2 are the resistances of the diode-connected transistors Q_8 and $Q_{11,}$ while I_{B1} and I_{B2} are the DC bias currents shown by I_{B2} and I_{B3} in Fig. 2. Therefore, the expression for filter parameters ω_o and Q becomes:

$$\omega_o = \frac{q}{kT}\sqrt{\frac{I_{B1} I_{B2}}{C_1 C_2}} \quad (18)$$

$$Q = \sqrt{\frac{C_1 I_{B2}}{C_2 I_{B1}}} \quad (19)$$

If $C_1 = C_2 = C$, then Eqs. (15) and (16) become:

$$\omega_o = \frac{q}{kT}\frac{1}{C}\sqrt{I_{B1} I_{B2}} \quad (20)$$

$$Q = \sqrt{\frac{I_{B2}}{I_{B1}}} \quad (21)$$

Also, if $I_{B1} = I_{B2} = I_{B,}$ then Eqs. (20) and (21) become:

$$\omega_o = \frac{q}{kT}\frac{I_B}{C} \quad (22)$$

$$Q = 1 \quad (23)$$

4 Simulation Results

The proposed current-mode Tow-Thomas-equivalent biquad has been tested through SPICE simulation carried out by using standard set of bipolar process parameters with transistors NR100N and PR100N for which the parameters are listed here in Table 1 [20–23]. The power supply voltage V_{CC} was chosen to be 1.5 V.

The circuit of Fig. 2 has been verified in PSPICE with a cut-off frequency of 3.06 MHz at $I_B = 10\ \mu A$ which gives $R \cong 2.6\ k\Omega$, $Q = 1$.

Table 1 Transistor process parameter

Transistor type	Process parameter
NR100N	TR = 0.5E−8 EG = 1.206 XTB = 1.538 XTI = 2.0 IS = 121E−18 BF = 137.5 VAF = 159.4 IKF = 6.974E−3 ISE = 36E−16 NE = 1.713 BR = 0.7258 RBM = 25 RC = 50 CJE = 0.214E−12 VJE = 0.5 MJE = 0.28 CJC = 0.983E−13 VJC = 0.5 MJC = 0.3 XCJC = 0.034 CJS = 0.913E−12 VJS = 0.64 MJS = 0.4 FC = 0.5 TF = 0.425E−8 VAR = 10.73 IKR = 2.198E−3 RE = 1 RB = 524.6
PR100N	TR = 0.610E−8 EG = 1.206 XTB = 1.866 XTI = 1.7 IS = 73.5E−18 BF = 110 VAF = 51.8 IKF = 2.359E−3 ISE = 25.1E−16 NE = 1.650 BR = 0.4745 VAR = 9.96 IKR = 6.478E−3 RE = 3 RB = 327 RBM = 24.55 RC = 50 CJE = 0.18E−12 VJE = 0.5 MJE = 0.28 CJC = 0.164E−12 VJC = 0.8 MJC = 0.4 XCJC = 0.037 CJS = 1.03E−12 VJS = 0.55 MJS = 0.35 FC = 0.5 TF = 0.610E−9

Table 2 Adjusted and nominal values of capacitors shown in Fig. 2

Capacitors	Adjusted	Nominal
C_1	9.1 pF	10 pF
C_2	39.5 pF	40 pF

In the above design example, it has been assumed that the gains of all the current mirrors are chosen to be 1.0 by taking into account that all transistors have the same emitter areas. The nominal values of capacitors are given in Table 2.

The values of the capacitor must be adjusted in order to absorb the parasitic capacitors. In Table 2, we have listed the adjusted area of the transistors used in Fig. 2, which makes the current gain of each current mirror unity (Fig. 3 and Table 3).

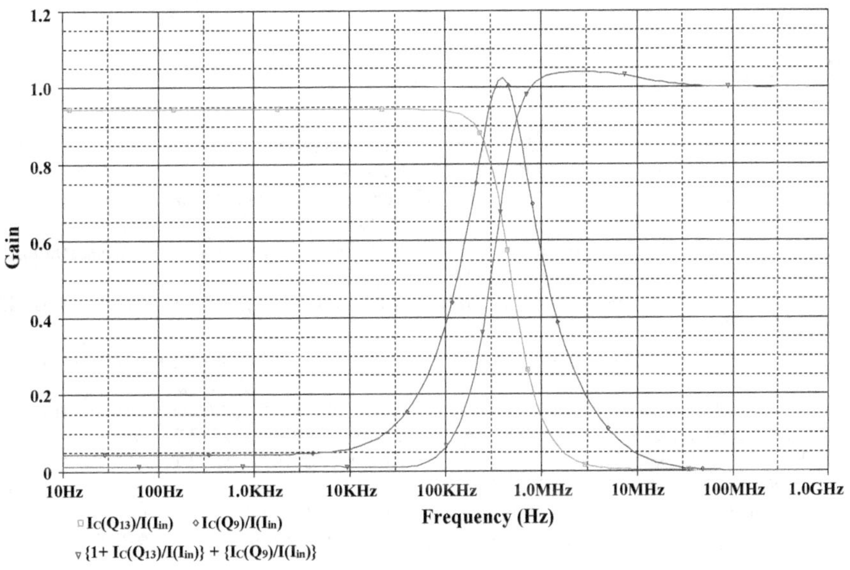

Fig. 3 Frequency response of current-mode Tow-Thomas-equivalent biquad

Table 3 Values of the transistor emitter areas as shown in Fig. 2

Emitter areas	Values	Emitter areas	Values
A_2	1.02	A_4	1.03
$A_{10} = A_{12}$	1.07	$A_8 = A_9$	1.04
$A_6 = A_7$	1.04		

5 Conclusions

The paper presents high-frequency current-mode Tow-Thomas-equivalent biquad filter suitable for operation at high frequencies till 100 MHz and can operate at low voltage of 1.5 V. Also, it can be concluded that either by changing the value of the bias current or the capacitor value, the gain and the cut-off frequency of the filter increase. The cut-off frequency of the proposed filter circuit can be electronically tuned through a single DC bias current. The proposed circuits was tested using SPICE, and the simulated result thus obtained confirms the theoretical results.

Acknowledgements This work was performed at Electronics CAD Lab of ECE Department, I.T.S. Engineering College, Greater Noida, India.

References

1. J.C. Ahn, N. Fujii, Current-mode continuous-time filters using complementary current mirror pairs. IEICE Trans. Fundam. **E79-A**(2), 168–175 (1996)
2. R. Angulo, M. Robinson, E.S. Sinencio, Current-mode continuous-time filters: two design approaches. IEEE Trans. Circuits Syst. **39**(5), 337–341 (1992)
3. R.W.J. Barker, Accuracy of current mirrors, in *IEE Colloquium on Current Mode Analogue Circuits*, London, vol. 25, paper 2 (1989)
4. B.L. Hart, R.W.J. Barker, Negative current-mirror using n-p-n transistors. Electron. Lett. **13**, 311–312 (1977)
5. B.L. Hart, R.W.J. Barker, Modified current mirror with a voltage-following capability. Electron. Lett. **18**, 970–972 (1982)
6. S.S. Lee, R.H. Zele, D.J. Allstot, G. Liang, CMOS continuous-time current-mode filters for high-frequency applications. IEEE J. Solid-State Circuits **28**(3), 323–329 (1993)
7. F.J. Lidgey, Looking into current mirrors. Wireless World **85**, 57–59 (1984)
8. R. Schaumann, M.S. Ghausi, K.R. Laker, *Design of Analog Filters: Passive, Active RC, and Switched Capacitor* (Prentice Hall, Englewood Cliffs, NJ, 1989)
9. R. Senani, A.K. Singh, A new universal current-mode biquad filter. Frequenz J. Telecommun. (Germany) **56**(1/2), 55–59 (2002)
10. A.M. Soliman, Mixed-mode biquad circuits. Microelectron. J. **27**(6), 591–594 (1996)
11. C. Toumazou, F.J. Lidgey, P.Y.K. Cheung, Current-mode analogue signal processing circuits—a review of recent developments, in *IEEE International Symposium on Circuits and Systems*, Portland, USA, vol. 3 (1989), pp. 1572–1575
12. C. Toumazou, F.J. Lidgey, D.J. Haigh, *Analogue IC Design: The Current-Mode Approach* (Peter Peregrinus Ltd, 1990)
13. Y.P. Tsividis, Integrated continuous-time filter design—an overview. IEEE J. Solid-State Circuits **29**(3), 166–176 (1994)
14. T. Tsukutani, M. Ishida, S. Tsuiki, Y. Fukui, Current-mode biquad without external passive elements. Electron. Lett. **32**(3), 197–198 (1996)
15. T. Voo, C. Toumazou, High-speed current mirror resistive compensation technique. Electron. Lett. **31**(4), 248–249 (1995)
16. G. Wegmann, E.A. Vittoz, Very accurate dynamic current mirrors. Electron. Lett. **25**, 644–646 (1989)
17. B. Wilson, F.J. Lidgey, C. Toumazou, Current mode signal processing circuits, in *IEEE Symposium on Circuits and Systems*, Helsinki, vol. 3 (1988), pp. 2665–2668

18. P. Wu, R. Schaumann, A high-frequency GaAs transconductance circuit and its applications. *IEEE International Symposium on Circuits and Systems*, New Orleans, LA, USA (1990), pp. 3081–3084.
19. X.R. Meng, Z.H. Yu, CFA based fully integrated Tow-Thomas biquad. Electron. Lett. **32**(8), 722 (1996)
20. A. Bhandari, A.A. Husain, M.S. Chadha, A. Gupta, Lossy and lossless current-mode integrators using CMOS current mirrors. Int. J. Eng. Res. Dev. **9**(3), 34–41 (2013)
21. A.A. Husain, M.S. Chadha, A. Gupta, A. Bhandari, Design of high frequency current-mode continuous-time filter using CMOS current-mirrors. IOSR J. VLSI Signal Process. **3**(6), 58–62 (2013)
22. M.S. Chadha, A. Gupta, A. Bhandari, A.A. Husain, Third-order current-mode filter realization using CMOS current-mirror. IOSR J. Eng. **3**(12), 32–39 (2013)
23. A. Gupta, M.S. Chadha, A. Bhandari, A.A. Husain, A novel approach for realization of higher order filter using bipolar and MOS current-mirrors. IOSR J. Electron. Commun. Eng. (IOSR-JECE) **9**(1), 99–104 (2014)

Continuous-Time High-Frequency Current-Mode Kerwin–Huelsmann–Newcomb (KHN)-Equivalent Biquad Filter Using MOS Complementary Current-Mirror

Ashish Gupta

Abstract Design of integrated circuits is possible both in voltage and current- mode type of analog signal processing. Designing analog integrated circuits is a state of the art that has received incredible advancement from the viewpoint of development of VLSI technology. The application of current-mode designs now dominates conventional voltage-mode designs. Numerous circuits and techniques are presented in the literature for designing wide variety of voltage and current-mode processing circuits which can be suitably implemented in CMOS and Bi-CMOS technology. Current-mode signal processing being a striking approach because of the ease of performing mathematical operations like addition, subtraction, multiplication and the circuits can be operated at high bandwidth than their voltage-mode counterparts. This paper presents the design of a current-mode continuous-time high-frequency biquadratic filter suitable for implementation in VLSI technology.

Keywords Continuous-time filters · Biquadratic filters · Complementary current-mirrors · Kerwin–Huelsmann–Newcomb (KHN) biquad · Current-mode signal processing

1 Introduction

Current-mode technique has been received commendable attention, as they offer following advantages: (i) high slew-rate, (ii) low-power consumption, (iii) operating frequency range is superior, (iv) enhanced accuracy and linearity. Interest in design of current-mode (CM) [1, 2] filters has developed, but for operation at higher frequencies (in MHz range) [3, 4], the main challenges faced during the design of analog filters are (i) consistent high-frequency performance, (ii) automatic on-chip availability of tuning and (iii) varying operating conditions.

Thus, for high-frequency applications, the output must be current, and it should be given as

A. Gupta (✉)
Department of ECE, I.T.S. Engineering College, 46, KP-III, Greater Noida, UP 201308, India
e-mail: ashishguptaas@its.edu.in

H. S. Saini et al. (eds.), *Innovations in Electronics and Communication Engineering*, Lecture Notes in Networks and Systems 107,
https://doi.org/10.1007/978-981-15-3172-9_37

$$I_{out} = g_m * V_{in} \tag{1}$$

where g_m is the transconductance of the active analog building block. For application as continuous-time filters, the transconductance (g_m) of the circuit should meet the following properties: linear, simpler and have wide frequency response, should have high output and input impedance to simplify circuit design, operate at low-voltage preserve power and to make it compatible with the digital technology. The g_m depends on DC bias voltage (V_B) or current (I_B) to make circuits electronically tunable against the environmental variation.

Depending on the choice of the technology, the operating frequency range of the g_m circuits [3] can be extended to be greater than 50 MHz (for CMOS technology), greater than 500 MHz (for bipolar technology) or even greater than 1 GHz (for GaAs technology) for designing high-frequency continuous-time telecommunication circuits.

As capacitors and the transconductors are the only components required for realizing filters, g_m-C [5, 6] filters can readily be implemented in fully integrated form [7], with desired technology. For active simulations, we consider identical transconductors and grounded capacitors for simple layout and processing. In addition to this, the large bandwidth of transconductance and reduced parasitic effects of device and circuit on filter performance results in higher working frequency range of the circuits.

2 Biquadratic Filter

An electric filter is a two-port network which shapes the frequency band of the input in order to obtain an output with the preferred frequency. Thus, it has pass band and stop band, respectively, in which the frequencies are transmitted and rejected at the output.

2.1 Current-Mode KHN-Equivalent BiQUAD

It can be implemented by cascading two lossless integrators [8–10] to obtain five filter functions namely: low-pass, high-pass, band-pass, band-elimination and all-pass responses.

Figure 1 shows the block schematic representation of current-mode KHN-equivalent biquad , and the routine analysis has been carried out to find the current transfer functions. Figure 2 shows the CMOS realization of the proposed CM KHN-equivalent biquad using complementary current-mirror pair [11].

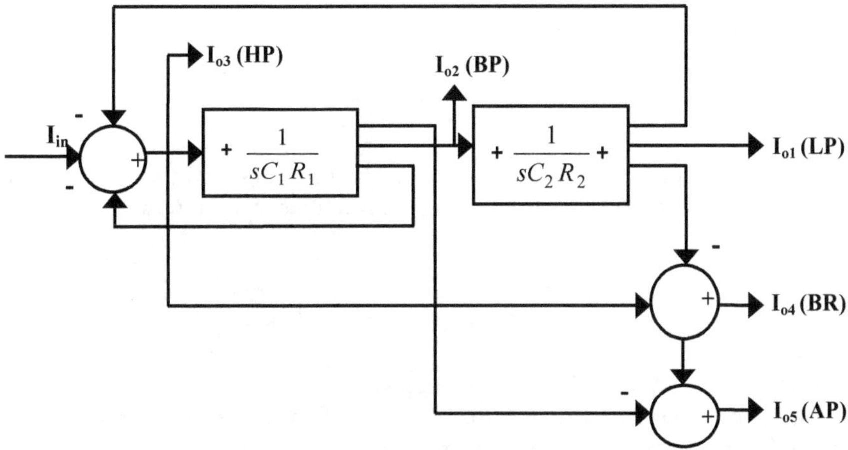

Fig. 1 Block diagram of current-mode KHN-equivalent biquad

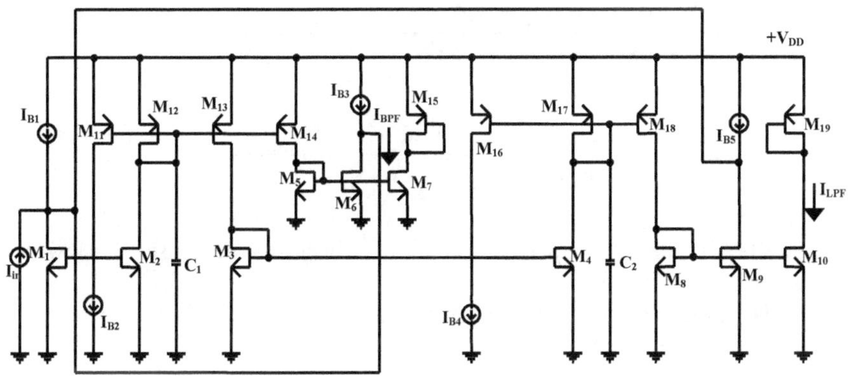

Fig. 2 Circuit diagram of CM KHN-equivalent biquad using MOS complementary current-mirror pairs

2.2 *Mathematical Analysis*

From Fig. 1, we obtain the following three equations:

$$I_{o3} = I_{in} - I_{o2} - I_{o1} \tag{2}$$

$$I_{o1} = \frac{I_{o2}}{sC_2R_2} \tag{3}$$

$$I_{o2} = \frac{I_{o3}}{sC_1R_1} \tag{4}$$

From Eqs. (3) and (4), we have

$$I_{o1} = \frac{I_{o3}}{s^2 C_1 C_2 R_1 R_2} \tag{5}$$

Now, substituting the value of I_{o1} and I_{o2} from Eqs. (4) and (5) in Eq. (1), we obtain an expression for the current I_{o3} in terms of input current I_{in} as

$$\frac{I_{o3}}{I_{in}} = \frac{s^2}{s^2 + s\left(\frac{1}{C_1 R_1}\right) + \left(\frac{1}{C_1 C_2 R_1 R_2}\right)} \tag{6}$$

Thus, Eq. (6) represents a high-pass response.

Similarly, when the value of I_{o3} is substituted from Eq. (6) in Eqs. (4) and (5), respectively, we obtain the expression for the currents I_{o1} and I_{o2} in terms of input current I_{in} as follows:

$$\frac{I_{o2}}{I_{in}} = \frac{s\left(\frac{1}{C_1 R_1}\right)}{s^2 + s\left(\frac{1}{C_1 R_1}\right) + \left(\frac{1}{C_1 C_2 R_1 R_2}\right)} \tag{7}$$

$$\frac{I_{o1}}{I_{in}} = \frac{\left(\frac{1}{C_1 C_2 R_1 R_2}\right)}{s^2 + s\left(\frac{1}{C_1 R_1}\right) + \left(\frac{1}{C_1 C_2 R_1 R_2}\right)} \tag{8}$$

Thus, Eqs. (7) and (8), respectively, represent a band-pass response and a low-pass response.

Similarly, the currents I_{o4} can be written as

$$I_{o4} = I_{o3} + I_{o1} \tag{9}$$

Now, substituting the value of the currents I_{o3} and I_{o1} from Eqs. (6) and (8), respectively, in Eq. (9), we obtain an expression for the current transfer function I_{o4}/I_{in}.

$$\frac{I_{o4}}{I_{in}} = \frac{s^2 + \left(\frac{1}{C_1 C_2 R_1 R_2}\right)}{s^2 + s\left(\frac{1}{C_1 R_1}\right) + \left(\frac{1}{C_1 C_2 R_1 R_2}\right)} \tag{10}$$

Thus, Eq. (10) represents a notch response.

Similarly, for the current I_{o5}, we have the following expression:

$$I_{o5} = I_{o4} - I_{o2} \tag{11}$$

Now, substituting the value of the currents I_{o4} and I_{o2} from Eqs. (10) and (7), respectively, in Eq. (11), we obtain the expression for I_{o5}/I_{in}.

$$\frac{I_{o5}}{I_{in}} = \frac{s^2 - s\left(\frac{1}{C_1 R_1}\right) + \left(\frac{1}{C_1 C_2 R_1 R_2}\right)}{s^2 + s\left(\frac{1}{C_1 R_1}\right) + \left(\frac{1}{C_1 C_2 R_1 R_2}\right)} \tag{12}$$

Thus, Eq. (12) represents an all-pass response.

The expressions for the filter parameters namely the cut-off frequency (ω_o) and the quality factor (Q) can be obtained from the above current transfer function, and they are given by:

$$\omega_o = \frac{1}{\sqrt{C_1 C_2 R_1 R_2}} \tag{13}$$

$$Q = \sqrt{\frac{C_1 R_1}{C_2 R_2}} \tag{14}$$

where

$$R_1 = \frac{1}{g_{m1}} \text{ and } R_2 = \frac{1}{g_{m2}} \tag{15}$$

where g_{m1} and g_{m2} are the transconductance of the diode-connected transistors Q_{12} and Q_{17}, respectively. Also, I_{B1} and I_{B2} are the DC bias currents shown as I_{B2} and I_{B4} in Fig. 2.

But the transconductances g_{m1} and g_{m2} are directly related to the square root of the bias currents I_{B2} and I_{B4} and are given as

$$g_{m1} = \sqrt{2\mu_n C_{ox}\left(\frac{W}{L}\right)_{12} I_{B2}} \text{ and } g_{m2} = \sqrt{2\mu_n C_{ox}\left(\frac{W}{L}\right)_{17} I_{B4}} \tag{16}$$

Therefore, the expression for filter parameters ω_o and Q becomes

$$\omega_o = \sqrt{\frac{I_{B2} I_{B4}}{C_1 C_2}} \tag{17}$$

$$Q = \sqrt{\frac{C_1 I_{B4}}{C_2 I_{B2}}} \tag{18}$$

If $C_1 = C_2 = C$, then Eqs. (17) and (18) become

$$\omega_o = \frac{1}{C}\sqrt{I_{B2} I_{B4}} \tag{19}$$

$$Q = \sqrt{\frac{I_{B4}}{I_{B2}}} \tag{20}$$

Also, if $I_{B2} = I_{B4} = I_B$, then Eqs. (19) and (20) become

$$\omega_o = \frac{I_B}{C} \tag{21}$$

$$Q = 1 \tag{22}$$

3 Simulation Results

The proposed circuit of Fig. 2 was verified using SPICE with 0.5 μm CMOS process parameters provided by MOSIS (AGILENT). These parameters are listed in Table 1 [12–16].

For the circuit shown in Fig. 2, the analysis was carried out with DC bias current $I_{B1} = I_{B3} = I_{B5} = 24$ μA, $I_{B2} = I_{B4} = 15$ μA, $C_1 = 0.01$ pF, $C_2 = 0.1$ pF, $(W/L)_P$ ratio = 1 μm/1 μm, $(W/L)_N$ ratio = 1 μm/1 μm and supply voltage $V_{DD} = 1.5$ V.

The simulated value of the cut-off frequencies for low-pass, high-pass and band-pass response is found to be: $(f_0)_{LPF} = 51.67$ MHz, $(f_0)_{HPF} = 95.404$ MHz and $(f_0)_{BPF} = 88.444$ MHz, respectively. The band-pass filter has a bandwidth of 186.894 MHz.

The simulated results are very well in agreement with the theoretical results that were found to be: $(f_0)_{LPF} = 52$ MHz, $(f_0)_{HPF} = 95$ MHz, $(f_0)_{BPF} = 88$ MHz and BW = 185 MHz, respectively. The SPICE simulation for the proposed circuit has been shown in Fig. 3.

Figures 4 and 5, respectively, represent the variation in the cut-off frequency

Table 1 CMOS process parameters

Transistor	Process parameters
nMOS	LEVEL = 3 UO = 460.5 TOX = 1.0E−8 TPG = 1 VTO = 0.62 JS = 1.08E−6 XJ = 0.15U RS = 417 RSH = 2.73 LD = 0.04U VMAX = 130E3 NSUB = 1.17E17 PB = 0.761 ETA = 0.00 THETA = 0.129 PHI = 0.905 GAMMA = 0.69 KAPPA = 0.10 CJ = 76.4E−5MJ = 0.357 CJSW = 5.68E−10 MJSW = 0.302 CGSO = 1.38E−10 CGDO = 1.38E−10 CGBO = 3.45E−10 KF = 3.07E−28 AF = 1 WD = 0.11U DELTA = 0.42 NFS = 1.2E11
pMOS	LEVEL = 3 UO = 100 TOX = 1.0E−8 TPG = 1 VTO = 0.58 JS = 0.38E−6 XJ = 0.10U RS = 886 RSH = 1.81 LD = 0.03U VMAX = 113E3 NSUB = 2.08E17 PB = 0.911 ETA = 0.00 THETA = 0.120 PHI = 0.905 GAMMA = 0.76 KAPPA = 2 CJ = 85E−5MJ = 0.429 CJSW = 4.67E−10 MJSW = 0.631 CGSO = 1.38E−10 CGDO = 1.38E−10 CGBO = 3.45E−10 KF = 1.08E−29 AF = 1 WD = 0.14U DELTA = 0.81 NFS = 0.52E11

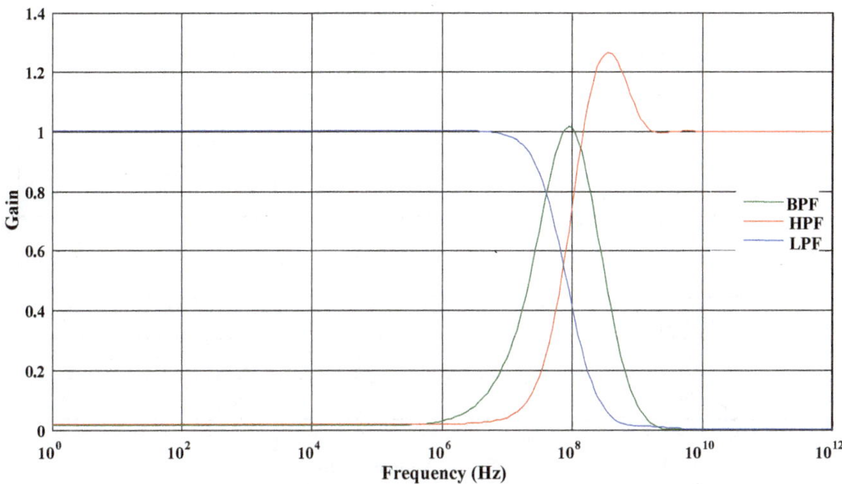

Fig. 3 Simulated response of current-mode KHN-equivalent biquad

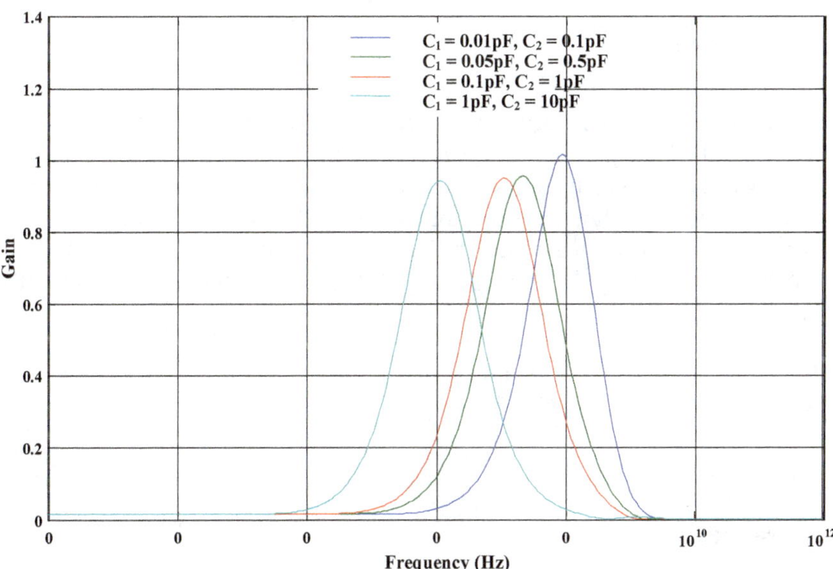

Fig. 4 Variations in cut-off frequency of band-pass response with capacitance value for current-mode KHN-equivalent biquad

and gain of the band-pass and the low-pass responses of the current-mode KHN-equivalent biquad circuit shown in Fig. 2.

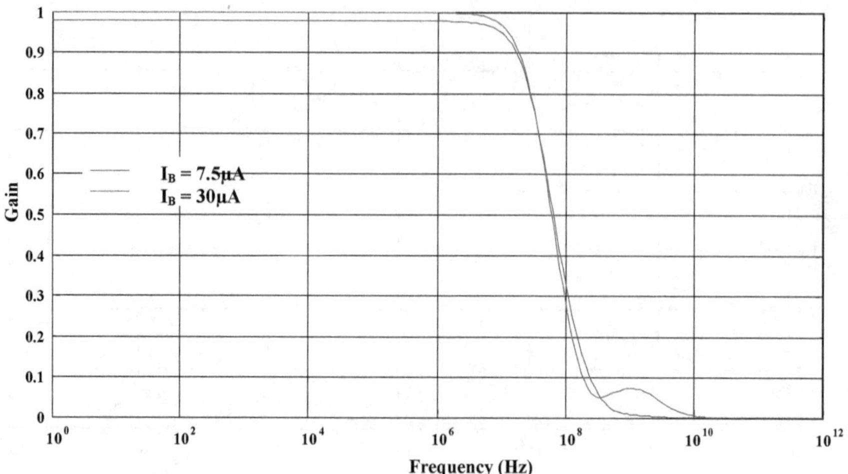

Fig. 5 Variations in gain of low-pass response with bias current for current-mode KHN-equivalent biquad

4 Conclusions

This paper presents high-frequency CM KHN-equivalent biquad filter suitable for operation at high frequencies they can operate at a low voltage of 1.0 V [17–24]. It can also be concluded that by varying the capacitor or the bias current, improves the gain and the operating frequency of the filter. Therefore, frequency of the filter can be controlled through a single DC bias current, thus providing good electronic tunability. The circuit was verified using SPICE, and simulation results confirm theoretical results.

Acknowledgements This work was performed at Electronics CAD Lab of ECE Department, I.T.S. Engineering College, Greater Noida, India.

References

1. J.C. Ahn, N. Fujii, Current-mode continuous-time filters using complementary current mirror pairs. IEICE Trans. Fundam. **E79-A**(2), 168–175 (1996)
2. R. Angulo, M. Robinson, E.S. Sinencio, Current-mode continuous-time filters: two design approaches. IEEE Trans. Circuits Syst. **39**(5), 337–341 (1992)
3. R. Schaumann, M.S. Ghausi, K.R. Laker, *Design of Analog Filters: Passive, Active RC, and Switched Capacitor* (Prentice Hall, Englewood Cliffs, NJ, 1989)
4. C. Toumazou, F.J. Lidgey, D.J. Haigh, *Analogue IC Design: The Current-Mode Approach* (Peter Peregrinus Ltd, 1990)
5. A. Bhandari, A.A. Husain, M.S. Chadha, A. Gupta, Lossy and lossless current-mode integrators using CMOS current mirrors. Int. J. Eng. Res. Dev. **9**(3), 34–41 (2013)

6. A.A. Husain, M.S. Chadha, A. Gupta, A. Bhandari, Design of high frequency current-mode continuous-time filter using CMOS current-mirrors. IOSR J. VLSI Signal Process. **3**(6), 58–62 (2013)
7. S.S. Lee, R.H. Zele, D.J. Allstot, G. Liang, CMOS continuous-time current-mode filters for high-frequency applications. IEEE J. Solid-State Circuits **28**(3), 323–329 (1993)
8. R. Senani, V.K. Singh, KHN-equivalent biquad using current conveyors. Electron. Lett. **31**(8), 626–628 (1995)
9. A.M. Soliman, Kerwin-Huelsman-Newcomb circuit using current conveyors. Electron. Lett. **30**(24), 2019–2020 (1994)
10. A. Toker, S. Ozouguz, C. Acar, Current-mode KHN-equivalent biquad using CDBAs. Electron. Lett. **35**(20), 1682–1683 (1999)
11. M.S. Chadha, A. Gupta, A. Bhandari, A.A. Husain, Third-order current-mode filter realization using CMOS current-mirror. IOSR J. Eng. **3**(12), 32–39 (2013)
12. R.W.J. Barker, Accuracy of current mirrors, in *IEE Colloquium on Current Mode Analogue Circuits*, London, vol. 25, paper 2 (1989)
13. B.L. Hart, R.W.J. Barker, Negative current-mirror using n-p-n transistors. Electron. Lett. **13**, 311–312 (1977)
14. B.L. Hart, R.W.J. Barker, Modified current mirror with a voltage-following capability. Electron. Lett. **18**, 970–972 (1982)
15. F.J. Lidgey, Looking into current mirrors. Wireless World **85**, 57–59 (1984)
16. R. Senani, A.K. Singh, A new universal current-mode biquad filter. Frequenz J. Telecommun. (Germany) **56**(1/2), 55–59 (2002)
17. A.M. Soliman, Mixed-mode biquad circuits. Microelectron. J. **27**(6), 591–594 (1996)
18. C. Toumazou, F.J. Lidgey, P.Y.K. Cheung, Current-mode analogue signal processing circuits—a review of recent developments, in *IEEE International Symposium on Circuits and Systems*, Portland, USA, vol. 3 (1989), pp. 1572–1575
19. Y.P. Tsividis, Integrated continuous-time filter design—an overview. IEEE J. Solid-State Circuits **29**(3), 166–176 (1994)
20. T. Tsukutani, M. Ishida, S. Tsuiki, Y. Fukui, Current-mode biquad without external passive elements. Electron. Lett. **32**(3), 197–198 (1996)
21. T. Voo, C. Toumazou, High-speed current mirror resistive compensation technique. Electron. Lett. **31**(4), 248–249 (1995)
22. G. Wegmann, E.A. Vittoz, Very accurate dynamic current mirrors. Electron. Lett. **25**, 644–646 (1989)
23. B. Wilson, F.J. Lidgey, C. Toumazou, Current mode signal processing circuits, in *IEEE Symposium on Circuits and Systems*, Helsinki, vol. 3 (1988), pp. 2665–2668
24. A. Gupta, M.S. Chadha, A. Bhandari, A.A. Husain, A novel approach for realization of higher order filter using bipolar and MOS current-mirrors. IOSR J. Electron. Commun. Eng. (IOSR-JECE) **9**(1), 99–104 (2014)

A Technical Shift in Monitoring Patients Health Using IoT

Bhamidi Rama and I. V. Subba Reddy

Abstract The patient monitoring system (PMS) plays a crucial role in early detection, diagnosis, and decision support. A lot of technological shift is observed in every area of life in this world, especially in the area of telemedicine. Due to the use of communication and information transfer, it is required to enhance the current patient monitoring system. Reports indicate that before a cardiac or respiratory arrest, 84% of the patients have physiological problems (Franklin and Mathew in Crit Care Med 22(22):244–247, 1994, [1]). Moreover, sending the measured data using Internet of Things (IoT) is an advanced way of wireless communication in the present scenario. During the shift duties of nurses, sometimes the data collected by another person may be out of sight which makes it difficult for the doctor to follow up the patient's case. Healthcare monitoring system designed in the present work will drastically reduce hospitalization, waiting time, consultation time, burden on medical staff, and overall health cost. The vital signs of patient are measured and sent to the personal computer (PC) of the doctor and on mobile phone, and to achieve this, Arduino-based microcontroller unit (MCU) is used with ESP8266 for Wi-Fi connection. Sensors for body temperature, heart rate, blood pressure, and tremors are connected to the MCU, and the measured readings are sent to the ThingSpeak server for web-based visualizations and to ThingView for mobile-based visualization.

Keywords Patient monitoring system · Healthcare solutions · ESP8266 · Arduino · Internet of Things · ThingSpeak · ThingView

---

B. Rama
Department of Electronics Technology, Loyola Academy, Old Alwal, Secunderabad, Telangana 500010, India
e-mail: ramakovur@gmail.com

I. V. Subba Reddy (✉)
Department of Physics, GITAM University-Hyderabad Campus, Rudraram, Medak, Telangana 502329, India
e-mail: isubbareddy@rediffmail.com

H. S. Saini et al. (eds.), *Innovations in Electronics and Communication Engineering*, Lecture Notes in Networks and Systems 107,
https://doi.org/10.1007/978-981-15-3172-9_38

1 Introduction

A smart way of connecting physical objects that are used in day-to-day life with Internet to communicate between two machines is called IoT [2]. These devices are linked to cloud platform through ThingSpeak Web services on which captured data is stored and analyzed using MATLAB analytics. It has been found that there will be 37.6% growth in healthcare IoT industry between the years 2015 and 2020. IoT helps people to enjoy personal attention for their health requirements [3]. The devices are tuned to remind them of their appointments. Due to the intervention of IoT, clinicians find the patients who are recuperating the post-anesthesia care unit easily, due to real-time monitoring. The medical staff can see the readings pertaining to patient health parameters on cloud as long as they are stored there.

US Bureau Census shows that an expenditure of $4 trillion is expected on health care in the next 10 years or 20% GDP on healthcare services [4]. An increase in 19.8% GDP is anticipated in comparison to 17.6% in 2010. Health care by 2020 is slated to have a cost of US $4.64 trillion [5, 6]. Hence, there is an absolute necessity of health services which can reduce the hospital expenses, save time, alleviate the workload of the staff in hospitals, and the overall cost. A major challenge in health care today is an increase in the population of senior citizens across the world. It has been found that the old aged people (more than 65 years) have drastically increased in the last 20 years and will still further rise to 1.2 billion by 2020 [7]. Current trends of vital parameters in HMS, its connectivity, types, places, and categories are studied [8–11].

The sensors used in this work are DS18B20 (temperature sensor), heart rate sensor, BP sensor, and MEMS (ADXL335) sensor which are very efficient and easy to program. This work discusses the design procedure of how patient's health status is continuously measured, monitored, and the information is sent to the concerned doctor before reaching the hospital. The early detection of vital signs using different methods helps the physicians to diagnose the problems of people more efficiently. Patient health parameters (PMS) are measured using sensors and the data is sent to the MCU. In the intensive care unit (ICU), patient is monitored using different devices with wired connections. The designed system helps in continuous monitoring of health parameters and transfers the data to a personal computer (PC) and to a mobile app.

2 Methodology

In medical applications, data is collected from various sensors, processed through MCU, and sent to devices having Wi-Fi. The block diagram of proposed system used to measure the physiological parameters using sensors, viz. body temperature, heart rate, blood pressure (BP), and tremor is represented in Fig. 1. The Arduino Uno development MCU board used in this work is a user-friendly, open-source MCU

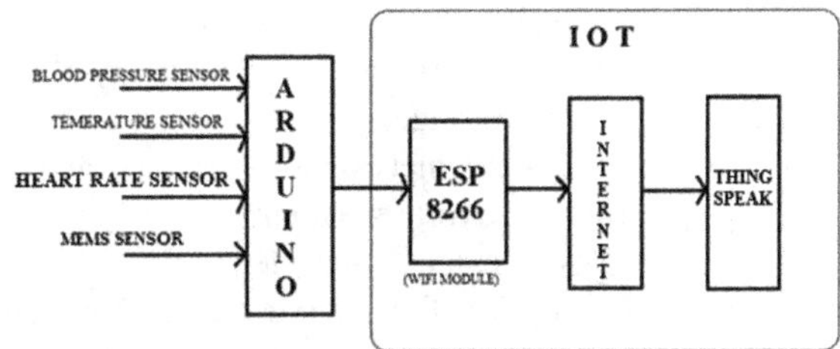

Fig. 1 Block diagram of PMS using IoT

with a clock frequency of 16 MHz [12]. Temperature sensor DS18B20 used here has 3 input pins VCC, GND, and DQ. The temperature sensor is interfaced with Arduino and is programmed using one wire and Dallas temperature libraries. The heart rate sensor is connected to the digital pin 8 of Arduino along with VCC and GND. The BP sensor used in this work is a serial communication device developed by Sunrom Technologies connected to the Analog pin 3. When the Analog pin 3 is low, BP sensor is 'ON' and when its high BP sensor is 'OFF.'

ESP8266 Wi-Fi microchip operates at 3.6 V and is used to interface between Arduino and IoT. It sends data using serial communication, at a rate of 115,200 baud rate, and has 1 MB disk size with system on chip. It has 4 pins VCC, GND, transmitter (Tx), and receiver (Rx). It helps in accessing Wi-Fi using the MCU. It has a pre-programmed attention (AT) commands that help in acting as an interface between Arduino and ThingSpeak Web site. The AT commands of ESP8266 make it easier to use this Wi-Fi (Ethernet Shield) microchip to transform it into IoT solutions.

The following steps are used to transfer the measured readings from sensor to ThingSpeak Web site using IoT

- ESP8266 microchip has 3 modes of operation, Mode 1—host, Mode 2—client, and Mode 3—both. In this work, Mode 3 is used. The AT command is AT + CW MODE = 3. The next step is to find whether Wi-Fi connection is there or not.
- ESP8266 identifies the SSID using the AT command AT + CWJAP 'SSID,' to find which Wi-Fi is connected and once it gets connected, it will ask for the password.
- To interact with Internet ESP8266, user needs ID and password. So, an intermediate interface is used, which is ThingSpeak Web site.
- The first thing required is the Internet protocol address (IP address). After getting the address, all the values are delivered to this address.
- The IP address used by ThingSpeak Web site is given in the program using AT command AT + CIP START = <type><address><port>. Hence, the IP address for ThingSpeak is 184. 106. 153. 149. 80. This address connects the hardware board to Internet using TCP/IP protocol.
- Identification is the next step in which the user will have the authority to store in that Web site. So, a password is obtained using API key. The ThingSpeak Web site gives this password (API key) to enter into that Web site after logging in through the Internet user ID and password.

- The API key obtained is copied and written in the program (16-bit key). Once the API key is sent, it will ask how much quantity of data is going to be sent and what is the length of the string.
- To upload data, a channel has to be created and it has to be given a name. The name given for this work is patient monitoring system (PMS). There are 8 fields where in 8 sensors can be connected.
- In this work, only five fields are used, i.e., heart beat (field 1), temperature (field 2), systole (field 3), diastole (field 4), and tremors (field 5).
- String name for field 1 'heartbeat.'
- String name for field 2 'temperature.'
- String name for field 3 'systole.'
- String name for field 4 'diastole.'
- Spring name for field 5 'tremors' (0 = no tremors and 1 = tremors).

Once the PMS using IoT gets connected to ThingSpeak, the string of data is sent to field 1 'heartbeat,' reading 83, field 2 'temperature,' reading 97, and so on. ThingSpeak Web site supports an android app (ThingView) which has to be logged in using channel ID to see the obtained readings.

3 Results

The study helps to monitor the physiological parameters of patients on day-to-day basis using Internet of Things (IoT). The measured physiological parameters are recorded and displayed using ThingSpeak Web site. The photographs of PMS using IoT developed evaluate the health parameters like heart rate, temperature, blood pressure (BP) along with nervous disorders caused in hands and the readings are displayed on the LCD as shown in Fig. 2.

The temperature sensor DS18B20 is a digital sensor with 9–12 bit resolution. This sensor is interfaced with Arduino and programmed using one wire and Dallas temperature libraries. Heart rate sensor works on the principle of photoplethysmograph. The heart rate sensor measures the flow of blood for 10 beats using an inbuilt

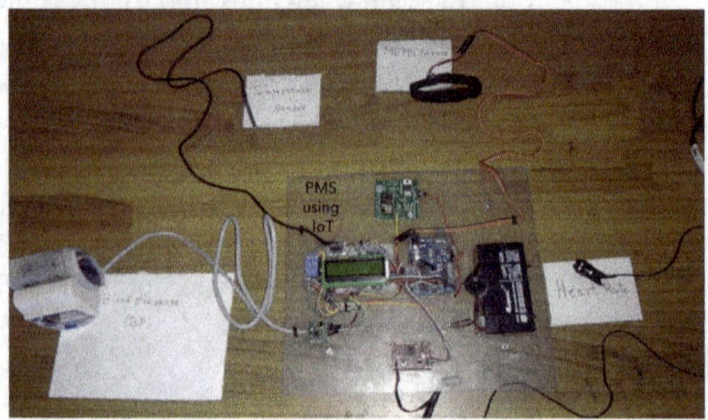

Fig. 2 Photograph of sensors interfaced to PMS using IoT

timer present on Arduino board and using the formula heart rate is calculated as 60 * 1000/rate, where Rate (1 beat) = Rate/10. The blood pressure sensor is a serial communication device developed by Sunrom Technologies that reads data of 15 bytes and transmits at a baud rate of 9600. BP sensor gives 13 byte ASCII code readings after measuring the systole and diastole values. ADXL335 is a 3D accelerometer MEMS sensor used to monitor tremors in hand. Analog read instruction is used to read *X*, *Y* values of MEMS sensor for 50 ms and the obtained value is multiplied by 100 using the program, for validation of tremors. If there is a shake in the hand for 5 s, then it is identified as tremor, else it is not tremor.

Figure 3 shows the heart rate reading on the LCD obtained using clip-type heart rate sensor and the graphical representation of the heart rate reading is obtained along with time of measurement in ThingSpeak Web site as shown in Fig. 4. Figure 5 shows the temperature reading obtained using DS18B20 temperature sensor and the graphical representation obtained using MATLAB analytics is displayed on ThingSpeak Web site along with time of measurement as shown in Fig. 6. Figure 7 displays the systole and diastole readings of blood pressure measured using BP sensor along with

Fig. 3 Photograph of measurement of heart rate using sensor

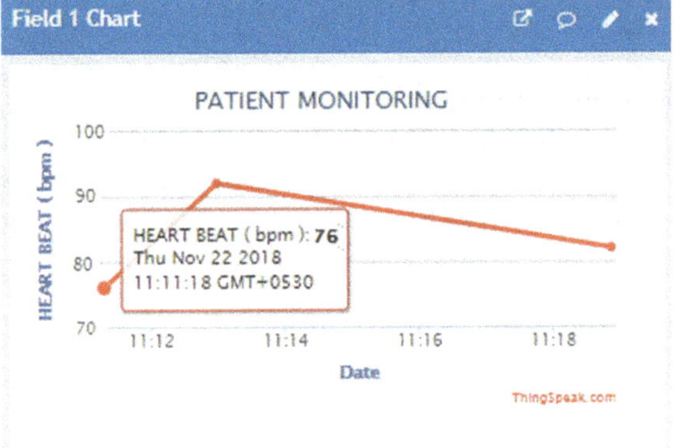

Fig. 4 Reading of the heart rate sensor displayed on ThingSpeak Web site

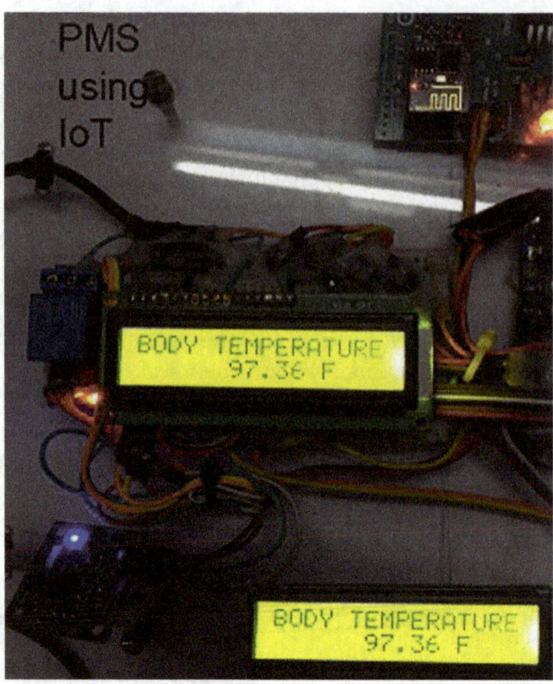

Fig. 5 Photograph of measurement of temperature using DS18B20 sensor

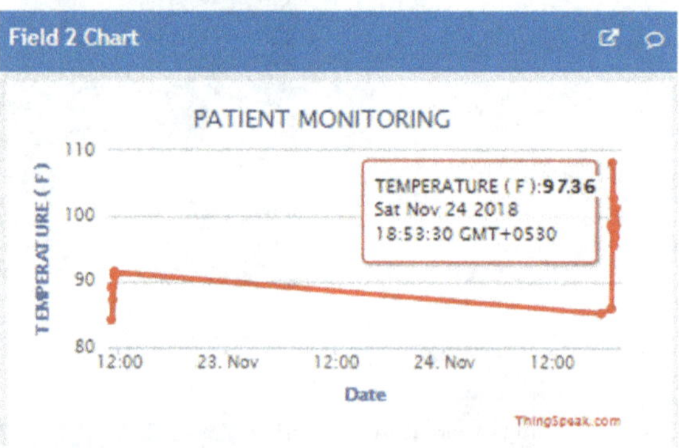

Fig. 6 Readings of temperature displayed on ThingSpeak Web site

its graphical representation as shown in Figs. 8 and 9. Figure 10 displays the tremor measured using ADXL335 MEMS sensor and its graphical representation shows a value 1 shown in Fig. 11. The non-occurrence of tremors is displayed on the LCD as shown in Fig. 12 and its value as 0, which is shown in Fig. 13 (Fig. 14). Figure 15 shows the display obtained on the mobile phone using ThingView mobile app.

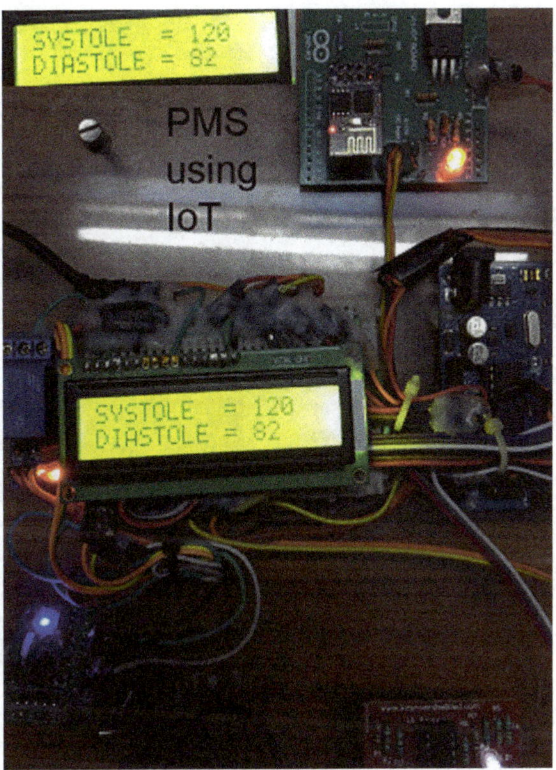

Fig. 7 Photograph of measurement of blood pressure using BP sensor

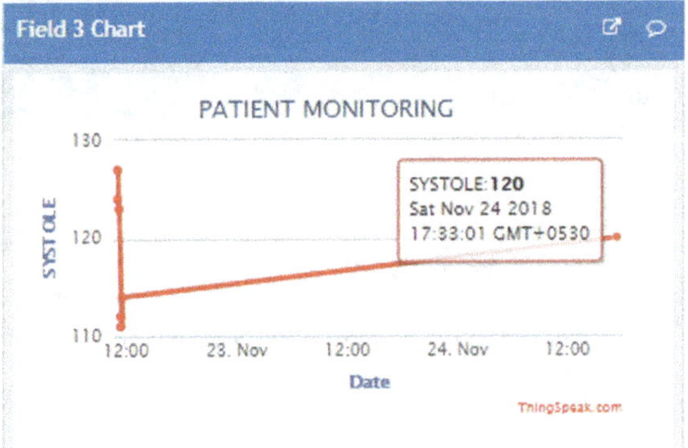

Fig. 8 Blood pressure systole readings displayed on ThingSpeak Web site

4 Conclusions

This work is useful for doctors to know the patient's health parameters from their desk only, instead of checking the charts written by nurses pertaining to the patient parameters throughout the day in the hospital. It can be used by the patient's relatives

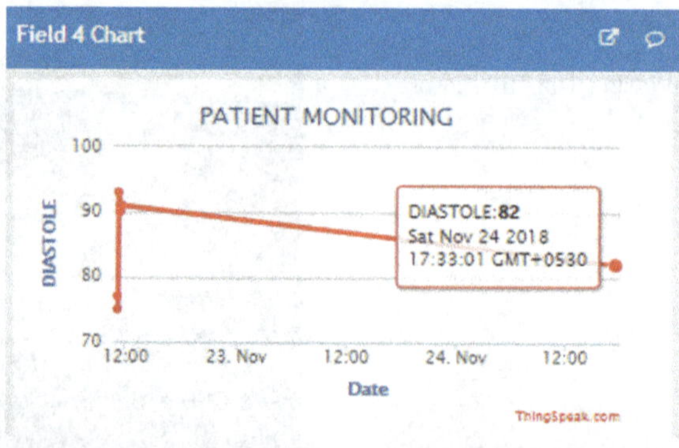

Fig. 9 Blood pressure diastole readings displayed on ThingSpeak Web site

Fig. 10 Photograph showing tremor occurrence

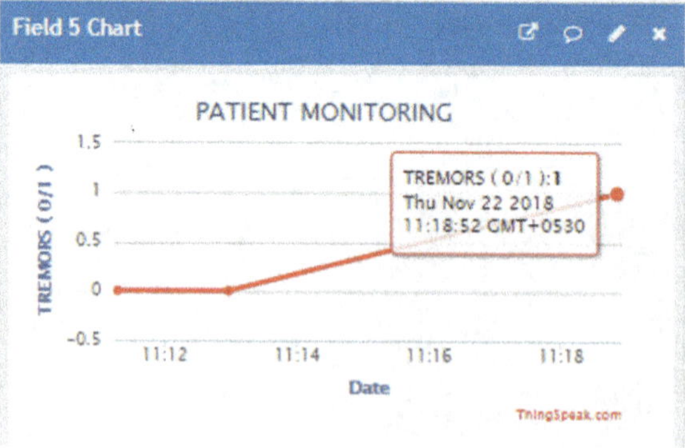

Fig. 11 Tremor occurrence reading displayed on ThingSpeak Web site

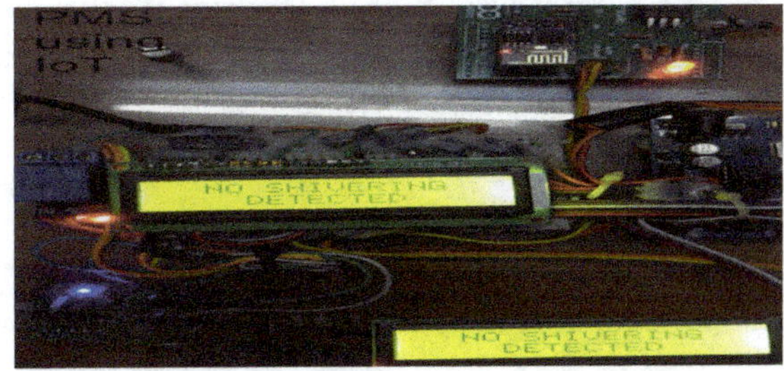

Fig. 12 Photograph of tremor not occured

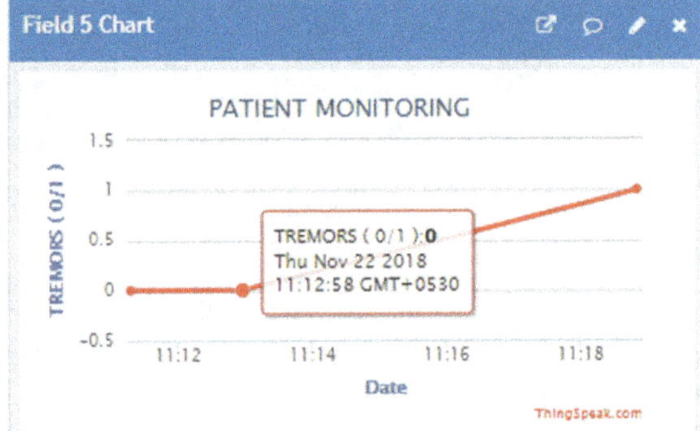

Fig. 13 Normal tremor reading displayed on ThingSpeak Web site

Fig. 14 Photograph of PMS sending data to ThingSpeak Web site

who are not in the vicinity and it can be achieved by choosing the option public in ThingSpeak web site and also on ThingView mobile app. All the readings of the parameters are uploaded into the Web site, and by using the MATLAB analytics, the graphical representation of the physiological parameters are displayed along with

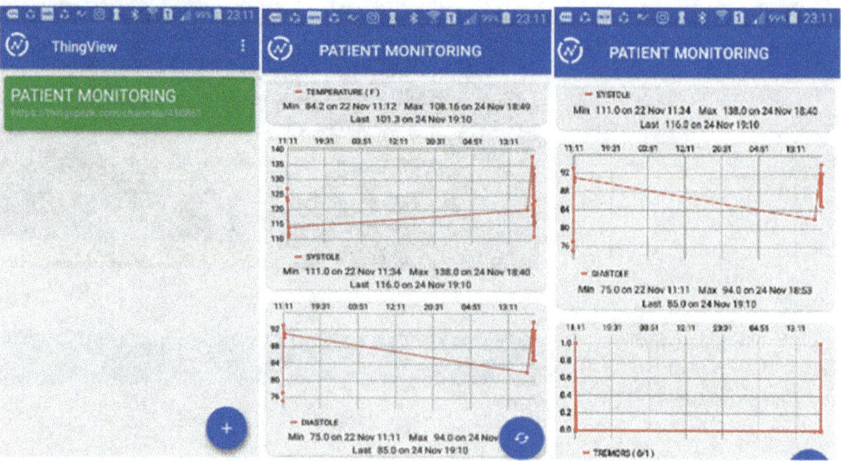

Fig. 15 Photograph of reading displayed on ThingView mobile app

time of measurement and the values at that time. Generally, nurses measure the patient parameters three times in a day. So, these readings can be remotely monitored by the doctor and accordingly they can monitor the patient. This work helps the specialists to screen patients who are in coma and doctors going for vitreo-retinal surgeries, as gentle developments in the body or hands are distinguished and sent on to the specialists personal computer.

An efficient PMS is designed to measure physiological parameters to diagnose patients at an early stage using IoT. The device is fully effective at several points of health care (places), in secondary care (hospitals), in an emergency (ambulance), in primary care (medical centers), in homes and for aged people. The physiological parameters of more number of patients can be measured in less time and patients in coma are monitored continuously.

The above conclusions are related to real-time physiological information, parameters representation, and interpretation methods. The overall system is developed in such a way that it integrates itself for the well-being of patients. The proposed system is sensible, accurate, and reliable in predictions.

References

1. C. Franklin, J. Mathew, Developing strategies to prevent in hospital cardiac arrest is to analyzing responses of physicians and nurses in the hours before the event. Crit. Care Med. **22**(22), 244–247 (1994)
2. C.A. Hussian, K. Vuha, M. Rajani, J.M. Vineeth, Smart health care monitoring using Internet of Things and android. Int. J. Adv. Res. Electron. Commun. Eng. **6**(3), 101–104 (2017)
3. M.M. George, N.M. Cyriac, S. Mathew, T. Antony, Patient health monitoring system using IoT and android. J. Res. **02**(01), 102–104 (2016)

4. S.J. Coons, J.A. Johnson, The United States spends more on health care than any other nation in the world. Soc. Behav. Asp. Pharma. Care, 279 (2010)
5. S.P. Keehan, A.M. Sisko, C.J. Tuffer, J.A. Poisal, G.A. Cuckler, A.J. Madison, J.M. Lizonitz, S.D. Smith, National health spending projections through 2020: economic recovery and reform drive faster spending growth. Health Aff. **30**(8), 1594–1605 (2011)
6. C.J. Truffer, S. Keehan, S. Smith, J. Cylus, A. Sisko, J.A. Poisal, J. Lizonitz, M.K. Clemens, Health spending projections through 2019 the recessions impact continues. Health Aff. **29**(3), 522–529 (2010)
7. M. Young, *The Technical Writer's Handbook* (University Science, Mill Valley, CA, 1989)
8. P.S. Pandian, K. Mohanavelu, K.P. Safeer, T.M. Kotresh, D.T. Shakunthala, P. Gopal, V.C. Padaki, Smart vest wearable multi parameter remote physiological monitoring system. Med. Eng. Phys. **30**(4), 466–477 (2008)
9. T.H. Tan, C.S. Chang, Y.F. Huang, Y.F. Chen, C. Lee, Development of a portable Linux-based ECG measurement and monitoring system. J. Med. Syst. **4**, 559–569 (2011)
10. L. Pollonini, N. Rajan, S. Xu, S. Madala, C. Dacso, A novel handheld device for use in remote patient monitoring of heart failure patients—design and preliminary validation on healthy subjects. J. Med. Syst. **36**(2), 653–659 (2012)
11. R.-G. Lee, K.-C. Chen, C.-C. Hsiao, C.-L. Tseng, A mobile care system with alert mechanism. IEEE Trans. Inf. Technol. Biomed. **11**(5), 507–517 (2007)
12. S. Gupta, Jeevan Rakshak patient monitoring system using Matuino. Int. J. Res. Eng. Technol. **4**(3), 31–34 (2016)

IoT-Based State of Charge and Temperature Monitoring System for Mobile Robots

Rameez Raja Chowdhary, Manju K. Chattopadhyay and Raj Kamal

Abstract This paper presents the Internet of Things (IoT)-based state of charge (SOC) and temperature monitoring system for battery of mobile robots. It uses robots (RBs) of Orchestration of Robotic Platform (ORP). Our system monitors the temperature of battery and terminal voltage at regular interval of time. The SOC is determined with the help of proposed re-modified extended Coulomb counting method. A robotic server is designed for collecting, storing and analysing the data. The server sends necessary messages to robotic electric vehicle (REV), based on the status of the readings. These messages are used to prevent overheating of battery and improve the operating cycle of battery.

Keywords IoT · Temperature measurement · SOC · SOH · Robotic electric vehicle

1 Introduction

IoT is a buzzword in vehicular industry [1]. It empowers the new possibilities for perceptional computing, advanced analytics and new technologies in vehicular industry [2]. The industry is expeditiously adopting this perceptional technique to take modern vehicles to the next level. Moreover, it assists the vehicle-maker to provide highly

R. R. Chowdhary
Department of Electronics and Telecommunication Engineering, Institute of Engineering and Technology, Devi Ahilya University, Indore 452001, India

R. R. Chowdhary (✉) · M. K. Chattopadhyay
School of Electronics, Devi Ahilya University, Indore 452001, India
e-mail: rameez.chowdhary@gmail.com

M. K. Chattopadhyay
e-mail: manju.elex@gmail.com

R. Kamal
Department of Electronics and Telecommunication Engineering, Prestige Institute of Engineering and Science, Indore 452001, India
e-mail: dr_rajkamal@hotmail.com

H. S. Saini et al. (eds.), *Innovations in Electronics and Communication Engineering*, Lecture Notes in Networks and Systems 107,
https://doi.org/10.1007/978-981-15-3172-9_39

customized services to commuters and owners. The subscribers prefer personalized and interactive vehicle.

Modern vehicles are smart vehicles [2–4]. It is always connected with a network. The vehicle periodically updates the sensor data to a server. The vehicle-maker or the service centre designs the server [1]. The following key areas have limitless possibilities in modern vehicular industry [2, 5].

Vehicle Maintenance: Smart vehicles are competent to upload equipped sensor's data at dedicated server. The makers or service provider uses the sensor's data for vehicle service [6]. Similarly, they also use these data to improve the performance of the vehicle.

Safety: Smart vehicles are competent to inform any undesired event to rescue team. The vehicle sends current information about the situation. It reduces the reaction time of rescue service that helps in saving life.

Navigation: Smart vehicles are equipped with GPS. The vehicles regularly upload the path, fuel consumption, traffic and parking data at server. The holder can monitor the fuel consumption and locate the vehicle from anywhere using these data.

Infotainment: All vehicle-makers give infotainment system. This provides accessibility to online services such as email, music, video streaming and news information through the dashboard to user.

Smart Vehicle Services: Vehicle-makers provide facility of keyless locking and unlocking of car, vehicle tracking, ignition detection, parking assistance and automatic climate control. Some vehicle-makers like Mercedes Benz, Google car, Tesla motor and BMW are working on driverless car.

The vehicle-makers are also focusing on environment-friendly vehicles [7]. Hybrid electric vehicle (HEV) and electric vehicle (EV) are energy-efficient [3]. EV releases fewer greenhouse gases that make it eco-friendly [8–10]. HEVs use dual fuel, petrol and battery to power the vehicle. This technique makes HEV energy-efficient [3, 4].

A battery is an important device for EVs and HEVs. The performance of EV depends on the health of the battery and other parameters. The surrounding temperature of the battery is one of the important parameters. It is a function of the ambient and internal temperature of the battery [11, 12]. The Joule heating effect introduces the internal temperature of the battery [13]: Temperature also affects the charging ability, power supply ability and electrochemistry of battery.

A battery delivers maximum energy to the vehicle when it is healthy. Accurate SOC calculation preempts the battery from overcharging and discharging. Similarly, accurate SOH calculation decreases the overall operating cost of the battery. Our system uses re-modified extended Coulomb counting (RECC) algorithm for SOH and SOC calculation of the lead–acid battery. Lead–acid battery is a cost-effective energy storage device with good safety. It finds application in smart grid, UPS and vehicle [11]. The advantage of our system is that it precisely calculates the SOC of the battery without any knowledge of its internal dynamic and electrochemistry. Additionally, the system updates the calculated readings on a server for live monitoring of battery. The server analyses the data and sends alert messages accordingly.

1.1 Related Work

Rahimi-Eichi et al. [14] presented an algorithm. The algorithm is a combination of Coulomb counting and open-circuit voltage (VOC) algorithm. The algorithm uses an adaptive parameter-updating technique to decrease the SOC determination error. They introduced a resistive capacitor equivalent module of the battery. The module explains the nonlinear connection between SOC and VOC for the battery. Zackrission et al. [15] presented a life cycle estimation method for EV in their study. They also discussed the environmental effect in life cycle estimation of the battery. Weng et al. [16] proposed a VOC method to calculate SOH and SOC for the Li-ion battery. They compared the method with the existing method. Their experimental result shows the effectiveness of the proposed method.

Zanella et al. [17] offered an IoT model for a smart city. The model incorporates a large number of diverse models. The model gives access to a particular set of records for the evolvement of an opulence of digital service. Joshua [18] proposed a system. The author uses system to explain the consequence of temperature on battery performance. The cold and hot start performance on the electric range for power train is also demonstrated by the author. Vore et al. [6] presented a data acquisition system for plug-in hybrid electric vehicle (PHEV); they composed the driving and charging data of PHEV. The data help in vehicle design, policy analysis and operator feedback. Yamamoto et al. [19] presented a vehicle-monitoring model. The authors implemented the model for achieving a better bus operation. The model logs diverse raw data such as battery temperature, accumulated travelling time, power consumption, SOC and mileage. The model uplinks this data to cloud using a 3G network for live monitoring.

2 System Design

We present IoT-based battery's SOC monitoring and alert system. The system improves the effectuation of the battery. Experiments are conducted using ORP platform designed by Chowdhary et al. [20–22]. The system is divided into two parts, data acquisition unit and a server. A data acquisition unit regularly computes current, terminal voltage and surface temperature of the battery. The data acquisition unit is appended with REV that obtains battery's reading at regular interval. The robotic node (RN) of ORP works as the REV.

The RECC algorithm uses this reading and calculates SOH and SOC. The REV transmits these readings, namely SOH, SOC and surface temperature of the battery to the server. The robotic Orchestrator (ORCH) is configured as a server. It uses a micro-SD card for the reading storage. The server further processes these reading using SOC and temperature analysis algorithm. It sends the essential messages for maximizing the safety of the battery. Therefore, the system helps to enhance the operating cycle of the battery. The IoT server and REV use MQTT protocol for information exchange.

API of MQTT is lightweight. It needs minimal computation on a device. IoT devices profusely use MQTT due to its low power consumption property. Chowdhary et al. [20] explained the hardware and software implementation of ORP in their work.

3 Coulomb Counting Algorithm

The SOC estimation is a sophisticated job. It depends on the type of battery and the application where the battery is being used. The Coulomb counting algorithm is a plausible algorithm for calculation of the battery's SOC. It provides precise result for both Li-ion and Lead-acid battery. On the other hand, VOC provides precise SOC calculation only for lead–acid battery. However, the method does not consider nonlinearity and output current of battery [22]. The following equation gives mathematical representation of Coulomb counting algorithm [23]:

$$\mathrm{SOC}(t) = \mathrm{SOC}(0) + \frac{1}{C_{\mathrm{new}}} \int_{t_0}^{t} I_{\mathrm{c}} \mathrm{d}t \tag{1}$$

where SOC(0) denotes initial value of SOC at t_0 (when we put a new battery in service), SOC(t) denotes present value of SOC, C_{new} denotes the total capacity of battery. $\int_{t_0}^{t} I_{\mathrm{c}} \mathrm{d}t$ denotes integral discharge current during the discharge process.

The algorithm faces drawbacks like accumulation error, discharging losses and charging losses during self-discharging of battery. Hence, these parameters should be considered for much precise calculation of SOC. Therefore, the SOC must be recalibrated periodically. Hence, we designed a RECC algorithm based on our system requirements.

A 100% charged battery has maximum dischargeable capacity (C_{max}) that cannot be same as the estimated capacity since it decreases with time. Therefore, SOH can be calculated by the following equation.

$$\mathrm{SOH}(t) = C\mathrm{max}(t)/C_{\mathrm{new}}(t_0) \quad \forall t \geq t_0 \tag{2}$$

$C_{\mathrm{max}}(t)$ represents maximum instantaneous dischargeable capacity. C_{new} represents capacity of new battery at time t_0.

The depth of discharge (DOD) can be defined as the amount of capacity that has been released with respect to estimated capacity.

$$\mathrm{DOD} = C_{\mathrm{discharging}}/C_{\mathrm{new}} \tag{3}$$

The difference of the DOD in a working interval (Δt) can be determined by our modified equation for a discharging and charging current of battery. It reduces the computation overhead.

$$\Delta\text{DOD} = -\left((t_0 - (t_0 + \Delta t))\left(\frac{I_c(t_0) + I_c((t_0 + \Delta t))}{2}\right)\right)/C_{\text{new}}, \quad \Delta t \geq 0 \quad (4)$$

where I_c is positive for charging and negative for discharging cycle. Therefore, the DOD can be computed by

$$\text{DOD}(t) = \text{DOD}(t_0) + \Delta\text{DOD} \quad (5)$$

Hence, the SOC can be computed by [24] as

$$\text{SOC}(t) = \text{SOH}(t) + \text{DOD}(t) \quad (6)$$

Equation (6) evaluates the SOC value for $t > t_0$.

3.1 RECC Algorithm

Algorithm 1 presents the pseudocode of the RECC algorithm. A new lead–acid battery is used for the experiments. SOH is 100% for the new battery. The initial reading of SOC is calculated using either the open-circuit voltage or loaded voltage. It also relies on the initial state of the battery. The algorithm stores the calculated reading in memory. Then, it measures the battery voltage V_c and current I_c for further calculation of the SOC and SOH. If $I_c < 0$, it means that the battery operates in discharging mode. When $I_c > 0$, it means that the battery operates in charging mode; $I_c = 0$ means that the battery operates in open-circuit mode.

The algorithm uses Eq. (5) to calculate DOD. During the open-circuit mode of battery, the SOC is calculated from relationship between the SOC and VOC. During discharging mode of the battery, the SOC is calculated using Eq. (6). The voltage V_c is less than lower threshold voltage V_{LT} when battery is fully charged. On the other hand, when the battery is fully charged, it means V_c is greater than upper threshold V_{UT}. The value of V_{LT} and V_{UT} is 11.4 V and 12.7 V, respectively. A fully charged battery is used for each experiment.

Algorithm 1: RECC Algorithm

```
if (new battery) then
    estimate DOD(0), SOC(0) and SOH = 100%.
    else
            Read reading from memory.
    end else
  end if
  while
    Get the value of I_b and V_b
    if (I_c < 0 && V_c > V_LT) then
            estimate DOD(t) and SOC(t).
```

```
        else
                SOH = DOD
        end else
      end if
      if ($I_c > 0$ && $V_b \geq V_{UT}$) then
                SOH = DOD
        else
                estimate DOD(t) and SOC(t).
        end else
      end if
      if ($I_c = 0$) then
                estimate open circuit DOD and SOC
      end if
    goto Get the value
  end while
```

Algorithm 2: SCTA Algorithm

```
While
      Read the value of temp, SOC and SOH.
      if (temp > 56) then
              Send master alert msg.
              Send SOC & temp value.
      end if
      if (temp > 35 || temp < 56) then
              Send Warning msg.
              Send SOC & temp value.
      end if
      if (temp > 0 || temp < 35) then
              Send SOC & temp value.
      end if
      if (SOH < 80%) then
              Send battery replacement msg.
        else
              Send SOC & temp value.
        end else
      end if
end while
```

The calculation of SOC can be done using SOC and charging voltage curve. The constant current mode (specified in manufacturer datasheet) is used for charging the battery and to plot the relationship between charging voltage and SOC.

3.2 State of Charge and Temperature Analysis (SCTA) Algorithm

We have designed a SCTA algorithm. Algorithm 2 gives the pseudocode of the SCTA algorithm. Our server uses SCTA algorithm to analyse the temperature and SOC data of REV. The algorithm divides the temperature reading into three groups, namely low, medium and high. The low temperature range for battery is 20–35 °C, a medium range is 36–55 °C and high range is beyond 56 °C. The algorithm does not send any alert message to REV if the temperature reading of battery lies under the range of low group. When temperature reading lies under the range of medium category or last five reading lies under this category, the algorithm sends an alert message to REV. When continuous five temperature readings lie under the range of critical category, the algorithm sends "Stop REV" message. The algorithm also analyses the SOC and SOH readings. This reading helps to monitor the life cycle of the battery. The battery needs to be replaced when the SOH of battery degrades beyond 80% [25]. When the SOH value goes less than 80%, then the algorithm sends the battery replacement message.

4 Experimental Process

One RBs of ORP is used to perform experiments. The REV configures as a node and ORCH as a server in the proposed information exchange model.

Figure 1 shows the REV moving in environment of experiments. Forty five experiments have been performed in the duration of one and half months. Each experiment

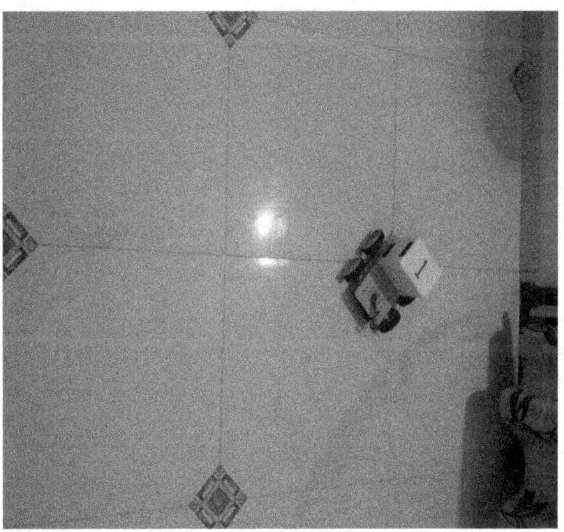

Fig. 1 REV moving in arena of experiment

lasted approximately two and half hours. The RBs run circularly in a given environment with three different speeds viz. slow, medium and fast. The REV uses function of SPFN class of ORP platform for speed control of motor.

The REV measures battery current, voltage and temperature data with sampling rate of one minute. The REV saves this information internally. The RECC algorithm uses these readings for calculation of the SOH and SOC. Thus, the REV has some processing capability. The REV sends these calculated readings in every 5 min to the server. The server further processes these readings. It reduces the data processing and information exchange overhead at the server end. The server collects these readings regularly and sends the alert messages based on the analyses of SCTA algorithm.

The maximum recorded temperature in the experiment is 65 °C. We have divided the temperature range into three groups based on this experimental temperature range. However, the battery can safely handle much higher temperature that is defined in SCTA. These temperature ranges of SCTA are only for demonstration purpose. The SCTA algorithm sends alert messages based on the temperature reading in the experiments.

5 Investigation and Discussion

The SOC for lead–acid battery is calculated using implementation of RECC. The Amptek lead–acid battery is used to power the REV. The operating voltage for this battery is 12 V. The recommended charging and discharging cut-off voltages for the battery are 11.4 V and 12.7 V, respectively.

SOH is assumed to be 100% for new battery. Our system regularly collects the reading of voltage, current and surface temperature of battery.

Fig. 2 Charging curve of battery

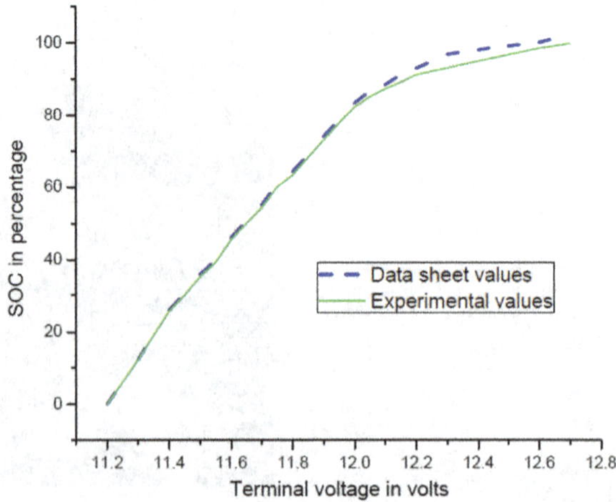

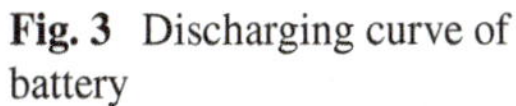

Fig. 3 Discharging curve of battery

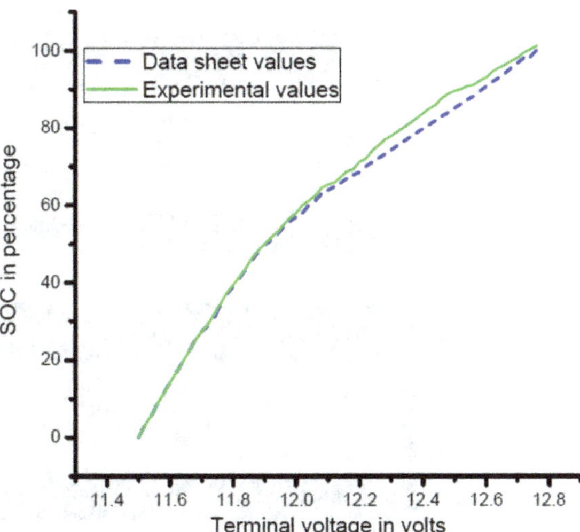

Figure 2 shows a charging SOC curve with respect to charging time. We have compared the datasheet charging and discharging curve with experimentally estimated SOC values. The red dashed curve shows the SOC reading of the datasheets in Fig. 2. The green curve shows the experimentally calculated SOC reading using RECC algorithm in Fig. 2.

The experimental SOC curve follows the reference SOC curve for initial readings, but after some readings, it deviates. The SOC calculation error is less than 3%. The calculation error is due to improper determination of the initial SOC and temperature variation. Hence, the accurate estimation of initial SOC can reduce the considerable amount of error in experimental SOC.

The discharging curve of SOC with respect to the terminal voltage is shown in Fig. 3. The SOC values obtained from datasheet is shown by red curve. The experimentally estimated SOC readings are shown by green curve. The green and red curves are almost identical for lower SOC readings. However, the experimental curve shows variation for higher SOC values. During the SOC estimation of a discharging cycle, the maximal obtained error is less than 4%. This error is due to the temperature variation and improper digitization.

The server uses the SCTA algorithm to analyse the data. When the temperature falls in the medium range, the robotic server sends alert message to REV. The server sends a message "Overheating" to warn the REV. It means that the REV requires cooling the battery. Figure 4a shows the message on a LCD connected on the REV. Additionally, the server sends master alert message to REV when the temperature reading exceeds the medium range. Figure 4b shows the master alert message. When REV receives master alert message, it stops itself. The server regularly gives the SOC and SOH reading to let the RB know how much time the RB runs in arena. The RB runs with a constant and variable speed in a circular path. Figure 5 shows the server data. REV sends these data to server.

Fig. 4 Server messages to REV. T, C and H represent the temperature reading, SOC value and SOH value in both figures, respectively

We do not have any battery replacement message from the server in experiments, because SOH reading is greater than 80%. The minimum SOH reading obtained during the forty-fifth experiment is 99%. The REV can send the data on any open-source IoT server. We have designed our own server and algorithms to analyse the readings due to the length of experiments.

We have implementing Eq. (4) using numerical integration method, instead of the definite integration method. It reduces the computation overhead. This allows us to implement the proposed RECC algorithm using the 8-bit microcontroller. The algorithm provides accurate SOC and SOH calculation compared to the other researchers. Our algorithm provides maximum SOC calculation error less than 3% for charging cycle and 4% for discharging cycle. On the other hand, the method given by Rahimi-Eichi et al. [14] gives the calculation error less than 5%. Rahimi-Eichi et al. used large hardware circuit and high power processor (e.g. laptop) for SOC calculation, compared to our circuit. In other words, our system uses less hardware and gives better accuracy compared to other researchers.

6 Conclusion

The SOH and SOC are calculated using RECC algorithm. Our algorithm decreases computation overhead compared to definite integral. Therefore, implementation of the algorithm is less complex. Additionally, the hardware requirement to measure V_c and I_c are less pricey. It also simplifies circuit design. The ORCH works fine as a robotic server. According to the status of the battery, it sends all necessary

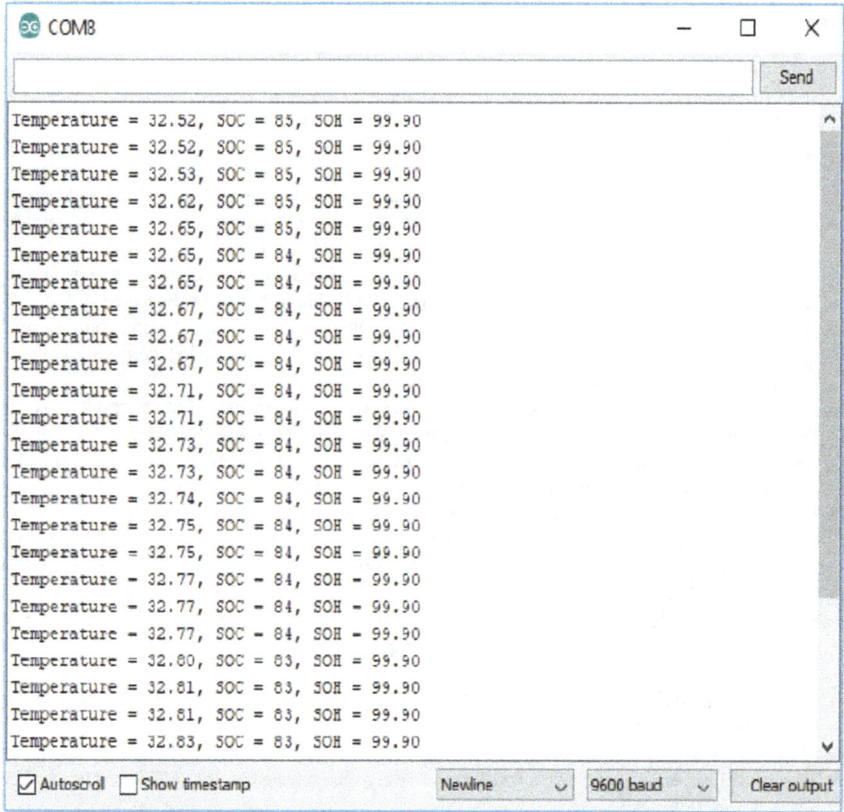

Fig. 5 Readings stored at server database

messages to REV. The proposed IoT system processes the SOH and SOC data at REV side. Therefore, it reduces the computation and message transmission overhead on the server side. The server calculates the real-time SOH. Hence, server may avoid the battery from overloading and make sure it is protection. Additionally, our projected system design is squashed and can be run on low power processor due to less computational overhead. The system gives better precision compared with other researcher's system. In future work, we will study the temperature effect on battery of REV. The proposed systems find application in data acquisition system.

References

1. X. Krasniqi, E. Hajrizi, Use of IoT technology to drive the automotive industry from connected to full autonomous vehicles, in *Proceeding of International Federation of Automatic Control* (2016), pp. 269–274
2. New opportunities are on the way with cognitive technology for cars (2018), http://www.ibm.co/watson-iot-cognitive-solutions-ww-flyer-wwf12354usen-20180103(1).pdf. Accessed 15 Mar
3. R. Trigui, F. Mensing, E. Bideaux, Eco-driving: potential fuel economy for post-manufactured hybrid vehicles. Int. J. Electr. Hybrid Veh. **8**(4), 321–334 (2016)
4. R.R. Chowdhary, M.K. Chattopadhyay, R. Kamal, IoT model based battery temperature and health monitoring system using electric vehicle like mobile robots. J. Adv. Robot. (2019)
5. B. Deshpande, 7 application IoT connected vehicle which may soon be common (2014). http://www.simafore.com/blog/bid/207243/7-applications-of-IoT-connected-vehicle-which-may-soon-be-common
6. S. Vore, M. Kosowski, Z. Wilkins, T.H. Bradley, Data management for geographically and temporally rich plug-in hybrid vehicle big data. World Electr. Veh. J. **8**, 293–304 (2016)
7. S. Boyd, D. Howell, An overview of the hybrid and electric systems R&D at the U.S.–DOE (FY 2015–2016). World Electr. Veh. J. **8**, 461–472 (2016)
8. V. Smil, Electric vehicles: not so fast. IEEE Spectrum **54**(12) (2017)
9. F. Zhang, X. Zhang, M. Zhang, A.S.E. Edmonds, Literature review of electric vehicle technology and its applications, in *5th International Conference on Computer Science and Network Technology (ICCSNT)*, Changchun (2017)
10. J.Y. Yong, V.K. Ramachandaramurthy, K.M. Tan, N. Mithulananthan, A review on the state-of-the-art technologies of electric vehicle, its impacts and prospects. J. Renew. Sustain. Energy Rev. **49**, 365–385 (2015)
11. J. Wang, I. Besselink, H. Nijmeijer, Battery electric vehicle energy consumption modelling for range estimation. Int. J. Electr. Hybrid Veh. **9**(2), 79–102 (2017)
12. L. Chen, W. Li, X. Xu, S. Wang, Energy management optimisation for plug-in hybrid electric sports utility vehicle with consideration to battery characteristics. Int. J. Electr. Hybrid Veh. **8**(2), 122–138 (2016)
13. A.A. Pesaran, Battery thermal models for hybrid vehicle simulations. J. Power Sources **110**, 377–382 (2002)
14. H. Rahimi-Eichi, M.Y. Chow, Adaptive parameter identification and state-of-charge estimation of lithium-ion batteries, in *IECON 2012-38th Annual Conference on IEEE Industrial Electronics Society*, Montreal, QC (2012)
15. M. Zackrission, L. Avellan, J. Orienius, Life cycle assessment of lithium ion batteries for plug-in-hybrid electric vehicle—critical issues. J. Clean. Prod. **18**(15), 1519–1529 (2016)
16. C. Weng, J. Sun, H. Peng, A unified open-circuit-voltage model of lithium-ion batteries for state-of-charge estimation and state-of-health monitoring. J. Power Sources **258**, 228–237 (2014)
17. A. Zanella, N. Bui, A. Castellani, L. Vangelista, M. Zorzi, Internet of things for smart cities. IEEE Internet Things J. **1**(1), 22–32 (2014)
18. L. Joshua, Effect of temperature on lithium-iron phosphate battery performance and plug-in hybrid electric vehicle range. M.S. thesis, Department of Mechanical Engineering University of Waterloo, Ontario, Canada, 2013
19. A. Yamamoto, M. Fukuda, H. Utsumi, Vehicle management and travel data analysis of E-Bus adopted in JR Kesennuma line. World Electr. Veh. J. **8**, 122–130 (2016)
20. R.R. Chowdhary, M.K. Chattopadhyay, R. Kamal, Orchestration of robotic platform and implementation of adaptive self-learning neuro-fuzzy controller. J. Electron. Des. Technol. **8**(3), 17–29 (2017)
21. R.R. Chowdhary, M.K. Chattopadhyay, R. Kamal, Study of an orchestrator for centralized and distributed networked robotic systems, in *Proceeding of the International Conference on Information Engineering, Management and Security*, Chennai (2015)

22. R.R. Chowdhary, M.K. Chattopadhyay, R. Kamal, Comparative study of orchestrated, centralised and decentralised approaches for orchestrator based task allocation and collision avoidance using network controlled robots. J. King Saud Univ. Comput. Inf. Sci. (2018)
23. Y. Zhihao, H. Ruituo, X. Linjing, State-of-charge estimation for lithium-ion batteries using a Kalman filter based on local linearization. Energies **8**, 7854–7873 (2015)
24. M. Murnane, A. Ghazel, A closer look at state of charge (SOC) and state of health (SOH) estimation techniques for batteries (2017). http://www.analog.com/media/en/echnical-documentation/technical-articles/A-Closer-Look-at-State-Of-Charge-and-State-Health-Estimation-Techniques
25. Fundamentals in battery testing, BU-901. http://www.batteryuniversity.com/learn/article/difficulties_with_testing_batteries

Signal and Image Processing

Optimized Segmentation of Oil Spills from SAR Images Using Adaptive Fuzzy K-Means Level Set Formulation

Kalyani Chinegaram, Kama Ramudu, Azmeera Srinivas and Ganta Raghotham Reddy

Abstract With the increasing amount and complexity of remote sensing image data and the difficulties faced in processing the data, the development of large-scale image segmentation analysis algorithms could not keep pace with the need for methods that improve the final accuracy of object recognition. So, the development of such methods for large-scale images poses a great challenge nowadays. Traditional level set segmentation methods which are Chan-Vese (CV), image and vision computing (IVC) 2010, ACM with SBGFRLS and online region-based ACM (ORACM) were suffered with more amounts of time complexity, as well as low segmentation accuracy due to the large intensity homogeneities and the noise. The robust region-based segmentation is impossible in remote sensing images is a tedious task because due to lack of spatial information and pixel intensities are non-homogenous. For this reason, we proposed a novel hybrid approach called adaptive particle swarm optimization (PSO)-based Fuzzy K-Means clustering algorithm. The proposed approach is diversified into two stages: in stage one, pre-processing the input image to improve the clustering efficiency and overcome the obstacles present in traditional methods by using particle swarm optimization (PSO) and Fuzzy K-Means clustering algorithm. With the help of PSO algorithm, we get the "optimum" pixels values that are extracted from the input SAR images; these optimum values are automatically acted as clusters centers for Fuzzy K-Means clustering instead of random initialization from original image. The pre-processing segmentation result improved the clustering efficiency but suffers from few drawbacks such as boundary leakages and outlier's even particle swarm optimization is used. To overcome the above drawbacks, post-processing is necessary to facilitate the superior segmentation results by using

K. Chinegaram (✉) · K. Ramudu · A. Srinivas · G. R. Reddy
Department of ECE, Kakatiya Institute of Technology and Science, Warangal 506015, India
e-mail: kalyanichinegaram@gmail.com

K. Ramudu
e-mail: ramudukama@gmail.com

A. Srinivas
e-mail: srinivas_azmeera@yahoo.com

G. R. Reddy
e-mail: grrece9@gmail.com

H. S. Saini et al. (eds.), *Innovations in Electronics and Communication Engineering*, Lecture Notes in Networks and Systems 107,
https://doi.org/10.1007/978-981-15-3172-9_40

level set method. Level set method utilizes an efficient curve deformation is driven by external and internal forces in order to capture the important structures (usual edges) in an image as well as curve with minimal energy function is defined. The combined approach of both pre-processing and post-processing is called as Adaptive Particle Swarm Optimization-based Fuzzy K-Means (AFKM) clustering via level set method. The proposed method is successfully implemented on large-scale remote sensing imagery, and the dataset are taken from the open source NASA earth observatory database for segmenting the oil slicker creeps, oil slicker regions, etc. So here in this, the proposed new hybrid method had feasibility and the efficiency which could attain the high accurate segmentation results when compared with traditional level set methods ...

Keywords Image segmentation · Remote sensing images · Adaptive Fuzzy K-Means clustering · Fast level set method

1 Introduction

Among many segmentation methods, mostly we come across is clustering analysis. Many modified clustering algorithms had been introduced and proven to be effective for image segmentation. Normally, images and its segmentation consists of some specific noise, and clustering accuracy can be specifically upgraded or developed by the process known as a "particle swarm optimization-PSO". In 1995, PSO [1] was introduced by Russell Eberhart along with James Kennedy. Originally, these two initiated an emerging computer simulations software of a birds flocking around the food sources, and then earlier comprehended in what way their algorithms are going to get worked on optimization issues. The method can be applied to optimization issues of large dimensions, often producing eminent solutions more rapidly than alternative approaches, and it has a low convergence rate in the iterative process [2–5].

Mostly, the significant difficulties in image analysis are image segmentation. For different applications, numerous segmentation algorithms have been presented and created. Now and again, inadmissible outcomes have been experienced for number of existing segmentation algorithms such as thresholding region growing and clustering techniques. Here, we proposed a novel hybrid clustering method, which is called Optimized Fuzzy K-Means (OFK) clustering for improving the clustering efficiency, but these clustered results suffers with few drawbacks even PSO algorithm is Introduced such as boundary or edge leakages and Outliers which are present during the clustering process. In order to overcome these difficulties, we introduced a new level set formulation for accurate as well as robust segmentation in the post-processing [6, 7].

This paper is systematized as follows: A detailed explanation regarding materials and methods is used in Sect. 2 such as PSO algorithm and Adaptive Fuzzy K-Means clustering are discussed. A modified version of level set method is proposed based on OFKM clustering; Sect. 3 gives a detailed explanation of the experimental results and discussions, and finally, conclusions are presented in Sect. 4.

2 Materials and Methods

In this chapter, a complete study on existing level set methods such as IVC 2010, Chan-Vese (CV) model, Active Contour Model (ACM) with SBGFRLS, Online Region-Based ACM (ORACM). These methods suffered with few limitations such as slow and complex, weak edge leaking and poor segmentation results with illumination changes. Improper image acquisitions like inadequate illumination conditions can also lead to poor image segmentation. The Geodesic ACM (GAC) having improved segmentation results for the SAR images which are having sharp edges, but it suffers with the few drawbacks such as unstable evolution, requires periodic reinitialization, balloon or pressure force causes boundary leakage, slow evolution due to smaller time step.

The performance of active contour models can be improved further by introducing selective additional pre-processing techniques such as one possible way is the combined approach introduced that is Particle Swarm Optimization (PSO) and Fuzzy K-Means (FKM) clustering algorithms, which is named as Optimized Fuzzy K-Means (OFKM) clustering and Here, Post-Processing is necessary to overcome the drawbacks in the Pre-processing such as Outliers, Boundary and Edge leakages. A new energy function is defined for evolving level sets in post-processing based on the OFKM Clustering. With this, accurate image segmentation is possible even for the images captured under any illumination conditions by using the proposed method. The proposed method which is called Optimized Fuzzy K-Means via level set evolution is explained in the following sections.

2.1 Calculate Optimum Object Pixels by Using PSO Algorithm

The particle swarm optimization (PSO) was introduced in the year 1995 [8] by Kennedy along with Eberhart for global optimum solution to all engineering applications. In this process, PSO algorithm is used in image processing for different applications such as image segmentation, denoising, enhancement for getting optimum output. The pixels considered in the input images are called "particles." Every particle has its own position as well as velocity. This approach is an automatic iterative process to get optimum pixels by using the below Eqs. (1) and (2)

$$V_{\text{updated}}(t+1) = W * V_i(t) + C_1 * r_1(p_i(t) - x_i(t)) + C_2 * r_2(g_i(t) - x_i(t)) \quad (1)$$

in which $v_i(t)$ indicates the value of current velocity of dimension of *ith* particle in *tth* iteration. A variable $x_i(t)$—indicates current location of the particle if *ith* particle in *t*th iteration. Variable w—is weight of inertia or inertial weight, c_1—self-cognition acceleration coefficient and c_2—social-cognition acceleration coefficient.

$$x_i(t+1) = x_i(t) + v_i(t+1) \tag{2}$$

Equation (2) represents new location of every particle refreshed utilizing a first position along with the new velocity by a condition (1) in which r_1 and additionally r_2 was produced independently. Range of uniform distributed random numbers are (0, 1). The flowchart of PSO algorithm is explained in the below figure.

The parameters inertial weight (W), acceleration coefficients C_1 and C_2 and r_1 and r_2 are used in Eq. (1). The proper values are chosen for better results. The standard PSO algorithm utilizes the C_1 and C_2 are equal to 2. For better segmentation results, changes are made in the inertial weight (W) and acceleration coefficients according to the specific application, which is called "Adaptive PSO." In this work, we considered the following parameter values for optimum pixel calculation from the input image.

2.2 An Adaptive Fuzzy K-Means Clustering Algorithm (AFKM)

The method called Adaptive Fuzzy K-Means (AFKM) clustering for image segmentation could also be applicable on images which are general. This method is also suitable for medical, remote sensing as well as microscopic images which are captured by the electronic products. This algorithm gives better quality and the adaptive clustering procedure when compared with the other conventional methods or clustering methods.

The K-Means (KM) [7, 9] clustering is widely used due to its simplicity. The K-Means algorithms' objective is o minimize the function of an object in order to allocate the group of data to its center. Due to the weakness of KM algorithm, its performance is limited. Therefore, to overcome the drawbacks of KM clustering here, Fuzzy C-Means (FCM) [10–13] had been introduced.

2.2.1 Algorithm Steps for Fuzzy K-Means Clustering Algorithm

Here in this section, we propose AFKM clustering algorithm which incorporates both the K-Means and Fuzzy C-Means clustering algorithm. The implementation of the proposed method is explained by considering an image which is a digital one with $R * S$ pixels (R—represents number of rows; S—represents number of columns) which had to be clustered into n_c clusters or regions. Let us consider $f(x, y)$ as the pixel in the image and c_j as a *j*th cluster center. For K-Means clustering, depending upon the Euclidean distance, a group of data was assigned to the closer center. Each new position of the cluster center is calculated by the equation given below

$$c_j = \frac{1}{n_{c_j}} \sum_{x \in c_j} \sum_{y \in c_j} f(x, y) \tag{3}$$

where $x = 1, 2, 3, \ldots, s$; $y = 1, 2, 3, \ldots, R$; $j = 1, 2, 3 \ldots, n_c$.

Similarly, the FCM, allocating the data member simultaneously to greater than one class this process is depended on a membership function given below.

$$M^m jf(x, y) = \frac{1}{\sum_{k=1}^{n_c} \left(\frac{d_{jf(x,y)}}{d_{kf(x,y)}}\right)^{2/(m-1)}}; \text{ if } d_{kf(x,y)} > 0, \quad \forall j, f(x, y) \tag{4}$$

$$\begin{cases} M^m kf(x, y) = 1 \\ M^m jf(x, y) = 0; \text{ for } f(x, y) \neq k \\ \text{if } d_{kf(x,y)} = 0. \end{cases} \tag{5}$$

From the above Eq. (4), $d_{jf(x,y)}$ is the distance from the pixel point (*x*, *y*) to the present jth cluster center, $d_{kf(x,y)}$ is the distance from the point (*x*, *y*) to the other *k* cluster centers, n_c is the number of cluster centers, *m* is the integer and $m > 1$ is the degree of fuzziness.

In this process, a new hybrid clustering algorithm is introduced in pre-processing, which is called Adaptive Fuzzy K-Means (AFKM) algorithm. The speciality of proposed hybrid clustering is to esteem all centers were initialized. So to ensure a better clustering of an image, Eq. (3) will be no more in order to update the cluster center. Hence, due to this, we utilize the fuzziness and its belonging concepts in the proposed method. So, here in this, the membership function $M^m jf(x, y)$ is determined with the help of Eqs. (4) and (5). To attain a good process of clustering, some changes were made in AFKM algorithm. The degree of belongingness B_j is calculated for every cluster after specifying the membership to all the individual data. When the degree of belongingness is higher, then it shows a strong relationship in between the center and its members where it ensures an improved data clustering. The equation for degree of belongingness is

$$B_j = \frac{c_j}{M^m_{jf(x,y)}} \tag{6}$$

To improve the clustering method, we need to update the degree of membership. It is optimized where it is depended on the degree of belongingness to ensure that the process of reallocating member places the data to its appropriate cluster. The esteem of $M^m_{jf(x,y)}$ is modified in an iteration accordingly as

$$(M^m_{jf(x,y)})' = M^m_{jf(x,y)} + \Delta M^m_{jf(x,y)} \tag{7}$$

Here, $(M^m_{jf(x,y)})'$ represents the new updated membership. $\Delta M^m_{jf(x,y)}$ is well-defined as:

$$\Delta M^m_{jf(x,y)} = \alpha(c_j)(e_j) \tag{8}$$

α—is a constant designed with a value in between 0 and 1. e_j value is calculated by using the following equation

$$e_j = B_j - \hat{B}_j \tag{9}$$

$\hat{B}_j$ represents normalized esteem for the degree of belongingness.

Lastly, the new center positions of total existing clusters were calculated depending on optimized or new membership function as follows:

$$c_j = \frac{\sum_{x \in c_j} \sum_{y \in c_j} (M^m_{jf(x,y)})' f(x, y)}{\sum_{x \in c_j} \sum_{y \in c_j} (M^m_{jf(x,y)})'} \tag{10}$$

This process is repeated until the values of all the centers were no longer variation.

3 Implementation to Level Set Segmentation

The proposed method is diversified into two stages, which are called pre-processing and post-processing an input image. So here, in pre-processing, we used a new hybrid approach by using particle Swarm Optimization (PSO) and Adaptive Fuzzy K-Means clustering, which is called Optimized Adaptive Fuzzy K-Means (OAFKM) clustering for improving the clustering efficiency. The detailed description of pre-processing is already discussed in Sect. 2. In the later stage, which is called post-processing by using level set method, it is necessary to remove the outliers, boundary leakages and to get the smoothed segmentation image. A brief explanation and mathematical implementation to level sets are given in below sections.

In post-processing, we use level set method in order to overcome the obstacles which are present in pre-processing. It is necessary to initialize the contour on AFKM clustered image instead of original input image. In traditional level set methods, with the help of two components which are called as data term as well as regularization term, curve evolution is controlled [14]. The data term attracts the curve towards boundary at the same time second one is to regulate the regularity of a curve. Now, let we discuss the proposed level set method based on the PSO and AFKM clustering; here, a new energy function is derived based on the Optimized FKM clustering. The modified energy function of the CV model is given in Eq. (11).

$$E(C_1, C_2, C) = \mu \cdot \text{Length}(c) + \lambda_1 \int\limits_{\text{inside}(c)} |I_{\text{Ofkm}} - C_1|^2 + \lambda_2 \int\limits_{\text{outside}(c)} |I_{\text{Ofkm}} - C_2|^2 \mathrm{d}x\mathrm{d}y \quad (11)$$

From, Eq. (11), the term "Length(c)" is used as regularizing term, which is mainly for smooth evolving of the level curve "C." To effectively control of the level set curve, we consider weighting coefficient parameters that are mu (μ) ≥ 0 and lambda (λ_1, λ_2) > 0. Similarly, the mean intensities, C_1 and C_2, are to be calculating for pre-processing image "IOafkm," which evolve the level curve inside as well as the outside of an image, respectively. The modified level set formulation (LSF) of the energy function is given in Eq. (11) [15–26]:

$$\frac{\partial \phi}{\partial t} = \delta_{\varepsilon}(\phi)\left[\mu \cdot \text{div}(\nabla\phi/|\nabla\phi|) - \lambda_1(I_{\text{ofkm}} - C_1)^2 + \lambda_2(I_{\text{ofkm}} - C_2)^2\right] \quad (12)$$

Equation (12) could be re-written as follows:

$$\frac{\partial \phi}{\partial t} = \delta_{\varepsilon}(\phi)[\lambda(C_1 - C_2)(2I_{\text{ofkm}} - C_1 - C_2)] \quad (13)$$

In $(C_1 - C_2)(2I_{\text{OAfkm}} - C_1 - C_2)$, from Eq. (13), right-hand side is the real effective part and the data term $(2I_{\text{OAfkm}} - C_1 - C_2)$ at which the sign can be utilized to regulate the direction of propagating level curve. The term $(C_1 - C_2)$ could be eliminated due to the insignificant contribution. In addition, it is feasible to substitute $\delta\varepsilon$ (φ) with $|\Delta\varphi|$. By utilizing, all these procedures, the following were very much simpler region-based level set formula which could be achieved as:

$$\frac{\partial \phi}{\partial t} = \lambda(2I_{\text{ofkm}} - C_1 - C_2)|\nabla\phi| \quad (14)$$

In Eq. (14), larger esteem of a λ could expedite an evolution of a level curve. However, to decrease the number of constraints, we speed up this prototype by utilizing a larger time step "Δt" instead of larger λ. Therefore, we remove the λ directly in Eq. (14).

The definition of c_1' and c_2' is same as that in Eq. (11). To attain the stable results, we normalize the data term $(2I_{\text{OAfkm}} - C_1 - C_2)$; thus, the proposed modified level set formula could be written as given in Eq. (15)

$$\frac{\partial \phi}{\partial t} = \frac{2I_{\text{optimumfkm}} - C_1 - C_2}{\max(\left|2I_{\text{optimumfkm}} - C_1 - C_2\right|)}|\nabla\phi| \quad (15)$$

Here, C_1 and C_2 are the mean intensities, which are given in the following equation:

$$C_{1(\text{inside})} = \frac{\int_\Omega I_{\text{Oafkm}}(x, y) \cdot H(\phi)\text{d}x\text{d}y}{\int_\Omega H(\phi)\text{d}x\text{d}y}$$
$$C_{2(\text{outside})} = \frac{\int_\Omega I_{\text{Oafkm}}(x, y) \cdot (1 - H(\phi))\text{d}x\text{d}y}{\int_\Omega (1 - H(\phi))\text{d}x\text{d}y} \tag{16}$$

Equations (14) and (15) are the final level set equations, which utilize only one parameter that is time step "Δt," which gives faster curve evolution on optimized adaptive Fuzzy K-Means clustered images. The time step and number of iterations are inversely proportional to each other. If the larger time step is considered, it will give the faster convergence and less number of iterations to get the converged final contour.

4 Simulation Results and Discussions

The proposed algorithm presented here in this paper is compared with the models which were widely used see Fig. 1, such as Chan-Vese (CV), ACM with SBGFRLS, Online Region-Based ACM (ORACM) methods in terms of an Area covered, Effective Area, less CPU time and Area error from remote sensing images. Similarly, Fig. 1f shows the final level set contour evolution by proposed method and its segmenting regions, respectively. The proposed level set method is segmenting the regions accurately and effectively when compared with conventional methods. The database was taken from NASA https://earthobservatory.nasa.gov/images. These images were rescaled to 256 × 256 for contour evolution of the both proposed and conventional methods for preprocessed images.

The performance parameters of conventional models were tabulated in Table 1, and the proposed model was tabulated in Tables 2 and 3, respectively. A pixel-based quantitative evaluation approach is used. In this evaluation, the proposed approach made a comparison between the final segmented image "Z" and ground truth image "U." The performance parameters used in this paper such as: Dice Similarity Index, Jaccard Similarity Index, True Positive Fraction (TPF), True Negative Fraction (TNF), False Positive Fraction (FPF) and False Negative Fraction (FNF) can be defined as

$$\text{Dice} = \frac{2|Z \cap U|}{|Z| + |U|} \quad \text{Jaccard} = \frac{Z \cap U}{Z \cup U}, \quad \text{TPF} = \frac{Z \cap U}{U} \tag{17}$$

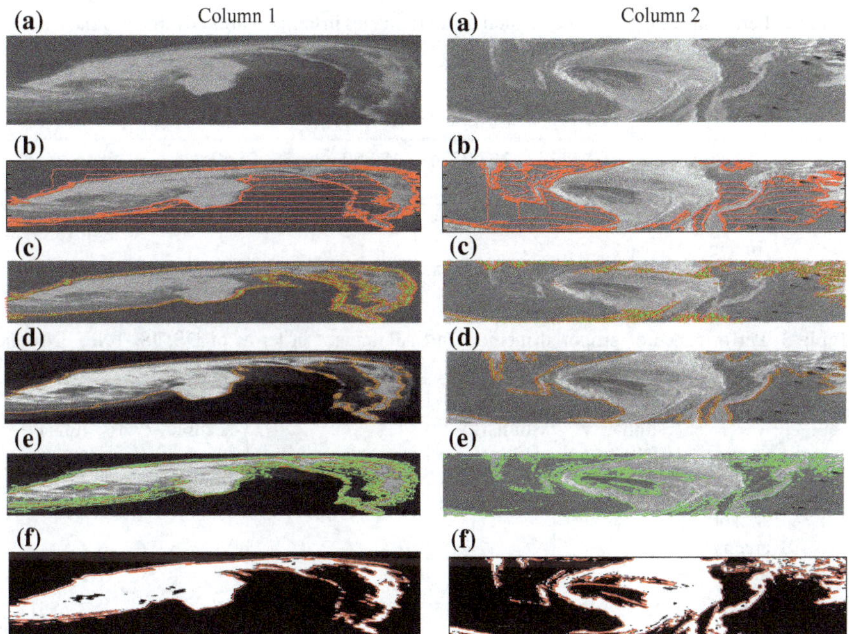

Fig. 1 Simulation results on different SAR images, first row shows original SAR images such as column 1 and column 2 are oil slicker creeps and oil spill images, respectively. Similarly, second row, third row, fourth row and fifth row depicts the simulation segmentation results of existing level set methods such as IVC 2010, Chan-Vese (CV), ACM with SBGFRLS, Online Region-Based ACM (ORACM) model, respectively, and finally, the last row depicts the proposed method segmentation results

Table 1 Performance of conventional methods on SAR images in terms of DS, JS, TPF, FNF and elapsed time

S. no.	Images	Method	DS	JS	TPF	FNF	SA (%)	Elapsed time (s)
1	Oil slicker creeps image	IVC 2010	0.88	0.79	0.791	0.209	88	82.4546
		Chan-Vese	0.21	0.19	0.196	0.080	21	50.9249
		ACM with SBGFRLS	0.44	0.39	0.398	0.602	44	90.2614
		ORACM	0.39	0.35	0.355	0.645	39	1.8835
2	Oil slick image	IVC 2010	0.70	0.63	0.633	0.366	70	21.605709
		Chan-Vese	0.0008	0.0004	0.0004	0.999	0.08	42.7014
		ACM with SBGFRLS	0.31	0.27	0.277	0.722	31	157.8172
		ORACM	0.46	0.41	0.418	0.581	46	1.4528

Table 2 Performance of proposed method on SAR images in terms of area covered by final contour, area error and final cluster centers, respectively

S. no.	Images	Final cluster centers		Area covered (mm^2)	Actual area (mm^2)	Area error (mm^2)
		Cc1	Cc2			
1	Oil slicker creeps image	102.1113	167.2374	45,850	55,696	9846
2	Oil slick image	138.6437	185.7085	42,649	55,696	13,047

Table 3 Performance of proposed method on SAR images in terms of DS, JS, TNF, FNF and elapsed time

S. no.	Images	Dice similarity (DS)	Jaccard similarity (JS)	TPF	FNF	Segmentation accuracy (%)	Elapsed time (s)
1	Oil slicker creeps image	0.91	0.82	0.823	0.176	91	8.0777
2	Oil slick image	0.85	0.76	0.765	0.23	85	10.5511

5 Conclusion

Optimized Adaptive Fuzzy K-Means clustering using level set method is presented in this paper for efficient segmentation of remote sensing images such as oil slicker, oil spills images. This proposed method improves the clustering efficiency by incorporating the Particle Swarm Optimization (PSO) and Fuzzy K-Means (FKM) clustering in the pre-processing, which is called Optimized Fuzzy K-Means Clustering. In the post-processing, level set method is used to refine the segmentation results and also for rejection of outliers, edge and boundary leakage problems in the pre-processing. The performance of the proposed method is analyzed based on the segmentation accuracy (SA), dice dimilarity (DS), jaccard similarity (JS), true negative fraction (TNF) and false negative fraction (TNF). The proposed method got the accuracy of 91 and 85% of two different oil spills images over existing level set methods. The proposed method is superior, maximum area is covered and more accurate segmentation of oil spills regions over existing level set methods.

References

1. N.A. Mat-Isa, M.Y. Mashor, N.H. Othman, Comparison of segmentation performance of clustering algorithms for Pap smear images, in *Proceedings of International Conference on Robotics, Vision, Information and Signal processing (ROVISP2003)* (2003), pp. 118–125

2. F. Gibou, R. Fedkiw, *A Fast Hybrid K-Means Level Set Algorithm for Segmentation*, Tech. Rep. (Stanford University, Stanford, CA, USA, 2002)
3. R. Ronford, Region based strategies for active counter models. Int. J. Comput. Vis. **3**(2), 229–251 (1994)
4. M. Fatih Talu, ORCAM: online region-based active contour model. Expert Syst. Appl. **40**, 6233–6240 (2013). www.elsevier.com/locate/eswa
5. G. Gan, C. Ma., J. Wu, *Data Clustering: Theory, Algorithms, and Applications* (Society for Industrial and Applied Mathematics, 2007)
6. F. Cui, L. Zou, B. Song, Edge feature extraction based on digital image processing techniques, in *IEEE International Conference on Automation and Logistics* (2008), pp. 2320–2324
7. N.A. Mat-Isa, Automated edge detection technique for Pap smear images using moving k-means clustering and modified seed based region growing algorithm. Int. J. Comput. Internet Manag. **13**, 45–59 (2005)
8. B. Bhanu, J. Peng, Adaptive integrated image segmentation and object recognition. IEEE Trans. Syst. Man Cybern. **30**, 427–441 (2000)
9. T. Kanungo, D. Mount, N. Netanyahu, C. Piatko, R. Silverman, A.Y. Wu, An efficient k-means clustering algorithm: analysis and implementation. IEEE Trans. Pattern Anal. Mach. Intell. **24**(7) (2002)
10. R.L. Cannon, J.V. Dave, J.C. Bezdek, Efficient implementation of the fuzzy c-means clustering algorithm. IEEE Trans. Pattern Anal. Mach. Intell. **8**, 248–255 (1986)
11. M.Y. Mashor, Hybrid training algorithm for RBF network. Int. J. Comput. Internet Manag. **8**(2), 50–65 (2000); C. Mao, S. Wan, A water/land segmentation algorithm based on an improved Chan-Vese model with edge constraints of complex wavelet domain. Chin. J. Electron. **24**(2), 361–365 (2015)
12. W. Cai, S. Chen, D. Zhang, Fast and robust fuzzy c-means clustering algorithms incorporating local information for image segmentation. Pattern Recogn. **40**, 825–838 (2007)
13. S. Krinidis, V. Chatzis, A robust fuzzy local information c-means clustering algorithm. IEEE Trans. Image Process. **19**, 1328–1337 (2010)
14. C. Li, C. Xu, C. Gui, M.D. Fox, Distance regularized level set evolution and its application to image segmentation. IEEE Trans. Image Process. **19**(12), 3243–3254 (2010)
15. S. Ahmadi, M.J.V. Zoej, H. Ebadi, H.A. Moghaddam, A. Mohammadzadeh, Automatic urban building boundary extraction from high resolution aerial images using an innovative model of active contours. Int. J. Appl. Earth Obs. Geoinf. **12**(3), 150–157 (2010)
16. K. Kim, J. Shan, Building roof modeling from airborne laser scanning data based on level set approach. ISPRS J. Photogramm. Remote Sens. **66**(4), 484–497 (2011)
17. M. Cote, P. Saeedi, Automatic rooftop extraction in nadir aerial imagery of suburban regions using corners and variational level set evolution. IEEE Trans. Geosci. Remote Sens. **51**(1), 313–328 (2013)
18. C. Li, C.Y. Kao, J.C. Gore, Z. Ding, Minimization of region scalable fitting energy for image segmentation. IEEE Trans. Image Process. **17**(10), 1940–1949 (2008)
19. V.P. Dinesh Kumar, T. Thomas, Clustering of invariance improved Legendre moment descriptor for content based image retrieval, in *IEEE International Conference on Signal Processing, Communications and Networking* (2008), pp. 323–327
20. L.D. Cohen, I. Cohen, Finite-element methods for active contour models and balloons for 2-D and 3-D images. IEEE Trans. Pattern Anal. Mach. Intell. **15**(11), 1131–1147 (1993)
21. T.F. Chan, L.A. Vese, Active contours without edges. IEEE Trans. Image Process. **10**(2), 266–277 (2001)
22. Y. Shi, W.C. Karl, A real-time algorithm for the approximation of level-set-based curve evolution. IEEE Trans. Image Process. **17**(5), 645–656 (2008)
23. K. Zhang, L. Zhang, H. Song, D. Zhang, Reinitialization-free level set evolution via reaction diffusion. IEEE Trans. Image Process. **22**(1), 258–271 (2013)
24. L. Bertelli, S. Chandrasekaran, F. Gibou, B.S. Manjunath, On the length and area regularization for multiphase level set segmentation. Int. J. Comput. Vis. **90**(3), 267–282 (2010)

25. E. Brown, T. Chan, X. Bresson, Completely convex formulation of the Chan-Vese image segmentation model. Int. J. Comput. Vis. **98**(1), 103–121 (2012)
26. K. Karantzalos, N. Paragios, Recognition-driven two-dimensional competing priors toward automatic and accurate building detection. IEEE Trans. Geosci. Remote Sens. **47**(1), 133–144 (2009)

Moving Object Tracking Using Optimal Adaptive Kalman Filter Based on Otsu's Method

Ravi Pratap Tripathi and Ashutosh Kumar Singh

Abstract Object tracking is the most challenging task, especially when object is deformable and moving. Kalman filter is widely used mathematical tool for tracking and prediction, but to use Kalman filter it is more important that noise variance must be well defined. In real situation, noise variances may be undefined. In this paper, a method is proposed to track object using Otsu's threshold method based on Kalman filter. Otsu's method is used for segmentation purpose, and Kalman filter is used for the prediction of various parameters. For noise variance estimation, we fuse maximum likelihood estimation along with memory attenuation method. In this paper, Kalman filter is applied on various parameters such as centroid, axis lengths, and elevation angles of segmented image for better tracking of deformable object.

Keywords Kalman filter · Innovation · Maximum likelihood · Otsu's thresholding · Memory attenuation

1 Introduction

Object tracking is an important subject of medical field; image processing, computer vision, RADAR, and missile guidance fields. Object tracking uses the concept of image processing, probabilistic theory, and state-space techniques [1, 2]. In tracking of object, we use video which is nothing but sequence of frames. From each and every frame, we try to find the location and orientation of object. Using this information, we locate trajectory which will be the tracking path of the object [3]. Tracking of object can be accomplished by two methods—(1) feature-based modal matching and (2) random motion parameter estimation [5–8]. This iterative matching matches the current image and template image and based on the maximum similarity, we try to track object. The major problem in this method is if the target color is similar to the background then the tracking of the object becomes difficult. In random motion parameter estimation method based on object motion constraints, in this, we estimate

R. P. Tripathi (✉) · A. K. Singh
Department of Electronics and Communication Engineering, Indian Institute of Information Technology Allahabad, Allahabad, India
e-mail: rse2018510@iiita.ac.in

H. S. Saini et al. (eds.), *Innovations in Electronics and Communication Engineering*, Lecture Notes in Networks and Systems 107,
https://doi.org/10.1007/978-981-15-3172-9_41

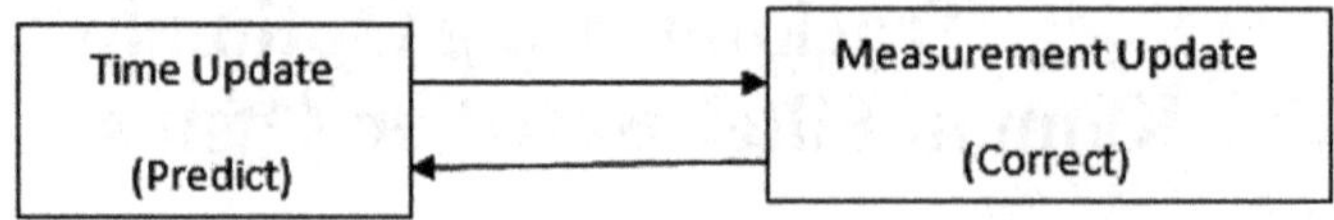

Fig. 1 Process of Kalman filters

the position of object. In this paper, Otsu's thresholding method is used with Kalman filter. The method is used to find the various parameters (centroid, axis length, elevation angle, etc.) of the target object. This measured value of various parameters is used by Kalman filter in estimation of the next position of deformable object. This algorithm also works very accurately when data is missing between the frames.

The flow of paper is divided as follows: Sect. 2 explains Kalman filter, Otsu's method explained in Sect. 3. In Sect. 4, the proposed method is given, simulation and results are discussed in Sect. 6 and gives the concluding remarks.

2 Kalman Filter

Kalman filter is mathematical tool, which is used for estimation of state from the noisy data [4]. Consider a dynamic system is represented as (Fig. 1):

$$x_k = A_{k-1}x_{k-1} + B_{k-1}u_{k-1} + \lambda_{k-1}w_{k-1} \tag{1}$$

$$z_k = H_k x_k + v_k \tag{2}$$

In the above equation,

$$E[W_k] = 0 \quad \mathrm{Cov}\left[W_k w_j\right] = \mathrm{Cov}[W_k w_k] = Q_k \delta_{kj}$$

$$E[v_k] = 0 \quad \mathrm{Cov}\left[v_k v_j\right] = \mathrm{Cov}[v_k v_k] = R_k \delta_{kj}$$

$$E[W_k v_k] = \mathrm{Cov}\left[w_k v_j\right] = 0$$

3 Optimal Adaptive Kalman Filter (OAKF)

In OAKF, we combine the characteristics of innovation adaptive estimation (IAE), maximum likelihood estimation (MLE) with memory attenuation (MA) [10]. In real-time applications, it may be possible that we are unable to find the exact value of Q and R, so in this case, we use IAE to find variances. In some cases, it may possible that the state of the system not well defined, so we use MA algorithm. In OAKF, for error covariance prediction, a scaling factor β is used. The flowchart of OAKF loop is given in Figs. 2 and 3.

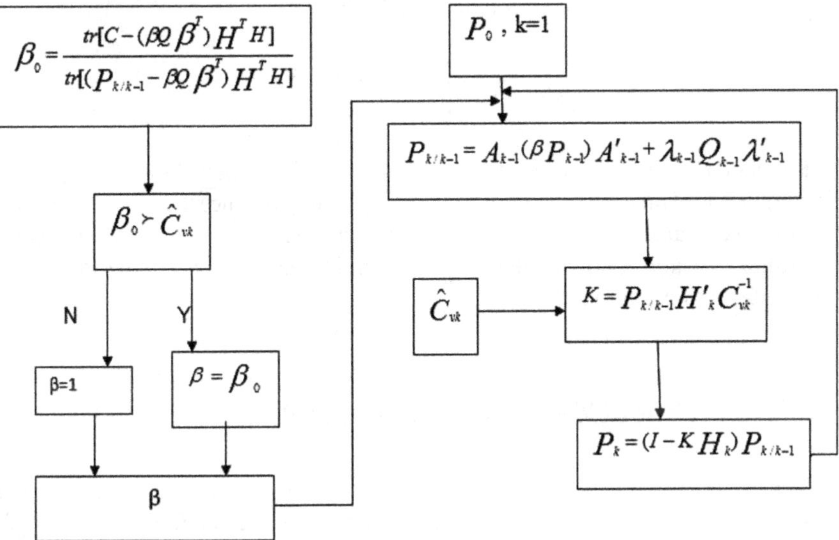

Fig. 2 Structure of OAKF gain loop

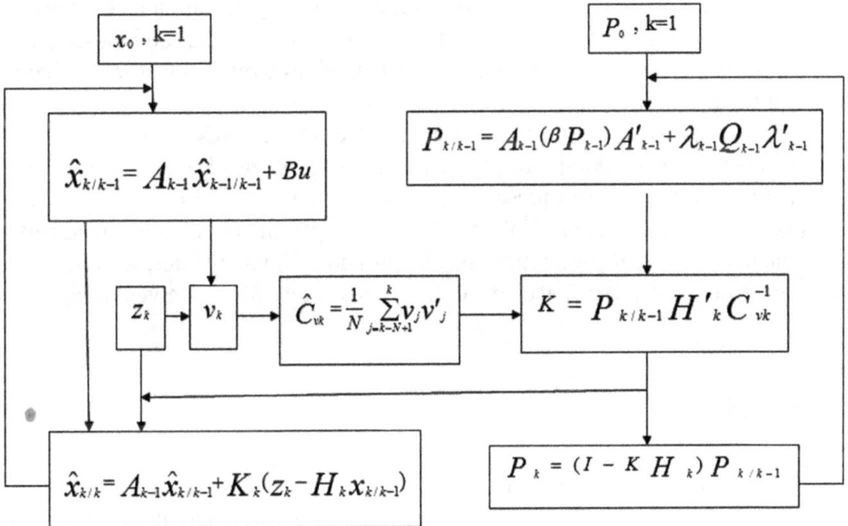

Fig. 3 Structure of OAKF

4 Otsu's Thresholding Method

Thresholding is the easiest way of image segmentation. In the image, each pixel has different intensity values. The pixel is divided on the basis of their intensity level [9]. For segmentation, we use a threshold value T given as:

$$g(x, y) = \begin{cases} 1 \text{ if } f(x, y) > T \\ 0 \text{ if } f(x, y) < T \end{cases} \tag{3}$$

For choosing the proper value of T, we use Otsu's method. In this, we perform clustering-based image thresholding. Otsu's algorithm assumes that images have two types of pixel value one is foreground and second is background pixel. On using this algorithm, our aim is to calculate such threshold value which is able to separate foreground and background.

5 Otsu's Thresholding-Based Kalman Filter

In this section, a novel method is proposed for tracking of moving and deformable object. In this algorithm, we fused Otsu's method with Kalman filter. Generally, for tracking of particular part from any video/frames, we require two techniques: (1) segmentation of particular part from image or videos and (2) prediction of state. For segmentation purpose, we use Otsu's thresholding method which is adaptive and fast. By using this method and some other image segmentation techniques, we find the characteristics parameters (color, centroid, axis length, and elevation) of the segmented images. The obtained segmented values work as measured value for Kalman filter.

In the second part, we predict the next value on the basis of measured value obtained from segmentation. For prediction purpose, we use Kalman filter. It takes the measured value, process noise error covariance matrix, and measurement error covariance matrix. We apply Kalman filter on all parameters of segmented image individually on X- and Y-direction movement and estimate the correct value from noisy measured data. The flowchart of the proposed algorithm is given in Fig. 4.

6 Simulation

All the simulation results are validated on robotic fish video taken from Teledyne DALSA. In this video, a robotic fish moving with constant acceleration 5 m/s^2. This fish organ (black spot) which is deformable is tracked using the above-proposed method. The measurement and process noise are considered independent to each other. Their mean and variance are 0 and 2.5. The state transition matrices for both

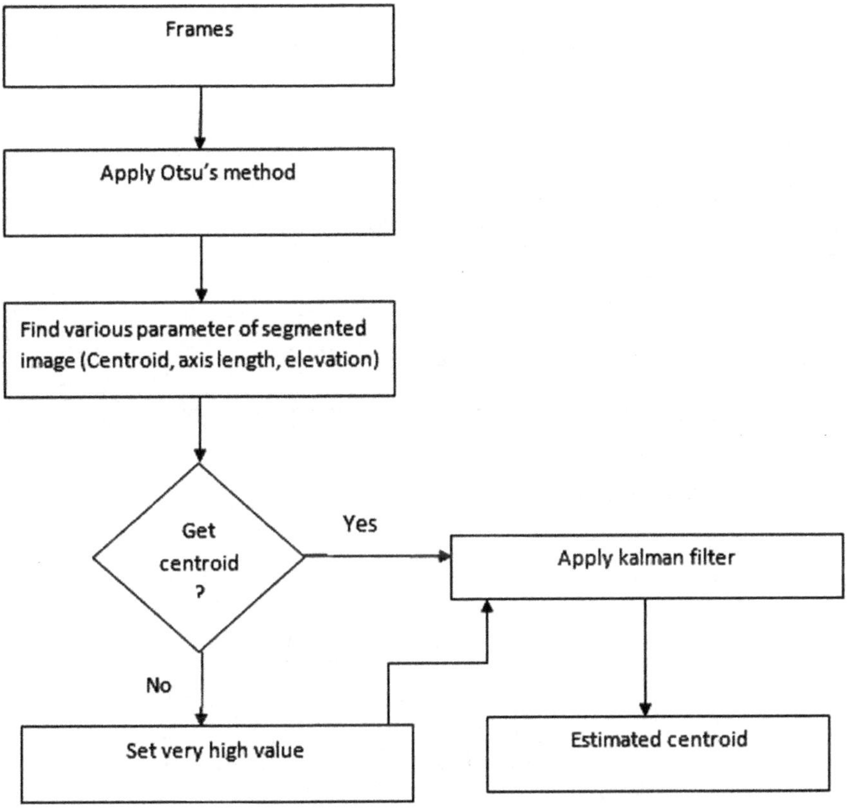

Frames
Apply Otsu's method
Find various parameter of segmented image (Centroid, axis length, elevation)
Get centroid ?
Yes
No
Apply kalman filter
Set very high value
Estimated centroid

Fig. 4 Structure of the proposed method (Otsu's thresholding-based Kalman filter)

measured
estimated
Y-axis movement
sample time
350
300
250
200
150
100
0
10
20
30
40
50
60
70
80

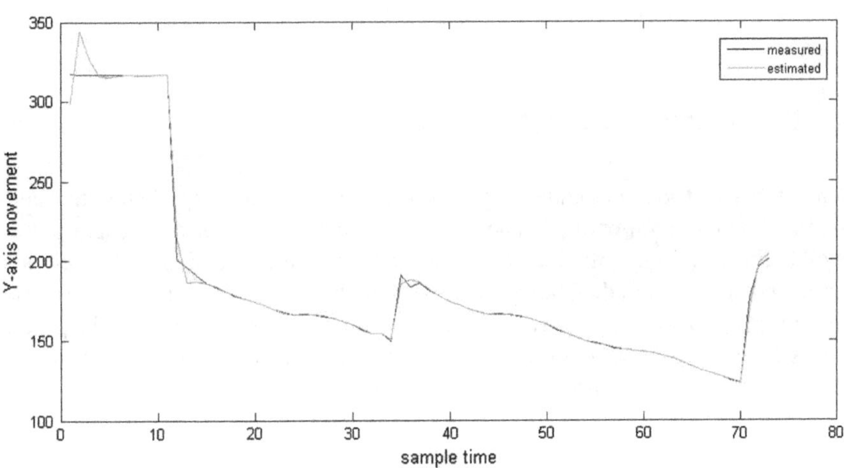

Fig. 5 Tracking in *X*-direction

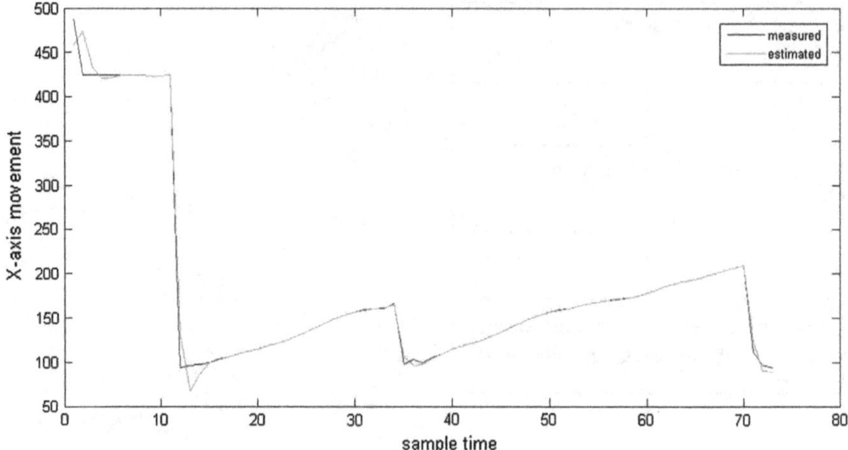

Fig. 6 Tracking in *Y*-direction

X- and *Y*-axis are as follow:

$$A = \begin{pmatrix} 1 & 0.1 \\ 0 & 1 \end{pmatrix} \quad B = \begin{pmatrix} 1 \\ 1 \end{pmatrix}$$

$$\lambda = \begin{pmatrix} 0.0075 \\ 0.175 \end{pmatrix} \quad B = (1 \quad 0)$$

$$P_0 = \begin{pmatrix} 0.1 & 0 \\ 0 & 0.1 \end{pmatrix} \quad x_0 = \begin{pmatrix} 0 \\ 0 \end{pmatrix} \quad y_0 = \begin{pmatrix} 0 \\ 0 \end{pmatrix}$$

7 Result and Discussion

On the basis of given condition, we track fish black spot using Otsu's thresholding method based on optimal adaptive Kalman filter when noise variances are not well defined. In Fig. 7, the red circle represents measured value and green circle is predicted value. From the above result, we can observe the advantages of proposed algorithm in tracking. It is observed that the algorithm works well when the process states are well defined and noise variances are inaccurate. It also performs well when there are errors in both process and measurement states (Figs. 5 and 6).

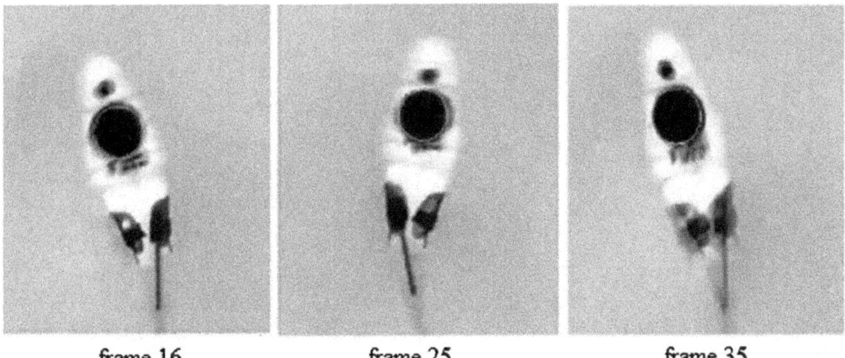

Fig. 7 Tracking of fish black organ (courtesy video data is taken from Teledyne DALSA)

8 Conclusion

In this paper, we have combined the characteristics of Otsu's thresholding method with Kalman filter which gives the superior tracking result in real-time application. We have seen that tracking of fish organ is achieved when system states and process noises are not well defined. Its accuracy, stability, and robustness make it one of the best methods of tracking-related problem. The work presented is inspired by medical imaging concept. In the future, we are trying to apply this similar type of concept on human lungs on tracking of cancerous tissue.

References

1. Y. Bar-shalom, X.R. Li, T. Kirubarjan, Estimation with application to tracking and navigation: theory algorithm and software (Wiley, New York, 2001)
2. Z.M. Durovic, B.D. Kovacevic, Robust estimation with unknown statistics. IEEE trans. automat. contr. **44**(6) (1999)
3. L. Luling, S. Dai, Research of objective tracking algorithm applied in passenger flow statistics in public traffic (2009) IEEE
4. M.S. Grewal, A.P. Andrews, *Kalman filtering theory and practice using Matlab*, 3rd edn. John Wiley & Sons, Hoboken, New Jersey (2001)
5. R.K. Mehra, On the identification of variances and adaptive Kalman filtering. IEEE Trans. Autom. Control. Ac-**15**, 175–184 (1970)
6. M.A. Gandhi, L. Milin, Roboust Kalman filter based on a generalized maximum likelihood type estimator. IEEE Trans. Signal Process. **58**(5), 2509–2520 (2010)
7. K. Myers, B. Tapley, Adaptive sequential estimation with unknown noise statistics. IEEE Trans. Autom. Control. **21**(4), 520–523 (1976)
8. Y. Xiaokun, Y. Jianping, Based on the maximum likelihood criterion adaptive Kalman filter algorithm. J. Northwest. Polytech. Univ. **23**(4), 469–474 (2005)
9. G.Y. Zhang, G.Z. Liu, H. Zhu, B. Qiu, Ore image thresholding using bi-neighbourhood Otsu's approach. Electron. Lett. (2010)
10. X.U. Fuzhen, S.U. Yongqing, L.I.U. Hao, Research of optimized adaptive Kalman filtering (2014) IEEE

Portable Camera-Based Assistive Device for Real-Time Text Recognition on Various Products and Speech Using Android for Blind People

Sandeep Kumar, Sanjana Mathew, Navya Anumula and K. Shravya Chandra

Abstract In this paper, we propose a model to help visually impaired persons in reading product labels on day-to-day products. This method serves as an eye for the blind. The user is asked to display the product in front of the camera which captures the region of interest. The extracted object is pre-processed in which the RGB object transformed to grayscale. Then the grayscale image is binarized with the help of Otsu's algorithm. If there are any small objects in the binary image, it is removed producing another binary image. The resultant image is filtered with the help of the median filter to remove salt-and-pepper noise. The pre-processing step is followed by detection or segmentation using Gabor's algorithm. Then the text of the image is extracted; it is given to a speaker which gives the audio output.

Keywords OTSU segmentation · Median filter · Gabor feature · Portable camera · Blind people

1 Introduction

A camera-based text reading framework is a mechanism wherein visually impaired person is made capable to read texts labeled on products which are used in daily life. This method serves as an eye for the blind [1–3]. The user is asked to display the product in front of the camera which captures the region of interest. The extracted object is pre-processed in which the RGB object is transformed into a grayscale image. Then the grayscale image is binarized with the help of Otsu's algorithm

S. Kumar (✉) · S. Mathew · N. Anumula · K. S. Chandra
Department of ECE, Sreyas Institute of Engineering and Technology, Hyderabad, Telangana, India
e-mail: er.sandeepsahratia@gmail.com

S. Mathew
e-mail: sanjana.m@sreyas.ac.in

N. Anumula
e-mail: navya.anumula@gmail.com

K. S. Chandra
e-mail: shravya.chandra1@gmail.com

H. S. Saini et al. (eds.), *Innovations in Electronics and Communication Engineering*, Lecture Notes in Networks and Systems 107,
https://doi.org/10.1007/978-981-15-3172-9_42

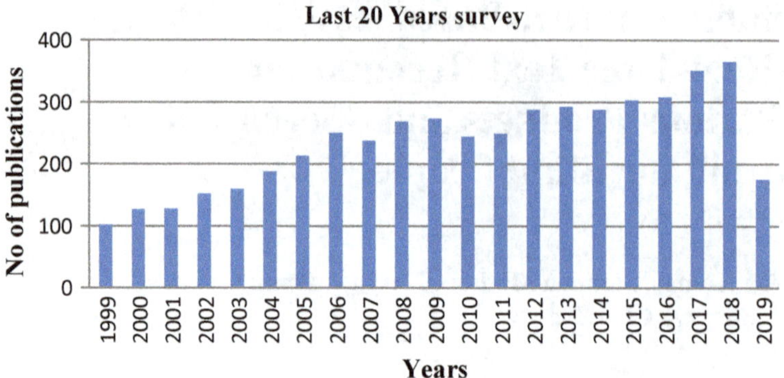

Fig. 1 Showing statistics of the research on portable camera-based assistive devices

[4, 5]. If there are any small objects in the binary image, it is removed producing another binary image. The resultant image is filtered with the help of the median filter to remove salt-and-pepper noise [6]. The pre-processing step is followed by detection or segmentation using Gabor's algorithm. Then the text of the image is extracted; it is given to a speaker which gives the audio output [7, 8] (Fig. 1).

Reading is one of the integral parts of the twenty-first century. We can find written texts in all walks of life be it certificates, statements of various banks, hotel menu's, classroom notices, packaging of products, etc. [9, 10]. There are various devices which assist the blind people such as optical aids, video magnifier, and screen readers to identify objects. Also, there are few devices that can help a visually challenged individual to identify day-to-day used objects such as packaging of the products and printed text on objects such as doctor prescriptions, medication bottles, etc. [11–13]. The ability of a visually challenged person to read printed labels will enhance independent living and will encourage economic and social growth; hence, this paper proposes a camera-based system that can be useful to blind people. Speech and text are an essential means of human communication [14]. In order to analyze the information in a text, human vision is very important. But blind as well as people with poor vision can access information from the text by converting the required text to speech or audio [15–19]. Hence to extract text details from difficult backgrounds, this methodology introduces detection of text followed by the recognition that incorporates detection and extraction of printed text patterns. The text characters consist of various fonts, scales, and colors, different orientations and alignments. In order to detect such text patterns, this system will help blind users to easily locate the object and get the necessary information from it. All the letters are recognized from text through OCR technique [20]. Text detection serves as a helping hand for the visually challenged person in order to become competent, confident, and economically independent. This methodology easily detects text from the captured image and the detected texts are given as audio/speech output to a blind person [21, 22]. However, this paper focuses more on text detection algorithms. It proposes text reading with the aid of a USB camera helping visually challenged person in reading and understanding the text present on the captured image and identifying the object [23–26].

2 Literature Work

Sr. no.	Author name	Year	Methodology	Remarks
1	R. Singh et al. [10]	2014	Text binarization and recognition using Otsu's algorithm of optical character recognition Haar cascade classifier algorithm to extract text regions	Datasets: ICDAR 2003 and ICDAR 2011 robust reading sets dataset collected using 10 blind persons
2	Yi et al. [2]	2014	OCR Cascade AdaBoost classifier for text stroke orientation and distribution of edge pixels	Datasets: ICDAR 2003 and ICDAR 2011 Robust reading sets dataset collected using 10 blind persons
3	P. Radeva et al. [21]	2015	OCR Translator Text to speech	Tesseract OCR engine and leptonica image processing libraries were tested
4	Wiwatcharakoses et al. [8]	2015	MSER components are used for detecting text areas Short circuit evaluation to reduce the complexity of the attributes extraction process	A multi-language dataset with large variation in text alignment and camera views The precision of 70.16% and recall of 93.06% To identify text regions, double thresholding is used cascade classifier
5	Jeong et al. [20]	2016	Electronic travel aid uses seven ultrasonic sensors for obstacle detection Tactile actuators for stimulating the area of fingers consisting of nine phalanges	In the first waking test, obstacle avoidance was 90% In the second walking test, it was 88.6%

(continued)

(continued)

Sr. no.	Author name	Year	Methodology	Remarks
6	Deshpande et al. [1]	2016	Maximally stable external regions for text detection Text localization and binarization using OCR	Evaluated on a dataset of 50 captured images Resolution of range 640 * 480 to 1600 * 1200. Logitech 5MP camera is used
7	Rajesh et al. [3]	2017	Text extraction using Tesseract OCR Convert text to speech using the e-speak tool Text binarization using Otsu's method	Raspberry Pi is used to provide portability OpenCV used for fast processing
8	Chang et al. [4]	2017	Identification involves image acquisition, pre-processing of OCR, pattern training, image recognition	Five different tags are compared Recognition performed using LabVIEW An LC SLM combined with plane mirror and a normal white paper
9	Rizdania et al. [5]	2017	Phases for text detection are MSER, canny edge detection, region filtering, OCR for the text recognition process	Testing data are taken from the grocery store in Malang city The previous method with two classifier accuracy-69% and proposed method accuracy-80.88%.
10	S. P. F. Joan et al. [14]	2017	Stroke width transformations for text detection Image correlation for text recognition	MSARA database with 85% average rate of recognition and average speed 5.03 s KAIST database with average rate of recognition 93% and average speed 2.11 s

(continued)

(continued)

Sr. no.	Author name	Year	Methodology	Remarks
11	Holanda et al. [17]	2018	OCR used for text and pattern recognition as well as enhance the image Pre-processing to reduce noise using median or Gaussian filters and Otsu's method Segmentation similarity using region growing and watershed. Segmentation discontinuance using connected objects	The segmentation using connected contours had accuracy 99% Segmentation using region growing had accuracy 92% Pattern recognition using MLP network had efficiency 99% Connected contours Pattern recognition using *k*-nearest neighbor algorithm and SVM
12	Joan et al. [16]	2018	Text information extraction systems Color-, edge-, texture-based features	ICDAR CHARS74K, VACE KAIST scene text SVT, IIIT 5 K word
13	Arakeri et al. [11].	2018	Raspberry Pi with a compatible camera to capture Sensor to notify the nearest object distance at eye level	Accuracy of OCR found to be 84% Accuracy of object recognition algorithm 93.6% and average time 15 s Raspberry Pi, No IR camera module v2, 8MP camera
14	Elgendy et al. [24]	2019	Shopping preparation using a CV, OCR, and STT	General methods for PVI for shopping

3 Problem Statement

Some of the existing approaches detect the background area rather than the actual size of the text when the background is complex and other outdoor scenery which can add to the clutter. Certain systems are not compatible with all the lighting conditions such as bright light, dim light, dark, and so on. Certain hybrid systems increase the complexity of the infrastructure as well as time consumption.

4 Proposed Methodology

With the help of proposed methodology, we overcome the few of the problems that are defined in the problem statement and also to help the visually challenged persons to read the texts from complex backgrounds; we have to use a camera-based framework which only extracts the required texts on the product and reads aloud the text information. Figure 2 shows the flowchart of the proposed algorithm.

The system framework is broadly classified into three parts namely: capture the image, pre-processing the image followed by edge detection, extraction of the text and speech output (Fig. 3).

Step 1: The image is captured with the help of a portable USB camera as shown in Fig. 4b. The image captured is in the RGB format. It undergoes pre-processing stages which include: Fig. 5

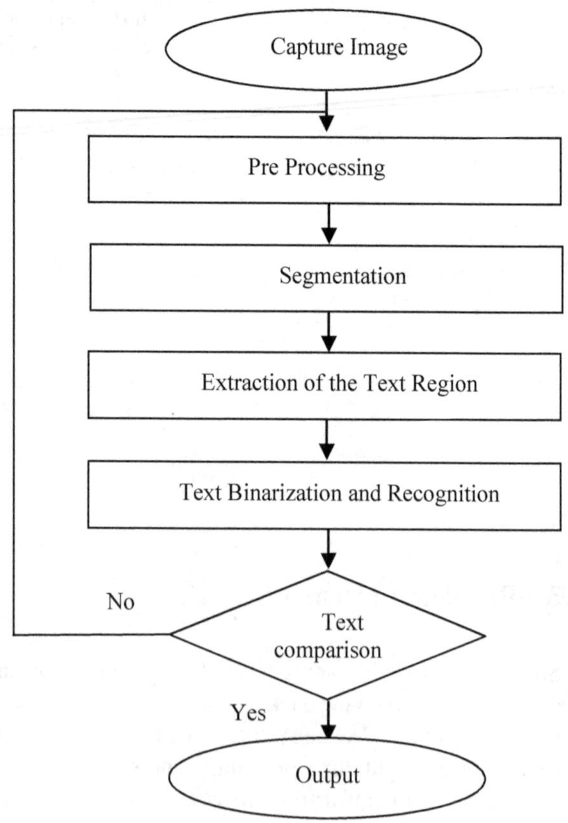

Fig. 2 Flowchart of the proposed methodology

Fig. 3 Snapshot demo system including scene capture, data processing, output display

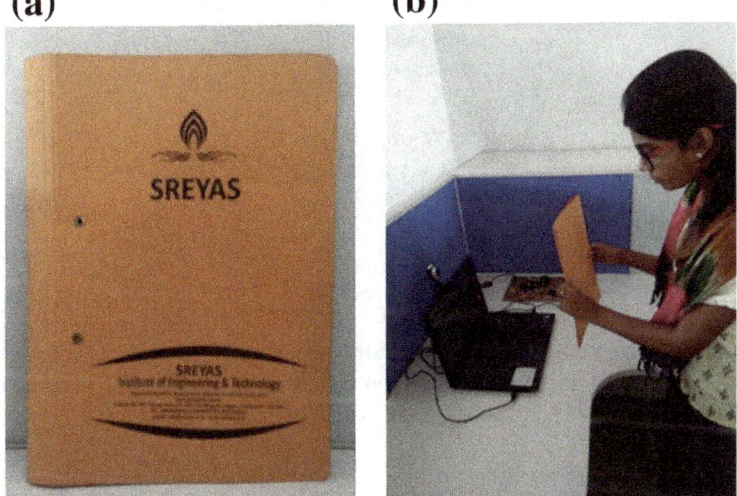

Fig. 4 **a** Represents the object taken for the experiment is a college file and. **b** Represents the object which is held in front of the USB camera

Pre-Processing Algorithm

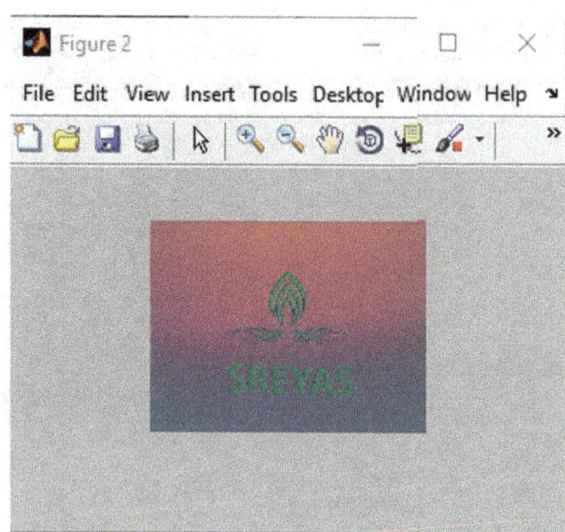

Fig. 5 Captured RGB image

Step 1: Read the color image.
Step 2: Convert color image into gray image.
Step 3: Convert the intensity image to binary image using Otsu's algorithm.
Step 4: Remove any irrelevant objects from the binary image-producing another binary image.
Step 5: Filter the binary image using median filter to remove noise.

In order to convert the grayscale image into the binary form, we incorporate Otsu's algorithm which is a thresholding method which is used worldwide in order to find the optimal threshold level based on the distribution of pixels.

Steps in Otsu's Algorithm

Step 1: Compute the histogram and probability of each intensity level
Step 2: Set the initial value of $v_i(0)$ and $v_j(0)$
Step 3: Increment through all the possible thresholds from $t = 1$ to the maximum intensity level.

(i) Increment the value of $v_i(0)$ and $v_j(0)$
(ii) Compute $\alpha_b^2(t)$

Step 4: Optimum threshold corresponds to the maximal α_b^2(t)

Otsu's Algorithm

$$\alpha^2 v(t) = V_0(t)\alpha_0^2(t) + V_1(t)\alpha_1^2(t)$$

V_0 and V_1 are the probability of the two classes.
$\alpha_0^2(t)$ and $\alpha^{12}(t)$ are the individual variances
$V_0(t) = \sum_{i=0}^{t-1} p(i)$
$V_1(t) = \sum_{i=t}^{K-1} p(i)$
The class probabilities are computed from the *K* bins of the histogram
$\alpha_b^2(t) = \alpha^2 - \alpha_w^2(t) = V_0(\mu_0 - \mu_T)^2 + V_1(\mu_1 - \mu_T)^2$
where $\mu_{0,1,T}$ are the class means which can be computed as follows: $\mu_0(t) = \sum_{i=0}^{t-1} ip(i)/V_0(t) \& \mu_1(t) = \sum_{i=t}^{K-1} ip(i)/V_1(t) \& \mu_T = \sum_{i=0}^{K-1} ip(i)$

Gabor Filter Algorithm

Step 1: Each colored (RGB) pixel is converted to a single value attribute:

$$\text{Pixel } s = \text{pixel } R + 2 * \text{pixel } G + 3 * \text{pixel } B \tag{4}$$

Step 2: The input data set $A = \{a_1, a_2, \ldots, a_n\}$ is the set of transformed pixels.
Step 3: Convolution of the input data with Gabor filter function is done at every pixel.
Step 4: Decimation of the Gabor filter frequencies reduces the pixel size. The phase information gives the details of edge locations.
Step 5: Various features can be extracted such as amplitude response, maximum energy, etc.

Step 2: After the image is passed through the pre-processing stage, it is subjected to detection or segmentation using the Sobel operator for edge detection. This operator is used in the processing of images for edge detection (Fig. 6).

Step 3: The edge detection is followed by extraction of the text region which uses Gabor's algorithm. We obtain Gabor filter function on multiplying a Gaussian window with a sinusoid function. This filter, when applied to an image, performs convolution operation and yields the edges of the image.

Step 4: The extracted text undergoes comparison by computing the correlation between the template and the extracted text. The trained template consists of letters, symbols, and other special characters. The window size of the extracted text is taken as 44 * 26 pixels. Finally, the output text is displayed as audio helping the blind people/visually impaired person.

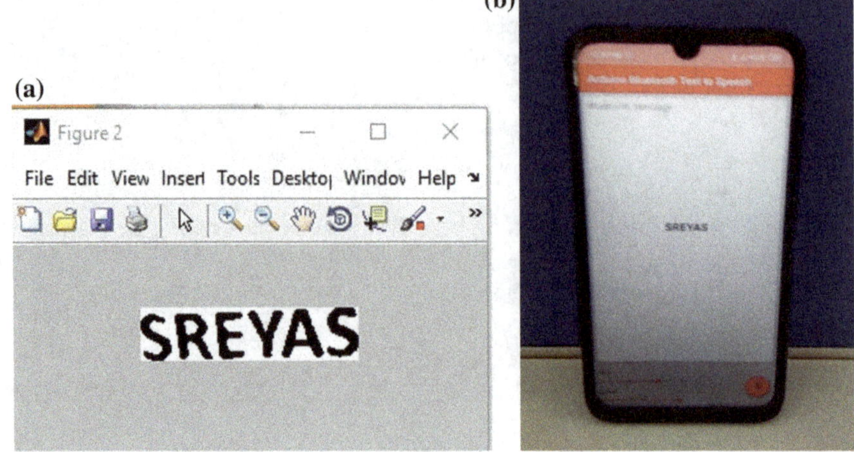

Fig. 6 **a** Extracted text that shows the name of the file whereas. **b** Displayed text as audio output with the help of Android Bluetooth text to speech

5 Result

In the proposed methodology, we have overcome the limitations of various lightning conditions. Our proposed system can work on any lighting conditions. Also, it overcomes hardware complexity because of its portable characteristics. All advantages achieved with the proposed systems (Figs. 7 and 8).

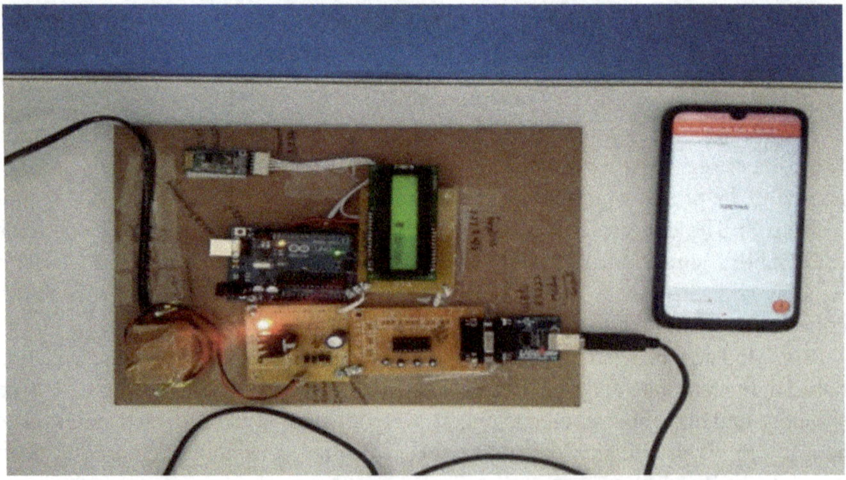

Fig. 7 The demo setup showing the text converted to speech

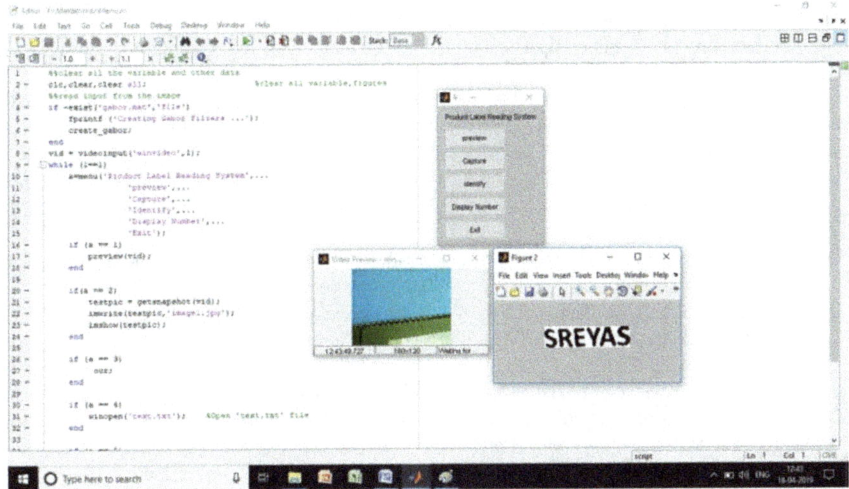

Fig. 8 The output window on screen

6 Acknowledgement

The source of the image in Fig. 4b is of the author and has been included in the paper with the consent of the author.

7 Conclusion

This paper mainly focuses on a camera-based assistive system which can be used by blind persons to help them identify the name of the handheld object. It gives superior performance under any lighting conditions as a USB camera is employed with adjacent lights in order to capture the image in low light. The proposed system is capable to extract the required text from the image from complex and cluttered backgrounds and read aloud the text on it.

References

1. S. Deshpande, R. Shriram, Real-time text detection and recognition on handheld objects to assist blind people, in *International Conference on Automatic Control and Dynamic Optimization Techniques (ICACDOT)*, (2016), pp. 1020–1024
2. X. Rong, C. Yi, Y. Tian, Unambiguous text localization and retrieval for cluttered scenes, in *IEEE Conference on Computer Vision and Pattern Recognition (CVPR)*, (2017), pp. 3279–3287
3. M. Rajesh, B.K. Rajan, A. Roy, K.A. Thomas, A. Thomas, T.B. Tharakan, C. Dinesh, Text recognition and face detection aid for visually impaired person using Raspberry PI, in *International Conference on Circuit Power and Computing Technologies (ICCPCT)*, (2017), pp. 1–5

4. C.-C. Chang, Y.-C. Yang, S. Ming-Chun, J.-C. Tsai, Tunable micro cat's eye array in an optical identification system and comparison of different ID tags. Sel. Top. Quantum Electron. IEEE J. **21**(4), 130–136 (2015)
5. Rizdania, F. Utaminingrum, Text detection and recognition using multiple phase method on the various product label for visually impaired people, in *International Conference on Sustainable Information Engineering and Technology (SIET)*, (2017), pp. 398–404
6. K.-M. Lee, M. Li, C.-Y. Lin, Magnetic tensor sensor and way-finding method based on geomagnetic field effects with applications for visually impaired users. Mechatron. IEEE/ASME Trans. **21**(6), 2694–2704 (2016)
7. Z. Fei, E. Yang, H. Hu, H. Zhou, Review of machine vision-based electronic travel aids, in *23rd International Conference on Automation and Computing (ICAC)*, (2017), pp. 1–7
8. C. Wiwatcharakoses, K. Patanukhom, MSER based text localization for multi-language using the double-threshold scheme, in *1st International Conference on Industrial Networks and Intelligent Systems (INISCom)*, (2015), pp. 62–71
9. R.Y. Acharya, C. Zhang, Y. Tian, Learning alphabets and numbers from hand motion trajectories for visually impaired children, in *IEEE 7th Annual International Conference on CYBER Technology in Automation Control and Intelligent Systems (CYBER)*, (2017), pp. 1356–1360
10. R. Singh, R. Borse, Alternative product label reading and speech conversion: an aid for blind person, in *International Conference on Computing Communication Control and Automation (ICCUBEA)*, (2017), pp. 1–6
11. M.P Arakeri, N.S. Keerthana, M. Madhura, A. Sankar, T. Munnavar, Assistive technology for the visually impaired using computer vision, in *International Conference on Advances in Computing Communications and Informatics (ICACCI)*, (2018), pp. 1725–1730
12. X. Rong, C. Yi, Y. Tian, Computer vision—ECCV 2016 workshops, (2016), pp. 99–109
13. X. Rong, B. Li, J.P. Muñoz, J. Xiao, A. Arditi, Y. Tian, *Advances in Visual Computing*, (2016), pp. 100–111
14. S.P.F. Joan, S. Valli, An enhanced text detection technique for the visually impaired to read the text. Inf. Syst. Front. (2016)
15. D.N. Aiguo Song, L. Tian, X. Xu, D. Chen, A walking assistant robotic system for the visually impaired based on computer vision and tactile perception. Int. J. Soc. Robot. (2015)
16. S.P.F. Joan, S. Valli, A survey on text information extraction from born-digital and scene text images, in *Proceedings of the National Academy of Sciences, India Section A: Physical Sciences* (2018)
17. G.B. Holanda, J.W.M. Souza, D.A. Lima, L.B. Marinho, A.M. Girão, J.B.B. Frota, P.P. Rebouças Filho, Development of OCR system on Android platforms to aid reading with a refreshable braille display in real time. Measurement (2018)
18. S. Wang, L. Huang, J. Hu, Text line detection from rectangle traffic panels of natural scene, in *Journal of Physics: Conference Series*, (2018), pp. 960–967
19. S.A. Angadi, M.M. Kodabagi, A light weight text extraction technique for hand-held device. Int. J. Image Graph. 15–22 (2015)
20. G.Y. Jeong, K.H. Yu, Multi-section sensing and vibrotactile perception for walking guide of visually impaired person. Sensors, 16–28 (2016)
21. P. Radeva, A. Verikas, D.P. Nikolaev, W. Zhang, J. Zhou, H. Jabnoun, F. Benzarti, H. Amiri, in *Ninth International Conference on Machine Vision (ICMV 2016)*, (2017), pp. 10341, 1034123
22. A. Alnasser, S. Al-Ghowinem, *Intelligent Computing*, (2019), pp. 231–858
23. N. Nicoletta, L. Giuliani, J. Sosa-Garciá, L. Brayda, A. Trucco, F. Odone, *Multimodal behavior analysis in the wild*, (2019), p. 79
24. M. Elgendy, C. Sik-Lanyi, A. Kelemen, Making shopping easy for people with visual impairment using mobile assistive technologies. Appl. Sci. **9**, 1061 (2019)
25. S. Kumar, S. Singh, J. Kumar, Live detection of face using machine learning with multi-feature method. Wirel. Pers. Commun. Springer J. (SCI). https://doi.org/10.1007/s11277-018-5913-0
26. S. Kumar, S. Singh, J. Kumar, Automatic live facial expression detection using genetic algorithm with haar wavelet features and SVM. Wirel. Pers. Commun. Springer J. (SCI). https://doi.org/10.1007/s11277-018-5923-y

Color Image Quality Assessment Based on Full Reference and Blind Image Quality Measures

P. Ganesan, B. S. Sathish, K. Vasanth, M. Vadivel, V. G. Sivakumar and S. Thulasiprasad

Abstract The degradation of the image at the acquisition and transmission severely affects the various stages of the image processing and leads to the unexpected results. So, it is necessary to be aware of the source and reason of the various noises that are incorporated with the original image. The image quality measures compute the quality of the corrupted or degraded image with or without reference (input) image. The comparison of the degraded and reference image produces numerical score that decides the quality of the image. The proposed work tests the images incorporated the various noises such as salt and pepper, Poisson, Gaussian and speckle. Blind (no reference) and full reference quality measures investigate the amount of degradation in images. Full reference quality measures are very simple to compute but do not correlate with human perception, whereas blind reference measures prove its supremacy in this aspect.

Keywords Image quality · Noise · PSNR · SSIM · BRISQUE · NIQE · PIQE

P. Ganesan (✉) · K. Vasanth · M. Vadivel · V. G. Sivakumar · S. Thulasiprasad
Department of Electronics and Communication Engineering, Vidya Jyothi Institute of Technology, Aziz Nagar, C.B.Road, Hyderabad, Telangana , India
e-mail: gganeshnathan@gmail.com

K. Vasanth
e-mail: vasanthecek@gmail.com

M. Vadivel
e-mail: drmvadivel79@gmail.com

V. G. Sivakumar
e-mail: sivakumarvg2004@gmail.com

S. Thulasiprasad
e-mail: stprasad123@yahoo.co.in

B. S. Sathish
Department of Electronics and Communication Engineering, Ramachandra College of Engineering, Eluru, Andhra Pradesh, India
e-mail: subramanyamsathish1@yahoo.co.in

H. S. Saini et al. (eds.), *Innovations in Electronics and Communication Engineering*, Lecture Notes in Networks and Systems 107,
https://doi.org/10.1007/978-981-15-3172-9_43

1 Introduction

In image processing, it is necessary to test the efficiency of imaging system using the image quality parameters or measures. The quality of the image demeans by the various degradation factors such as artifacts, noise and blurring [1]. The image quality metrics can be applied to assess the subjective and objective measure of the image excellence [2]. Most importantly, image quality metric should be correlated well with the human visual perception to produce faithful result. The quality of the output image can simply be evaluated without the precise use of the input image in the case of univariate measures. Pixel count and defocus blur are examples for univariate measures. The univariate measures have limited application in analyzing the quality of the images. In the case of bivariate measures, the quality of the output image can be determined using the comparison of corresponding image pixels between input and output images. The image quality measures compute the quality of the corrupted or degraded image with or without reference (input) image. The comparison of the degraded and reference image produces numerical score that decides the quality of the image. The proposed work tests the images incorporated the various noises such as salt and pepper, Poisson, Gaussian and speckle. Blind (no reference) and full reference quality measures investigate the amount of degradation in images.

2 Review on Image Quality Measures

Based on the reference (model), the quality measures are broadly categorized into two groups as full reference (bivariate) and no reference (blind). For full reference metrics, an output image is necessary to weigh against reference one to exploit the quality of the image. For example, in image compression, the uncompressed (original) image can be utilized as a reference one to measure the quality of the image after compression. Similarly, to assess the quality of the image degradation process, the degraded image is the valuable reference to compare with the original image [3]. The basis for the PSNR is mean squared error. It is a measure of the peak value of the error. Larger the PSNR score is the greater the image quality and vice versa [4]. SSIM is another important quality measure frequently utilized in image processing applications to compute the similarity or dissimilarity between the input and output images [5]. This metric is a combination o luminance, contrast and image structure (pattern of pixel intensities) [6]. This quality measure is very easy to compute and also correlates well with human visual system. It is wise to go with blind (or) no reference measures if reference image is not accessible. In this case, the quality of the image can be estimated using the image statistical features [7]. The example for this category includes blind/reference-less image spatial quality evaluator (BRISQUE), natural image quality evaluator (NIQE) and perception-based image quality evaluator (PIQE). BRISQUE model is well instructed on distortion image database, and it is restricted to estimate the superiority of the image with the same class of degradation [8]. The least value of this measure indicates better quality

Table 1 PIQE score and image quality based on live image quality assessment

S. no.	PIQE score	Image quality
1	0–20	Excellent
2	21–35	Good
3	36–50	Fair
4	51–80	Poor
5	81–100	Bad

of the image. So, generally the degraded image has the highest BRISQUE score than the original undistorted image. NIQE can compute the image quality with random distortion. This opinion-unaware measure does not employ subjective quality scores. The lowest value of this measure indicates better quality of the image. The PIQE, opinion-unaware and unsupervised, score does not necessitate a trained model for computation. The smallest value of PIQE score means the superior quality of the image. Based on the experiment conducted on the LIVE dataset, the PIQE score and image quality are given in Table 1 [9].

The BRISQUE and the NIQE determine the superiority score of an image with computational competency, once the model is trained. PIQE is computationally not competent; however, it offers both global quality score and local gauge of eminence [10].

3 Experimental Research

Figure 1 illustrates the reference (input) image. Image noise is arbitrary deviation of intensity or chrominance information in images. It is an undesirable consequence of image acquisition that unintelligible the original information. These unwanted electrical fluctuations may be due to the sensor and other factors. The salt and pepper, also called as impulse noise, originated by the spiky and abrupt turbulences in the

Fig. 1 Reference image. Image courtesy: Mathworks Inc

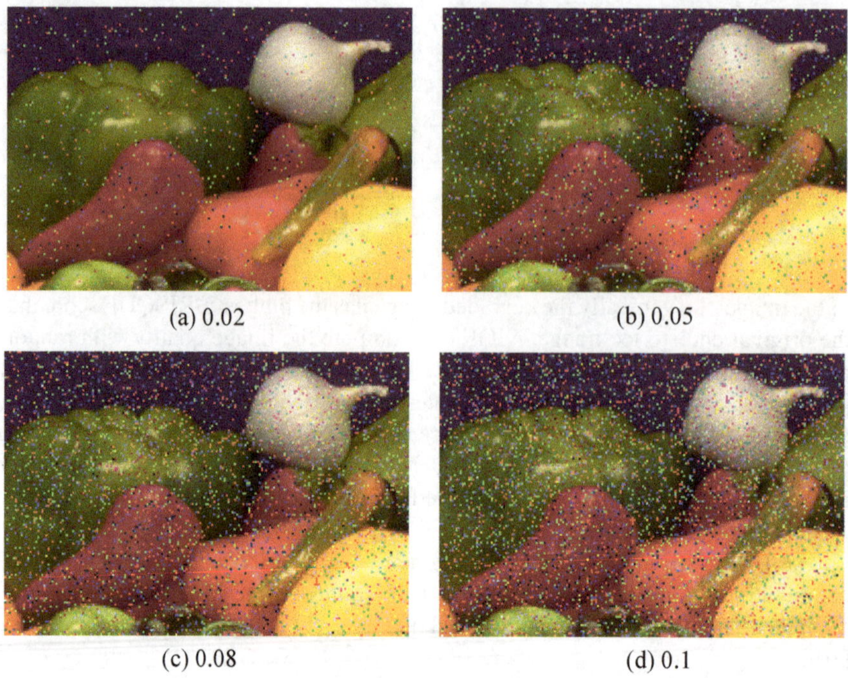

(a) 0.02 (b) 0.05 (c) 0.08 (d) 0.1

Fig. 2 Reference image corrupted by salt and pepper noise. The numerical value indicates the density of the noise

image signal [5]. The noise appears as a collection of white (salt) and black (pepper) pixels to degrade the quality of the image. To test the efficiency of the various image quality measures, salt and pepper noise of different density is added to the reference image. In this work, noise density of 0.02, 0.05, 0.08 and 0.1 is added to the reference image as shown in Fig. 2.

The Poisson noise, also known as shot noise, can be mathematically described by Poisson distribution process. Figure 3 illustrates the reference (input) image with added Poisson noises.

The speckle is a coarse noise that intrinsically subsists in and corrupts the eminence of the images like SAR, optical coherence tomography or ultrasound. Mostly, the image acquisition surfaces are awfully uneven on the extent of the wavelength. Images acquired from these rough surfaces by SAR or ultrasound experience a universal observable fact called speckle. This noise is the effect from the arbitrary variations in the return signal from an object. This process raises the mean dynamic intensity level of an image that adds white and black dots in the image. To test the efficiency of the various image quality measures, speckle noise of different variances is added to the reference image. In this work, noise variance of 0.02, 0.05, 0.08 and 0.1 is added to the reference image as depicted in Fig. 4.

Fig. 3 Reference image corrupted by Poisson noise

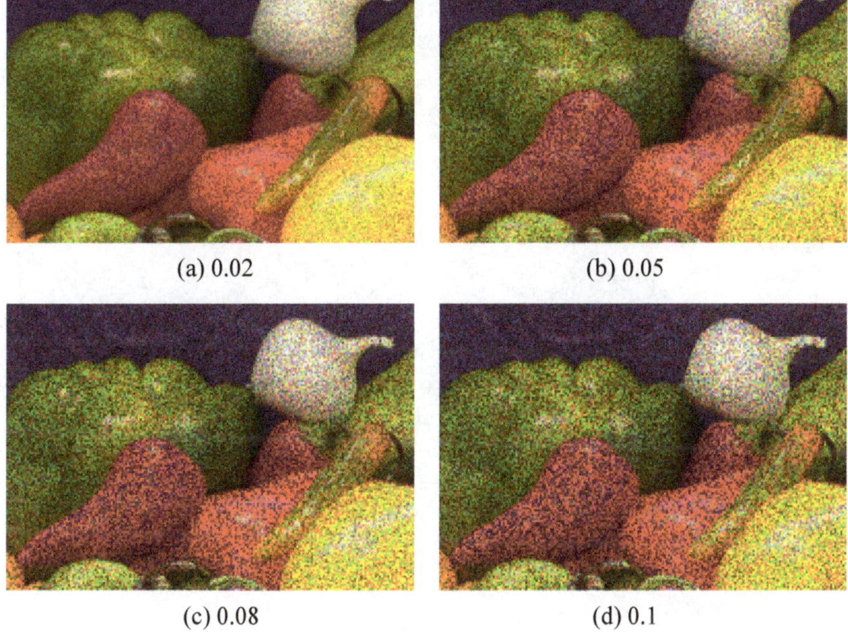

Fig. 4 Reference image corrupted by speckle noise. The numerical value indicates the variance of the noise

The Gaussian noise is a statistical noise based on the Gaussian (normal) distribution. In images, the major causes of this unnecessary noise are sensor (poor illumination or high temperature) and faulty circuits. This type of noise can be removed by the spatial filters such as mean, median or Gaussian smoothing filtering. To test the efficiency of the various image quality measures, Gaussian noise of different

(a) 0.02
(b) 0.05
(c) 0.07

Fig. 5 Reference image corrupted by Gaussian noise. The numerical value indicates the variance of the noise

variances is added to the reference image. In this work, noise variance of 0.02, 0.05 and 0.07 is added to the reference image as shown in Fig. 5.

The end result of the proposed approach for the full reference image quality measures is tabulated in Table 2. The result is graphically displayed in Fig. 6.

The upshot of the proposed approach for the no reference image quality measures is tabulated in Table 3. This clearly shows that the degree of the image degradation is varied according to the amount of the noise added to it. The result is graphically displayed in Fig. 7.

Table 2 Outcome of the proposed method for full reference image quality measures

S. no.	Noise type	SSIM	SNR	PSNR
1	Salt and pepper (0.02)	0.8400	14.6451	21.6520
2	Salt and pepper (0.05)	0.7001	10.6612	17.6680
3	Salt and pepper (0.08)	0.6059	8.6406	15.6475
4	Salt and pepper (0.1)	0.5581	7.6927	14.6996
5	Poisson	0.9538	21.7438	28.7507
6	Gaussian (0.02)	0.8723	13.6729	20.3701
7	Gaussian (0.05)	0.6980	9.8923	16.9824
8	Gaussian (0.07)	0.5606	7.6543	14.7525
9	Speckle (0.02)	0.9195	17.6000	24.6069
10	Speckle (0.05)	0.8360	13.7266	20.7335
11	Speckle (0.08)	0.7740	11.8111	18.8179
12	Speckle (0.1)	0.7401	17.9263	10.9195

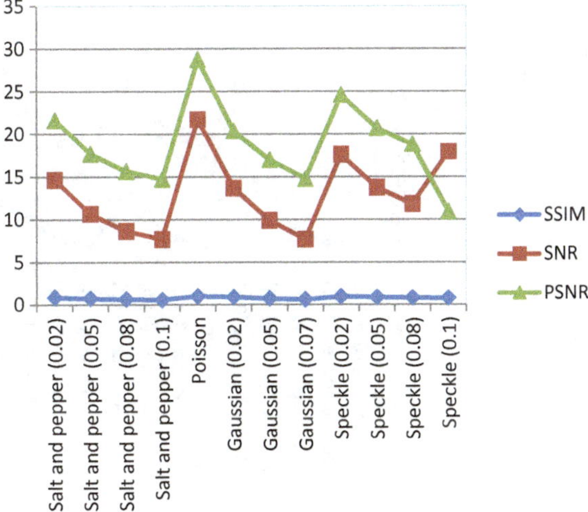

Fig. 6 End result of the proposed approach for the full reference image quality measures

Table 3 Outcome of the proposed method for no reference image quality measures

S. no.	Noise type	NIQE	PIQE	BRISQUE
1	No noise (original image)	1.7281	14.9026	17.6792
2	Salt and pepper (0.02)	8.2189	63.9025	46.2801
3	Salt and pepper (0.05)	14.2671	71.8969	58.9008
4	Salt and pepper (0.08)	19.2426	79.9158	71.8912
5	Salt and pepper (0.1)	31.5721	86.0046	83.8914
6	Poisson	7.7191	59.6201	41.9012
7	Gaussian (0.02)	11.6721	60.7293	49.2781
8	Gaussian (0.05)	17.2718	69.0047	62.9104
9	Gaussian (0.07)	26.8192	83.9412	79.5619
10	Speckle (0.02)	7.0123	51.9102	39.6707
11	Speckle (0.05)	10.2571	58.2503	51.8923
12	Speckle (0.08)	19.0271	73.9512	68.3278
13	Speckle (0.1)	29.0163	89.0127	77.2105

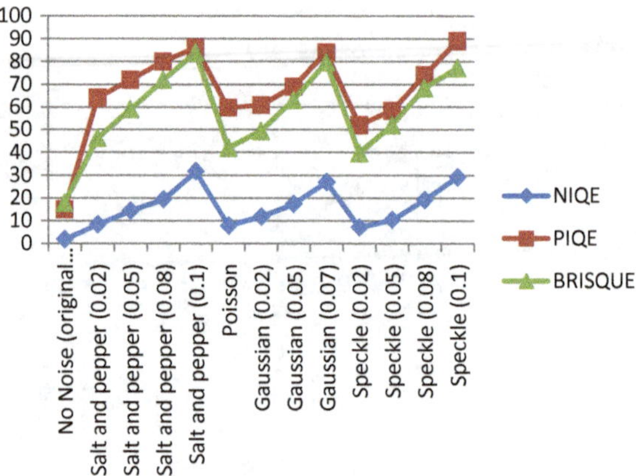

Fig. 7 End result of the proposed approach for the no reference image quality measures

4 Conclusion

The image quality assessment based on blind and full reference quality measures is proposed. The proposed work investigated the impact of various noises such as salt and pepper, Poisson, Gaussian and speckle on the color images. The image quality measures computed the quality of the corrupted or degraded image with or without reference (input) image. The quality of the image is estimated using the numerical score obtained by the comparison of the degraded and reference image. Blind (no reference) and full reference quality measures investigated the amount of degradation in images. Though full reference quality measures are very simple to compute, no reference measures are the optimal choice because of its correlation with human perception.

References

1. P. Ganesan, V. Rajini, Assessment of satellite image segmentation in RGB and HSV color space using image quality measures, in *2014 International Conference on Advances in Electrical Engineering (ICAEE)*, (2014), pp. 1–5
2. H.R. Sheikh, M.F. Sabir, A.C. Bovik, A statistical evaluation of recent full recent full reference image quality assessment algorithms. IEEE Trans. Image Process. **15**(11), 3441–3456 (2006)
3. P. Ganesan, K.B. Shaik, HSV color space based segmentation of region of interest in satellite images, in *2014 International Conference on Control, Instrumentation, Communication and Computational Technologies (ICCICCT)*, (2014), pp. 101–105
4. V. Kalist, P. Ganesan, B.S. Sathish, J.M.M. Jenitha, Possiblistic-fuzzy C-means clustering approach for the segmentation of satellite images in HSL color space. Procedia Comput. Sci. **57**, 49–56 (2015)
5. I. Avcıbas, B. Sankur, K. Sayood, Statistical evaluation of image quality measures. J. Electron. Imaging **11**(2), 206–223 (2002)
6. Z. Wang, A.C. Bovik, H.R. Sheikh, E.P. Simoncelli, Image quality assessment: from error visibility to structural similarity. IEEE Trans. Image Process. **13**(4) (2004)
7. A.K. Moorthy, A.C. Bovik, Blind image quality assessment: from scene statistics to perceptual quality. IEEE Trans. Image Process. **20**(12), 3350–3364 (2011)
8. A.K. Mittal, R. Soundararajan, A.C. Bovik, Making a completely blind image quality analyzer. IEEE Signal Process. Lett. **22**(3), 209–212 (2013)
9. https://in.mathworks.com/help/images/image-quality.html
10. A. Mittal, A.K. Moorthy, A.C. Bovik, No-reference image quality assessment in the spatial domain. IEEE Trans. Image Process. **21**(12), 4695–4708 (2012)

Performance Investigation of Brain Tumor Segmentation by Hybrid Algorithms

M. Vadivel, V. G. Sivakumar, K. Vasanth, P. Ganesan and S. Thulasiprasad

Abstract Brain tumor is now the leading tumor root of demise in industrialized world. A miserably short cure rate mostly imitates the tendency of Brain tumor to present as clinically superior tumors. Most Brain tumors are revealed tardy during their medical track, at that time the choice for efficient healing intrusion are inadequate. The early detection of Brain tumor is a exigent crisis, owing to the structure of the tumor cells and deformation, where the majority of the cells are overlie with all other. In recent years Brain tumor detection is most popular problems in spatial image identification due to 2D dimensional datasets. This paper presents two existing segmentation methods namely SVM and FCM methods, for fragmenting images to discover the Brain tumor in its premature stages. A cluster based hybrid improved algorithm is proposed to overcome training and learning models.

Keywords Tumor · SVM · FCM · Medical · Segmentation · Clustering

1 Introduction

A brain tumor is described as curious enlargement of cells interior to the brain and inmost spinal duct. A few tumors can be cancerous thus they need to be detected and cured in time. The precise root of brain tumors is not lucid and neither is accurate

M. Vadivel (✉) · V. G. Sivakumar · K. Vasanth · P. Ganesan · S. Thulasiprasad
Department of ECE, Vidya Jyothi Institute of Technology, Aziz Nagar, C.B.Road, Hyderabad, Telangana, India
e-mail: drmvadivel79@gmail.com

V. G. Sivakumar
e-mail: sivakumarvg2004@gmail.com

K. Vasanth
e-mail: vasanthecek@gmail.com

P. Ganesan
e-mail: gganeshnathan@gmail.com

S. Thulasiprasad
e-mail: stprasad123@yahoo.co.in

H. S. Saini et al. (eds.), *Innovations in Electronics and Communication Engineering*, Lecture Notes in Networks and Systems 107,
https://doi.org/10.1007/978-981-15-3172-9_44

set of indications described. Hence public may be anguish from it lacking recognize the risk. Main brain tumors can be either spiteful or benign.

Brain tumor occurred, if the cells were separating and rising unusually. It is emerge to be a hard heap when it is analyzed with analytical health imaging methods. Primary brain tumor and metastatic brain tumor are the two major class of brain tumor. Main brain tumor is the situation when the tumor is produced in the brain and be likely to reside there. The metastatic brain tumor is the tumor that is produced in a different place in the body and extends throughout the brain.

The indication [1] of brain tumor relies on the position, dimension and category of the tumor. It happens, if the tumor condensing the neighboring cells and provides out pressure. In addition, it arises, if the tumors chunk the liquid that flows all over the brain. The general indications are having headache, sickness, vomiting, difficulty in balancing and walking. Brain tumor can be detected by the diagnostic [2] imaging techniques namely MRI and CT scan. The pair of the techniques has benefit in detecting, relies on the spot type and the idea of assessment needed.

Segmentation [3, 4] is a vital step towards the analysis phase in several image processing chore. It allows quantification and visualization of the objects of interest. It has been decided that segmentation of therapeutic images is fixed a hard task and completely automatic segmentation measures are distant from satisfying in numerous practical situations. If the concentration or configuration of the object varies appreciably from the surroundings, segmentation is noticeable. In all supplementary circumstances manual discovering of the object boundaries by an expert only seems to be the valid truth. But it is undeniably a very time overriding job.

2 Research Problem

All research papers aims to develop such a system which predict and detect the tumor in its premature stages [5]. Furthermore it is attempted to develop the exactness of the premature guess and revealing system through preprocessing, segmentation aspect extraction and categorization techniques of extorted database.

For instance, the segmentation of brain tissue may not have the same requirements as the bifurcation of the liver and also the parameters differ. General imaging [6] relic such as sound, motion and partial volume effects also have significant effects on the attainment of segmentation algorithms. The substance to be disjointed from medical images is exact anatomical configuration, which are frequently non-stiff and intricate in outline and reveal substantial variability from person to person. Furthermore, there are no clear outline models yet available that fully confines the abrasion in anatomy. Magnetic timbre images are more intricate due to the restrictions in the imaging utensils that guide to a non-linear gain relic in the images. Also, the signal is tainted by movement artifacts due to voluntary or involuntary displacement of the patient through the scanning practice.

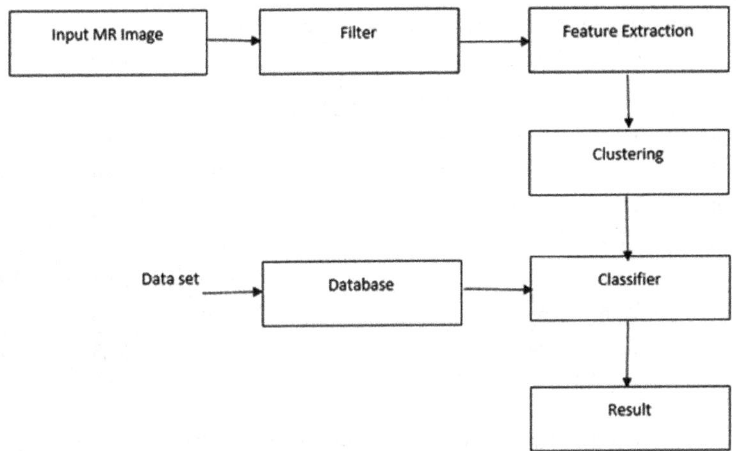

Fig. 1 Building blocks of proposed system

3 System Model of Proposed Method

Figure 1 shows the block diagram of proposed system. It composed of different stages in proposed system namely filtering, clustering and classification.

4 Proposed Hybrid Algorithm

In this paper, hybrid algorithm [7, 8] beside with the edge protection is utilized for enhanced categorization and discovery of the brain tumor [9]. Hybrid algorithm is the simplest classification algorithm which provides highly competitive results and can also be used for regression problems as well. *K* nearest neighbor algorithm classifies new cases by storing all cases available and situated on a resemblance measure for instance, distance functions. Hybrid algorithm is a non-parametric method which is used in geometric evaluation and pattern identification [10]. It is easy for interpretation and calculation time is low. This algorithm works for every constraint of deliberation.

Hybrid algorithm presume that the data is in a trait space that is, in a measure space. The information can either be scalars or multidimensional vectors. As already mentioned, the spot are in attribute space and hence they have a concept of space which is the Euclidean distance. The training data comprise a collection of vectors and class tag related with every vector. In the transparent type, it is assumed to be either positive or negative classes. It will be able to work uniformly well with random number of classes.

'*k*' is the number, which decides the number of neighbors which is recognized based on the span metric and persuade the classification. It is typically an odd quantity

Table 1 Performance analysis of existing and proposed methods

Type	Cases	Accuracy		MSE	PSNR	SSIM
		SVM	Hybrid algorithm			
Normal	Case 1	85.8	95.9	0.058	60.5	0.996
Normal	Case 2	74	92.8	0.0917	58.5	0.992
Normal	Case 3	73.8	92.7	0.0493	61.2	0.996
Tumour	Case 4	70.1	91.6	0.0236	64.4	0.999
Tumour	Case 5	71.4	92	0.19	55.4	0.98
Tumour	Case 6	68.6	96.1	0.00903	68.6	1
Tumour	Case 7	74.4	92.9	0.386	52.3	0.953

if the numeral classes are 2. Class probabilities can be calculated as the normalized frequency of samples that belong to each class. For instance, in binary categorization crisis, it is possible to have two classes namely class 0 and class 1. In this paper, initially the k value which is the number of nearest neighbor is calculated and the Euclidian distance between the query occurrences and every training samples are determined. After which the distances calculated are sorted and then nearest neighbor is determined based on k-th minimum distance.

Then, assign the majority class among the nearest neighbor in the grey and determine the class using those neighbors and hence the segmented image with brain abnormalities are formed. Edge safeguarding method fresh up the image noise in the consistent areas, but maintain all image formation like edges or curves. The hybrid algorithm is not just pertinent to grayscale images, it can be extended to color images also. They are plain, quick and excellent at safeguarding edges and slim architectural particulars in images. Boundaries and curve of unreliable disparity that may be there in real images have been replicated.

These nonlinear methods are used to determine the refined gray value in reliance of the pleased of a described neighborhood. From the list of the surroundings pixels, only few pixels are appropriated for the averaging, it may have analogous gray values contrast to the pixel in deliberation. Thus, with this Hybrid approach better performance like accuracy, MSE and PSNR values are achieved.

5 Simulation Results

Simulation is done for tumor image [11, 12] as well as normal image under different stages using Matlab software which is shown in Figs. 2 and 3.

The new approach provides the improving results and enhanced performance compared to the current approaches [13] and the comparison results are shown here. The simulation outcomes of various stages of the processing are displayed comparing

(i) Simulation results of tumor Image

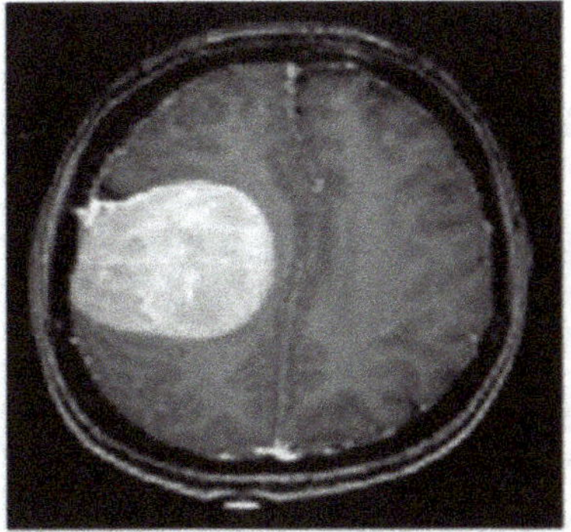

(a) Original Image

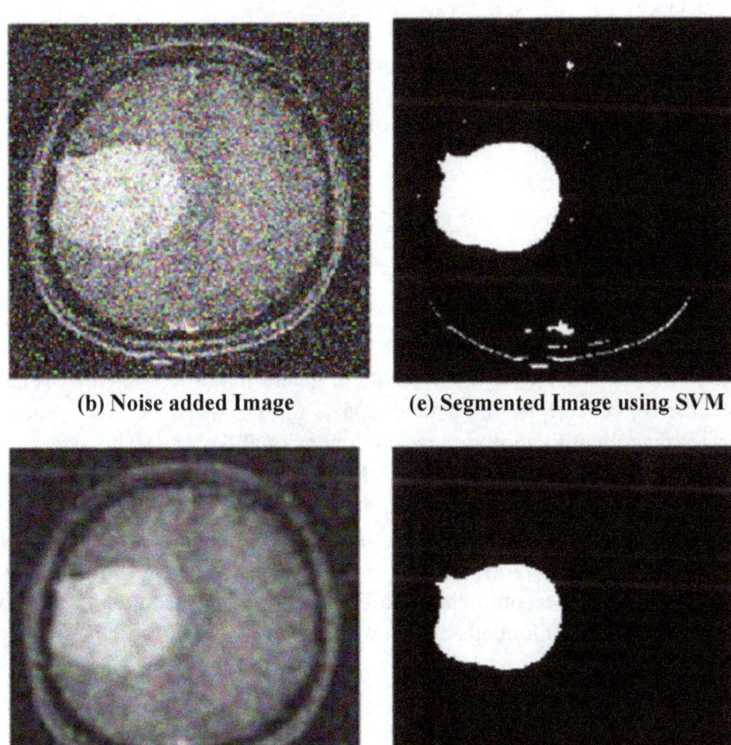

(c) Noise Filtered Image

(f) Segmented Image using Hybrid Algorithm

Fig. 2 (**a–g**) simulation result of tumor image

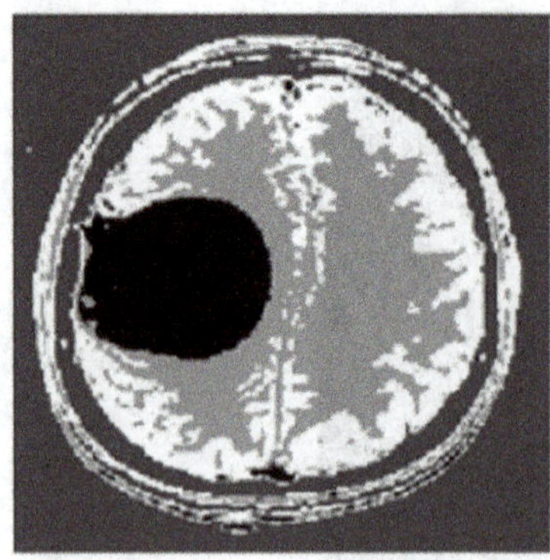

(d) Fuzzy C-Means clustered Image

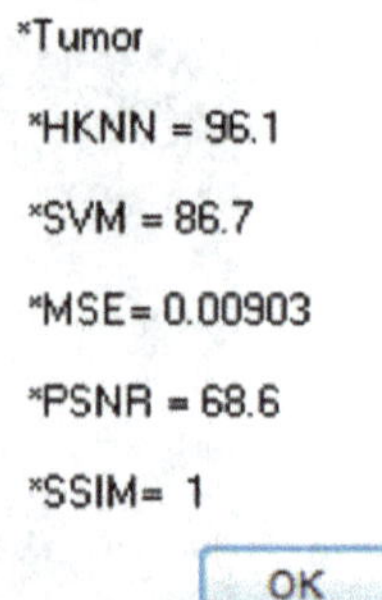

(g) Message dialog box to display the detection and parameter details

Fig. 2 (continued)

the existing and new approach. It is evident that the Hybrid algorithm provides better performance and discovery of the brain tumor as comparing to previous approaches (Table 1).

6 Conclusions

The early detection of Brain tumor is a demanding crisis, owing to the structure of the tumor cells structure and deformation, where majority cells are overlie with each other. In recent years tumor detection is the utmost imperative problems in structural image identification due to 2D dimensional datasets. This paper exhibits two segmentation techniques namely SVM and FCM algorithm, for slicing Brain cells classification to detect tumor at its early stages. A cluster based hybrid algorithm is proposed to conquer real time problem in training and edification models. In this paper we apply adaptive hybrid density estimation and block based Kernel density estimation, to group the comprehensive data. In this approach, Hybrid algorithm donates in finding the tumor cells at the earliest as compared to existing approaches.

(ii) Simulation results of normal image

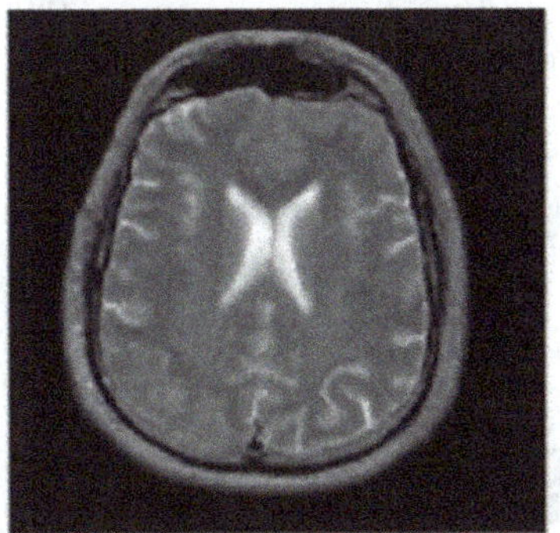

(a) Original Image

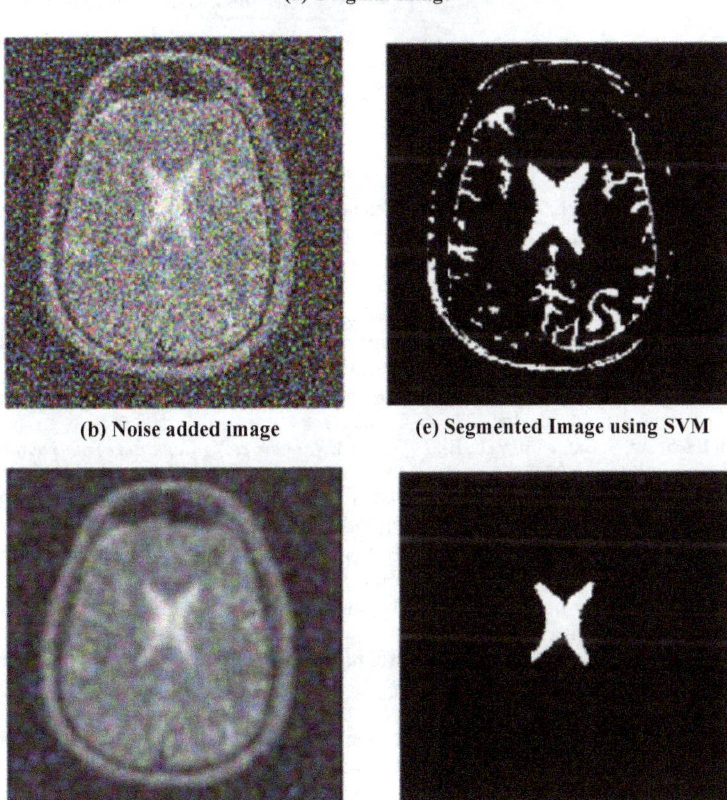

(b) Noise added image
(e) Segmented Image using SVM
(c) Noise Filtered Image
(f) Segmented Image using Hybrid Algorithm

Fig. 3 (**a**–**g**) simulation result of normal image

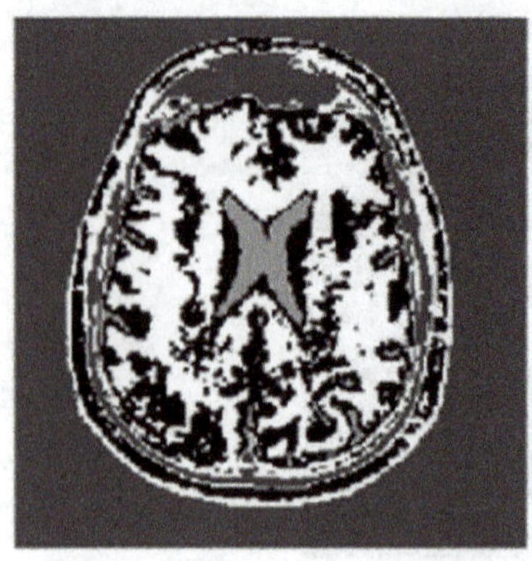

(d) Fuzzy C-Means clustered image

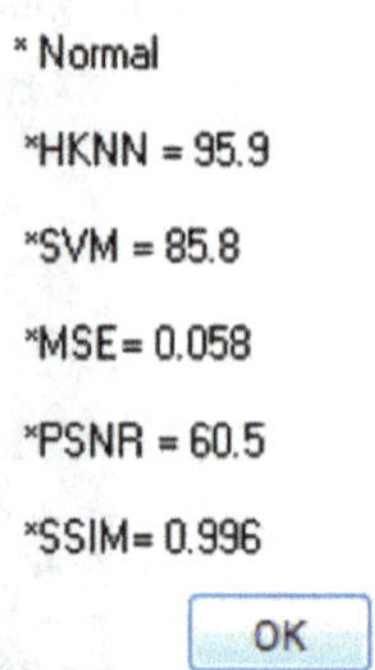

(g) Message dialog box to display the detection and parameter details

Fig. 3 (continued)

References

1. Y. Sun, E.J. Delp et al., A comparison of feature selection methods for the detection of breast cancer in mammograms: adaptive sequential floating search versus genetic algorithm, in *IEEE 27th Annual International Conference in Medicine and Biology Society*, (2005), pp. 6532–6535
2. R.N. Al-Okaili, J. Krejza, J.H. Woo et al., Intracranial brain masses: MR imaging-based diagnostic stratergy—initial experience. Radiology **243**(2), 539–550 (2007)
3. P. Sinthia, M. Malathi, MRI brain tumor segmentation using hybrid clustering and classification by back propagation algorithm. Asian Pac. J. Cancer Prev. **19**, 3257–3263 (2018)
4. M. Angulakshmi, G.G. Lakshmi Priya, Automated brain tumour segmentation techniques—a review. Wiley Periodicals, Inc., **27**, 66–77 (2017)
5. J.M. Baehring, S. Bannykh, W.L. Bi, *Diffusion MRI in the early diagnosis of malignant glioma* (Springer, 2006), pp. 221–225
6. N.I. Weisenfield, S.K. Warfield, Normalization of joint image—intensity statistics in MRI using the Kullback—Leibler divergence, in *IEEE International Symposium on Biomedical Imaging: Nano to Macro*, vol. 1(15) (2004), pp. 101-104
7. E.A.A. Maksoud et al., MRI brain tumor segmentation system based on hybrid clustering techniques. Springer book series in Communications in Computer and Information Science, vol 488 (2014), pp. 401–412
8. K.M. Iftekharuddin et al., Automatic brain tumor detection in MRI: methodology and statistical validation, in *Proceedings of the SPIE Symposium and Medical Imaging*, vol. 5747 (2005), pp. 2012–2022
9. A. Islam et al., Automatic brain tumor detection in MRI: methodology and statistical validation, in *Proceedings of the SPIE Symposium and Medical Imaging*, vol. 5747 (2005), pp. 2012–2022
10. T. Adah et al., A feature based approach to combine functional MRI, structural MRI and EEG brain imaging data, in *IEEE, EMBS Annual International Conference* (2007)
11. S. Ahmeed, K.M. Iftekharuddin, Efficacy of texture, shape, and intensity feature fusion for posterior-fossa tumor segmentation in MRI. IEEE Tran. Inf. Tech. Biomed. (2011)
12. K.M. Iftekharuddin, R. Marsh, W. Jia, Fractal analysis of tumor in brain MR images. Mach. Vis. Appl. **13**, 352–362 (2003)
13. A. Islam, K.M. Iftekharuddin, R. Ogg, F.H. Laningham, B. Sivakumar, Multifractal modeling, segmentation, prediction and statistical validation of posterior fossa tumor, in *SPIE Medical Imaging Conference* (2008)

Liveness Detection and Recognition System for Fingerprint Images

Munish Kumar and Priyanka Singh

Abstract In recent years, for verification and identification, uses of biometrics information are increased rapidly because they are more secure and reliable. Biometrics (face, finger, iris, palm, etc.) recognize the person based on human traits that are physiological and behavioral traits. Fingerprint recognition systems are widely used biometric for verification and identification due to its universality in nature and easiness. But they are various types of attacks are present that affect the performance of the fingerprint recognition system like spoofing attacks, displacement error, and physical distortion, etc. In this proposed system, work is carried out to overcome these types of errors and enhances the accuracy of the system. For spoofing detection, supervised learning with minutiae extraction method is used, for displacement error, alternating direction method multiplier (ADMM) is used and enhance the accuracy of the system by a technique that uses crossing number for minutiae extraction, for feature extraction gray-level difference method, discrete wavelet transforms, and feature matching using hamming distance. For learning and classification, support vector machine is used. In this fingerprint verification competition (FVC) 2002, FVC2004, FVC2006, and ATVS are considered for testing purpose and calculation of accuracy.

Keywords Fingerprint · ACC · TAR · Liveness detection

1 Introduction

Today security is the main issues in all the fields due to various types of threats. Password or PIN that may be without difficulty cracked by using cybercriminal so there is a requirement of the more secure system that protects the systems from various types of attacks and that can be achieved by biometrics [1, 2]. Biometrics secure the system by measuring the life sign that is by finger, face, palm, and iris, etc. In this study,

M. Kumar (✉) · P. Singh
ECE Department, Deenbandhu Chhotu Ram University of Science and Technology, Murthal, India
e-mail: engg.munishkumar@gmail.com

P. Singh
e-mail: priyankaiit@yahoo.co.in

H. S. Saini et al. (eds.), *Innovations in Electronics and Communication Engineering*, Lecture Notes in Networks and Systems 107,
https://doi.org/10.1007/978-981-15-3172-9_45

the fingerprint is considered due to their uniqueness in nature. In this study database, FVC 2000–2006 and ATVS are considered [3–5]. Earlier accuracy is calculated on FVC 2000, FVC 2002, and FVC2004 database, these databases are made from student's finger and FVC 2006 is more complex than the previous database. In this database, the fingerprint is taken from students, workers, and employees, etc. [6–8]. From the previous study, a conclusion is taken that biometric systems are drastically used for protection purposes and offer huge scope to become aware of/authenticate the users, thereby removing lots of burden from the person's end. Fingerprints are extensively utilized in biometric systems for authentication due to their uniqueness. Any people cannot have the same fingerprints. There is a burden on a human being to recall all the facts/statistics for his or her to get access to any particular system. If these statistics/data is stolen, then the unauthorized person can effortlessly access the machine and without having any problems misuse the information/facts. The huge difficulty is the safety of the device to protect any unlawful entry to any system [9–12].

2 Problem Statements

But still, there are few regions that want to be taken care in this area, which is considered as challenges to using fingerprints in biometrics system. The problem identification can be summarized pointwise as given below:

1. At the time of fingerprint scanning, if there is displacement and due to this there may be mismatch then this is a challenge to be considered for fingerprint recognition systems.
2. By using dummy printing making fingerprint from the impression left by using plasticity or clay to get unlawful access to any system. This is also a problem.
3. Distortion in fingerprints due to any work or reason may cause a problem in matching.
4. There is various types of frauds in bank transaction or any online service, which can be avoided by using fingerprint recognition (FPR) system.

3 Proposed Method

A complete FPR system is proposed for live detection and verification purpose. The proposed FPR system is designed to overwhelm the problem in the system, and to achieve this, the objectives are summarized pointwise as given below:

1. Displacement error is considered and reduce it.
2. To overcome the problems of fake fingerprint matching.
3. To decrease the recognition time for fingerprint by classification into a real or fake fingerprint.

4. Mismatch due to physical distortion is reduced.
5. To make a graphic user interface of the fingerprint recognition system.

The whole system is divided into two modules 'A' and 'B'. In the module, 'A' spoofing detection is performed and the displacement error is removed. In module 'B', identification of fingerprint is carried out and based on that resulting system confirm the authenticity of the person. In the first step, it works on module 'A' then on module 'B'. If the output of module 'A' is real, then only it goes to module 'B' otherwise it terminates the system for further processing and displays that the input finger is fake. By doing this, it protects the whole system from spoofing attacks and increases the performance of the system by classification of a real and fake finger. In module 'B', further processing is performed on the fingerprint image to enhance FPR system performance.

The complete FPR system for live detection and authentication purpose is shown in Fig. 1. The whole system is divided into two modules.

Module 'A': It is a fingerprint spoofing detection system that comprises of acquisition, pre-processing, feature extraction, loading database, recognition, and performance.

Acquisition:

1. Select the input image.
2. Show the test image.
3. Convert color image to gray-scale image.

Pre-processing:

1. Reducing displacement error from the input image.

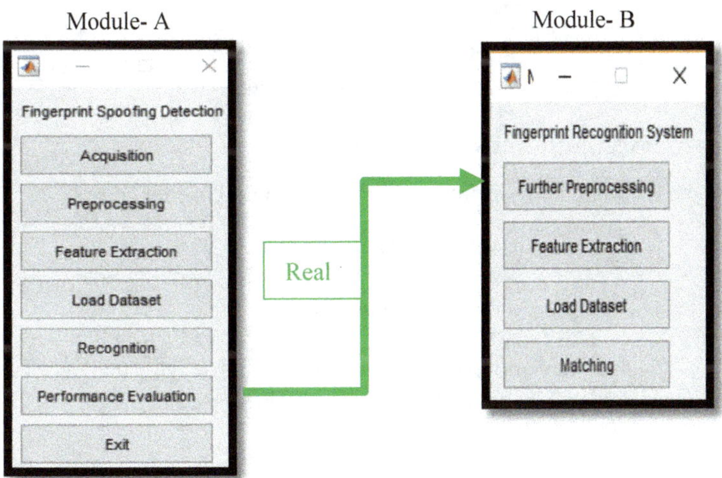

Fig. 1 Fingerprint recognition system

2. Ridge flow detection is performed.
3. Thinning process is done.
4. Minutiae extraction is carried out.
5. Finding ridge ends.
6. Finding bifurcation.

Feature extraction: In this step, features like homogeneity, contrast, energy, entropy, mean, and last histogram are extracted from the image.
Load database:
In this step, database mat file is loaded for the system.
Recognition:

1. Select training database path.
2. For classification, SVM is used, and classification is done by utilizing one against one approach.
3. Recognition of the finger is performed and decide whether the input finger is real or fake.

Performance:
In this step, if the finger is real, then go for further processing, and calculation is computed and based on that the performance of the system is decided.
Exit:
In this step, if the detected finger is fake then it exits the system to staring phase.
Module 'B': It comprises of noise removing, feature extraction, loading database, recognition, and performance.
Noise removing: Wavelet shrinkage with a soft threshold is used for removing noise from a distorted image.

$$\omega_\delta = \begin{cases} \operatorname{sign}(\omega)(|\omega| - \delta) & , |\omega| \geq \delta \\ 0 & , |\omega| < \delta \end{cases} \tag{1}$$

where ω—wavelet coefficients, δ—threshold.

Threshold: It computed by Donoho and Johnstone proposed model.

$$\delta = \sigma\sqrt{2\log_e N} \tag{2}$$

where σ—standard deviation of the noise, N—signal length, unidentified image noise standard deviation is estimated using the following equations:

$$\sigma_{iHH} = \frac{\text{Median}(|w_{iHH}|)}{0.6745} \tag{3}$$

$$\sigma_{iLH} = \frac{\text{Median}(|w_{iLH}|)}{0.6745} \tag{4}$$

$$\sigma_{iHL} = \frac{\text{Median}(|w_{iHL}|)}{0.6745} \tag{5}$$

Ridge Flow: Directions of ridges are shown by ridge flow and ridges are unfair in case of distorted fingerprint, whereas ridges are uniform in normal fingers. Ridge flow is calculated by the gradient-based approach.
The region of the Interest: More information-based area is extracted to provide a more accurate result.
Thinning: In this process, pixels of the image are widened by one pixel without any change in minutiae orientation and location.
Minutiae Extraction: For minutiae extraction, the crossing number method is used.
Feature Matching: Features are extracted using gray-level difference method (GLDM) and match feature using hamming match algorithm. A predefined threshold value is fixed if the matching score is greater than that value then fingers are matched. GLDM is used for feature extraction, i.e., entropy, mean, energy, homogeneity, and contrast.

$$\text{hom} = \sum \frac{P}{\text{greydiff}^2 + 1} \tag{6}$$

where greydiff = [0:1], P—probability function and hom—homogeneity

$$\text{con} = \sum (PX\,\text{greydiff}^2 + 1) \tag{7}$$

where con—the contrast of an image

$$\text{eng} = \sum (P)^2 \tag{8}$$

where eng—the energy of an image

$$\text{ent} = -\sum PX\log(P + \text{eps}) \tag{9}$$

where ent—the entropy of an image

$$B1 = \sum PX\text{greydiff} \tag{10}$$

where $B1$ denotes the mean of an image.

Feature Classification: From FVC2006, 960 fingerprints are taken for testing. Owner identity is found by the recognition process. Support vector machine (SVM)

is used to reduce misclassification error in training data and test data. SVM separates the normal or distorted finger image.
Accuracy: It is computed by using this formula:

$$\mathrm{ACC} = 100 \times ((\mathrm{TP} + \mathrm{TN})/N) \tag{11}$$

$$\mathrm{TAR(Recall)} = \frac{\mathrm{TP}}{\mathrm{TP} + \mathrm{FN}} \tag{12}$$

$$\mathrm{TRR(Specificity)} = \frac{\mathrm{TN}}{\mathrm{TN} + \mathrm{FP}} \tag{13}$$

$$\mathrm{FAR(Miss\ Rate)} = 1 - \mathrm{TAR} \tag{14}$$

$$\mathrm{FRR(Fall - out\ Rate)} = 1 - \mathrm{TRR} \tag{15}$$

where TP—true positive, TN—true negative, N—datasets size, ACC—accuracy, FRR—false rejection rate, FAR—false acceptance rate, TRR—true rejection rate, TAR—true acceptance rate.

The complete FPR system is shown in Fig. 1 and it consists of two modules 'A' and 'B'. The workflow is shown in Fig. 2. With the help of this proposed work, the accuracy of the fingerprint recognition system is calculated.

4 Result

The proposed FPR system is tested on standard datasets, i.e., FVC and ATVS. For verification purpose, 40 features are taken and cross-validation 10 K fold is used. Support vector machine (SVM) is used for classification and recognition. Module 'A' results are shown in Table 1 and module 'B' results are shown in Table 2. The overall accuracy of the system is computed and found to be 95.70% for complete FPR system. Table 3 gives a comparison of the proposed FPR system with existing methods for module 'B'. By combing both the module, the complete system graphic user interface is created to overcome the problem in fingerprint recognition systems.

Comparison of accuracy between proposed FPR systems with base is shown in Fig. 2. The complete system performs live detection and recognition that are shown in three different steps that are live detection with authorizing, live detection with unauthorized, and fake finger. Process flow for authorizing user is shown in Fig. 3a, b. There will no warning sound after that operation and the user can access the system. Process flow for unauthorized user is shown in Fig. 4.

In this process, a warning is produced to warn the staff that an unauthorized user wants to access the system. And the last for fake user wants to access the system by

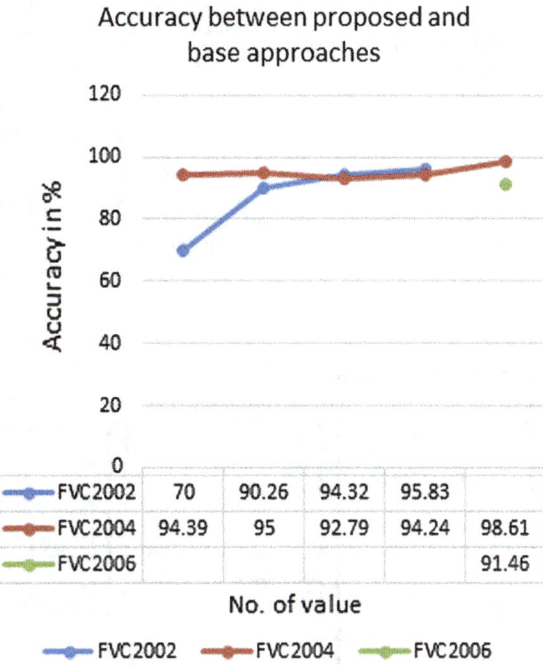

Fig. 2 Accuracy comparison between proposed and base method

Table 1 Proposed FPR result for real/fake detection by module 'A'

Databases	Proposed method accuracy (%)
FVC2002	93.51
FVC2004	95.37
FVC2006	96.80
ATVS-FFp	98.58

Table 2 Proposed FPR result for recognition by module 'B' on FVC2006

FVC2006 database	Proposed method accuracy (%)
DB1_A	91.94
DB2_A	90.22
DB3_A	88.55
DB4_A	88.94
DB1_B	95.38
DB2_B	91.93
DB3_B	92.81
DB4_B	91.93

Table 3 Analysis of accuracy between proposed and existing approaches module 'B'

Authors	Datasets	Accuracy (%)
Sangram Bana et al.	FVC2002	65–70
Maheshwai et al.	FVC2004	94.39
Le Hoang Thai et al.	FVC2004	95
Gowthami et al.	FVC2002, FVC2004	90.26,92.79
Gowthami et al.	FVC2002, FVC2004	94.32, 94.24
Proposed approach	FVC2002, FVC2004	95.83,98.61
Proposed approach	FVC2006	91.46

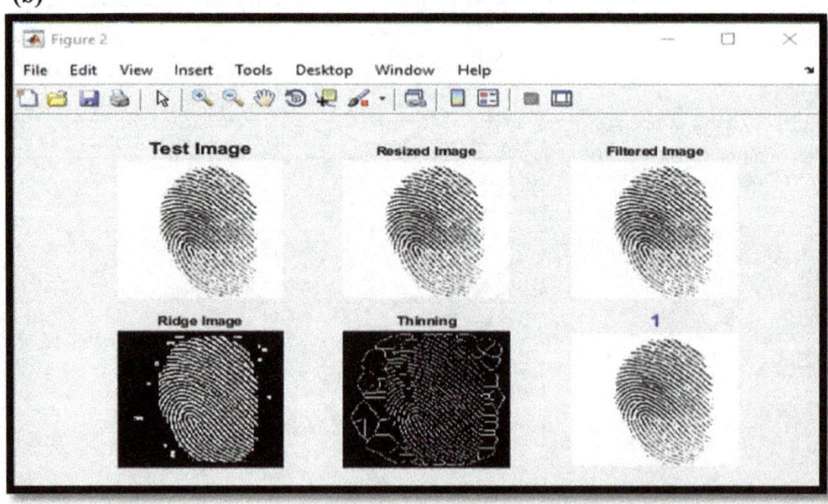

Fig. 3 **a** Process flow for authorize user, **b** process flow for authorizing user

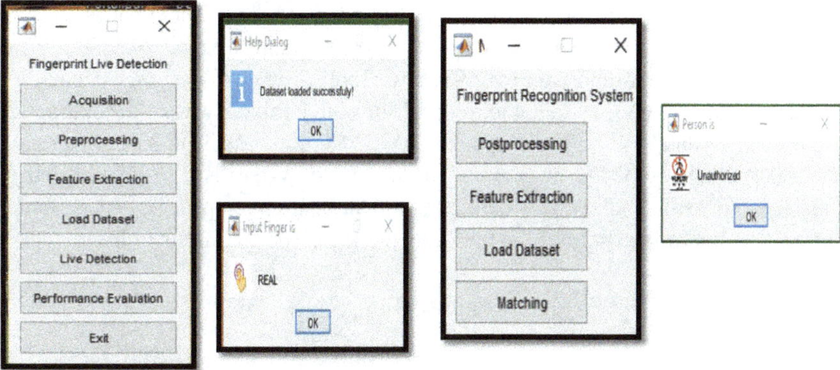

Fig. 4 Process flow for unauthorized user

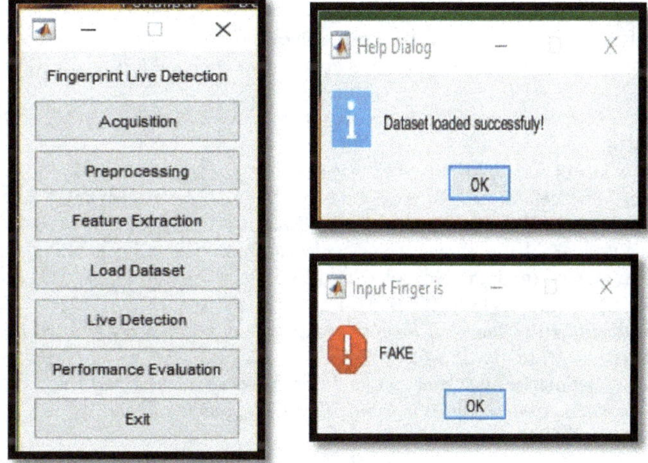

Fig. 5 Process flow for fake user

fake finger shown in Fig. 5. In this, the staff is alert by a warning sound that a fake user wants to access the system.

5 Conclusion

The proposed FPR system gives better performance in spoofing attacks and in the recognition process. FPR system is implemented with two modules 'A' and "B'. In module 'A' the accuracy is computed based on the spoofing detection. Module 'A' is tested on standard datasets, i.e., FVC and ATVS and average accuracy are calculated as 96.06%. And module ‘B’ accuracy is dependent on the authentication process. The proposed FPR system has been tested on the standard datasets FVC2000–2006 and the precision is found well over 95.30%. Table 2 gives the average accuracy of the system for the FVC2006 database is 91.46% and Table 3 gives the accuracy for the FVC2004 database is 98.61% and the FVC2002 database is 95.83%. The overall accuracy with module ‘A’ and ‘B’ of the proposed system is 95.70% with all the databases for real or fake finger detection and authentication process. Also, Table 3 shows a comparative study of the previous work carried out in this field.

References

1. S.S. Kulkarni, H.Y. Patil, Survey on fingerprint spoofing, detection techniques, and databases. IJCA Proc. Natl. Conf. Adv. Comput. NCAC **2015**(7), 30–33 (2015)
2. A. Al-Ajlan, Survey on fingerprint liveness detection, in *International Workshop on Biometrics and Forensics*, (2013), pp. 1–5
3. G. Arunalatha, M. Ezhilarasan, Fingerprint liveness detection using probability density function, in *International Conference on Communication and Signal Processing*, (2016), pp. 6–8, IEEE
4. Y.H. Baek, *The Fake Fingerprint Detection System Using a Novel Color Distribution (ICTC)*, IEEE (2016)
5. https://en.wikipedia.org/wiki/Fingerprint_Verification_Competition
6. Q. Huang, S. Chang, C. Liu, B. Niu, M. Tang, Z. Zhou, An evaluation of fake fingerprint databases utilizing SVM classification. Pattern Recognit. Lett. **60–61**, 1–7 (2015)
7. M. Kumar, Priyanka, Various image enhancement and matching techniques used for fingerprint recognition system. Int. j. inf. technol. (Springer Singapore Print, ISSN 2511–2104), (2017). https://doi.org/10.1007/s41870-017-0061-4
8. A.T. Gowthami, H.R. Mamatha, *Fingerprint Recognition Using Zone-Based Linear Binary Patterns*, VisionNet’15, pp. 552–557
9. H.S. Brar, V.P. Singh, Fingerprint recognition password scheme using BFO, in *IEEE International Conference on Advances in Computing, Communications and Informatics (ICACCI)*, (2014), pp. 1942–1946
10. D. Fang, X. Lv,Bin Lei, *A Novel InSAR Phase Denoising Method via Nonlocal Wavelet Shrinkage*, IEEE (2016)
11. H. Fronthaler, K. kollreider, J. Bigun, Local features for enhancement and minutiae extraction in fingerprints. IEEE Trans. Image Process. **17**(3), 354–363 (2008)
12. M. Kumar, P. Singh, FPR using machine learning with multi-feature method. IET Image Process. (2018). https://doi.org/10.1049/iet-ipr.2017.1406. IET Digital Library, https://digital-library.theiet.org/content/journals/10.1049/iet-ipr.2017.140

Munish Kumar completed his schooling at Holy Child Public School, Rewari, Haryana, India. He completed his Bachelor of Technology (2008) from The Technological Institute of Textile and Sciences, Bhiwani, India, and Master of Technology (2012) degrees from D.C.R. University of Science and Technology, Murthal, Sonipat, India. Currently, he is Research Scholar in Electronics and Communication Engineering Department at D.C.R. University of Science and Technology, Murthal, Sonipat, India. His current areas of interest include biometrics, image processing, and medical images.

Dr. Priyanka Singh is working as Professor at D.C.R. University of Science and Technology, Murthal, Sonipat, India. Her current areas of interest are signal processing, image processing, multimedia communication, and SAW filter design. Her highest qualification is a Ph.D. in Electronics Engineering from Indian Institute of Technology, Delhi, India. She has published several papers in refereed journals including IEEE Transactions and Conferences.

Character Recognition for ALPR Systems: A New Perspective

Sahil Khokhar and Pawan Kumar Dahiya

Abstract The automatic license plate recognition (ALPR) systems are utilized to locate vehicles' license (or number) plates and extract the information it contains from the image or video. The paper presents a new method of computing the recognition efficiency in which a successful recognition is only considered if the whole license plate is correctly recognized instead of focusing on individual characters, as it is more useful to consider the license plate as a whole. The recognition efficiency of template matching algorithm and SVM-based feature matching algorithm was determined to be 76.36% and 80%, respectively.

Keywords Automatic license plate recognition (ALPR) · Feature matching · Support vector machine (SVM) · Optical character recognition (OCR)

1 Introduction

The automatic license plate recognition (ALPR) systems are used to extract a vehicle's license (or number) plate information from an image or video. Some uses of the system include its use at toll roads payment collection, vehicle parking, etc., for more rapid service; it may be used in control of traffic to observe the flow of traffic around the road network [1] and at traffic signals to identify the vehicles that jump the red light [2].

The automatic license plate recognition (ALPR) system has five stages. The very first step is to obtain the video or image of a vehicle. Secondly, the image is pre-processed, so that the relevant details are preserved, whereas the unnecessary details and noise are removed from the image [3]. Then, the next step comprises of locating the license plate present in the image. The fourth step comprises of segmenting

S. Khokhar (✉)
Department of ECE, GJUS&T, Hisar, India
e-mail: khokhar.sahil0809@gmail.com

P. K. Dahiya
Department of ECE, DCRUST, Murthal, Sonepat, India
e-mail: pawan.dahiya@gmail.com

H. S. Saini et al. (eds.), *Innovations in Electronics and Communication Engineering*, Lecture Notes in Networks and Systems 107,
https://doi.org/10.1007/978-981-15-3172-9_46

each character that exists in the license plate [4, 5]. The segmented character is then recognized during the fifth and final step [4, 6–8].

The paper's purpose is to compare the character recognition techniques proposed in the literature and the stand-alone template matching and feature matching using SVM techniques for the recognition of the characters in Indian license plates.

The rest of the paper is structured in the following manner. In Sect. 2, the related work is reviewed. Section 3 explains the detailed implementation of the work. In Sect. 4, the results of the recognition techniques are discussed. In Sect. 5, the conclusions of the work that is done is discussed.

2 Related Work

The ALPR process comprises of five steps as shown in Fig. 1 [10, 16]. Firstly, the image that contains the license plate is acquired. The image is then preprocessed based on the demand for the detection and recognition algorithms. Some common preprocessing steps are resizing, grayscale conversion, contrast enhancement, and noise removal.

After enhancing the image, the region of interest, i.e., the license plate, is located within the image. The features of the license plate were used for its localization. In [4, 9, 17], an analysis of high-density regions and their color information is carried out to localize the license plate. In [14], the dimensional features of the plates were utilized for the purpose of localization. The relative positioning of the plates with respect to the tail lights of the vehicle was used for zeroing in on the plates in [11]. There are many more techniques that can be used for the localization of plates such as connected-component analysis (CCA) [6], morphological operations [12], genetic algorithms [5], and neural networks [15].

Classifiers can be used for the segmentation of characters as done in [6], where SVM is used for analyzing each connected component. In [4], the projection plots of the characters are used for their segmentation.

The fifth and final step of the ALPR system is the recognition of characters. In [6], a two-level SVM classifier is used to recognize characters, in which the second

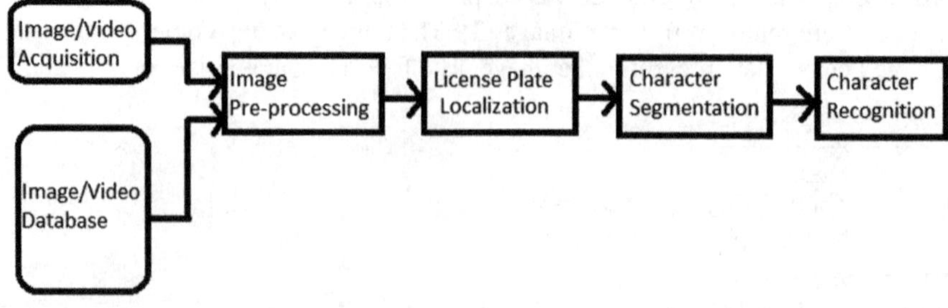

Fig. 1 Steps involved in the ALPR process

SVM is used to differentiate between similar-looking characters. In [4], a hybrid of the decision tree classifier and the SVM classifier is proposed. The recognition was done by a sparse network of winnows (SNoW) classifier that used successive mean quantization transform (SMQT) features in [8]. In [7], the recognition of Chinese license plates was carried out with a combination of convolutional neural network (CNN) and extreme learning machines (ELM).

3 Proposed Work

The preprocessing for the acquired images consisted of resizing, grayscale conversion, and noise removal. The license (or number) plate localization was concluded in mainly has four steps (viz. edge detection, separation of the objects from the background, searching for the connected components, and the selection of the candidate). For detection of the edges, the morphological processes erosion and dilation are used. The region of interest is separated from background by eliminating the edges that have a low probability of being the license plate. The region of interest is then carefully chosen based on the properties of the characters. Most of the noise from the license plate was also eliminated in the process and the characters were segmented using connected-component analysis (CCA).

The techniques used for recognition of the character can be generally classified into two types, i.e., the template matching and the feature matching [13]. In the template matching technique, the correlation between the character extracted from the image and the available character templates is measured and the character template which has the highest correlation coefficient with the extracted character is selected as the target. While, in the feature matching technique, extracted character's predefined features are extracted and then it is recognized with the help of a classifier, such as an artificial neural network (ANN) and support vector machines (SVM).

To improve the accuracy of the process, the information about the location (Fig. 2) of the alphabets and digits is utilized. Hence, the alphabets are only compared with

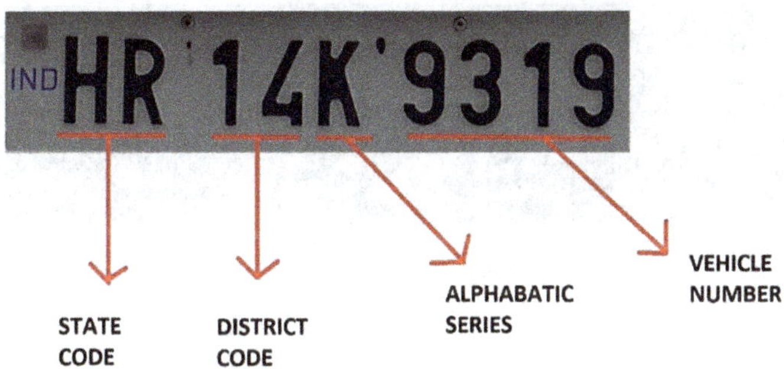

Fig. 2 Location of digits and alphabets in a license plate

Fig. 3 Image of license plate used for template matching

alphabets and digits with digits to eliminate the probability of error in recognizing two similar-looking numbers and alphabets such as (0 & O) and (8 & B).

3.1 Template Matching

In template matching, each segmented character's correlation with the respective templates (26 alphabets or 10 digits) is computed and the template with which the correlation coefficient is highest is chosen as the recognized character (Fig. 3).

3.2 Feature Matching Using SVM

In this technique, the features of a character are used instead of comparing the character on a pixel-by-pixel basis. For feature extraction of the image, the skeletal image (Fig. 4) of the character is used and its Euler number, eccentricity, extent, and orientation are taken as four features that are used. Besides these four features, other features are extracted by first dividing the given character into 9 equal parts. Then, for each part of the image, the filled area, the number of the horizontal, right diagonal, vertical and left diagonal lines, and their respective normalized lengths were used as features. Hence, there are a total of 85-dimensional features associated with each character. The SVM utilized RBF (Gaussian radial basis function) kernel.

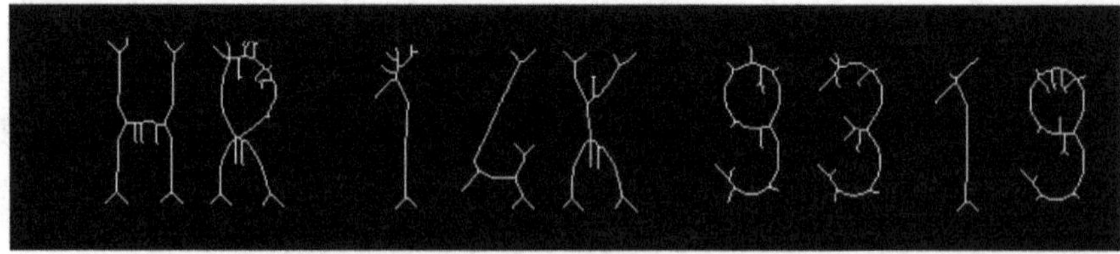

Fig. 4 Skeletal image of the license plate shown in Fig. 2

Table 1 Success rate of the proposed work

Recognition technique	Success rate (%)
Template matching	76.36
SVM	80

4 Results

The success rate of recognition techniques for ALPR systems is computed as the ratio of successfully recognized characters to total number of characters to be recognized. According to this, the success rate of template matching was 96.7% and that of SVM was 97.2%. However, this method may work in other OCR applications as if any character is misrecognized it can still be interpreted correctly based on its context, but this is not the case in ALPR systems. Even if a single character is erroneously recognized in the ALPR systems, then the recognition of the whole number plate should be considered incorrect. Therefore, the recognition efficiency of the system in this paper is computed as the ratio of the number of license plates that are successfully recognized to the total number of license plates. A successful recognition is one in which each and every character in the license plate is recognized correctly.

The success rate of license plate recognition given by the template matching is 76.36% and that given by the SVM classifier is 80%. These results are also shown in Table 1. The license plates that were recognized had 9 characters.

The efficiency of the previous works considers success rate based on individual characters. This information can be utilized to determine an approximate success rate for the whole license plate. If the probability of successful recognition of a character is p and the license plates have a total of n characters, then the probability of successful recognition of the plate can be computed as

$$P(n) = p^n \tag{1}$$

Using Eq. (1), an estimated success rate (with the plate consisting of n characters) of license plate recognition is presented in Table 2 and the comparison is shown between the proposed technique and the recognition methods presented in the literature. The highest accuracy as seen from the Table 2 is given by the SNoW classifier [8]. However, it had only 80% yield, i.e., the 20% characters with the lowest confidence value did not qualify for the recognition stage.

5 Conclusion

The recognition efficiencies of the proposed algorithm were on par with some of the more complex algorithms presented in the literature, yet as can be seen from the results, the ALPR systems do not have high enough accuracy that they can

Table 2 Comparison of the proposed work to the recognition methods in literature

Recognition method	Reported accuracy (p) (%)	Expected accuracy ($P(n)$) (%)			Reported plate accuracy ($n = 9$)
		$n = 8$	$n = 9$	$n = 10$	
Decision tree—SVM hybrid [4]	95.85	71.2	68.2	65.5	–
2 level SVM [6]	97.6	82.3	80.3	78.4	–
CNN—ELM hybrid [7]	96.38	74.5	71.7	69.2	–
SNoW [8]	99	92.2	91.3	90.4	–
Proposed template matching	**96.7**	76.5	74	71.5	**76.36%**
Proposed SVM	**97.2**	79.7	77.4	75.3	**80%**

be implemented without any human intervention. There is still a need for ample research before a truly automated ALPR system can be implemented. In addition, the recognition techniques for different regions might provide different results. So, the recognition technique should be carefully chosen.

References

1. B. Tian, Y. Li, B. Li, D. Wen, Rear-view vehicle detection and tracking by combining multiple parts for complex. IEEE Trans. Intell. Transp. Syst. **15**(2), 597–606 (2014)
2. I.S. Ahmad, B. Boufama, P. Habashi, W. Anderson, T. Elamsy, Automatic license plate recognition: a comparitive study, in *IEEE International Symposium on Signal Processing and Information Technology (ISSPIT)*, (Abu Dhabi, UAE, 2015), pp. 635–640
3. K.S. Raghunandan, P. Shivakumara, H.A. Jalab, R.W. Ibrahim, G.H. Kumar, U. Pal, T. Lu, Riesz fractional based model for enhancing license plate detection and recognition. IEEE Trans. Circuits Syst. Video Technol. **28**(9), 2276–2288 (2018)
4. A.H. Ashtar, M.J. Nordin, M. Fathy, An Iranian license plate recognition system based on color features. IEEE Trans. Intell. Transp. Syst. **15**(4), 1690–1705 (2014)
5. G. Abo Samram, F. Khalefah, Localization of license plate number using dynamic image processing techniques and genetic algorithms. IEEE Trans. Evol. Comput. **18**(2), 244–257 (2014)
6. R. Panahi, I. Gholampour, Accurate detection and recognition of dirty vehicle plate numbers for high-speed applications. IEEE Trans. Intell. Transp. Syst. **18**(4), 767–779 (2017)
7. Y. Yang, D. Li, Z. Duan, Chinese vehicle license plate recognition using kernel-based extreme learning machine with deep convolutional features. IET Intel. Transport Syst. **12**(3), 213–219 (2018)
8. O. Bulan, V. Kozitsky, P. Ramesh, M. Shreve, Segmentation-and annotation-free license plate recognition with deep localization and failure identification. IEEE Trans. Intell. Transp. Syst. **18**(9), 2351–2363 (2017)
9. M.R. Asif, Q. Chun, S. Hussain, M.S. Fareed, Multiple licence plate detection for Chinese vehicles in dense traffic scenarios. IET Intell. Transp. Syst. **10**(8), 535–544 (2016)

10. S. Rani, P.K. Dahiya, A review of recognition technique used automatic license plate recognition system. Int. J. Comput. Appl. (0975–8887) **121**(17) (2015)
11. H. Kuang, L. Chen, F. Gu, J. Chen, L. Chan, H. Yan, Combining region-of-interest extraction and image enhancement for nighttime vehicle detection. IEEE Intell. Syst. **31**(3), 57–65 (2016)
12. J. Yepez, S.B. Ko, Improved license plate localisation algorithm based on morphological operations. IET Intel. Transport Syst. **12**(6), 542–549 (2018)
13. D. Shan, M. Ibrahim, M. Shehata, W. Badawy, Automatic license plate recognition (ALPR): a state-of-the-art review. IEEE Trans. Circuits Syst. Video Technol. **23**(2), 311–325 (2013)
14. Q. Li, A geometric framework for rectangular shape detection. IEEE Trans. Image Process. **23**(9), 4139–4149 (2014)
15. L. Xie, T. Ahmad, L. Jin, Y. Liu, S. Zhang, A new CNN-based method for multi-directional car license plate detection. IEEE Trans. Intell. Transp. Syst. **19**(2), 507–517 (2018)
16. S. Khokhar, P.K. Dahiya, A review of recognition techniques in ALPR systems. Int. J. Comput. Appl. (0975–8887) **170**(6), 30–32 (2017)
17. Y. Yuan, W. Zou, Y. Zhao, X. Wang, H. Xuefeng, N. Komodakis, A robust and efficient approach to license plate detection. IEEE Trans. Image Process. **26**(3), 1102–1114 (2017)

A System for Disease Identification Using ECG and Other Variables

Amana Yadav and Naresh Grover

Abstract Cardiovascular disease is a major health problem in every part of the world, causing increase in death rate. The noninvasive efficient diagnosis of heart is possible through the study of electrocardiogram (ECG) of the patient under observation. A single system is necessary to know about few diseases based upon ECG data of the person and other parameters. Here, a new technique has been proposed which will measure kidney and heart disease depending on creatinine and urea nitrogen levels. This technique will detect the diseases in prior to reach the severe condition of their health. The MIT-BIH database has been considered as a reference for ECG signals and MATLAB version 8.2 was used to display the results.

Keywords Electrocardiogram (ECG) · R peak detection · QRS complex · Cardiac arrhythmia · MIT-BIH · MATLAB

1 Introduction

Electrocardiogram (ECG) is an indispensable clinical means that describes the electrical activity of heart registered by the skin electrode fixed on the body surface [1]. Depending on various factors morphology of the heart, age, size, relatively body weight, ECG of every person is different [2, 3]. ECG signal can be used to detect cardiac arrhythmias and is used by many of the physicians for different applications [4]. The normal heart rate is about 70 beats/min. Any change in the heart rate is called arrhythmia, and it is widely classified in two categories based on R-R interval, i.e., bradycardia and tachycardia [5].

Each beat consists of three different features of ECG waves like P wave, QRS and finally T wave (ventricular repolarization) are represented in Fig. 1.

From the past few years, many methods have already been developed for ECG arrhythmia classification. Few of them are based on nearest neighbor classifier based

A. Yadav (✉) · N. Grover
Faculty of Engineering and Technology, Manav Rachna International Institute of Research and Studies, Faridabad, India
e-mail: amana.fet@mriu.edu.in

H. S. Saini et al. (eds.), *Innovations in Electronics and Communication Engineering*, Lecture Notes in Networks and Systems 107,
https://doi.org/10.1007/978-981-15-3172-9_47

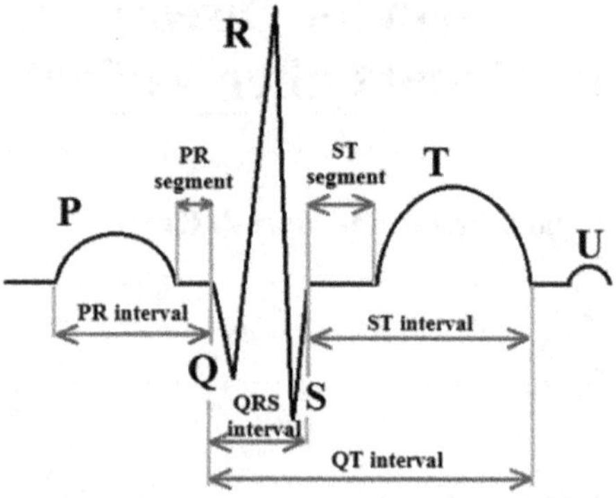

Fig. 1 Normal ECG signal consists of a P wave, a QRS complex and a T wave [6]

on a local fractal [7], a method for cardiac abnormality detection based on electrocardiogram (ECG) with Gaussian mixture model (GMM) [8] and a new approach for detection of the rhythm of heart using assimilation of orthogonal polynomial decomposition (OPD) based on Hermite and classification of support vector machines (SVMs) [9]. Few methods for ECG beat classification have also been developed depending on the multiclass SVM implementation in combination with the fault improving output codes [10] and a method of ECG printout with support vector machines and principle components analysis for ECG beat classification [11].

The existing techniques for ECG beat detection used R peak detection which is accurate and take longer time to detect QRS complex. In few of the techniques, real-time detection method is proposed but it depends on a suitable choice of the bandwidth for the bandpass filter.

The main purpose of the proposed work is to find out the kidney disease and heart disease using ECG feature and creatinine and urea nitrogen levels. The proposed system performs diagnosis of disease in two-stage ECG features have been classified, and in the second stage, disease has been identified using the arrhythmia type and with the level of urea nitrogen and creatinine. For reference point of view of ECG signal, MIT-BIH database has been used. The features which have been extracted from ECG signals were given to decision circuit with other few variables.

This paper is organized as follows.

Section 2 represents the disease finding technique in detail. Here, proposed methodology has been described. In Sect. 3, the experimental results are explained which represent the quality of the work. MIT-BIH arrhythmia database was used to assess the proposed method. Finally, Sect. 4 shows the conclusion of our work and also presents future scope.

2 Methodology

Several steps have been used in the proposed method to evaluate ECG signal, these steps are data acquisition then preprocessing after that peak detection with feature extraction and finally decision circuit. In data acquisition, MIT-BIH database was used to get ECG signal. The ECG signals that are produced by body are mixed with some noise and artifacts [12]. So, FIR filter was used to eliminate this interference in preprocessing step. Now, ECG features have been obtained like mean, SD (standard deviation) and variance of the ECG signal. These features and values of creatinine and urea nitrogen levels are given to decision circuit as mentioned in Fig. 2. The decision circuit will find the type of disease based on these variables. Fig. 2 shows the block diagram of disease's detection system in the human beings.

Along with the ECG signal analysis, we have measured the systolic pressure, diastolic pressure, kidney disease and heart diseases. The normal range of urea nitrogen level in the human body is 7–20 mg/dL. The normal range of the creatinine is 0.6–1.1 mg/dL in males and its value in females is 0.5–1.1 mg/dL [13]. We are keeping the constant values for creatinine and urea nitrogen levels, depending on these two and blood pressure, QRS intervals, identifying the kidney disease and heart diseases. If urea nitrogen level and creatinine are above normal range means, it is identified as the kidney disease [14]. By considering RR interval, QQ interval, SS interval, frequency-domain features of the ECG signal, identifying the heart diseases.

The detailed discussions of each stage are as follows.

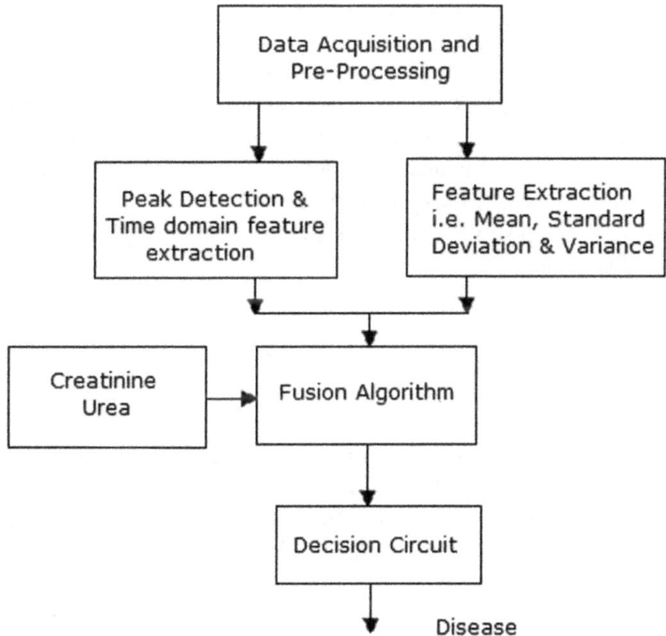

Fig. 2 Block diagram of disease detection system

2.1 Preprocessing and Filtering

Please check that the lines in line drawings are not interrupted and have a constant several noises or undesired interferences mixed with the ECG signals and corrupt it, such as influence of the muscle noise, the baseline drift, the interference due to power line, frequency noise impacts and electro mayo gram (EMG) [15]. So, before applying any algorithm for detection, the preprocessing of the signal should be done; FIR filter technique is used for removal of baseline wander.

2.2 R Peak Detection

After the preprocessing and filtering steps, R peaks have been found using four stages approach, as first-order forward differencing which highlights the QRS complex, amplitude normalization, Smooth Shannon energy envelope extraction and a peak finding logic which uses thresholding [16].

2.3 Feature Extraction

In this stage, attribute of ECG waves like P, Q, R, S, T peaks, QRS interval, RR distance, ST distance and PR distance has been calculated. The approximation coefficients of wavelet of the obtained ECG signal were used to represents the signals. Then few characteristics like mean, standard deviation (SD) and variance of the coefficient of wavelet are observed. These are used to represent the frequency and time sharing of the ECG signals.

2.4 Additional Parameter

After finding ECG signal features, we need few variables like creatinine and urea nitrogen levels to identify the kidney disease. The normal range of urea nitrogen level in the human body is 7–20 mg/dL. The normal range of the creatinine in males is 0.6–1.1 mg/dL and its value in females is 0.5–1.1 mg/dL.

2.5 Fusion Algorithm

To find out the disease depending on the arrhythmia information and additional variables information, an algorithm has been designed. This algorithm will classify

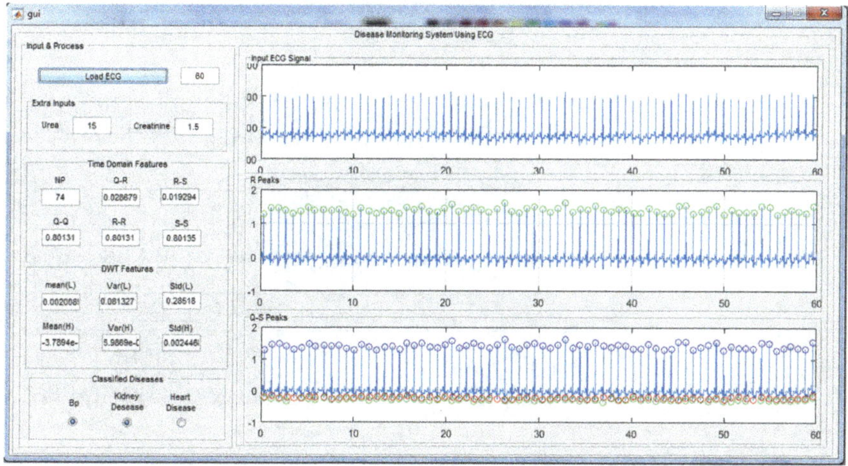

Fig. 3 Kidney disease identification in MATLAB

the diseases whether it is kidney disease and heart diseases. If urea nitrogen level and creatinine are above normal range means, it is identified as the kidney disease. By using QRS interval and frequency-domain features of the ECG signal, identifying the heart diseases.

3 Results and Discussion

To examine the proposed method for disease detection, MIT-BIH arrhythmia database has been used. These data contain 48 half an hour of two-channel ECG recordings at sampling frequency 360 Hz [17]. The given algorithm of disease detection was using MATLAB version 8.2 and it was verified on different ECG signals taken from the database of the MIT-BIH arrhythmia.

Depending on creatinine and urea nitrogen levels and blood pressure, QRS intervals diseases of few patients have been identified. The results of detected disease on different patients are given (Figs. 3, 4, 5, 6 and 7).

4 Conclusion

To diagnose the kidney and heart diseases, it is very important to find the ECG parameters, as it is used for classification and arrhythmia detection. Hence, we have proposed a method to diagnose R peak and feature detection of ECG signals. After that, this method used few additional parameters like creatinine and urea nitrogen levels were used to identify the kidney and heart diseases of different patients. The

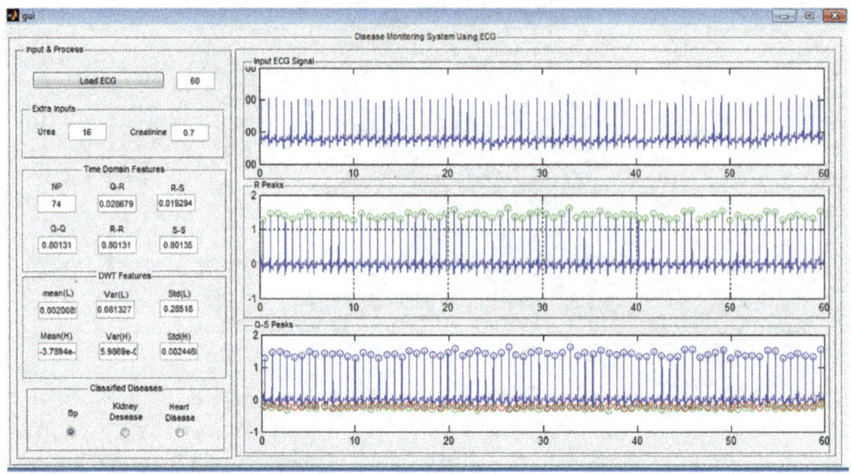

Fig. 4 BP identification in MATLAB

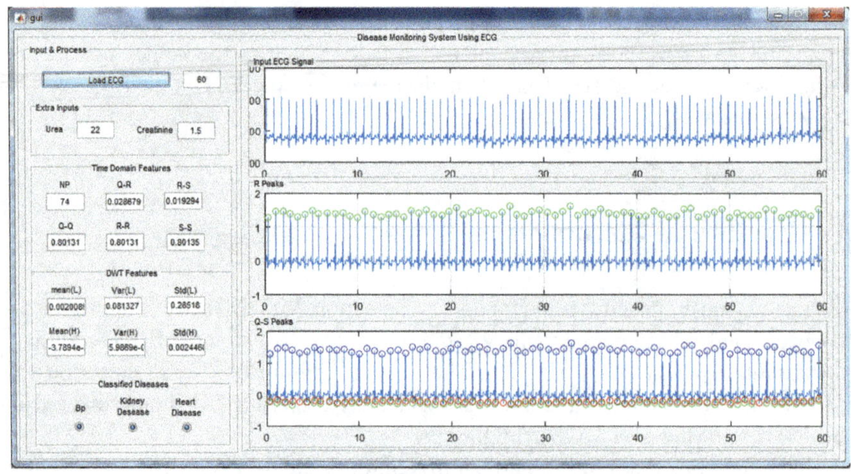

Fig. 5 Kidney disease and heart disease identification in MATLAB

results received from the given method have been shown and discussed. This proposed method has been tested and verified by using the standard MIT-BIH arrhythmia database.

For prior and accurate identification of more disease at the output, this algorithm can be enhanced further. And, this proposal can be used further to observe the disease for real-time ECG signals in future.

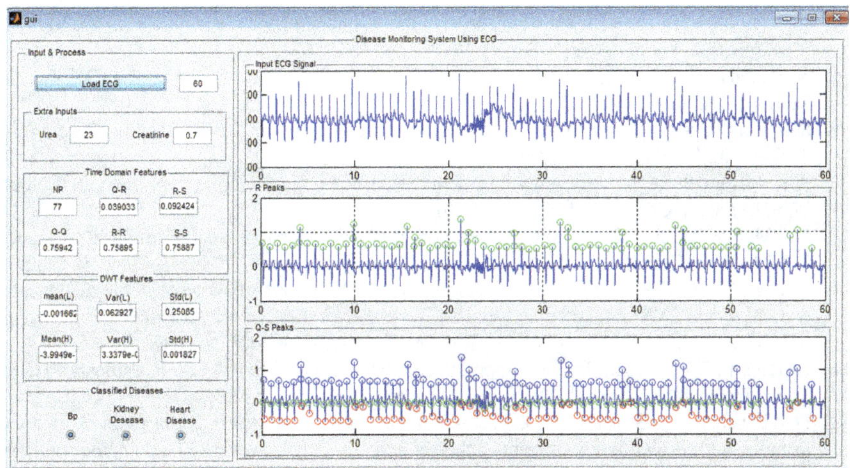

Fig. 6 Kidney disease and heart disease identification in MATLAB

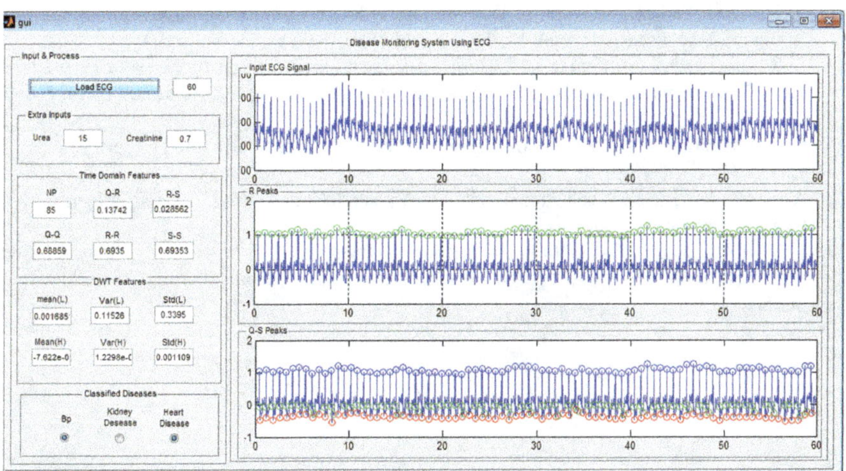

Fig. 7 Heart disease identification in MATLAB

References

1. N. Yun-Hon, J.H. Yap, D.U. Jeong, Implementation of the abnormal ECG monitoring system using heartbeat check map technique, in *Proceedings of International Conference on IT Convergence and Security*, (Macao, 2013), pp. 1–4, December 16–18
2. C.P. De, M. Dwyer, R.B. Reilly, Automatic classification of heartbeat using ECG morphology and heart beat interval features. IEEE Trans. Biomed. Eng. **51**, 1196–1206 (2004)
3. A.E. Zadeh, A. Khazaee, V. Ranaee, Classification of electrocardiogram signal using supervised classifier and efficient features. Comput. Methods Programs Biomed. **99**, 179–194 (2010)

4. R.M. Korurek, B. Dogan, ECG beat classification using particle swarm optimization and radial basis function neural network. Expert. Syst. Appl. **33**, 7563–7569 (2010)
5. N. Saxena, S. Kshitij, Extraction of various features of ECG signal. Int. J. Eng. Sci. Emerg. Technol. **7.4**, 707–714 (2015)
6. S.T. Prasad, S. Varadarajan, Heart rate detection using Hilbert transform. Int. J. Res. Eng. Technol. **02**(8), 508–513 (2013)
7. A.K. Mishra, S. Raghav, Local fractal dimension based ECG arrhythmia classification. Biomed. Signal Process. Control **5**, 114–123 (2010)
8. R.J. Martis, C. Chakraborty, A.K. Ray, A two stage mechanism for registration and classification of ECG using Gussian mixture model. J. Pattern Recognit. Arch. **42**(11), 2979–2988 (2009)
9. W.K. Lei, M.C. Dong, J.Shi, B.B. Fu, Automatic ECG interpretation via morphological feature extraction and SVM inference nets, in *IEEE Asia Pacific Conference on Circuits and Systems (APCCAS)*, (2008). https://doi.org/10.1109/apccas.2008.4746008
10. I. Guler, E.D. Ubeyli, Multiclass support vector machines for EEG-signals classification. IEEE Trans. Inf Technol. Biomed. **11**(2), 117–126 (2007)
11. D. Thanapatay, C.Suwansaroj, C. Thanawattano, ECG beat classification method for ECG printout with principle components analysis and support vector machines, in *International Conference on Electronics and Information Engineering (ICEIE)*, (2010), pp. VI 72–75. https://doi.org/10.1109/iceie.2010.5559841
12. S. Das, M. Chakraborty, QRS detection algorithm using Savitzky-Golay filter. Int. J. Signal Image Process. **03**(1) (2012)
13. S. Shafi, M. Saleem, R. Anjum, W. Abdullah, T. Shafi, ECG abnormalities in patients with chronic kidney disease. J. Ayub Med. Coll. Abbottabad **29**(1), 61–64 (2007)
14. D. Aronson, M.A. Mittleman, A.J. Burger, Elevated blood urea nitrogen level as a predictor of mortality in patients admitted for decompensated heart failure. Am. J. Med. **116**(7), 466–73 (2004)
15. P. Phukpattaranont, QRS detection algorithm based on the quadratic filter. Expert. Syst. Appl. **42**(11), 4867–4877 (2015)
16. A. Yadav, N. Grover, A robust approach for R-peak detection. Int. J. Inf. Eng. Electron. Bus. (IJIEEB) **9**(6), 43–45 (2017)
17. MIT-BIH Database distribution, Massachusetts Institute of Technology, 77 Massachusetts Avenue, Cambridge, MA 02139 (1998). http://www.physionet.org/physiobank/database/mitdb/

Image Quality Analysis Based on Texture Feature Extraction Using Second-Order Statistical Approach

V. Kalist, P. Ganesan, L. M. I. Leo Joseph, B. S. Sathish and R. Murugesan

Abstract When the image is of very larger in size to be analyzed and contains redundant pixels, the image is transformed into a cluster of feature representations as intensity, color or texture. Texture is one of the most important attribute employed to segment the input images into number of meaningful regions of interest (ROI) and to categorize those regions. It is exemplified by the spatial allocation of intensity levels in a region (neighborhood). The segmentation of the texture is the process of finding out of the edge between texture clusters of an image. The classification of texture is the procedure to determine a particular texture pattern from a group of texture classes. The salient feature of the feature extraction is to reduce the amount of data needed to illustrate an image accurately. It is very easy to compute the first-order features, whereas higher-order computation is very difficult to implement due to its interpretation and computational cost. In this work, GLCM is constructed to extract the second-order statistical features of image texture. Here, the texture features are

V. Kalist
Department of Electronics and Communication Engineering, Sathyabama Institute of Science and Technology, Chennai, India
e-mail: kalist.v@gmail.com

P. Ganesan (✉)
Department of Electronics and Communication Engineering, Vidya Jyothi Institute of Technology, Aziz Nagar, C.B.Road, Hyderabad, India
e-mail: gganeshnathan@gmail.com

L. M. I. Leo Joseph
Department of Electronics and Communication Engineering, S.R. Engineering College, Warangal, Telangana, India
e-mail: leojoseph@srecwarangal.ac.in

B. S. Sathish
Department of Electronics and Communication Engineering, Ramachandra College of Engineering, Eluru, Andhra Pradesh, India
e-mail: subramanyamsathish@yahoo.co.in

R. Murugesan
Department of Electronics and Communication Engineering, Malla Reddy College of Engineering and Technology, Secunderabad, India
e-mail: rmurugesan61@gmail.com

H. S. Saini et al. (eds.), *Innovations in Electronics and Communication Engineering*, Lecture Notes in Networks and Systems 107,
https://doi.org/10.1007/978-981-15-3172-9_48

calculated from the geometric distribution of gray level of each pixel with respect to neighborhood pixels of the image.

Keywords Feature extraction · Segmentation · GLCM · Texture

1 Introduction

There are many ways to extract the necessary features from the input image [2]. These include spatial (an image is defined by the spatial distribution of its gray level), transform (provides information in the frequency domain for edge and boundary detection), color (an image is a combination of three different color components), shape (based on its physical profile and structure and mainly applied for object recognition) and texture (image is a collection of replicated pattern at regular intervals) [6]. The image attribute texture defines the features and exterior of an object using shape, size and arrangement [9]. The structural methods provide more image information which is useful for image synthesis. Model-based methods base on the organization of an image that can be used for depicting texture. The feature extraction based on models purely depends on the neighboring image pixels and noise. Markov model is the best example for this type of feature extraction. Transform methods, based on any one of the transforms as Fourier or wavelet, extracting the features in frequency domain. The statistical-based approach based on the linear rapport between the image pixels gray levels. They are applied to investigate the spatial allocation of gray values by calculating the confined features at each pixel in the image and extracting necessary details from the local features. According to the order (number of pixels), statistical approaches categorized into first (single pixel), second (two pixels) and higher (more than two pixels) order. The first-order approach computes the simple quality measures such as mean and variance of the individual image pixel without considering the spatial relationship among the pixels [3]. The second or higher-order approaches compute the quality measures such as energy, correlation based on the relationship among the image pixels. The major advantage of the statistical approach is that it presents excellent outcomes with lesser execution time. This approach considers only the location of the gray level of the image pixels and their relationship. The numerical 2D array is constructed based on the pixels values which represent the texture of the image. This texture pattern mainly depends on the orientation of a particular pixel and its immediate neighbors. GLCM is the best example for the statistical texture analysis. It approximates image features using second-order statistics. The salient feature of the feature extraction is to reduce the amount of data needed to illustrate an image accurately. In this work, GLCM is constructed to extract the second-order statistical features of image texture. Here, the texture features are calculated from the geometric distribution of gray level of each pixel with respect to neighborhood pixels of the image. It is very easy to compute the first-order features, whereas higher-order computation is very difficult to implement due to its interpretation and computational cost.

2 Texture in Image Processing

In image processing, texture is defined as the intensity deviation of the image surface. It presents the image information such as regularity, coarseness and smoothness. Texture is a replicating pattern of neighborhood deviation in image intensity. So, it cannot be described for a single pixel [5]. For instance, consider an image of 60% white and 40% black pixels distribution. For this intensity information, number of different binary images can be created with dissimilar textures. It is a collection of texels, i.e., texture elements or primitives. It may be fine or coarse according to texel size. Fine texture consists of smaller texels and the intensity (tone) variation between texels also very large. The coarse texture includes larger texels, i.e., more pixels. Texture operators may be simple as range (variation between low and high intensity values in a neighborhood) and variance. These operators are very easy to compute but do not present any intimation about the replicating character of texture. In computer vision, it is necessary to classify or cluster the textures for automatic recognition or inspection. The natural texture images are very difficult to compress because of their high-frequency components. So, texture analysis is indispensable to represent the images in very convenient and compact form [10].

3 Computation of GLCM

A gray-level co-occurrence matrix (GLCM) is an appropriate statistical-based approach for extorting the second-order texture details from the images [1]. It is a 2D array (matrix) which consists of the amount of rows and columns, i.e., one and the same to the amount of different pixel (gray) values of the image [7]. It also explains the rate of the occurrence of a particular gray level in a stipulated spatial orientation with other gray levels of an area of interest [8]. The two major parameters to be considered to compute GLCM are the spatial distance between the two image pixels (d) and their relative direction (angle) as 0°, 45°, 90° and 135°. The computation of GLCM from the input image is explained using the following illustration 1. GLCM has zero value at (0, 1). This is due to there are no occurrences in the test image where two straight nearby pixels with the values 0 and 1. In this fashion, all the values of GLCM matrix are computed (Fig. 1).

4 Materials and Methods

The texture feature extraction for the image analysis using the second-order statistical approach is explained in this section. To extract the necessary second-order statistical features, GLCM is constructed. In this, the texture features are calculated from the geometric distribution of gray level of each pixel with respect to neighborhood pixels

Fig. 1 GLCM computation **a** input image matrix **b** GLCM matrix **c** generalized form of GLCM matrix

(a)

3	4	0
1	0	3
3	0	2
3	0	3

(b)

0	0	1	2	0
1	0	0	0	0
0	0	0	0	0
2	0	0	0	0
1	0	0	0	0

(c)

Gray Level	0	1	2	3	4
0	(0,0)	(0,1)	(0,2)	(0,3)	(0,4)
1	(1,0)	(1,1)	(1,2)	(1,3)	(1,4)
2	(2,0)	(2,1)	(2,2)	(2,3)	(2,4)
3	(3,0)	(3,1)	(3,2)	(3,3)	(3,4)
4	(4,0)	(4,1)	(4,2)	(4,3)	(4,4)

of the image. Figure 2 illustrates the test images to investigate the competence of the proposed method. These are the standard test images frequently used for image processing applications.

Let i and j the co-occurrence matrix coefficients
$P(i, j)$ the element of the matrix at (i, j)
N the co-occurrence matrix dimension.

The second-order statistical measures to analyze the image based on the texture features are elucidated as follows [4].

Contrast ($C1$) is the standard deviation which assesses the intensity differences between a reference and its adjacent pixels. The largest value of contrast means that more intensity deviation between them.

Fig. 2 Test images to investigate the competence of the proposed method. Image courtesy: Mathworks

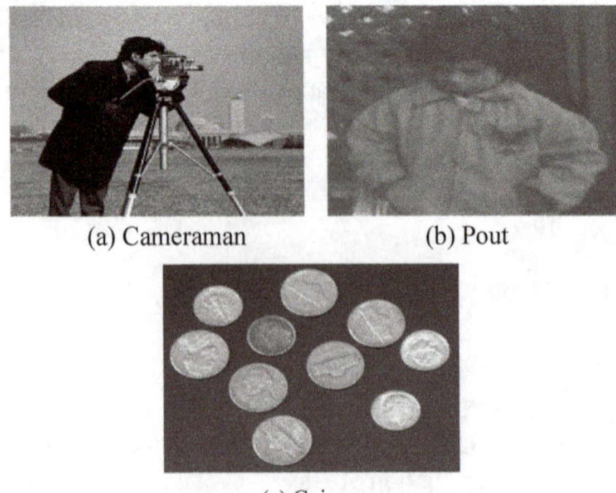

(a) Cameraman (b) Pout

(c) Coins

$$C1 = \sum_i \sum_j (i - j)^2 P(i, j) \tag{1}$$

Homogeneity (H) explains the closeness of GLCM member's distribution to its diagonal. H and $C1$ are inversely proportional to each other. It achieves its maximum when the majority of the elements of GLCM are accumulated close to the major diagonal. It is inversely proportional to GLCM contrast.

$$H = \sum_i \sum_j \frac{1}{1 + (i - j)^2} P(i, j) \tag{2}$$

Entropy calculates the image disarray (randomness) of the image. If all the elements of matrix P are same, entropy is the largest one. It should be least when GLCM consists of unequal elements. The value of entropy is large when all the pixels of an image or region have unique magnitude and low for random variation in their magnitude.

$$E1 = -\sum_i \sum_j P(i, j) * \ln P(i, j) \tag{3}$$

The parameter energy ($E2$) computes the consistency or regularity of the image. The energy is maximum when the uniform or periodic distribution of the image intensity level.

$$E2 = \sum_i \sum_j \sqrt{P^2(i, j)} \tag{4}$$

The linearity of the image is computed in terms of the correlation ($C2$). For linear structured images, the correlation is very high.

$$C2 = \sum_i \sum_j P(i, j) \frac{(1 - \mu_x)(1 - \mu_y)}{\sigma_x \sigma_y} \tag{5}$$

μ_x, μ_y are means, and $\sigma_{x,} \sigma_y$ are standard deviations of P_x and $P_{y.}$
The autocorrelation is given by

$$\text{AC} = \sum_i \sum_j (i, j) * P(i, j) \tag{6}$$

The dissimilarity of GLCM is computed by

$$D = \sum_i \sum_j (i - j) * P(i, j) \tag{7}$$

The calculation of cluster shade is presented as

$$\mathrm{CS} = \sum_i \sum_j (i + j - \mu_x - \mu_y)^3 * P(i, j) \tag{8}$$

The cluster prominence is represented as

$$CP = \sum_i \sum_j (i + j - \mu_x - \mu_y)^4 * P(i, j) \tag{9}$$

Maximum probability of GLCM is illustrated as

$$\mathrm{MP} = \mathrm{Max}\{P(i, j)\} \tag{10}$$

The variance of GLCM matrix is portrayed as

$$V = \sum_i \sum_j (i - \mu)^2 * P(i, j) \tag{11}$$

The sum average of GLCM is given by

$$\mathrm{SA} = \sum_{i=2}^{2N_g} i P_{x+y}(i) \tag{12}$$

The sum entropy is mentioned as

$$\mathrm{SE} = -\sum_{i=2}^{2N_g} P_{x+y}(i) * \log\{P_{x+y}(i)\} \tag{13}$$

The sum variance of GLCM is indicated as

$$\mathrm{SV} = \sum_{i=2}^{2N_g} (1 - \mathrm{SA})^2 P_{x+y}(i) \tag{14}$$

The difference variance is represented as

$$\mathrm{DV} = \mathrm{Var}\{P_{x-y}\} \tag{15}$$

The difference entropy is computed as

$$\mathrm{DE} = -\sum_{i=0}^{N_{g-1}} P_{x-y}(i) * \log\{P_{x-y}(i)\} \tag{16}$$

Table 1 Extracted features from the texture analysis

Feature	Cameraman	Pout	Coins
Correlation	0.9000	0.8968	0.9081
Contrast	0.7565	0.1436	0.5672
Energy	0.1685	0.2534	0.2326
Homogeneity	0.8672	0.9340	0.8884
Cluster prominence	436.93	13.779	304.21
Cluster Shade	−40.286	0.9044	32.852
Dissimilarity	0.3522	0.1347	0.2865
Autocorrelation	21.472	11.643	17.076
Entropy	2.3694	1.5954	2.1211
Max. probability	0.3106	0.3128	0.4183
Variance	21.7844	16.372	17.219
Sum average	8.4969	7.9533	7.5623
Sum variance	56.6713	44.481	44.435
Sum entropy	1.9989	1.4850	1.8390
Difference variance	0.75,654	0.1436	0.5622
Information measure of correlation	−0.5294	−0.6116	−0.5739
Inverse difference	0.9646	0.9851	0.9708

5 Experimental Result and Discussion

The extracted features from the second-order statistical texture analysis of the test images are tabulated in Table 1. The experiment is conducted on different images to extract the features. Only the result of three images (cameraman.tif, pout.tif, coins.png) is illustrated here.

The graphical illustration of the outcome for the image, cameraman, is shown in Fig. 3. Here, only four quality measures (correlation, contrast, energy and homogeneity) are illustrated with respect to horizontal offset.

Figures 4 and 5 depicted the result for the images, pout and coins, respectively. Here, only four quality measures (correlation, contrast, energy and homogeneity) are illustrated with respect to horizontal offset.

6 Conclusion

The proposed work explained the texture feature extraction for the image analysis. The texture features are calculated from the geometric distribution of gray level of each pixel with respect to neighborhood pixels of the image. The first-order approach

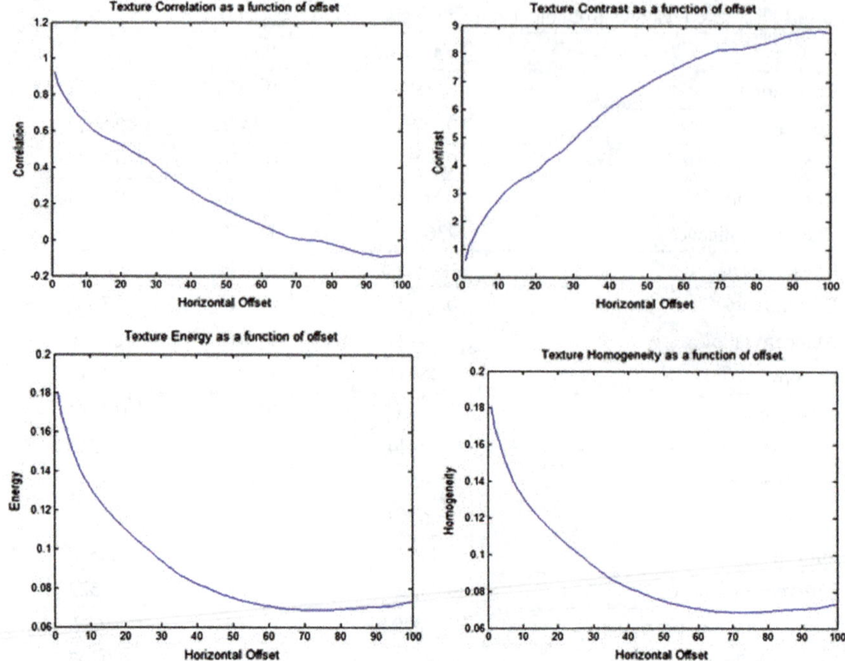

Fig. 3 Graphical illustration of extracted features for cameraman test image

computes the simple quality measures such as mean and variance of the individual image pixel without considering the spatial relationship among the pixels. In this work, the second or higher-order approaches, compromised between first and higher order, computed the quality measures such as energy, correlation based on the correlation among the pixels.

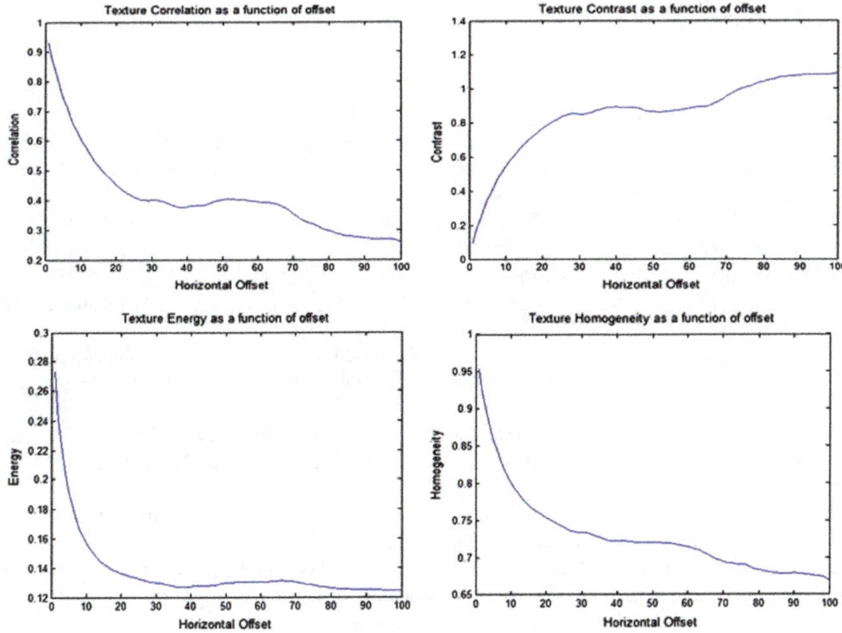

Fig. 4 Graphical illustration of extracted features for pout test image

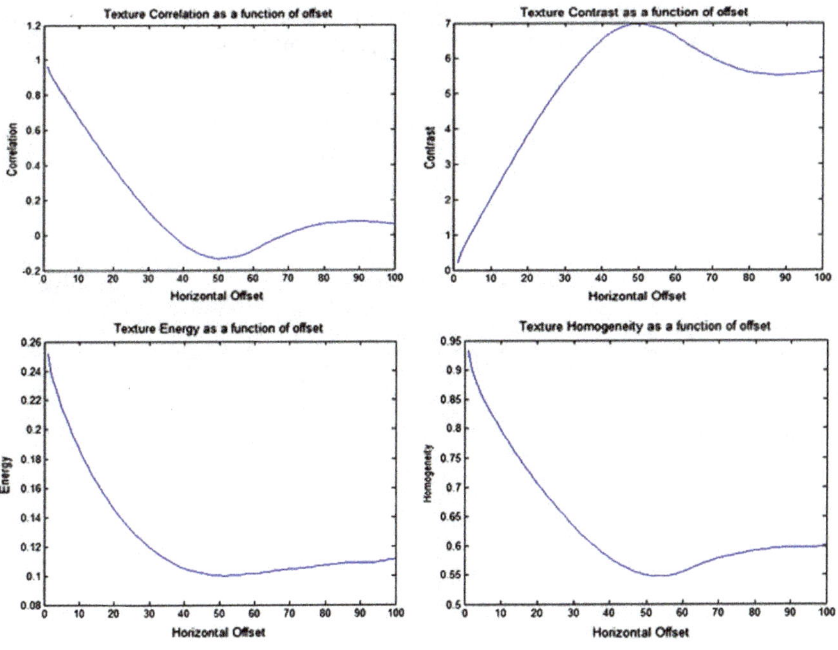

Fig. 5 Graphical illustration of extracted features for cameraman test image

References

1. D.A. Clausi, An analysis of co-occurrence texture statistics as a function of grey level quantization. Can. J. Remote Sens. **28**(1), 45–62 (2002)
2. P. Ganesan, V. Rajini, Assessment of satellite image segmentation in RGB and HSV color model using image quality measures, in *International Conference on Advances in Electrical Engineering (ICAEE)*, (2014), pp. 1–5
3. B. Gao, X. Li, W.L. Woo, T.Y. Tian, Physics-based image segmentation using first order statistical properties and genetic algorithm for inductive thermograph imaging. IEEE Trans. Image Process. **27**(5), 2160–2175 (2018)
4. R.M. Haralick, K. Shanmugam, I. Dinstein, Textural features of image classification. IEEE Trans. Syst. Man Cybern. **3**(6), 610–623 (1973)
5. X. Huang, W. Yang, Classification of remotely sensed imagery according to the combination of gray scale and texture features based on the dynamic windows. J. Geomat. Sci. Technol. **32**, 277–281 (2015)
6. V. Kalist, P. Ganesan, B.S. Sathish, J.M.M. Jenitha, Possiblistic-fuzzy *C*-means clustering approach for the segmentation of satellite images in HSL color space. Procedia Comput. Sci. **57**, 49–56 (2015)
7. H. Kekre, S.D. Thepade, T.K. Sarode, V. Suryawansh, Image retrieval using texture features extracted from GLCM, LBG and KPE. Int. J. Comput. Theory Eng. **2**, 695–700 (2010)
8. C.N. Rao, S.S. Sastry, K. Mallika, H.S. Tiong, K. Mahalakshmi, Co-occurrence matrix and its statistical features as an approach for identification of phase transitions of mesogens. Int. J. Innov. Res. Sci. Eng. Technol. **2**, 4531–4538 (2013)
9. G. Sajiv, P. Ganesan, Comparative study of possiblistic fuzzy *C*-means clustering based image segmentation in RGB and CIELuv color space. Int. J. Pharm. Technology. **8**(1), 10899–10909 (2016)
10. L. Soh, C. Tsatsoulis, Texture analysis of SAR sea ice imagery using gray level co-occurrence matrices. IEEE Trans. Geosci. Remote Sens. **37**(2), 780–795 (1999)

Frequency Impact Analysis with Music-Evoked Stimulated Potentials on Human Brain

Shidhartho Roy, Monira Islam, Md. Salah Uddin Yusuf and Tanbin Islam Rohan

Abstract The most common stimulation of brain is sound or music consisting of multiple frequency ranges which affect differently on the EEG bands. This paper shows an innovative approach to indicate the mental states due to music-evoked stimulated potentials. Different Bengali, English and Tamil songs are categorized as bass, medium and treble music genre according to their frequency ranges. From PSD analysis, the activation of alpha and beta bands is evaluated from which the mental state of brain is determined. Due to the increment of time duration, alpha activity of human brain is decreasing and beta activity is on the rise so mental state moves from relax to stress. The β/α ratio is found 0.507 for relax state and it is found that after 3 h brain relaxation is obtained for the three frequency levels. Finally, brain relaxation is modeled with the β/α band power ratio applying least square algorithm.

Keywords EEG bands · Stress analysis · Music · Band power ratio · Relaxation modeling

1 Introduction

Music is one of a modest group of human cultural universals that evoke different feelings, from excitement to relaxation, happiness to sorrow, dread to convenience or

S. Roy · M. Islam · Md. S. U. Yusuf (✉) · T. I. Rohan
Department of Electrical and Electronic Engineering, Khulna University of Engineering and Technology (KUET), Khulna, Bangladesh
e-mail: ymdsalahu2@gmail.com

S. Roy
e-mail: swapno15roy@gmail.com

M. Islam
e-mail: monira_kuet08@yahoo.com

T. I. Rohan
e-mail: tanbinislam009@gmail.com

H. S. Saini et al. (eds.), *Innovations in Electronics and Communication Engineering*, Lecture Notes in Networks and Systems 107,
https://doi.org/10.1007/978-981-15-3172-9_49

even combinations. Neurosurgeons use it in order to increase concentration, to coordinate motions and strengthen collaboration [1], to improve attention and vigilance, and to increase concentration or for motivation athletes use them [2].

Perceptual, behavioral and physical reactions to continued emotional activities can be predicted to cause stress. The assessment of these stress reactions includes subjective estimates and views in distinct experiment environments. One of the most common techniques for measuring pressure intensity is self-report questionnaires [3]. A subjective technique is the assessment of cognitive pressure through questionnaires [4], and stress is quantifiable through human bio-signals [5]. The EEG signal is subdivided into several spectral analysis bands consisting of alpha bands (8–13 Hz), beta bands (14–30 Hz), delta bands (0.5–4 Hz) and theta bands (4–8 Hz). Increased EEG activity in alpha waves has been linked to increased relaxation; otherwise, it declines during stress [6].

Previously, many works have been conducted on stress indication. Johannes Sarnthein et al. experimented on persistent patterns of brain activity and relate a positive impact on brain for short duration music exposure [7]. Petsch et al. investigated on the relation of brain activity changes for music stimuli via EEG signal [8]. Rideout and Laubach studies are based on the positive impact of classical music (low-frequency music) on human brain on short interval [9]. Another big term of this research arena is the Mozart effect.

This paper analyzes the impact of music at different frequencies from beta and alpha activation using PSD, and the relaxation analysis is performed from the band power ratio (BAR). The block diagram of the proposed research is given in Fig. 1. The rest of the paper is organized as follows. Section 2 shortly introduces this proposed work methodology. Section 3 reports the experimental findings and debate justifying the PSD assessment for pressure indication and brain soothing mathematical modeling. Finally, the conclusion is given in Sect. 4.

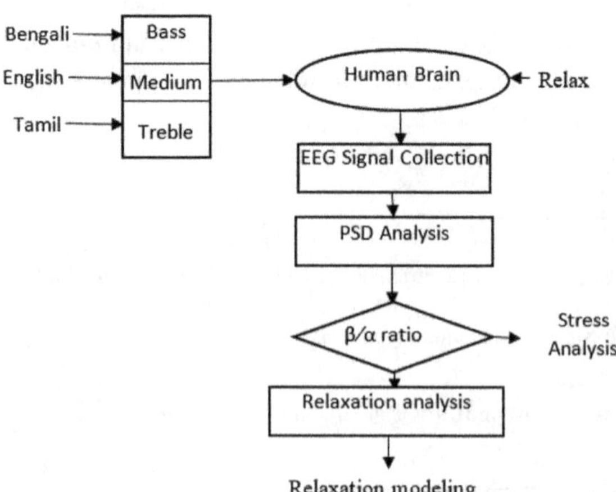

Fig. 1 Flow diagram of proposed method

2 Proposed Methodology

2.1 *Categorization of Song Stimuli*

The mental states and brain relaxation due to music-evoked stimulated potentials are evaluated in this research. Stress is determined through a range of songs in the test environment at different frequencies. Bengali, English and Tamil songs are thus considered to stimulate stress to subjects. Songs of three different languages are further subdivided into three frequency bands like bass (20 Hz–1 kHz), medium (1–7.5 kHz) and treble (7.5–17 kHz).

2.2 *EEG Signal Acquisition*

EEG signals are recorded via the BIOPAC MP36 system in BME Laboratory, KUET. The purchase of the signal requires BIOPAC electrode guide array (SS2L), disposable rubber electrodes (EL503), information storage device (MP36 and MP150) with cable and power. Figure 2a, b shows the data acquisition protocol and the raw EEG signal, respectively.

2.3 *PSD Analysis*

Power spectral density (PSD) analysis is significant to identify the brain activation. The PSD value of any band shows the average power of the individual band in terms of frequency. The spectral analysis is performed to extract the PSD of the EEG bands. The spectral analysis is performed to extract the PSD of the EEG bands. In this analysis, Hamming window is taken with zero padding and 1024 data samples are taken for extracting the value of PSD. Equation 1 represents the power spectral density function of alpha/beta band, respectively, where the average power is in the time range (t_1, t_2).

$$\int_0^F \mathrm{PSD}_{\mathrm{alpha/beta}}(k)\mathrm{dk} = \int_0^F \frac{2|X(k_{\mathrm{alpha}})|^2}{(t_2 - t_1)} \mathrm{d}k \tag{1}$$

The method FFT analysis gives us the opportunity to analyze alpha and beta band. It is given by Eq. 2.

$$X_{\mathrm{alpha/beta}}(k) = \{X(0), X(1), \ldots, X(n-1)\} \tag{2}$$

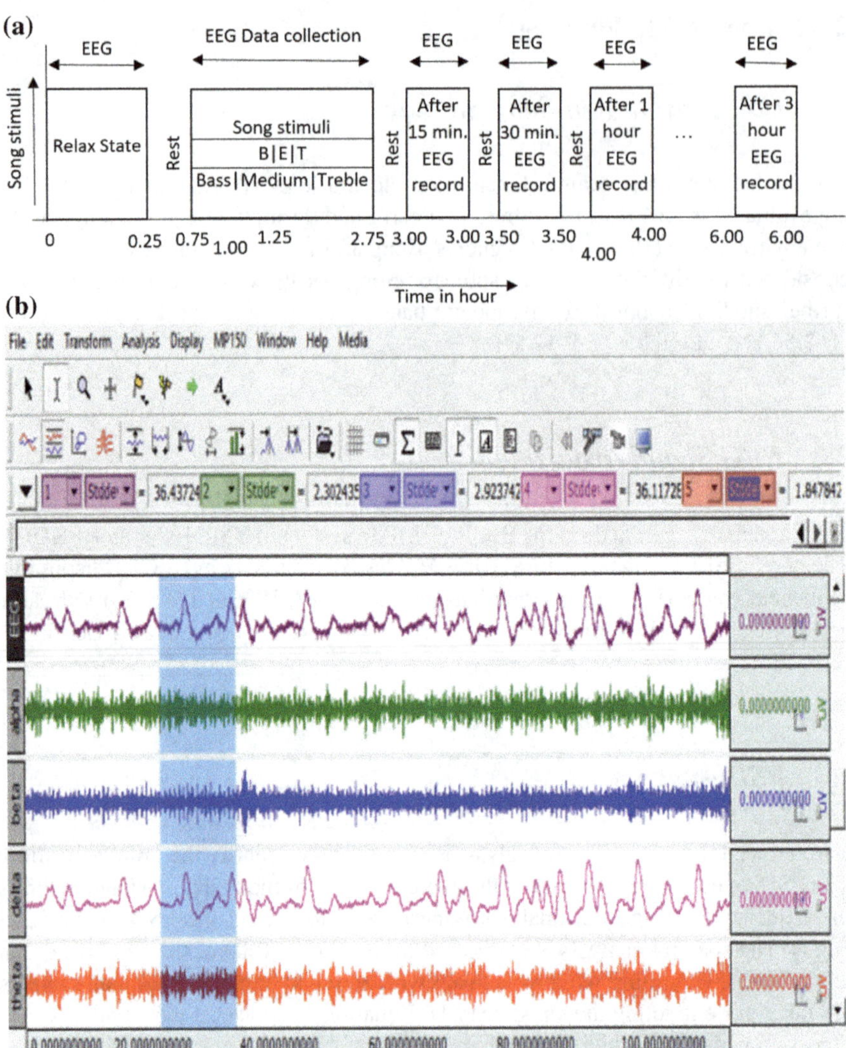

Fig. 2 **a** Data acquisition protocol; **b** raw EEG signal

3 Results Analysis and Discussion

3.1 Stress Detection from PSD

Table 1 exhibits the beta to alpha ratio (BAR) during hearing the song stimuli duration and the percentage of the changes in mental states at different times. When the BAR is greater than 1 it indicates stress but less than 1 indicates more alpha activity and it indicates relax. At relax state BAR is 0.507 and when the percentage change in BAR

Table 1 Stress identification from BAR of human brain for different songs stimuli

Song	Frequency	Relax		After 15 min.		After 30 min.		After 2 h	
		BAR	BAR changes	BAR	BAR changes (%)	BAR	BAR changes (%)	BAR	BAR changes (%)
Bengali	Bass	0.507	0	0.471	−6.92	0.508	0.36	1.402	176.68
	Medium			0.472	−6.79	0.512	1.09	1.604	216.51
	Treble			0.480	−5.25	0.518	2.27	1.661	227.70
English	Bass			0.478	−5.62	0.514	1.58	1.646	224.76
	Medium			0.478	−5.52	0.521	2.90	1.739	243.22
	Treble			0.484	−4.48	0.534	5.37	2.219	337.82
Tamil	Bass			0.510	0.65	0.530	4.58	2.408	375.22
	Medium			0.521	2.78	0.547	8.07	3.028	497.53
	Treble			0.526	3.91	0.549	8.33	3.213	533.87

is negative it indicates more relax but the more positive changes in BAR indicate more stress.

3.2 Brain Relaxation Analysis and Modeling

Brain relaxation is analyzed with the β/α band power ratio. After song stimulation, EEG is collected at different intervals and when BAR reaches nearly at 0.507 then relaxation is obtained. Table 2 shows the brain relaxation analysis with BAR. The BAR is plotted over different time intervals and modeled with least square algorithm. At different frequency levels of different songs, brain relaxation is modeled with fourth-order polynomial as shown in Eq. 3 and the coefficients are shown in Table 3. Figure 3 shows the brain relaxation modeling for Bengali song at bass level.

$$Y_{\text{Relaxation}} = P_0 + P_1 x_{\beta/\alpha}(t) + P_2 x^2_{\beta/\alpha}(t) + P_3 x^3_{\beta/\alpha}(t) + P_4 x^4_{\beta/\alpha}(t) \quad (3)$$

Table 2 Brain relaxation analysis with beta/alpha ratio

Songs	Frequency	After hearing song stimuli (with beta/alpha ratio)								
		Immediately	15 min	30 min	1 h	1.5 h	2 h	2.5 h	3 h	Residual stress in %
Bengali	Bass	1.40	1.10	0.94	0.81	0.74	0.58	0.51	0.504	−0.22
	Medium	1.60	1.18	1.01	0.85	0.79	0.61	0.51	0.509	0.25
	Treble	1.66	1.29	1.01	0.92	0.84	0.64	0.51	0.511	0.48
English	Bass	1.64	1.31	1.12	0.91	0.82	0.67	0.57	0.514	0.78
	Medium	1.73	1.41	1.14	0.92	0.84	0.70	0.56	0.517	1.09
	Treble	2.21	1.78	1.32	1.02	0.89	0.71	0.59	0.518	1.17
Tamil	Bass	2.40	2.01	1.54	1.12	0.89	0.72	0.57	0.523	1.61
	Medium	3.02	2.54	1.62	1.26	0.93	0.74	0.57	0.527	2.05
	Treble	3.21	2.62	1.70	1.34	0.94	0.75	0.62	0.539	3.22

Table 3 Coefficients of the mathematical model along with norm of residual

Songs	Frequency	P_4	P_3	P_2	P_1	P_0	Norm of residuals
Bengali	Bass	0.09	−0.6	1.4	−1.4	1.4	0.0487
	Medium	0.12	−0.83	1.9	−1.9	1.6	0.5684
	Treble	0.14	−0.89	2	−2	1.7	0.0712
English	Bass	0.072	−0.5	1.3	−1.5	1.6	0.0312
	Medium	0.081	−0.58	1.5	−1.8	1.7	0.0357
	Treble	0.09	−0.69	1.9	−2.5	2.2	0.0951
Tamil	Bass	0.048	−0.4	1.3	−2.3	2.4	0.0908
	Medium	0.099	−0.79	2.4	−3.6	3.1	0.2955
	Treble	0.1	−0.83	2.5	−3.8	3.3	0.2714

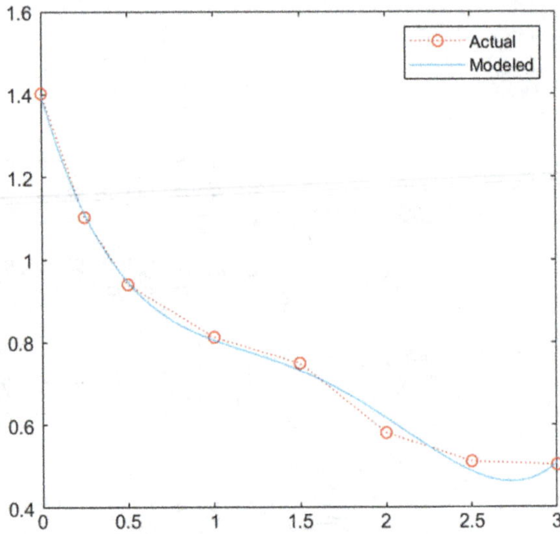

Fig. 3 Brain relaxation modeling for Bengali song at bass level

4 Conclusion

In this paper, music with different languages and frequencies are taken as stimuli to determine mental state of human brain. Experimental findings show that the short duration of music provides relief from stress but when subjects were exposed to music for a long period then the stress is on increasing state. For the increasing time duration, Tamil song for treble frequency shows more stress and Bengali song with bass frequency shows more relax than the others. It can be decided that the native language with lower frequency shows a positive response on human brain. Finally, an approach of brain relaxation modeling is implemented using least square algorithm although after 3 h residual stress is still present for high frequency level.

Acknowledgements This work is supported by the Department of EEE, Khulna University of Engineering and Technology (KUET), Bangladesh. Special thanks are due to the subjects who have given their kind consent for data collection.

References

1. D. McNeil, *Keeping Together in Time: Dance and Drill in Human History*. Harvard University Press (1995)
2. P.C. Terry, C.I. Karageorghis, A.M. Saha, S. D'Auria, Effects of synchronous music on treadmill running among elitetriathletes. J. Sci. Med. Sport **15**, 52–57 (2012)
3. T.H. Holmes, R.H. Rahe, The social readjustment rating scale. J. Psychosom. Res. **11**, 213–218 (1967)
4. T.K. Liu, Y.P. Chen, Z.Y. Hou, C.C. Wang, J.H. Chou, Noninvasive evaluation of mental stress using by a refined rough set technique based on biomedical signals. Artif. Intell. Med. **61**, 97–103 (2014)
5. R. Zheng, S. Yamabe, K. Nakano, Y. Suda, Biosignal analysis to assess mental stress in automatic driving of trucks: palmar perspiration and masseter electromyography. Sensors **15**(3), 5136–5150 (2015)
6. A.A. Saidatul, M.P. Paulraj, S. Yaacob, Y. Mohd Ali, R. Fadzly, M. Fauziah, Spectral density analysis: theta wave as mental stress indicator. J. Commun. Comput. Inf. Sci. **260**, 103–112 (2011)
7. J. Sarnthein, A. vonStein, P. Rappelsberger, H. Petsche, F. Rauscher, G. Shaw, Persistent patterns of brain activity: an EEG coherence study of the positive effect of music on spatial-temporal reasoning. Neurol. Res. **19**(2), 107–116 (1997)
8. H. Petsche, K. Linder, P. Rappelsberger, The EEG: an adequate method to concretize brain processes elicited by music. Music Percept. Winter **6**(2), 133–160 (1988)
9. B.E. Rideout, C.M. Laubach, EEG correlates of enhanced spatial performance following exposure to music. Percept. Mot. Skills **82**, 427–432 (1996)

Simulation of a Robust Method for of License Plate Recognition Using Block Process

Nimmagadda Satyanarayana Murthy

Abstract In the previous decades, the telematics business has expanded in size and extension. The extent of vehicle telematics is by all accounts expanding wherein VIN gives data on where and how individuals are driving, what toll streets they are utilizing and what highlights are adding to their autos. This data could be utilized as an incredible well-being asset. The reason that VIN is prominent for telematics objects is that every vehicle's VIN is novel and that it very well may be decoded to give fundamental data concerning the vehicle. In this paper, vehicle license plate number (VLP) utilizing area-based image examination has been proposed. Character recognition is achieved by OCR and SIS based on block processing to get the corresponding gray image, and then, it is segmented by symbol analysis (SA). The results showed to build up an accurate and automatic license plate acknowledgment framework.

Keywords Gray scale · Vehicle license plate number (VLP) · Thresholding · Image segmentation

1 Introduction

There are divergent applications where programmed license plate acknowledgment can be utilized. The term "complete license plate recognition system" does not have a distinct significance out of the setting of the explicit application. Diverse applications may mean rather extraordinary license plate recognition systems as far as format, equipment and innovation, and notwithstanding for similar applications producers give LPR frameworks with similar usefulness however very unique structure. Presumably the most widely recognized LPR application is stopping and access control. In the extent of these applications, we would already be able to characterize a typical kind of license plate acknowledgment framework with a run of the mill equipment arrangement and framework design. For instance, of a license plate acknowledgment

N. Satyanarayana Murthy (✉)
ECE Department, Velagapudi Ramakrishna Siddhartha Engineering College, Vijayawada, Andhra Pradesh, India
e-mail: nsmmit@gmail.com

H. S. Saini et al. (eds.), *Innovations in Electronics and Communication Engineering*, Lecture Notes in Networks and Systems 107,
https://doi.org/10.1007/978-981-15-3172-9_50

framework, we will present here an extremely straightforward, still run of the mill LPR framework for access control.

It is exceptionally valuable for some traffic management systems. VLP acknowledgment requires some complex undertakings, for example, VLP location, division and acknowledgment. These assignments turn out to be increasingly advanced when managing plate pictures taken in various slanted edges or plate pictures with commotion. Since this issue is generally utilized in real-time systems, it requires exactness as well as quick handling. Most VLP acknowledgment applications reduce the multifaceted nature by setting up some compels on the position and separation from the camera to vehicles and the slanted edges. By that way, the acknowledgment rate of VLP acknowledgment frameworks has been improved significantly.

2 Literature Survey

Anagnostopoulos et al. [1] proposed another calculation that a PC vision and character acknowledgment algorithm for license plate acknowledgment. Jiao et al. proposed [2] another technique for multi-style LP acknowledgment by speaking to the styles with quantitative parameters, i.e., plate turn point, plate line number, character type and arrangement. Chen et al. [3] proposed a novel strategy to perceive license plates in which a division stage finds the license plate inside the picture utilizing its remarkable highlights and the character recognizer separates remarkable highlights of characters and uses layout coordinating administrators to get a hearty arrangement. Gonzalez and Woods [4] recommended that an image pre-preparing approach that incorporates estimation of the histogram, finding the total number of zeniths and smothering immaterial apexes. Finlayson et al. proposed [5] another color invariant picture portrayal dependent on a current dim scale picture upgrade method: histogram evening out.

Ridler and Calvard [6] proposed a procedure for determining a threshold where to section the image that has been illustrated, the iterative methodology talked about gives a straightforward programmed strategy to ideal threshold determination. D. Liu et al. proposed the target capacity of Otsu strategy is proportionate to that of K-implies technique in staggered thresholding [7]. They are both dependent on an equivalent model that limits the inside class change. Gasparini and Schettini [8] proposed a strategy for recognizing and evacuating a color cast from a computerized photograph with no from the earlier information of its semantic substance. Petr Cika et al. proposed the influence of error correction codes on digital watermarking systems which work in the frequency domain [9]. The scheme is based on the well-known watermarking scheme, which uses the discrete wavelet transform (DWT). G. Adorni et al. present a model [10] that depends on both uniform and nonuniform cell automata and is equipped for perceiving license plates for moderate moving vehicles under any ecological conditions.

Auty et al. [11] displayed an invention that gives an object observing framework including camera implies for checking movement and decide a procurement time.

Davies et al. [12] presented a compact stand-alone automatic system. The essential components of the gear are a strobe light source, a covered camcorder and a picture processor. Eikvil and Huseby et al. [13] depicted the advancement of a video-based framework for traffic reconnaissance that works progressively on a standard PC-stage. Martín et al. [14] proposed License plate area depends on numerical morphology and character acknowledgment is actualized utilizing Hausdorff remove. Results are tantamount to those acquired with different strategies.

J.-M. Shyu et al. proposed a contraption and strategy [15] for naturally perceiving the characters on a license plate which is settled on a vehicle. The A sensor/controller gadget is used to identify whether vehicle has achieved a foreordained position for picture inspecting and to discharge a trigger flag when the vehicle has achieved the foreordained position. Deb et al. [16] proposed a technique for vehicle permit plate acknowledgment dependent on sliding concentric windows and counterfeit neural system. This paper incorporates techniques like dark scale transformation, binarization, distinguishing competitor area, sifting shading, tilt amendment, division and acknowledgment.

3 Proposed Work: Algorithm for Character Segmentation

The planned framework is made out of four principle modules, comparing to the four fundamental computational advances, which incorporate the license plate zone module, the pre-preparing module, the acknowledgment module and the investigation of probable license number.

3.1 Pre-processing Algorithms

A license plate acknowledgment framework chiefly comprises three noteworthy parts: license plate discovery, character division and character acknowledgment. Because of the assorted variety of parameters associated with vehicle pictures, the initial step, i.e., license plate location, is the most essential errand among these means. The pre-processing techniques used in this paper are as shown in Fig. 1.

- Image enhancement
- Luminance transformation
- Thresholding
- Image segmentation into regions
- Evaluation of regions
- Optical character recognition
- Comparing recognized license plates with a database.

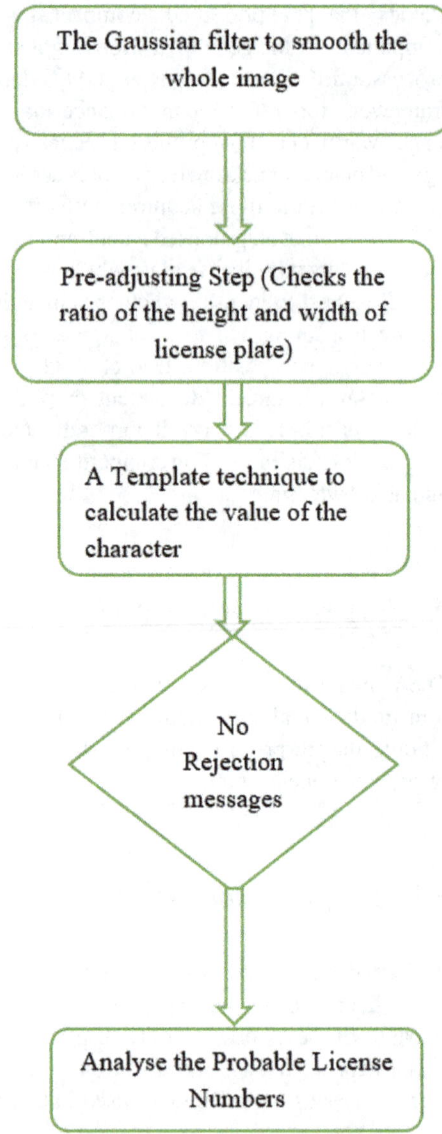

Fig. 1 Flowchart of license plate recognition using block process

3.2 *Luminance Transformation*

The luminance part (dark scale part) is reasonable for the protest recognition in advanced picture. All the standard cameras and camcorders catch the video in the red, green, blue (RGB) shading model. This model is inadmissible for picture preparing. Much of the time, the picture is changed from RGB into *YUV* shading model, where the *Y* is the luminance part (dark scale part) and *U*, *V* are the chrominance part of the first picture. Just the luminance part of the picture is required for the following procedure. To get the Y part from the RGB display, the condition is utilized. $Y = 0.299\text{R} + 0.587\text{G} + 0.114\text{B}$ Chrominance parts are unimportant for next processing.

3.3 Thresholding

Thresholding is one of the main parts used during image segmentation. It is used to transform the gray-scale image to binary image. A threshold level T is set as first before the global thresholding.

3.4 Image Segmentation into Regions

Division alludes to the way toward dividing an advanced picture into numerous districts (sections). Pixels in the locale are comparable concerning some trademark or processed property, on account of license plates to luminance level. The guideline of a straightforward division technique is as per the following:

- Single pixels are red in raster filter.
- Segments containing white pixels are sought.

In the event that the pixel with the asked for shading is discovered, all area pixels are contrasted and one another. On the off chance that where the pixels have a similar force, the protest is recognized. This procedure is rehashed for the entire picture. The fundamental thought is to discover a vehicle license plate. From the rundown of discovered questions, just the articles with rectangular shape are chosen. Every one of these items is tried for square shape and elliptical. The square shape is surrounded over the chose protest in the primary stage. The picture is turned from 0 to 90 degrees in light of the fact that the question inside the picture can be pivoted as well. The base and greatest in x-and y-hub are recognized in each point of turned picture. The base and most extreme in x-and y-pivot are distinguished in each edge of turned picture. The base separation in x-pivot decides the protest width and in y-hub the question stature as appeared in Fig. 2.

A locale is extricated from the picture in the event that if the size equivalents to the explicit angle proportion. Every one of the locales in the chose protest that estimate is littler or more prominent than the predefined image measure is disposed of. On the off chance that the tally of the rest images is between six and nine, the selected object declares the license plate Fig. 3.

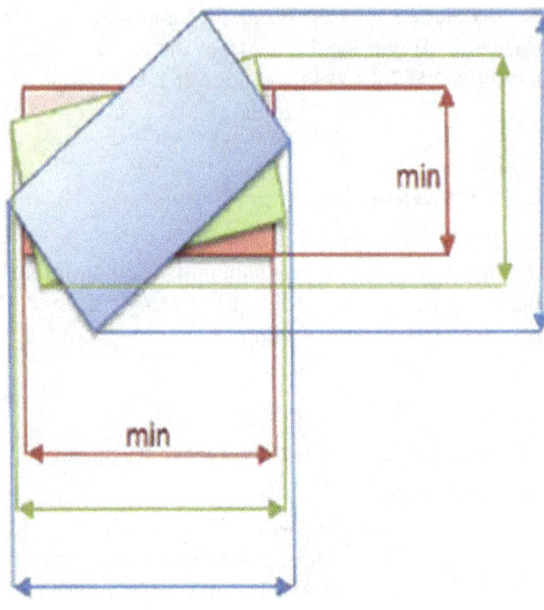

Fig. 2 Rectangle and oblong parameters

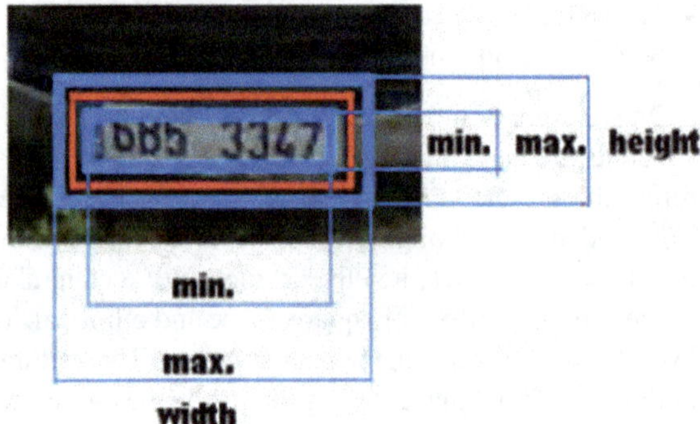

Fig. 3 Region selecting

3.5 *Optical Character Recognition in Vehicle License Plate*

The last advance in vehicle license plate location and acknowledgment is perusing of single characters and numbers. This progression is imperative for instance at the passage to vehicle leave or for the police for stolen autos seek. Single components on license plate must be fragmented and broke down. The investigation is called as optical character recognition (OCR) technique.

We looked over two variations.

- comparison with a predefined display
- analysis of image area.

4 Result Analysis

The input image from a camera, and with this image, we can proceed to other steps of results by applying different algorithms as shown in Fig. 4.

The input image is then contrast stretched so that the image is clearer than the previous one after that luminance transformation is done. The input image is converted from RGB to *YUV* display where *Y* is the luminance part that is dark scale part and *UV* are the chrominance part which is unsatisfactory for picture handling activities as shown in Fig. 5.

After complexity extending and luminance change, the edge location calculation is connected, so the limit of the license plate is effectively perceived. To perform this, angle of the license plate must be required. Since there can be various conditions in different countries. After the boundary is detected, the analysis of symbols in the number plate is done, and if the characters are between 0 and 9, then it is detected as a number plate as shown in Fig. 6.

The last advance in vehicle license plate discovery and acknowledgment is perusing of single characters and numbers. Single components on license plate must be portioned and broke down. Optical character recognition (OCR) is done by two methods.

- Comparison with predefined models
- Analysis of symbol region.

Finally, the result is shown in Fig. 7.

Fig. 4 Input image

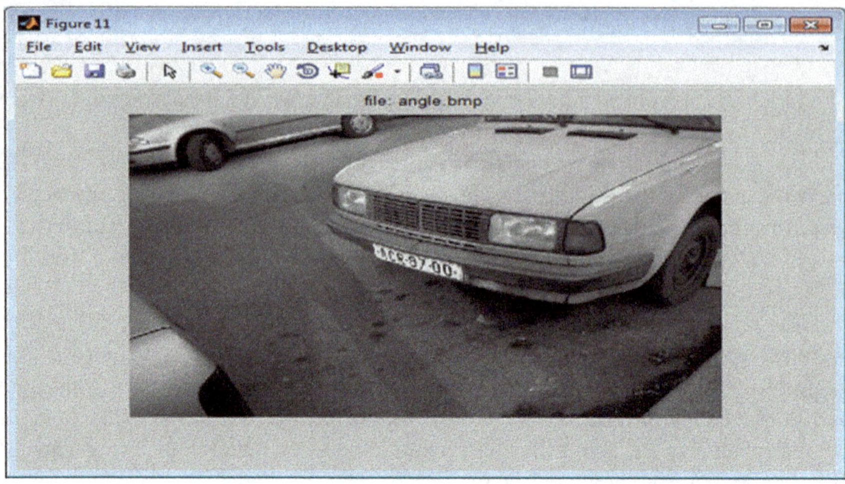

Fig. 5 Picture after difference extending and luminance change

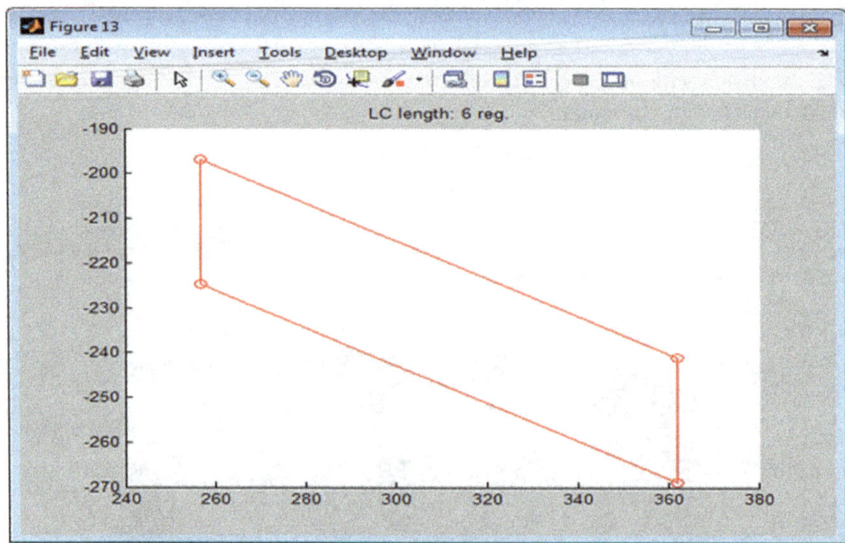

Fig. 6 Angle of the number plate

5 Conclusion

The procedure of vehicle number plate recognition requires a high level of precision when we are chipping away at an extremely bustling street or leaving which may not be conceivable physically as an individual will in general get exhausted because of dreary nature of the activity and they cannot monitor the vehicles when there are

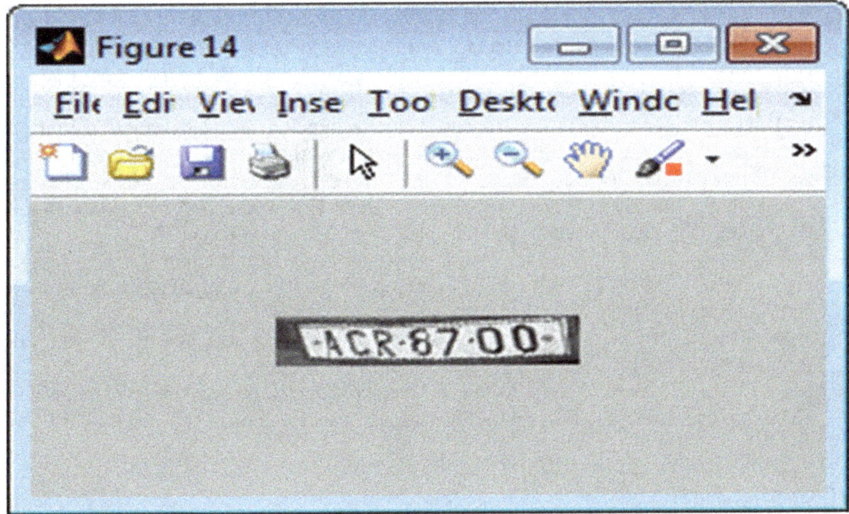

Fig. 7 Final result

various vehicles are going in a brief span. To defeat this issue, numerous endeavors have been made by the scientists over the globe for last numerous years. A comparable exertion has been made in this work to build up a precise and programmed number plate acknowledgment framework.

References

1. C. Anagnostopoulos, I. Anagnostopoulos, V. Loumos, E. Kayafas, A license plate-recognition algorithm for intelligent transportation system applications. IEEE Trans. Intell. Transp. Syst. **7**(3), 377–392 (2006)
2. J. Jiao, Q. Ye, Q. Huang, A configurable method for multi-style license plate recognition. Pattern Recognit. **42**(3), 358–369 (2009)
3. Z.-X. Chen, C.-Y. Liu, F.-L. Chang, G.-Y. Wang, Automatic license plate location and recognition based on feature salience. IEEE Trans. Veh. Technol. **58**(7), 3781–3785 (2009)
4. R.C. Gonzalez, R.E. Woods, *Digital Image Processing*, 3rd edn. (Prentice-Hall Inc, Upper Saddle River, NJ, USA, 2006)
5. G. Finlayson, S. Hordley, G. Schaefer, G.Y. Tian, Illuminant and device invariant colour using histogram equalisation. Pattern Recogn. **38**(2), 179–190 (2005)
6. T.W. Ridler, S. Calvard, Picture thresholding using an iterative selection method. IEEE Trans. Syst. Man Cybern. **8**(8), 630–632 (1978)
7. D. Liu, J. Yu, *Otsu method and k-means*. HIS '09 Ninth International Conference on Hybrid Intelligent Systems, vol. 1, (Aug 2009), pp. 344–349
8. F. Gasparini, R. Schettini, Color balancing of digital photos using simple image statistics. Pattern Recogn. **37**(6), 1201–1217 (2004)
9. P. Cika, *Watermarking scheme based on discrete wavelet transform and error-correction codes*. IWSSIP 2009 16th International Conference on Systems, Signals and Image Processing (June 2009)

10. G. Adorni, F. Bergenti, S. Cagnoni, *Vehicle license plate recognition by means of cellular automata*. IEEE International Conference on Intelligent Vehicles (1998), pp. 689–693
11. G. W. Auty, P. I. Corke, P. A. Dunn, I. B. MacIntyre, D. C. Mills, B. F. Simons, M. J. Jensen, R. L. Knight, D. S. Pierce, P. Balakumar, Vehicle monitoring system, Patent US5809161, 1998
12. P. Davies, N. Emmott, N. Ayland, License Plate Recognition Technology for Toll Violation Enforcement. Proceedings of IEE Colloquium on Image Analysis for Transport Applications, vol. 035, (1990), pp. 7/1–7/5
13. L. Eikvil, R. B Huseby, *Traffic Surveillance in Real-Time Using Hidden Markov Models*. Proceedings 12th Scandinavian Conference on Image Analysis (Bergen, 2001)
14. F. Martín, M. García, J. L. Alba, *New Methods for Automatic Reading of VLP's (Vehicle License Plates)*. IASTED International Conference Signal Processing, Pattern Recognition, and Applications (2002)
15. J.-M. Shyu, I.-M. Chen, T.-Q. Lee, Y.-C. Kung, Method and apparatus for automatically recognizing license plate characters, Patent US4878248, 1989
16. K. Deb, I. Khan, A. Saha, K.H. Jo, An efficient method of vehicle license plate recognition based on sliding concentric windows and artificial neural network. Procedia Technol. **4**, 812–819 (2012)

Recognition of Handwritten Digits with the Help of Deep Learning

Sunita S. Patil, V. Mareeswari, V. Chaitra and Puneet Singh

Abstract Handwritten numerical recognition is becoming the most interesting topic in research area today due to great growth in artificial intelligence and its different learnings and computer visual perception algorithms. This project shows the comparison of digit recognition among machine learning algorithms like support vector machine (SVM), *K*-nearest neighbor (KNN), random forest classifier (RFC) and with deep learning algorithm like multilayer convention neutral network (CNN) using Keras (Keras is a high-level neural networks library written in Python which is simple enough to be used. It works as a protector to low-level and high-level libraries like TensorFlow or Theano) with Theano and Tensorflow (An open source software library which provides high performance numerical computation. Its architecture is flexible in such a way that it is easily deployed across various platforms like (CPUs, GPUs, TPUs), form desktop to clusters of servers to smart handsets devices). Further looking and comparing for the accuracy produced by above-mentioned algorithms, the results appear to be like this: The accuracy of digit recognition is 98.69% in convolutional neural network, 97.90% in support vector learning (SVM), 96.67% using *K*-nearest neighbor (KNN) and 96.89% using random forest classifier (RFC) which clearly shows that convolutional neural network produces more accurate prediction with better results comparatively.

Keywords *K*-nearest neighbor (KNN) · Support vector machine (SVM) · Modified National Institute of Standards and Technology (MNIST)

S. S. Patil (✉) · V. Mareeswari · V. Chaitra · P. Singh
ACS College of Engineering, Bangalore, India
e-mail: sunita.chalageri@gmail.com

V. Mareeswari
e-mail: mareesh.prasanna@gmail.com

V. Chaitra
e-mail: chaitrav568@gmail.com

P. Singh
e-mail: Puneetsingh1427@gmail.com

H. S. Saini et al. (eds.), *Innovations in Electronics and Communication Engineering*, Lecture Notes in Networks and Systems 107,
https://doi.org/10.1007/978-981-15-3172-9_51

1 Introduction

Handwritten digit recognition is the computer ability in predicting the handwritten digits as an input which varies from individual in the sources like papers, emails, bank cheque, envelopes, pictures, written messages, etc. The topic has become an inquisition for many years where few of the experimental environments are postal address, envelopes, bank cheque processing, application forms, etc. Many classification techniques have been developed using machine learning, and this paper uses *K*-nearest neighbors, support vector machine classifier, random forest classifier, etc., though these algorithms show 96% of accuracy and still is not sufficient for the application in legitimate world. Considering an example, if 'y' person is willing to credit an amount to 'x' person whose amount number is 789,654 and you write it has 789,054, the transaction is definitely kept on hold until the proper verification is made, this is because your handwritten digit '6' is understood has '0.' Here comes the role of deep learning for better accurate result. The artificial neural networks are same as human brain reacts and responds for all the changes. Here, it plays a major role in image processing.

2 Related Work

DI NUOVO, Alessandro (2017): The given paper deals with the study of human behaviors for mathematical gestures and uses them to create a model, which would eventually recognize digits out of the scenario. This model has been achieved by integrating three main components: first one for processing real-life data, another to control the finger gestures and one to build a relationship between the digits and its respective hand gestures as a network [1]. This model has huge potential in robotics because that is how human babies develop initial skills, using their tiny fingers and legs. This is the first sign of intelligence and learning which can be used in robotics too. Various studies show that behavioral gestures are really important in learning mathematics [2]. The given paper uses supervised deep learning to investigate and train the dataset [3]. This is a much generalized approach in order to train robots in cheap manner [4].

2018, S M Shamim, Mohammad Badrul Alam Miah, Angona Saker, Masud Rana and Abdullah Al Jobair: Recognizing digits which are written by humans can vary to a very high extent, the problem associated with this can be found in mailing systems, bank cheques, etc. This paper deals with a solution to such problems by introducing machine learning algorithms like random forest classifier, support vector machines, J48, etc. The highest accuracy reached by this paper is just more than 90% which has already been surpassed by supervised deep learning algorithm whose accuracy ranges up to 98% using TensorFlow. There has been immense work in attempt to research field of recognizing digits using data mining and machine learning [5]. The research takes into study, a variety of techniques like feature extraction based on structures and pixels [6]. The ideal algorithm will use the dataset in the form of training and testing dataset by segregating the dataset in the parent phase. As an output to this paper, perceptron using multiple layers gives the highest accuracy of all machine learning algorithms.

Saeed AL-Mansoori Int. Journal of Engineering Research and Applications: The given paper tends to use neural networks over other algorithms to accurately predict the particular mathematical digit individually. It also uses the predefined dataset made by Americans, Modified National Institute for Standards and Technology (MNIST). MNIST comprises 50,000 training and 10,000 testing or validation dataset. The training of this neural network algorithm is achieved with back propagation technique in gradient descent along with testing algorithms. After this, comparison of results is done with specific attributes. The present system gives accuracy beyond 99%. The image input is achieved using off-line tools like camera or scanner [7].

3 Database Modeling and Structure

A. MNIST dataset

MNIST dataset is the subset of the large dataset called NIST dataset. This dataset consists of 70,000 handwritten digits, in these 70,000 datasets, 60,000 data is used to train the machine along with the label, and rest 10,000 data is used for testing the machine for its evaluation. MNIST dataset array is represented in the form of 28*28 value image along with their labels.

B. Structure

Prior-processing: A multiple function conducted on given picture scanned, which improves the picture interpreting efficiently which is relevant to deterioration. It plays an important role in capturing the interesting part in the image scanned. Usually filter in noise, lubricating and equalizing is conducted in each step. The definition for pre-processing can also be a dense illustration of the interested part captured. Binarization is used to commute a gray-scale picture into ones and zeros pictures.

Deterioration: In deterioration stages, a picture of series of digits is deteriorating into small images of single alphabets. The prior-processed provided picture is segregated into separate digits, where the number is assigned to each digit using labeling process. This process provides the count of digits in the image. Each individual digit is equally resized into pixels.

Equalization: The extracted digit needs to be normalized to the quantity of the digits because of the large variations in the sizes of each digit, method to normalization in the size is necessary.

Extrication of feature: Extrication of feature is considered as interesting slice in the image. It includes information of shape, motion, color, the texture of the handwritten digit scanned image. It extracts the important and meaningful information from

Fig. 1 System architecture

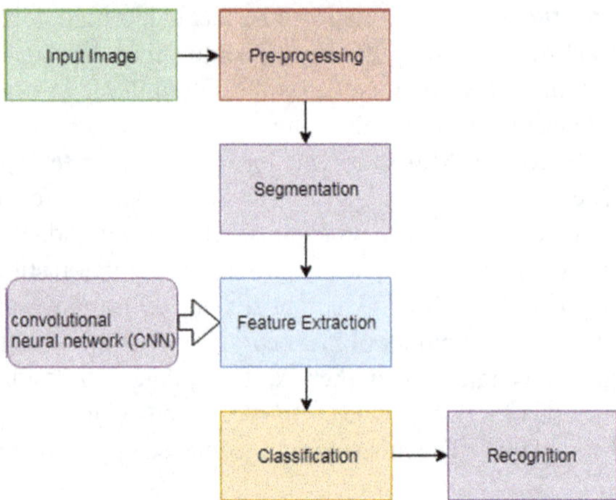

image. Feature extracted image contains less information when compared to the original image which helps in the storage purpose.

Classification: Classification stage performance just like animals recognizing the objects and classifying what kind of object is that, which takes minimal efforts. Artificial neural network works same as the animal and human brain which first classifies based on shapes and then based on features of the object and finally comparing the images extracted and previously classified to the original image which provides the accurate and efficient result.

The system architecture which includes all the above functionalities is given in Fig. 1.

4 Machine Learning Classification Algorithm

To compare the accuracy obtained by various algorithm, the below-mentioned algorithm is used.

1. *K*-nearest neighbors (KNN)
2. Supervised vector machine (SVM)
3. Random forest classifier (RFC).

A. *K*-Nearest Neighbor Classification (KNN)

K-nearest neighbor (KNN) is the simple and good classifier among the other classifier which uses Euclidian or Hamming distance for the nearest neighbor prediction [8]. The classifier is provided with the input in which the trained classifier searches for its desired number of the nearest neighbor in predicting which classification it belongs to.

B. **Supervised Vector Machine Classification (SVM)**

SVM—Supervised vector machine is one of the important classifiers. It is used for the supervised learning for classification and regression [9]. This classifier is trained with some datasets so that whenever the input is given to the classifier, the classifier classifies the input based on the features recognized during the training period and then categories it to one of the clusters.

C. **Random Forest Classifier Classification (RFC)**

RFC is a concept used for both regression and classification. In this algorithm, the huge number decision tree results are formed, i.e., for each input, the new matrix is formed, finally, the overall matrix for n inputs is collected, the majority of the similar results are collected and classified to one cluster, another similar result matrixes are classified to another cluster and so on.

To find out the average prediction once the machine is trained by the individual decision trees, the below-mentioned formula is used:

5 Classification Utilizing Deep Learning

A Convolutional Neural Network

[CNN] A convolutional neural network is one among the forward feeding artificial neural network. This concept is inspired by the biological neural system [10].

Convolutional neural network is same as the organization of animal neural system in which it consists of neurons with the weights and the biases. Figure 2 shows the basic layout of neural network which are input layer, hidden layer (which can be multiple) and the output layer. Each neuron is fed with the input to the next neuron, i.e., the output of the previous neuron is served as the input to the next neuron, by this feature of biological concept, the highest accurate and efficient result is obtained because of the multiple layer between the input and the output layer in which the most refined features are extracted to obtain the correct output.

Input Layer: The input image is turned into its corresponding pixel values.

Hidden layer: NumPy array gets conceded into hidden layer to specify the number of filters (3*3). Each filter generates receptive fields. Convolutional produces a feature map that characterizes how pixel values are improved.

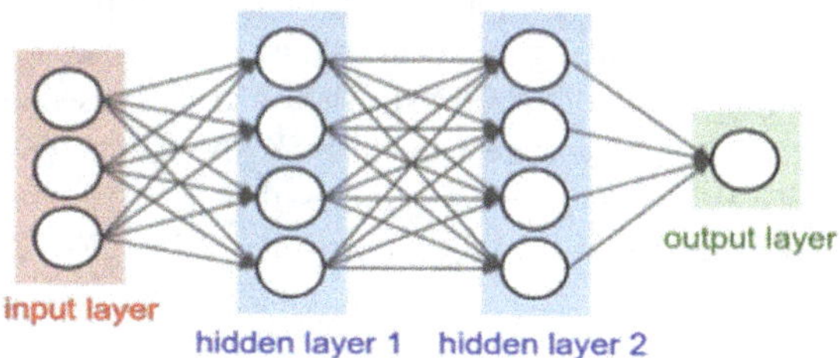

Fig. 2 Convolutional neural network basic layout

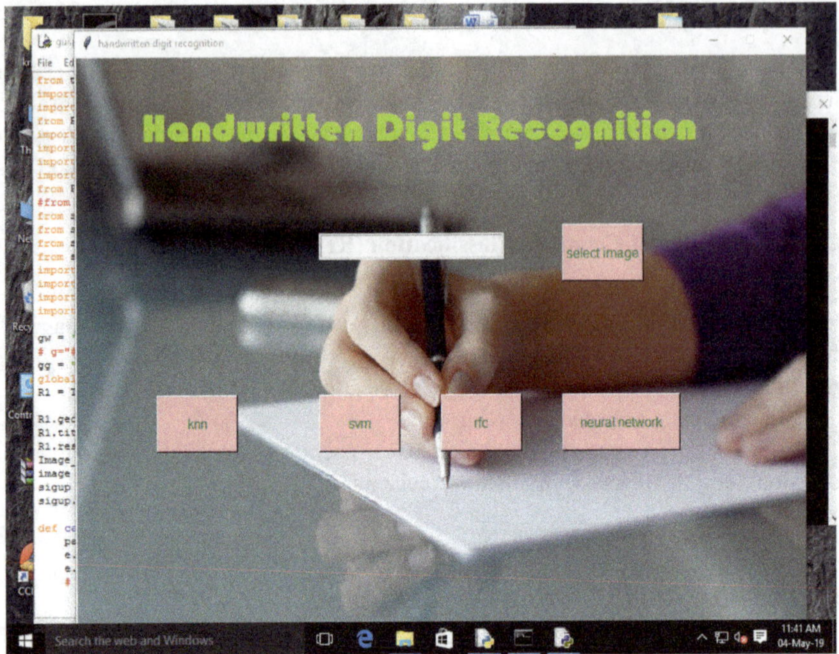

Fig. 3 GUI screen for selecting the algorithm performing for the image given as an input

Output layer: The output represents itself as a probability for each digit class. The model is able to demonstrate the feature conformation of the digit in the image

6 Results

See Figs. (3, 4, 5 and 6).

7 Conclusion

For the real-world application, the accurate number identification is a great thirst for knowledge. The figure given below gives the accurate result compared in machine learning methods like KNN, SVM and RFC used. The result shows that the digit recognition in K-nearest neighbor (KNN) is 96.67%, support vector machine (SVM) is 97.91%, and random forest classifier (RFC) is 96.86. And comparatively, we see that the convolutional neural network (CNN) gives the highest accuracy of 98.72%.

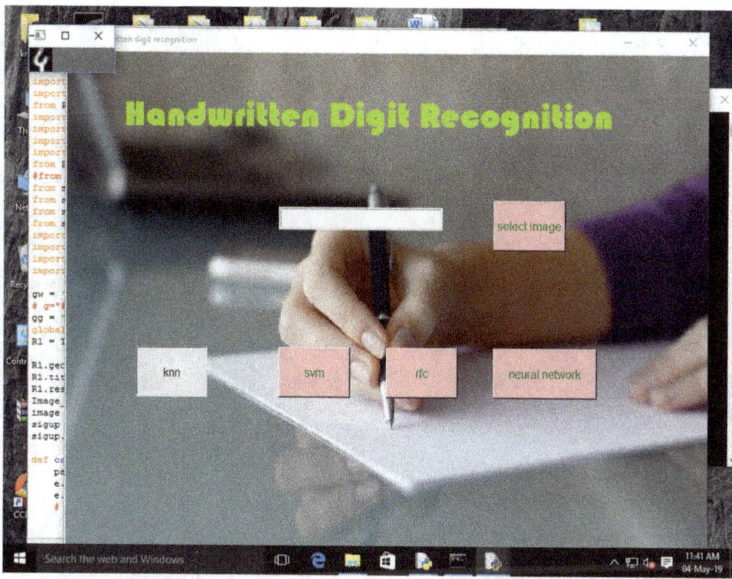

Fig. 4 Image of digit 4 on the left top corner is the random pick for which KNN algorithm has to predict

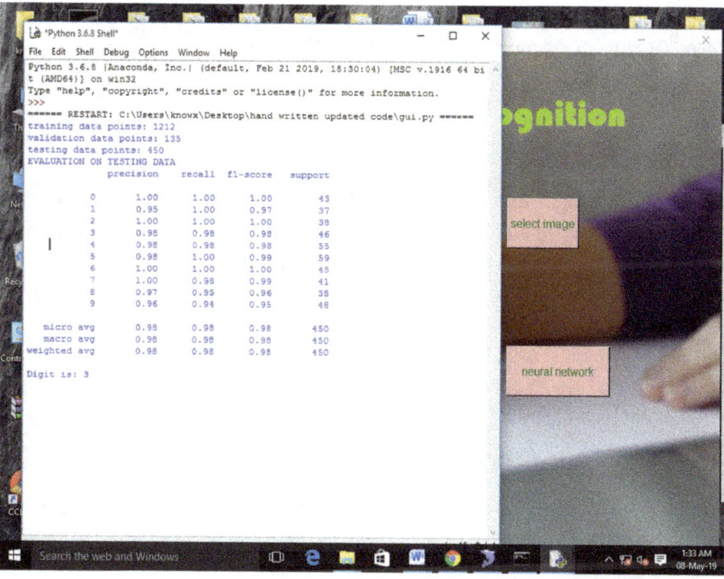

Fig. 5 Number prediction for the random digit input

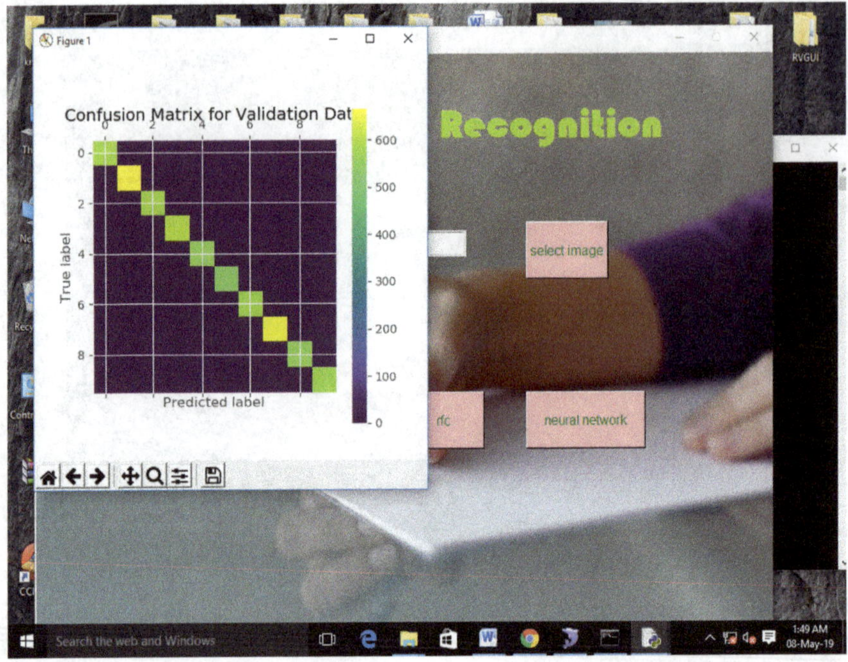

Fig. 6 Confusion matrix against true label and predicted label which gives the accurate value of prediction made through the intensity of the colors in the scale

8 Future Scope

Using convolutional neural network for future studies provides the best results, and it is currently applied for handwritten digit recognition. There are many more areas like recognizing language of the text, recognition of characters, objects, online digits and character and in image sectionalisation for which CNN can be applied for better results and accuracy.

References

1. R. Pfeifer, J. Bongard, S. Grand, How the Body Shapes the Way we Think: A New View of Intelligence, (MIT press, 2007)
2. S. Goldin-Meadow, S. C. Levine, S. Jacobs, *Gesture's role in learning arithmetic*, ed. by L. D. Edwards, F. Ferrara, D. Moore-Russo, (Information Age Publishing, 2014)
3. N. Tschentscher, O. Hauk, M.H. Fischer, F. Pulvermüller, You can count on the motor cortex: Finger counting habits modulate motor cortex activation evoked by numbers. Neuroimage **59**(4), 3139–3148 (2012)
4. A. Droniou, S. Ivaldi, O. Sigaud, Deep unsupervised network for multimodal perception, representation and classification. Rob. Auton. Syst. **71**, 83–98 (2015)
5. W. Kloesgen, J. Zytkow, Handbook of Knowledge Discovery and Data Mining

6. R. Tokas, A. Bhadu, A comparative analysis of feature extraction techniques for handwritten character recognition. Int. J. Adv. Technol. Eng. Res. **2**(4), 215–219 (2012)
7. R. Plamondon, S.N. Srihari, On-line and off- line handwritten character recognition: A comprehensive survey. IEEE Trans. Pattern Anal. Mach. Intell. **22**(1), 63–84 (2000)
8. C-L. Liu, K. Nakashima, H. Sako, Handwritten Digit Recognition Using State-of-the-Art Techniques, 2016
9. M. M. A. Ghosh, A. Y. Maghari, A Comparative Study on Handwriting Digit Recognition Using Neural Networks, 2016
10. C. Ma, H. Zhang, Effective Handwritten Digit Recognition Based on Multi-Feature Extraction and Deep Analysis, 2015

Image-Based Localization System

Omsri Kumar Aeddula and Irina Gertsovich

Abstract The position of a vehicle is essential for navigation of the vehicle along the desired path without a human interference. A good positioning system should have both good positioning accuracy and reliability. Global Positioning System (GPS) employed for navigation in a vehicle may lose significant power due to signal attenuation caused by construction buildings or other obstacles. In this paper, a novel real-time indoor positioning system using a static camera is presented. The proposed positioning system exploits gradient information evaluated on the camera video stream to recognize the contours of the vehicle. Subsequently, the mass center of the vehicle contour is used for simultaneous localization of the vehicle. This solution minimizes the design and computational complexity of the positioning system. The experimental evaluation of the proposed approach has demonstrated the positioned accuracy of 92.26%.

Keywords Automatic indoor positioning system · Center of mass · Gradient positioning system · SLAM

1 Introduction

Due to increased demand in autonomous systems for applications in closed spaces, the deployed indoor positioning systems have gained researcher's attention and wide range research and development in sensor networks and the Internet of Things [1]. The demand in use of location services and ability to develop intelligent systems led to the design of various positioning systems using but not limited to radar, laser, Global Positioning System (GPS), and computer vision [2].

O. K. Aeddula (✉) · I. Gertsovich
Blekinge Institute of Technology, Karlskrona, Sweden
e-mail: omsri.kumar.aeddula@bth.se

I. Gertsovich
e-mail: irina.gertsovich@bth.se

H. S. Saini et al. (eds.), *Innovations in Electronics and Communication Engineering*, Lecture Notes in Networks and Systems 107,
https://doi.org/10.1007/978-981-15-3172-9_52

GPS employs triangulation process to determine the physical location and can be affected negatively by the surrounding buildings [3]. Infrared and ultrasound-based methods estimate the position of the vehicle under the line-of-sight and radio-frequency signals are prone to reflection, diffraction, and multi-path fading [4]. Ultrasound tracking was an accurate positioning system developed by AT&T Cambridge. Users and the vehicle were tagged with ultrasonic identifiers named as "bats," used for position estimation. The system required a large number of receivers across the ceiling [5].

Simultaneous localization and mapping (SLAM) methods estimate the position of the vehicle and map it to the unknown environment to keep track of the vehicle location [6]. The review of state-of-the-art designs of SLAM positioning systems describes the systems that map the moving object location based on landmarks estimated by the moving object in its environment [2]. The environment landmarks are represented by features, which require feature extraction and matching procedures to be implemented in a SLAM system.

In this work, we propose an image-based localization system for estimation of the position of a moving vehicle in a static environment and subsequently tracking the path of the vehicle. Instead of extracting and mapping the landmarks in the environment, we use the vehicle itself as the detected feature for a vehicle localization. This paper presents a method of image gradient-based vehicle positioning system in real-time scenario using a static camera with a single lens. The work is carried out under an assumption of the constant illumination, shadow-free environment, and a static background. Such environmental conditions can be partly found in the road tunnels or indoor sites. The described system requires reduced hardware and computation complexity as it is intended to operate in the constrained environment.

2 Materials and Methods

The theory of the computer vision techniques that are the parts of the proposed gradient-based image positioning system is shortly presented in the following subsections. The performance evaluation methodology, including experimental setup and evaluation parameter, is described in the end of this section. The overview of the gradient-based positioning system is shown in Fig. 1. As a preprocessing step, a static background is discarded in each acquired video frame, thus leaving the foreground information, generated by a vehicle, presented in the frame.

2.1 Image Gradient

To detect edge pixels of the object in the image, an image gradient technique can be used to discover the directional change in the pixel intensity values. In the first step of the proposed method, the gradient magnitude in the image I is approximated in this work as

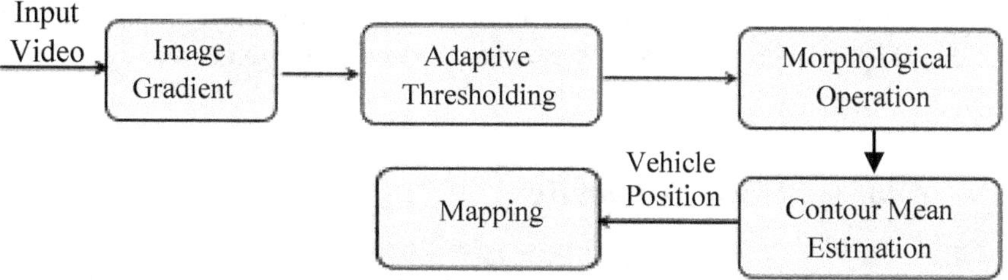

Fig. 1 Overview of gradient positioning system

$$\|\nabla\|G = \sqrt{(G_x^2 + G_y^2)}, \tag{1}$$

where horizontal G_x and vertical G_y gradients are computed by convolving I with a Sobel filter in the corresponding direction, respectively [7].

The pixels processed by the image gradient estimation may include the pixels that are not related to the edge of the vehicle in the image. Such erroneous pixels can possibly reduce the precision of the vehicle location. As a median filter is highly effective in removing salt and pepper noise from the video frames, it was used in the proposed system to remove erroneous pixels from the edge information [8].

2.2 Adaptive Thresholding

To remove the uncertainties in the video frames to further simplify vehicle contour estimation in the image, the edge information evaluated in the previous step is binarized using an adaptive thresholding [8]. Adaptive thresholding uses a different threshold value for different regions of the video frames, where the threshold level, denoted by t is determined by the mean of the neighborhood area. An adaptive threshold operation is given according to

$$\theta(I, t) = \begin{cases} 1 \text{ if } I(x, y) \geq t \\ 0 \text{ otherwise}, \end{cases} \tag{2}$$

where $I(x, y)$ is a video frames pixel intensity [7].

2.3 Morphological Operations

Morphological operations can be performed on the images to remove the uncertainties due to, e.g., noise pixels [7]. Dilation, erosion, and closing operations with a rectangular structuring element were performed on the binary images to close the holes or to remove noise pixels. These procedures allow creating a continuous shape of the vehicle, which is useful for subsequent contour detection in this work.

2.4 Contour Mean Estimation

The contour of the vehicle can be extracted by tracing the boundary of the vehicle and joining all the connected points along the vehicle outline. A contour approximation method described in [9] was employed by proposed system to extract the boundary points along the edge of the vehicle. This procedure became a major contributor to the accuracy of the estimated vehicle position in this work. The center of the area, limited by contour, can be established as an arithmetic mean, which refers to a central value of the discrete points [7]. For the contour data set, mean value of x- and y-coordinates is computed according to [7]

$$[x', y'] = \left[\frac{1}{p}\sum_{i=0}^{p-1} x_i, \frac{1}{p}\sum_{i=0}^{p-1} y_i\right] \forall (x_i, y_i) \in c, \tag{3}$$

where x' and y' are the x-coordinate and y-coordinate of the contour mean, c are the contour pixels with x- and y-coordinates, and p is the number of x- or y-coordinates of the pixels in the contours. Estimated mean coordinates of the contour points provide the information about the vehicle position in an image.

2.5 Experimental Setup

The experimental work was carried out at a demo site designed by Volvo CE with a vehicle prototype at Blekinge Institute of Technology. The demo site is shown in Fig. 2a, where the center object is the vehicle prototype that moves from one end to another end of the test site. To evaluate the performance of the presented methodology, a marker line was drawn on the site along the desired path of the vehicle, see Fig. 2a. A Microsoft Lifecam studio camera, shown in Fig. 2b was set up at a height 1.8 m above the ground, to identify the vehicle and track the motion in real time.

2.6 Performance Evaluation

An expert in driving the vehicle was instructed to drive a vehicle on the marked path drawn on the site. The image of the site with the marked line was used to evaluate the performance of the proposed system. The position estimated by the

(a) A demo site for evaluation of the proposed indoor positioning system.

(b) Camera used for vehicle tracking.

Fig. 2 Equipment used for testing

evaluated system was considered to be correct if the estimated position belonged to the marked line. The performance accuracy in this work shows the percentage of correctly estimated points R in the total number of estimated points T. It is calculated as $\text{Accuracy} = \frac{R}{T} \times 100\%$.

3 Results and Discussions

The aim of this work is to evaluate the suitability of the proposed gradient-based image positioning system to detect a position of a vehicle in real-life scenario. Figure (3a–c) show the results of the image gradient magnitude extraction, image adaptive thresholding, and vehicle contour extraction, respectively. Figure 3d shows the vehicle movement in real-time scenario, where the positional coordinates of the vehicle were mapped to the environment to trace the vehicle movement by a series of black spots.

Performance evaluation of the proposed positioning system in several trials allowed estimating the average accuracy of 92.26%, thus indicating that the vehicle can be localized with a great degree of accuracy.

In an attempt to relate the performance of the proposed system to the current state of the art, the performances of localization systems, found in the literature, have been studied. The presented performance methodologies differ between themselves which made the comparison of the performance results between different systems unfeasible.

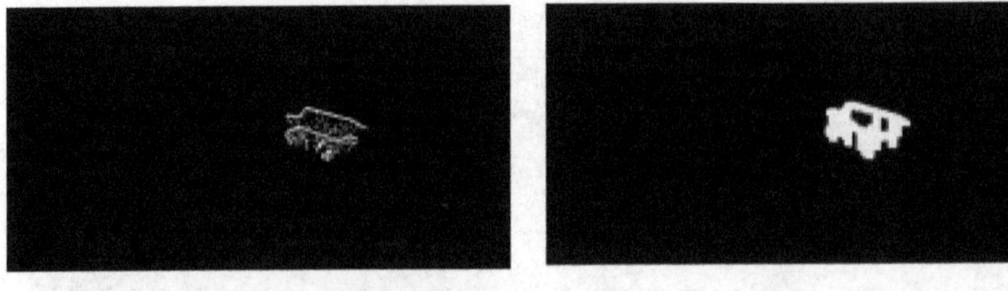

(a) Gradient magnitude image.

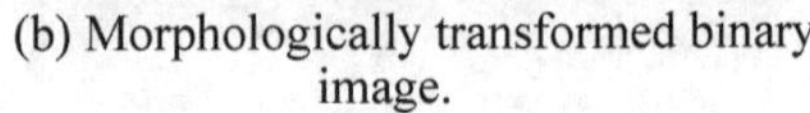

(b) Morphologically transformed binary image.

(c) Vehicle contour image.

(d) Estimated position of a vehicle in a frame.

Fig. 3 Generated results at various stages

4 Conclusion

The presented method had successfully located the position of the vehicle in the indoor environment using a static camera. An image gradient-based positioning system was developed to trace the path of a vehicle using the located position. The experimental work was carried out in a real-time scenario to test the performance of the positioning system on a specially designed Volvo CE mini-site. The system had achieved the positional average accuracy of 92.26% and found to be reliable under the assumptions of shadow-free environment, static background, and usage of single vehicle.

As a future work, two directions are considered. One direction is to improve the proposed system performance under relaxed constraints regarding the vehicle environment. And secondly, we consider developing the performance evaluation framework where different localization system can be objectively and consistently compared.

Acknowledgements This work has been supported by Product Development and Research Laboratory, Karlskrona, Sweden and Volvo CE, Eskilstuna, Sweden.

References

1. B. Lin, Z. Ghassemlooy, C. Lin, X. Tang, Y. Li, S. Zhang, An indoor visible light positioning system based on optical camera communications. IEEE Photonics Technol. Lett. **29**(7), 579–582 (2017). https://doi.org/10.1109/LPT.2017.2669079
2. G. Kim, J. Kim, K. Hong, Vision-based simultaneous localization and mapping with two cameras. 1671–1676 (2005). https://doi.org/10.1109/IROS.2005.1545496
3. A.H. Lashkari, B. Parhizkar, M.N.A. Ngan, WIFI-based indoor positioning system. 76–78 (2010). https://doi.org/10.1109/ICCNT.2010.33
4. X. Ye, WiFiPoz, an Accurate Indoor Positioning System, n.d., 84
5. M. Hazas, A. Hopper, Broadband ultrasonic location systems for improved indoor positioning. IEEE Trans. Mob. Comput. **5**(5), 536–547 (2006). https://doi.org/10.1109/TMC.2006.57
6. A.R. Khairuddin, M.S. Talib, H. Haron, Review on Simultaneous Localization and Mapping (SLAM). 2015 IEEE International Conference on Control System, Computing and Engineering (ICCSCE). 85–90 (2015). https://doi.org/10.1109/ICCSCE.2015.7482163
7. R.C. Gonzalez, R.E. Woods, digital image processing (Pearson, New York, NY, 2018)
8. R. Szeliski, Computer Vision: Algorithms and Applications, n.d., 979
9. OpenCV: Contours: Getting Started [Online]. https://docs.opencv.org/3.2.0/d4/d73/tutorial_py_contours_begin.html. Accessed 21 June 2018

NN_Sparsity_Based Model for Semiautomatic Road Extraction from High-Resolution Satellite Images Using Adaptive Multifeature

Sreevani Srungarapu and S. Nagaraja Rao

Abstract In this exploration, a technique is proposed to make additionally refresh guides in urban–rural zone utilizing high-goals satellite pictures. "Street cover" is characterized in this exploration as a veil of street pixels, which are segregated from others utilizing business remote detecting programming MATLAB. "Street seed" is characterized in this examination as a guiding point, demonstrating that a street is going over the topic beside the course. Street image sources are separated commencing edge pixels. Street line training is led in a self-loader route through melding together street veil and street seeds. Trials are led utilizing road pictures of adjacent Delhi city, in, with India a variable goal of 1 m, along with four groups, for example red, green, blue, as well as close infrared. Here with the assistance of NN-sparsity model increase the completeness of the image. Test results demonstrate that the strategy is legitimate in extricating fundamental streets in high thick structure territory and all streets in farmland effectively.

Keywords NN-sparsity (neural network sparsity model) · High-resolution · Multifeature

1 Introduction

Suppositions on a geometric-stochastic street exemplary are unmistakably recorded as pursues.

(1) Road breadth difference is little and also street thickness alteration is probably going to exist moderate.
(2) Road course variations present are probably going to remain moderate.
(3) Street neighbourhood normal dim dimension is probably going to differ just gradually.
(4) Grey dimension variety among street and foundation is probably going to be huge.

S. Srungarapu (✉) · S. Nagaraja Rao
E.C.E Department, G. Pulla Reddy Engineering College (Autonomous), Kurnool, A.P, India
e-mail: srivani997@gmail.com

H. S. Saini et al. (eds.), *Innovations in Electronics and Communication Engineering*, Lecture Notes in Networks and Systems 107,
https://doi.org/10.1007/978-981-15-3172-9_53

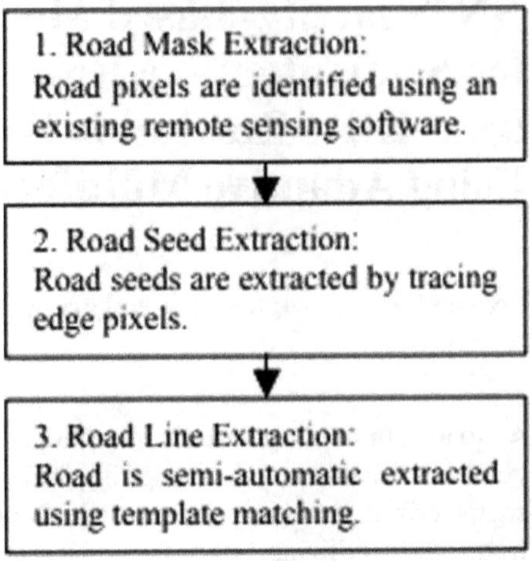

Fig. 1 Road extraction flow

1. Road Mask Extraction:
Road pixels are identified using an existing remote sensing software.

2. Road Seed Extraction:
Road seeds are extracted by tracing edge pixels.

3. Road Line Extraction:
Road is semi-automatic extracted using template matching.

(5) Roads are probably not going to be short.

Anyway, they are not valid in every case, as street pictures differ a ton with ground goals, street type, and thickness of encompassing items, etc. A particular street model just as a street removal technique is essential on behalf of extricating street outlines utilizing high-goals satellite picture, for example street and HR pictures, where a street model is desirable over have as less yet conventional suppositions as possible [1, 2].

In this work, we propose a self-loader technique for street withdrawal starting urban/rural scene utilizing towering-goals satellite pictures, which contain a pounded goal of around 1 m, also four groups, for example red, green, blue, in addition to close ultraviolet. The technique comprises three stages as shown in Fig. 1. A "street veil" is characterized in this examination as a cover of street pixels, which are produced through arranging street pixels of a multi-range cable picture utilizing business accessible detecting software [3, 4]. A "Street Seed" is characterized in this exploration as a indicator point, demonstrating that a street is going concluded the point beside the course. Street nodes are located extricated by following edge pixels, as a extended edge mark through just a moderate alter of course recommend a street or waterway going finished. Street streaks are presented extricated in a self-loader method, where assumed a beginning stage, a street stripe is followed by iteratively coordinating a street layout through mutually street veil in addition to street seeds, regulator focuses demonstrating the correct headings are relegated through administrator if the street line became gone.

2 Existing Work

In the accompanying work, we address each progression in detail. Trials are led utilizing a street picture close by Delhi city, India. Exploratory outcomes and discourses are given accordingly, where street extraction and guide age are contemplated in detail at both thick structure territory and field.

A pixel-based street cover is separated utilizing business remote detecting programming MATLAB. Given the preparation estimations of undergrowth, soil, construction (solid), liquid, prompt expressway, principle street as well as little street, picture pixels of a multi-range satellite broadcasting picture are arranged hooked on various gatherings utilizing greatest probability technique. As the multi-range D-SAT picture utilized in this exploration has IV groups, for example red, green, also blue in addition to close infrared, characterization outcome is significantly additional dependable than individuals on three groups, for example red, green as well as blue, or fewer. Picture pixels named direct roadway, fundamental street and little street are situated misused in this exploration to create a "street mask" [5, 6]. A street mask is a binary image, where white pixel suggest that a street like object is most probably over there. The street cover is abused in street line mining [7–9] (Fig. 2).

Edge pixels extricated by semantic channel reflect not just the limits of streets, structures, streams, or shadows, yet in addition some other neighbourhood fast changes of photometric attributes. Albeit particularly in focal urban, authority pixels of structures, vegetation in addition to shadows are located blended through that of streets, so street limits be present not as pure as to be extricated, a extended superiority stripe through just a moderate alter in course prepare emphatically recommend the limit of a street otherwise a waterway. In this research as street seed (see Fig. 3), and

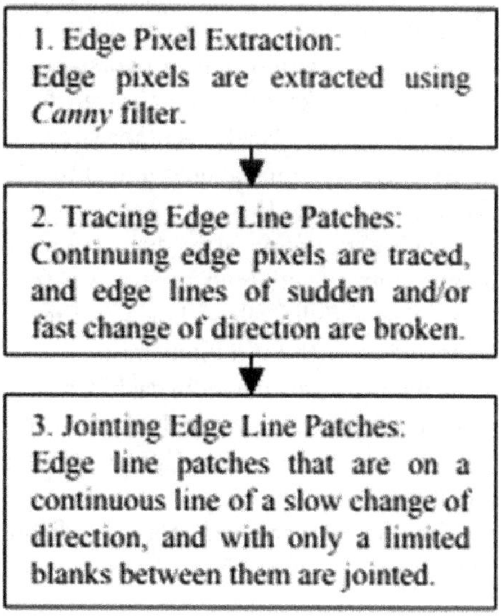

Fig. 2 Flow of extracting road seed

1. Edge Pixel Extraction:
Edge pixels are extracted using *Canny* filter.

2. Tracing Edge Line Patches:
Continuing edge pixels are traced, and edge lines of sudden and/or fast change of direction are broken.

3. Jointing Edge Line Patches:
Edge line patches that are on a continuous line of a slow change of direction, and with only a limited blanks between them are jointed.

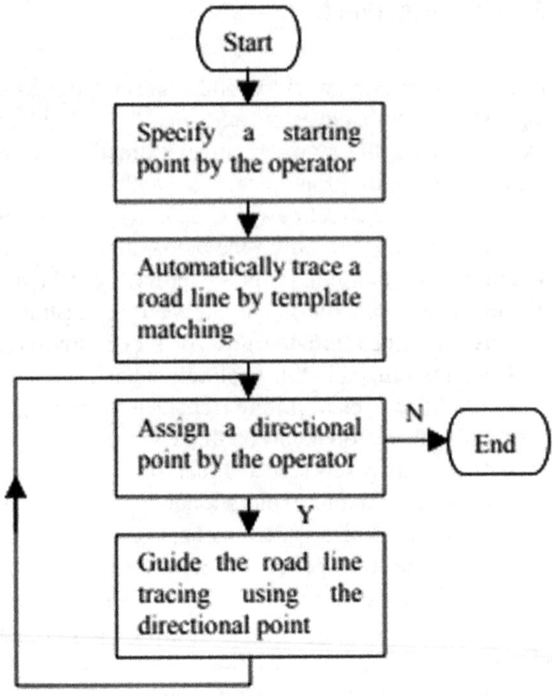

Fig. 3 Semi-automatic road line extractions

exploited as the complimentary information of "street veil" in street line extraction [10, 11].

Applying semantic channel on satellite broadcasting picture consequences in a paired picture of control pixels, where boundaries are diminished to a width of solitary pixel. Withdrawal of lengthy superiority positions commencing the twofold picture of edge pixels is directed in two stages, following also jointing superiority stripe patches. Street kernels are produced for every edge pixels on the superiority line covers, where two headings of the street seed, showing two different conducts of the street track, are situated determined by utilizing the extraneous line of the edge line fix at the edge pixel.

In this exploration, street lines are removed in a self-loader path by coordinating a street format with both street cover (M) and street seeds-(S). The progression of street line withdrawal is shown in Fig. 2. Towards the start, a beginning stage is indicated through the administrator. A street line is extended from the starting point in an iterative way, where in each iteration, a quadrilateral street layout (T), breadth (B) and distance (d) is rotated at one of the terminal point o the street line and matched with the integrated image of street mask (street stripe) and street seed to find the next street line point (Fig. 3).

The method breaks while coordinating outcome decreases to a dimension lower than a given limit. Now, the administrator looks at whether the street line is right or raced to a misguided course or misfortune its way in a nearby greatest. A control point is doled out at whatever point fundamental by the administrator to direct the street line extraction. In this examination, coordinating expense of the street format at a picture point p with a turn edge α is characterized as pursues [12] (Figs. 4 and 5).

$$f_T(\alpha, p) = \sum_{(i,j)\subset T} \left((i', j') \in M\right) | \left((i', j') \in S\right)$$

$$\begin{pmatrix} i' \\ j' \end{pmatrix} = \begin{pmatrix} \cos\,\alpha & -\sin\,\alpha \\ \sin\,\alpha & \cos\,\alpha \end{pmatrix} \begin{pmatrix} i \\ j \end{pmatrix} + \begin{pmatrix} p \cdot x \\ p \cdot y \end{pmatrix}$$

Given a steering switch opinion q, the street line is altered and further reached out as appeared previously. The separation from any street direct kp towards q is determined as pursues.

Expanding street stroke after q is directed similarly as tended to in past area, while following the street line from p to q is diverse at the accompanying binary residences [13].

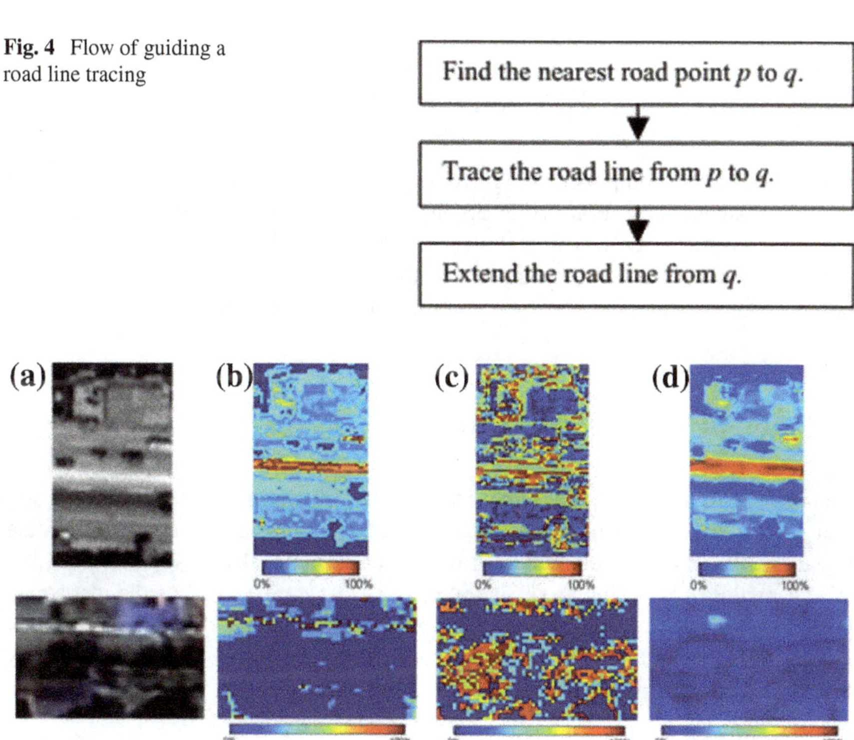
Find the nearest road point p to q.
Trace the road line from p to q.
Extend the road line from q.
(a)
(b)
(c)
(d)
0%
100%

Fig. 4 Flow of guiding a road line tracing

Fig. 5 Per-pixel reactions of highlights, weighted by histogram esteem. **a** Duplicate. **b** Colour include. **c** Local entropy highlight. **d** HSC include

(1) every cycle, the extensional bearing of street incurable point pk 'allotted by pk q'
(2) Corresponding expense of street format at a picture opinion '*p*' with a turn edge is characterized equally as pursues.

3 Proposed Work (Sparse Representation)

The essential guideline of meagre portrayal is that sign '*x*' contains spoken to as a straight mix of a couple of premise courses in gathered collection *D* by upgrading the l1-regularized least-squares issue. However, street extraction is not simply an issue of remaking. The educated target library limits the remaking blunder for the meagre portrayal; at the same time, more significantly, it requires a solid discriminative capacity to isolate the street focus from the foundation. Luckily, a straightforward yet powerful coding plan called area obliged direct codes had been presented [14], which processes the portrayal with the nearby format covers (premise) in '*D*' that are like the applicant test '*x*'.

$$\min_{\alpha}\left\| x - \sum_{i=1}^{K} D_i\alpha_i \right\|^2 + \lambda\|d \odot \alpha\|^2$$

where o indicates component insightful duplication, λ controls the sparsity, and *d* is the Euclidean separation vector among *x* and the premise vectors in *D*. Moreover, a word reference learning strategy called the *K*-closest neighbour technique is acquainted with point of confinement, the space traversed by straightforwardly choosing the *K* information vectors from the informational index [11].

3.1 Multifeature Sparse Model

The contrast between a street target and a general following target lies in the way that not exclusively does the street target not have rich surface and shading data; however, it is likewise deficient with regard to an unmistakable and stable profile. We characterize $\alpha*$ as the arrangement of (2) and x∗ *I*, $I = 1{:}N$, as the inadequate portrayal of the picture patches focused at pixel position ci inside the hopeful window focused at *y*. The street target appearance is portrayed by the joint sparsity-based histogram circulation of the shading, neighbourhood entropy, and HSC highlights.

(a) Colour Feature:

It has been demonstrated that shading data is strong to pivot, bending, and incomplete impediment [15]. The shading histogram can be processed as

$$h_j^I(y) = C_I \sum_{i=1}^{N} K\left(\left\|\frac{y - c_i}{h}\right\|^2\right)\delta\left[b\left(x_i^*(c_i)\right) - j\right]$$

where CI is a standardization steady, h is the data transmission parameter, $K()$ is the piece work, and $b(\bullet)$ is the capacity to figure which canister pixel $x* I$ (ci) is in.

(1) Native 'Entropy' Article: 'Local entropy' data is an alluring supplement to shading data on account of variations in enlightenment and foundation mess. The district entropy histogram container is processed below

$$h_j^E(y) = C_E \sum_{i=1}^{N} K\left(\left\|\frac{y - c_i}{h}\right\|^2\right)|\delta[b(E_i) - j]$$

(2) HSC Feature: The HSC depends on basic attributes [16] compared with the features that are more likely to change, such as gradient and the texture, the HSC is more robust for a street target that is susceptible to surface objects. The HSC can be figured as (Fig. 6)

$$h_j^S(y) = C_S \sum_{i=1}^{N} K\left(\left\|\frac{y - c_i}{h}\right\|^2\right)\left|\alpha_{ij}^*\right|$$

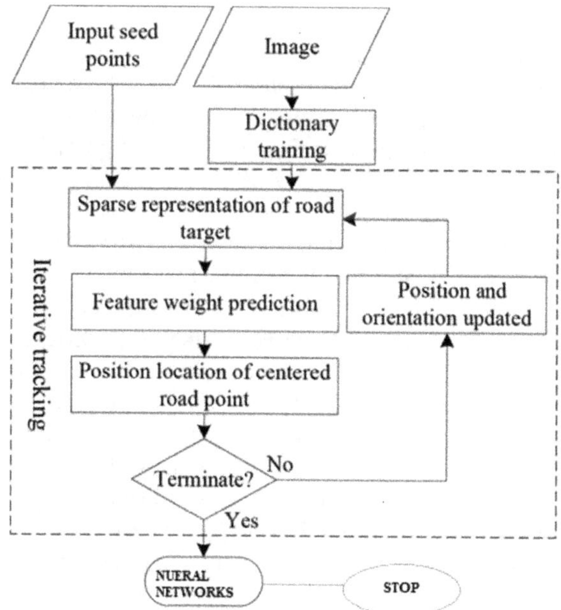

Fig. 6 Proposed work

where CS is a standardization steady and $\alpha * Ij$ is e jth meagre coefficient of the Ith picture fix. In Fig. 1, imagine three highlights utilizing the overwhelming loads in every pixel. Plainly, the circulation of the loads of each component differs along the street surface and its region. Consequently, using the correlative attributes of the three unique highlights enables us to complete street target acknowledgement.

(1) Initialization: human administrator is expected to acquire the beginning stage, the course, and the width of the street via naming stone emphases on the information picture.
(2) Wordlist Exercise: word reference '*D*' is found out by the preparation tests by the *K*-closest 'neighbour' strategy.
(3) Objective Illustration: 'sparsity-based' element histogram of the objective area and the hopeful district are acquired by the projected multi-highlight inadequate classical.
(4) Future Mass Calculation: to gauge the capacity of every component to recognize a contrast among the street goal and foundation, we initially make *L* target perceptions (close to the street target) and *B* foundation perceptions (along the vertical street course). The discriminative presentation of each element would then be able to be evaluated by the Fisher separate paradigm [16].

3.2 NN-Neural Networks

From a numerical perspective, explore on the guess abilities of feedforward neural systems has focussed on two angles: all comprehensive estimation on smaller info circles and estimate in a incomplete set. Various researchers consume examined the all complete deduction abilities of normal multilayer feedstuff 'NEURAL-NETWORKS' [1–3]. In honest applications, the *N*-systems are prepared in incomplete making set. For capacity supposition in a restricted making usual, Huang and Babri [4] demonstrated that a single-shrouded layer feedforward neural system (SLFN) with at record *n* concealed neurons and practically some nonlinear enactment competence can learn *N* particular perceptions with zero blunder. It ought to be noticed that the info loads (linking the material coating to the primary shrouded layer) would be stable in all these past imaginary study roles just as in practically all viable education intentions of feed frontward neural systems. After the info loads and the covered sheet inclinations are elected self-assertively, SLFNs tin are principally careful as a straight framework and the yield loads (connecting the concealed layer to the yield layer) of SLFNs container be rationally obvious finished frank gathered up opposite task of the shrouded layer yield networks.

In the light of this idea, this paper proposes a basic learning calculation for SLFNs called extraordinary learning machine (ELM) whose learning pace can be a large number of times quicker than customary feedforward arrange learning calculations like back-engendering calculation while getting better speculation execution. Unique in relation to customary learning calculations, the proposed learning calculation not just will in general achieve the littlest preparing mistake yet additionally the littlest standard of loads. Bartlett's hypothesis on the speculation execution of feedforward neural networks [9] states that for feedforward neural systems achieving littler preparing blunder, the littler the standard of loads is, the better speculation execution the systems will in general have. Along these lines, the proposed learning calculation will in general have better speculation execution for feedforward neural systems.

4 Results

In our analyses, two informational indexes are applied to assess the presentation of the future practice. The primary informational index, as appeared in 'Fig. 3', remained gathered by the GF-2 settlement on the zone of Delhi, India, with a goal of 0.80 M/PIXEL. This picture, with a size of 4000.0 × 3500.0 'pixels', incorporates various kinds of clamours, for example impediment of automobiles, drooping plants, and structure shadows. The second informational index, as appeared in Fig. 5, was gathered by the proposed satellite image on the zone of Delhi, India, with a 0.60 m goals band. This picture, with a size of 1600.0 × 1200.0 pixels, incorporates an assortment of street settings, for example street factual alteration, impediment of vehicles, as well as a strident bend in the highway (Fig. 7).

Figure 8 shows the image observed from the database which is sent by satellite; image contains no information, but we need to reduce noise and get information from this obtained image.

Figure 9 shows that histogram equalization of image, and this contains resolution and pixel intensity and blurring brightness, but we have to reduce and increase these properties.

Figure 10 explains about red green, red, blue models of our selected image; this image gives more information for future extraction.

Figure 11 explains that comparison of three methods; the final one got no errors compared to the remaining.

Figure 12 shows different filter methods on image.

Enter a number: case 1: Kim method, case 2: Miao method, case 3: proposed method 3

The number of seed points = 51

Total road length pixel = 33,239

completeness = 96.61

RMSE (PIXELS = 2.25

Figure 13 shows that final full information image; in this image, all the information is available like a perfect image, and this is possible with NN-sparse model (Table 1).

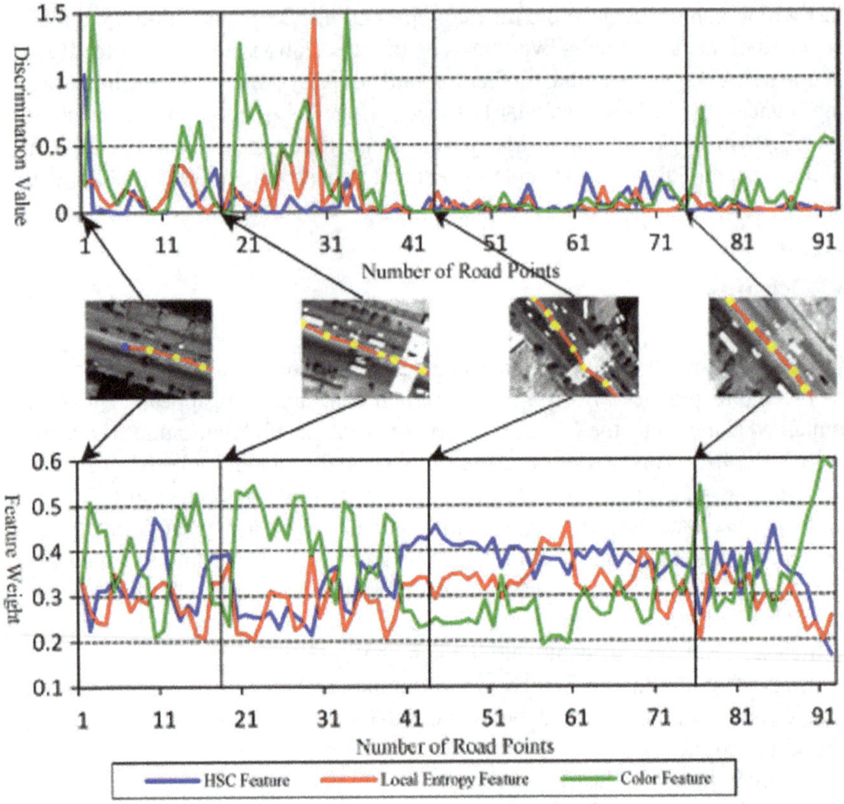

Fig. 7 Graphical analyses

Fig. 8 Satellite image

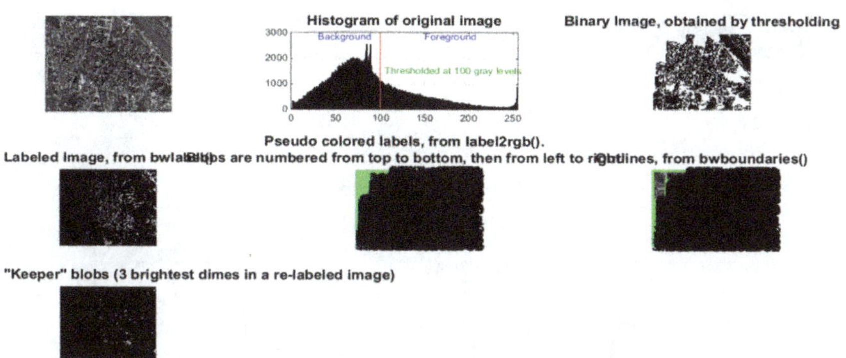

Fig. 9 Histogram equalization

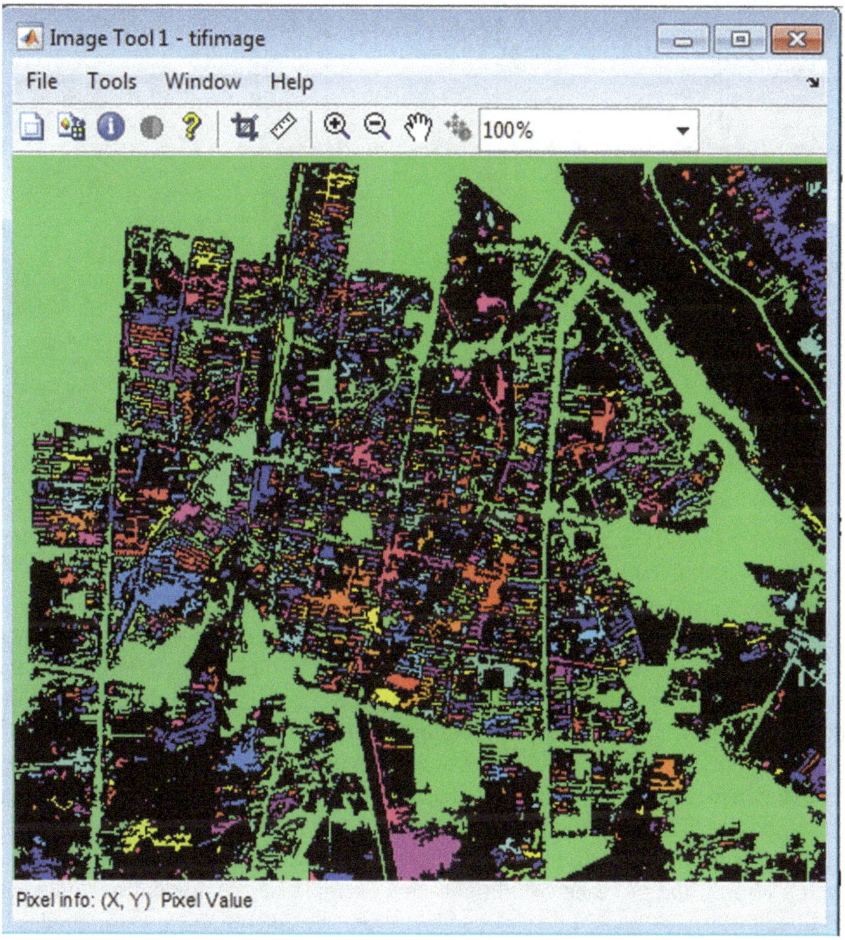

Fig. 10 RGB model

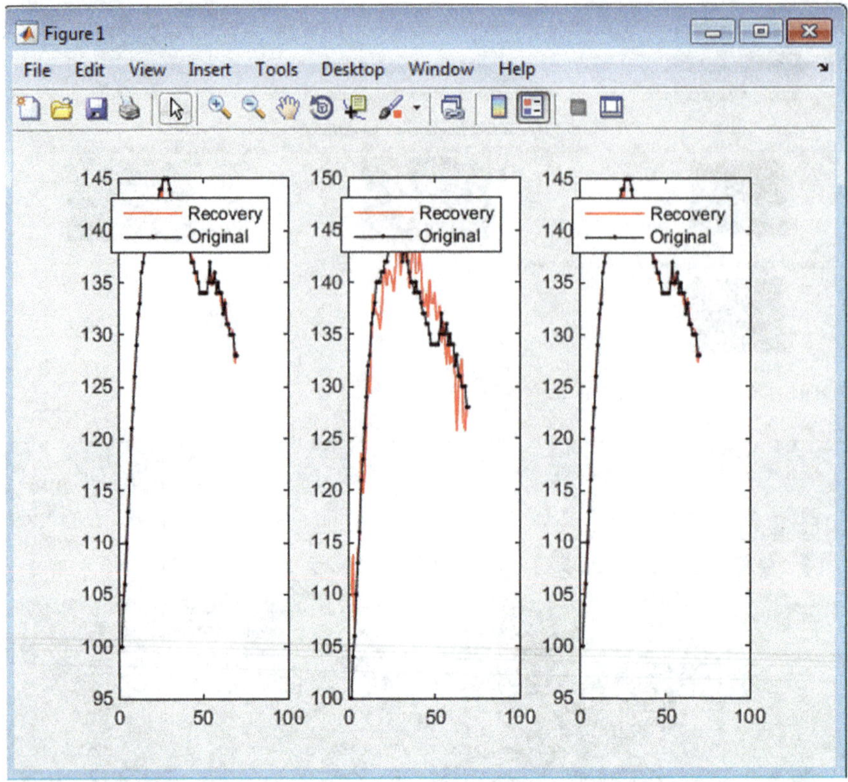

Fig. 11 Graphical comparisons

5 Conclusion

In this work, a self-loader strategy for street extraction from HR satellite pictures has been exhibited. In particular, neighbourhood inadequate portrayal is performed with a versatile weighted blend of shading, nearby entropy, as well as HSC highlights to demonstrate the presence of the street aim, which gives a superior answer for street withdrawal beneath composite circumstances. The test consequences affirm that the suggested technique beats the present best in class arrangements. Inferable from the distinction of the multifeatures includes between the street target and its neighbourhood, we can viably utilize the complementation of the multi highlights, just as keep up the dynamic parity in street target demonstrating. Extra investigations found that the proposed model may not give adequate qualification, which is brought about by the absence of item highlight data. In these conditions, a conceivable arrangement would join all the more new highlights, for example morphological qualities and local co-event highlights. In our upcoming effort, we determine variety upgrades to the programmed determination of the info limitations, which impact the separated streets.

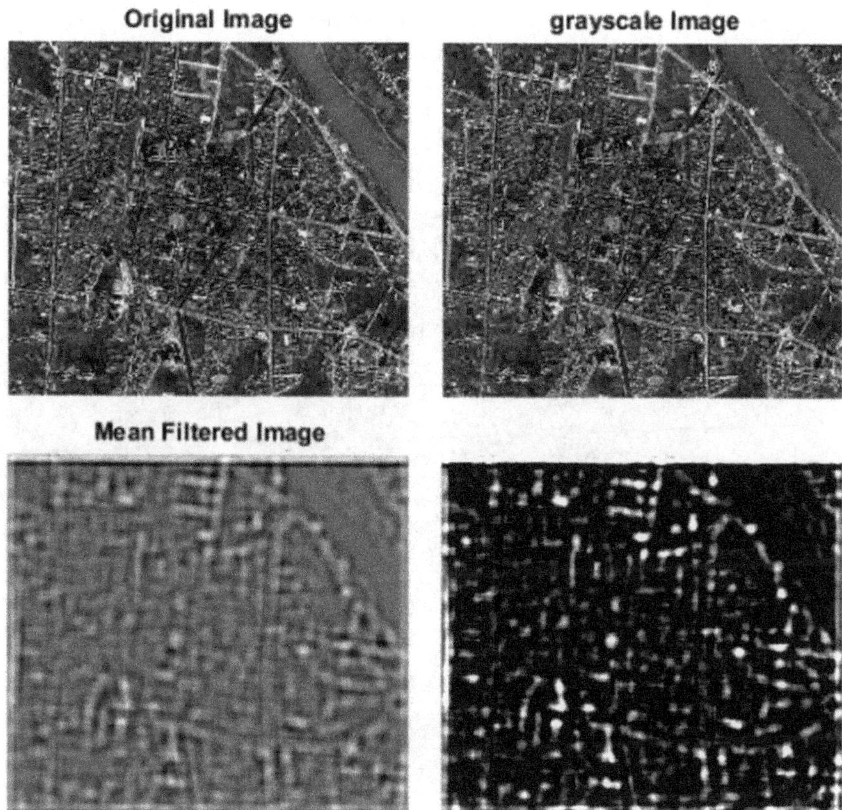

Fig. 12 Filter methods on image

Fig. 13 Proposed output

Table 1 Comparison of work

	Kim et al.'s method	Miao et al.'s method	Proposed method
Number of seed points	57	48	51
Total road length (pixels)	33,131	33,922	33,239
Completeness (%)	93.58	96.51	96.61
RMSE (pixels)	3.34	3.77	2.25

References

1. W. Shi, Z. Miao, J. Debayle, An integrated method for urban mainroad centerline extraction from optical remotely sensed imagery. IEEE Trans. Geosci. Remote Sens. **52**(6), 3359–3372 (2014)
2. B. Liu, J. Huang, L. Yang, C. Kulikowsk, Robust tracking using local sparse appearance model and k-selection. Proc. IEEE CVPR, 1313–1320 (Sep. 2011)
3. Y. He, M. Li, J. Zhang, J. Yao, Infrared target tracking based on robust low-rank sparse learning. IEEE Geosci. Remote Sens. Lett. **13**(2), 232–236 (2016)

4. G. Cheng, Y. Wang, F. Zhu, C. Pan, Road extraction via adaptive graph cuts with multiple features. Proc. ICIP, 3962–3966 (2014)
5. J. Wang, J. Yang, K. Yu, F. Lv, T. Huang, Y. Gong, Locality constrained linear coding for image classification. Proc. IEEE CVPR, 3360–3367 (Apr. 2010)
6. Y. Mingfeng, B. Yuming, Z. Gaopeng, Z. Weijun, *Adaptive Blockfusion Multiple Feature Tracking in a Particle Filter Framework*. IEEE International Conference on Cyber Technology in Automation, Control and Intelligent Systems (CYBER), (Sep. 2013), pp. 400–404
7. X. Ren, D. Ramanan, Histograms of sparse codes for object detection. Proc. IEEE CVPR, 3246–3253 (Apr. 2013)
8. J.B. Mena, State of the art on automatic road extraction for GIS update: A novel classification. Pattern Recognit. Lett. **24**(16), 3037–3058 (2003)
9. X. Lin, J. Zhang, Z. Liu, J. Shen, M. Duan, Semi-automatic extraction of road networks by least squares interlaced template matching in urban areas. Int. J. Remote Sens. **32**(17), 4943–4959 (2011)
10. Y. Zang, C. Wang, Y. Yu, L. Luo, K. Yang, J. Li, Joint enhancing filtering for road network extraction. IEEE Trans. Geosci. Remote Sens. **55**(3), 1511–1525 (2016)
11. Y. Zang, C. Wang, L. Cao, Y. Yu, J. Li, Road network extraction via aperiodic directional structure measurement. IEEE Trans. Geosci. Remote Sens. **54**(6), 3322–3335 (2016)
12. J. Zhou, W.F. Bischof, T. Caelli, Road tracking in aerial images based on human—computer interaction and Bayesian filtering. ISPRS J. Photogram. Remote Sens. **61**(2), 108–124 (2006)
13. X. Hu, Z. Zhang, C. Vincent, C.V. Tao, A robust method for semi-automatic extraction of road centerlines using a piecewise parabolic model and least square template matching. Photogram. Eng. Remote Sens. **70**(12), 1393–1398 (2004)
14. T. Kim, S.-R. Park, M.-G. Kim, S. Jeong, K.-O. Kim, Tracking road centerlines from high resolution remote sensing images by least squares correlation matching. Photogram. Eng. Remote Sens. **70**(12), 1417–1422 (2004)
15. Z. Miao, B. Wang, W. Shi, H. Zhang, A semi-automatic method for road centerline extraction from VHR images. IEEE Geosci. Remote Sens. Lett. **11**(11), 1856–1860 (2014)
16. S. Das, T. Mirnalinee, K. Varghese, Use of salient features for the design of a multistage framework to extract roads from high resolution multispectral satellite images. IEEE Trans. Geosci. Remote Sens. **49**(10), 3906–3931 (2011)

Detection of Brain Tumor in MRI Image Using SVM Classifier

Rudrapathy Bhavani and Kishore Babu Vasanth

Abstract Biomedical imaging plays an important role in diagnosis and early detection of tumor. Brain Tumor is an uncontrolled tissue growth found in any part of the brain. Early stage tumor detection makes the treatment easier. The important imaging techniques are X-ray, computed tomography, magnetic resonance imaging, ultrasound, etc. Magnetic resonance imaging (MRI) is the best technique to detect the tumor portion in the brain. The existing system uses fusion technique of image which is a method of joining complementary information and multi-modality images of the patient. Moreover, the early stages of tumor cannot be detected effectively. In the proposed system, SVM classifier is used in the detection of tumor affected portion. Noise in the acquired image is removed using Gabor filter. The function of Gabor filter is edge detection, feature extraction and noise removal. The morphological function such as dilation and erosion will be applied through the filtered image. Then, the enclosed region will be spitted out separated by the SVM classifier. Using SVM classifier, the early stages of tumor can be easily detected. Also, it is used to segment the tumor portion. This paper also aims at sending notifications to the guardians, about the patient's condition and the medications to be taken by the patients by means of both mails and messages.

Keywords Medical image · Tumor · MRI · SVM classifier · Morphological function · Feature extraction

1 Introduction

Biomedical imaging has developed rapidly in recent years. Recent researches and their experiments really helped in advancing diagnostic tools for medical field [1].

R. Bhavani (✉)
Research Scholar, Sathyabama Institute of Science and Technology, Chennai, India
e-mail: bhavanirudra@gmail.com

K. B. Vasanth
Professor, Vidya Jyothi Institute of Technology, Aziz Nagar, C.B.Road, Hyderabad, Telangana, India
e-mail: vasanthecek@gmail.com

H. S. Saini et al. (eds.), *Innovations in Electronics and Communication Engineering*, Lecture Notes in Networks and Systems 107,
https://doi.org/10.1007/978-981-15-3172-9_54

The various types of biomedical imaging technologies are magnetic resonance imaging, computed tomography scan, ultrasound, SPECT, PET and X-ray. Brain tumor is the most common disease in medical science. Detection of brain tumor in early stages can improve the prevention mechanism to higher level. The MRI technique is the most effective technique for brain tumor detection. MRI is better than X-rays which results in high-quality images [2].

2 Existing Methodology

Image fusion is a method of joining complementary information and multi-modality images of the same patient into an image. Therefore, the obtained image consists of more informative than the individual images alone. In feature level fusion, source images are partitioned into regions and features like intensities of pixel, edges or texture are used for fusion technique [3].

Feature level fusion between images is a challenging problem of inter-image variability such as pixel mismatches (scale, rotations and shifts), missing pixels, image noise, resolution and contrast [4]. The inaccuracies in feature representation can lead to poor fusion performance and lesser robustness of the feature representation. In addition, this also means that wrong feature representation can lead to wrong conclusion that reduces the reliability of medical image analysis in clinical settings [5].

Region-based image fusion of feature level would be highly efficient when compared to the pixel-based fusion methods. The fusion method has multi-modal images which are partitioned into regions using automatic segmentation process [3], and the image fusion is performed based on the rules of region-based fusion. The major disadvantage of the existing system is that fusion system passes information within each decomposition level so that the source image details are preserved expressing the artifacts [6, 7].

3 Proposed Methodology

The proposed system has several advantages that overcome the disadvantages of the existing feature level fusion system.

1 Sensitivity to noise and blurring effects can be achieved.
2 The use of SVM classifier aids in detecting the tumor in the early stage.
3 The proposed method uses the kernel trick, so we can build in expert knowledge about the problem (Fig. 1).

In this work, support vector machines (SVM) are supervised learning models that analyze data used in classification and regression analysis. In addition to performing linear classification, SVMs can efficiently perform a nonlinear classification using

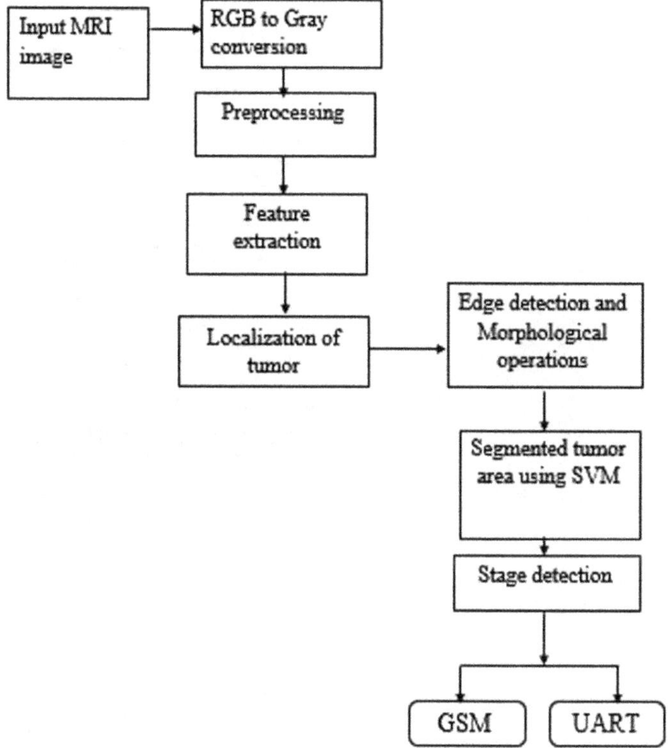

Fig. 1 Block diagram of the proposed system

the kernel trick, by mapping their inputs into high-dimensional feature spaces [8]. An efficient classification method is proposed to recognize normal as well as abnormal MRI brain images.

SVM classifier is implemented to segment the affected portion of cancer. To segment the portion, first, we have to filter out the noise in the acquired image based upon the masking methodology [9]. The morphological function will be applied throughout the filtered image. Using morphological bounding, box will be drawn over the detected portion. Hence, by means of SVM classifier, the region enclosed by bounding box will be spitted out separately (Fig. 2).

The input images are scanned using MRI and will be stored in MATLAB. These stored images are displayed as a grayscale image of size 256*256. The colored images are converted into grayscale image using RGB to gray conversion. The obtained image may contain noise. White Gaussian noise is the most commonly occurring noise in the MRI images. Hence, the noise removal is mandatory for the tumor detection from magnetic resonance images [3]. There are many types of filters used for image noise removal. The images are preprocessed to filter out the noise from a grayscale converted image. Edge detection is the most vital part in tumor detection. A Gabor filter is a linear filter which is used for edge detection [6, 10].

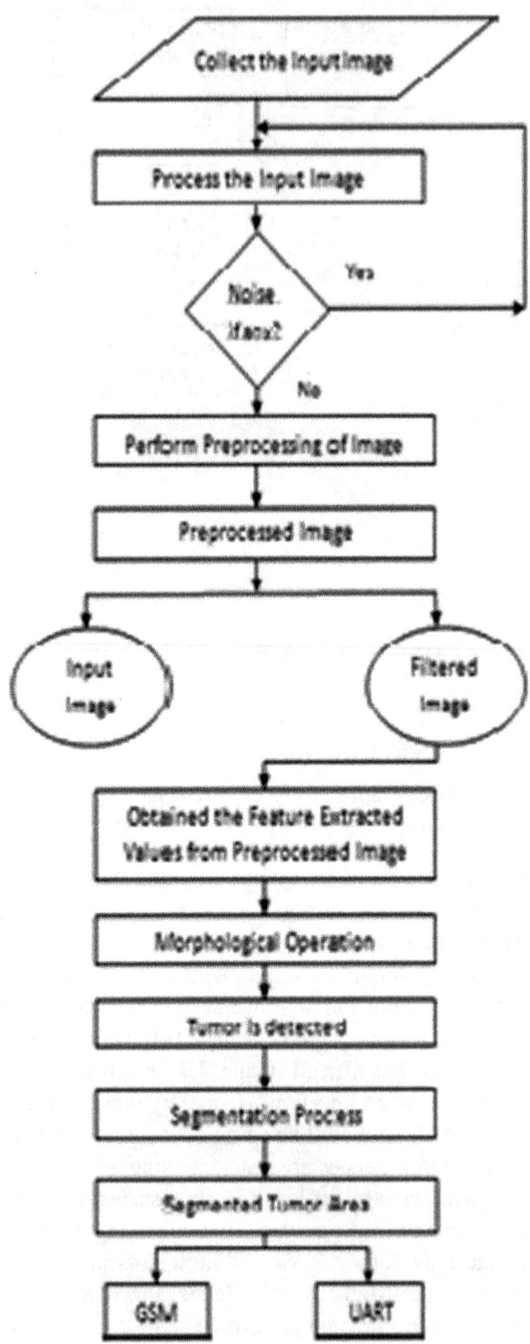

Fig. 2 Workflow of the proposed system

The next process will be the morphological operation where the boundary of the tumor part is approximately sized out in red. During this process, adding and removing of pixels take place through which the border of the tumor is detected. During the process of segmentation, two output portions are appeared, namely the tumor area is indicated as green boundary and the second image shows the segmented tumor portion. After detecting and segmenting the tumor area, a message is forwarded to the doctor. Also, a message is sent to the guardian's mobile phone through global system for mobile communication (GSM) which points out the corresponding stage of the tumor and suggests the respective medicines to be taken [11].

4 Results and Discussion

In this paper, the support vector machine is applied to segment the detected portion of cancer. To segment the portion, first, we have to filter the acquired image based upon the masking methodology. The morphological function will be applied and extracted throughout the filtered image. With the help of morphological bounding, the box will be drawn over the detected portion. Then, the region enclosed by bounding box will be taken out separately by means of SVM classifier.

Copy the collected images for the initial process to the current folder and then execute it. If any noise occurs, it will be completely cleared. Get the input image from the collected data and apply it for preprocessing. The resultant image will be preprocessed image, i.e., input image and filtered image. The input MRI image is obtained from https://www.insight-journal.org/midas/gallery for research work and simulation purpose (Figs. 3 and 4).

Perform the feature extraction and obtain the corresponding feature extracted values corresponding to correlation, energy, contrast and homogeneity from the preprocessed image (Fig. 5).

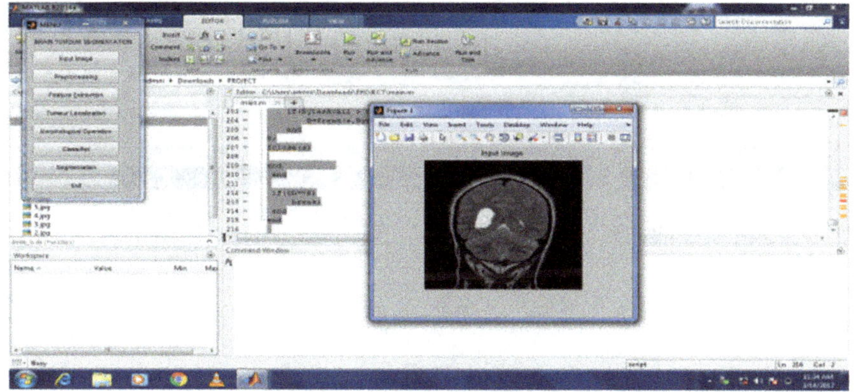

Fig. 3 Window displaying input image [12]

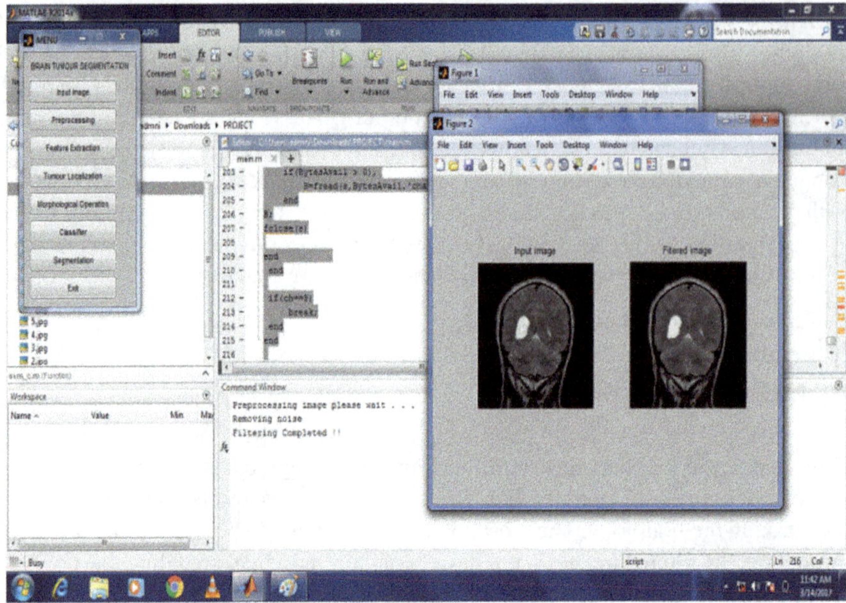

Fig. 4 Window displaying preprocessed image

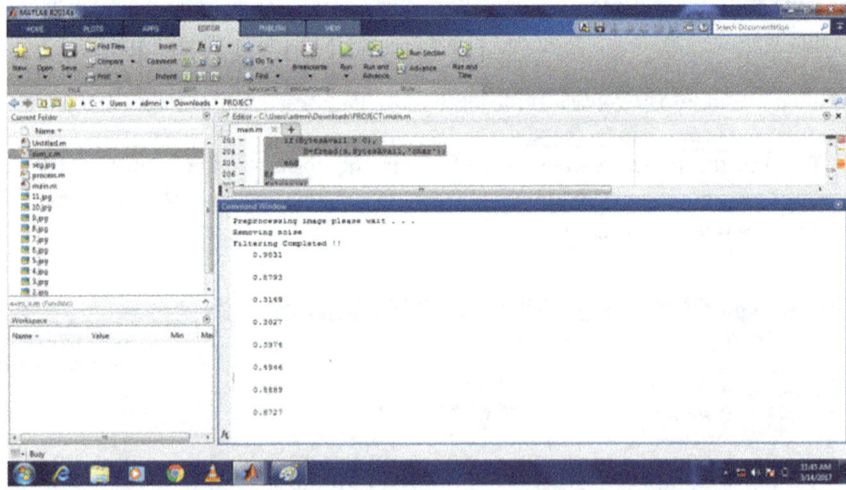

Fig. 5 Window with feature extraction values

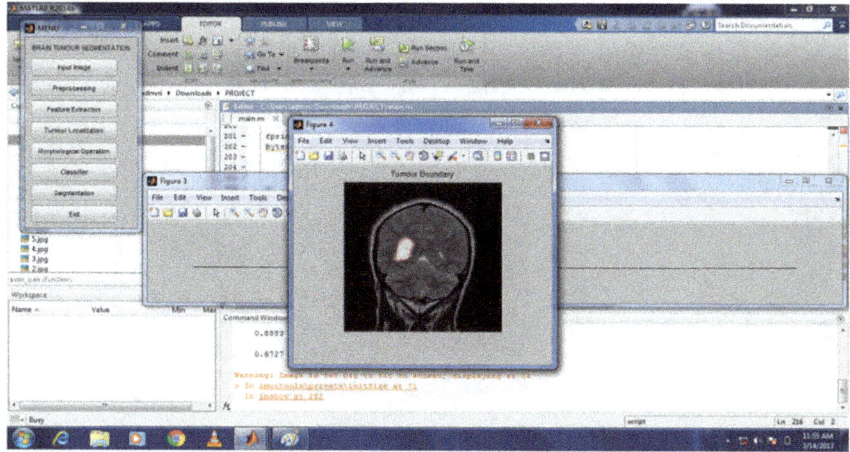

Fig. 6 Image after morphological operation

The next process will be the morphological operation where the boundary of the tumor part is approximately sized out in red. During this process, adding and removing of pixels take place through the border of the detected tumor (Fig. 6).

The image after morphological operation is appeared on the screen in which unwanted dots and lines are removed (Fig. 7).

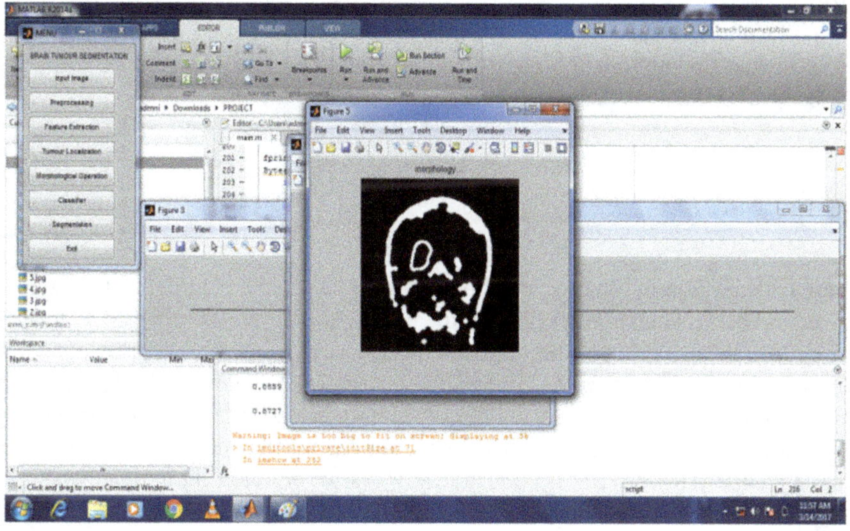

Fig. 7 Window showing classified image

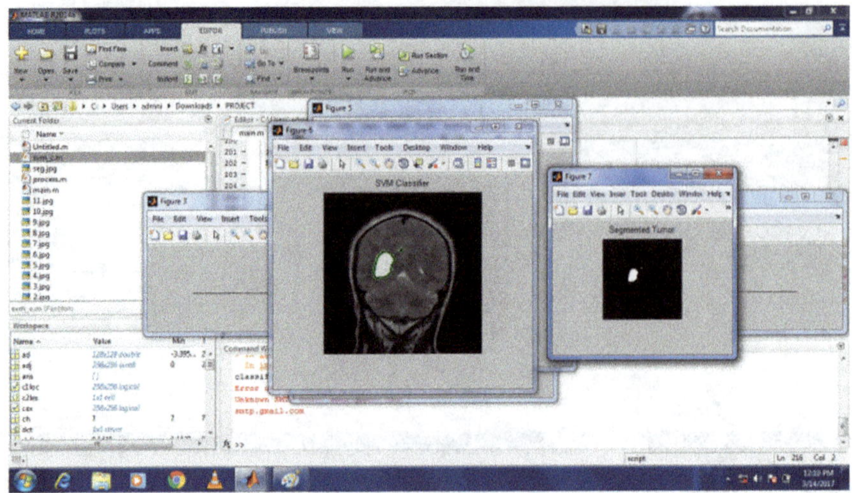

Fig. 8 Display of segmented tumor area

During the process of segmentation, two output portions are appeared, namely the tumor area which is indicated as green boundary and the second image shows the segmented tumor portion (Fig. 8).

After the process of segmenting the tumor area, it is being mailed to the doctor. The stage of the tumor (early or advanced) is indicated on the screen (Figs. 9 and 10).

Following the mail notification, a message is also sent to the guardian's mobile phone which points out the corresponding stage of the tumor and suggests the respective medicines to taken (Fig. 11).

5 Conclusion

Initial stage of brain tumor is a major challenge in the medical field. Brain tumor detection is a tedious job because of the complex structure of brain. MRI images provide an easier method to detect the tumor and also to perform the surgical approach for its removal. The existing image fusion feature level technique has limitation in accuracy, exactness and ability to detect the tumor earlier. To overcome the existing system limitations, SVM classifier is used for early detection of tumor and stage classification. This method is comparatively best when referring to other available algorithms.

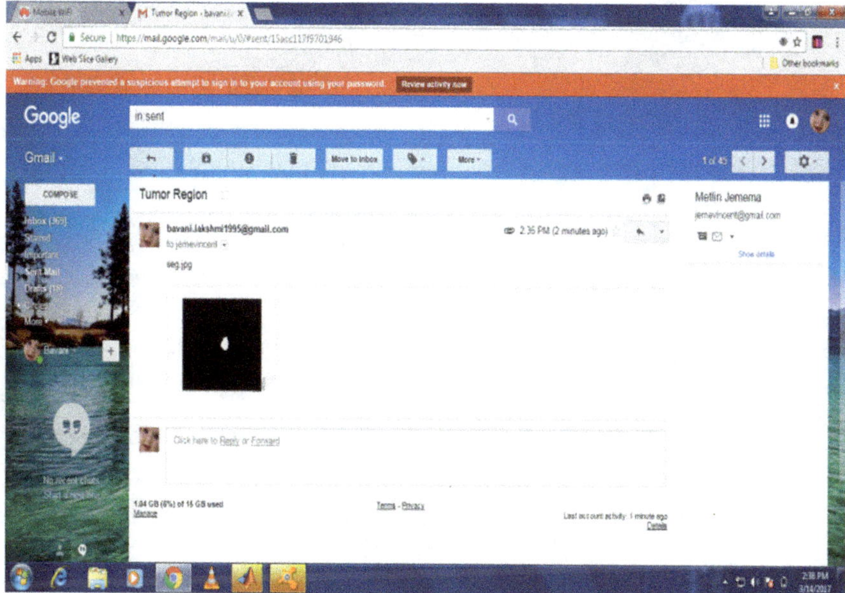

Fig. 9 Screenshot of sender's mail page with resultant image

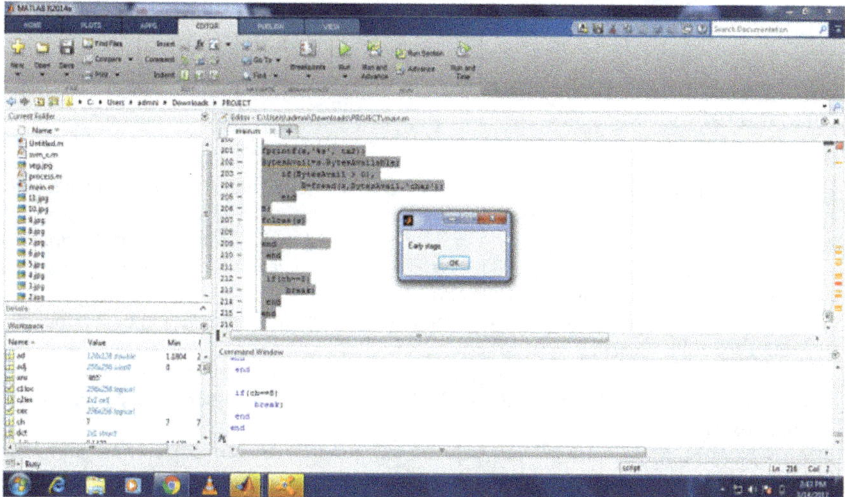

Fig. 10 Window displaying tumor stage

Fig. 11 Screenshot of receiver's message

References

1. M.A. Balafar, A.R. Ramli, M.I. Saripan, S. Mashohor, Review of brain MRI image segmentation methods. Artif. Intell. Rev. **33**(3), 261–274 (2010)
2. M. G. Linguraru, W. J. Richbourg, J. Liu, J. M. Watt, V. Pamulapati, S. Wang, R. M. Summer, Tumor burden analysis on computed tomography by automated liver and tumor segmentation. IEEE Trans. Med. Imaging **31**(10), October (2012)
3. K. K. Manoj, S. Yadev, Brain tumor detection and segmentation using histogram thresholding. Int. J. Eng. Adv. Technol. (April 2012)
4. M. Shasidhar, V. S. Raja, B. V. Kumar, MRI brain image segmentation using modified fuzzy C-means clustering algorithm. IEEE Int. Conf. Commun. Syst. Netw. Technol. **32**(6), 234–239 (June 2011)
5. M. N. Nobi, M. A. Yousuf, Anew method to remove noise in magnetic resonance and ultrasound images. J. Sci. Res. (2010)
6. N. Rőhrl, J. R. Iglesias-Rozas, G. Weidl, *Computer Assisted Classification of Brain Tumors*, (Springer, Berlin Heidelberg, 2008), pp. 55–60
7. S. Datta, M. Chakraborty, *Brain Tumor Detection from Pre-Processed MR Images using Segmentation Techniques*, IJCA Special Issue on 2nd National Conference-Computing, Communication and Sensor Network, CCSN (2011)
8. M. C. Clark, L. O. Hall, Automatic tumor segmentation using knowledge based techniques. IEEE Trans. Med. Imaging **17**(2) (Apri1 998)
9. F. J. Galdames, F. Jaillet, C. A. Perez, An accurate skull stripping method based on simplex meshes and histogram analysis in magnetic resonance images. Rapport de recherche RR-LIRIS (2011)

10. R. Preetha, G. R. Suresh, Performance analysis of fuzzy C means algorithm in automated detection of brain tumor, (IEEE CPS, WCCCT, 2014)
11. R. B. Dubey, M. Hanmandlu, S. Vasikarla, Evaluation of three methods for MRI brain tumor segmentation. (IEEE computer society, 2011)
12. https://www.insight-journal.org/midas/gallery

Power Quality Event Recognition Using Cumulants and Decision Tree Classifiers

M. Venkata Subbarao, T. Sudheer Kumar, G. R. L. V. N. S. Raju and P. Samundiswary

Abstract The main intention of this article is to introduce machine learning (ML) algorithms to recognize the power quality (PQ) events. In this paper, a new pattern recognition (PR) algorithm is introduced for PQ detection using decision tree classifiers (DTCs). To train the classifier, a set of higher-order cumulants has been extracted from all classes of PQ signals. Further, principle component analysis (PCA) is carried out to reduce prediction time by removing the redundant features. Performance of the proposed DTCs is tested under different training and testing rates. The performance comparison of DTCs with that of the existing techniques is done in terms of classification accuracy.

Keywords PQ disturbances · Gini index · Feature extraction · Cumulants · Pattern recognition

1 Introduction

In recent years, usages of sensitive electronic devices in smart homes, buildings and cities are growing in an exponential manner. These sensitive devices are easily affected by PQ disturbances, such as sags, surges and interruptionsetc. In real-time applications, failure of these sensitive devices may cause serious damage, especially in smart applications. The quality of the power signal may degrade mainly due to internal elements in the transmission system, unbalancing loads and also due to other external factors such as weather. In the past, several researchers came with different solutions to analyze these PQ disturbances such as discrete wavelet based [1–5], rough set [6], S-transform [7–9], neural network [10, 11] and wavelet packet [11]. For PQ signal classification, wavelet methodologies [1–4] have been widely used. But in this approach, there is a considerable degradation in the performance in case

M. Venkata Subbarao (✉) · T. Sudheer Kumar · G. R. L. V. N. S. Raju
Department of ECE, Shri Vishnu Engineering College for Women (A), Bhimavaram, Andhra Pradesh, India
e-mail: mandava.decs@gmail.com

P. Samundiswary
Department of EE, Pondicherry University, Pondicherry, India

H. S. Saini et al. (eds.), *Innovations in Electronics and Communication Engineering*, Lecture Notes in Networks and Systems 107,
https://doi.org/10.1007/978-981-15-3172-9_55

of lower SNR. To overcome this, Liao [5] introduced a modified wavelet approach with noise suppression. S. Dalai et al. [6] developed a PQ classification scheme with a minimal cross-correlation feature set using rough set theory. Further, Zhao et al. [7] introduced an S-transform (ST)-based classification algorithm for combination of single and complex PQ signals. Gargoom et al. [8] developed an hybrid algorithm by combination of Parseval's theorem and ST to monitor the PQ events. Dash et al. [9] developed an algorithm for detection of short duration PQ events based on Kalman filter and hybrid ST. Mishra et al. [12] developed a hybrid algorithm by combination of ST features and PNN to detect and classify PQ disturbances. Lee et al. [10] developed a probabilistic neural network (PNN)-based classification approach which is developed for classification of complex PQ events. Tong et al. [11] introduced a classification algorithm using wavelet packet transform (WPT) and support vector machine (SVM). Biswal et al. [13] developed an algorithm with a combination of fuzzy *C*-means and particle swarm optimization (PSO) for classification of PQ events. The classification process in all the existing algorithms involves three stages, applying the transform to the disturbance, extracting the features from the transform output and finally a decision making with the classifier. The computational complexity of all these algorithms is more because of the involvement of transform.

In order to reduce the complexity and to improve the efficiency of classification, a new attempt has been made in this paper. A new set of features is extracted directly from signal, and they are further applied to classifier for training and testing. The paper is organized as follows: The system model is discussed in Sect. 2. A detailed description of the proposed DTCs is presented in Sect. 3. Classification results of PQ events using the proposed DTCs are presented in Sect. 4. Finally, Sect. 5 concludes the paper.

2 System Model

The framework of the proposed PQ event recognition is shown in Fig. 1. It involves feature extraction, training and testing. In this paper, a new set of statistical features such as moments and cumulants is extracted directly from each PQ disturbance without applying any transform.

Further, these extracted features are used for training the proposed DTCs and to distinguish the unknown disturbance.

The multi-order moments of a PQ signal $x(n)$ are given by

$$M_{ab} = E\left[x(n)^{a-b}x^{*}(n)^{b}\right] \tag{1}$$

These moments are further used to derive the multi-order cumulants as follows

$$C_{20} = E\left[x^{2}(n)\right] \tag{2}$$

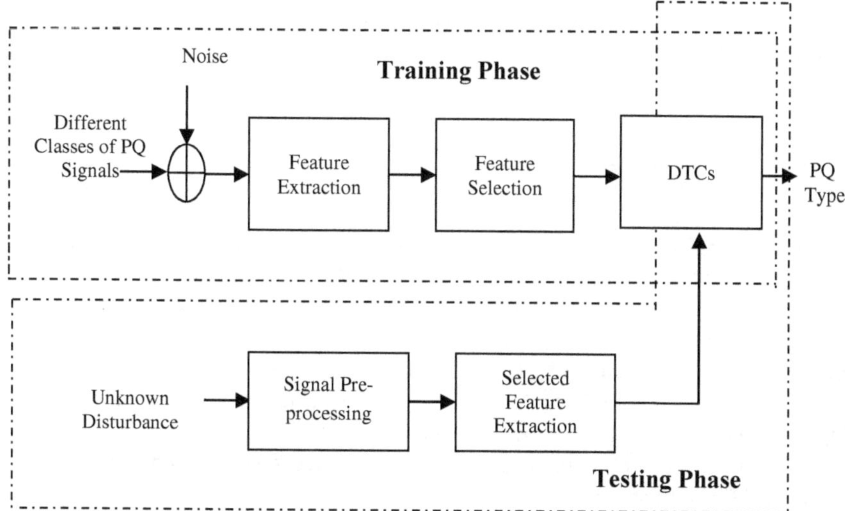

Fig. 1 Proposed approach for PQ detection

$$C_{21} = E\left[|x(n)|^2\right] \tag{3}$$

$$C_{40} = M_{40} - 3M_{20}^2 \tag{4}$$

$$C_{41} = M_{40} - 3M_{20}M_{21} \tag{5}$$

$$C_{42} = M_{42} - 2M_{21} - |M_{21}|^2 \tag{6}$$

$$C_{60} = M_{60} + 30M_{20}^3 - 15M_{20}M_{40} \tag{7}$$

$$C_{61} = M_{61} + 30M_{20}^2M_{21} - 10M_{20}M_{41} \tag{8}$$

$$\begin{aligned} C_{62} = M_{62} &- 6M_{20}M_{42} + 24M_{21}^2M_{22} - 8M_{21}M_{41} \\ &+ 6M_{20}^2M_{22} - M_{22}M_{40} \end{aligned} \tag{9}$$

$$\begin{aligned} C_{63} = M_{63} &+ 18M_{20}M_{21}M_{22} + 12M_{21}^3 - 9M_{21}M_{42} \\ &- 3M_{20}M_{43} - 3M_{22}M_{41} \end{aligned} \tag{10}$$

Here, C_{20} and C_{21} are second-order cumulants, C_{40}, C_{41} and C_{42} are the fourth-order cumulants and C_{60}, C_{61}, C_{62} and C_{63} are sixth-order cumulants.

To reduce the training and testing time of the proposed DTCs, redundant features are removed from the training set based on PCA. A set of three noise robust features is selected without degrading the performance of the classifier, and they are C_{20}, C_{40}

and C_{60}. Further, these selected features are used for training the proposed PRC and to test the unknown signal.

3 PQ Classification Using DTCs

Decision trees (DT) are nonparametric supervised learning classifiers. These are simple to understand and fast in classification or prediction. The complexity of these classifiers is less so that they require less memory. Increase in the splits of the trees leads to improvement in the classification performance. DTCs are binary trees, and the recognition is carried out by moving from the root node to the leaf node. Towing rule is considered to split the nodes, and the prediction accuracy at any node of DTCs depends on cross-entropy and Gini's diversity index.

If a feature set F contains n classes of modulated signals, then Gini index of feature set F is defined as

$$G(F) = 1 - \sum_{i=0}^{n} P_i^2 \tag{11}$$

where P_i is the relative probability of class i in feature set F.

If feature set F is split on D into F_1 and F_2 subsets, then Gini index is given by

$$G_D(F) = \frac{|F_1|}{|F|} G(F_1) + \frac{|F_2|}{|F|} G(F_2) \tag{12}$$

To split the node, the largest difference between $G(F)$ and $G_D(F)$ is to be considered.

Based on the number of splits, the DTCs are categorized into coarse tree (CT), fine tree (FT) and medium tree (MT). FT has more splits, so it provides more accuracy than MT and CT. FT is suitable for classification of large class dataset, whereas MT and CT are suitable for small class problems. Table 1 shows the characteristics of different DTCs.

Table 1 Characteristic of decision trees

Classifier	Prediction speed	Classification accuracy	Flexibility	Max. no. of splits
CT	Fast	Medium	Low	4
MT	Fast	High	Medium	20
FT	Fast	High	High	100

4 Results and Discussions

The classification performance of the proposed DTCs is tested with seven classes of PQ signals which are listed in Table 2. For the simulations, 1000 copies of each PQ disturbance are considered under varying noise conditions. For each copy of the PQ signal, a set of three selective features is extracted for training and testing which are listed in Table 2. Initially, the performance measures are carried out with 90% of training and 10% testing. Further, the simulations are carried out with 50–85% training cases.

Table 2 shows the numerical values of the three selected features for all classes of PQ disturbances which are considered for simulation. These selected features are used for training the proposed DT classifiers.

The scatter plot and confusion matrix (CM) of the proposed CT classifier are shown in Figs. 2 and 3, respectively. In CM, diagonal elements denote the true classification rates and off-diagonal elements represent false-negative rates. From Figs. 2 and 3, it is clear that the CT is unable to distinguish the spikes, swell from sag. Therefore, CT has poor performance in classification of PQ events, and its classification accuracy is 71.4%.

The classification of PQ events using CT is shown in Fig. 4. It depicts how the CT differentiates the different PQ disturbances with the use of cumulants. Due to limited number of splits, CT is unable to differentiate the sag with swell and sag with spikes.

The CM and decision tree of the proposed MT classifier are shown in Figs. 5 and 6, respectively. From Fig. 5, it is clear that the MT is perfectly distinguishing all classes of PQ events. The overall classification accuracy of MT is 100% with 90% training.

The performance of the proposed FT classifier with 90% is shown in Fig. 7. At 90% training, 100 copies of each PQ event are applied to FT classifier for classification. The classification performance of the proposed FT classifier in terms of observations

Table 2 Cumulants of PQ events

PQ signal	Label	Features					
		C_{20}		C_{40}		C_{60}	
		Min	Max	Min	Max	Min	Max
Sag	$S1$	0.00	0.39	−1.42	−1.32	8.30	9.20
Swell	$S2$	0.03	0.10	−1.45	−1.43	9.37	9.51
Outage	$S3$	0.04	0.14	−0.83	−0.81	2.60	2.81
Transient	$S4$	0.02	0.07	5.30	5.50	65.7	71.0
Flicker	$S5$	0.03	0.14	−1.24	−1.21	6.90	7.18
Harmonics	$S6$	0.03	0.12	−1.30	−1.28	7.60	7.86
Spikes	$S7$	0.02	0.10	−1.46	−1.44	9.50	9.67

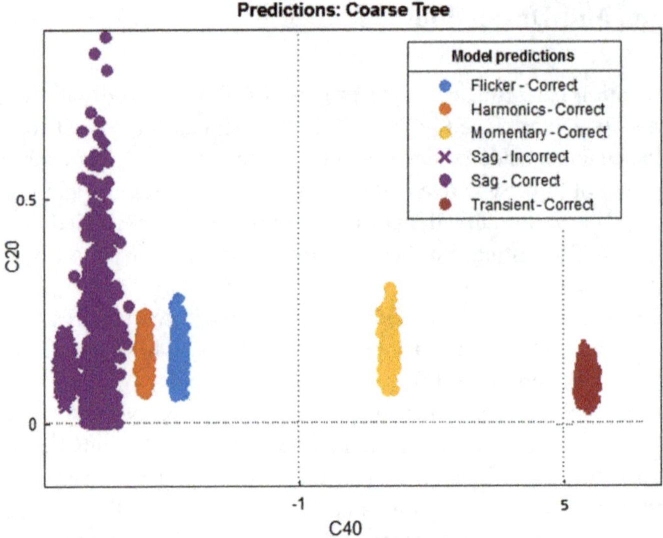

Fig. 2 Scatter plot of CT

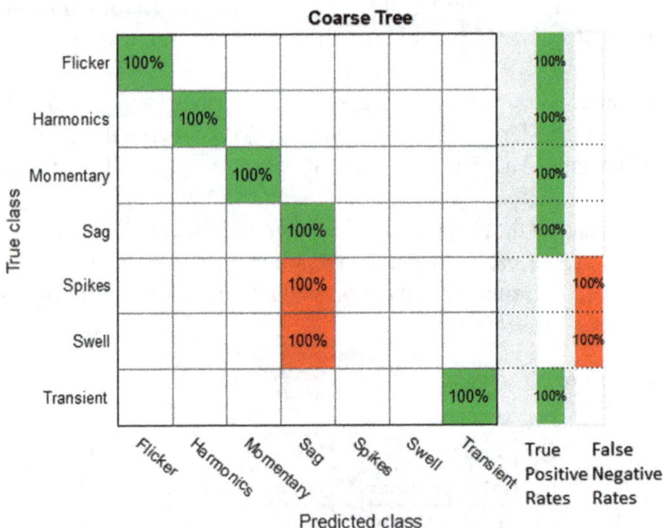

Fig. 3 Confusion matrix of CT

and percentages is represented in Fig. 7a, b, respectively. The overall classification accuracy of FT is 100% with 90% training.

The performance of the proposed FT classifier with 50% is shown in Fig. 8. At 50% training, 500 copies of each PQ event are applied to FT classifier for classification. The classification performance of the proposed FT classifier in terms of observations

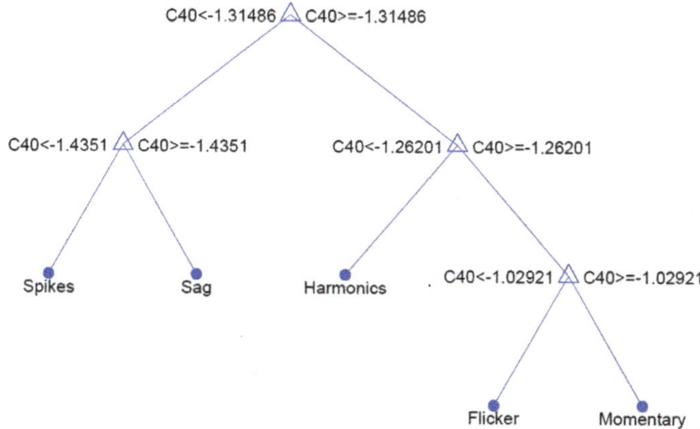

Fig. 4 Classification of PQ events using CT

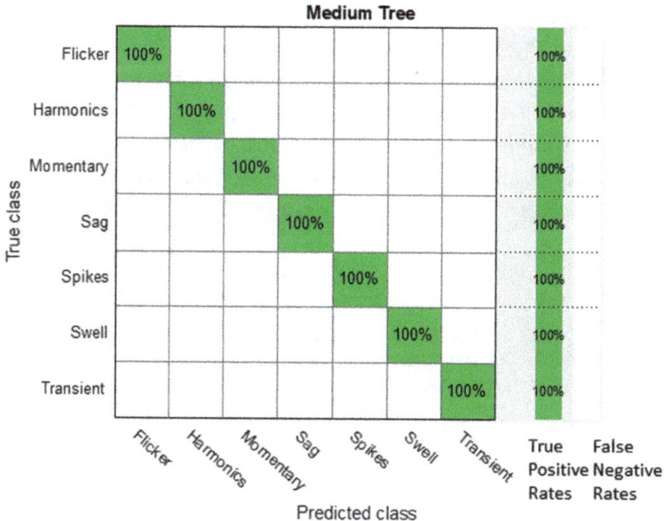

Fig. 5 Confusion matrix of MT at 90% training

and percentages is represented in Fig. 8a, b, respectively. The overall classification accuracy of FT is 99.8% with 90% training.

The performance of the proposed DTCs with different training rates is shown in Table 3, and it is clear that even with less training, the proposed DT classifiers achieve optimal classification accuracy. The performance comparison of various PQ classification algorithms is listed in Table 4. The performance of the proposed DTCs is superior to all the existing approaches in terms of accuracy and in terms of training time.

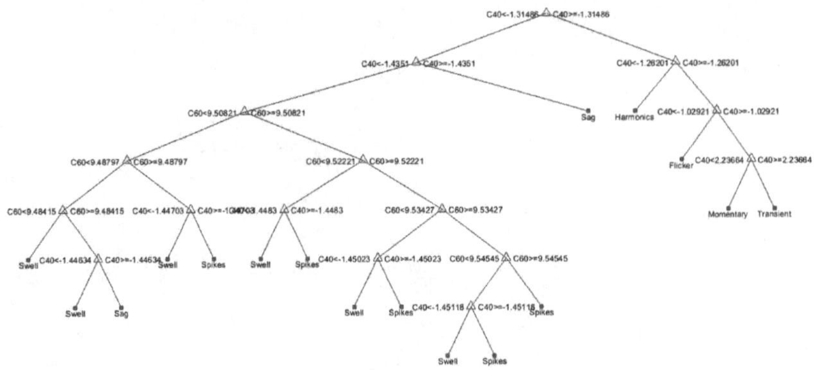

Fig. 6 Classification of PQ events using MT

5 Conclusion

In this paper, a new PQ event classification approach is developed with the use of cumulants and decision tree classifiers. Extraction of different higher-order statistical features for each PQ event and selection of appropriate futures using PCA are presented. The performance of the proposed DT classifiers is analyzed with various training rates. From the simulation results, it is proved that even with less training the proposed DT classifiers achieve more classification accuracy than that of the existing approaches.

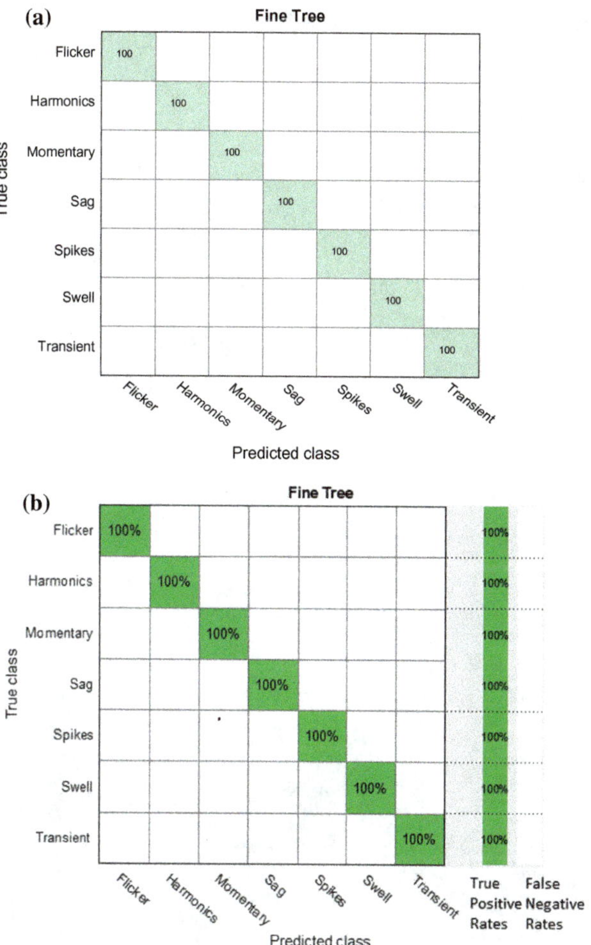

Fig. 7 Performance of FT at 90% training. **a** No. of observations. **b** Confusion matrix

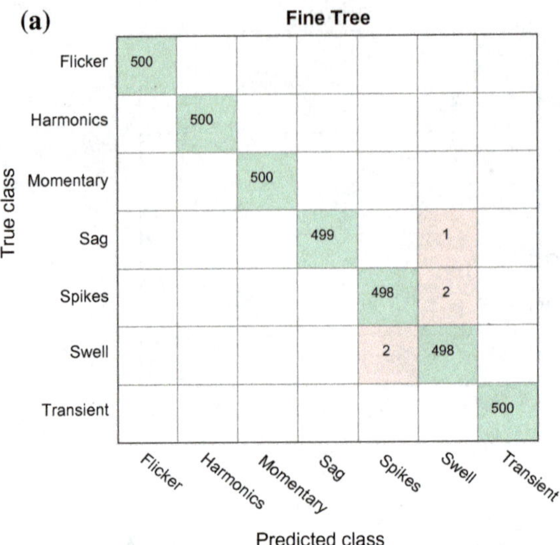

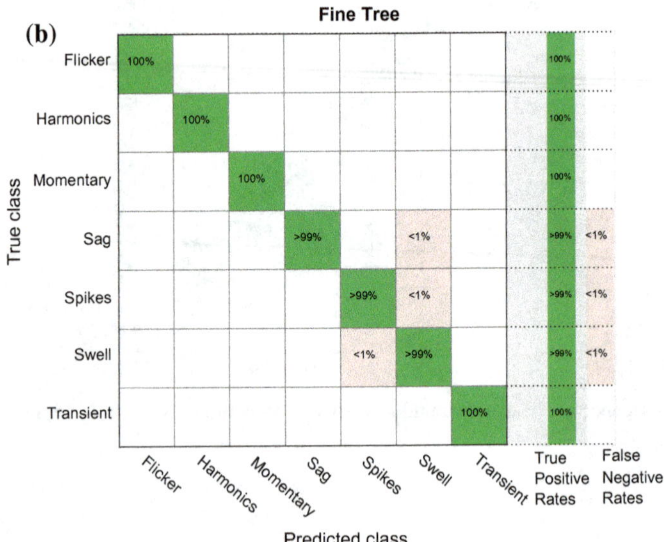

Fig. 8 Performance of FT at 50% training. **a** No. of observations. **b** Confusion matrix

Table 3 Performance of the proposed DTCs with different training cases

Classifier type/% of training	% of classification accuracy							
	90	85	80	75	70	65	60	55	50
FT	100	100	99.9	99.9	99.9	99.9	99.9	99.8	99.8
MT	100	100	99.9	99.9	99.9	99.9	99.9	99.8	99.8
CT	71.4	71.4	71.4	71.4	71.4	71.4	71.4	71.4	71.4

Table 4 Performance comparison of different methods

Method classification	No. of features used for classification	Training time (sec)	Accuracy (%)
WPT and ANN [11]	7	31.26	95.25
ST and PNN [12]	3 4	0.9 (CPU)	95.91 97.40
ST and PNN [10]	7	–	96.10
ST, fuzzy *C*-means and APSO [13]	8	–	96.33
ST, fuzzy *C*-means and GA [13]	8	–	96.45
WPT and SVM [11]	7	27.68	97.25
DWT [2]	8	–	97.81
Proposed FT, MT ≥85% training 60–80% training 50–55% training	3	2.12–2.41	100 99.9 99.8

References

1. H. He, J.A. Starzyk, A self-organizing learning array system for power quality classification based on wavelet transform. IEEE Trans. Power Deliv. **21**(1), 286–295 (2006)
2. M.A.S. Masoum, S. Jamali, N. Ghaffarzadeh, Detection and classification of power quality disturbances using discrete wavelet transform and wavelet networks. IET Sci. Meas. Technol. **4**(4), 193–205 (2010)
3. Z.-L. Gaing, Wavelet-based neural network for power disturbance recognition and classification. IEEE Trans. Power Deliv. **19**(4), 1560–1568 (2004)
4. S. Santoso, E.J. Power, W.M. Grady, A.C. Parsons, Power quality disturbance waveform recognition using wavelet-based neural classifier. I. Theoretical foundation. IEEE Trans. Power Deliv. **15**(1), 222–228 (2000)
5. C.C. Liao, Enhanced RBF network for recognizing noise-riding power quality events. IEEE Trans. Instrum. Meas. **59**(6), 1550–1561 (2010)
6. S. Dalai, B. Chatterjee, D. Dey, S. Chakravorti, K. Bhattacharya, Rough-set-based feature selection and classification for power quality sensing device employing correlation techniques. IEEE Sens. J. **13**(2), 563–573 (2013)
7. F. Zhao, R. Yang, Power-quality disturbance recognition using S-transform. IEEE Trans. Power Deliv. **22**(2), 944–950 (2007)
8. A.M. Gargoom, N. Ertugrul, W.L. Soong, Automatic classification and characterization of power quality events. IEEE Trans. Power Deliv. **23**(4), 2417–2425 (2008)
9. P.K. Dash, M.V. Chilukuri, Hybrid S-transform and Kalman filtering approach for detection and measurement of short duration disturbances in power networks. IEEE Trans. Instrum. Meas. **53**(2), 588–596 (2004)
10. C.Y. Lee, Y.X. Shen, Optimal feature selection for power-quality disturbances classification. IEEE Trans. Power Deliv. **26**(4), 2342–2351 (2011)
11. W. Tong, X. Song, J. Lin, Z. Zhao, Detection and classification of power quality disturbances based on wavelet packet decomposition and support vector machines. In Proceedings 8th International Conference on Signal Processing, IEEE Xplore, 1–4 (2006)

12. S. Mishra, C.N. Bhende, B.K. Panigrahi, Detection and classification of power quality disturbances using S-transform and probabilistic neural network. IEEE Trans. Power Del. **23**(1), 280–287 (2008)
13. B. Biswal, P.K. Dash, B.K. Panigrahi, Power quality disturbance classification using fuzzy *C*-means algorithm and adaptive particle swarm optimization. IEEE Trans. Ind. Electron. **56**(1), 212–220 (2009)

Foreground Segmentation Using Multimode Background Subtraction in Real-Time Perspective

Veerati Raju, E. Suresh and G. Kranthi Kumar

Abstract Nowadays foreground segmentations are becoming more complex in videos and images while capturing at distinct backgrounds. In this work, we addressed the multimode background suppression in video change detection, where it has many challenges to handle like illumination changes, different backgrounds, camera jitter and moving cameras. The framework contains different inventive systems in background modeling, displaying, order of pixels and use of separate shading spaces. This framework firstly allows numerous background scene models that are pursued by an underlying foreground/background used to estimate the probability for each pixel. Next, the image pixels are merged to form megapixels which are used to spatially denoise the underlying probability assessments to generate paired shading spaces for both RGB and YCbCr. The veils formed during the processing of these information pictures are then merged to separate the foreground pixels from the background. A comprehensive assessment of the suggested methodology on freely available test arrangements from either the CDnet or the ESI dataset indexes shows prevalence in the implementation of our model over other models.

Keywords Foreground segmentation · Video change detection · Computer vision · Background modeling · Background subtraction · Color shading spaces · Pixel classifiers

1 Introduction

Background subtraction (BS) is a standout topic, which has gained wide interest in computer vision perspective. It is a important step in advance video preprocessing and has various applications including video surveillance, movement checking, detection human body, motion acknowledgment and so forth. BS method typically generates a foreground (FG) binary mask for a given input picture and a background (BG) model. BS has become difficult task due to the account of the assorted variety of scenes foundation and the progressions started in the camera itself. Scene varieties can

---

V. Raju (✉) · E. Suresh · G. Kranthi Kumar
Department of ECE, Kakatiya Institute of Technology and Science, Warangal, India
e-mail: rajureddyv@gmail.com

H. S. Saini et al. (eds.), *Innovations in Electronics and Communication Engineering*, Lecture Notes in Networks and Systems 107,
https://doi.org/10.1007/978-981-15-3172-9_56

be in numerous structures, for example, dynamic foundation, brightening changes, irregular question movement, shadows, features, cover and in addition a large number of natural conditions like snowfall and daylight change. Moreover, these progressions are because of camera-related and sensor problems.

Current BS frameworks can solve few of these problems and a substantial segment of them are unanswered because of the diversity in background of the picture, camera development and ecological circumstances. Most of the methodologies developed till now unable handle all the key challenges simultaneously with good performance. So we presented a BS system which produces better results for variety of issues faced while processing the picture. This method uses Background Model Bank (BMB) which contains various BG models. To isolate FG pixels from changing BG pixels induced by scene or camera variation itself, we use megapixel (MP) spatial denoising for estimating the pixel probability of different color spaces in order to produce various FG masks, These FG masks then were used together to produce a final FG mask.

The prime contribution to this study is a universal BG subtraction system called multimode background subtraction (MBS) with many significant innovations like MP spatial denoising of measures, Background Model Bank (BMB), the combination of several binary masks and multi-color spaces for BS. Feasibility study findings of using similar model to manage changes in lighting and camera movement were discussed in [1] and [2], respectively. Enhancements to this work include:

- A thorough assessment of the fusion of suitable color spaces for BS.
- A new update mechanism for a model.
- A new MP spatial denoise based and a dynamic selection system that considerably decreases the variety of variables and computing nature.

Background subtraction is very much investigated area in computer vision perspective, so we illustrate MBS efficiency by offering a extensive comparison with 15 other cutting edge BS algorithms on a collection of freely accessible difficult sequences across over 12 distinct classifications, totaling to 56 video sets. To maintain a strategic distance from inclination in our assessments, we have embraced indistinguishable arrangements of measurements from prescribed by the CDnet 2014 [3]. The broad assessment of our framework shows better FG segmentation and prevalence of our framework in examination with existing best in class approaches.

The paper is organized as follows: Literature work is presented in Sect. 2. Proposed work is discussed in Sect. 3. Algorithm used in the work and its implementation is discussed in Sect. 4. Section 5 presents discussion of the results. Finally Sect. 6 concludes the work.

2 Literature Review

The author Ghobadi et al. presented how the complexity of 2-D/3-D pictures is eliminated simply by means of characterizing a quantity of intrigue, its utilization for hand detection and also for signals recognition. Harville et al. related the standard technique of basis demonstrating by means of Gaussian blends for shading and profundity recordings.

They use complete measured intensity images, so that there is no gripping object to deal with the numerous resolutions. Bianchi et al. reported a rather straightforward way to address vanguard area for 2D/3D images that relies upon on locale growing and shuns screening, though in Leens et al. a primary pixel-based totally basis demonstrating approach referred to as visual background extractor (ViBe) is applied for shading and profundity measurements independently. The following frontal face covering is blended with the assistance of paired image sports, for example, disintegration and widening. An extra targeted method for melding, shading and profundity is—sided by isolating the input, which is applied. Crabb et al. explained how preparatory leading edge is added by way of a separating aircraft in area and a reciprocal channel is attached to pick up the remaining consequences. The method is shown on profundity expanded alpha tangling, which is moreover the focus by Wang et al.

Schuon et al. presented the potential of two-side isolation for manipulating geometric gadgets. The difficulty of combining the profundity and shading measurements and taking care of their different nature is also examined throughout profundity upscaling. To that give up a fee capacity or volume is characterized by Yang et al. that portrays the fee of in principle every viable refinement of the profundity for a shading pixel. Again a respective channel is connected to this extent and after subpixel refinement a proposed profundity is picked up. The development is completed iteratively to perform the last profundity outline. The consolidation of a second see is moreover mentioned.

Bartczak and Koch discussed a similar method utilizing duplicate views. An approach running with one shading picture and several profundity snapshots is portrayed by Rajagopalan et al. Here the data mixture is described in a real manner and validated using Markov Random Fields on which a power minimization strategy is attached. Another propelled approach to join profundity and shading information is by Lindner et al. It relies upon nerve-racking safeguarding biquadratic upscaling and plays out an unusual treatment of invalid profundity estimations below refinements of the profundity for a shading pixel. Again a reciprocal channel is hooked up to this volume and after sub-pixel refinement a proposed profundity is picked up. The advancement is achieved iteratively to accomplish the closing profundity delineate. The consolidation of a second see is likewise examined.

R. H. Evangelio presents element Gaussians in blend display (SGMM) for foundation extraction. Gaussian blend fashions widely utilized as part of the space of reconnaissance. Because of low memory necessity, this model applied as part of the non-stop software. Split and union calculation gives the association if primary mode extends and that reasons weaker movement trouble. SGMM represents criteria of willpower of modes for the example of basis subtraction. SGMM gives better foundation fashions as far as low dealing with time and low memory stipulations; for that reason it's miles undertaking reconnaissance location.

L. Maddalena and A. Pestrosino addressed the self-organizing background subtraction (SOBS) for discovery of moving images in view of neural foundation display. Such model produces self-sorting out model naturally with out in advance studying approximately covered example. This flexible model basis extraction with scene containing constant enlightenment range, transferring foundations and conceal can include into shifting article with basis display shadows cast and accomplishes recognition of various varieties of video taken via desk bound camera. The presentation of spatial rationality out of spotlight display refresh techniques activates the SC-SOBS calculation that offers inspire vigor toward false popularity. L. Maddalena and A. Pestrosino talk approximately large exploratory results of SOBS and SC-SOBS in mild of development place challenges.

A. Morde, X. Ma, S. Guler discussed an idea about for traffic spot. Change place or frontal area and foundation department has been extensively utilized as a part of picture managing and PC imaginative and prescient, as it's miles important increase for disposing of motion facts from video outlines. Chebyshev's likelihood disparity-based totally foundation shows a strong regular basis/frontal area department approach. Such version upheld with fringe and repetitive motion identifiers. The framework utilizes identification of moving item shadows and complaints from extra multiplied quantity protest following and question grouping to refine the further division precision. In this method show-off, exploratory final results on great sort of test recordings show approach superior with camera jitter, dynamic foundations, and heat video and additionally forged shadows. Pixel-based adaptive segmenter (PBAS) is one of the strategies for recognizing shifting article within the video define using basis division with complaint.

Martin Hofmann, Philipp Tiefenbacher and Gerhard Rigoll communicate the unconventional approach for recognition of protest, i.e., for frontal vicinity division. This versatile division machine takes after a non-parametric basis displaying worldview and the foundation is composed by way of as of late watched pixel history. The choice part plays a vital in pixel-based totally versatile department for taking closer view preference. In this method, studying the model used to refresh foundation of the question. The learning parameter provides dynamic controllers for each considered pixel region and attribute it to be dynamic. Segmented adaptive based pixel is ideal in the techniques of refinement.

3 Proposed Method

Subtraction of background mainly classified into five-advance process: preprocessing, modeling of background, preface detection, validation of data and update of model. Preprocessing includes straight forward video processing on input video, such as converting formats and resizing images for subsequent steps. The modeling of the background is accountable for the creation of a statistical scene model followed by the pixel classification in the first phase of foreground recognition step. In this step, false detected preliminary pixels are eliminated in the data verification, forming the final phase foreground mask [4]. The last stage is if required to update the model. The last advance is to refresh the models if important. Our developments principally fall in the utilization of various color specs, background model relying on process of background modeling, formation of MP and foreground label correction, and new model update procedures. We outline every one of these developments in the following sections.

A. **BS with many color spaces**

For the precise segmentation of the foreground, the selection of color space is critical. For background subtraction, numerous color spaces have been used including RGB, YCbCr, HSV, HSI, Lab2000 and normalized-RGB (rgb). HSV, HSI, are utilized for foundation subtracting. Among of these color spaces, we consider the top four since these are broadly utilized color spaces: HSV, RGB, YCbCr, HSI. For several reasons, RGB is a reliable option:

- Luminosity and color statistics are equally scattered in every one of the three color channels.
- Powerful against both ecological and camera motion [5].
- It is most cameras' output format and its direct BS utilization prevents color conversion calculation costs [2].

The utilization of another color spaces: HSI, YCbCr and HSV are inspired from visual system of humankind. Characterizing color observation in HVS is that it has a tendency for assigning out reliable color even under varying illumination over a quite a few time or gap. The color spaces isolate the luminosity and color data in YCbCr coordinate direction and though HSI and HSV on polar directions. As color consistency influences BS to act stronger against brightness alterations, shadow and features. Comparative color space studies [6] have demonstrated that YCbCr outperforms the colors RGB, HSI and HSV and is regarded as the most appropriate color space to foreground segmentation [7].

Because of its autonomous color channels, YCbCr is minimal insubstantial to noise, brightness variations and shadow. RGB is found to be succeeding after HSV or HSI at the base as their mapping coordinates are very inclined to camera motion [1]. Moving from RGB to YCbCr saves hardware and software cost compared to HSV or HSI. In brightness of the correlation, YCbCr acts as a decision for division. Be that as it may, [4]. On the basis of the comparison above, YCbCr offers a natural segmentation option. But earlier techniques do recognize possible issues with YCbCr color space [8–10]. If the present picture includes very dark pixels, the likelihood of error rises as dark pixels are near the origin of RGB space. Similar circumstances do not happen exactly when illumination is less, yet moreover occurs when part of the image ends in darkness. It is customary in indoors with complex scene dimensions and light sources. Object's shadow is one such case were prohibitive usage of YCbCr in such condition realizes a decrease in precision.

In view of computer vision, two color spaces YCbCr and RGB are used to manage illumination conditions. Then we select the channels suitable for the given scene. The difference is that all channels and only one color are used in all current methods. Under poor light environment, Y and RGB channels are used because color information which is spread transversely and Y use the channels (Cr and Cb) of YCbCr to assemble foreground region division accuracy. In the midst of direct lighting conditions, both above color spaces supplement one another in giving an effective FG or BG classification.

To justify the various color spaces, a point by point quantitative examination is displayed in segment by differentiating division accuracy across more than 12 particular arrangements using each color space autonomously, two color spaces together, and by with dynamic picking of color channels.

B. **To Model a Background**

BG modeling is the basic step in BS procedure and the Model's efficiency and accuracy effects the segmentation. Many BG models use either a version of a statistical background model that is multimodal in the pixel sense. Such a strategy has two issues: First, the amount of methods used to model the distribution of pixel probability is hard to determine. Secondly, the dependencies between pixels are ignored and result in bad segmentation.

Recollecting the genuine goal of the BG, we put forward Background Model Bank (BMB), to incorporate various background models. In creating BMB, every BG image is overseen as a background model up with picked color channel merged to form a vector [11, 12]. The vital parts of background models are then joined into various common BG models utilizing an iterative consecutive sequential clustering framework. Two background average models (p and q alive and well) to measure more detectable than estimated corr_th are then combined. The measurement can be calculated by

$$\operatorname{corr}(p, q) = \left(\frac{(p-\mu_p)(q-\mu_q)'}{\sqrt{(p-\mu_p)(p-\mu_p)'}\sqrt{(q-\mu_q)(q-\mu_q)'}} \right) \tag{1}$$

Where μ_p and $\mu_q = \frac{1}{|X|}\Sigma_j q_j$

$$\mu_p = \frac{1}{|X|}\sum_j P_j \text{ and } \mu_q = \frac{1}{|X|}\sum_j q_j \tag{2}$$

This procedure proceeds repetitively unless there are not any normal background models with Corr > corr_th. Utilization of frame level cluster overcomes limitations of dimensions of scene. Normally genuine (real life) scenes consist of various kinds of objects. The diversity in settings and relationships between various kinds of subjects and items produces very complex and infinite geometry in the scene. Illustrations incorporate varieties as a result of sudden variations in light and shake in the camera. This assorted variety makes it hard to precisely catch and model the scene.

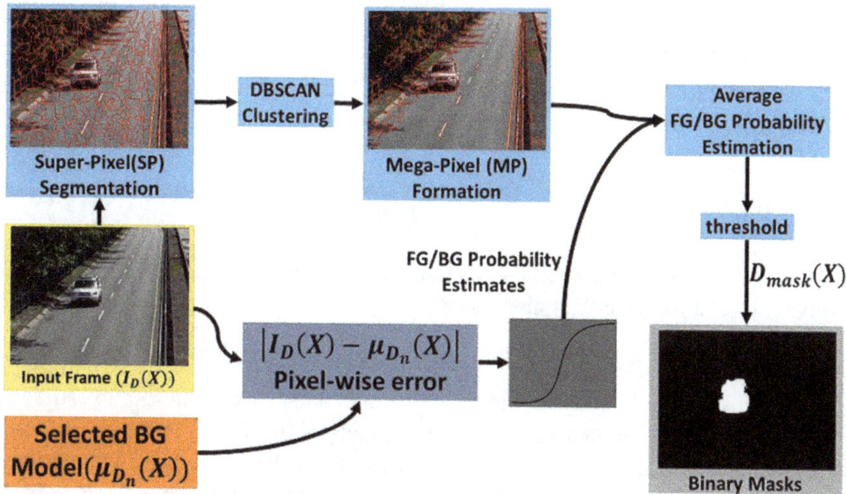

Fig. 1 Block diagram of proposed binary classification and mask generation

Utilization of various background models enables us to capture a scene more correctly. Other preferred standpoint of BMB is that it is mathematically easier than other ways. Only one Model is selected from BMB and rest is left. The experimental data illustrates how various background models could capture scene precisely. Contrasting with more unpredictable multi-modular, this approach gets equivalent or high-quality outcomes, with the use of simple binary classifier for the arrangement of pixels, which makes the application effective.

C. **Double Categorization**

In this section, for each color channel chosen, we address the generation of binary masks. The process carried out in four steps: Activation/disabling of the color channel, probability estimate of pixel level, MP preparation and estimate of mean probability. Figure 1 shows the design process of the binary mask.

(1) Color Channels Activation/disabling:

In this progression, activate or deactivate color channels like Cb and Cr. The two channels are used if the average intensity of input data image is larger than specifically chosen parameter channel_th, otherwise for the most part they are not used. In the event that the information of image strength is more significant than the decided parameter channel_th then one channel otherwise we can utilize both color channels.

(2) Estimation of Pixel-Level Probability:

Pixel wise error, error $D(X)$ is calculated in each of color channel YCbCr and RGB and based on the values, we chose one BG model.

$$\text{err}D(X) = |I_D(X) - \mu D_n(X)| \tag{3}$$

where D implies the color channel under test, $I_D(X)$ is the input image data, and μDn (X) is the mean of BG model. Initially, we determine the error for the individual pixel, then we calculate a initial probabilities I_p of all pixels by subjecting through a sigmoid function.

$$i_p(\mathrm{err}D(X)) = \frac{1}{\left(1 + e^{-\mathrm{err}D(X)}\right)} \tag{4}$$

This method of analysis behind this transform is that the larger the error, the more probability that the pixel belongs FG.

(3) Formation of Megapixel:

The aim of this step is to demonstrate spatial denoising by considering ip estimates and color data information of the surrounding pixels within the Super-Pixels (SP). SPs give advantages, when it comes to capture the local context and reduce computational level of complexity substantially. These algorithms merge nearby pixels into a single pixel based on a measure of resemblance. The graph partitioning problem is used for formulating SP segmentation. The goal for a graph $G = (V, E)$ and M of the SPs is to locate a subset of the edges $A \subseteq E$ to estimate a graph $G = (V, A)$ with at least M sub-graphs connected. There are two elements of the clustering objective function: entropy rate H in a random direction and balance term.

$$\begin{aligned} &\mathrm{maxAH}(A) + \lambda B(A), \\ &s.t. A \subseteq E \text{ and } N_A \geq M \end{aligned} \tag{5}$$

where NA is the quantity of associated parts in G. A substantial entropy is related to smaller and similar groups, though the adjusting term empowers compact with relative clusters. To improve over segmentation, SPs are consolidated to frame substantially greater MPs utilizing density-based spatial clustering for applications with noise clustering (DBSCAN).

DBSCAN clustering algorithms are based on density and clusters are characterized as denser areas, while the sparse areas are considered as outliers or boundaries to separate clusters. Two SP are combined to megapixels (MP) with below-mentioned criteria:

$$\mathrm{MP} = \begin{cases} 1 & \text{dist} \leq \text{threshold of color} \cap \text{SPs are bordering} \\ 0 & \text{dist} > \text{threshold of color} \cup \text{SPs are not ordering} \end{cases}$$

For any two neighboring SPs, distance range function depends on average lab color deviation and is characterized as:

$$\mathrm{dist} = \left|\mu_y^L - \mu_z^L\right| + \left|\mu_y^a - \mu_z^a\right| + \left|\mu_y^b - \mu_z^b\right| \tag{6}$$

$$\mu_y^{\mathrm{ch}} = \frac{1}{Y}\sum_{\mathrm{np}=1}^{Y} \mathrm{ch}(\mathrm{np}) \tag{7}$$

(4) Labeling and Estimation of Average Probability:

Next stage is to find mean probability of a MP y, denoted as AP y, with a sum of Y pixels: In the following stage we are figuring the normal probability of

$$\mathrm{AP}_y = \frac{1}{Y}\sum_{\mathrm{np}=1}^{Y} \mathrm{ip}(\mathrm{np}) \tag{8}$$

np represents pixel index and ip is the elemental estimated probability of BG/FG of all pixels. The Average Probability (AP) is allocated to all pixels that belongs megapixel. To get Binary Mask *D*mask (*X*) for every colored *D*, the AP is rounded with use of empirically estimated parameter threshold prob_th. Usage of MP and its specific AP empower us to assign out a mean probability to each pixel having a place with a comparable inquiry and in this way fabricates the segmentation accurateness.

For example, each one of the pixels that belong to the road in Fig. 2 is supposed to be BG. obviously, in Fig. 2, as we move from left to right, pixels of road having incorrect calculation of probability estimates the center out using neighboring pixels by methods SPs or MP, in like manner upgrading the segmentation precision.

The normal probability of a MP addresses a comparative inquiry for every pixel or SPs. Model update is indeed an important element for an algorithm to address scene modifications over time. After several frames or time periods, the classical method of model updates is to replace older values well into the model with newer

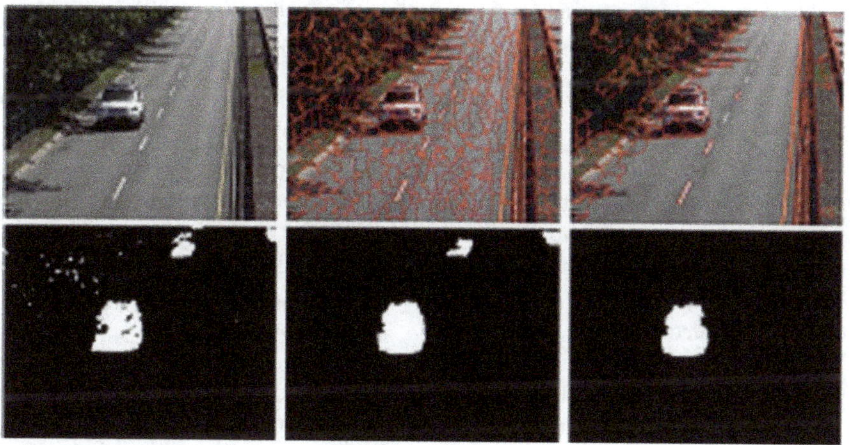

Fig. 2 Correlation of segmentation with mean probability measurement of every pixel, normal movement mean probability estimation (center) based on SP, and normal movement probability estimation based on MP (right)

ones. Such processes for upgrading can be a problem because the update frequency is difficult to be determined.

For instance, a man sitting at rest in a scene may be part of a BG if update frequency is fast. Another situation could be baggage that is forgotten, during which issue arises as to whether it would ever become a background or be a part of background? Two issues should be addressed through an update system. First, is updating the model necessary? Secondly, what is the update frequency? We contend that a shift rate of the FG pixels can help activate the update of the model and set an adequate update rate. The amount of FG pixels in a typical supervision scene fluctuates comparatively slightly and substantial changes can lead to a departure from the conventional BG model.

$$\text{model update} = \begin{cases} 1 \text{ if rate of change} \geq \text{th} \\ 0 \qquad\quad \text{otherwise} \end{cases}$$

where th is a precisely chosen parameter that suggests an adequately essential change in support of model update. The rate of change is figured in perspective of the departure of the amount of foreground pixels in present packaging from the average. Formally, we describe it as:

$$\text{rate of change} = \frac{\sum_{x \in X} O_t(X) - \frac{1}{h}\left(\sum_{i=t-h-1}^{t-1} \sum_{x \in X} O_i(X)\right)}{\frac{1}{h}\left(\sum_{i=t-h-1}^{t-1} \sum_{x \in X} O_i(X)\right)} \tag{9}$$

where $O_t(X)$ represents the o/p binary mask at moment of t for a given current i/p image.

Upon triggering the model update system and calculating the rate of change, the update feature f can be used to plot the pace of transition to identify the suitable change rate U and define it as:

$$U = f(\text{rate of change}) \tag{10}$$

To know the need of an update rate feature f, we need to know first how and what sort of modifications inside a scene can actually occur. BG changes can be made at various rates from slow to abrupt. The gradual light shift from the beginning of sunrise to end of sunset shows that a slowly evolving BG needs a slow update rate. There can be startling changes caused by sudden light changes in domestic conditions or camera related problems.

Unable to choose a reasonable update rate can become conscious over to the number of false positives. Therefore, it is essential to determine the correct update rate for altering BG dynamically. The choice of the update rate feature f from easy linear functions to complicated features is distinct. Two applicants are a simple or nonlinear function based on the ease and efficiency of parameters.

An immediate limit gives an unmistakable direct link between advance rate and rate of update. An exponential limit is used for greater sensitivity. This limit can suit to adjust during unexpected change in light. So we used a less complex function:

$$U = m * \text{rate of change} \tag{11}$$

where m = slope and $0 < m < 1$. For instance, when $m = 0.25$ and rate of advance = 1, the processed revive rate would be 0.25, implies less weightage to old BG and more weightage to present one. After selecting the update rate, a model is then revived as takes after:

$$\mu_n(X) = (1 - U).\mu_n(X) + U.I_t(X) \tag{12}$$

where an It (X) address current data diagram at time t and (µn (X)) is the picked BG appear for current edge and is being updated.

The dynamic model update framework grants to give provisions to various circumstances in which standard approaches make an impact. For example, no model revive will be associated when there is no FG in the scene or FG is not changing as the rate of advance is close to zero.

Eventually, at whatever point there is a change in BG, it can continuously choose revive rate and a while later update BG appear efficiently.

Because of the randomly generated initial starting points, *K*-means algorithm is, however, hard to reach the global optimum; instead, it produces one of the local minimum, leading to improper clustering results. Barakbah and Helen performed that the error ratio of *K*-means is more than 60% for well-separated datasets. To prevent this, we use our prior work for the optimization of initial clusters for K-means with pillar algorithms. The Pillar algorithm is very robust and superior for optimizing initial centroid for *K*-means by placing all centroids far apart in the data distribution.

4 Algorithm Development

In this sub-section, we outline how individual works are grouped in our structure. The proposed structure includes five phases which is showed in Fig. 3. Each and every stage is explained below.

Stage 1: Selecting the Background Model

The first stage will be to choose the suitable BG model for the incoming image and to define the model BG in the BMB which enhances the correlation with that of input image $I(X)$:

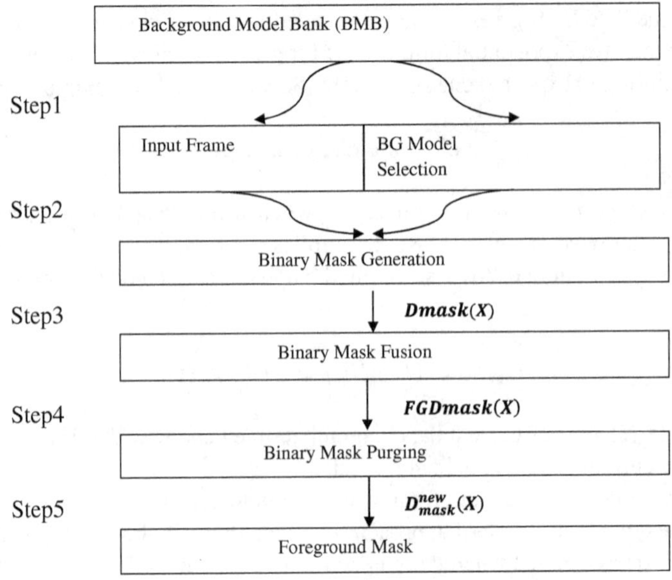

Fig. 3 Universal multimode subtraction system

$$\text{corr} = \arg\max n = 1, \ldots, N \times \left(\frac{(1-\mu_i)(\mu_n-\mu)'}{\sqrt{(1-\mu_i)(1-\mu_i)'}\sqrt{(\mu_n-\mu)(\mu_n-\mu)'}} \right) \tag{13}$$

Stage 2: Generation of Binary Mask (BM)

Here, the information picture and the picked BG exhibit are utilized to evaluate a initial probability of each pixel. The input information image is passed with same time as MP module, which partitions the image in individual number of MPs. Avg probability estimates will be computed for every MP utilizing pixel-level probability estimates and from there thresholded to produce binary masks (BM) for every color channel. We demonstrate the BM for color channel D as Dmask (X).

Stage3: Binary Masks Aggregation/Fusion

The BMs will then be used to create foreground detection (FGD) masks for color spaces RGB and YCbCr. For the color of YCbCr, the FG DY YCbCr mask will be decreased only to Y channel BM if Cb and Cr channels are disabled. Finally, both FGD masks are merged to obtain the actual FGD mask with a logical AND between distended versions.

Stage 4: Purging of Binary Masks

The FGD mask is implemented in step 3 is applied to the individual BMs. This eliminates all the erroneous foreground areas and improves consistency in the final phase in the classification of FG and BG pixels.

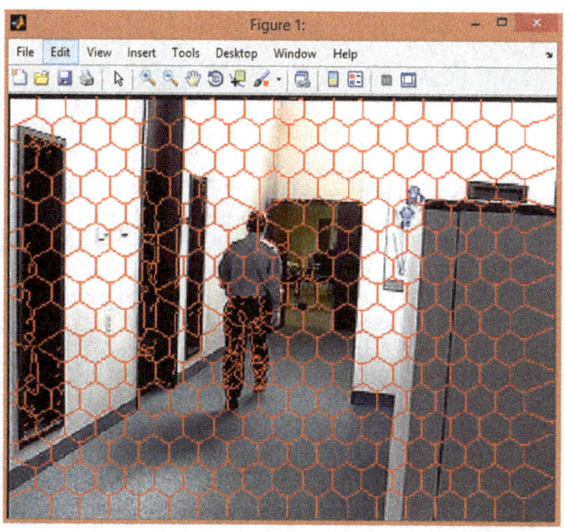

Fig. 4 Super-pixels segmented input image

Stage 5: Foreground Mask

In the last stage of the method, FG mask is derived from the logical OR of all the *D*mask new (*X*) masks.

5 Results

(See Figs. 4, 5, 6, 7, 8 and 9).

6 Conclusion

In this paper, we proposed a method where we can segment the foreground and background and we can improve the efficiency. To do this, the pixel-level comparison is done and probabilities are estimated by which spatial denoising occurs. Based on the illumination conditions, low light vision, RGB and Y color channels bright light CbCr are used to get the foreground segmentation. In this, we are using *K*-means method by using pillars algorithm. *K*-means algorithm gives more compatible results compare with DBSCAN clustering due to the large size of data and small number of variables. However, *K*-means algorithm gives accurate results when we are using large dataset. In shadow suppression and in moving camera categories, Mbs will give the better performance results. And the proposed method implementation is done by using the MATLAB software. And in the future, we can develop by using the C and C++ algorithms.

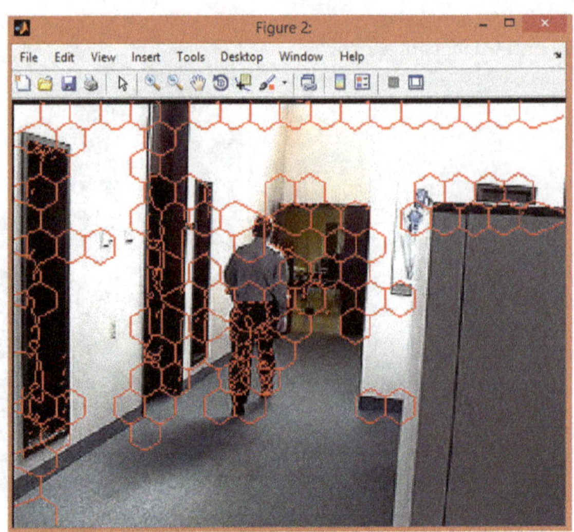

Fig. 5 Image after DBSCAN clustering

Fig. 6 Binary mask image output

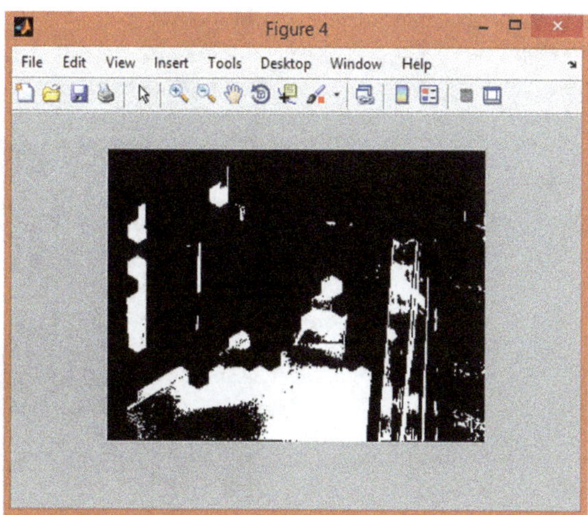

Fig. 7 Binary mask after thresholding

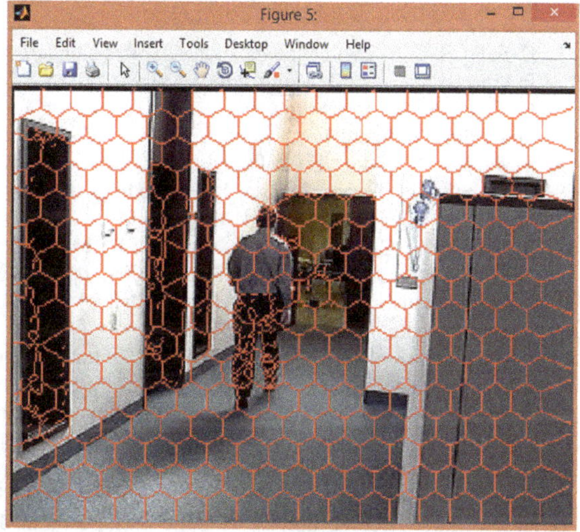

Fig. 8 Image after applying pillar *K*-means

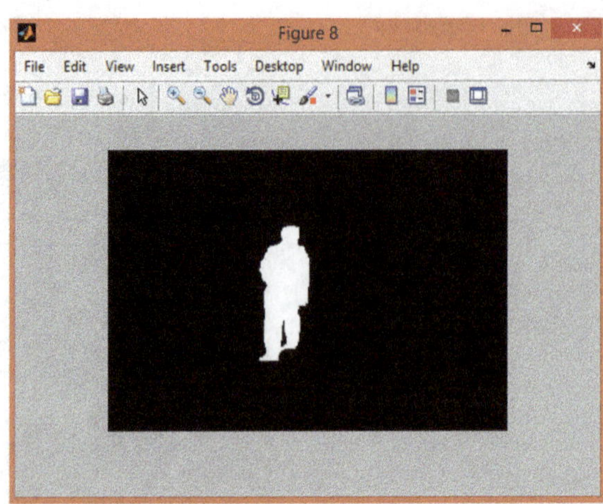

Fig. 9 Detected foreground by proposed work

References

1. H. Sajid, S. C. S. Cheung, Foundation subtraction underneath unexpected brightening change. In Proceedings of IEEE Multimedia Signal Process (MMSP) (Sep. 2014), pp. 1–6
2. H. Sajid, S.- C. S. Cheung, Foundation subtraction for static moving digital camera. In Proceedings of International Conference on Picture Process, (Sep. 2015), pp. 4530–4534
3. Y. Wang, P. -M. Jodoin, F. Porikli, J. Konrad, Y. Benezeth, P. Ishwar, CDnet 2014: An prolonged trade region benchmark dataset. In Proceedings of Computer Vision Example Recognition Workshops (CVPRW), 387–394 2014
4. S. C. S. Ching, C. Kamath, Strong structures for basis subtraction in urban rush hour gridlock video. In Proceedings of Electron Image,(2004), pp. 881–892
5. Change popularity Dataset, got to on [Online] Accessible 15 Dec. 2016: https://www.Changedetection.Internet
6. L. P. Vosters, C. Shan, T. Gritti, Background subtraction beneath sudden illumination changes. In Proceedings of AVSS, (Sep. 2010), pp. 384–391
7. S. Brutzer, B. Hoferlin, G. Heidemann, Evaluation of historical past subtraction techniques for video surveillance. In Proceedings of CVPR, (Jun. 2011), pp. 1937–1944
8. K. Toyama, J. Krumm, B. Brumitt, B. Meyers, Introvert: Principles and habitual on the subject of basis help. In Proceedings of ICCV, pp. 255–261 Sep. 1999
9. T. Bouwmans, Recent superior statistical background modeling for foreground detection—A systematic survey. Recent. Pats Comput. Sci. **4**(3), 147–176 (2011)
10. C. Stauffer, W. E. L. Grimson, Adaptive heritage mixture fashions for real-time monitoring. In CVPR, (1999)
11. A. Elgammal, D. Harwood, L. Davis, Non-parametric model for heritage subtraction. In Proceedings of ECCV, (2000), pp. 751–767
12. P.D.Z. Varcheie, M. Sills-Lavoie, G.-A. Bilodeau, A multiscale location-primarily based motion detection and background subtraction algorithm. Sensors **10**(2), 1041–1061 (2010)

A Comparative Study on LSB Replacement Steganography

C. Sailaja and Srinivas Bachu

Abstract Steganography is an approach of communication where sender and receiver can expect the hidden message. Major goal of steganography is to hide data and should not be susceptible, with less MSE, High PSNR and also with maximum capacity to hold the secret data. In this paper, a comparative analysis is made to evaluate the performance of the LSB replacement algorithm for different number of LSBs. The performance of the LSB replacement steganographic method has been evaluated by computing capacity, mean square error (MSE), peak signal-to-noise ratio (PSNR) and robustness.

Keywords Steganography · LSB replacement technique · MSE · PSNR · Robustness · Capacity

1 Introduction

Steganography is method for correspondence in mystery way that just the sender and the proposed recipient can estimate about the message. As it conceals the private or delicate data inside something that has all the earmarks of being nothing out of the standard thing. Generally, there is disarray among steganography and cryptology in light of the fact that the two are comparative in the way that they both are utilized to secure imperative data. The main distinction between two is that steganography includes concealing data so that it shows up as no data is covered up by any means. In the incident that the individual or people see the item in which the data is covered up and the person has no clue that there is any data holed hidden behind, at that point, the individual will not endeavour to decode the data.

The original version of this chapter was revised: The author's "S. Bachu" affiliation has been updated. The correction to this chapter is available at https://doi.org/10.1007/978-981-15-3172-9_75

C. Sailaja (✉)
Gurunank Institutions Technical Campus, Hyderabad, India
e-mail: sailaja.cheruvupalli@gmail.com

S. Bachu
Marri Laxman Reddy Institute of Technology and Management, Hyderabad, India

H. S. Saini et al. (eds.), *Innovations in Electronics and Communication Engineering*, Lecture Notes in Networks and Systems 107,
https://doi.org/10.1007/978-981-15-3172-9_57

Typically, a secret word is utilized for encryption when a data is covered up inside the document. Steganography is one of the basic ways that helps the people to hide the information. Data is imperative to human endeavours to be sure computerized data offers awesome chances and enhancements to individual's life, specifically with the presentation of Web. Presently, multi day's Web has turned into the most famous wellspring of data, which offers boundless channels to exchange data starting with one station then onto the next. Verified correspondence is a vital issue for quite a while. At first, cryptography was made for keeping the correspondence verified [1].

In cryptography, the message content is kept secret. But keeping information secret is not enough in most of the cases and to maintain the information unsusceptible is necessary. Steganography is developed to for this purpose. In cryptography, secret message is encoded, cipher text is generated, encoded message is unreadable, the message can be decoded only through the secret key, whereas steganography is a method to hide the information in such way that it cannot be detected by any one. Steganography is an art of invisible communication and has been used in various forms for 2500 years. Steganography is the word derived from the Greek language, the words "stegos" and "grafia". "stegos" means "cover" and "grafia" means "writing". In steganography, a cover source such as digital images, text files, videos and TCP/IP packets is used to hide the message. Both the methods steganography and cryptography are focused on same thing, i.e. avoiding an intruder to detect the secret information from intruder. But a single technology cannot be perfect. If the existence of hidden message is detected/suspected, the purpose of steganography is somewhat conquered. In some of the methods to increase the strength of invisible communication, both steganography and cryptography were combined. These techniques have many applications in communications and other related fields. They are used to protect e-mails, credit card information, corporate data, personal files, military messages, etc. There are different types of steganographic techniques available which are as follows:

- Text Steganography: In this method, secret text file is hidden in the cover source.
- Image Steganography: In this method, secret image is hidden in the cover source.
- Audio/Video Steganography: In this method, secret audio/video file is hidden in the cover source.
- TCP/IP packets: In this method, TCP/IP packets are hidden in the cover source.

In recent days, the use of Internet is increased a lot, transferring images also increased, and steganography has become a burning research area [2]. The major goal of digital steganography is to hide data as Internet is a popular communication media [3]. In steganography, the cover source may be digital images, audio files or video to hide the secret information; in this paper, digital image is considered as the cover material to hide text information. After hiding the secret data into cover image, the output image is called stego-image, which is transmitted through Internet, and at the receiver side, the stego-image is searched for the secret data by using the key.

2 Related Work

To design helpful steganography algorithm, it is imperative that the stego-image does not have any visual curio and it is measurably like regular images. In the event that an outsider or spectator has some doubt over the stego-image, steganography algorithm ends up futile. Amid the most recent decade, numerous steganographic algorithms for advanced images have been proposed [2]. Then again, numerous steganalysis techniques, whose objective is to break down an image to distinguish the nearness of a shrouded message in it, are additionally proposed. The algorithms of image steganography can be divided broadly into two types by its embedding method: spatial domain embedding method, which works directly on the pixels of image and frequency domain embedding technique, which converts the image into frequency domain and then performing operations. LSB replacement strategy is one of the well-known spatial domain embedding methods, which replaces least significant bits of the image by a secret message or its encoded rendition. The concealing limit of this procedure is straightforwardly related with the image measure.

Major advantages of the LSB replacement embedding technique are simple execution, imperceptibility of hidden message to human visual system and secret data can be embedded at a high rate.

A. **LSB Replacement Technique**

LSB replacement is a typical basic way to deal with inserting data in an image, but it is powerless against even slight image control. Changing over image from an arrangement like GIF or BMP to JPEG and back could demolish the data covered up in LSBs [4]. One of the most widely recognized techniques for usage is least significant bit insertion, in which the least noteworthy piece of each byte is adjusted to shape the bit-string speaking to the implanted document. Modifying the LSB will just purpose minor changes in shading and subsequently is generally not observable to the human eye. While this method functions admirably for 24-bit colour image documents, steganography has not been as effective when utilizing a eight-bit colour image record because of limitations in shading varieties and the utilization of a shading map. LSB strategy is a simple basic technique for concealing information. In any case, stego-images can draw doubt or be effectively identified from measurable investigation.

3 Major Goals of Steganography

Three significant objectives of image steganography to accomplish are capacity, robustness and imperceptibility [3]. These goals dependent of each other, and are not free of one another, somewhat in the vast majority of the cases it moves towards becoming exchange off among these objectives for considering any steganographic method. Other than these fundamental perspectives, some functional angles like simplicity of usage (complexity in design) and time complexity of the method are also important. Basic goals are briefly discussed in the next.

A. **Capacity**

The capacity of a cover image can be defined as the maximum secret data that can be hidden in the image. For basic LSB substitution, 12.5% is the maximum capacity that can be achieved. The property of cover image and the algorithm used to embed the secret data decides the maximum capacity. Many algorithms were proposed to enhance the capacity of cover image. Instead of using a single bit, more number of bits can be used to increase the capacity, and it is the simplest technique in LSB replacement technique. The capacity and robustness are tradeoff. If the capacity is increased, then the image may become susceptible to the viewer (up to two bits, LSB change will not affect the image, but more than two bits replacement, the image will be much noticeable.

In general to calculate capacity, the following equation is used,

Capacity = Number of pixels in the image × Number of LSB bits replaced.

The above equation cannot be adaptable to all the steganographic methods. Because now many steganographic methods are introduced by compressing the data and cover image. For these methods, the capacity ratio varies according to the compression method, and any function cannot correctly define embedding capacity of a cover image.

B. **Imperceptibility**

The goal of any steganographic method is that intruder should not suspect the cover image or communicating image for secret data. To provide better imperceptibility for a stego-image, it should be of higher dependability. The stego-image should be maintained similar to cover image so that intruder cannot identify the difference with naked eye [5].

PSNR, MSE are the most common metrics used to evaluate any steganographic method.

To calculate MSE of a cover image,

$$\mathrm{MSE} = \frac{1}{\mathrm{mn}} \sum_{i=0}^{m-1} \sum_{j=0}^{n-1} [I(i, j) - K(i, j)]^2$$

where

$I(i, j)$ = an original cover image,
$K(i, j)$ = a reconstructed cover image or stego-image,

m, n are the number of rows and columns correspondingly.
To calculate PSNR,

$$\mathrm{PSNR} = 10.\log_{10}\left(\frac{\mathrm{MAX}_I^2}{\mathrm{MSE}}\right)$$

Or

$$\text{PSNR} = 20.\log_{10}\left(\frac{\text{MAX}_I}{\sqrt{\text{MSE}}}\right)$$

where

MAX_I is the maximum deviation in the original image $I(i, j)$.

Ideally, MSE of a cover image should be very minimum, and PSNR should be maximum.

C. **Robustness**

Robustness can be defined as the resistance of stego-image towards different steganalysis techniques. Steganalysis is the countermeasure of steganography. It checks the image to detect the secret message [6].

The only aim of any steganalysis technique is to detect the image for secret data, and the steganalysis algorithms calculate statistical difference between input image (stego) and cover input (output). So, any method used to embed data in an image should defeat the steganalysis techniques and should transfer the information successfully without making the image susceptible to anyone. Robustness of any steganography method can be measured using histogram plot of image [7].

Histogram is a graphical representation of colour distribution in any digital picture. It plots the quantity of events of pixel intensity in a picture [8]. In the event of LSB replacement steganography, as the LSB bits are changed, the intensity esteems change, and consequently, the histogram of that picture likewise changes. These progressions can be utilized to distinguish nearness of mystery message in steganalysis. So, it is beneficial if the histogram contrasts among spread and stego-picture are lesser as that makes it progressively resistive towards identification.

4 Implementation

Algorithm for LSB Replacement

Step1: Read the cover image and message to be hiding.
Step2: Check whether the cover image is enough to hide the given message or not if not display error.
Step3: Convert cover image and message to bits.
Step4: Identify the number of LSB bits (right most bit) of each pixel in cover image to be replaced.
Step5: Replace LSB bit(s) with message bits in every pixel.
Step6: Display cover image and stego-image.
Step7: Calculate MSE, PSNR for both cover image and stego-image.
Step8: Plot histogram of both the images.
Step9: Repeat the steps from 4 to 8 for LSB two bits, three bits and four bits.

Algorithm for LSB Replacement:

Step1: : Number of bits to be replaced should be grouped and shifted as least significant bits using reshape().
Step2: Perform complement operation of the shifted byte and AND operation with the pixel value. Hence, least significant bits are cleared in each pixel.
Step3: Perform OR operation with the data to be hides. Then, LSBs are replaced with the message bits.
Step4: It results n bits of message embedded into the pixel LSB.

Algorithm for Retrieval process

Step1: Read the stego-image.
Step2: Collect the key, i.e. number of bits embedded in the image pixels.
Step3: Collect all the least significant bits from image pixels.
Step4: Step4: Reshape them into ASCII values and then to characters.

If the data to be hidden is "HELLO SAILAJA", then each character of text data is converted to ASCII then to binary.

Character	ASCII value	Binary
H	72	1001000
E	101	1100101
L	108	1101100
L	108	1101100
O	111	1101111
S	83	1010011
A	65	1000001
I	73	1001001
L	76	1001100
A	65	1000001
J	74	1001010
A	65	1000001

The character "H" is 1001000 can embedded in the pixels as follows:
For one-bit LSB replacement

1010101**1**	0101010**0**	1100111**0**	0011001**1**
1110001**0**	0001110**0**	1011011**0**	0110110

For two-bit LSB replacement

101010**10**	010101**01**	110011**00**	001100**00**

For three-bit LSB replacement

10101**100**	01010**100**	11001**110**

For four-bit LSB replacement

1010**1001**	0101**1000**

5 Results and Discussions

To implement the above algorithm, MATLAB Version R2016b was used. The figures show stego-images for different bit LSB replacement (Figs. 1, 2, 3, 4 and 5).

The figures show histogram plots of stego-image and original image histogram plots (Figs. 6, 7, 8, 9 and 10).

Fig. 1 Original test image

Fig. 2 Bit LSB replacement

Fig. 3 Bit LSB replacement

Fig. 4 Bit LSB replacement

Fig. 5 Bit LSB replacement

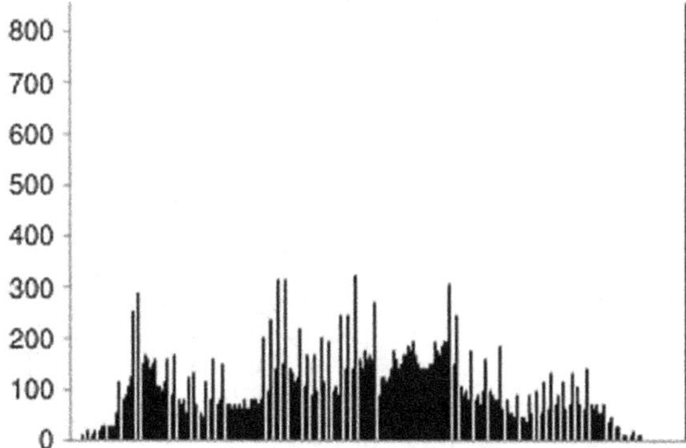

Fig. 6 Original image histogram

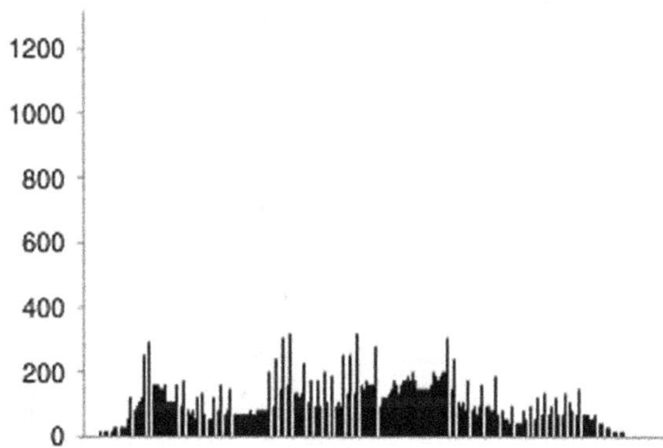

Fig. 7 Bit LSB replacement histogram

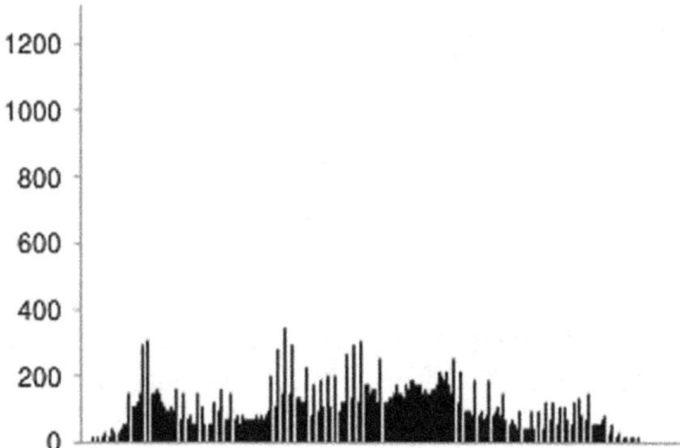

Fig. 8 Bit LSB replacement histogram

From the above stego-images and their histogram plots, it is observed that one-bit and two-bit LSB replacement does not affect the image or its histogram plot, whereas three-bit and four-bit LSB replacement shows some difference.

Using the above formulae, the MSE and PSNR for different LSB bits substitution are calculated (Table 1).

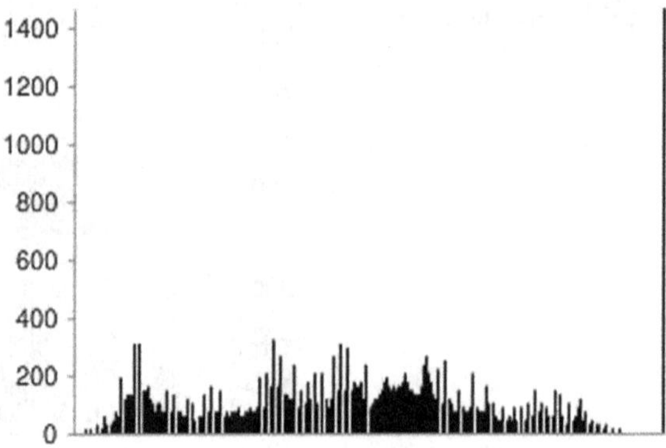

Fig. 9 Bit LSB replacement histogram

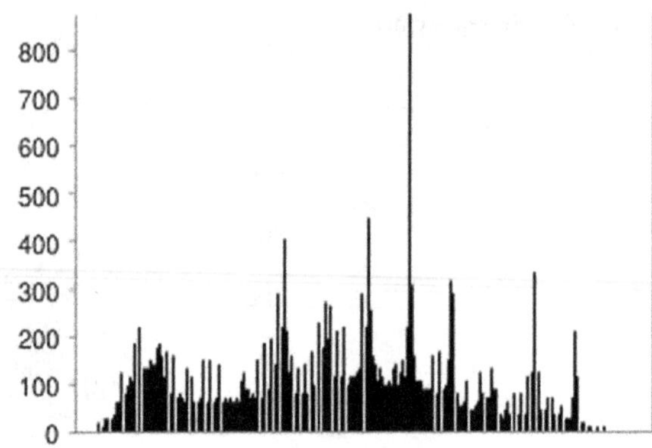

Fig. 10 Bit LSB replacement histogram

Table 1 MSE and PSNR between cover and stego-images

Number of bits replaced	MSE	PSNR (db)
1	17.85742	35.3005
2	14.32685	36.2571
3	9.392198	38.0910
4	6.601758	39.6221

From the table, it is observed that as number of LSB bits (capacity) are increasing, MSE value is decreasing, and PSNR value is increasing.

6 Conclusion

In this paper, the performances of different LSB replacement steganography techniques are analysed. The major goals of text steganography are embedding capacity, MSE, PSNR and robustness. From the simulation results, it is observed that embedding capacity and robustness, MSE, PSNR are trade off. The LSB replacement up to two bits doesnot affect the MSE, PSNR and robustness, but the embedding capacity is less. Three- and four-bit LSB replacement increases the capacity, but MSE, PSNR and robustness are poor. This can be overcome by introducing new methods to embedding more data without making the image suspicious.

References

1. J. S. Priya, et al., Ensuring Security in Sharing of Information Using Cryptographic Technique International Conference on Intelligent Computing and Applications, pp. 33–39
2. A. Hernandez-Chamorro, A. Espejel-Trujillo, J. Lopez-Hernandez, M. Nakano-Miyatake, H. Perez-Meana, *A Methodology of Steganalysis for Images CONIELECOMP*. 09 Proceedings of the 2009 International Conference on Electrical, Communications, and Computers. pp. 102–106
3. G. Maji, S. Mandal, S. Sen, N. C. Debnath, Dual Image based LSB Steganography, (2018). https://doi.org/10.1109/sigtelcom.2018.8325806
4. Z. Zhao, F. Liu, X. Luo, X. Xie, L. Yu, *LSB Replacement Steganography Software Detection Based on Model Checking*, ed. by Y. Q. Shi, H. J. Kim, F. Pérez-González. The International Workshop on Digital Forensics and Watermarking 2012. IWDW 2012. Lecture Notes in Computer Science, vol 7809 (Springer, Berlin, Heidelberg, 2013)
5. A. K. Gulve, M. S. Joshi, Image steganography algorithm with five pixel pair differencing and gray code conversion. Int. J. Image, Graph. Signal Process. (IJIGSP) **6**(3), Feb. (2014)
6. W.C. Kuo, S.H. Kuo, L.C. Wuu, Multi-bit data hiding scheme for compressing secret messages. Appl. Sci. **5**, 1033–1049 (2015). https://doi.org/10.3390/app5041033
7. L. Chen, Y. Q. Shi, P. Sutthiwan, X. Niu, *A Novel Mapping Scheme for Steganalysis*. The International Workshop on Digital Forensics and Watermarking, (2012) pp. 19–33
8. K. B. Raja, S. Siddaraju, K. R. Venugopal, L. M. Patnaik, *Secure Steganography using Colour Palette Decomposition*. 2007 International Conference on Signal Processing, Communications and Networking, Chennai, pp. 74–80 (2007). https://doi.org/10.1109/icscn.2007.350699
9. V. S. Burepalli, Rao, Digital watermarking for relational databases using traceability parameter. Int. J. Comput. Appl. Technol. (2009)

C. Sailaja M.Tech., working as Assistant Professor, Gurunanak Institutions Technical Campus, Hyderabad. Her area of interest is image processing. Published papers on image processing. Member of IAENG and IRED.

Srinivas Bachu Associate Professor, Department of ECE, Marri Laxman Reddy Institute of Technology and Management. His research area of interest is Image and Video Processing. He published more that ten papers in international journals and eight conferences. He has 13 years of teaching experience.

VLSI

Comparative Analysis of Partial Product Generators for Decimal Multiplication Using Signed-Digit Radix-10, -5 and -4 Encodings

Dharamvir Kumar and Manoranjan Pradhan

Abstract This paper presents the comparative analysis of partial product generators (PPG) for decimal multiplication using signed digit (SD) radix-10, radix-5 and radix-4 encoding schemes. PPG is the first step in obtaining the multiplication of two given numbers. For recoding of the BCD numbers, we have used 4221, 5211 and 5421 redundant BCD codes. The performance is compared based on the synthesis results obtained. Verilog is used as the hardware description language for the design. Synthesis and simulation is performed using Xilinx Vivado 2016.1.

Keywords Decimal multiplication · Partial product generation · Signed-digit (SD) high radix encoding · Verilog

1 Introduction

Hardware implementation of decimal arithmetic has attracted attention of many authors in recent past. IEEE standard for floating-point arithmetic, IEEE 754–2008, which came in the year 2008 is one of the reasons behind it. IBM had dedicated hardware unit for performing only decimal arithmetic in its eServer, POWER6 and z10 processor [1]. Decimal arithmetic is used in many areas such as financial, banking, scientific, stock market and Internet-based applications [1]. It becomes imperative in those situations where precision is required. For example, decimal number 0.2 does not have exact binary representation. It is expected that other microprocessor manufacturers will also include dedicated hardware units for floating point decimal arithmetic in their products.

Among arithmetic operations, multiplication is one of the most frequently performed operation in digital signal processing and other application areas. It is also, at the same time, more complex than addition and subtraction. So, a very wide range of

D. Kumar (✉) · M. Pradhan
Department of Electronics and Telecommunication Engineering, VSSUT, Burla Sambalpur, Odisha, India
e-mail: dkumar_etc@vssut.ac.in

M. Pradhan
e-mail: mpradhan_etc@vssut.ac.in

H. S. Saini et al. (eds.), *Innovations in Electronics and Communication Engineering*, Lecture Notes in Networks and Systems 107,
https://doi.org/10.1007/978-981-15-3172-9_58

work had been done on it. Being specific, decimal multiplication is more complex to achieve with the same type of digital hardware which is used for performing binary multiplication. This is because of the requirement of greater number of multiplicand multiples and inefficiency of representing decimal numbers in binary number system [2].

Basically, multiplication process involves three steps:

(i) Generation of required number of partial products(PPs),
(ii) Reduction of partial products by using carry-save adder(CSA) tree,
(iii) Final product formation by a carry-propagate addition of the last two partial products.

This paper describes the design, synthesis and simulation results of PPGs for decimal multiplication based on SD radix-10, -5 and -4 encodings. Number of partial products generated has an impact on the speed of calculation of the final product. For example, to multiply two 32-bit numbers, generally, thirty-two numbers of PPs are required. There are various schemes proposed by different authors to reduce the number of PPs formed, e.g. [3–6]. In this paper, PP generation using SD radix-10, -5 and -4 encoding is considered. And for the representation of BCD numbers, recoders based on redundant BCD codes 4221, 5211 and 5421 are used.

The rest of the paper is organized as follows: Sect. 2 describes PP generation for the three different schemes mentioned; Sect. 3 discusses the comparative results obtained; and Sect. 4 presents the conclusion of the work done.

2 Partial Product Generation

The multiplicand (X) and multiplier (Y) are assumed to be unsigned decimal integer p-digit BCD numbers. So, we can write,

$$X = \sum_{i=0}^{p-1} X_i \cdot 10^i \tag{1}$$

and,

$$Y = \sum_{i=0}^{p-1} Y_i .10^i \tag{2}$$

Then, the final product of X and Y will be a non-redundant $2p$ digit BCD number. Normally, for each multiplier digit a corresponding multiplicand multiple is generated in each partial product. The multiplier can be recoded in different ways to reduce the number of multiplicand multiples generation. In this paper, for recoding of multiplier SD radix-10, -5 and -4 encodings are used. All the partial products are generated in parallel as each recoded multiplier will generate a partial product by selecting a proper multiplicand multiple.

2.1 Multiplicand Multiples Generation

Figure 1 shows the block diagram for the generation of required multiplicand multiples for all the recoding schemes. For minimally redundant radix-10 scheme, the required multiplicand multiples are ($-5X$, $-4X$, $-3X$, $-2X$, $-X$, 0, X, $2X$, $3X$, $4X$, $5X$). It produces $p + 1$ number of partial products but also requires the generation of complex multiples like $3X$ and $-3X$ [7]. For SD radix-5 recoding scheme, we need to generate ($-2X$, $-X$, 0, X, $2X$) multiplicand multiples. This method generates $2p$ number of PPs. Although the architecture of SD radix-4 is simpler than SD radix-10, the number of PPs generated is more. And finally, for SD radix-4 multiplicand multiples ($-2X$, $-X$, 0, X, $2X$, $4X$ and $5X$) are generated. It also generates $2p$ number of PPs [7].

For representing decimal digits, BCD format is generally used, where each digit is represented by a 4-bit binary number with weights of 8, 4, 2 and 1 from most

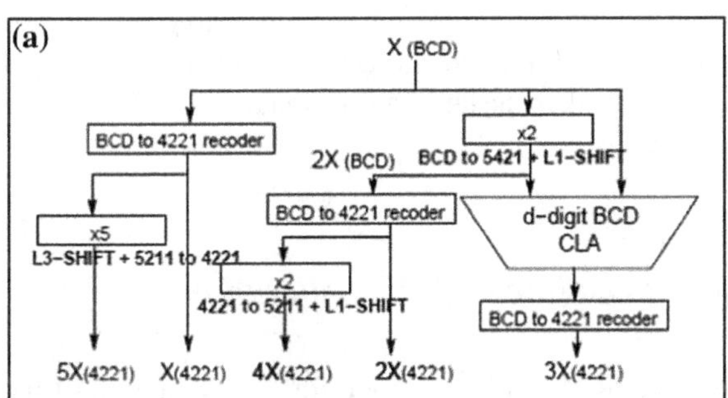

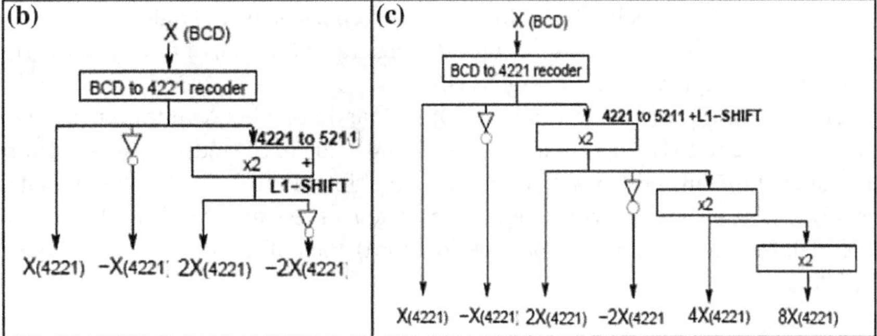

Fig. 1 Multiplicand multiples generation for **a** SD radix-10, **b** SD radix-5 and **c** SD radix-4

Table 1 Redundant BCD representations

Digit	BCD-8421	BCD-4221	BCD-5211	BCD-5421
0	0000	0000	0000	0000
1	0001	0001	0001–0010	0001
2	0010	0010–0100	0100–0011	0010
3	0011	0011–0101	0101–0110	0011
4	0100	1000–0110	0111	0100
5	0101	1001–0111	1000	1000–0101
6	0110	1010–1100	1001–1010	1001–0110
7	0111	1011–1101	1100–1011	1010–0111
8	1000	1110	1101–1110	1011
9	1001	1111	1111	1100

significant to least significant bit. This encoding for digits is unique in its representation as each digit will have only a specific BCD representation. There exist other representations (Table 1) of decimal digits which are redundant in nature. By redundancy, we mean a digit can be represented by more than one encoding. For example, in BCD-4221 encoding a decimal digit '4' can be represented by either 1000 or 0110 [8]. There are several advantages of using such encodings. The use of such coding can simplify carry-save addition, obtaining 9's complement, etc. [8]. In Table 1, for each such cases we have taken the left-side portion for obtaining the logical equations through *k*-map reduction.

2.2 *Multiplier Recoding and PP Generation*

Figures 2, 3 and 4 show the block diagram for the generation of required PPs for SD radix-10, -4 and -5, respectively. The right block in each figure represents the multiplier recoding block. The multiplier, *Y*, is recoded using SD radix-10, -5 and -4 schemes, respectively. There is one bit extra generated for each of the schemes. This extra bit is the sign bit denoted as ys_i [7].

The recoding process for multiplier, *Y*, is done in one-hot coding format for all SD radix schemes. The decision of the recoding value in SD radix-10 is decided by the value of the just previous digit. For example, for Yi 0010, i.e. 2 if the sign bit of previous digit is 1(i.e. previous digit is negative and less than 5), then SD radix-10 digit will be 3 and if sign bit of previous digit is 0, then SD radix-10 digit is taken as 2 [7, 9].

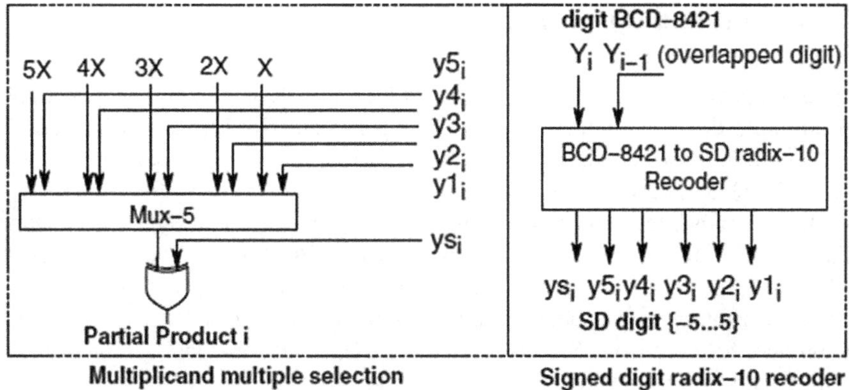

Fig. 2 Partial product generation for SD radix-10

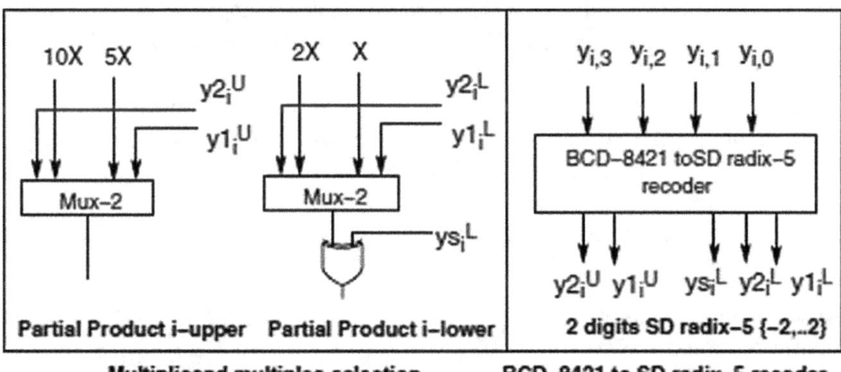

Fig. 3 Partial product generation for SD radix-5

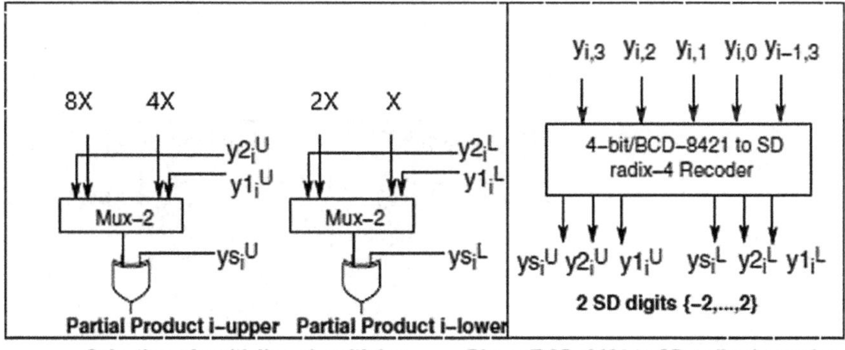

Fig. 4 Partial product generation for SD radix-4

3 Results and Discussion

Here, we present the results obtained for 64 bit (16 digit) partial product generators using these three techniques, i.e. SD radix-10, -5 and -4. The design is coded in Verilog. Synthesis and simulation is performed using Xilinx Vivado 2016.1. The Artix-7 architecture was chosen with the part number xc7a35tcpg236-3.

Table 2 shows the comparison of various recoders used in the scheme. It shows that the maximum area and delay is for the 64-bit BCD adder module. It uses 95 LUTs and shows a delay of 19.234 ns. It was expected as it involved rippling of the carry bits from one 4-bit adder to the next 4-bit adder till it reaches to the last block to produce final carry out. So, we can expect that the path for recoding scheme using this adder would be the critical path.

Apart from this out of the four recoders that were designed recoder 8421–4221 is consuming least area with 16 LUTs. Rest three recoders use 32 LUTs each with approximately similar delay figures. The results look in compliance with the logical equations for the recoders. The last column shows the ratio of all modules with respect to 8421–4221 recoder.

Table 3 shows the comparison of the three schemes used. The delay comes out to be maximum for the SD radix-10 scheme which was expected. It shows a delay of 13.399 ns that is approximately twice than the values for other two recoding schemes. It also consumes most number of LUTs at 2306. It can be said that there is more relative difference between the respective delays than the number of LUTs used.

Table 2 Synthesis result of 64-bit recoders for and BCD adder

Building blocks of the circuit	No. of LUTs used	Delay (ns)	Ratio LUT (delay)
64 bit recoder 8421–4221	16	5.824	1 (1)
64 bit recoder 8421–5421	32	6.199	2 (1.06)
64 bit recoder 4221–5211	32	5.824	2 (1)
64 bit recoder 5211–4221	32	6.186	2 (1.06)
64 bit BCD adder	95	19.234	5.94 (3.45)

Table 3 Synthesis results of 64-bit PPGs for SD radix-10, -5 and -4 encoding schemes

Scheme	No. of LUTs used	Delay (ns)	Ratio LUT (delay)
SD radix-10	2306	13.399	1.03 (1.86)
SD radix-5	2239	7.212	1 (1)
SD radix-4	2238	7.366	1 (1.02)

4 Conclusion

In this paper, we have compared PPGs for decimal multiplication based on the three different encoding schemes. The encoding schemes used are SD radix-10, -5 and -4. From the synthesis results, we found that SD radix-5 and -4 are at par. SD radix-5 and -4 show similar results in terms of area and delay. Also, SD radix-5 and -4 show better results than SD radix-10 encoding scheme in terms of both area and delay.

References

1. G. Jaberipur, A. Kaivani, Improving the speed of parallel decimal multiplication. IEEE Trans. Comput. **58**(11), 1539–1552 (2009)
2. M. A. Erle, E. M. Schwarz, and M. J. Schulte, Efficient Partial Product Generation. Int. Bus. (2005)
3. A. Nannarelli, “A Radix-10 Combinational Multiplier,” Electr. Eng., pp. 313–317, 2006
4. A. Vázquez, E. Antelo, P. Montuschi, A new family of high—Performance parallel decimal multipliers. Proc. Symp. Comput. Arith. 195–204 (2007)
5. A. Vazquez, E. Antelo, J.D. Bruguera, Fast radix-10 multiplication using redundant BCD codes. IEEE Trans. Comput. **63**(8), 1902–1914 (2014)
6. A. Vazquez, E. Antelo, P. Montuschi, Improved design of high-performance parallel decimal multipliers. IEEE Trans. Comput. **59**(5), 679–693 (2010)
7. A. Vazquez, High-performance decimal floating-point units : Ph.D. dissertation, (Jan. 20092014)
8. A.A. Wahba, H.A.H. Fahmy, Area efficient and fast combined binary/decimal floating point fused multiply add unit. IEEE Trans. Comput. **66**(2), 226–239 (2017). https://doi.org/10.1109/TC.2016.2584067
9. S. R. Carlough, E. M. Schwarz, Decimal Multiplication Using Digit Recoding. US 7,136,893 B2, (Nov. 14, 2006)

SCSGFRA: Sine and Cosine Signal Generation for Fixed Rotation Angle

S. Shabbir Ali, K. V. Suresh Kumar and Srinivas Bachu

Abstract There are several ways to define trigonometric signals; CORDIC is the more efficient algorithm to perform trigonometric operations for generating sine and cosine waveforms. By using rotation and vectoring modes in the coordinate system implementation in hardware becomes easy, and also different scaling factors and its compensated techniques can also be calculated further. The proposed system can generate sine and cosine signals in signal processing applications. Also, we can perform hyperbolic and exponential calculations. It can be further enhanced by extending for extended hyperbolic and linear coordinates for the trigonometric functions and inverse trigonometric functions. The proposed project is implemented in Xilinx ISE 13.2, and the output waveforms are simulated in Questasim 10.0b. The design implementation of CORIDC algorithm thus reduces the hardware implementation by using shift registers and adders, thereby increases its speed.

Keywords CORDIC algorithm · Coordinate systems · Vector modes

1 Introduction

In the field of electronics, the need for high speed increases day by day. Although, the speed increases but still, optimizing the size and power would be a challenging task. The need for multipliers has increased drastically in the applications of Communication Systems and Signal Processing. Optimization of these multipliers can be done using various techniques in allocating different lookup tables. Also, generating

S. Shabbir Ali
Department of ECE, Anantha Lakshmi Institute of Technology and Science, Anantapur, India
e-mail: shabbir.ali46@gmail.com

K. V. Suresh Kumar · S. Bachu (✉)
Department of ECE, Marri Laxman Reddy Institute of Technology and Management, Hyderabad, India
e-mail: bachusrinivas@gmail.com

K. V. Suresh Kumar
e-mail: kvskumar29@gmail.com

H. S. Saini et al. (eds.), *Innovations in Electronics and Communication Engineering*, Lecture Notes in Networks and Systems 107,
https://doi.org/10.1007/978-981-15-3172-9_59

the analogous signal using multipliers would consume more power and area which can be reduced by using the CORDIC algorithm.

CORDIC is nothing but shift and add operations that perform the mathematical functions like trigonometric, hyperbolic and logarithmic signals. In this paper, the algorithm is simulated that occupies the minimum area that is implemented in the SPARTAN 3E FPGA family that consumes very few lookup tables and slices [1].

Coordinate Rotational Digital Computer (CORDIC) is a well-known algorithm used for performing trigonometric, hyperbolic and exponential operations. These functions are computed more often in scientific and engineering applications [2–4].

2 CORDIC Algorithm

Two vectors are defined as $[X_0, Y_0]$ and $[X_n, Y_n]$. This algorithm is designed for 'n' number of iterations which access both vectoring mode and rotation mode [2, 5].

In the first mode, i.e., mode of rotation, the components of vector and rotational angle are set to the novel vectors of the coordinate components for which the rotations are allowed to pass through the given angle for computation. In another mode, i.e., vectoring mode, the scale of angular movement of the initial vectors can be calculated by the components of the coordinates in a vector [5].

The conventional method, implementation of 2D vector rotation, is done using a coordinate system which is shown in Eq. (1).

$$\begin{bmatrix} \mathrm{Cos}\,\theta & -\mathrm{Sin}\,\theta \\ \mathrm{Sin}\,\theta & \mathrm{Cos}\,\theta \end{bmatrix} \begin{pmatrix} X \\ Y \end{pmatrix} = \begin{pmatrix} X' \\ Y' \end{pmatrix} \tag{1}$$

where (X', Y') is one of the coordinate for a unity circle with a 'θ' rotation angle (Fig. 1).

The matrix operation can be expressed in Eqs. (2) and 3) as follows:

$$X' = X\,\mathrm{Cos}\,\theta - Y\,\mathrm{Sin}\,\theta \tag{2}$$

$$Y' = Y\,\mathrm{Cos}\,\theta + X\,\mathrm{Sin}\,\theta \tag{3}$$

Equations (2) and (3) are written in another form as follows:

$$X' = \mathrm{Cos}\,\theta\,[X - Y\tan\theta] \tag{4}$$

$$Y' = \mathrm{Cos}\,\theta\,[Y + X\tan\theta] \tag{5}$$

Assume that the angle of rotation is fixed to $\tan\theta = 2^{-i}$.

This is performed by shifting the x and y variables to right. The expression $\mathrm{Cos}\,\theta$ can be expressed in terms of $\tan\theta$ as shown in Table 1, i.e.,

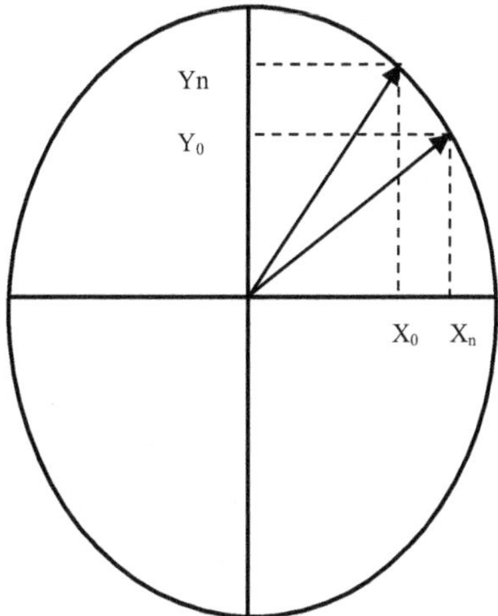

Fig. 1 Coordinate points on the unity circle with a 'θ' rotation angle

$$\left.\begin{aligned} \operatorname{Cos}\theta &= \frac{1}{\sqrt{1+\tan^2(\theta)}} \\ &= \frac{1}{\sqrt{1+(\pm 2^i)^2}} \\ &= C_i \end{aligned}\right\} \tag{6}$$

$$\left.\begin{aligned} &X_n = C_i\left[X_0 - Y_0 \cdot I \cdot 2^{-i}\right] \\ &\text{and} \\ &Y_n = C_i\left[Y_0 + X_0 \cdot I \cdot 2^{-i}\right] \end{aligned}\right\} \tag{7}$$

where $I = \pm 1$, the positive unit refers to the positive iterations, and the negative unit refers to the negative iterations.

Assume the rotational angle is 45° as input to the CORDIC algorithm. By considering the series of consecutive basic rotations, the required angle of rotation can be obtained.

where $i = 0, 1, 2 \ldots n - 1$.

For $\tan\theta = 2^{-i}$, the angle of rotations is shown in Table 2.

From this calculation, we can confirm that the CORDIC algorithm has the advancement of high precision that is capable of generating continuous analogous signals. The decision at each iteration i is the direction to rotate rather than whether or not to rotate because eventually to remove the C_i term from the expressions.

$$C_i = \frac{1}{\sqrt{1+(\pm 2^i)^2}}$$

$$\left.\begin{aligned} &X_{i+1} = C_i\left[X_i - Y_i \cdot I \cdot 2^{-i}\right] \\ &\text{and} \\ &Y_{i+1} = C_i\left[Y_i + X_i \cdot I \cdot 2^{-i}\right] \end{aligned}\right\} \tag{8}$$

Table 1 Different iteration values from 0 to 9

i	0	1	2	3	4	5	6	7	8
$\tan\theta = 2^{-i}$	1	0.5	0.25	0.125	0.0625	0.03,125	0.0156	0.0078	0.0039
$\theta = \text{arc}(\tan(2^{-i}))$	4.5	26.565	14.036	7.125	3.576	1.79	0.895	0.448	0.224

Table 2 Desired angle of rotation for successively smaller elementary rotations

i	$\tan\theta = 2^{-i}$	$\theta = \text{arc}\ \tan\left(2^{-i}\right)$	Z_i	Rotation θ	Final angle
0	1	45	20	−45	−25
1	0.5	26.565	−25	26.565	1.565
2	0.25	14.036	1.565	−14.036	−12.47
3	0.125	7.125	−12.471	7.125	−5.346
4	0.0625	3.576	−5.346	3.576	−1.77
5	0.03125	1.79	−1.77	1.79	0.020
6	–	–	–	–	–
.	–	–	–	–	–
.	–	–	–	–	–
19	–	–	0.000195	–	–

$$Z_{i+1} = \left[Z_i - I \cdot \tan^{-1}(2^{-i})\right] \tag{9}$$

In the hardware, X_i is fixed to 16 bit wide, and if the limit exceeds then the term $X_i \cdot I \cdot 2^{-i}$ moves to 0.

For example, if the design of CORDIC is 4 bit, then iterations beyond 4 bit would lead to a garbage memory. In order to restrict to the limit of its iterations, we move for avoiding arc tangent calculations. To yield the shift and add algorithms in vectoring rotation, delete the magnitude constant. The term C_i is able to be used as a fraction of a gain of the system for further process.

To obtain the accurate gain, not including C_i term, A_n will depend on I (Iterations) as shown in Eq. (10).

$$A_n = \sqrt{1 + (2)^{-2i}} \tag{10}$$

For the sinusoidal function limits 0 to π and $i = 0$ to $n - 1$. The design of sine and cosine outputs can be implemented in digital hardware by setting the variables at

$$\begin{gathered} Y_0 = 0 \quad \text{and} \quad X_0 = \frac{1}{A_n} \\ X_n \approx \text{Cos}\,(Z_0) \quad \text{and} \quad Y_n \approx \text{Sin}\,(Z_0) \end{gathered} \tag{11}$$

This algorithm is limited to the rotational angle between $-\frac{\pi}{2} \leq Z_0 \leq \frac{\pi}{2}$. To improve the rate of convergence of all rotational angles, $|Z_0| \leq 2\pi$. Also, angles outside the range must be pre-rotated as below:

$$\left.\begin{aligned} X_0' &= -IY_0 \\ Y_0' &= IX_0 \\ Z_0' &= Z_0 - I\tfrac{\pi}{2} \end{aligned}\right\} \tag{12}$$

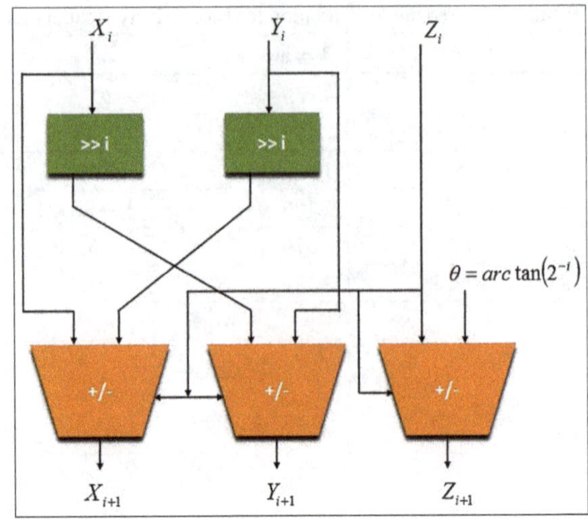

Fig. 2 Block diagram of CORDIC iterations

where $I = \begin{cases} +1 \text{ for } Y_0 > 0 \\ -1 \text{ elsewhere} \end{cases}$.

Multiple iterations are used in unit angle operation. The way by which every iteration found is according to the results of the previous iteration (Fig. 2).

3 Results and Discussion

The design is simulated in Questasim 10.0b version, and the output is verified using Xilinx ISE 13.2. CORDIC algorithm generates the sine and cosine waveforms which are shown in Fig. 3.

CORDIC algorithm device utilization summary report and RTL schematic are shown in Figs. 4 and 5. A total of 4.52 ns of delay is being observed in the synthesis report generated by SPARTAN 3E (XA3S1200E) FPGA as shown in Fig. 6.

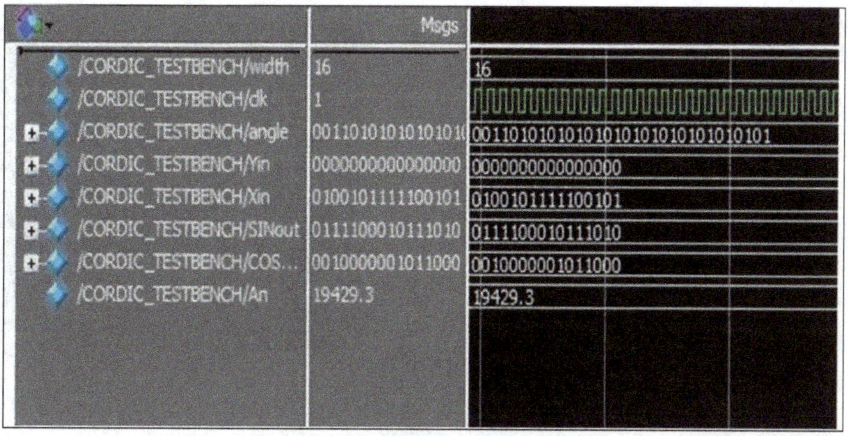

Fig. 3 CORDIC algorithm output (sine and cosine waveforms)

Device utilization Summary				
Logic Utilization	**Used**	**Available**	**Utilization**	**Note(s)**
Number of slice Flip Flops	955	17,344	5%	
Number of 4 input LUTs	1,420	17,344	8%	
Number of occupied Slices	935	8,672	10%	
Number of Slices containing only related logic	935	935	100%	
Number of Slices containing unrelated logic	0	935	0%	
Total Number of 4 input LUTs	1,828	17,344	10%	
Number used as logic	1,389			
Number used as a route-thru	408			
Number used as Shift registers	31			
Number of bonded IOBs	97	304	31%	
Number of BUFGMUXs	1	24	4%	
Average Fanout of Non-Clock Nets	1.83			

Fig. 4 Internal block of IOBs

4 Conclusion

The implementation of CORDIC algorithm for a fixed rotation angle calculates using number of iterations to generate sine and cosine waveforms using the Questasim 10.0b simulator and synthesized by using Xilinx ISE 14.5 Design Suite. The proposed method thereby reduces the number of LUTs by 31.6% and delay by 1.85 ns which in turn reduces the hardware implementation. This design module is much more efficient and consumes less area than the conventional approach of designing

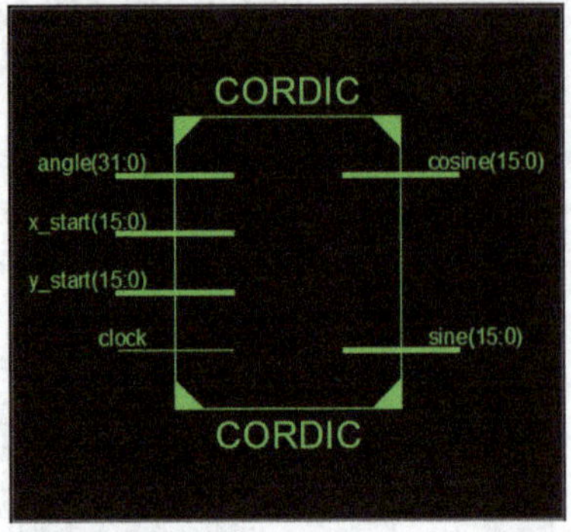

Fig. 5 RTL schematic view of CORDIC algorithm

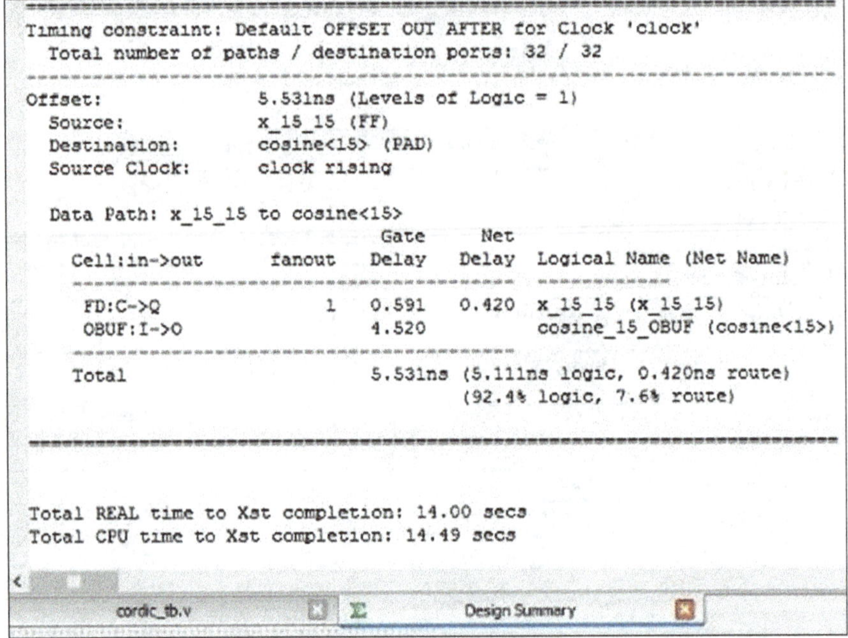

```
Timing constraint: Default OFFSET OUT AFTER for Clock 'clock'
  Total number of paths / destination ports: 32 / 32
---------------------------------------------------------------------------
Offset:              5.531ns (Levels of Logic = 1)
  Source:            x_15_15 (FF)
  Destination:       cosine<15> (PAD)
  Source Clock:      clock rising

  Data Path: x_15_15 to cosine<15>
                                Gate     Net
    Cell:in->out      fanout   Delay   Delay  Logical Name (Net Name)
    ----------------------------------------  ------------
     FD:C->Q               1   0.591   0.420  x_15_15 (x_15_15)
     OBUF:I->O                 4.520          cosine_15_OBUF (cosine<15>)
    ----------------------------------------
    Total                      5.531ns (5.111ns logic, 0.420ns route)
                                       (92.4% logic, 7.6% route)

===========================================================================


Total REAL time to Xst completion: 14.00 secs
Total CPU time to Xst completion: 14.49 secs
```


Fig. 6 Design summary: delay calculations

hardware. The CORDIC algorithm can be further implemented for different variable rotational angles.

References

1. J. Li, J. Fang, Study of CORDIC algorithm based on FPGA, in *IEEE Xplore—Control and Decision Conference (CCDC)* (2016). (Chinese)

2. M. Garrido, P. Kallstom, CORDIC II: a new improved CORDIC algorithm. IEEE Trans. Circ. Syst. II, **63**(2) (2015)
3. C. Haettich, *YAC—Yet Another CORDIC Core* (2014)
4. A. Kaivani, G. Jaberipur, Decimal CORDIC rotation based on selection by rounding: algorithm and architecture. Comput. J. **54**(11) (2011)
5. F. Elguibaly, α-CORDIC: an adaptive CORDIC algorithm. Can. J. Electr. Comput. Eng. **23**(3) (1998)

Power Efficient Router Architecture for Scalable NoC

Nivya Varghese and Swaminadhan Rajula

Abstract A power efficient router architecture to increase the performance in a NoC interconnect is suggested in this paper. Static power reduction by using power-gating techniques has been extensively discussed in the past. More recent papers implement router architectures to resolve the issues in power gating such as early wake-up latency and energy dissipation, thereby achieving high performance and reduction in static power. The work proposed in this paper modifies on a recent router architecture that remedied the early wake-up latency overhead and achieved significant power savings. The proposed LPSQ router architecture provides an additional reduction in dynamic power and area overhead by 20.2% and 24.5%, respectively, without degrading the overall system performance.

Keywords Power gating · Network latency · Wake-up latency · Shared queue · VC router

1 Introduction

Network-on-chip (NoC) has emerged as the fundamental communication paradigm for many core architectures. Future applications call for thousands of cores on a single chip. While NoC is a scalable solution for the emerging multi-core architectures, static power consumption is becoming a prominent issue in this field. The entire chip's power budget is being consumed in the NoC communication fabric. As the transistor size continues to shrink, the static power consumption is exacerbated.

To overcome these, power-gating techniques have emerged as the ultimate solution. In VLSI, power gating [1] is used to reduce static power consumption by turning off the idle circuit blocks. The concept is borrowed to NoC in [2], wherein the routers

N. Varghese · S. Rajula (✉)
Department of Electronics and Communication Engineering, Amrita School of Engineering, Amrita Vishwa Vidyapeetham, Bengaluru, India
e-mail: r_swaminadhan@blr.amrita.edu

N. Varghese
e-mail: nivyarosev@gmail.com

H. S. Saini et al. (eds.), *Innovations in Electronics and Communication Engineering*, Lecture Notes in Networks and Systems 107,
https://doi.org/10.1007/978-981-15-3172-9_60

are powered-off during consecutive idle time. However, power-gating techniques suffer from large wake-up latencies when the packets have to wait for the powered-off router to turn on and become fully functional. When multiple routers are powered-off in a packet's path, it needs to wait multiple times. This will add up to a cumulative delay in the overall operation of the chip.

Power punch [3] is a technique to hide the wake-up latency and improve network latency. The Power punch signals are sent from an active NI to a powered-off router which is present in the packet's path to reach destination router. This will provide sufficient time for the powered-off router to wake-up before the packet reaches it. However, since the sleep time of router is getting reduced, there are not much power savings. NoRD [4] provides a better solution to increase the length of idle periods. It also addresses the disconnection problem of powered-off routers as well as the wake-up latency. Here, the sleepy routers are bypassed by allowing a unidirectional ring to operate within the network. This bypass ring connects all the nodes via NI of routers in the chip and is responsible for forwarding the flits to next router's NI or to the local node. The ring topology provides limited scalability. In MP3 [5], authors opted for an indirect network topology, namely Clos network which has a higher path diversity so that more routers can be powered-off in its path and there is reduced critical path. The higher link per mesh cost reduces the feasibility of the design.

Routers can be powered-off during less traffic to reduce the static power dissipation. When there is less traffic, it is more convenient and power effective to perform a simple switching technique that is achieved through the EZ Pass with VC router architecture [6]. By using this simple switching technique, there is no need to completely wake-up the powered-off router. The architecture proposed in our paper is borrowing the concept of router design from [6] which deals with the elimination of wake-up latency to an extent while a packet encounters a powered-off router. Here, the concept of shared buffer [7] is also incorporated which increases the overall throughput of chip network. This can help in achieving lower zero-load latency and also contributes toward the reduction in dynamic power component. The overall router design has 20.2 and 24.5% reduction in power and area, respectively.

The rest of the paper is organized as follows. Section 2 gives an overall briefing of the working of NoC routers. Section 2.1 gives the effects of power gating. Section 3 gives the proposed design which modifies the EZ-Pass architecture with the shared buffer under high traffic. Section 4 gives the comparison results, and Sect. 5 concludes the paper.

2 Background and Motivation

2.1 NoC Routers

A multi-core system is illustrated in Fig. 1, in which the cores (IPs) are communicating through a mesh network of routers. There are four input and output ports

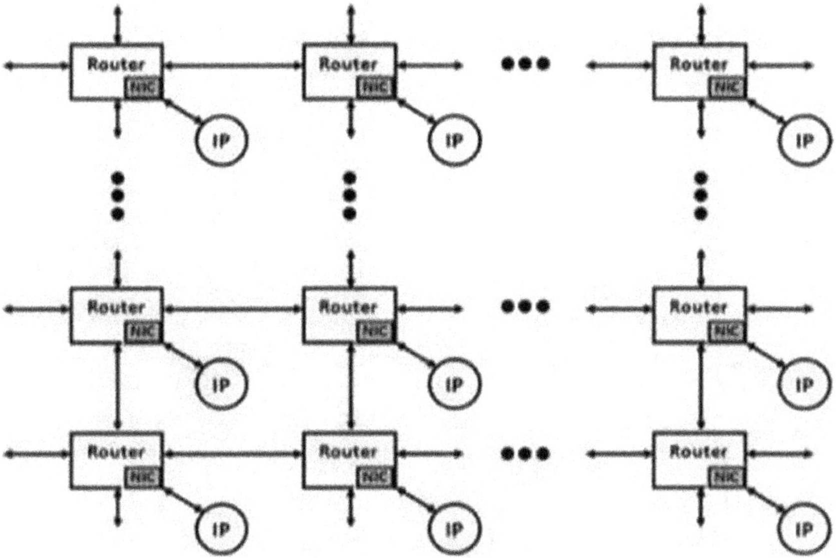

Fig. 1 Basic structure of NoC

for communication within adjacent routers. The local port receives and transfers messages between core and router. The network interface (NI) is located between its processor and router and is responsible for converting the messages from core to flit for circulating between routers and vice versa. Figure 2 shows a four-stage, five-input port, wormhole NoC router. Segmentation of a single packet is as follows—a header flit, body flits, and a tail flit. In any typical wormhole router, there are four stages of processing a flit—virtual channels (VC) for storing arriving packets, router computation (RC), wherein the address of each flit is extracted and the output port is identified, virtual channel allocation (VA) for flow control and switch allocation (SA) for establishing the connection between the input port and the desired output port. Finally, switch traversal (ST) takes place.

2.2 *Power Gating of NoC Routers*

The idle circuit blocks in a device are a source of leakage currents due to the continuous supply of source voltage into it. This is the major cause for static power dissipation. To mitigate its effects, power-gating techniques were devised by designers to keep the static power at bay. Figure 3a shows the use of a transistor T1 that controls the circuit block to turn on/off according to the circuit block being in either operational mode or idle. So, when the block is idle, the T1 cuts off the supply to the block. Similarly, to reduce the static power in NoC routers, power-gating techniques are applied. In NoC's, the main component that has an impact on the static power is

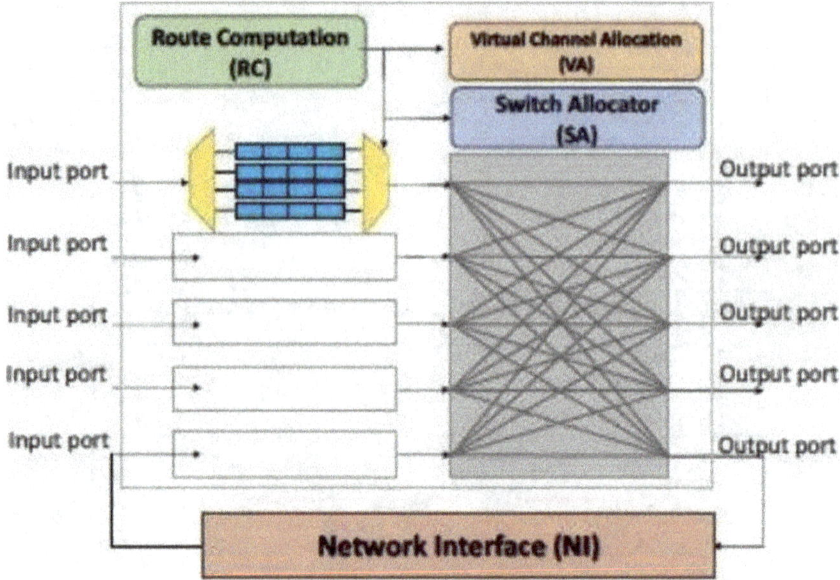

Fig. 2 Basic router architecture [6]

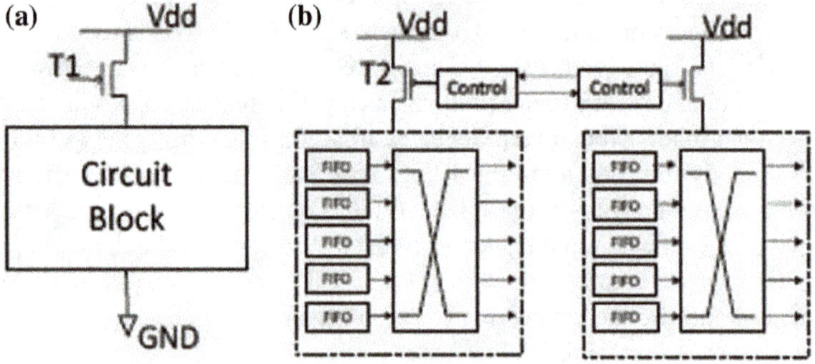

Fig. 3 Power-gating techniques [6]

the routers. However, they can be either operational or idle depending on the traffic patterns in the network. In Fig. 3b, the control blocks and the transistors perform the power gating by turning the transistors on/off according to the idle/busy states.

While power-gating offers reduction in static power, there is some degradation in system performance due to the following reasons:

1. Minimum idle period—to take the full benefits of energy saving, these idle blocks/routers must remain in its sleep mode for a longer duration of time in order to reduce the associated energy overhead. Breakeven time (BT) is defined as the minimum number of consecutive clocks cycles the router must remain in

its sleep mode. Prior researches [2] have confirmed this breakeven time to be at least 10 clock cycles. This in most cases may not be possible since heavy traffic patterns compel these idle routers to turn on to forward the flits causing a major reduction in energy savings.

2. Large wake-up latency—the delay incurred by the router while turning on from its powered-off mode is wake-up latency. This delay gets cumulatively added up in NoC of powered-off routers with a greater number of cores. The flit needs to wait for the router to be fully functional before it can forward it. This wake-up latency must be either fully/partially reduced to increase the performance.
3. Connectivity issues—each node can send or receive packets via NI depending on the on/off status of the router. This is known as node-router dependency. Once the routers are powered-off, it will be completely disconnected from the rest of the network causing connectivity issues.

Under high traffic, the router is fully utilized, and therefore, it is not required to power off the router. The factor that determines whether to turn off the router is detection time (DT). It is defined as the time the router remains idle for consecutive clock cycles. According to prior works, [1] it is 4 clock cycles. Once DT is detected, router goes to powered-off mode. Under powered-off condition, there are various factors that deteriorate the system performance as mentioned above due to low traffic in the system. A recent approach to reduce the static power under powered-off condition is proposed by the authors of paper [6] and the router architecture is known as EZ-Pass router architecture. The proposed design in this paper offers a modified design to the one proposed in [6] by further reducing the power by 20.2% and the overall area by 24.5%.

3 LPSQ Router Architecture

Figure 4 modifies the EZ-Pass router architecture by replacing the conventional VC router with a shared queue router in [7] so that there is an increased advantage of reducing the power and increasing the throughput additional to the benefits of eliminating wake-up latency and providing static power savings.

Outside the router, there are some single flit latches associated with each input ports, multiplexers, de-multiplexers, and a control unit. The incoming flits are directly stored inside single-flit latch one at a time and the control unit analyze the status of the router. Under powered-off status of a router, the incoming flit is routed through the NI toward the designated output port. This is the bypass route followed so that the router is not woken up. The router is powered-on when the traffic to single input port increases to three flits at a time. While router is getting powered-on, simultaneously, the flit already inside the latch can get switched toward the desired output port through NI. This way, the wake-up latency is masked and there is considerable saving in energy while allowing the router to remain idle for a longer time.

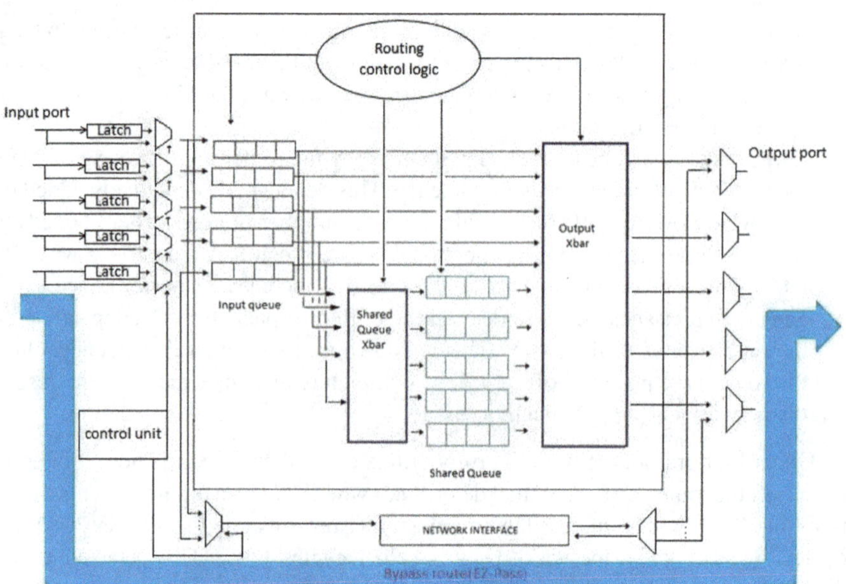

Fig. 4 Proposed LPSQ router architecture

Hence, under low traffic mode, the NI processes the flit and switches it forwards to the next router in its destination path. Under high traffic mode, the router is turned on and the flits enter inside the router. The incoming flits are initially entering their respective input queues. The routing logic identifies the output port and if its available, the output crossbar switch will establish the connection and switch traversal (ST) takes place. If the required output port is busy, the flits get buffered into the shared queue via shared crossbar switch. When the desired output port opens up, these flits from shared queue move toward the output port.

Using a shared buffer router, there is maximum utilization of buffer space. The accumulation of flits at one input port alone can be avoided as was the case in VC routers. Rather, the flits from all the input ports are accommodated inside the shared queue till it receives the grant signal to advance toward output port. This in turn increases the throughput, and the number of buffer queues is reduced causing an overall reduction in router area compared to VC routers. Due to the reduction in area and increase in throughput of router, the dynamic component of power is getting considerably reduced.

4 Experimental Results

Simulations for the base and proposed router architecture are performed using Xilinx Vivado and synthesize is done in Cadence RTL compiler using TSMC 45 nm technology. The results of LPSQ are compared with EZ-Pass router architectures [6].

We calculated the average latency of each flit entering the powered-on router by analyzing the simulation waveform from Vivado HLS. Since single stage router is proposed here, the latency is measured from the time the head flit enters the router to the time its tail flit takes to completely exit through the output port. It was observed to be on an average of 12 ns. To demonstrate the throughput of shared queue in LPSQ architecture, we increased the number of flits entering the north input port such that traffic at the north port was high. The graph in Fig. 5 demonstrates the throughput—it is more for our proposed LPSQ compared to EZ Pass.

Figure 6 shows the area and power comparisons of the proposed and baseline design. Figure 6a plots the power of router alone for LPSQ and EZ Pass and also the

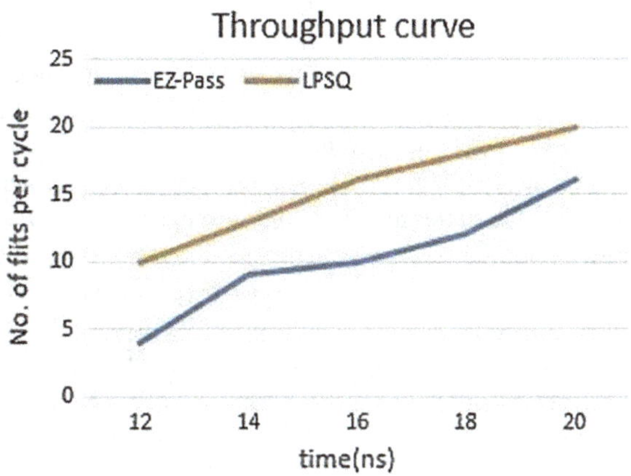

Fig. 5 Throughput comparison curve

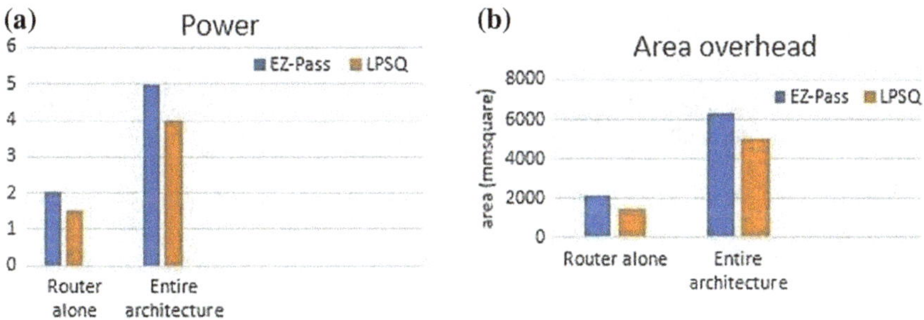

Fig. 6 Power and area comparisons of EZ Pass and LPSQ

Table 1 Slice logic of base and proposed routers

Slice type	Used	Available	Utilization %
Slice LUT			
Base	296	47,200	0.36
Proposed	218	94,400	0.23
Slice registers			
Base	375	47,200	0.4
Proposed	349	94,400	0.4

Table 2 Comparison of parameters between base and proposed approaches

Router architecture	Power (mW)	Area (μm^2)	Delay (ps)
EZ-pass router (Base)	6281	4.989	990
LPSQ (Proposed)	5043	3.979	1000

overall power consumed by the proposed LPSQ router and EZ-Pass router architectures. The shared queue router had 25% less power consumption compared to VC router in our design. It was noticed that there was an overall reduction of power by 20.2% in our proposed LPSQ router architecture.

In Fig. 6b, the first comparison shows the area overhead of the general VC router from Fig. 2 used in EZ-Pass architecture and the shared queue router in Fig. 4 from LPSQ architecture excluding the components outside their routers. This shows that a shared queue router requires 32% lesser area than a conventional VC router due to the lesser number of buffer queues required for the proposed router design. The second part of the comparison in Fig. 6b shows the entire router architecture area comparison of the proposed LPSQ and EZ-Pass router. Here, there is a 24.5% area reduction in the overall router design from the EZ-Pass router architecture.

We obtained the slice logic after running the implementation on Artix-7 FPGA in Vivado HLS. The overall area utilized by the base and proposed router on the FPGA after implementation is shown in Table 1. A slice comprises of lookup tables (LUTs) and flip-flops. It is noted that our LPSQ router architecture has 0.17% less area utilization on Artix-7 FPGA after implementation.

Table 2 shows the parameter comparisons of the base and proposed router architectures. The proposed LPSQ router architecture achieves a 20.2 and 24.5% reduction in power and area, respectively, compared to EZ-Pass router architecture.

5 Conclusion

This work gives a new router architecture for eliminating the issues in power-gating technique. Due to the sharing of buffer queues inside the router during high traffic, the utilization of buffer space is increased which had a positive impact on the throughput of the system. Under low traffic, the router remains powered-off for a longer duration due to the alternate path followed by the incoming flits in turn increasing the energy savings. The proposed LPSQ router architecture achieves a 20.2 and 24.5% reduction in power and area, respectively, compared to EZ-Pass router architecture. This router architecture can be extended too many stages and connected in any topologies along with appropriate routing algorithms to analyze the effectiveness of the design.

References

1. Z. Hu, A. Buyuktosunoglu, V. Srinivasan, V. Zyuban, H. Jacobson, P. Bose, Microarchitectural techniques for power gating of execution units (2004), pp. 32–37
2. H. Matsutani, M. Koibuchi, D. Ikebuchi, K. Usami, H. Nakamura, H. Amano, Ultra fine-grained run-time nn/c gating of on chip routers for cmps, in *Networks-on-Chip*. IEEE (2010), pp 61–68
3. L. Chen, D. Zhu, M. Pedram, T.M. Pinkston, Power punch: towards non-blocking power-gating of noc routers, in *High Performance* (2015)
4. T. Pinkston, NoRD: node-router decoupling for effective power-gating of on-chip routers, in *International Symposium on Microarchitecture*
5. L. Chen, T.M. Pinkston, Mp3: minimizing performance penalty for power-gating of clos network-on-chip, in *International Symposium on High-Performance Computer Architecture (HPCA)* (2014)
6. H.Z.A. Louri, EZ-Pass: an energy and performance-efficient power-gating router architecture for scalable NoCs. IEEE Comput. Archit. Lett. **17**(1), 88–91 (2018)
7. A.T.T.B.M. Baas, Achieving high-performance on-chip networks with shared-buffer routers **22** (2014)

PLEADER: A Fast and Area Efficient Hardware Implementation of Leader Algorithm

Payel Banerjee, Tapas Kumar Ballabh and Amlan Chakrabarti

Abstract Leader algorithm is a clustering algorithm generally used as a pre-clustering step in various accelerated techniques for clustering large datasets. The algorithm while clustering large datasets can be computationally expensive because of its high convergence time. In this paper, a fast and area-efficient hardware implementation of leader algorithm is proposed which utilizes an efficient parallelism technique to speed up the algorithm without much resource consumption. It uses a master–slave architecture model where the slave processes after comparing a data with the existing clusters send the result to the master process which then updates the clusters. The algorithm is capable of automatically choosing the required number of slave processes and instead of increasing its number reutilizes them repeatedly making the process an area-efficient technique. Experiments are conducted with various datasets, and the result confirms that the proposed algorithm outperforms the original method in terms of speed.

Keywords Leader algorithm · Data mining · Parallelism · Hardware · FPGA

1 Introduction

Clustering is a tool that group objects into clusters such that the bonding is strong among the objects of same cluster and is weak among the objects of different clusters. It is a vital task of data mining and finds application in various fields like machine learning, pattern recognition, image processing, spatial data analysis, information retrieval, data compression, etc. [1]. Clustering algorithm can be further classified into two types, viz. partitional and hierarchical clustering based on the method of

P. Banerjee (✉) · T. K. Ballabh
Department of Physics, Jadavpur University, Kolkata, India
e-mail: payelbanerjee54@gmail.com

T. K. Ballabh
e-mail: tkb@phys.jdvu.ac.in

A. Chakrabarti
Calcutta University, Kolkata, India
e-mail: achakra12@yahoo.com

H. S. Saini et al. (eds.), *Innovations in Electronics and Communication Engineering*, Lecture Notes in Networks and Systems 107,
https://doi.org/10.1007/978-981-15-3172-9_61

formation. Partitional clustering techniques break or partition the dataset into non-overlapping set of clusters. Hierarchical clustering produces a set of nested clusters that are organized as a tree called dendrogram. Both these algorithms can further be classified into two main types, viz. distance-based and density-based depending on the criteria of clustering. Distance-based method uses a suitable distance metric to find distance between patterns and is the main criterion for clustering. On the other hand, density-based methods take clustering decision depending on the density information of the patterns. The leader algorithm is an incremental, distance-based, partitional clustering technique that using a single scan assigns the data as a leader or as a follower for a given threshold. This algorithm mainly finds its application in the pre-clustering phase of various well-known data mining techniques such as *K*-means [2], DBSCAN [3], average linkage [4], and single linkage [5] method and its main purpose is to partially cluster the data points using a single scan making the overall clustering technique easier and faster.

The main contributions of our algorithm are:

a. Uses a parallelism scheme to reduce the convergence time of the algorithm to a much lower value.
b. The master–slave architecture model is an area-efficient technique where the master process can reutilize the slave processes again and again to make simultaneous comparisons. This method removes the need of increasing the number of slave processes to enhance speedup and thereby causes lower resource consumption.

Both of these above factors make the algorithm highly suitable for clustering large datasets where high convergence time and resource consumption act as serious bottlenecks in the path of efficient clustering.

The paper is organized as follows. Section 2 describes the background and the related work, Sect. 3 explains the proposed algorithm. Section 4 describes the hardware-implemented design of the algorithm, Sect. 5 reports the time and space complexity of the algorithm, Sect. 6 gives the challenges in the hardware implementation of the algorithm, and Sect. 7 discusses the experimental-based results. Section 8 finally presents the conclusions and the future scopes of the research.

2 Leader Clustering Algorithm

Leader algorithm [6] is an incremental partitional clustering algorithm that for a given threshold ζ produces a set of leaders $L = \{l_1, l_{2,} \ldots\}$. For every pattern '$x$', if there is a leader $l_i \in L$ such that $\|x - l_i\| <= \zeta$ then 'x' will be assigned as a follower of $l_{i.}$ If the condition is not satisfied for any of the leaders, then it will be assigned as a new leader. A data once assigned as a follower of leader l_i cannot be assigned to a leader l_j for $i \neq j$. The process of scanning continues unless and until a data is assigned either as a leader or as a follower.

2.1 Steps of the Conventional Leader Algorithm

Let $l_i \in L$ $(i = 1)$ be the first scanned pattern, 'ζ' being the user-defined threshold, D being the set of input patterns.

1. Assign the first scanned pattern as the first leader 'l_i' for i = 1.
2. For any unscanned pattern x $\in D$, do
3. Compute dist = $\|x-l_i\|$.
4. If 'dist' <= ζ, go to step 5 else go to step 6.
5. Assign 'x' as follower of l_i. Go to step 8.
6. If all leaders are checked go to step 7 else repeat from 3 with next unchecked leader $l_i \in L$ for i = i + 1.
7. Assign 'x' as a new leader. Go to step 8.
8. If any pattern is left to be scanned repeat from step 2 with i = 1 else go to 9.
9. Output 'L' with its followers.

The time complexity of this algorithm for 'n' data points is $O(mn)$ with 'm' being the number of leaders provided by the algorithm. The space complexity is however $O(m)$ if only leaders are stored else $O(n)$ for storing all the followers along with the leaders.

An accelerated approach of the leader algorithm is discussed in [4] using triangle inequality. This technique is highly beneficial as it eliminates the redundant distance computations making the algorithm to converge faster without affecting the cluster quality. The disadvantage of this algorithm is that it is data dependent as depending on the datasets used, the distance computation may or may not be avoided using the triangle inequality. Our algorithm is totally a different approach which shows the hardware implementation of leader algorithm along with the parallelism scheme used to speed up the method. This method accelerates the algorithm irrespective of the type of datasets and that also without much resource consumption.

3 PLEADER-Parallel Leader Algorithm

The proposed algorithm is a parallel version of the leader algorithm and is called as PLEADER. This method uses a divide and conquer rule where it divides a single clustering process into multiple steps and applies parallelism technique in each step to speed up the process. It uses a master–slave architecture model where the slave processes compare a data to the existing leaders and send the result to the master process which then takes a clustering decision. Consider that there are 'N' slave processes and one master process. Initially, the first data of the dataset will be assigned as the first leader. At the next stage, the second data of the dataset will be compared with this first leader. If the threshold criterion is satisfied, then both of them will be merged into the same cluster else will be split into two different clusters. This comparison is done only by the first slave process. Now if comparing first and second data gives a new leader then the second slave process gets active along with the first process. Now the third data is compared with the first and second leaders by the first and the second slave processes, respectively. This data will be either assigned as follower to any of these leaders if the threshold criterion is satisfied for that leader else will be treated as a newly found leader. So the next step is to compare the fourth data with all the existing leaders. Now if we have three leaders then three slave processes will be active to do the simultaneous comparisons. But it is not possible to have '*Z*' number of slave processes for '*Z*' number of clusters for a very high value of '*Z*' because of resource limitation. So depending on the resource availability of the system we have to fix the number of slave processes.

So when the number of clusters exceeds the number of slave processes, then the slave processes must repeat them unless and until a clustering decision is taken by the master process. Say, if at any stage of clustering, we have four clusters and three slave processes and then the three slave processes firstly compare the data with the first three out of four leaders. If the data lies within a threshold distance of any of these three leaders, then the data is assigned as a follower to that leader. At the next stage, the comparison again starts with the next data of the dataset. In case, the data is not found in threshold to any of these three leaders, the first slave process repeats itself once again to perform the comparison with the fourth leader. During this process, the other slave processes will remain inactive. If at any stage of clustering, we have '*N*' slave processes and '*Z*' clusters, and '*Z*' is greater than '*N*' then firstly '*N*' slave processes will get active to perform the comparison with the '*N*' clusters. If the data cannot be merged with any of these '*N*' clusters, then at the next stage (*Z*–*N*) number of slave processes get active to complete the remaining comparisons provided (*Z*–*N*) is less than or equal to *N*. If we find (*Z*–*N*) still greater than '*N*', then again '*N*' process will be active to complete the comparisons with '*N*' clusters out of (*Z*–*N*) number of remaining clusters. So now we are left with (*Z*–2*N*) number of clusters. If at any of these stages, the data is found to lie within the threshold distance of any of these clusters then it will be assigned to that cluster. If the threshold criterion is not met even after comparing the data with all the existing clusters, then the data will be treated as a newly formed leader. After taking a clustering decision, the comparison begins with the next data of the dataset. This process repeats unless and until all the data points are either assigned as followers or as leaders. Thus, here parallelism techniques are used in every iteration unless and until a clustering decision is taken.

The steps of the proposed algorithm are summarized below:

a. Activate the master process and assign the first object of the set as the first leader. Assign the number of leaders to be compared as unity.
b. Deactivate the master process and activate the required number of slave processes depending on the number of leaders to be compared and compare the next object of the set with the existing leaders.
c. Deactivate the slave processes and starts the master process. If the data comes in threshold with any of these compared leaders, then assign the data as follower to the leader and go to step 'g' else go to step 'd'.
d. If no leaders are left to be compared, then go to step 'f' else go to step 'e'.

e. Deactivate the master process and activate the required number of slave processes depending on the number of leaders left to be compared and compare the data with the unchecked leaders. Go to step ‘c’.
f. Assign the data as a new leader. Update the number of leaders and go to step ‘g’.
g. Repeat from step ‘b’ unless and until all the objects are clustered.

Thus here the master process, at every iteration controls the activation and deactivation of slave processes and reutilizes them again and again unless a clustering decision is reached.

4 Hardware Implementation of the Proposed Algorithm

The entire functional block of the algorithm is divided into four major Components:

(a) Distance calculation unit;
(b) Comparing unit;
(c) Clustering unit;
(d) Control unit.

Figure 1 shows the functional block diagram of the design. Memory ‘M’ contains all the input patterns to be clustered and memory ‘C’ contains all the output clusters, ‘C_i’ denotes each cluster block (the entire memory ‘C’ has been broken into ‘i’ blocks where each block represents a cluster, ‘i’ ranging from 1 to the maximum number

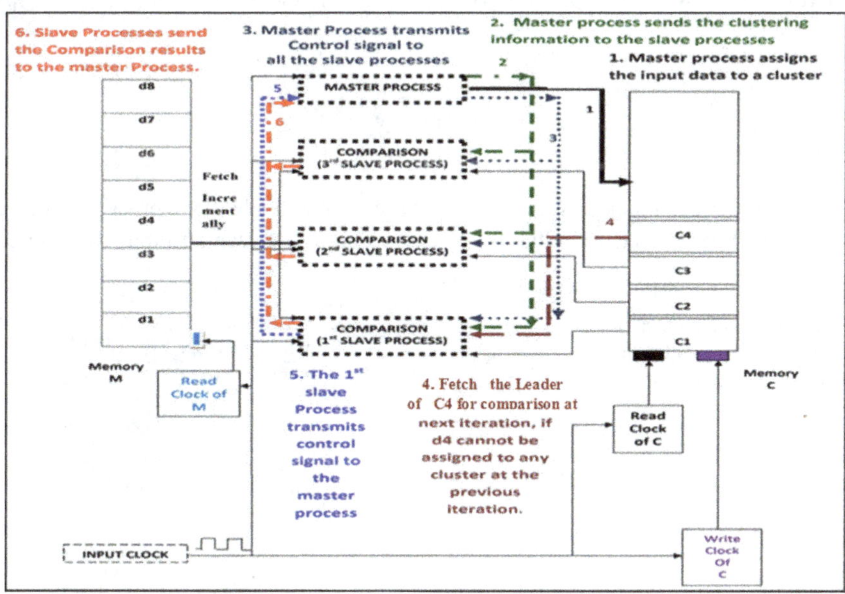

Fig. 1 Functional block diagram of the proposed algorithm hardware design

of clusters that can be produced), 'd_j' corresponds to the data at the jth location of memory 'M'. All the signals are marked with numbers, and the description of the signals has been explained with the corresponding numbers.

4.1 Distance Calculation Unit

Here $l_1, l_2, l_3, \ldots, l_N$ is the series of leaders according to their order of formation. In other words, 'l_1' is the first formed leader and l_N is the most newly formed leader. Choice of distance metric is a crucial task of data mining applications which highly depends on the dimensionality of the input data. Different measures give different amount of proximity to a given query point specifically at higher dimensions. Even the performance of the distance metric degrades as dimensionality increases. The behavior of L_k norm [7] shows that the choice of distance metric in high dimensionality depends on the value of k. This means that the Manhattan distance metric or L_1 norm is much more preferable than the Euclidean distance metric or L_2 norm and maximum metric or $L\infty$ norm for high dimensional data mining applications [7, 8]. Implementation of Manhattan metric is also beneficial than Euclidean metric in terms of resource consumptions as Euclidean distance metric involves performing square roots which is very expensive to be performed on field-programmable gate array (FPGA). Due to these advantages of Manhattan metric over Euclidean metric, the proposed algorithm has used the former metric for calculating the similarity between clusters. For the Manhattan metric of dimension 'd' in a plane, the Manhattan distance between the point $P1$ $(x_1, x_2, \ldots, x_d)$ and the point $P2$ at $(y_1, y_2, \ldots, y_d)$ is given by

$$\sum_{i=1}^{d} |x_i - y_i|$$

This functional block as shown in Fig. 2 is executed by the slave processes where each slave process performs the distance calculation between an input pattern and

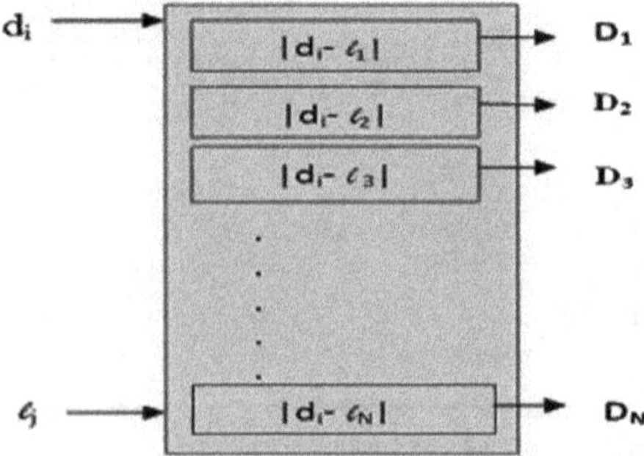

Fig. 2 Distance calculation between an input pattern and a leader

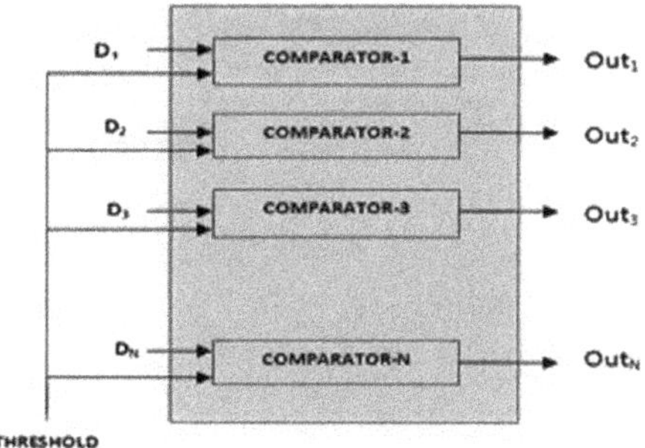

Fig. 3 Comparison of the distances with threshold

the corresponding leaders simultaneously. They get a data pattern 'd_i' from the input memory and compare it with the leaders 'l_j' fetched from the cluster memory. All the distances are then applied to the input of the comparators which then compare the distance with threshold.

4.2 Comparing Unit

Figure 3 shows the comparing unit where the comparators compare the distances provided by the distance calculation unit and the user-specified threshold. The output of the comparator goes high whenever the distance exceeds threshold else goes low. These output signals are then transferred to the master process which then takes a clustering decision. All the comparisons are done simultaneously by the slave processes.

4.3 Clustering Unit

The clustering unit as shown in Fig. 4 is performed by the master process. The master process after getting the results of the comparison from the comparing unit simply makes a vector of signal with the bits at the output of the comparing unit. In case, there are more than one leaders lying within the threshold distance to the input pattern, a priority rule is followed for assigning the data to a cluster according to their order of formation. So if there are three such leaders and all are formed following the order (first → second → third) then the first leader will add the pattern as its follower. The MSB of the signal holds the bit of highest priority and is followed by the bits with

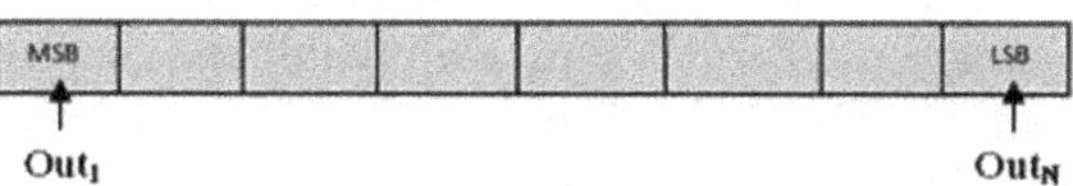

Fig. 4 Vector of signal assigning priority to leaders

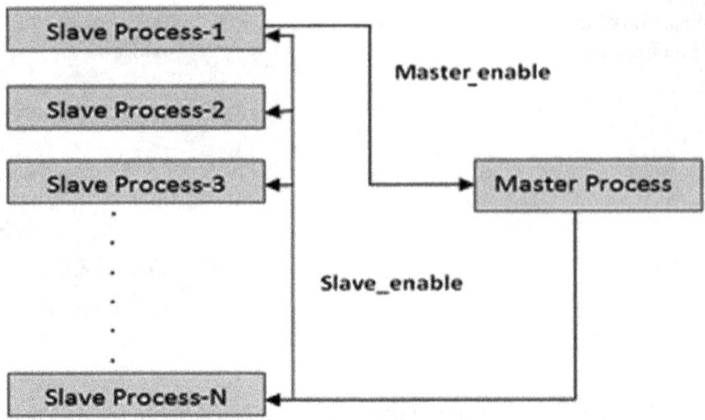

Fig. 5 Control signals generated by slave and master process

decreasing order of priority as one move toward the LSB. Here bit of highest priority refers to the output of the comparator which compares the distance of the pattern and the first formed leader with threshold. So the clustering unit firstly checks the MSB bit of the vector and if found low, the data is assigned as the follower to that leader else the bit with next higher priority will be checked. Thus, all the bits are checked sequentially starting from the MSB and continue unless and until a bit with low state is reached.

4.4 Control Unit

The control unit shown in Fig. 5 is the backbone of the entire design which controls the activation and the deactivation of the master and the slave processes by tracking the completion of their corresponding functions. The master and the slave processes are clocked processes. Hence at every rising edge of the clock, both of them try to get activated but the control signals used in the design prevent both the master and the slave processes to work simultaneously. There are two control signals namely 'slave_enable' and 'master_enable'. The slave processes get activated only after the initial cluster assignment step. The slave process works only if the slave_enable signal is high and the master_enable signal is low. Similarly, the master process works only if the slave_enable signal is low and the master_enable signal is high. At any stage of operation, all the activated slave processes work simultaneously and perform same operations. Hence all the slave processes complete their corresponding operations at the same clock period. The slave process that gets activated for at least a single existing cluster will activate for any higher value of clusters. Therefore, the master_enable signal generated by this slave process indicates the completion of operations by all the slave processes. Hence the master_enable signal generated by this slave process can be used as a control signal for activating the master process. At the rising edge of the clock after the initial cluster assignment step, the slave processes activate depending on the number of clusters and after performing its functions make the master_enable signal high. At the same rising edge, the master process although cannot perform its operations but makes the slave_enable signal low so that at the next rising edge the slave processes cannot get activated again. At the next rising edge, getting the slave_enable signal low and the master_enable signal high, the master process starts and after completing its functions makes the slave enable signal high. At the same rising edge although the slave processes cannot get activated to perform its functions, it makes the master enable signal low so that at the next rising edge the master process cannot get activated again. It should be noted from Fig. 1, that read operation from memory '*M*' and '*C*' is done by the slave processes and the write operation to memory '*C*' is done by the master process. As the slave and master processes cannot get activated at the same clock, hence read and write operations of the cluster memory (memory *C*) are not occurring simultaneously.

5 Time and Space Complexity of the Algorithm

(a) The first input data of the dataset is assigned as the first leader. This requires two clock pulses. The first clock pulse is used for initial data read operation and the second clock pulse is used for writing the data to the cluster memory.
(b) At the next rising edge, the slave process gets activated and then compares the second data in the dataset with the existing leader. At the end of the process, the master_enable signal is generated which at the next clock cycle will activate the master process.
(c) The fourth clock is spent by the master process to take a clustering decision. After the end of its operations, the master process generates the slave enable signal that in the next clock activates the slave processes.

So thus we see that starting of a slave process to its restart requires two clock cycles. If after '*X*' number of iterations (starting of a slave process to its restart), we find the number of clusters left to be compared is less than or equal to the number of slave processes, then it needs just one more iteration to assign the data to a new cluster. Thus, in total $(X + 1)$ number of iterations are required to complete the task. So, for '*Z*' existing clusters and '*N*' number of slave processes, if after '*X*' iterations we have $(Z - XN)$ number of clusters left to be compared, which is less than or equal to the number of slave processes then we can find the total number of iterations for assigning a data to a new cluster.

For,

$$Z - XN < N$$

Or,

$$\frac{Z}{N} < X + 1$$

For,

$$Z - XN = N$$

Or,

$$\frac{Z}{N} = X + 1$$

** If $(X + 1)$ is a fraction, then it will always be rounded up to the next higher integer and that will give the exact number of iterations to assign a data to a cluster. Each iteration takes two clock cycles, so $2 \times (X + 1)$ number of clock cycles are required to assign a data to a new cluster. Depending on the resource availability, the value of 'N' has to be adjusted. To find the maximum number of clock pulses required to create 'm' leaders using these 'N' slave processes, we have made two assumptions.

(a) Firstly 'm' leaders are formed with no follower of each.
(b) After formation of 'm' leaders, remaining '$n - m$' number of data points will be assigned as follower to the mth leader.

Using the above two assumptions, the maximum number of required clock pulses are,

$$2 + 2\left[\frac{1}{N} + \frac{2}{N} + \frac{3}{N} + \cdots + \frac{m-1}{N}\right] + 2 \times (n - m)\frac{m}{N}$$
$$= 2 + 2 \times \sum_{z=1}^{m-1} \frac{Z}{N} + 2 \times (n - m)\frac{m}{N} \quad (1)$$

In Eq. 1, '2' has been added because that includes the clocks required to assign the first data of the dataset as the initial cluster.

Here the series $\frac{1}{N} + \frac{2}{N} + \frac{3}{N} + \cdots + \frac{m-1}{N}$ is the total number of iterations required to create 'm' leaders.'$\frac{m}{N}$' is the number of iteration required to assign each of the 'n-m' data points as the follower of the mth leader.

Simplifying equation '1' gives the time complexity of the proposed algorithm as $O\left(\frac{mn}{N}\right)$. Thus, the time required by the algorithm decreases with increasing number of slave processes, but increases with the increase in data elements in the set.

The space complexity of the algorithm will remain same as $O(m)$ if only leaders are stored else $O(n)$ for storing all the patterns. After retrieving the leaders from the patterns, the followers can be assigned to the leaders at any later stage without storing the followers and thereby decreasing the space complexity to $O(m)$.

6 Challenges in Hardware Implementation

a. The design utilizes look-up tables (LUTs) to implement the memories. In case of large memories, many LUTs are required, which can consume a lot of resource of the field-programmable gate array (FPGA) board and hence in that case Block RAMs are more preferable over LUTs. Block RAMs are dedicated large memories embedded in FPGA and one must check the parameters of the Block RAM from the datasheet of the target FPGA before selecting it for implementing the design.
b. In this design, the slave processes are reading the leaders from the cluster memory simultaneously. It should be noted that this function requires the implementation of multiporting techniques of Block RAM [9] to allow simultaneous read operations from the memory by the slave processes.
c. As Manhattan metric is supported by L_k norm for higher dimensions, our algorithm has employed the same for the design. Increase in dimensions of input data will also result in increase of resource consumption as shown in Table 5. This is because increase in dimensions increases the co-ordinates of the vectors or points in a datum which increases the load of performing addition between the corresponding co-ordinates while computing the distance between the two data. This is however not the drawback of the design or metric but an unavoidable problem of high dimensional distance computations.

7 Experimental Results

The full design of the pre-clustering algorithm has been checked and verified using Xilinx ISE design suite 14.4 and implemented in Kintex 7 (KC705) using the Xilinx Vivado platform and VHSIC Hardware Description Language (VHDL) to create the design code. The clock used for the design is a 200 MHz differential clock. Table 1 gives the datasets on which the algorithm has been tested to check the performance of the algorithm. Figure 6 compares the convergence time of the pre-clustering algorithm with increasing number of slave processes as well as data size. This is the time taken by the algorithm when all the clusters are singletons, i.e., none of the objects

Table 1 Experimental dataset

Data		No. of instances	No. of attributes	Attribute type
Random		100	1	Integer
Pendigits	Testing	3498	16	Integer
	Training	7494	16	Integer
Letter		20,000	16	Integer
Shuttle		58,000	9	Integer

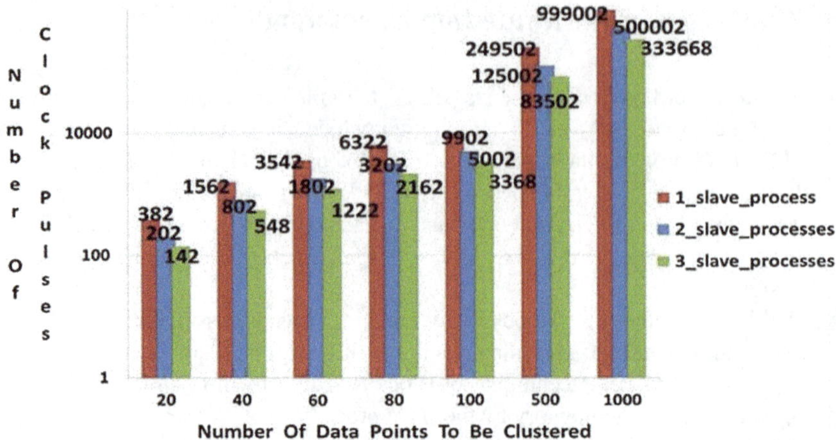

Fig. 6 Variation of convergence time of the algorithm with varying slave processes and data size

are within the threshold distance of each other. This is the worst case that can hardly happen and therefore shows the maximum possible number of clock pulses required by the method. It is shown that increasing slave processes remarkably decreases the use of clock pulses for a clustering action, thereby making the algorithm suitable for clustering large data volume. To check the performance of the design on hardware, the algorithm has been implemented on a small one-dimensional random dataset [10]. Tables 2 and 3 respectively show the speedup achieved and the post-implementation resource utilization with increasing slave processes when applied on the one-dimensional dataset. To check the performance of the algorithm on multivariate real-word datasets, we conducted experiments with datasets as obtained from UCI machine learning repository [11] and the improvement in speed is presented in detail in Table 4. To check the effect of the proposed method on the clustering result, we have applied Rand Index [12] and the R.I of '1.0' in all cases implies that the clustering result does not get affected by increasing slave processes. Figure 7 shows the speeding up of the algorithm with variable data size and the number of clocks saved by our algorithm is found to increase with increase in data size. Thus, our algorithm is really beneficial for big data analysis which is a serious challenge envisaged by experts in the field of data science. We have not implemented the clustering of

Table 2 Speedup achieved versus number of slave processes for one-dimensional dataset

Data	Threshold	Total No. of clocks required for distance computations			R.I.
		1 slave process	3 slave process	5 slave process	
Random	0.5	4870	1694	1078	1.0
	2.5	1736	668	452	1.0
	4.5	1202	490	348	1.0
	6.5	988	420	314	1.0

Table 3 Post-implementation resource utilization report for increasing slave processes for one-dimensional clustering

Resource	Available	Utilized	Utilized %	Utilized	Utilized %	Utilized	Utilized %
		One slave process		Three slave processes		Five slave processes	
LUT	203,800	18,458	9.06	47,590	23.35	72,949	35.79
FLIPFLOP	407,600	8354	2.05	9585	2.35	9995	2.45
IO	500	11	2.20	11	2.20	11	2.20
BUFG	32	1	3.12	1	3.12	1	3.12

Table 4 Speedup achieved versus number of slave processes for large real datasets

Data		Threshold	No. of clocks (in Millions) required for distance computations			R.I.
			1 slave process	3 slave process	5 slave process	
Pendigits	Testing	10	12.22	4.06	2.44	1.0
		20	12.12	4.04	2.42	1.0
		30	11.46	3.82	2.28	1.0
		40	9.96	3.32	1.98	1.0
	Training	10	56.14	18.72	11.22	1.0
		20	55.90	18.64	11.18	1.0
		30	53.12	17.70	10.62	1.0
		40	45.52	15.16	9.1	1.0
Letter		3	216.66	72.22	43.34	1.0
		5	125.86	41.96	25.18	1.0
		7	65.72	21.92	13.16	1.0
		9	34.54	11.52	6.92	1.0
Shuttle		5	296.80	98.96	59.40	1.0
		7	158.88	53.00	31.82	1.0
		9	92.42	30.84	18.52	1.0
		11	57.24	19.12	11.48	1.0

these multivariate data points in FPGA and have only shown the speedup that can be reached by our algorithm when applied on these datasets. It is quite obvious that increase in dimensions will increase resource utilization as shown in Table 5 and for clustering a data with very large number of attributes an advanced version of FPGA with higher resource availability will be preferred.

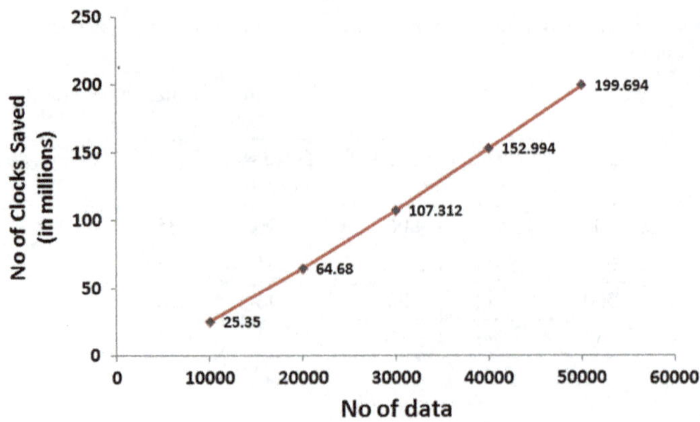

Fig. 7 No. of clocks saved for shuttle dataset with increasing data size. # Slave processes = 5, Threshold = 5

Table 5 Post-implementation resource utilization report of Manhattan metric at one and four dimensions

LUT			Flip flops		
Utilization		Availability	Utilization		Availability
1D	4D		1D	4D	
3	10	203,800	4	13	407,600

8 Conclusion and Future Scope

The paper presents a hardware-implemented version of leader algorithm called PLEADER that uses parallelism scheme using master–slave architecture model to speed up the convergence of the algorithm. The method instead of using large number of slave processes reutilizes the existing slave processes on every iteration parallely thereby optimizing the resource consumption along with providing high speed. The algorithm has been implemented on FPGA for one-dimensional dataset with maximum of five slave processes and the increase in speed is found to be remarkable even without consuming half of the FPGA resources. The paper has also shown an unavoidable effect of increase in dimensions on the resource utilization of the design. Experiments are conducted on large real multivariate datasets to show the speedup achieved by our algorithm on these data and the result is found to be highly beneficial specially for large dataset. Future work may include a more area-efficient approach of the implementation of multivariate clustering in FPGA and also to check the performance of the algorithm for streaming data clustering.

References

1. M.H. Dunham, *Data Mining: Introductory and Advanced Topics* (Prentice Hall, New Delhi, India, 2003)
2. T. Hitendra Sarma, P. Viswanath, B. Eswara Reddy, A hybrid approach to speed-up the k-means clustering method. Int. J. Mach. Learn. Cybern. (IJMLC). https://doi.org/10.1007/s13042-012-0079-7
3. A. Amini, T.Y. Wah, LeadenStream: a leader density-based clustering algorithm over evolving data stream. J. Comput. Commun. **1**, 26–31 (2013)
4. B.K. Patra, N. Hubballi, S. Biswas, S. Nandi, Distance based fast hierarchical clustering method for large datasets, in *Proceedings of the 7th international Conference on Rough Sets and Current Trends in Computing, RSCTC'10* (Springer, Berlin, 2010), pp. 50–59. https://doi.org/10.1007/978-3-642-13529-3_7
5. B.K. Patra, S. Nandi, P. Viswanath, A distance based clustering method for arbitrary shaped clusters in large datasets. Pattern Recogn. **44**(12), 2862–2870 (2011)
6. J.A. Hartigan, *Clustering Algorithms* (Wiley Inc, New York, NY, USA, 1975)
7. C.C. Aggarwal, A. Hinneburg, Daniel A. Keim, On the surprising behavior of distance metrics in high-dimensional space First publication, in *ICDT 200, 8th International Conference Database theory*, London, UK, ed. by J. Van den Bussche. (Springer, Berlin, 2001), pp. 420–434. (Lecture notes in computer science, 1973)
8. M. Steinbach, L. Ertoz, V. Kumar, The challenges of clustering high dimensional data, in *New Vistas in Statistical Physics—Applications in Econophysics, Bioinformatics, and Pattern Recognition* ed. by L.T. Wille (Springer, Berlin, 2003)
9. C. Eric LaForest, J. Gregory Steffan, *Efficient Multi-ported Memories for FPGAs*. Department of Electrical and Computer Engineering, University of Toronto {laforest, steffan} @eecg.toronto.edu
10. https://goo.gl/RVtSBS
11. C.L. Blake, C.J. Merz, *UCI Repository of Machine Learning Databases* (1998). Available from: http://www.ics.uci.edu/~mlearn/MLRepository.html
12. W.M. Rand, Objective criteria for evaluation of clustering methods. J. Am. Stat. Assoc. **66**(336), 846–850 (1971)

Analytical Comparison of Power Efficient and High Performance Adders at 32 nm Technology

Imran Ahmed Khan, Md. Rashid Mahmood and J. P. Keshari

Abstract In this paper, analytical comparison of full adders has been presented on the basis of power, delay and PDP. All simulations are performed using SPICE in 32 nm CMOS technology. Full adder is the basic block of an arithmetic logic unit (ALU) so the power consumption and delay of an ALU are reduced by optimizing full adder. Simulation results show that for input to output carry FA-Tung has the highest speed and the lowest PDP. While for input to output sum, FA-Goel has the highest speed and the lowest PDP. FA-Conventional has the highest power consumption while FA-Tung has the lowest power consumption. For multi-bit adders, FA-Tung has the best performance.

Keywords Arithmetic circuit · DSP · Performance · VLSI · Multiplexer · Integrated circuit

1 Introduction

The arithmetic circuits are very important in electronic systems. If the characteristics of the arithmetic circuit are good, then overall performance of the electronic system will be improved drastically. Adder is a fundamental arithmetic circuit that is used in several VLSI systems, for example, application-specific digital signal processing (DSP) architectures and microprocessors. This is the core of many arithmetic operations such as addition/subtraction, multiplication, division and address generation [1]. The arithmetic logic unit is a digital circuit that performs arithmetic operations and logical operations; full adder significantly affects the performance of ALU [2].

I. Ahmed Khan (✉)
Jamia Millia Islamia, New Delhi, India
e-mail: imran.vlsi@gmail.com

Md. Rashid Mahmood
Guru Nanak Institutions Technical Campus, Hyderabad, India

J. P. Keshari
ABES Engineering College, Ghaziabad, India

H. S. Saini et al. (eds.), *Innovations in Electronics and Communication Engineering*, Lecture Notes in Networks and Systems 107,
https://doi.org/10.1007/978-981-15-3172-9_62

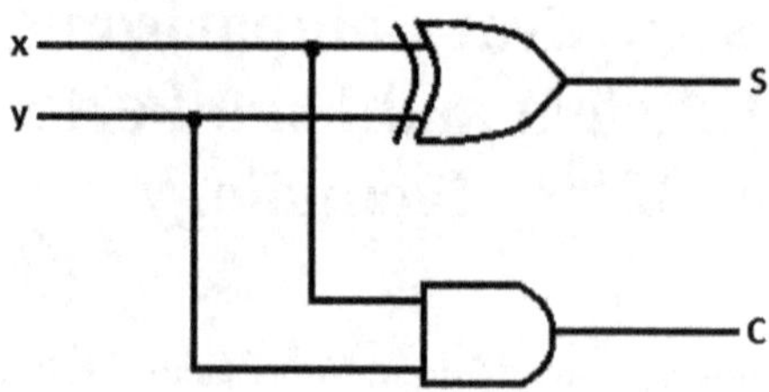

Fig. 1 Half adder using Ex-OR and AND gate

A half adder is a circuit that accepts two inputs and gives two outputs. The two inputs are single-bit binary values. The two outputs are sum and carry. A half adder uses two logic gates, an AND gate, and an Exclusive OR gate. A half adder is represented by Fig. 1. It has a two inputs x and y and two outputs *S* and *C* [3, 4]. The logical expressions for the sum and carry may be written as follows:

$$s = sy' + x'y \tag{1}$$

$$c = x \cdot y \tag{2}$$

The full adder performs the arithmetic sum of three bits. It has three inputs and two outputs. Three inputs involve two input bits '*x*' and '*y*' and an extra bit for an incoming carry '*z*'. Two outputs are sum '*S*' and carry out '*C*'. The incoming carry input '*z*' is used for cascading full adders to make *N*-bit adder. The realization of the full adder using half adders is shown in Fig. 2.

Full adder is the crucial building block used to implement multiplier, adder-subtractor, microprocessor, digital signal processor (DSP), ALU, etc. So, a full adder design with smaller chip area, low power consumption and short delay is required for IC design. Most of the complex computational circuit needs full adder. The power consumption of computational block can be reduced by using low power full adder. Several full adders have been proposed in past, among them full adders having less transistor using pass transistor logic are widely used because of lesser

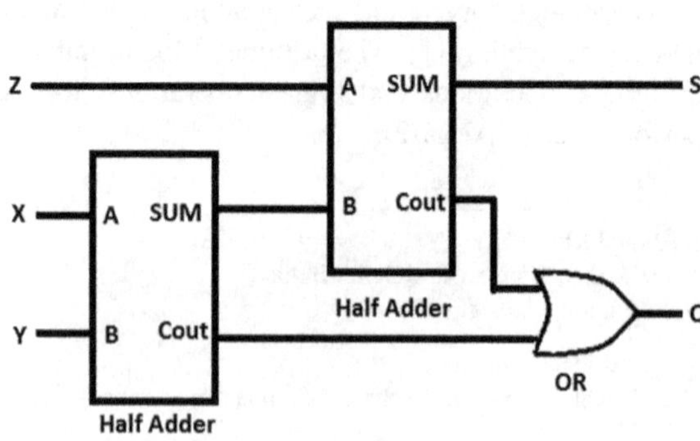

Fig. 2 Full adder realized with half adders

Table 1 Truth table of full adder

Inputs			Outputs	
x	y	z	s	c
0	0	0	0	0
0	0	1	1	0
0	1	0	1	0
0	1	1	0	1
1	0	0	1	0
1	0	1	0	1
1	1	0	0	1
1	1	1	1	1

power consumption [5–7]. But these designs suffer from output signal degradation and cannot be employed in low voltage operations [8].

The truth-table of a full adder is shown in Table 1. The logical expressions for the sum and carry bits can be written as follows:

$$s = x'y'z + x'yz' + xy'z' + xyz \tag{3}$$

$$c = xy + xz + yz \tag{4}$$

2 Circuits of Full Adder

A. Conventional Full Adder

Three inputs of conventional full adder involve two input bits 'A' and 'B' and an extra bit for an incoming carry 'C_i'. Two outputs are sum 'S_o' and carry out 'C_o'. The incoming carry input 'C_i' is used for cascading full adders to make N-bit adder. The realization of the conventional full adder is shown in Fig. 3 [4].

B. Full Adder Proposed by C. K. Tung et al.

Multiplexer-based full adder was proposed by Tung et al. [9]. FA-Tung composed of three modules: XOR–XNOR module, sum module and carry module to generate circuit, sum and carry, respectively. Figure 4 represents FA-Tung.

If $H = 0, C_o = A$;
and if $H = 1, C_o = C_i$.

Hence, the propagation delay from input carry (C_i) to output carry (C_o) is equal the delay of only an NMOS transistor. So, this adder has fast carry propagation.

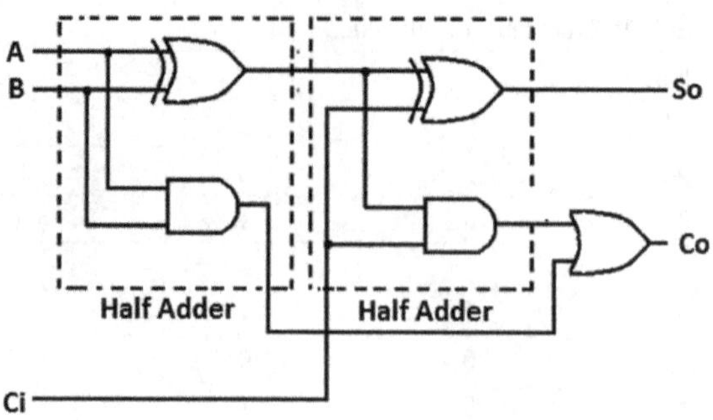

Fig. 3 Full adder realized with basic gates

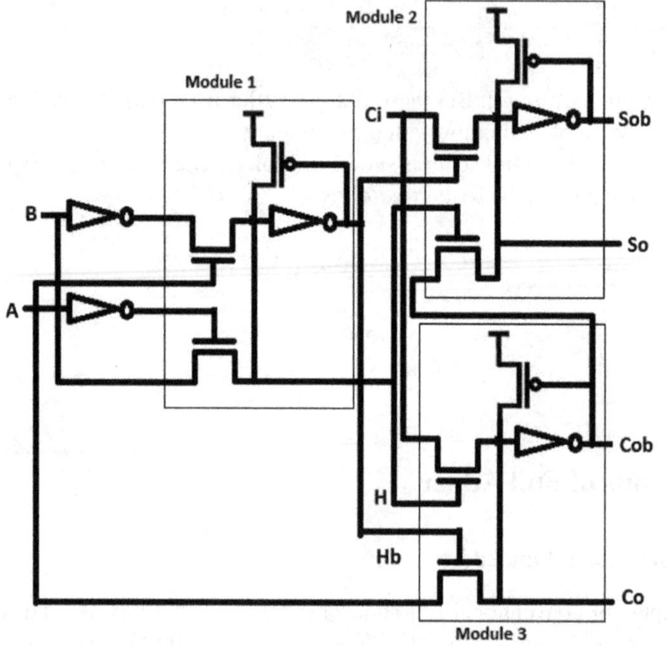

Fig. 4 Schematic diagram of full adder by C. K. Tung

Equation (5) shows the Boolean expression for sum output S_o and Eq. (6) shows the Boolean expression for carry output C_o.

$$S = C_i H' + C_o' H \tag{5}$$

$$C_o = C_i H + H' A \tag{6}$$

where $H = A$ XOR B and $H' = A$ XNOR B

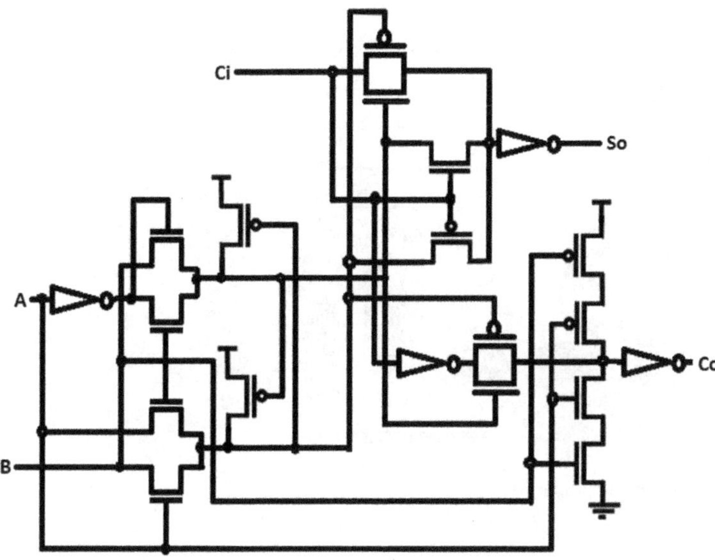

Fig. 5 Schematic diagram of full adder by S. Goel et al.

C. **Full Adder Proposed by S. Goel et al.**

Figure 5 represents hybrid-CMOS full adder (FA-Goel) [10]. This full adder uses pass transistor logic as well as static CMOS circuit techniques. This has three logic modules: XOR–XNOR, sum and carry modules. Hybrid-CMOS full adder generates XOR and XNOR outputs simultaneously by using only eight transistors. XOR and XNOR outputs are used to generate sum and carry outputs. Through static CMOS techniques, the carry module of FA-Goel obtains improved performance and increased driving capability. The number of transistor of FA-Goel is twenty-four so it has lesser chip area.

3 Simulation Results and Comparative Analysis

The full adder circuits have been simulated with a 32 nm CMOS technology with nominal supply voltage 0.9 V. Rise time and fall time of inputs are 50pS. Power consumption, different types of delays and PDPs have been considered. The simulation has been done using SPICE software. In full adder, there are four types of delay:

(i) Input carry C_i to output carry C_o
(ii) Input carry C_i to output sum S_o
(iii) Input data A to output carry C_o
(iv) Input data A to output sum S_o.

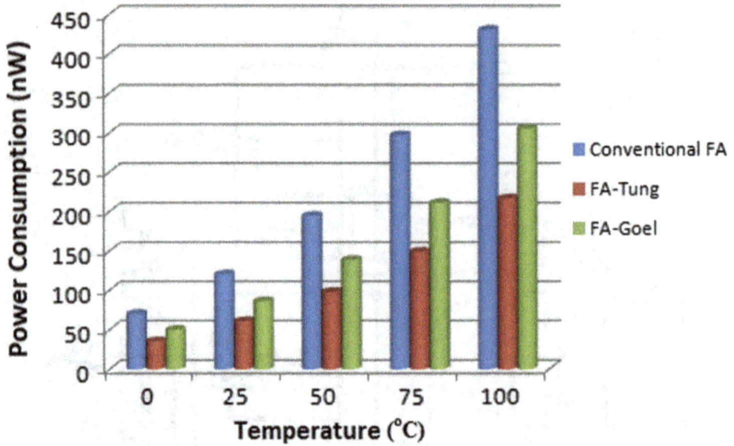

Fig. 6 Power consumption (nW)

For fair comparison of full adders, all four types of delay have been considered in this paper.

Figure 6 shows that for all temperatures, FA-Conventional has the highest power consumption while FA-Tung has the lowest power consumption. In FA-Tung, the propagation of carry signal from C_i to C_o requires the delay time for passing only an NMOS transistor. Therefore, this adder offers very fast carry propagation. Figure 7 shows that FA-Tung has the shortest delay at all temperatures. FA-Goel has the longest delay at 0, 50 and 75 °C while for remaining temperatures FA-Conventional has the longest input carry C_i to output carry (C_o) delay.

Figure 8 shows that FA-Conventional has the highest PDP at all temperatures while FA-Tung has the lowest PDP at all temperatures. The PDP exhibited by the

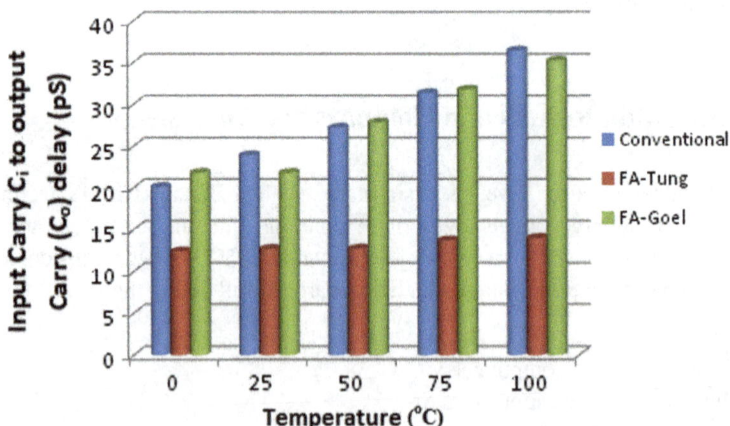

Fig. 7 Input carry C_i to output carry (C_o) delay (pS)

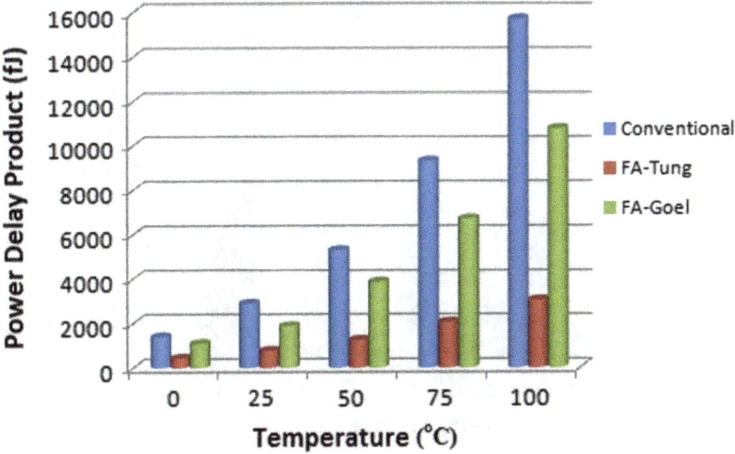

Fig. 8 PDP for input carry C_i to output carry (C_o) (fJ)

full adder would affect the system's overall performance [11]. So, FA-Tung has the best performance among the considered full adders when input carry C_i to output carry (C_o) delay is more important. Figure 9 shows that FA-Goel has the shortest delay at all temperatures while FA-Conventional has the longest input carry C_i to output sum (S_o) delay for all temperatures. Figure 10 shows that FA-Conventional has the highest PDP at all temperatures while FA-Goel has the lowest PDP at all temperatures. FA-Goel has the best performance among the considered full adders while input carry C_i to output sum (S_o) delay and respective PDP is important.

Table 2 represents delay from input A to output carry (C_o). It is clear from the table that FA-Conventional has the shortest delay at all temperatures except 75 and

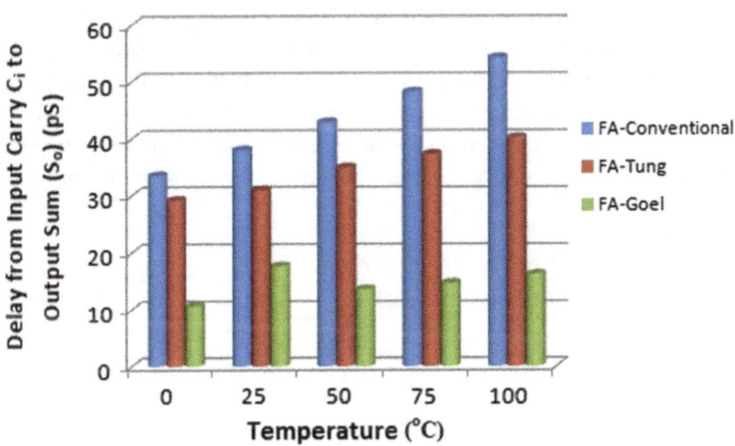

Fig. 9 Delay from input carry C_i to output sum (S_o) (pS)

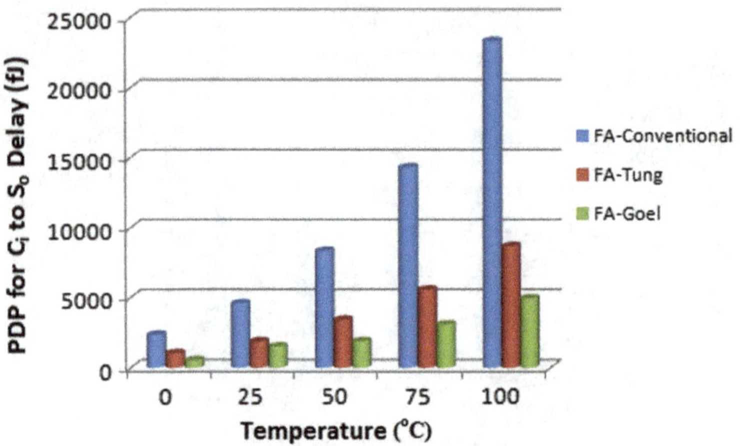

Fig. 10 PDP for input carry C_i to output sum (S_o) (fJ)

Table 2 Delay from input A to output carry (C_o)

Temperature (°C)	FA-Conventional (pS)	FA-tung (pS)	FA-goel (pS)
0	17.27	36.39	26.22
25	20.86	27.45	28.92
50	24.20	25.36	31.51
75	27.87	25.21	34.32
100	31.36	25.29	36.92

100 °C, at these two temperatures FA-Tung has the shortest delay. At all temperatures, FA-Goel has the longest delay except 0 °C, at this temperature FA-Tung has the longest delay. Table 3 represents PDP from input A to output carry (C_o). It is clear from the table that FA-Tung has the lowest PDP at all temperatures except 0 °C, at this temperature FA-Conventional has the lowest PDP. At all temperatures, FA-Conventional has the highest PDP except 0 °C, at this temperature FA-Goel has the highest PDP. The PDP exhibited by the full adder would affect the system's overall performance [11]. So, FA-Tung has the best performance among the considered full

Table 3 PDP for input A to output carry (C_o) delay

Temperature (°C)	FA-Conventional (fJ)	FA-Tung (fJ)	FA-Goel (fJ)
0	1209.07	1282.02	1306.02
25	2516.13	1665.67	2485.38
50	4713.43	2484.27	4371.38
75	8275.16	3763.85	7245.98
100	13,515.85	5479.08	11,281.64

adders when input data to output carry (C_0) delay is more important (Figs. 11 and 12). Figure 13 represents input (A) to output sum (S_0) delay. It can be observed from the figure that FA-Conventional has the longest delay while FA-Goel has the shortest delay for all temperatures.

Table 4 shows that FA-Goel has the shortest delay at all temperatures while FA-Conventional has the longest input A to output sum (S_0) delay for all temperatures. Figure 14 shows that FA-Conventional has the highest PDP at all temperatures while FA-Goel has the lowest PDP at all temperatures. FA-Goel has the best performance

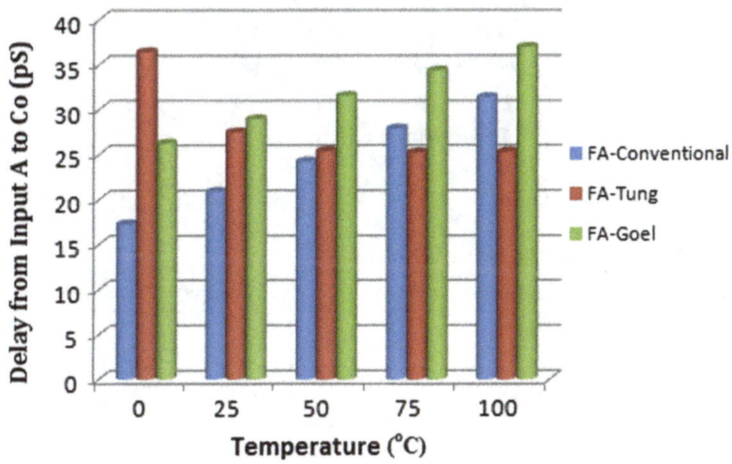

Fig. 11 Delay from input A to output carry (C_0)

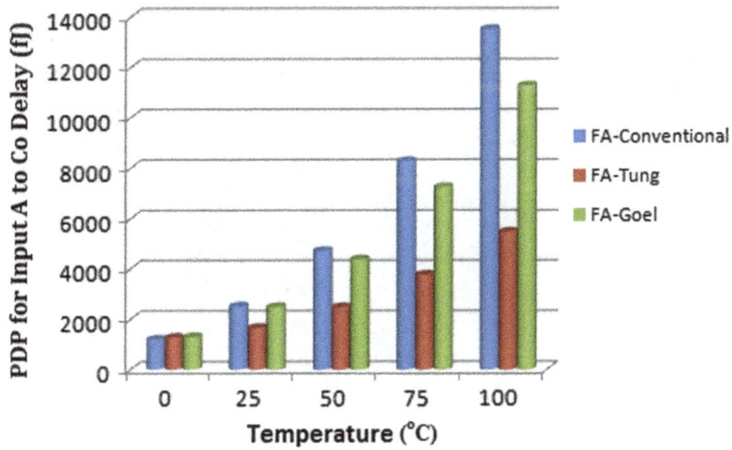

Fig. 12 PDP for input A to output carry (C_0) delay (fJ)

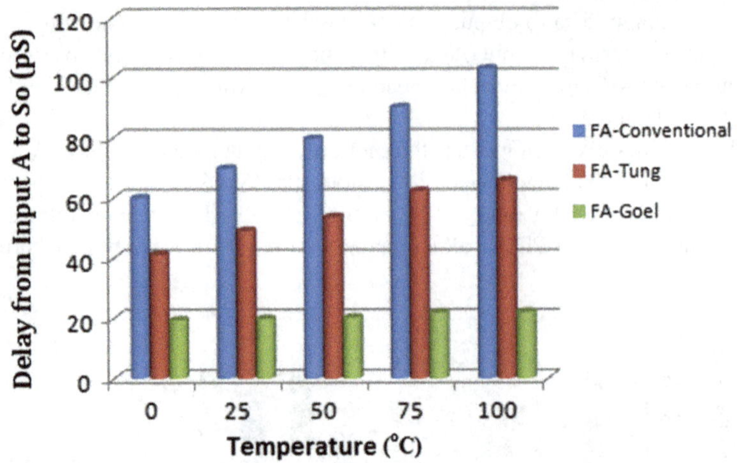

Fig. 13 Delay from input A to output sum (S_o) (pS)

Table 4 Delay from input A to output sum (S_o)

Temperature (°C)	FA-Conventional (pS)	FA-Tung (pS)	FA-Goel (pS)
0	59.99	41.16	19.39
25	69.74	49.02	19.79
50	79.62	53.58	20.19
75	90.16	62.22	21.74
100	102.90	65.76	21.96

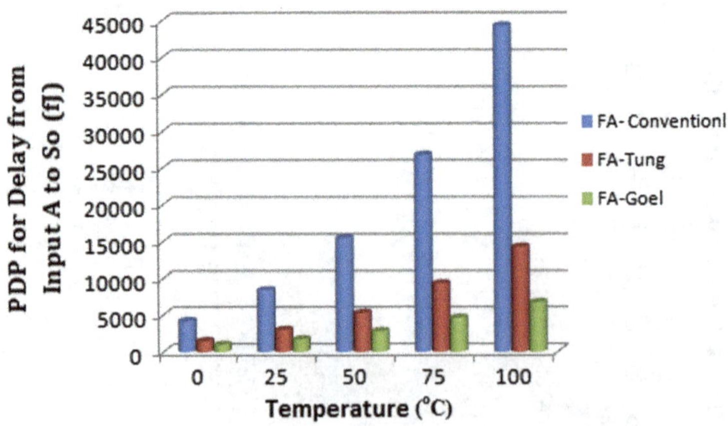

Fig. 14 PDP for delay from input A to output sum (S_o) (fJ)

among the considered full adders while Data Input to Output Sum (S_o) delay and respective PDP is important.

4 Conclusion

In this paper, a comparative analysis of full adders has been done. The full adder circuits have been simulated with a TSMC 32 nm CMOS technology. Power consumption, different types of delays and PDPs have been considered. The simulation has been done using SPICE software. From the results, it is found that for all temperatures, FA-Conventional has the highest power consumption while FA-Tung has the lowest power consumption. FA-Tung has the best performance among the considered full adders when input carry C_i to output carry (C_o) delay is more important. For all temperatures, FA-Conventional has the longest delay and the highest PDP while input carry C_i to output sum (S_o) delay and respective PDP is considered. FA-Goel has the best performance among the considered full adders while input carry C_i to output sum (S_o) delay and respective PDP is important. FA-Tung has the best performance among the considered full adders when input data to output carry (C_o) delay is more important. FA-Goel has the best performance among the considered full adders while Data Input to Output Sum (S_o) delay and respective PDP is important while FA-Conventional has the longest delay and the highest PDP for these conditions.

References

1. A.M. Shams, M. Bayoumi, Performance evaluation of 1-bit CMOS adder cells. in *Proceedings of IEEE ISCAS,* Orlando, FL, (vol. 1, 1999), pp. 27–30
2. Q.A. Al-Haija, H. Al-Amri, M. Al-Nashri, S. Al-Muhaisen, C of 4-bit special purpose microprogrammed processor, in *The 4th International Conference on Emerging Ubiquitous Systems and Pervasive Networks (EUSPN-2013),* Elsevier Procedia Computer Science (vol. 21, 2013), pp. 512–516
3. S. Parmar, K.P. Singh, Design of high speed hybrid carry select adder, in *3rd IEEE International Advance Computing Conference (IACC),* Ghaziabad (2013), pp. 1656–1663
4. M.M. Mano, M.D. Ciletti, *Digital Design,* 5th edn. (Pearson Education, 2012)
5. J.-F. Lin, Y.-T. Hwang, M.-H. Sheu, C.-C. Ho, A novel high speed and energy efficient 10-transistor full adder design. IEEE Trans. Circuits Syst. I **54**(5), 1050–1059 (2007)
6. Y. Jiang, A. Al-Sheraidah, Y. Wang, E. Sha, J.G. Chung, A novel multiplexer-based low-power full adder. IEEE Trans. Circuits Syst. II. Analog Digit. Signal Process **51**, 345–348 (2004)
7. H.T. Bui, Y. Wang, Y. Jiang, Design and analysis of low-power 10-transistor full adders using XOR XNOR gates. IEEE Trans. Circuits Syst. II, Analog and Digital Signal Processing **49**(1), 25–30 (2002)
8. D. Wang, M. Yang, W. Cheng, X. Guan, Z. Zhu, Y. Yang, Novel low power full adder cells in 180 nm CMOS technology, in *4th IEEE Conference on Industrial Electronics and Applications* (2009), pp. 430–433
9. C.K. Tung, S.H. Shieh, C.H. Cheng, A regularly modularized multiplexer-based full adder for arithmetic applications. Appl. Math. Inf. Sci. Int. J. **8**(3), 1257–1265 (2014)

10. S. Goel, A. Kumar, M.A. Bayoumi, Design of robust, energy-efficient full adders for deep-submicrometer design using hybrid-CMOS logic style. IEEE Trans. Very Large Scale Integr. VLSI Syst. **14**, 1309–1321 (2006)
11. M.A. Hernandez, M.L. Aranda, CMOS full-adders for energy-efficient arithmetic applications. IEEE Trans. Very Large Scale Integr. VLSI Syst. **19**(4), 718–721 (2011)

Miscellaneous

Optimize Generation Scheduling with Real-Time Power Management System for Isolated Hybrid Microgrid

Kuldip Singh, Satyasis Mishra and Demissie J. Gelemecha

Abstract Isolated microgrid with renewable energy sources had vital role for rural electrification from in India. The major challenge in village electrification is generation scheduling from renewable energy sources with respect to load demand on isolated hybrid microgrid. The optimize or scheduling of electrical utilization will not only affect the demand on distribution generation system but also alter the loading of generation system. The RPMS is used for optimizing the generation schedule with respect to load at utilization end for balancing the load radial distribution system. In this paper, the study is carried out for optimize generation scheduling for isolated microgrid with RPMS with respect to time and priority of load utilization. The power management system is communication hub between generation from different renewable energy sources and utilization for improvement of reliability and stability of isolated hybrid microgrid for rural electrification.

Keywords Isolated hybrid microgrid · RPMS · Renewable sources · Optimize · Time scheduling · Priority

1 Introduction

The electrical power generation from renewable energy sources in India for rural electrification has been increased due to huge energy demands and environmental protection requirements. This pressures for clean and environmentally friendly rural distributed generation system, while ensuring stability and reliability of distributed

K. Singh (✉) · D. J. Gelemecha
Centurion University, Paralakhemundi, R.Sitapur, Odisha, India
e-mail: ksmann3@gmail.com

D. J. Gelemecha
e-mail: gelmechad@gmail.com

S. Mishra
Adama Science and Technology University, Adama, Ethiopia
e-mail: satyasismishra@gmail.com

H. S. Saini et al. (eds.), *Innovations in Electronics and Communication Engineering*, Lecture Notes in Networks and Systems 107,
https://doi.org/10.1007/978-981-15-3172-9_63

generation system from solar, biomass and fuel cells. The distributed generation system isolated hybrid microgrid approach was designed as a bridge technology toward the smart or real-time-based system with small-scale energy generation and local distribution system along with minimum energy cost [1, 2].

Due to instability in power generation from renewable energy sources with local distribution system has motivated the integration of different renewable energy sources for increasing the power generation capacity [2]. The major challenge faced to implement the isolated hybrid microgrid concept for village electrification is the fact that the integration of renewable energy sources with distribution generation/loads, which substantially increase the complexity of control, communication in demand management system. It is reflecting on power quality, protection, stability, reliability, and efficiency of isolated hybrid microgrid [2]. For improving the reliability and stability of isolated hybrid microgrid an innovative demand management system, smart grid technology is expected to revolutionize the solution of integration and optimization between sources and loads [1].

To enhance the power management or demand management for effective power dispatch and load scheduling with respect to time and priority, the real-time power management system (RPMS) has been deployed. With the installation of isolated hybrid (PV-FC-Biomass) microgrid, it would add extra values to the households by contributing a control of electrical utilization based on time and priority system, which is contribution for cost saving and energy saving. The RPMS is implemented with optimize technique by considering both technical and economical parameters. The system is built with customized controller for generation and load scheduling at each household based on time and priority [3]. The main objective of study is optimizing the power utilization based on the generation scheduling from renewable energy sources with respect to time and Ppriority for improving the reliability and stability of isolated hybrid microgrid for village electrification. The paper is organized as follows: In Sect. 2, electric load demand model based on each household connected load is explained by mathematical approach, Sect. 3 explains the basic algorithm for load scheduling and integration of sources with respect to time and load demand from each home, and Sect. 4 explains simulation results based on the demand and generation from renewable energy sources, and finally, the paper is concluded.

2 Electrical Load Demand Model

Consider the total electrical load demand in village with a set of $P_D = \{1, 2, 3 \ldots N\}$ houses that shared with isolated hybrid microgrid by integration of solar PV plant, fuel cell, and biomass. For each house η, let L_η^h total load demand with respect to appliance in each house and total loads demand at time $t\,\varepsilon\,\mathcal{T} = \{1, 2, 3 \ldots T\}$. In this, for daily operation of isolated hybrid microgrid, the time slot is taken as one hour, and total time is $H = 24$. The total daily load demand for user η is denoted by $l_\eta \triangleq \left[L_\eta^1, L_\eta^2 \ldots L_\eta^H\right]$ [4]. The total load for all users during a time t can be describe as follows:

$$L_{Dt} \triangleq \sum_{\eta \in \mathrm{N}} L_{\eta}^{t} \tag{1}$$

Daily peak load demand

$$L_{\mathrm{Dpeak}} = \max_{t\varepsilon\mathcal{T}} L_{Dt} \tag{2}$$

Average load demand

$$\boldsymbol{L}_{\mathbf{DAvg}} = \frac{1}{\boldsymbol{H}} \sum_{t\varepsilon\mathcal{T}} \boldsymbol{L}_{\boldsymbol{Dt}} \tag{3}$$

Peak to average ratio of load demand is

$$\boldsymbol{P}_{\mathbf{Drms}} = \frac{\boldsymbol{L}_{\mathbf{Dpeak}}}{\boldsymbol{L}_{\mathbf{DAvg}}} = \frac{\mathbf{H}\max_{t\varepsilon\mathcal{T}} \boldsymbol{L}_{\boldsymbol{Dt}}}{\sum_{t\varepsilon\mathcal{T}} .\boldsymbol{L}_{\boldsymbol{Dt}}} \tag{4}$$

We define $\mathcal{T}$ as the set of operating time slot of house $\eta \in \mathrm{N}$ and we restrict the minimum and maximum power consumption for each house $\eta \in \mathrm{N}$ in operating time slot

Let $P_{Dm_{\eta}}$ = pre-determine total daily load demand in isolated hybrid microgrid from $\eta \in \mathrm{N}$ houses.

The energy balance equation for isolated hybrid microgrid is

$$\sum_{t\varepsilon\mathcal{T}} P_{D\eta}^{t} = P_{Dm_{\eta}} \tag{5}$$

Now consider the generation system for isolated hybrid microgrid for fulfill the electricity demand for each household.

The solar *PV* plant generation profile with respect to light intensity is λ

$$P_{\mathrm{PVGen}} = \left[P_{\lambda}^{1}, \ldots P_{\lambda}^{t} \ldots P_{\lambda}^{\mathcal{T}}\right] \tag{6}$$

where P_{λ}^{t} = maximum generation from solar power plant

The generation from fuel cells is

$$P_{\mathrm{FCGen}} = \left[P_{fc}^{1}, \ldots P_{fc}^{t} \ldots P_{fc}^{\mathcal{T}}\right] \tag{7}$$

The generation from biomass with respect to biomass availability is B_{mt}

$$P_{\mathrm{BMGen}} = \left[P_{BM_{Bmt}}^{1}, \ldots P_{BM_{Bmt}}^{t} \ldots P_{BM_{Bmt}}^{\mathcal{T}}\right] \tag{8}$$

The total demand in isolated microgrid is shared by three different sources

$$\sum_{t\varepsilon T} P^{t}_{D\eta} = \sum_{t\varepsilon T} P_{\text{PVGen}} + P_{\text{FCGen}} + P_{\text{BMGen}} \tag{9}$$

3 Scheduling and Integration of Sources With Respect To Time

In the above section, we discuss the load demand from each household with respect to time and the sharing of total load demand with three renewable energy sources, namely solar, fuel cell, and biomass generation system for village electrification. The basic model diagram is shown in Fig. 1.

The scheduling and integration of sources in isolated hybrid microgrid is optimized by real-time power management system (RPMS) as shown in Fig. 1. The maximum optimize utilization of each sources is scheduling algorithm based on time.

Time-Based Algorithm for Generation Scheduling With RPMS

1. //Check total load demand by Eq. (2)
2. //Initialize the total number of hours H=24
3. **for (t=1; t<=H; t++)**
4. **if** (t=>22.01 & t<04.01) **then**
5. Connect the load to fuel cell $\boldsymbol{L_{Dpeak} = P_{FCGen}}$
6. **if** $(L_{Dpeak} = P_{FCGen})$ **then**
7. go to 3
8. **else**

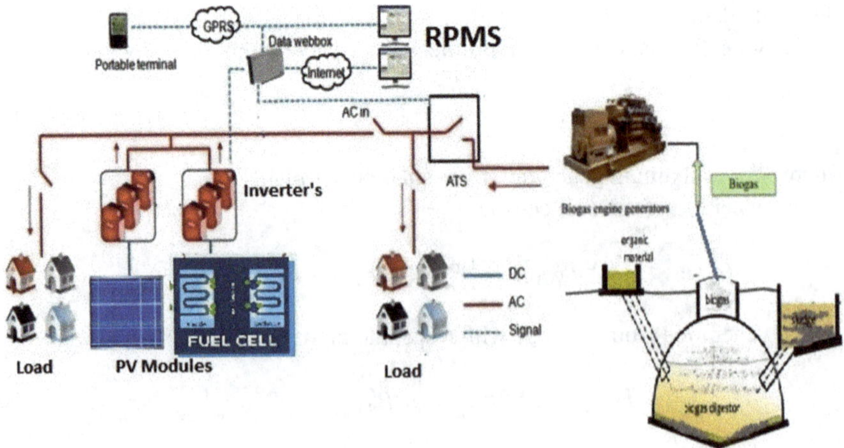

Fig. 1 Block diagram for isolated hybrid microgrid with RPMS

9. Integrate the biomass and fuel cell with RPMS system
10. $\boldsymbol{L_{Dpeak} = P_{FCGen} + P_{BMGen}}$
11. **end if**
12. **end if**
13. **if** ((t=>4.01 & t<7.01)|| (t=>17.01 & t<22.01)) then
14. connect the load to biomass generation $\boldsymbol{L_{Dpeak} = P_{BMGen}}$
15. **if** $(L_{Dpeak} = P_{BMGen})$ **then**
16. go to 14
17. **else**
18. go to 9
19. **end if**
20. **end if**
21. **if** (t=>7.01 & t<17.01) **then**
22. connect the load to solar PL plant $\boldsymbol{L_{Dpeak} = P_{PVGen}}$
23. **if** $(L_{Dpeak} = P_{PVGen})$ **then**
24. go to 21
25. integrate the solar PV plant with biomass
26. $\boldsymbol{L_{Dpeak} = P_{PVGen} + P_{BMGen}}$
27. **end if**
28. **end if**
29. return true
30. **end for**

4 Simulation Results and Discussion

As discussed in Sect. 3, the electrical demand in the village is shared by different renewable energy sources in isolated hybrid microgrid. The simulation study is carried out with Homer software for isolated hybrid microgrid with solar PV plant, fuel cell, and biomass integration based on load demand and priority of generation sources based on time. The load data and generation scheduling from different renewable energy sources with RPMS system is given in Table 1. The load profile is shown in Fig. 2 for village load on January 3 in 24 h.

Based on the Algorithm, the generation is scheduled based on time and priority of sources, the maximum load sharing on solar PV plant in daytime. The generation of solar power plant is shown in Fig. 3 with respect to time in hours.

Solar power generation depends on the solar radiations, which are not constant throughout the day or month or year. Due to uncertainty in solar power generation, load needs to integrate with multiple sources to fulfill the load demand. Figure 4 shows the load sharing with PV-FC-biomass for fulfilling the electric load demand with respect to time, and during excess load demand, more one source are integrating for load sharing on distribution system with real-time power management system. In this process, the un-interrupted power supply from renewable energy sources is

Table 1 Generation scheduling with RPMS system

Time (h)	Load (kW)	PV output (kW)	FC output (kW)	Biomass output (kW)
01:00	21.12272	0	22.23444	0
02:00	15.45492	0	16.26834	0
03:00	16.46071	0	17.32706	0
04:00	23.93764	0	25.19752	0
05:00	21.12502	0	0	22.23686
06:00	29.15772	0	0	30.69234
07:00	22.11621	11.45605	0	11.22122
08:00	120.793	206.5757	0	0
09:00	162.3238	209.8804	0	0
10:00	172.668	218.157	0	0
11:00	158.4356	255.7696	0	0
12:00	351.8632	272.2982	0	79.565
13:00	244.088	275.0422	0	0
14:00	262.952	274.7148	0	0
15:00	245.2233	262.9713	0	0
16:00	299.1845	233.9629	0	65.2216
17:00	286.2814	150.4747	0	135.8067
18:00	237.9118	0	0	237.9118
19:00	172.6553	0	0	181.7424
20:00	117.9621	0	0	124.1706
21:00	171.9285	0	0	180.9773
22:00	176.0125	0	0	185.2763
23:00	25.03068	0	26.34808	0
00:00	21.45582	0	22.58508	0

integrated with load demand, which improves the stability and reliability of isolated hybrid microgrid for village electrification.

The main comparison between demand management system and real-time power management system is that in demand management system the total load demand at grid side is control based on peak demand load, but in case of the RPMS system, the demand management is control based at load end for overcome the peak demand.

5 Conclusion

The future isolated hybrid microgrid with load optimization has been considered a complex and advanced distribution generation system with multiple renewable energy sources for village electrification in India. The integration and scheduling

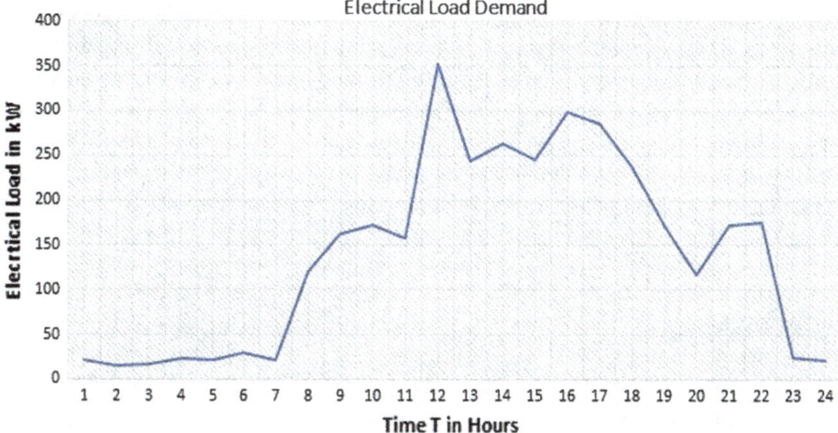

Fig. 2 Village load demand with respect to time. *Source* 3 Jan with Homer software

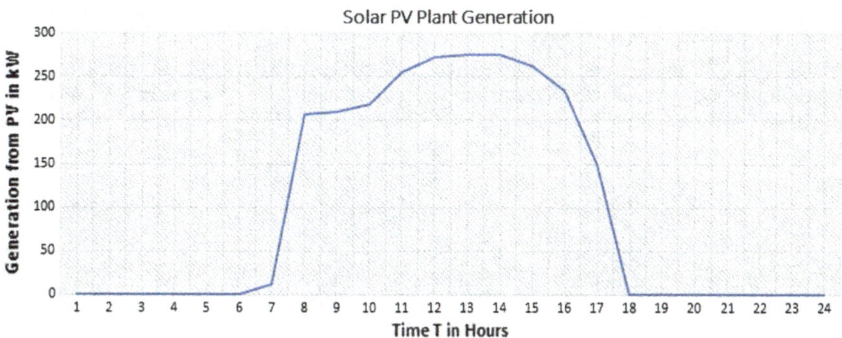

Fig. 3 Generation from solar power plant with respect to time. *Source* Homer software

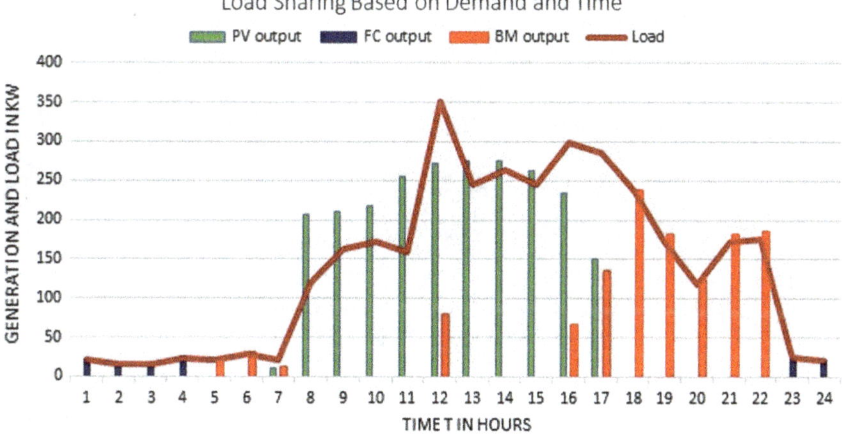

Fig. 4 Generation scheduling with respect to load demand and time with RPMS system

of generation with respect to load demand and time is major challenge for isolated microgrid for village electrification, which is directly related to stability and reliability of distribution generation system. In this study, real-time power management system is proposed for optimize generation scheduling by integration of different renewable energy sources with respect to time and load demand. The RPMS system improves the stability and reliability of the isolated hybrid microgrid for village electrification with real-time optimize conditions.

References

1. Y. Park, S. Kim, Game theory-based Bi-level pricing scheme for smart grid scheduling control algorithm. J. Commun. Netw. **18**(3), 484–492 (2016)
2. L. Trigueiro dos Santos, M. Sechilariu, F. Locment, Optimized load shedding approach for grid connected DC microgrid systems under realistic constraints. MDPI J. Build. **6**(50), 1–15 (2016)
3. F. Yang, X. Xia, Techno-economic and environmental optimization of a household photovoltaic-battery hybrid power system within demand side management. Renew. Energy J. **108**, 132–143 (2017)
4. B. Gao, W. Zhang, Y. Tang, M. Hu, M. Zhu, H. Zhan, Game-theroretic energy management for residential users with dischargeable plug-in electrical vehicle. MDPI J. Energ.S **7**, 7499–7518 (2014)

Churning of Bank Customers Using Supervised Learning

Hemlata Dalmia, Ch V S S Nikil and Sandeep Kumar

Abstract In the current challenging era, there is prominent competition in bank industry. To improve quality and level of service, bank concentrates on customer retention as well as customer churning. This paper discusses the classification problem of banking industry. It focuses on the customers of a bank concerns towards churning, predicting the departing customers from potential customers. Machine learning is the cutting edge technology that is practical and handy to solve such problems. Using supervised machine learning, a proprietary algorithm (a typical machine learning model) is created to forecast and inform the bank about the customers who are at the highest risk in leaving the bank. A customer churn prediction can be used here as churn and nonchurn customers are to be defined. Using ML, gap is to be resolved between churn and nonchurn customers. Different accuracy levels are achieved by classifiers using different data sheets. A novel approach K-nearest neighbor algorithm (KNN) is presented in which dataset is suitably grouped into training and testing models depending on weighted scales along with XGBooster algorithm for high and improved accuracy.

Keywords Customer churning · Machine learning · XGBooster · KNN

1 Introduction

Customers are surrounded by number of resources of information in today's digitized environment, and they have all resources at the tip of their fingers. Smartphones, e.g., provide instant access to various branded products, mbanking, comparative information. And customer perspectives demand based on advanced technology,

H. Dalmia (✉) · S. Kumar
ECE Department, Sreyas Institute of Engineering and Technology, Hyderabad, Telangana, India
e-mail: dalmiahemlata@gmail.com

S. Kumar
e-mail: er.sandeepsahratia@gmail.com

Ch V S S Nikil
Sreyas Institute of Engineering and Technology, Hyderabad, Telangana, India
e-mail: saisainikil@gmail.com

H. S. Saini et al. (eds.), *Innovations in Electronics and Communication Engineering*, Lecture Notes in Networks and Systems 107,
https://doi.org/10.1007/978-981-15-3172-9_64

convenience reasons, price sensitivity, service quality reasons and socio factors [1]. Due to availability of different options as a boon of advanced technologies, customers keep on changing from one service provider to another. That is why for companies, it is difficult to retain and attract the customers and loses their wealth due to switching action by their customers. The procedure of customers leaving their service providers is called churn [2].

For banking sector, this customer churn prediction [3] is the serious issue and gargantuan impact on the profit line of bankers. Thus, customer retention scheme can be targeted on high-risk customers who wish to discontinue their custom and switch to another competitor. To minimize the cost of bank sectors customer retention marketing scheme [4], an accurate and prior identification of these customers is hypercritical.

Customer churning [5, 6] is the estimate or analysis of degree of customers who turn to shift to an alternative. It is the most common problem witnessed in any industry. Banking is one such industry that focuses a lot on customer's behavior by tracking their activities. It is very extortionate to add a new customer to the bank when compared to retention [7]. Companies can raise their profits by handling these customers. Hence, there is a need to keep up the existing customers, which will be achieved only by understanding the customer's grievances of changing the bank. The paper presents a model to churn the bank customers using *k*-nearest neighbor (KNN) algorithm. This simple KNN algorithm is used to classify the customers into two classes, those who will leave the bank and those who will not leave. To enhance the accuracy, XGBooster algorithm is applied, whereas many research papers are available from various journals based on bank customer churn prediction, but techniques applied are decision tree [8], logistic regression [9], random forest [10], unsupervised learning [11], artificial neural network (ANN), data mining, [12, 13] neurocomputing [14]. Next section explains the literature survey based on different algorithms used in various papers.

2 Related Work

2.1 Literature Review

From the above discussion, it is clear that customer retention is important for a company and for its business strategy. Customer churning becomes business intelligence to know which customers will shift or who will get retained. To achieve customer churning, companies started adapting machine learning techniques for customer churn prediction models. In this section, a few techniques are compared considering churn prediction.

Data mining by author 'Sen K' aims to analyze large dataset by converting the sets of data into useful data. And a customer churn prediction model is developed and is measured using accuracy, sensitivity and specificity and Kappa's statistics [15].

The support vector machine (SVM) is the popular technique providing guide to the bank for customer strategy. SVM has larger probability of customer churns in the samples. With good number of plenty vectors, SVM provides good precision in predicting technique models. SVM gives high fitting accuracy rate of 0.59 by the author 'Zhao Jing' [16]

'Guoxun Wang' focuses on the comparison of all techniques used to build credit card holder churn model for the banks in China based on multi-criteria decision algorithm and constructing techniques using PROMETHEE and TOPSIS methods [17]. In MCDM algorithm, decision tree methods are implemented. 'Shaoying Cui' presents [18] improved FCM algorithm as data mining algorithm to facilitate the banks with a new idea for predicting customer churn. It achieved accuracy rate of 80% for high-value customers and 83% for low-value customers. 'Pradeep B' proposed to construct a model for churn prediction for a company using logistic regression and decision trees techniques. In Pradeep's approach there is a trial to retrieve the important factors of the customer churn that provides additional and useful knowledge which supports decision making [19].

Alisa Bilal Zorić applied a data mining technique 'neural network' in the software package ANN to predict churn in bank customer. Using this model, the reason of customer leaving the bank can be easily acquainted by entering the parameters [20].

'Abinash Mishra' proposed methodology of ensemble classifiers comprising bagging, boosting and random forest [21] to predict customer churn for telecom industry. Random forest achieves high accuracy of 96% with low specificity and high sensitivity and low error rate [21].

'Ning Lu' presented a paper in which an experimental evaluation proves that the boosting provides a good source of churn data, efficiently providing the customer churn model. The measures for churn prediction are calculated using a training set of customers over a period of six months [22].

'Hend Sayed' presented a methodology of decision tree in which two packages ML and MLib were conducted, to evaluate accuracy, model training and model evaluation. They got effective result with ML package [23].

2.2 *Data Acquisition*

Dataset used for this supervised prediction is acquired from an online source. The target dataset is subjected to churning of customers of bank containing information about 10,000 customers with 14 features for each customer. The customers of the bank are identified as churn or loyal based on the potential features like credit score, age, gender, estimated salary, etc. A user of the bank is classified as loyal if he/she is active and remains with the bank. Customers are classified as churners if they switch to another bank. The variable exited in the dataset gives the actual status of the customer if he/she had switched to another bank.

2.3 *Data Preprocessing*

The process to identify the required independent variables for predicting the exit status of a customer and to predict the binary dependent variable 'EXITED' using the independent variables is data preprocessing. The dataset used for predicting churning of customers of a bank contains information about 10,000 customers with 14 features for each customer. These features include row number, customer id, surname, credit score, geography, gender, age, tenure, balance, number of products, has cr card, is active member, estimated salary, exited [24].

To predict the churning of customers, dataset is split suitable for training and testing. At this instance, splitting has 80% training rate and 20% testing rate (Table 1).

The value of this attribute will be 1 if the customer has left the bank and 0 if remained there.

Feature scaling or data normalization is a technique used to standardize the range of independent variables in the dataset.

3 Methodology

In this paper, whole focus is using flexible technique to boost the accuracy in customer churning process. So, along with K-nearest neighbors (KNN) algorithm, XGBoost algorithm is implemented. The block diagram is represented below to describe the whole process (Fig. 1).

Table 1 Utilization of features of dataset

Serial No.	Features	Utilisation
1.	RowNumber	Unused attribute
2.	CustomerId	Unused attribute
3.	Surname	Unused attribute
4.	Credit score	Unused attribute
5.	Geography	Used as input after encoding
6.	Gender	Used as input after encoding
7.	Age	Used as input
8.	Tenure	Used as input
9.	Balance	Used as input
10.	NumOfProducts	Used as input
11.	HasCrCard	Used as input
12.	IsActiveMember	Used as input
13.	EstimatedSalary	Used as input
14.	Exited	Used as target

Fig. 1 Block diagram to build customer churn model

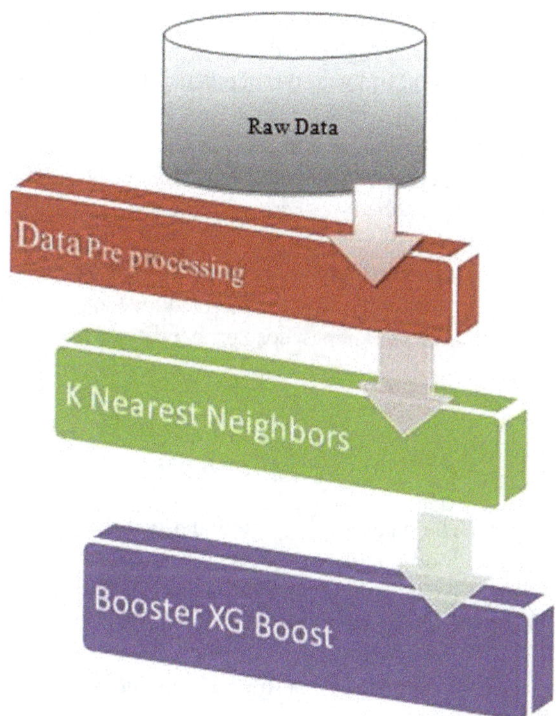

3.1 K-Nearest Neighbors

K-nearest neighbors is a machine learning and data mining algorithm used to address classification and regression problems. It uses Euclidean distance to find the similarity between the classes.

3.2 Boosting

The idea of using Boosting algorithm for customer churn prediction is to train a series of classifier simultaneously and keep updating the model accuracy for improving the performance of the classifier.

3.3 XGBoost

XGBoost is extreme gradient boosting. It is the mostly used predominant and advanced algorithm to solve machine learning problems. It dispenses support in most of the developing environments (C++, python, R). It is mainly used for increased performance and high speed.

Installing XGBoost: To implement XGBoost model, use XGBoost function from XGBoost package in RStudio by importing the package. Fit the model to the data and predict the churning customers using the function XGBoost by supplying the training dataset, dependent variable and the number of iterations to which the classifier is to be trained. Use the *k*-fold cross-validation technique to find the average accuracy of XGBoost classifier.

The customers are classified into two classes those who will leave the bank and who those who will not leave the bank; we apply an effective algorithm and predict the churning customers whose probability of leaving the bank is very high.

3.4 KNN Algorithm

Step 1: The dataset is imported and preprocessed. Preprocessing is needed to get good quality results. The dataset is split suitable for training and testing. We have 80% for training and 20% testing.

Step 2: Training data is fitted to the KNN classifier, and the exit status of customer is predicted as follows:
y_pred = knn (train = training set[, −11], test = test

Step 3: Step 3: Find the accuracy obtained using the KNN classifier. The *K* value can be identified by checking with multiple values. Accuracy is improved by tuning algorithm with different *K* values.
_set [, −11], cl = training set [, 11], $k = 5$, prob = TRUE)

The working of K-nearest neighbor is explained as follows:

To find the exit status of a customer (record) *X*, the Euclidean distance of the record with respect to all the other records X_i $i = \{1, 2, 3 \ldots n\}$ is calculated using the following equation

$$\text{Euclidean distance} = \text{Sqrt}\left(\sum (X_i - X)^2\right), \quad i = \{1, 2, 3 \ldots n\}$$

The distances obtained are arranged in ascending order. And the first *K* distances from the obtained distances ($K > 0$) are selected. The records (points) corresponding to the distances are identified, and the exit status for each of the records is observed. The exit status of record *X* based on majority voting has to be evaluated at last.

3.5 XGBoost Algorithm

Step 1: XGBoost classifier is installed from XGBoost package and fitted to the training set by specifying the maximum number of iterations.

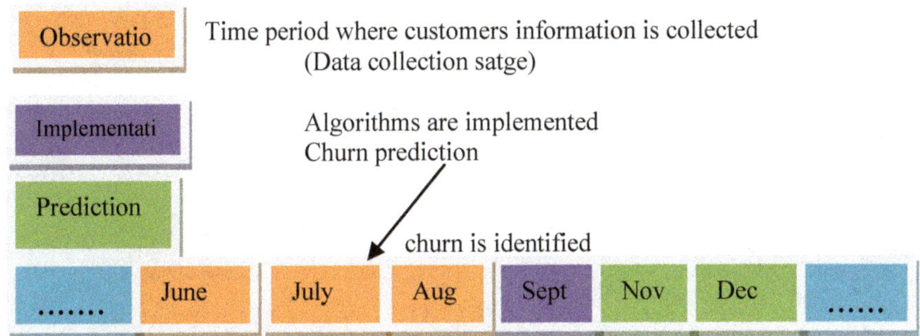

Fig. 2 Time line diagram for customers of bank

Step 2: Evaluate the XGBoost classifier accuracy using *K*-fold cross-validation by mentioning the number of folds. The ultimate accuracy of the XGBoost model is the mean of all the folds.

In training and testing, the training data is fit to the classifier. The data frame is constructed as matrix to pass it as an argument, and then, the maximum number of iterations is specified. In testing, the predict exit status of the customer is tested with the actual status to determine the accuracy of the classifier. Here, the confusion matrix shows an accuracy of 86.85%.

K-fold cross-validation method is applied by specifying number of folds (sample 10), and the average accuracy 88.07% is evaluated.

The performance is measured using the following parameters:

1. Accuracy 2. Specificity 3. Sensitivity 4. Error rate

Along with performance matrix, confusion matrix is also selected to prove the efficiency of model on the dataset for which the true values are familiar. The values for different classifiers are known using *R*Studio tool. Using confusion matrix, all the above-mentioned performance parameters are calculated (Fig. 2).

$$\text{Accuracy} = \frac{\text{TP} + \text{TN}}{\text{FP} + \text{FN} + \text{TP} + \text{TN}}$$

$$\text{Sensitivity} = \frac{\text{TP}}{\text{FN} + \text{TP}}$$

$$\text{Specificity} = \frac{\text{TN}}{\text{FP} + \text{TN}}$$

$$\text{Error Rate} = 1 - \text{Accuracy}$$

TP—True Positive TN—True Negative FP—False Positive FN—False Negative

The flowchart depicts the systematic approach for classifying the customers of a bank by implementing predictive machine learning algorithm 'K-nearest neighbor' and a booster 'XGBoost' (Fig. 3).

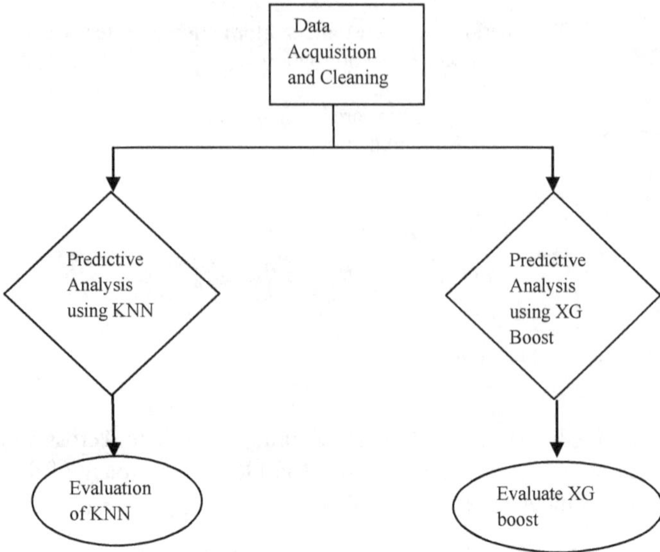

Fig. 3 Flowchart for customer churning system

4 Result

To assess the performance of the classifier model for churn prediction, bank data is trained for a specific period, and then, the customers of the bank are classified into loyal or churn based on their activities. This prediction gives useful insights to the bank officials regarding its customers and functioning of bank. The performance of the prediction model is the capability to identify customers exit status accurately. We use confusion matrix for evaluation (Figs. 4, 5, 6 and 7; Tables 2, 3 and 4).

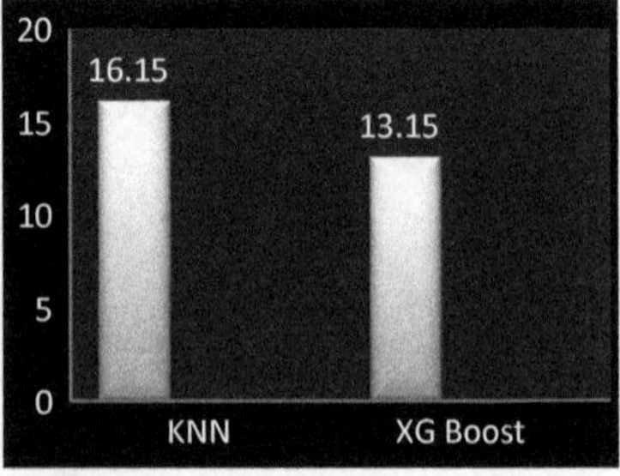

Fig. 4 Error rate

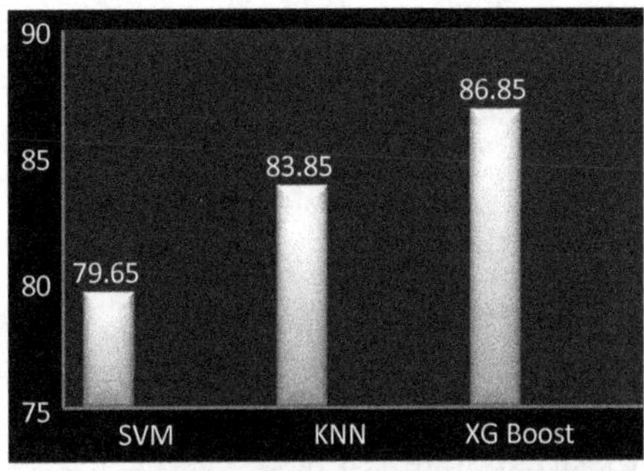

Fig. 5 Accuracy

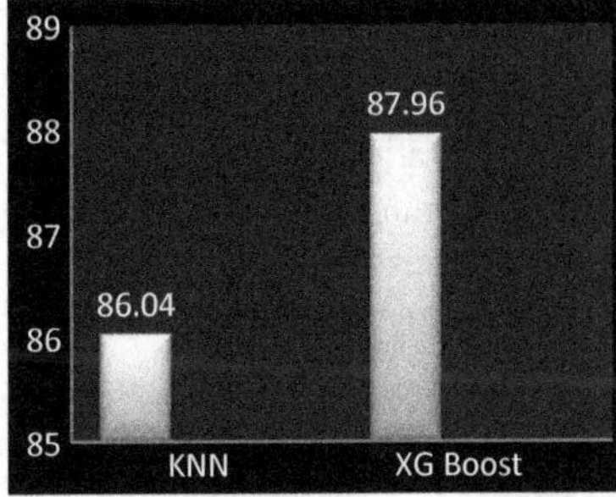

Fig. 6 Sensitivity

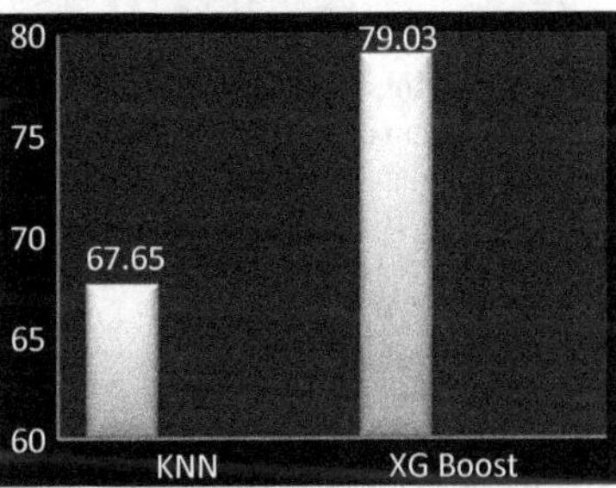

Fig. 7 Specificity

Table 2 Confusion matrix of KNN algorithm

Exit status	Predicted churn	Predicted retention
Actual churn	1516	77
Actual retention	246	161

Table 3 Confusion matrix of XGBoost algorithm

Exit status	Predicted churn	Predicted retention
Actual churn	1541	52
Actual retention	211	196

Table 4 Performance metrices

Classifier	% Accuracy	Error rate	Sensitivity	Specificity
KNN	83.85	16.15	86.04	67.65
XGBoost	86.85	13.15	87.96	79.03

5 Conclusion

In this paper, we propose an effective model of churn in bank industry. It combines the KNN with XGBoost algorithm to enhance the accuracy of the model; this proves the advantage of the technique used. XGBoost gives the best result in terms of accuracy, sensitivity and specificity. Boosting has given the increased accuracy of 86.85 with low error, high sensitivity and specificity.

Organizations periodically calculate customer churn in multiple aspects. Churning can be the number of customers lost, ratio or percentage of customers lost compared with total customers in bank. Churn can be calculated on quarter or annual basis. An accurate forecast can give insights on future using which a strategy can be formulated.

References

1. N. Hashmi, N.A. Butt, M. Iqbal, Customer churn prediction in telecommunication in a decade review and classification. Int. J. Comput. Sci. Issues (IJCSI) **10**(5), 271–281 (September 2013)
2. V. Mahajan, R. Mishra, R. Mahajan, Review of data mining techniques for churn prediction in telecom. JIOS **37**(2), 183–197 (2015)
3. L. Yan, R.H. Wolniewicz, R. Dodier, Predicting customer behavior in telecommunications. 1094-7167/04© 2004 IEEE Published by the IEEE Computer Society (2004)
4. B. Kaderabkora, P. Malecek, Churning and labour market flows in the new EU member states. Int. Inst. Soc. Econ. Sci. 372–378 (2015)
5. S.A. Qureshi, A.S. Rehman, A.M. Qamar, A. Kamal, *Telecommunication Subscribers' Churn Prediction Model Using Machine Learning* (IEEE, 2013), pp. 131–136
6. B. Mishachandar, K.A. Kumar, Predicting customer churn using targeted proactive retention. Int. J. Eng. Technol. **7**(2.27), 69–76 (2018)
7. K. Mishra, R. Rani, Churn prediction in telecommunication using machine learning, in *International Conference on Energy, Communication, Data Analytics and Soft Computing (ICECDS-2017)* (IEEE, 2017), pp. 2252–2257
8. E.M.L. Peters, G. Dedene, J. Poelmans, Understanding service quality and customer churn by process discovery for a multi-national banking contact center, in *Proceedings—IEEE 13th International Conference on Data Mining Workshops, ICDMW 2013* (2013), pp. 228–233. Art. no. 6753925
9. N. Wang, D.X. Niu, Credit card customer churn prediction based on the RST and LS-SVM, in *Proceedings of the 2009 6th International Conference on Service Systems and Service Management, ICSSSM '09* (2009), pp. 275–279. Art. no. 5174892
10. Y. Xie, X. Li, E.W.T. Ngai, W. Ying, *Customer Churn Prediction Using Improved Balanced Random Forests*. (Elsevier, Amsterdam, 2008), pp. 5445–5449
11. P. Spanoudes, T. Nguyen, Deep learning in customer churn prediction: unsupervised feature learning on abstract company independent feature vectors. arXiv:1703.03869v1 1–22 (2017)

12. W. Ying, X. Li, Y. Xie, E. Johnson, Preventing customer churn by using random forests modeling, in *IEEE International Conference on Information Reuse and Integration IEEE IRI-2008* (2008), pp. 429–434. Art. no. 4583069
13. Y. Chen, L. Zhang, Y. Shi, Post mining of multiple criteria linear programming classification model for actionable knowledge in credit card churning management, in *Proceedings—IEEE International Conference on Data Mining ICDM* (2011), pp. 204–211. Art. no. 6137381
14. A. Amin, S. Anwar, A. Adnan, M. Nawaz, K. Alawfi, A. Hussain, K. Huang, Customer churn prediction in telecommunication sector using rough set approach. Neurocomputing (2016). https://dx.doi.org/10.1016/j.neucom.2016.12.009
15. K. Sen, N.G. , in *Proceedings of 23rd Signal Processing and Communications Applications Conference, SIU* (2015), pp. 2384–2387. Art. no. 7130361
16. J. Zhao, X.H. Dang, Bank customer churn prediction based on support vector machine: taking a commercial bank's VIP customer churn as the example, in *2008 International Conference on Wireless Communications, Networking and Mobile Computing WiCOM 2008* (2008). Art. no. 4680698
17. G.Wang, L. Liu, Y. Peng, G. Nie, G. Kou, Y. Shi, Bayazit, Predicting credit card holder churn in banks of China using data mining and MCDM, in *Proceedings—2010 IEEE/WIC/ACM International Conference on Web Intelligence and Intelligent Agent Technology—Workshops, WI-IA 2010* (2010), pp. 215–218. Art. no. 5615798
18. S. Cui, N. Ding, Customer churn prediction using improved FCM algorithm, in *IEEE 3rd International Conference on Information Management (ICIM)*, (2017) Art. no. 16967234
19. B. Pradeep, S. Vishwanath Rao, & S. M. Puranik, Analysis of customer churn prediction in logistic industry using machine learning. Int. J. Sci. Res. Publ. **7**(11), 401–403 (2017)
20. A. Bilal Zorić, Predicting customer churn in banking industry using neural networks. 117–123 (2016)
21. A. Mishra, U.S. Reddy, A comparative study of customer churn prediction in telecom industry using ensemble based classifiers, in *Proceedings of the International Conference on Inventive Computing and Informatics (ICICI 2017)* (2017), pp. 721–725. Art. no. 17803488
22. N. Lu, H. Lin, J. Lu, G. Zhang, A customer churn prediction model in telecom industry using boosting. IEEE Trans. Ind. Inform. **10**(2), 1659–1665 (May 2014). https://doi.org/10.1109/TII.2012.2224355
23. H. Sayed, M.A. Abdel-Fattah, S. Kholief, Predicting potential banking customer churn using apache spark ML and MLlib packages: a comparative study. Int. J. Adv. Comput. Sci. Appl. (IJACSA) **9**(11), 674–677 (2018)
24. A.R.K. Ahmad, A. Jafar, K. Aljoumaa, Customer churn prediction in telecom using machine learning in big data platform. J. Bigdata. (2019). https://doi.org/10.1186/s40537-019-0191-6
25. A. Keramati, H. Ghaneei, S.M. Mirmohammadi, Developing a prediction model for customer churn from electronic banking services using data mining. Financial Innov. (2016). http://creativecommons.org/licenses/by/4.0/
26. J. Xiao, Y. Wang, S. Wang, A dynamic transfer ensemble model for customer churn prediction, in *2013 6th International Conference on Business Intelligence and Financial Engineering* (IEEE, 2014), pp. 115–119
27. S.F. Sabbeh, Machine-learning techniques for customer retention: a comparative study. Int. J. Adv. Comput. Sci. Appl. (IJACSA) **9**(2), 273–281 (2018)
28. B. He, Y. Shi, Q. Wan, X. Zhao, Prediction of customer attrition of commercial banks based on SVM model. Procedia Comput. Sci. **31**, 423–430 (2014)

Study and Analysis of Apriori and *K*-Means Algorithms for Web Mining

K. Ramya Laxmi, N. Ramya, S. Pallavi and K. Madhuravani

Abstract Emerging technologies such as cloud computing and data mining deal with issues such as scalability, security and efficiency. Web mining is broadly classified under data mining, refers to the resultant data combination obtained by assembling data available in the Web and information mining techniques. In general, mining can be defined as the process of extracting significant things from huge datasets. The applications of Web mining include the evaluation of the successful completion of a particular task, assessing the applicability of specific Web sites and comprehending client conduct. The proposed work is concerned with distributed networks and is aimed at improving the efficiency of data mining and cloud computing techniques. The main goal is to find a solution for generating various itemsets in each site. A well-known algorithm in data mining is the Apriori algorithm which discards infrequent items at the cost of useful data. A major limitation of this algorithm is its slowness, owing to increased transactions. The *K*-means segmentation algorithm is employed for increasing the efficiency by clustering the initial itemset.

Keywords Cloud computing · Apriori algorithm · *K*-means algorithm

1 Introduction

The process of extracting hidden information from vast dataset is termed as data mining. Data mining is a process of predictive information, which is useful for companies that are dealing with data warehouses in their day-to-day activities. Data

K. R. Laxmi (✉) · N. Ramya · S. Pallavi · K. Madhuravani
Department of CSE, Sreyas Institute of Engineering and Technology, Nagole, Hyderabad, India
e-mail: kunta.ramya@gmail.com

N. Ramya
e-mail: ramya.n@sreyas.ac.in

S. Pallavi
e-mail: pallavi.s@sreyas.ac.in

K. Madhuravani
e-mail: madhuravani.k@sreyas.ac.in

H. S. Saini et al. (eds.), *Innovations in Electronics and Communication Engineering*, Lecture Notes in Networks and Systems 107,
https://doi.org/10.1007/978-981-15-3172-9_65

mining allows the users to predict the future of modern world, thereby permitting the business to be knowledge driven and proactive. Data mining offers the automatic prospective analysis to prevent the past analysis of decision support systems by providing retrospective tools. Data mining finds application in various fields as they resolve the drawbacks of higher time consumption. It also aims at evacuating the hidden pattern database, determining expectation beyond predictive data.

Enormous number of data are refined and collected by many different companies. The implementation of hardware and software techniques of data mining enhances the values of information resources and is integrated by new systems and products through online. The implementation of parallel processing or client-server techniques with higher performance allows the data mining to ask questions like "Answer questions related to the client and the reason behind them to reasonably respond to the next promotion mail"?

Data mining is also known as knowledge-discovery and data mining or knowledge-discovery in databases (KDD). Many number of patterns can be easily identified by parameters like clustering, association rule mining, classification and so on. Data mining is a very difficult, and it consists of multiple core fields like computer science. And by adding the computational techniques from machine learning, pattern recognition, information retrieval, and statistics its value increases.

Information mining technique includes six normal classes of tasks:

Anomaly discovery—The recognizable proof of uncommon information records, that may be intriguing or information blunders that require further examination.

Association Rule Learning—This class connects the variables. Case in point, a store may accumulate information on client obtaining propensities. Utilizing affiliation standard taking in, the market can figure out which items are as often as possible purchased together and utilize this data for showcasing purposes.

Clustering—is the assignment of finding gatherings and structures in the information that are somehow "comparable", without utilizing referred to structures as a part of the information.

Process of summation the known structures to use to the new information. Case in point, an email project may endeavor to group an email as "genuine" or as "spam".

Regression—endeavors to discover a capacity which models the information with last error.

Outline—giving a more minimal representation of the information set, including perception and report generation.

2 Literature Survey

In analytical system, retail organization of market basket analysis is determined to analyze the placement of goods [1], to design the sales promotions, to improve the performance of supermarket and to quantify customer satisfaction.

For identifying the patterns, two techniques such as classification and association deal with data mining [2]. The set rules of classification and association techniques include *K*-nearest neighbor and Apriori algorithm and occur challenges from huge databases like inefficiency and time utilization.

In supermarket, the customers buy product from different categories; it means the loyalty of customer depends on different ways [3]. Cloud computing plays a vital role in Architecture, Engineering and Construction (AEC) sector. AEC sector is the process of connecting several organizations and professions with project-based industry, high fragmentation and data intension [4]. For *K*-means rule, the *K*-value is most significant and there is no applicable proof for the choice of *K* (number of cluster to generate), and sensitive to initial worth [5], for various initial worth, there is also totally different clusters generated. The *K*-means sector consists of clustering centers with high dependencies. In data mining, *K*-means approach is comprised of large amount of initial cluster centers. If these cluster centers are fully discarded from the available information, total cluster iterations will be raised to infinity. This in turn produces low-level enhancement to achieve optimal results, but the cluster outcomes remain improper.

Emergence of contemporary techniques for scientific knowledge [6] collection has resulted in accumulation of data pertaining to various fields. Typical information querying methods are inadequate to extract helpful info from huge knowledge banks. Cluster analysis is one among the foremost knowledge analysis strategies, and therefore, the *K*-means rule is popularly used for several kinds of applications.

Cluster analysis of information is a very important task [7] in knowledge discovery and data processing. Cluster analysis aims to cluster knowledge on the premise of similarities and dissimilarities among the info components. The method may be performed during a supervised, semi-supervised or unattended manner. Apriori formula contains some disadvantages despite being straightforward and clear [8]. The goal is to verify the datasets from the candidate set with many frequent itemsets, large datasets and low minimum support by wasting of time. For example, if frequent itemset contains 10^4 datasets, then the candidate set must contain 2-length of 10^7 datasets, and finally, they can be tested and accumulated. Moreover, to the frequent datasets of size one hundred (e.g.) $v_1, v_2 \ldots v_{100}$, generates the size of 2- one hundred candidate itemsets which exactly waste their time in generating the size of the pattern.

Efficient numbering of candidate sets in Apriori requires a tree along with breadth-first search. A sum of *K*-candidate itemsets is obtained from a total of $k - 1$ itemsets. The process of pruning is then carried out on candidates with rare sub-patterns [9]. The candidate itemsets consist of K-length sets-based downward closure lemma. The class association rule mining discovers the association and relation between huge datasets [10]. The association rule mining plays a vital role in the research of data mining, and also, it acts as a typical style for data mining. There are some important fields that value the itemsets in the database; they are information science, information retrieval, statistic, visible and artificial intelligence [11–13].

3 Proposed Methodology and Solutions

Despite, Apriori algorithm is have some weakness and it is simple, straight forward, easy to understand and very expensive to large number of datasets. The example for Apriori algorithm is 104 frequent dataset with 1-item sets in the pattern of frequent datasets must have 107 frequent 2-item sets in the candidate sets and are tested and accumulated. The frequent dataset pattern with size 100 datasets, such as $\{a_1, a_2 \ldots, a_{100}\}$, generates 2–100 candidate itemsets in total. In Apriori algorithm, implementation method is installed to inherit the cost of candidate generation. There are different types of problems, and they are identified as follows:

- Apriori algorithm verifies and scans the datasets in the database. Data mining is true by matching patterns; it is difficult to verify numerous datasets and to scan the database of candidate itemsets.
- The efficiency becomes lower, when the system I/O load, data services and limited memory capacity time are considered to be longer in the database.

Apriori algorithm does not have frequent *K*-itemset, so the transaction is being reduced to improve the efficiency of the algorithm. By scanning the database, the itemset which is potential to frequent in DB at least should be the partitions of frequent in DB. Data mining depends on the subsequent set of data, i.e., lower support threshold + method of determining the completeness.

3.1 Apriori Algorithm

Boolean association rule is used in the mining of frequent itemset of the Apriori algorithm. Bottom-up approach is used in the Apriori algorithm, where the frequent subsets are extended from one itemset at a time; i.e., a group of candidates and candidate generation is tested. The Apriori algorithm operates on the transaction database; the best example is the items collected by the customers.

Algorithm:

1. Identify frequent itemsets, that is, the itemset contains minimal support.
2. The subset of frequent itemset should also be a frequent. For example, consider $\{XY\}$ are the subset of frequent itemset then $\{X\}$ and $\{Y\}$ must be frequent itemset.
3. To find frequent itemset with cardinality which varies from 1 to *K*-itemset.
4. For generating association rules, frequent itemsets are used.

3.2 The K-Means Algorithm

In data mining, the *K*-means clustering technique is very popular for cluster analysis. The *K*-means clustering divides *n* observation into *K*-clusters, and each observation consists of cluster with nearest mean which acts as a prototype of cluster. The computational problem is very hard to find (NP-hard). Data mining consists of heuristic algorithm which is efficiently employed and converged to the local optimization. The *K*-means algorithm is same as the expectation–maximization algorithm by the combination of an iterative refinement approach and Gaussian distributions. Cluster centers are used for modeling the data [14, 15]. The expectation–maximization algorithm allows the cluster centers to have different shapes, whereas *K*-means clustering finds cluster centers of spatial extents.

Algorithm:

1. Choose a value for the total no. of clusters *K*.
2. Randomly select *K* data points in the dataset because these are initial cluster centers.
3. To assign the remaining datapoints of closet cluster center, Euclidean distance is used.
4. Each cluster consists of instances to calculate the new mean for every cluster.
5. According to the previous iteration, if the new mean values are similar to mean values, then the process is terminated. Else repeat the step 3–5 for new mean values of cluster centers.

The flow chart (Fig. 1) shows the overall work performed for the proposed methodology. According to flow chart, the transaction dataset is first initialized in the flow.

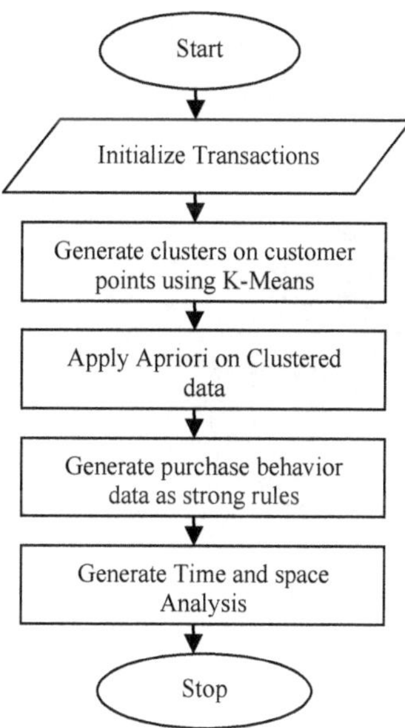

Fig. 1 Methodology of the proposed work

The clusters are then generated on the basis of customer points using *K*-means. The clustered data is then provided to Apriori algorithm for further analysis. The output of the Apriori algorithm is generation of strong rules which shows the purchasing behavior of customers. There, Apriori algorithm is executed twice, i.e., first for unclustered transactions and secondly, for clustered data. After the execution of algorithms, a graphical analysis is generated displaying the memory consumed in bytes by algorithm for clustered and non-clustered data. Another analysis is generated which displays the time consumed by algorithm in milliseconds.

Flow chart—Apriori algorithm

The flow chart (Fig. 2) shows the working of Apriori algorithm. According to flow chart, the transaction dataset is first initialized in the flow. Candidates are generated on the basis of the initial transactions. The candidate sets are further processed, and frequent item sets are filtered out. If all the transactions are covered, then the output of the Apriori algorithm is generation of strong rules which shows the purchasing behavior of customers. There, Apriori algorithm is executed twice, i.e., first for unclustered transactions and secondly, for clustered data.

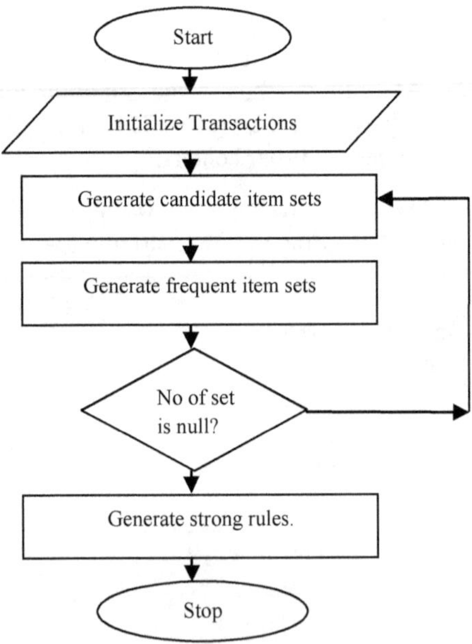

Fig. 2 Apriori algorithm

4 Results and Discussions

Space Consumption

The comparative analysis of space consumption among the algorithms is as given in Table 1.

The above space consumption reading clearly shows that the proposed approach has less space consumption as compared to traditional Apriori algorithm.

Space Analysis

The graphical analysis is more explanatory as it shows that the reading is much higher in case of improved Apriori, therefore the most efficient one. The *x*-axis shows the algorithm performance, and the *y*-axis shows memory space consumed (Fig. 3).

Time Analysis

The comparative analysis of running time among the algorithms is as given in Table 2.

The above execution time reading clearly shows that the proposed approach has less execution time as compared to traditional Apriori algorithm.

Table 1 Space consumption

	Apriori	Improved Apriori
Space	1,079,072	866,080

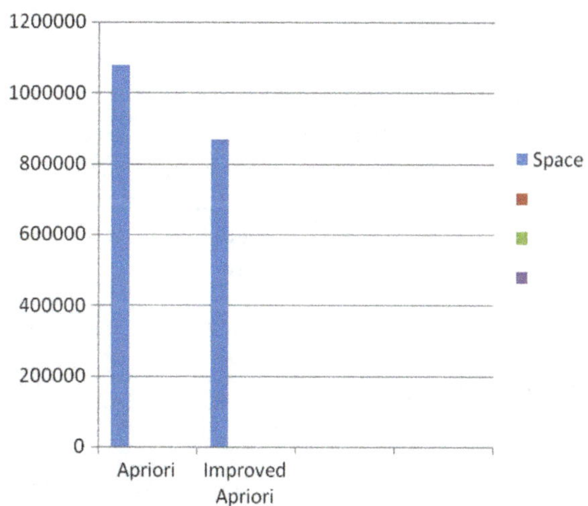

Fig. 3 Analysis for space consumption reading

Table 2 Execution time

	Apriori	Improved Apriori
Time	1936	284

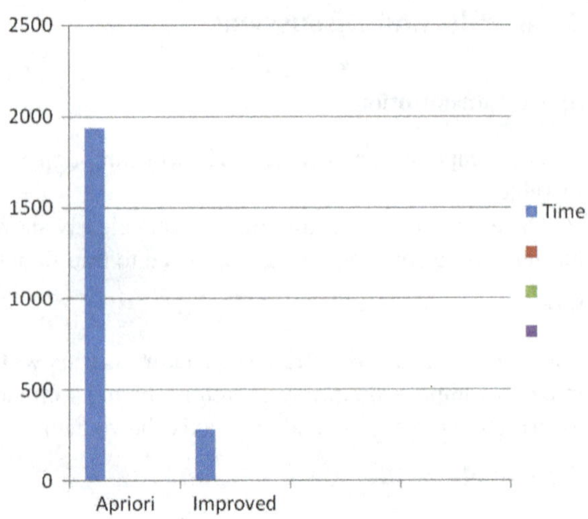

Fig. 4 Analysis for execution time

Execution Time Analysis

The graphical analysis is more explanatory as it shows that the reading is much higher in case of improved Apriori, therefore the most efficient one. The *x*-axis shows the algorithm performance, and the *y*-axis shows execution time in milliseconds (Fig. 4).

5 Conclusions

The focal objective of this present thesis work is to analyze the performance of Apriori algorithm which is one of the popular data mining algorithms in conjunction to clustering algorithms like *K*-means. The time and space performances are recorded, and appropriate analytical reports are generated. It is also concluded that clustered data helps in reducing the amount of transactions provided to Apriori in conditions where number of transactions are very large.

References

1. L.M.C. Annie, A.D. Kumar, Market basket analysis for a supermarket based on frequent item set mining. Int. J. Comput. Sci. (IJCSI) **9**(5), 257 (2012)
2. R. Agarwal, B. Kochar, D. Srivastava, A novel and efficient KNN using modified Apriori algorithm. Int. J. Sci. Res. Publ. **1**, 112–117 (2012)
3. M.A. Farajian, S. Mohammadi, Mining the banking customer behavior using clustering and association rules methods. Int. J. Ind. Eng. Prod. Res. **21**(4), 239–245 (2010)
4. C. West, S. MacDonald, P. Lingras, G. Adams, Relationship between product based loyalty and clustering based on supermarket visit and spending patterns. Int. J. Comput. Sci. Appl. **2**(2), 85–100 (2005)

5. T.H. Beach, O.F. Rana, Y. Rezgui, M. Parashar, Cloud computing for the architecture, engineering and construction sector: requirements, prototype and experience. J. Cloud Comput. Adv. Syst. Appl. **2**(1), 8 (2013)
6. R. Agrawal, T. Imieliński, A. Swami, Mining association rules between sets of items in large databases, in *Proceedings of ACM SIGMOD Conference*, Washington DC, USA, (1993)
7. R. Agrawal, R. Srikant, Fast algorithms for mining association rules, in *Proceedings of VLDB Conference* (Santiago, Chile, 1994)
8. S. Dhanaba, S. Chandramathi, A review of various k-nearest neighbor query processing techniques. Int. J. Comput. Appl. **31**(7), 0975–8887 (2011, October)
9. S. Nandagopal, S. Kartikh, Mining of meteorological data using modified Apriori algorithm. Eur. J. Sci. Res. **47**(2), 295–308 (2010) ISSN: 1450-216X (Euro Journals Publishing, Inc. 2010)
10. S.S. Weng, J.L. Liu, Feature-based recommendations for one-to-one marketing. Expert Syst. Appl. **26**, 493–508 (2004)
11. S. Pathak, R. Raja, V. Sharma, The impact of ICT in higher education. IJRECE **7**(1) (2019). ISSN: 2393-9028 (Print) ISSN: 2348-2281 (online) ISSN: 2393-9028 (Print). **7**(1), 1650–1656. (UGC Approved)
12. R. Raja, T.S. Sinha, R.K. Patra, S. Tiwari, Physiological trait based biometrical authentication of human-face using LGXP and ANN techniques. Int. J. Inf. Comput. Sec. **10**(2/3), 303–320 (2018). (Scopus Index)
13. R. Raja, T.S. Sinha, R.P. Dubey, Soft computing and LGXP techniques for ear authentication using progressive switching pattern. Int. J. Eng. Futur. Technol. **2**(2), 66–86 (2016). ISSN: 2455-6432
14. R. Raja, T.S. Sinha, R.P. Dubey, Orientation calculation of human face using symbolic techniques and ANFIS, Int. J. Eng. Futur. Technol. **7**(7), 37–50 (2016). ISSN: 2455-6432
15. R. Raja, T.S. Sinha, R.P. Dubey, Recognition of human-face from side-view using progressive switching pattern and soft-computing technique. Assoc. Adv. Model. Simul. Tech. Enterp. Adv. B. **58**(1), 14–34 (2015). ISSN:-1240-4543

A Study on Ontology Creation, Change Management for Web-Based Data

Nittala Swapna Suhasini, R. Mantru Naik, K. Uma Pavan Kumar and A. Ramaswamy Reddy

Abstract The research and development of artificial intelligence (AI) involve a generic representation known as ontology. Ontology is a generic representation of the domain data like medical ontology, manufacturing ontology, etc. The present work deals with the construction of the ontologies with Web pages like Naukri, Monster and Time Jobs Web sites, and the name of the ontology is job portals ontology. The reason behind the development of this work is to propose simple methods and models to create the ontology, to handle the changes happened to the source data and to apply the changes dynamically to the existing ontology. The concept of dynamic ontology creates more impact on the research of ontology models. We believe that the change management aspect is also helpful to the researchers so as to track the changes caused in the source data, in this case the Web portal data. The other dimension of the work is integration of recommender systems with ontology model so as to emit the resultant technologies which are not having sufficient resources in the job market. The proposed work involves the simple way of ontology creation, change management and dynamic ontology creation based on the changes of the source data. The outcome of the work is ontology creation, change management, dynamic ontology generation and a brief description of recommender systems usage.

Keywords AI · Ontology · Change management · Dynamic ontology · Recommender systems

N. S. Suhasini · R. M. Naik · K. Uma Pavan Kumar (✉) · A. Ramaswamy Reddy
Department of Computer Science and Engineering, Malla Reddy Institute of Technology, Hyderabad, Telangana, India
e-mail: dr.kethavarapu@gmail.com

N. S. Suhasini
e-mail: nittala_swapna@yahoo.com

R. M. Naik
e-mail: manthru440@gmail.com

A. Ramaswamy Reddy
e-mail: ramaswmyreddymail@gmail.com

H. S. Saini et al. (eds.), *Innovations in Electronics and Communication Engineering*, Lecture Notes in Networks and Systems 107,
https://doi.org/10.1007/978-981-15-3172-9_66

1 Introduction

Ontology is a generic representation of the knowledge related to various domains such as medical, manufacturing, education mechanical and many others. Ontology creation involves identification of the source data and mapping of the classes, attributes and methods based on the metadata observed. Once the creation is over, the changes happened to the source data tracking and applying the same to the existing ontology are tedious tasks in general [1–3].

The dynamic ontology work is very nominal from the literature and the current work address, a simple solution to that problem. The proposed framework provides a way to handle the ontology creation, change management and dynamic ontology generation, and all these activities are tagged as common flow of activities which makes the task of ontology users much simple.

The organization of work flows like in Sect. 2 and the existing methods in ontology creation were described; Sect. 3 involves the change management-related aspects, issues and research gaps in the existing works; Sect. 4 involves the proposed method of ontology creation and change management; Sect. 5 involves the conclusion and future scope of the work.

2 Existing Methods in Ontology Creation

The following are some of the observations from various ontology models and the corresponding methods (Table 1).

Ontology construction involves various categories like domain ontology and theory ontology to capture the ontology from the source the methods involve identification of the key concepts, class and subclass relations.

The ontology can be evaluated based on certain parameters like consistency, reusability, precision and recall [22, 23]. The ontology construction involves the support of the domain experts; the source data might be structured, semi-structured and unstructured in nature.

Each and every model follows a kind of source data, and they are all static in nature; there is no provision of updating the ontology dynamically based on the changes in the source data instantly. The majority of the methods is dependent on textual data, and the methodology is something like pattern matching and tagging-based approach [7, 24, and 25].

Table 1 Various ontology models creation with methodology [4–6]

Method	Source data	Methodology	Implementation	Remarks
InfoSleuth	Domain thesaurus	Pattern matching	POS tagged	Structural ambiguity
SKC (stanford)	Online dictionary	Page ranking	Arc rank	Misspelled head words
Ontology learning	Free text from web	Tokenizer	Preposition, verb	Noisy data
ECAI2000	Free text	Conceptual clustering	Description logic	Relation extraction
Inductive logic programming	A corpus of sentences	Slot fillers	Inductive logic programming	Deployment issues
Library science and ontology	Controlled vocabulary	Manual refined relations	Subject headings	Modeling issues
TERMINAE	Text	Automatic ontology	Concurrent semantic notations	New text cannot be inserted
Artequart	Artists and paintings	Dynamic knowledge extraction	Extraction by name	Overlapping data
SALT	Text data	Lexical analysis	Concurrent semantic notations	Static approach

3 Change Management in Ontologies

As the ontologies are commonly available source of the data, there is always a need to convey the changes happened to the ontology, and tracking of the changes must be in simple and logical way.

Most of the change management methods in ontology depend on two aspects, schema versioning and schema evolution. The methods cause the compatibility issues. The major drawback of the existing models is that tracking of the changes from the beginning may not be possible [8, 9].

Schema versioning is a way of tracking the changes in various versions of ontologies, and then, by merging all those versions, the final version can be expected, but the problem is in compatible merging of the versions [19].

Schema evolution is a kind of new mappings added to the schema which requires the embedding of the new aspects into the existing architecture and then publishes the change happened to the ontology.

A complex approach involves the steps like change capture, change representation, change of semantics, change implementation and verification and propagation of the change. The main issues related to this process are: the complexity is more in the implementation, and tracing of change is also tedious [20, 21] (Table 2).

The issues in the existing works are observed and presented here [13–15, 20, 21]

Table 2 Various change management techniques in ontologies [10–12, 18]

Approaches	Change request	Change representation	Conflict resolution	Change implementation	Change propagation	Working
L.S. Tojanovic, Ontology theory, management and design, IEEE Journal of Ocean Engineering, 2012	The complete change request is represented in formal representational format	These changes (due to business requirements) are specified by ontology engineer	Ontology engineer resolves all the inconsistencies due to requested changes by incorporating deduced changes	The requested changes (including deduced changes) are applied to the source ontology	Applied changes are propagated to dependent data,	User intervention required for system working
T. Gabel, et al., Ontology-based applications for enterprise system, 2012	Specified by ontology engineer logy engineer	Formal Formal representation of changes	Predefined strategies for conflict resolution	Provides interface for user interaction and also logs the changes	Propagation of changes to dependent artifacts	User intervention required for system working
P. Plessers, Understanding ontology evolution, Journal of Web Semantics, 2010	Different versions of ontologies are used in this approach	Changes among different versions are represented formally	After change implementation, it checks for inconsistencies and implements change recovery	First, it implements the change request and then checks for any conflicts	It does not	User intervention required for system working
A. M. Khattak, Mapping evolution of dynamic Web ontologies, Journal of IS: ACM, 2015	New changes such as (change in single concept, group of concepts and concepts in a hierarchical structure) are detected automatically using H-Match and WordNet	Change representation is provided by change history ontology (CHO)	For conflict resolution, KAON API is used with some suggested extensions	Changes are implemented atomically	Change propagation is not handled in this approach	This approach provides suggestions toward automation of the process

- The ontology creation, change management and dynamic ontology construction are not in a common process.
- There is no common framework to handle all the aspects as a common process.
- There is no specific procedure existing to track the changes happened to the source data.
- Most of the methods require the support of ontology engineers to observe the consistency and quality of the ontology.
- There is a need to use the tools, and the process is complex.

4 Proposed Method of Ontology Creation and Change Management

Automatic ontology construction has been proposed. The steps involved in the method are data extraction, data integration and generation of the common file to construct the ontology.

The proposed method involves the automatic ontology construction with the help of key value-based data extraction. Ontology change management is proposed with quantum elastic search approach (Figs. 1, 2 and 3).

The proposed method of ontology involves the Finite Space Access (FSA) by avoiding unnecessary capture of animations from the source data. The existing Natural language Processing (NLP) method follows the separation of the data based on the syntactic and semantic observations along with lexical analysis.

Pre-processing Phase - I

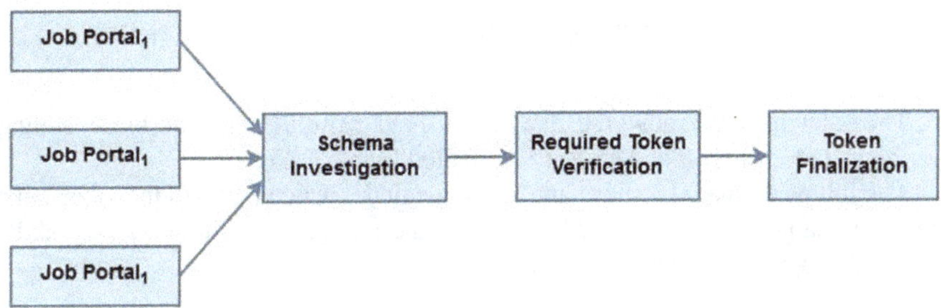

Input : Job portal data

Output: Tokens[Job name, skill, exp, location, company name]

Fig. 1 Preprocessing the Web portals data

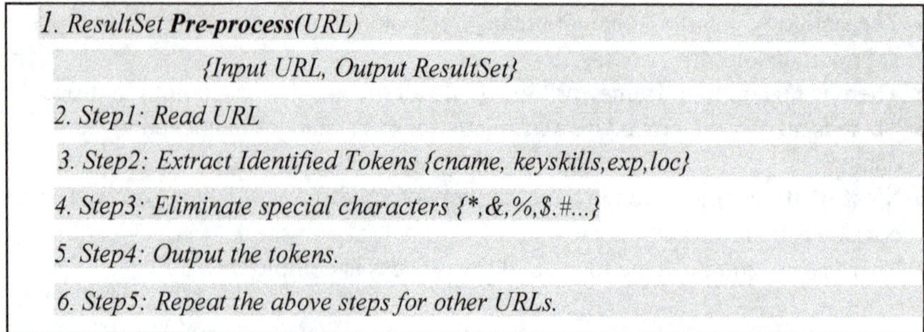

1. ResultSet **Pre-process(URL)**
{Input URL, Output ResultSet}
2. Step1: Read URL
3. Step2: Extract Identified Tokens {cname, keyskills,exp,loc}
4. Step3: Eliminate special characters {*,&,%,$.#...}
5. Step4: Output the tokens.
6. Step5: Repeat the above steps for other URLs.

Fig. 2 Algorithm to preprocess the Web data

Data Extraction Phase - II

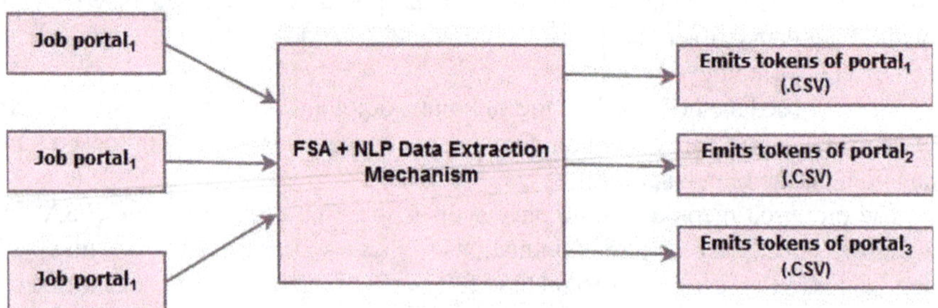

Input : Job Portal Data

Output : Tokens from each portal

Fig. 3 Generation of the .csv from the job portals data

The following table shows the time taken to read, write and total time by depending on the methods of NLP and NLP+FSA method (Table 3; Fig. 4).

The following diagram shows the ontology model which involves the tokens like company name, vacancy available, name of the job, place of the work, key skills required and experience for the job profile (Fig. 5).

Table 3 Time required to read and write data with the existing and proposed method

Method applied	Cumulative CPU time (s)	HDFS read	HDFS write	Total map reduce time (TMR)	Total time taken (TT)
NLP-based extraction	9.87	13,261	130	13.590	110.324
NLP+FSA method	6.49	13,261	123	10.300	97.068

In NLP method, time required to extract data is: 9.87 s
In combined method, time required to extract data is: 6.49 s
In NLP method, HDFS writes: 130 bytes
In combined method, HDFS writes: 123 bytes
In both methods, HDFS reads: 12,261 bytes
In NLP method, TMR time: 13.590 s
In combined method, TT: 97.068 s

1.Input: XML File

2.Output: Ontology

3.Step 1: Identify the generated ontology map file in the source.

4.Step 2: Open the file with the ontology editor

5.Step 3: Display the Ontology with classes and values

6.Step 4: With Ontograph the generation of required tokens along

with dependent data.

Fig. 4 Algorithm to identify ontology map file

5 Conclusion and Future Scope of the Work

The work explained the concept of ontology, the existing methods of ontology creation along with methodology and issues [16, 17]. The change management in ontology with the existing works has been explained along with the issues in the tracking of the changes. The work also described the proposed method of ontology creation with NLP+FSA-based approach along with the results in case of read and write of data into Hadoop distributed file system (HDFS). The work depicts the ontology model so as to represent the source data from the job portals in the tokens of company name, skills, experience, etc.

The future scope of the work is to describe the change management proposed approach and creation of the dynamic ontology based on the changes observed in the source data. The construction of ontology even is the main focus of the future scope. The change management with detailed implementation and a complete framework

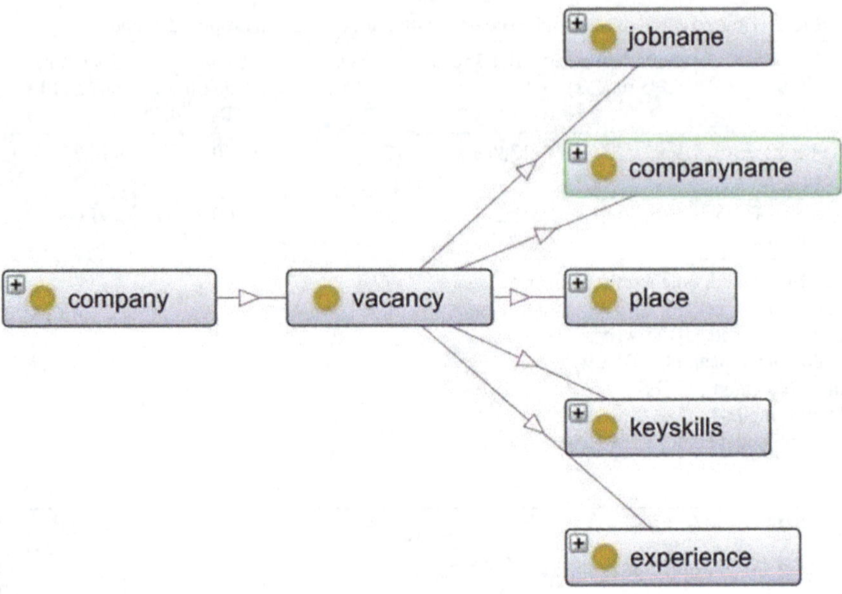

Fig. 5 Ontology for job portals data

to handle ontology creation, change management and dynamic ontology creation is the main focus of the future work.

References

1. N. Lasierra, Designing an architecture for monitoring patients at home: ontologies and web services for clinical and technical management integration. IEEE J. Biomed. Health Inform. **18**(3) (2014)
2. M. Roantree, Automating the integration of clinical studies into medical ontologies, in *47th Hawaii International Conference on System Science* (2014)
3. V. Jain, Architecture model for communication between multi agent systems with ontology. Int. J. Adv. Res. Comput. Sci. **4**(8) (2013)
4. P.A. Smirnov, Domain ontologies integration for virtual modelling and simulation environments. Procedia Comput. Sci. Elsevier **29**, 2507–2514 (2014)
5. S. Monisha, A framework for ontology based link analysis for web mining. J. Theor. Appl. Inf. Technol. **73**(2) (2015)
6. V. Jain, Mining in ontology with multi agent system insemantic web: a novel approach. Int. J. Multimed. Appl. (IJMA) **6**(5) (2014)
7. M. Fernández López, Overview of methodologies for building ontologies, in *Proceedings of the IJCAI-99 Workshop on Ontologies and Problem-Solving Methods (KRR5)*, Stockholm, Sweden, 2 Aug 2013
8. P. Gokhale, Ontology Development Methods. DESIDOC J. Libr. Inf. Technol. **31**(2), 77–83 (2011)
9. Y. Ding, Ontology research and development—a review of ontology mapping and evolving. J. Inf. Sci. **28**(5), 375–388 (2012)

10. H.S. Pinto, A methodology for ontology integration, in *K-CAP'01*, Victoria, British Columbia, Canada, 22–23 Oct 2001, Copyright 2001 ACM 1-58113-380-4
11. T.H. Duong, Complexity analysis of ontology integration methodologies: a comparative study. J. Univers. Comput. Sci. **15**(4) (2009)
12. M. Roantree, Automating the integration of clinical studies into medical ontologies. IEEE Comput. Soc. (2014)
13. Y. Huang, Using ontologies and formal concept analysis to integrate heterogeneous tourism information. IEEE Trans. Emerg. Top. Comput. (2015)
14. O. Iroju, State-of-the art: a comparative analysis of ontology matching systems. Afr. J. Comput. ICT. **5**(4) (2012). IEEE ISSN 2006-1781
15. Y. Huang, L. Bian, Using ontologies and formal concept analysis to integrate heterogeneous tourism information. IEEE Trans. Emerg. Top. Comput. (2015)
16. L. Zhao, Ontology integration for linked data. J. Data. Semant. (2014)
17. L. Zhao, Integrating ontologies using ontology learning approach. IEICE Trans. **E-96**(1) (2013)
18. T.H. Duong, Complexity analysis of ontology integration methodologies: a comparative study. J. Univers. Comput. Sci. **15**(4), 877–897 (2009)
19. S. Saraswathi, S. Venkataramanasam, Design of dynamically updated automatic ontology for mobile phone information retrieval system. Int. J. Metadata Semant. Ontol. **9**(3) (2014)
20. U.P. Kumar, D.S. Saraswathi, Automatic ontology generation for health care data with key-value extraction and combiner based integration, in *National Conference on ICT Solutions for Challenges and Issues in e-Health (NCICTEH'17) at Pondicherry Engineering College* 1 Sept 2017
21. U.P. Kumar, D.S. Saraswathi, Survey on techniques of ontology creation and integration. Int. J. Eng. Technol. Manag. Appl. Sci. **3**(Special Issue), 166–171 (2015). ISSN 2349-4476
22. U.P. Kumar, D.S. Saraswathi, Elastic search usage in automatic ontology based job recommendation system for reliable and accurate notifications. Int. J. Comput. Technol. Appl. [Scopus Indexed] **9**(3), 1679–1685 (2016)
23. U.P. Kumar, D.S. Saraswathi, Concept based dynamic ontology construction for job recommendation system. Procedia Comput. Sci. **85**, 915–921 (2016)
24. U.P. Kumar, D.S. Saraswathi, Ontology based job recommendation system with dynamic source updates by slowly changing source detection. Int. J. Knowl. Eng. Soft Data Parad. [-Inderscience-Publication]
25. U.P. Kumar, D.S. Saraswathi, FSA and NLP based un-supervised non template Web data extraction in the construction of dynamic ontology. Int. Conf. Inform. Anal. 112–118 (2016). (ACM Digital Library)

Multivariate Regression Analysis of Climate Indices for Forecasting the Indian Rainfall

S. Manoj, C. Valliyammai and V. Kalyani

Abstract Climate change and global warming are a reality. Extreme climate-related events are happening in various parts of the world. Rising temperatures, frequent droughts and floods are common now across many cities including new cities. Various climate indexes like Indian Ocean Dipole, Southern Oscillation Index, Arctic Oscillation, Pacific Decadal Oscillation and El Nino Southern Oscillation have a combined effect in Indian subcontinent. In this paper, monthly numbered climate indexes and how it influences the Indian rainfall have been studied. Regression analysis is performed on these climate indices for evaluating their impact on Indian rainfall.

Keywords Regression analysis · Indian rainfall · Indian Ocean Dipole · Southern Oscillation Index · Arctic Oscillation · Pacific Decadal Oscillation · El Nino Southern Oscillation

1 Introduction

Climate change is affecting many sectors including agriculture. In a developing country like India, for many people, agriculture is the main job. Indian agriculture depends on monsoon rainfall which includes south-east monsoon and north-east monsoon. Finding the climate indices which plays a key role in controlling the Indian rainfall will be useful for all stakeholders.

S. Manoj (✉)
Wipro Limited, Chennai, India
e-mail: haimanoj.a2z@gmail.com

C. Valliyammai · V. Kalyani
Department of Computer Technology, MIT Campus, Anna University, Chennai, India
e-mail: cva@mitindia.edu

V. Kalyani
e-mail: vkalyani29@gmail.com

H. S. Saini et al. (eds.), *Innovations in Electronics and Communication Engineering*, Lecture Notes in Networks and Systems 107,
https://doi.org/10.1007/978-981-15-3172-9_67

2 Related Work

Climate change is analysed by many researchers. It poses a variety of challenges with wide-ranging effects. It affects all sectors including agriculture. Agriculture is one of the prime jobs of India. Agriculture industry has to handle irregular rainfall patterns and irregular timing of the rainfall [1]. The irregular rainfall patterns lead to low agricultural production, and it leads to food insecurity. The people living in sea-shore side have to face many problems including sea-level rise [2].

2.1 Indian Monsoon

Monsoon is the main source of water for agriculture in India. From July to September month, India gets rainfall from south-west monsoon. During the last three months of every year, India gets rain through north-east monsoon. Indian Meteorological Department (IMD) measures the rainfall all over India throughout the year [3].

2.2 Precipitation Extremes

Many new regional climate extremes and unexpected local climate extremes are the result of interplay between various climate forces and local environmental factors. Dittus et al. [4] focus on temperature and precipitation events at less than 24 h timescale. Historical temperature extremes, precipitation extremes and its projected changes have been reviewed by [5].

2.3 Temperature Extremes

Intergovernmental Panel on Climate Change (IPCC) reports conclude that daily warm temperatures will increase across the world. Extreme climate events are devastating for economy. Many scientists have done work related to whether the length of the hurricanes season will increase or not. Heat waves will occur more frequently than ever before [6]. Dong et al. [7] discussed about a method to build accurate pattern-aided regression (PXR) models. Alessandro Tiesi et al. [8] proposed a statistically significant approach for rainfall prediction within 12–24 h time frame.

2.4 Sea Surface Temperature Influencing Indian Monsoon

Indian summer monsoon, along with many other factors, is also controlled by Indian Ocean sea surface temperature (SST) and tropical Pacific Ocean SST biases. When troposphere temperature gradient (TTG) changes from negative to positive, Indian summer monsoon is onset [9].

The winter precipitation that affects seasonal and monthly precipitation over north-western India (NWI) was analysed, and if enough precipitation is available in December month, then it will help to predict the individual monthly precipitation as well as December-January-February (DJF) seasonal precipitation. It is also noted that if December precipitation is good, then the following will be good: Individual precipitation of January, February, January-February and DJF seasonal precipitation [10]. Analysis of rain events in Himalayan mountain region revealed that the Western Himalayan (WH) region is predominantly inhabited by low-to-medium-level clouds and it impacts extreme orographic rain events in that region [11].

2.5 ENSO (EL Nino Southern Oscillation)

ENSO is a phenomenon which affects climate of many countries. ENSO affects many real-world economies including Australia, Southeast Asia, the Western Hemisphere and Sub-Saharan Africa. ENSO forecasts released for agricultural planning in many countries including Asia, Australia, Europe and USA. ENSO drives floods and hurricanes in US south-west coast, droughts in North-east India and Brazil and unusual winters in Canada and north-west of Alaska [12].

3 Proposed Methodology

3.1 System Architecture

The proposed framework is shown in Fig. 1 which extracts the climate data, discovers the available climatic indices and then performs regression analysis based on multivariate distribution to identify the important dependents of climate. It must weed out the bad data such as missing or duplicate data and then have to identify the relevant data using the training pattern. The training pattern is identified from the test patterns which are fabricated through pre-processed data.

The total Indian rainfall, various climate indices and their values from year 1950–2018 are used. The three rainfall events analysed in this study are overall rainfall in India, total rainfall in India on June, July, August and September (JJAS) and rainfall departure percentage in India on JJAS months.

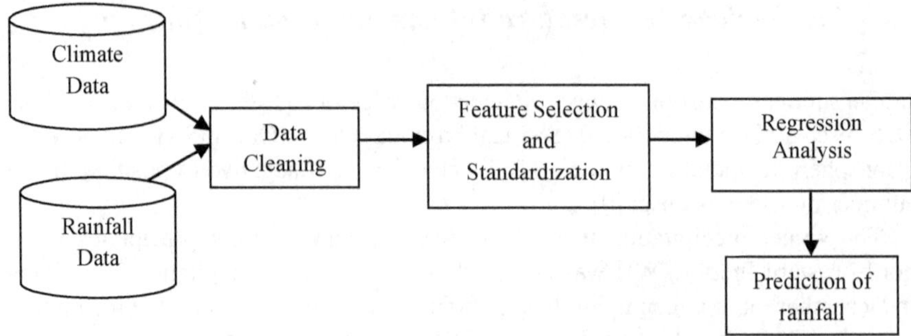

Fig. 1 System architecture

India gets most of the rainfall from June to December month. Hence, climate indices are also analysed for these seven months. Standardized procedure is applied to the rainfall data, so that regression factor values are easy to interpret. A table is generated using monthly numbers for all twelve months of every climate index for this study. For example, SOI index values for January month is SOI1 and for December month is SOI12. Table 1 describes climate indices and the corresponding monthly numbered information.

Table 1 Climate indices (monthly numbered) used in this study

Climate index	Climate index (monthly numbered)
IOD—Indian Ocean Dipole	IOD1, IOD2 …. IOD12 IOD1—January value of IOD in a year; IOD12—December value of IOD in a year
SOI—Southern Oscillation Index	SOI1, SOI2 … SOI12
AO—Arctic Oscillation	AO1, AO2 … AO12
PDO—Pacific Decadal Oscillation	PDO1, PDO2 … PDO12
NINO12—Nino Index for region 1 and 2	NINO12_1, NINO12_2 … NINO12_12
NINO3—Nino Index for region 3	NINO3_1, NINO3_2 … NINO3_12
NINO34—Nino Index for region 3 and 4	NINO34_1, NINO34_2 … NINO34_12

3.2 Regression Analysis

Hidden and significant relationship between sets of variables in data is analysed using multivariate regression analysis (MRA). Using mathematical tools, trends in sets of data are analysed. Simple linear regression (SLR) uses a single X variable for each dependent Y variable, whereas MRA uses multiple X variables for each dependent Y variable. Equation (1) captures the mathematical format of MRA which is used in this study.

$$Y = \beta_0 + \beta_1 X_1 + \beta_2 X_2 + \cdots + \beta_k X_k + e \tag{1}$$

where Y is predictand and $X_1, X_2, X_3 \ldots X_k$ are the predictors. 'e' is the error. And, $\beta_1, \beta_2, \ldots \beta_k$ are estimated using the training data.

4 Results and Discussion

4.1 Dataset

The data related to various climate indices is from Working Group on Surface Pressure [13]. The Indian rainfall data is downloaded from [14].

4.2 Analysis of Climate Data

MRA on following climate indices is helpful to identify the most influencing parameter for Indian rainfall. The indices used are IOD, SOI, AO, PDO, NINO12, NINO3 and NINO34. These indices are analysed against three important rainfall events in India, and the results are as follows:

Figure 2 shows that the total rainfall in India throughout the year is highly influenced by NINO34_7, IOD7, NINO3_6, NINO3_8, NINO12_9, NINO12_7, PDO8, PDO7, SOI6 and AO7.

Figure 3 shows that total rainfall in India on JJAS months is highly influenced by IOD7, IOD8, NINO12_9, NINO3_9, NINO12_7, NINO34_7, NINO34_8, AO9, PDO6 and NINO34_6.

Figure 4 shows that percentage of rainfall departure in India on June, July, August and September months is highly influenced by NINO34_7, IOD8, IOD7, NINO3_9, NINO12_7, NINO3_6, NINO12_9, PDO6, AO9 and PDO8.

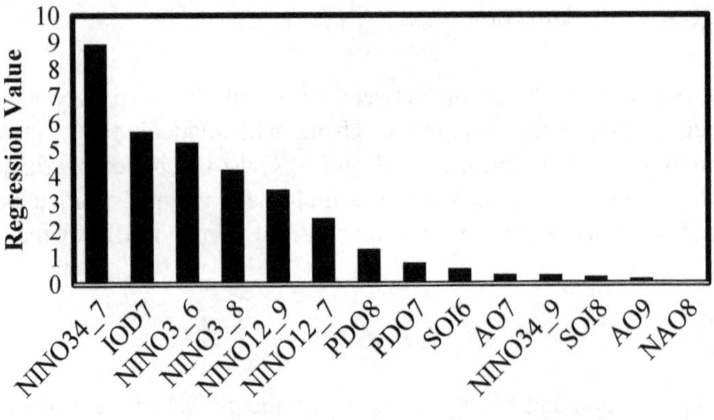

Fig. 2 Top climate indices (monthly numbered) influencing overall rainfall in India

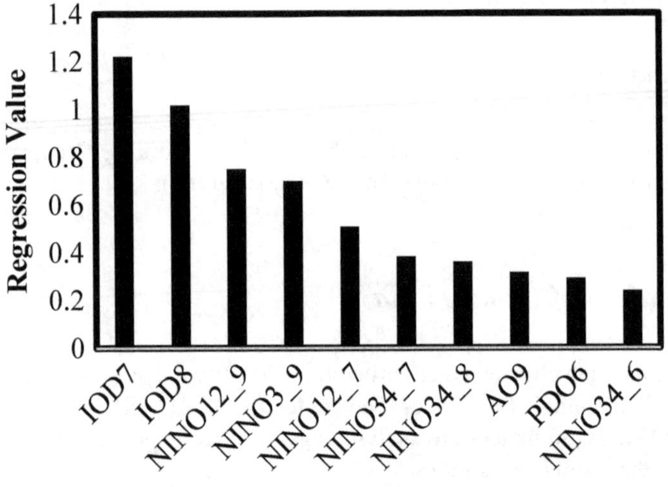

Fig. 3 Top climate indices (monthly numbered) influencing JJAS rainfall in India

5 Conclusion and Future Work

Climate change will lead to more extreme weather events, which include heat waves, drought and famine, strong winds, unexpected and heavy rainfall and flash floods. Monthly numbered climate indexes and their influence on Indian rainfall were analysed, and the results were discussed. Various climate indices like IOD, NINO3,

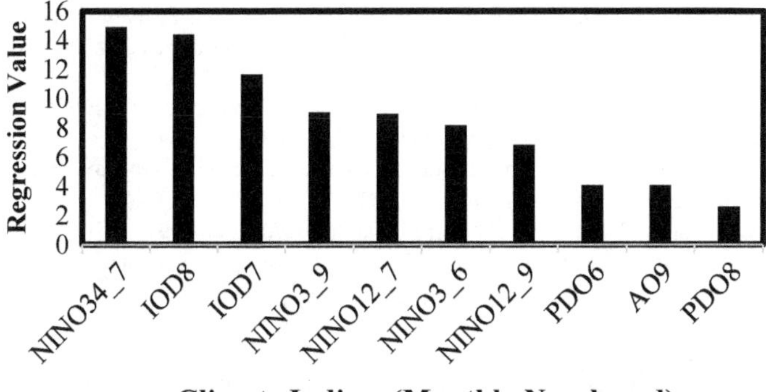

Fig. 4 Top climate indices (monthly numbered) influencing JJAS rainfall departure percentage in India

NINI12, NINO4, AO, PDO and ENSO have a combined effect in Indian rainfall. Along with these indices, other indices like Madden–Julian Oscillation and greenhouse gases including CO_2 have to be analysed.

References

1. A. Lopez, E.C. de Perez, J. Bazo, P. Suarez, B. van den Hurk, M. van Aalst, Bridging forecast verification and humanitarian decisions: a valuation approach for setting up action-oriented early warnings. Weather. Clim. Extrem. (2018) https://doi.org/10.1016/j.wace.2018.03.006
2. F. Iturbide-Sanchez, S.R.S. da Silva, Q. Liu, K.L. Pryor, M.E. Pettey, N.R. Nalli, Toward the operational weather forecasting application of atmospheric stability products derived from NUCAPS CrIS/ATMS soundings. IEEE Trans. Geosci. Remote. Sens. **56**(8), 4522–4545 (2018)
3. B.P. Shukla, C.M. Kishtawal, P.K. Pal, Satellite-based now casting of extreme rainfall events over western Himalayan region. IEEE J. Sel. Top. Appl. Earth Obs. Remote. Sens. **10**(5) (2017)
4. A.J. Dittus, D.J. Karoly, M.G. Donatb, S.C. Lewisb, L.V. Alexander, Understanding the role of sea surface temperature-forcing for variability in global temperature and precipitation extremes. J. Weather. Clim. Extrem. (2018)
5. M. Das, S.K. Ghosh, Detection of climate zones using multifractal detrended cross-correlation analysis: a spatio-temporal data mining approach, in *International Conference on Advances in Pattern Recognition* (2015)
6. Intergovernmental Panel on Climate Change Working Group Reports. http://www.ipcc.ch/report/ar5/wg2/
7. G. Dong, V. Taslimitehrani, Pattern-aided regression modelling and prediction model analysis. IEEE Trans. Knowl. Data Eng. **27**(9), 2452–2465 (2015)
8. A. Tiesi, M.M. Miglietta, D. Conte, O. Drofa, S. Davolio, P. Malguzzi, A. Buzzi, Heavy rain forecasting by model initialization with LAPS: a case study. IEEE J. Sel. Top. Appl. Earth Obs. Remote. Sens. **9**(6), 2619–2627 (2016)
9. S. Joseph, A.K. Sahai, S. Abhilash, R. Chattopadhyay, N. Borah, Development of evaluation of an objective criterion for the real-time prediction of Indian summer monsoon onset in a coupled model framework. J. Clim. 6234–6248 (2015)
10. C. Prodhomme, P. Terray, S. Masson, G. Boschat, T. Izumo, Oceanic factors controlling the Indian summer monsoon onset in a coupled model. Clim. Dyn. 978–1001 (2015)

11. M.M. Nageswararao, U.C. Mohanty, A.P. Dimri, K.K. Osuri, Probability of occurrence of monthly and seasonal winter precipitation over Northwest India based on antecedent-monthly precipitation. Theor. Appl. Climatol. **132**(3–4), 1247–1259 (2018)
12. K. Ashok, Z. Guan, T. Yamagata, A look at the relationship between the ENSO and the Indian Ocean dipole. J. Meteorol. Soc. Jpn. **81**(1), 41–56 (2003)
13. NOAA Earth System Research Laboratory's Physical Sciences Division. https://www.esrl.noaa.gov/psd/gcos_wgsp/Timeseries/
14. Open Government Data (OGD) Platform India. https://data.gov.in/

A Model-Based Analysis of Impact of Demographic and Other Factors on the Brand Preference of Three-Wheeler Automobile Drivers in Adama City

Mohd Arif Shaikh, U Deviprasad and Mohd Wazih Ahmad

Abstract The problem of consumer preference for three-wheeler brands was studied in the case of Adama City, Ethiopia. The demographic and non-demographic factors of the drivers and owners of the different three-wheeler brands available in the local market were obtained using a questionnaire. The impact of variables like age, marital status, driving experience, whether a driver is part-time or owner of the three-wheeler and possibility of accidents was studied on brand preference variable. Three different models, namely multivariate regression model, multinomial logit model (MLM) and random forest (RF) model, were used to learn the consumer preferences from the above data set in different settings. On experimentation, it was found that regression model fails to learn the exact consumer preferences from the above data set, while MLM model has performed in favor of a particular class. At the same time, different variants of RF models were created using various sampling schemes. It was found that using appropriate sampling scheme RF model outperforms the regression and MLM model in accuracy.

Keywords Consumer behavior · Oromia transport · Three-wheeler · Brand preference

1 Introduction

A band plays an important role in the competitive market environment and also in the mind of the customers [1]. Generally, the quality of the goods and services is differentiated by their associated brand. Influence of brand drivers such as brand self-similarity, brand distinctiveness, social benefits of brand and experiences that

M. A. Shaikh (✉) · U. Deviprasad
GITAM University, Hyderabad, India
e-mail: mdarifet@gmail.com

U. Deviprasad
e-mail: prasad_vungarala@yahoo.co.in

M. W. Ahmad
Intelligent Systems SIG, ASTU, Adama, Ethiopia
e-mail: wazihahmad786@gmail.com

H. S. Saini et al. (eds.), *Innovations in Electronics and Communication Engineering*, Lecture Notes in Networks and Systems 107,
https://doi.org/10.1007/978-981-15-3172-9_68

are memorable with relation to a brand has higher involvement on brand preference [2] Sometimes, brand becomes the name of good or service instead of the product. It is very crucial decision from customer point of view to purchase the new product unless they have the experience of it. In Ethiopian market, there are several imported products in the field of automobile sector, especially in the case of three-wheeler automobile, there are few companies providing their products to the end users through various dealers and distributors. In absence of manufacturing facility of three-wheelers in Ethiopia, the demand of three-wheelers is fulfilled by imported products only. The demand of three-wheelers is increasing in Ethiopia day by day as they are economically affordable and designed according to the terrain requirements of the country. This research has been focused on identifying the important factors and their impact on the choice of three-wheeler brand among available brands in Ethiopian market. The model developed in this research can be used to identify the existing brand impact on the market as well as the model and findings of this research can be used by different manufacturers and new entrants to learn the behavior of Ethiopian market as well as the choice of the drivers. The results of study can be utilized to understand why certain types of demographic factors of drivers and the owners of the existing three-wheelers have attracted them for that particular brand. These results can be used by companies, dealers and designers of three-wheelers to evolve the design of the product as per the requirement of customers as well as to improve their promotional, sales, marketing and after sales activities in the region. The data is collected from various sources like regional three-wheeler associations. Sampling methodology followed in this research was stratified sampling for selecting the associations and random sampling for selecting individual subjects for questionnaire. There are 17 associations of three-wheelers in Adama region, two associations for four-wheelers, three associations for seven-seater three-wheelers. In each association, there are 300 three-wheelers registered and some of the passenger three-wheelers running in the area do not belong to any of the associations. In this research, a total of 520 questionnaires have been distributed among which 500 were received and found correct. Totally, 53 variables were studied on the stratified sample of population of drivers. As per the research guidelines, a total of six demographic variables have been selected for model fitting in supervised settings. Two models, namely multivariate regression and multinomial classification, are used to learn the patterns in the data so that the impact of each demographic and non-demographic variable on the brand selection can be computed.

1.1 Background

This paper points out that the attitude of three-wheeler drivers is highly affected by not only the policies which are implemented in the form of rules and regulations at various levels but various demographic and socioeconomic factors contribute to perception and behavior of the drivers [3, 4]. A detailed study of three-wheeler passengers, drivers and various policies which effect the fare, safety and other issues like driver attitude, income, etc. is studied in [5] in the case of Mumbai city of India. The important factors like working hours and daily travel, vehicle purchase, and ownership and operation (including fuel economy and consumption, maintenance practices, costs related to renting, loan repayment, fuel and oil, maintenance, revenues and daily income) [6] have considerable impact on the operation of three-wheelers as well as on the behavior of the drivers. In [7], it has been identified that the major problem of three-wheeler drivers is related to the prices of petrol, gas and behavior of customers to expect lowest fares and increasing pressure of population on cities. This paper has studied different factors related to the passengers, drivers and the state bodies like RTO in order to evaluate the current state of the profession. Based on the exploratory data analysis on the responses obtained from various respondents, important recommendations are drawn. In [8], the characteristics of urban transport system are identified in relation of three-wheelers and different aspects like regulations, operational characteristics, profile of drivers and users, financial aspects and infrastructure are studied. This research uses a mix of qualitative and qualitative methods of data analysis. After identifying the problems and major challenges faced by three-wheeler drivers, author proposed various solutions for the identified problems in the form of training, enforcement, infrastructure and fleet services for drivers. In [9], the purchase behavior of drivers is studied in the case of passenger cars. The important factors that influence the selection of particular brand are fuel efficiency, price, safety, capacity and comfort. This article performed an in-depth review of various driving factors reported by researchers in brand preference like name of company, durability, driving comfort [8], popularity, spare parts availability, convenience and overall look. However, in the case of cars necessity, prestige, market resale value, advanced technology used, etc. also play major role in selecting the car brand but in the case of our research, since three-wheelers are commercial purpose vehicles, major factors which decide the selection of a brand are driver-centric like demographic factors [10], repair cost, sound, pollution, mileage, suspension, brake, resale value, etc. Therefore, this research has identified a total of 53 variables including demographic and brand-specific features of the existing three-wheelers. This paper has focused on five important demographic factors of drivers like marital status, age, whether the driver is part-time or full-time, whether the driver is renter or owner of the three-wheeler, driving experience.

The machine learning models are able to learn from past experience and fitted models can be used to infer or predict the values corresponding to unseen records of same type. The regression is a task in which predictor variables are mapped with response variable with the help of a linear model in general. If there are multiple predictor variables, the basic linear regression model is modified and called multivariate regression. The response variable in the case of regression model must be real valued while predictor variables can be real or ordinal. On the other hand, multinomial classification model uses a likelihood function and performs multiclass classification; this model can also be fitted for ordinal data.

1.2 Problem Statement

Three-wheeler is an important form of transport in Ethiopia where terrain of the city is highly non-uniform and the scope of large city buses is limited by the complexity of roads and curved surface. There are different imported brands of three-wheelers in the region, which are manufactured outside the country. The major problem for the importers and the consumers is to decide which brand of three-wheeler is preferable in such complex terrain conditions of Ethiopian roads, in terms of performance, after sales service requirements, availability of spare parts, etc. This research focuses on the study of driver's perception of various available brands according to their demographic attributes like age, years of experience, renter and owner drivers. The problem of finding the reason for particular brand preference is central to various companies; therefore, this research is performed on formulating above problem in terms of a regression and classification models in which various features taken from the experienced drivers of three-wheeler brands are studied in a model-based approach to identify the relation between above variables.

1.3 Objectives

The objective of this research was to identify the relation between different demographic and non-demographic factors on the brand preference of the three-wheeler out of four possible brands available in the market, i.e., Bajaj, Mahindra, TVS, Piaggio. The market of Ethiopia is stable for these brands since the year 2005 and there is frequent increment in the number of three-wheelers from every brand. However, this research shall provide a valuable input for newcomers in the market to understand and analyze the attributes of the heavily sold brands in the market so that the entry into the market shall be managed as per the requirement and behavior of the consumers. In addition to this, the new entrants can decide on the important design issues as well as formulate their marketing and promotional activities across the important factors analyzed in this research paper. Also, our research outcomes shall be usable for the existing companies to improve their sales as well as after sales services.

1.3.1 Research Questions

In this research, the main question was to identify and learn the impact of age, renter versus drivers, their marital status, part-time and full-time nature and experience of driving on the brand preference.

1.3.2 Hypothesis

In the view of above research question, the following hypothesis is constructed:

H0: There is no impact of the demographic and non-demographic factors (age, renter–owner driver, marital status, years of experience, number of accidents) on brand preference.
H1: The demographic and non-demographic factors (age, renter–owner driver, marital status, years of experience, number of accidents) have significant impact on the brand preference.

If above research supports hypothesis H1, then a model can be learned from the above data set in such a way that the patterns obtained from the past data can be used to predict the brand preferences of the future data. Therefore, we explore the answer of our research question in terms of three proposed models named multivariate regression model, multinomial logit model and random forest model. These models were selected based on their ability to capture the relation between the response variable and predictor variables of mixed type.

2 Conceptual Framework

The problem of predicting the brand choice, given the features $\mathbf{X} = \{\mathbf{x1}, \mathbf{x2}, \mathbf{x3} \ldots \mathbf{xn}\}$ where each of xi representing a demographic and non-demographic feature of the object in question, here, three-wheeler of different companies, can be modeled as a regression problem as well as a classification problem. This research has developed three models, namely multivariate regression model, multinomial logit model [11] and random forest model, for learning consumer preferences. The training of model was performed in supervised learning settings with brand preference being the target variable. The conceptual framework [4] for this research includes activities of model selection, training and testing along with data preprocessing and actual prediction on new data (Fig. 1).

3 Methodology

This research is based on questionnaires obtained from the owners and drivers of three-wheeler in Adama city, Oromia region, Ethiopia. The questionnaires were collected to understand the buying behavior of customers for specific brands of three-wheeler from existing four main brands. For performing model selection and training, the response of questionnaires was coded on Likert scale. The data collection and encoding were done based on a questionnaire distributed to various associations in Adama region who manages three-wheelers. Exploratory data analysis was performed on 53 variables and six important variables were selected for model

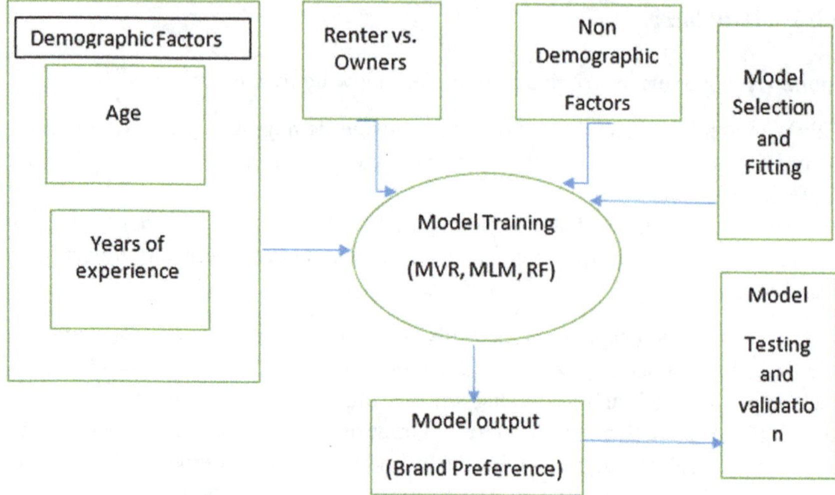

Fig. 1 Demographic model of brand preference

development. The data was divided in the ratio of 80 and 20 to perform training and testing of the models. The learned models were tested with hold out data and test results were generated for all the four brands. The comparative study of two models was performed and results were generated across the similar evaluation criteria like residual and accuracy. For performing analysis and developing the model, various libraries of R programming language were used.

The next section describes the mathematical form of the models used in this research in order to learn the relation between demographics and non-demographic factors with the brand preference. These models were selected based on experience of authors in similar cases of prediction problems and a deep literature survey.

4 Models

4.1 Multivariate Regression Model

A multivariate regression model learns the response variable from the predictor variables considering that the response variable is a continuous values. Following model equation was learned from the data considering target variable for brand preference as contentious variable:

$$Y = \text{Im}(\text{brand.pre} \sim \text{age} + \text{driver.part} \ldots + \text{expe})$$

Table 1 Coefficients of MV regression

Intercept	Age	Driver part	Marital	Rent/own	expe	acci
1.7146	0.056	−0.252	0.205	0.024	−0.129	−0.061

Table 2 Standard error in coefficients of MV regression

	Estimate	Std. error	t-value	Pr > t
Intercept	1.714	0.255	6.717	8.3e−11***
Age	0.056	0.077	0.725	0.468
Driver part	−0.252	0.108	−2.32	0.0207*
Marital	0.205	0.112	1.83	0.0677
Rent own	0.024	0.105	0.228	0.8198
expe	0.129	0.0812	−1.59	0.1120
Accident	−0.061	0.023	−2.56	0.018*

Here, we coded the Signif. codes as: 0 '***' 0.001 '**' 0.01 '*' 0.05 '.' 0.1 ' ' 1
In summary, the results on regression model are: residual standard error: 0.464 on 493 degrees of freedom, multiple R-squared: 0.0244, adjusted R-squared: 0.01252, F-statistic: 2.055 on 6 and 493 DF, p value: 0.0570

The coefficients of each variable are computed and the model was tested on hold out data. The results of this model are presented in Tables 1 and 2.

4.2 *Multinomial Logit Model*

A multinomial logit model is fitted in order to compute the probability of response variable **Y**, considering it as a multinomial variable. This aspect of multinomial logit model is represented by the following function:

$$\Pr(\text{brand.pre} = 1|X, Y, Z\ldots) = \frac{\exp(\beta_0 * \text{age} + \beta_1 * \text{expe} + \ldots \beta_2 * \text{driver.part})}{1 + \exp(\beta_0 * \text{age} + \beta_1 * \text{expe} + \ldots \beta_2 * \text{driver.part})}$$

The model is fitted with maximum likelihood estimate and the values of parameters and coefficients are reported in Tables 3, 4, 5 and 6.

4.3 *Random Forest Model with Probabilistic Sampling*

The given data set contains unbalanced classes [12] of brand preference. Therefore, the results of multivariate regression and multinomial logit model limited in accuracy and fails to correctly capture the true brand preference of the customers in Adama city.

Table 3 Coefficients of MLM model

	2	3	4
(Intercept)	0.383	−5.172	−2.238
age.L	−0.162	0.364	−0.537
age.Q	−8.303	−0.489	−0.564
age.C	0.335	−6.249	−0.312
driver.part2	0.635	−2.884	−0.004
marital2	0.551	−0.102	0.126
rent.own2	−0.474	0.012	0.232
expe2	0.084	−0.331	0.532
accident2	−0.532	0.274	−0.136

Table 4 Standard error of MLM model

	2	3	4
(Intercept)	0.729	1.027	0.957
age.L	0.531	0.279	0.699
age.Q	0.359	0.372	0.466
age.C	0.262	0.288	0.313
driver.part2	0.285	0.457	0.376
marital2	0.287	0.391	0.341
rent.own2	0.259	0.341	0.302
expe2	0.210	0.308	0.254
accident2	0.128	0.225	0.170

Table 5 Confusion matrix with MLM model

Pred.class	1	2	3	4
1	51	16	6	13
2	2	4	0	0
3	0	0	0	0
4	0	0	0	0

Table 6 Confusion matrix with RF model

Pred	Ref	
	0	1
0	7	2
1	38	96

Therefore, we have preprocessed the data again and the brand preference variable is converted into a binary variable with only two classes 0 and 1. In the new data set, 1 represents the preference of brand A while 0 represents brand preference of company B. Since most of the three-wheeler in Adama belongs to class A, class B was prepared by combining the frequency of all other company three-wheeler models. A random forest binary classification model is trained with probabilistic sampling on new dataset in order to investigate the problems of multivariate regression and multinomial logit model. The following probabilistic sampling methods used to train random forest model.

4.3.1 RF Model with Oversampling

In this method, model is trained after doing extra sampling with replacement from smaller class. At training time, the class balance was maintained by providing probabilistic selection of the data points from smaller class by doing resampling of data points with certain probability.

4.3.2 RF Model with Undersampling

The sample size from larger is restricted by the size of smaller class. The method works by doing undersampling from the class whose data points are more in the data set.

4.3.3 RF Model with Mixed Sampling

The sample size is derived from considering the size of smaller class as well as bigger class, so the sample finally selected represents the balanced classes.

4.3.4 RF Model with ROSE Sampling

Random oversampling examples are a method proposed in [13]; it works on the principle of bootstrap sampling and tries to balance the samples in both the classes. In this problem, the ROSE sampler was used to make balanced sampling of the two three-wheeler classes named A and B.

5 Results

The models were trained on 70% data and almost 30% data was used in testing the performance of each model. The following section describes the results of each model.

5.1 *Results on Regression*

Multivariate regression model considers the brand preference as continuous variable. Next sections describe the coefficients of predictors and standard error in estimation of each. The coefficients of different variables taken are as follows.

The value of coefficients represents the importance of each in multivariate regression model learned from this data.

5.1.1 Standard Error in Estimates

Table 2 depicts the standard error through fitted model in the various estimates.

5.2 *Results on Multinomial Logit Model*

In multinomial logit model, all four classes of the brand types are used in training the model. The results are shown in the following subsection.

5.2.1 Coefficients

The coefficients obtained by MLM model are shown Table 3.

5.2.2 Standard Error

Standard error in estimation of each variable is shown in Table 4.

5.2.3 Confusion Matrix

The confusion matrix of the MLM model is shown in Table 5. It is important to notice that the reference class was taken 1, 2, 3 and 4, respectively, but in each case single row confusion matrix is returned.

Table 7 Confusion matrix with RF-OVER model

Pred	Ref	
	0	1
0	14	21
1	31	77

Table 8 Confusion matrix with RF-UNDER model

Pred	Ref	
	0	1
0	15	32
1	30	66

5.3 Results on Random Forest Model with Probabilistic Sampling

The results on random forest model with binary brand preference data are shown in Table 6. The random forest model in this setting was trained on 240 instances of class A and 117 instances of class B.

5.3.1 Results with Oversampling

The table shows confusion matrix with RF with oversampling is shown in Table 7. The RF-OVER model was trained with 240 instances of each class.

5.3.2 Results with Undersampling

Results with RF-UNDER which perform undersampling are shown in Table 8. The total number of data points used from each class was 117.

5.3.3 Results with Both Sampling

The RF-BOTH model was trained with almost equal proportions of both classes. From class A, there were 134 data points taken, and from class B 151, data values were used (Table 9).

5.3.4 Results with ROSE Sampling

The results with RF-ROSE model with bootstrap sampling are given in Table 10.

Table 9 Confusion matrix with RF-BOTH model

Pred	Ref	
	0	1
0	26	49
1	19	49

Table 10 Confusion matrix with RF-ROSE model

Pred	Ref	
	0	0
0	15	11
1	30	87

Table 11 Comparison of RF models

	Accuracy	Sensitivity	Specificity
RF model	0.72	0.97	0.15
RF-OVER	0.63	0.78	0.31
RF-UNDER	0.56	0.67	0.33
RF-BOTH	0.52	0.5	0.57
RF-ROSE	0.71	0.88	0.33

ROSE model was trained on 500 data points in which 234 were belonging to class A and 266 were from class B.

5.3.5 Summary of RF Model

Table 11 shows the summary of various versions of RF model each with a sampling plan mentioned in the previous section.

6 Discussions

It was found that the multivariate regression model was unable to capture the true contribution of predictor variables on the response variable. This can be observed in the low value of *R*-square as well as the values of coefficients in Table 1. The reason behind such a poor performance of the regression model is the fact that the brand preference is not a continuous variable. Therefore, we decided to convert the above variable into a categorical variable with four possible categories. The multinomial logit model (MLM) was trained on the improved data set with categorical response and it was found that model was able to successfully capture the classes at training time. But, on test data, same model was unable to discriminate among various classes and the confusion matrix of the test data was found to be in single row as shown in Table 5. The problem of class imbalance was the main reason for poor performance of the MLM model, as the number of vehicles belonging to class 1 is preferred in Adama city as compared to other classes. Finally, a random forest model was trained on the same data set divided into training and test data set in the ratio of 80 and 20%. The base RF model was tested totally on 143 samples in which there were 96 class A and 7 class B predictions out of 98 and 45, respectively. The confusion matrix in Table 6 gives an idea that the model was able to predict class A with higher percent as compared to class B. In Table 11, the overall classification accuracy of the model is shown to be 72% which is better than the accuracy of a model which sequentially assigns test data points as class A, which is 68% due to the large proportion of class A data points in test data. In addition, the simple RF model has shown very high sensitivity of 97%, which shows the prediction abilities of this model for class A of the brand preference, as compared to specificity of 15% for class B.

RF-OVER model was developed to increase the specificity of the model or in other words, we wanted to improve the capability of model to capture class B of the brand preference. The results of RF-OVER model in Table 11 show its specificity is increased because of oversampling from class B at training time. The current level of specificity is 33% but it is obtained at the cost of model accuracy in predicting class A which is now 78% as compared to simple RF model. But in the confusion matrix of RF-OVER model in Table 7, we can observe that now 14 instances of class B are correctly classified as compared to the previous model in which only 7 instances of class B were correctly classified. On the other side, some data points belonging to class A are misclassified in this model. Similarly, the RF-UNDER, RF-BOTH and RF-ROSE models can be analyzed in terms of their sensitivity and specificity as well as the effect of sampling methodology can be seen on corresponding confusion matrix. In our finding, RF-BOTH is the best model in terms of its ability to predict both the classes, but at the same time the accuracy of this model is quite low on the test data. It depicts a kind of trade-off between the overall accuracy of the model with the ability of model to predict specific class. As per our research, the hypothesis H1 was found to be correct as the predictor variables like age, renter–owner driver, marital status, years of experience, number of accidents have significant impact on brand preference and that behavior of the customers can be learned in RF model.

7 Plots

See Figures 2, 3, 4, 5 and 6.

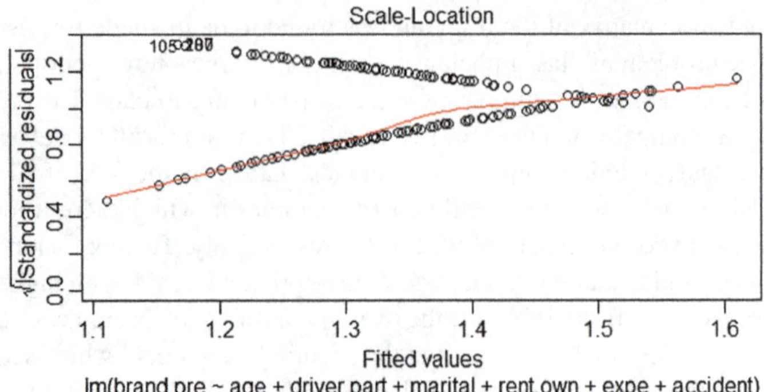

Fig. 2 A figure scale and location

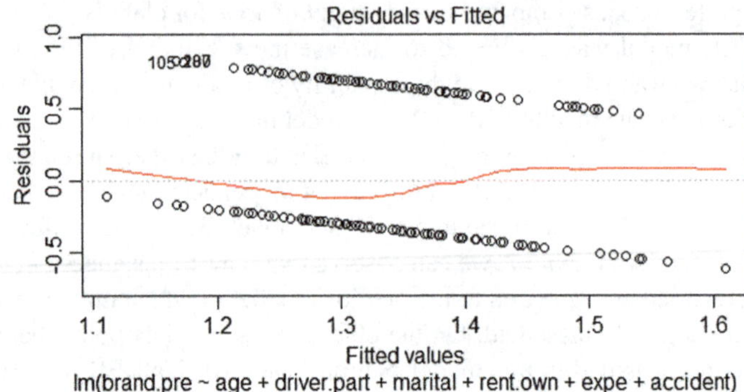

Fig. 3 It illustrate the residuals versus fitted

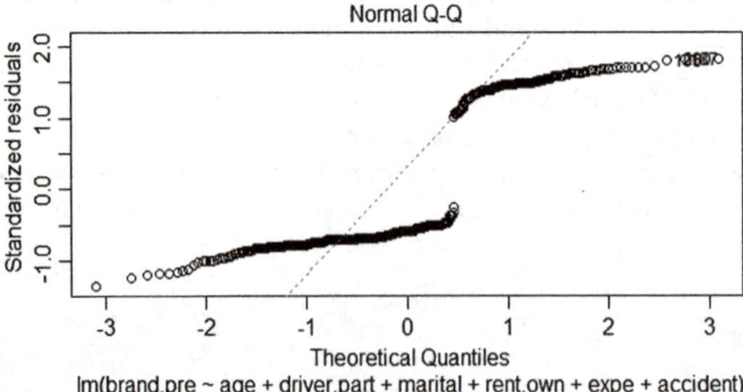

Fig. 4 It illustrate the Normal *Q*–*Q*

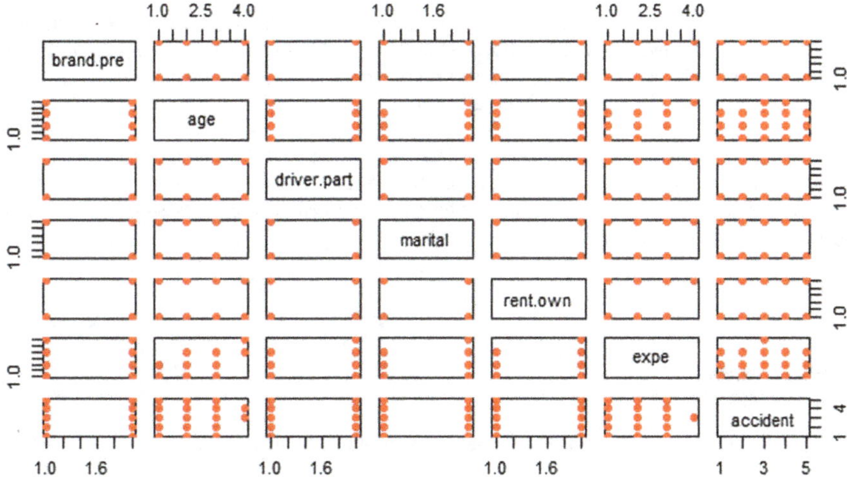

Fig. 5 Scree location plot

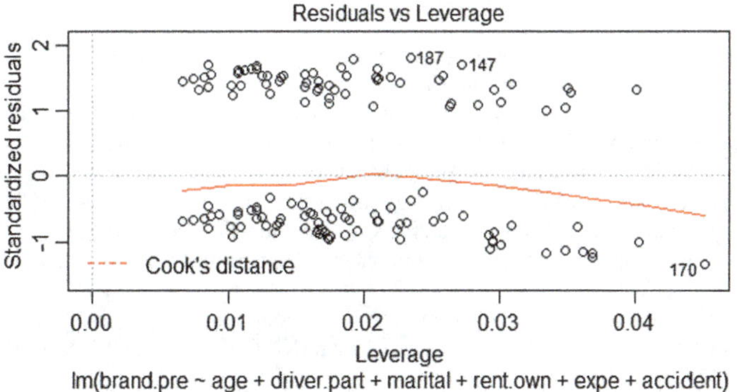

Fig. 6 Residuals versus leverage

8 Conclusions

The random forest model with proportional sampling method for each class can be used by individuals and companies to compute the likelihood of a particular instance of data for each available brand in the market. Also, multinomial logit model with four-class labeling of the customer preference fails to identify the instances when customer will like to buy other new brands. Similarly, multivariate regression model fails to learn the true response behavior from the given data set and needs to be modified for ordinal or multinomial response variables.

Acknowledgements The authors of this paper are thankful to the Intelligent Systems Special Interest Group, Department of Computer Science and Engineering, School of Electrical Engineering and Computing, Adama Science and Technology University Adama, Ethiopia, for providing a platform for performing this research. Special thanks go to Ato Tagel, who has helped our team in building the models and performing various experiments. Also, we acknowledge the help of Mr. Bhartendra Rai for his lecture on various variants of RF models used here and their implementation in R programming language.

Ethical Approval The questionnaires used in this research were distributed as per the sampling methodology among the respondents. Data obtained from various respondents was anonymized, and the data, as well as results, was approved for ethics by Intelligent Systems SIG, ASTU, Adama by the senior approval authority.

References

1. B.J. Bronnenberg, J.-P. Dub, The formation of consumer brand preferences. Annu. Rev. Econ. **9**, 353–382 (2017)
2. N. Stockberger-Sauer, S. Retneshwar, S. Sen, Drivers of consumer-brand identification. Int. J. Res. Mark. **29**(4), 406–418 (2012)
3. M. van Amerongen, Sri Lanka's three wheelers: regulation of the auto-rickshaw business in Galle. University of Amsterdam Graduate School of Social Sciences, Submitted on: 20 June 2011
4. C.B. Bhattacharya, S. Sen, Consumer–company identification: a framework for understanding consumers' relationships with companies. J. Mark. **67**(2), 76–88 (2003)
5. S. Harding, M.G. Badami, C. Reynolds, M. Kandlikar, Auto-rickshaws in Indian cities: public perceptions and operational realities. Transp. Policy **52**, 143–152 (2016)
6. E. Shlaes, A. Mani, Case study of Autorickshaw Industry in Mumbai, India, in *Transportation Research Record: Journal of the Transportation Research Board*, vol. 2416 (Transportation Research Board of the National Academies, Washington, DC, 2014), pp. 56–63
7. S.F. Quadari, Issues of Autorickshaw commuters in Aurangabad City: a case study. Bus. Dimn. **1**(1), 59–77 (2014)
8. T. Rajasekar, S. Rameshkumar, Determinants of preference towards passenger cars—the case of Madurai City in Tamilnadu. IOSR J. Bus. Manag. (IOSR-JBM) **17**(7), 84–88 (2015) (e-ISSN: 2278-487X, p-ISSN: 2319-7668, Ver. III)
9. A. Isik, M.F. Yasar, Effects of brand on consumer preferences: a study in Turkmenistan. Eur. J. Bus. Econ. **8**(16), 139–150 (2015)
10. H.H. Chovanováa, A.I. Korshunov, D. Babčanová, Impact of brand on consumer behavior, in *Business Economics and Management 2015 Conference, BEM2015*
11. C.M. Heilman, D. Bowman, G.P. Wright, The evolution of brand preferences and choice behaviors of consumers new to a market. J. Mark. Res. **37**(2), 139–155 (2000)
12. G. Menardi, N. Torelli, Training and assessing classification rules with imbalanced data. Data Min. Knowl. Disc. **28**, 92–122 (2014)
13. N. Lunardon, G. Menardi, N. Torelli, ROSE: a package for binary imbalanced learning. R J. **6**, 82–92 (2014)
14. A.S. Kumarage, M. Bandara, D. Munasinghe, Analysis of the economic and social parameters of the three-wheeler taxi service in Sri Lanka. Res. Transp. Econ. **29**, 395–400 (2010)

Cloud Computing Trends and Cloud Migration Tuple

Naim Ahmad, Shamimul Qamar, Nawsher Khan, Arshi Naim, Mohammad Rashid Hussain, Quadri Noorulhasan Naveed and Md. Rashid Mahmood

Abstract Cloud computing has come a long way to become an established platform to host the ICT infrastructure. Almost all of the Fortune 500 companies have cloud presence. And the latest announcement of Pentagon to offer a contract worth USD 10 billion to migrate to cloud shows the confidence level in the cloud infrastructure. Migration to cloud needs careful planning, strong business case, credible migration strategy, and robust migration frameworks. This paper reviews the area of cloud computing migration frameworks. It utilizes the co-word analysis technique of bibliometric or science mapping. The keywords of 299 research articles downloaded from the Web of science database have been analyzed. A 10 elements cloud migration tuple has been developed. Additionally, this paper gives cloud computing technology trends that have evolved to address the challenges of latency and jitter, context awareness, Internet of Things, voluminous data, and mobility support. The results of this paper will help in understanding the cloud computing and migration thereof.

N. Ahmad (✉) · S. Qamar · N. Khan · A. Naim · M. R. Hussain · Q. N. Naveed
College of Computer Science, King Khalid University, Abha 62529, Kingdom of Saudi Arabia
e-mail: nagqadir@kku.edu.sa

S. Qamar
e-mail: sqamar@kku.edu.sa

N. Khan
e-mail: nawsher@kku.edu.sa

A. Naim
e-mail: arshi@kku.edu.sa

M. R. Hussain
e-mail: humohammad@kku.edu.sa

Q. N. Naveed
e-mail: qnaveed@kku.edu.sa

Md. R. Mahmood
School of Engineering and Technology, Guru Nanak Institutions Technical Campus, Ibrahimpatnam, Telangana, India
e-mail: er.mrashid@gmail.com

H. S. Saini et al. (eds.), *Innovations in Electronics and Communication Engineering*, Lecture Notes in Networks and Systems 107,
https://doi.org/10.1007/978-981-15-3172-9_69

Keywords Cloud computing trends · Cloud migration tuple · Cloud native applications · Co-word analysis · Fog computing

1 Introduction

Cloud computing is relatively a recent paradigm to solve decades-old challenges of efficiency, effectiveness, and sustainability in the investments in the information and communication infrastructures (ICT). The cloud platform offers virtually unlimited compute, storage, and networking resources. Cloud computing has encapsulated all the previous standards for optimal computing as it is defined to consist of elastic computing, supporting variety of devices, running all present and past software, and compatible with range of network [1]. These features characterize it to be platform-independent, portable, and ubiquitous. The definition of cloud computing given by the National Institute of Standards and Technology (NIST) of USA is most cited in the literature. It focuses on the key features of cloud computing such as opportune and on-demand networked access, sharable and customizable collection of computing resources, quick and elastic allocation mechanism not requiring much interventions of providers or client organizations [2].

Cloud computing is highly discussed agenda in the ICT circles of corporations and being adopted at a very rapid pace. The revenue for public cloud services is expected to rise at a staggering percentage of 17.3 in 2019 to reach up to USD 206.2 billion [3]. In order of highest revenue share, these services are cloud application services (SaaS), cloud business process services (BPaaS), cloud system infrastructure services (IaaS), cloud application infrastructure services (PaaS), and cloud management and security services. And top 5 leading pubic cloud service providers along with their cloud services are Microsoft (SaaS, IaaS, and PaaS), Amazon (IaaS and PaaS), IBM (SaaS, IaaS, and PaaS), Salesforce (SaaS and PaaS), and SAP (SaaS, IaaS, and PaaS) [4], whereas cloud management and security services are the integral offerings of all cloud vendors. Leading traditional business process outsourcing (BPO) companies such as Capgemini are providing their services through cloud in BPaaS service model.

The objective of this paper is to illustrate upon the cloud computing trends and review of cloud computing migration frameworks through co-word analysis. The rest of the article is organized into sections of cloud computing trends, research methodology, data description, data analysis and results, conclusion, limitation and future research.

2 Cloud Computing Trends

Cloud computing is a culmination of numerous efforts in the different areas dating several decades back to distributed computing, grid computing, utility computing, etc. In its current formal form, it can be said that commercially cloud computing was offered by Amazon in 2002. It was through the launch of Amazon Web Services (AWS). Thereafter, other information technology companies Microsoft, Oracle, and SAP followed the suit [5]. Its exponential growth of adoption is due to recent accountability placed on ICT in the form of sustainability and high demand of ICT in latest innovations such as Smart city [6–8] and E-learning [9, 10]. Cloud computing itself is evolving very fast to cope with challenges namely low latency and jitter, context awareness, Internet of Things, voluminous data, and mobility support. This section will illustrate upon basics of cloud computing, container technology, fog computing, and edge computing.

Cloud computing is deployed through public clouds, private clouds, community clouds, and hybrid clouds. It is offered through majorly one of the service delivery models such as infrastructure as a service (IaaS), platform as a service (PaaS), and software as a service (SaaS). Many new prominent service delivery models have evolved such as business process as a service (BPaaS), security as a service, database as service (DBaaS), or anything as a service (XaaS). At the hardware and systems level, cloud computing utilizes virtualization to offer unified interface and services from the multiple hardware resources through virtual machine monitors (VMM) or hypervisors. Hypervisors are of two kinds. First, type-1 or bare metal hypervisors works straight on the hardware to manage virtual machines (VM) such as XEN, Microsoft Hyper-V, Oracle VM Server, VMware ESX and ESXi. Second, type-2 or hosted hypervisors execute on top of an operating system such as Redhat KVM and VMware Workstation [11]. Whereas, cloud services are provided by the means of Web services conceptualized on Web Services Description Language (WSDL) and Simple Object Access Protocol (SOAP), and Universal Description, Discovery, and Integration (UDDI) industry standards [12]. And looking at intra-cloud, these services are principally managed and administered through service-oriented architecture (SOA).

There is another virtualization technology, containers. Many large cloud providers such Google, Joyent, IBM/Softlayer, and others offer them [13]. Mostly containers utilize Linux kernel containment features, LXC. Containers sharing the same host OS, therefore, are lighter and hundreds of them can be run on single hardware machine, whereas on top of hypervisors, each VM can have its different OS. Containers provided just provide the view of underlying OS to developers unlike VM that has complete implementation of OS. Dockers help in the deploying Linux application inside containers. And Google's Kubernetes manages the cluster of docker containers. Cloud computing also provides a new approach to application development and deployment known as cloud native. Cloud native approach among others ensures the stateless computing so that elasticity of cloud can be achieved in real time and hence user traffic can be dynamically directed to any server regardless of their state of sessions. Cloud-native application (CNA) characterizes a distributed, elastic, and horizontal scalable system consisted of (micro) services that segregates state in a minimum of stateful components [14]. CNA may be developed with combination of best languages and managed through the DevOps processes.

The global cloud infrastructure is not always near to the consumers hence causes latency and jitters. Fog computing is deployed to solve this problem and the term was coined by Cisco in 2014 [15]. Fog computing utilizes edge entities to execute a significant amount of compute, store, and data transfer functions locally and transmitted over the Internet [16]. Fog computing is a standard that facilitates the edge computing and is highly desirable in the mobile computing and Internet of Things (IoT) applications to overcome the problems of latency and jitter, context awareness, and voluminous data. Irfan and Ahmad [17] have given three-layered IoT architecture in the medical field, namely things layer, intermediate layer, and backend computing layer. Intermediate layer or gateways are implemented using multi-agent, SAO, RESTful, or fog computing technologies, whereas backend computing or cloud computing layer is responsible for big data analytics and high-performance computing. This section has shed some light to explain the cloud computing trends that will help in planning the cloud deployment.

3 Research Methodology

Web of Science database was explored to collect the manuscripts in the domain of cloud computing migration framework; Web of Science database was used, as this database contains the high-quality peer reviewed research papers. The query for the keyword "cloud computing migration framework" resulted in 299 research articles. Their details are given in the data description section.

Bibliometric mapping or science mapping demonstrates the structural and dynamic aspects of scientific research in any domain [18]. Co-word analysis is a technique to analyze the significant words or keywords to study the conceptual structure of a research field [19]. In this article, the authors' keywords of the 299 articles have been studied to derive the trends in the field of cloud computing migration framework.

4 Data Description

The articles were between the time period of 2009–2018, Table 1. Maximum articles were published in the year of 2015. Table 2 shows the publishing sources having frequency count 5 or more. Articles in this domain are being cited at good rate showing the research concentration in this area Table 3. Most referred article had the Google citation count of 478 and Web of Science citation count of 247. Similarly, China leads the publication count with 23% articles among countries having 10 or articles, followed by USA, England, India, and Canada with 18%, 10%, 10%, and 6%, respectively.

Table 1 Year-wise publication frequency

Publication years	Article count	Bar chart
2018	37	
2017	46	
2016	55	
2015	60	
2014	34	
2013	29	
2012	24	
2011	12	
2009	2	

Table 2 High frequency sources of publication

Source of publication	Count
Lecture Notes in Computer Science	9
IEEE International Conference on Cloud Computing	7
Future Generation Computer Systems The International Journal of eScience	6
Journal of Network and Computer Applications	6
Communications in Computer and Information Science	5
IEEE Access	5
International Conference on Cloud Computing Technology and Science	5
Journal of Supercomputing	5

Table 3 Top five highly cited articles

Articles' field of study	Web of Science/Google citations
Wireless network virtualization [20]	247/478
Distributed application processing frameworks [21]	103/250
Resource management in clouds [22]	102/304
The cloud adoption toolkit [23]	92/301
Virtualisation-based high performance simulation platform [24]	77/73

5 Data Analysis and Results

There are a total of 1263 keywords in 299 articles, and their frequencies from 0 to 10 in the articles are shown in Table 4. Thirty-seven articles had no keywords and mostly articles had 3–7 keywords. Using Excel software, all the keywords were aggregated and duplicate entries were removed and resulted in 768 keywords.

There are multiple ways to analyze this data. One of the techniques is to form the clusters of related terms. Due to space limitation in this study, clusters having 10 or more related terms are being described. Ten such clusters have been identified that happen to contain 10 or more related terms. Keywords starting with cloud happen to be 71 but have not been included in the discussion as for its natural addition in many keywords. These terms are as follows virtual, service, data, multi, dynamic, mobile, performance, secure, application, and migration consisting of 30, 19, 18, 13, 11, 11, 11, 11, 10, and 10 related terms, respectively. These terms form a cloud migration tuple and are explained in the environment of cloud computing in the following paragraphs.

Virtual: This term is very integral of cloud computing as the whole technology works on virtualization of resources. Similarly, the virtual machine (VM)-related concepts and processes have also appeared in this cluster such as VM consolidation, migration, live migration, placement, provisioning, storage migration, scheduling, and security. Moreover, virtual network and virtual data center also happen to fall under this cluster.

Service: It is also a unique characteristics of cloud computing as it is delivered in the form of services. Some important terms related are service orientation, availability, innovation, migration, portability, and replication. Service-oriented architecture deserves a mention here as well. And service-level agreements (SLA) are also the most researched topic in this cluster, and some related terms are SLA assurance and monitoring.

Table 4 Keywords statistics

Number of keywords	Number of articles
0	37
1	0
2	4
3	30
4	79
5	83
6	43
7	14
8	5
9	2
10	2
Total key words in 299 articles	1263

Data: Data is the primary resource of information technology. Many important terms are mentioned with data such as locality, distribution, portability, migration, security, mining, and deduplication. Data centres (DC) are the basic infrastructural component of cloud computing and related terms were DC networking and management.

Multi: Multi-term has many-faceted implication in cloud computing. Most common of all is multi-tenant environment. Multi-agent system (system of multiple interacting intelligent system) [25] has also become significant in the perspective of cloud computing. This term is also associated with replication and redundancy such as multicast, multi-cloud, and multi-gateway system.

Dynamic: dynamic term is associated with real-time changes for instance cloud-native applications support dynamic resource allocation to ensure elasticity. This term is important for load balancing, consolidation, priority, resource migration, and scheduling. This term is also used as dynamic migration and consolidation of VMs, and dynamic structures.

Mobile: Support for mobile computing is a mandatory condition for all the leading cloud service providers. Mobile edge computing is also essential to solve the problems of jitter and latency. Additionally, the term is used in conjunction with devices, network, platform, and services.

Performance: This is also an important criterion for the success of cloud computing migration. This term has been given as a precursor to the migration such as performance matrix, modeling, prediction, and testing. Similarly, in the post migration stage, performance attributes are analyzed, evaluated and managed.

Secure: Security and privacy have been the most debated and stumbling block for migration to the cloud. Security is important in all the aspects such as migration, services, and data. Security needs careful planning. Security requirements, metrics, and measurements should be well defined. This term is also used in combination with compliance and transparency.

Application: In the traditional sense, software were said to be applications rather than services. Old valueable software is also termed as legacy application that is migrated to the cloud. Application migration and offloading are important concepts here. Post-migration terms will be application adaptation and optimization. Moreover, AppSpecCloudlet [26] framework for application-specific cloudlet selection of offloading is also mentioned. Some other important terms are application replication and application-aware allocation.

Migration: As the global information technology resources have been deployed on the premises, they now need to be migrated to cloud. Types of migration are an important topic such as re-host, re-platform, re-factor, or rebuild [27, 28]. Similarly, migration policy, methodology, and framework need to be developed well in advance before migration process execution. Migration patterns [29] also assist in managing the migration.

6 Conclusion, Limitation, and Future Research

Cloud computing has become a norm of the day and all the leading organization commercial or government of different size and domain are adopting it, recent being the US Defense. Public cloud service revenue is expected to reach 206.3 billion in 2019. Cloud computing technology is maturing day by day with hypervisors, containers, cloud-native applications, flog computing, edge computing, and cloudlets to become a robust platform to host complete range of information technology services. This paper has adopted the co-word analysis approach from the science or bibliometric mapping. The keywords of 299 articles have been analyzed and 10 clusters have been identified to derive a tuple of 10 elements in the context of cloud migration. These elements are virtual, service, data, multi, dynamic, mobile, performance, secure, application, and migration. These terms have been described in the previous section. In addition to awareness of cloud computing tends, this cloud migration tuple will help in understanding intricacies of migration to the cloud.

Web of Science database was considered only, and in future, more databases will be used such as ProQuest. Only bigger clusters have been explained, whereas smaller cluster can shed lights to more advancing frontiers. Moreover, specialized science mapping software will reveal further details of scientific and structural developments of cloud migration.

Acknowledgements We are thankful for all the support provided by the King Khalid University and motivational support of colleagues, family and friends.

References

1. N. Ahmad, Cloud computing: technology, security issues and solutions, in *2017 2nd International Conference on Anti-Cyber Crimes* (*ICACC*) (2017), pp. 30–35
2. P. Mell, T. Grance, *The NIST Definition of Cloud Computing*, http://nvlpubs.nist.gov/nistpubs/Legacy/SP/nistspecialpublication800-145.pdf
3. Gartner, *Gartner Forecasts Worldwide Public Cloud Revenue to Grow 17.3 Percent in 2019*, https://www.gartner.com/en/newsroom/press-releases/2018-09-12-gartner-forecasts-worldwide-public-cloud-revenue-to-grow-17-percent-in-2019
4. Forbes, The top 5 cloud-computing vendors: #1 Microsoft, #2 Amazon, #3 IBM, #4 Salesforce, #5 SAP, https://www.forbes.com/sites/bobevans1/2017/11/07/the-top-5-cloud-computing-vendors-1-microsoft-2-amazon-3-ibm-4-salesforce-5-sap/#45e113526f2e
5. A. Mohamed, A history of cloud computing. Comput. Wkly. **27** (2009)
6. N. Ahmad, R. Mehmood, Enterprise systems for networked smart cities, in *Smart Infrastructure and Applications* (Springer, Cham, 2020), pp. 1–33
7. N. Ahmad, R. Mehmood, Enterprise systems and performance of future city logistics. Prod. Plan. Control. **27**, 500–513 (2016)
8. N. Ahmad, R. Mehmood, Enterprise systems: are we ready for future sustainable cities. Supply Chain Manag. Int. J. **20**, 264–283 (2015)
9. N. Ahmad, N. Quadri, M. Qureshi, M. Alam, Relationship modeling of critical success factors for enhancing sustainability and performance in E-Learning. Sustainability **10**, 4776 (2018)

10. Q.N. Naveed, N. Ahmad, Critical success factors (CSFs) for cloud-based E-Learning. Int. J. Emerg. Technol. Learn. **14**, 140–149 (2019)
11. S. Campbell, M. Jeronim, *An Introduction to Virtualization* (2006)
12. D.S. Linthicum, *Cloud Computing and SOA Convergence in Your Enterprise: A Step-By-Step Guide* (Pearson Education, 2009)
13. D. Bernstein, Containers and cloud: from lxc to docker to kubernetes. IEEE Cloud Comput. **1**(2), 81–84 (2014)
14. N. Kratzke, P.-C. Quint, Understanding cloud-native applications after 10 years of cloud computing-a systematic mapping study. J. Syst. Softw. **126**, 1–16 (2017)
15. D. Linthicum, Edge computing vs. fog computing: definitions and enterprise uses (2018). https://www.cisco.com/c/en/us/solutions/enterprise-networks/edge-computing.html
16. Wikipedia, *Fog Computing*
17. M. Irfan, N. Ahmad, Internet of medical things: architectural model, motivational factors and impediments, in *2018 15th Learning and Technology Conference* (*L&T*) (2018), pp. 6–13
18. M.J. Cobo, A.G. López-Herrera, E. Herrera-Viedma, F. Herrera, Science mapping software tools: review, analysis, and cooperative study among tools. J. Am. Soc. Inf. Sci. Technol. **62**, 1382–1402 (2011)
19. M. Callon, J.-P. Courtial, W.A. Turner, S. Bauin, From translations to problematic networks: an introduction to co-word analysis. Inf. Int. Soc. Sci. Counc. **22**, 191–235 (1983)
20. C. Liang, F.R. Yu, Wireless network virtualization: a survey, some research issues and challenges. IEEE Commun. Surv. Tutorial. **17**, 358–380 (2015)
21. M. Shiraz, A. Gani, R.H. Khokhar, R. Buyya, A review on distributed application processing frameworks in smart mobile devices for mobile cloud computing. IEEE Commun. Surv. Tutorials. **15**, 1294–1313 (2013)
22. B. Jennings, R. Stadler, Resource management in clouds: survey and research challenges. J. Netw. Syst. Manag. **23**, 567–619 (2015)
23. A. Khajeh-Hosseini, D. Greenwood, J.W. Smith, I. Sommerville, The cloud adoption toolkit: supporting cloud adoption decisions in the enterprise. Softw. Pract. Exp. **42**, 447–465 (2012)
24. L. Ren, L. Zhang, F. Tao, X. Zhang, Y. Luo, Y. Zhang, A methodology towards virtualisation-based high performance simulation platform supporting multidisciplinary design of complex products. Enterp. Inf. Syst. **6**, 267–290 (2012)
25. S. Laghari, M.A. Niazi, Modeling the internet of things, self-organizing and other complex adaptive communication networks: a cognitive agent-based computing approach. PLoS One **11** (2016)
26. D.G. Roy, D. De, A. Mukherjee, R. Buyya, Application-aware cloudlet selection for computation offloading in multi-cloudlet environment. J. Supercomput. **73**, 1672–1690 (2017)
27. Gartner, *Gartner Identifies Five Ways to Migrate Applications to the Cloud*, https://www.gartner.com/newsroom/id/1684114
28. N. Ahmad, Q.N. Naveed, N. Hoda, Strategy and procedures for migration to the cloud computing, in *2018 IEEE 5th International Conference on Engineering Technologies and Applied Sciences* (*ICETAS*) (2018), pp. 1–5
29. P. Jamshidi, C. Pahl, N.C. Mendonça, Pattern-based multi-cloud architecture migration. Softw. Pract. Exp. **47**, 1159–1184 (2017)

Design of Super-Pipeline Architecture to Visualize the Effect of Dependency

Renuka Patel and Sanjay Kumar

Abstract Nowadays, pipelining is a very common phenomenon for getting speedup in processors. Super-pipeline architecture can issue more than one instruction in less than one clock cycle but dependency is major obstacle in super-pipeline architecture because dependency puts a stop in issuing the instruction in less than one clock cycle. In this paper, a simulator is designed for visualizing the effect of dependency on super-pipeline architecture. With the help of a simulator, we can make any previous instructions and instruction can be dependent on current instruction (dependency can be of any type like it can be of data dependency or control dependency or resource conflicts). Simulator also visualizes the effect of dependency for each instruction like how many numbers of stalls are encountered to handle dependency, in which cycle stall is encountered, how many clock cycles required to execute the instruction, and then finally, simulator calculates the performance parameters like total clock cycles, stalls, CPI, IPC, and speedup.

Keywords Pipelining · Super-pipeline · Dependency · CPI · IPC

1 Introduction

As the demand for processing increases, computer system architects and designers are forced to use techniques that result in high-performance processing and taking advantages of parallelism. Parallelism is one of the common methods for improving the performance of any processor. Parallelism allows the hardware to accelerate the execution speed of an application by executing multiple independent operations parallel. To achieve parallelism in an easy manner, pipelining is a widely used technique [1]. Pipelining is one form of embedding parallelism in a computer system [2]. Pipelining is a very popular and easy technique to increase the performance of

R. Patel (✉) · S. Kumar
Department of Computer Science, Pt. Ravishankar Shukla University, Raipur, Chhattisgarh 492010, India
e-mail: renu.access.me@gmail.com

S. Kumar
e-mail: sanraipur@gmail.com

H. S. Saini et al. (eds.), *Innovations in Electronics and Communication Engineering*, Lecture Notes in Networks and Systems 107,
https://doi.org/10.1007/978-981-15-3172-9_70

processors. This is because ideal pipeline can give the speedup that equals to the number of pipeline stages.

$$S_k = (n * k)/(k + (n - 1))$$

where k is the number of pipeline stages and n is the number of instructions.

But, practically it is very difficult to achieve the speedup S_k of ideal pipelining, because of the presence of dependencies, interrupts, branches, and other practical reasons [3, 4].

Pipelining is a technique in which one instruction is divided into number of independent steps called segments or stages. Each segment works on a particular part of the computation. The output of one segment becomes the input for the next segment [5, 6].

A typical instruction execution consists of a sequence of operations including load (L), decode (D), fetch (F), execute (E), and write (W) phases [7, 8]. Depending on the instruction type and processor/memory architecture used, each phase may require one or more clock cycles to execute [9]. Pipeline can be divided into five categories, namely scalar pipeline, super-scalar pipeline, super-pipeline, under-pipeline, and super-scalar super-pipeline [10, 11].

In this paper, the effect of dependency on super-pipeline architecture is studied.

2 Super-Pipeline Architecture

The super-pipelining is based on dividing the stages of a pipeline into several substages, and thus, it increases the number of instructions which are handled by the pipeline at the same time [12]. For example, by dividing each stage into two substages, a pipeline can perform at twice the speed in the ideal situation. Many pipeline stages may perform tasks that require less than half a clock cycle [13]. No duplication of hardware is needed for these stages, whereas in super-scalar architecture, duplication of hardware is needed [14]. Some example of super-pipeline processors are CDC 7600 (Control Data Computer 6600) [15], Cray 1, and MIPS R4000 [16].

In super-pipeline architecture instruction issue, latency is less than 1 clock cycle, i.e., more than one instruction can be issued in one clock cycle. In super-pipeline machine, effective cycle time is shorter than the baseline processor or scalar processor [17]. In other words, another instruction can be issued after a minor cycle. So, instead of waiting for the major cycle, instructions are issued after a minor cycle (here major cycle means 1 clock cycle and minor cycle means less than 1 cycle or half-cycle or 1/4th cycle). Super-pipeline architecture employs temporal pipeline. If a super-pipeline architecture can issue 1 instructions in 1/2 clock cycle and another instruction in another 1/2 clock cycle, then this architecture is known as super-pipeline architecture of degree two, which is shown in Fig. 1.

Figure 1 shows that the instruction load (L) is divided into two substages. Then the second stage, i.e., the instruction decode (D) stage and third stage fetch (F) stage are

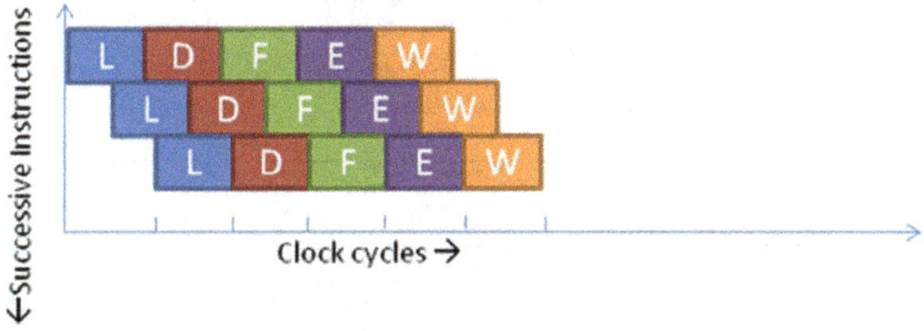

Fig. 1 Super-pipelime architecture of degree 2

also divided into two substages. Similarly, execution stage (E) and write back (W) stage are again divided into two substages. For studying the effect of dependency on super-pipeline architecture in this paper, super-pipeline architecture of degree 2 is chosen.

3 Experimental Setup

For an experiment, a simulator is designed in 'C' language. To evaluate the performance of super-pipelining architecture, simulator calculates the performance matrixes like total clock cycles to execute all the instructions, total stalls due to dependency, instructions per cycle (IPC), cycles per instruction (CPI), and speedup.

3.1 Terms Used for Designing of Super-Pipeline Architecture

- **Instruction issue latency** (**IL**)—IL means the interval between the two continuous instructions.
- **Instruction issue parallelism (IP)**—IP is the number of instructions which can issue at a time.
- **Operation latency** (**OL**)—OL is number of clock cycles to perform a particular task at the execute stage.
- **Machine parallelism** (**MP**)—MP is the number of instructions which can process simultaneously at a particular instant of time in any processor.

3.2 Design of Super-Pipeline Architecture

Figure 2 shows the timescale diagram of simple super-pipeline architecture of degree 2, where '*x*-axis' represents number of clock cycles and '*y*-axis' represents number

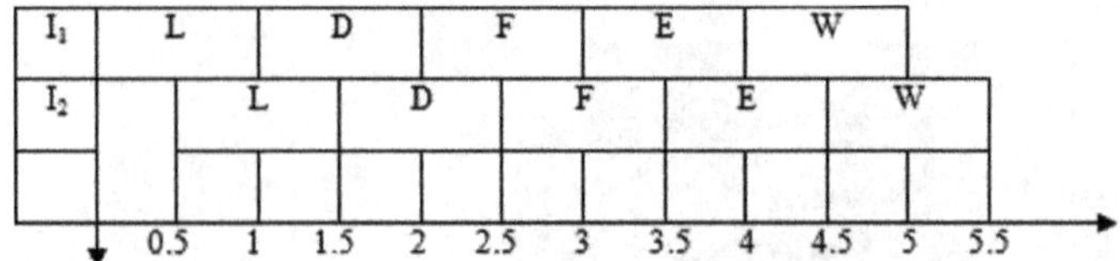

Fig. 2 Super-pipeline architecture without dependency

of successive instructions. In Fig. 2, IL is 0.5 cycle, IP is 1, MP is 1.5, and OL is 1 for both instructions I_1 and I_2.

3.2.1 Effect of Dependency on Super-Pipeline Architecture

There are different cases to stall the pipeline which is shown below.

Case 1: Stall due to data hazard

Let

$$I_1: a=b$$
$$I_2: c=a$$

Since I_2 is dependent on I_1 for value of variable 'a', that is why, this is the example of data hazard, fetching of instruction I_2 is possible only after the completion of write operation of instruction I_1. Therefore, fetch of I_2 is on clock cycle 6.5 instead of 3.5, this is shown in Fig. 3.

Case 2: Stall due to control hazard

Let

$$I_1: \text{if } a>b$$
$$I_2: \text{print a}$$
$$I_3: \text{print bye}$$

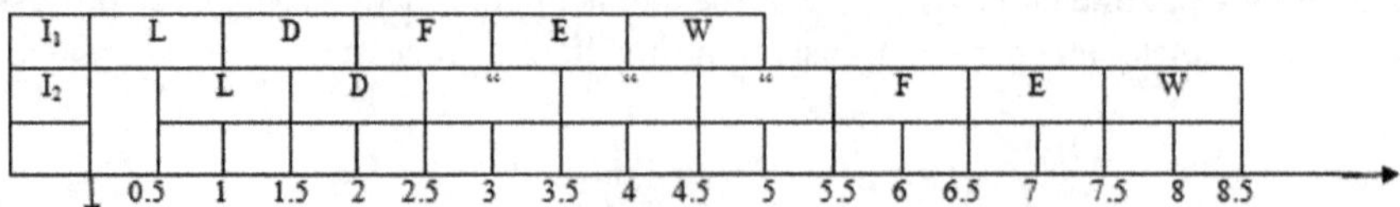

Fig. 3 Stall due to data hazard in super-pipeline architecture

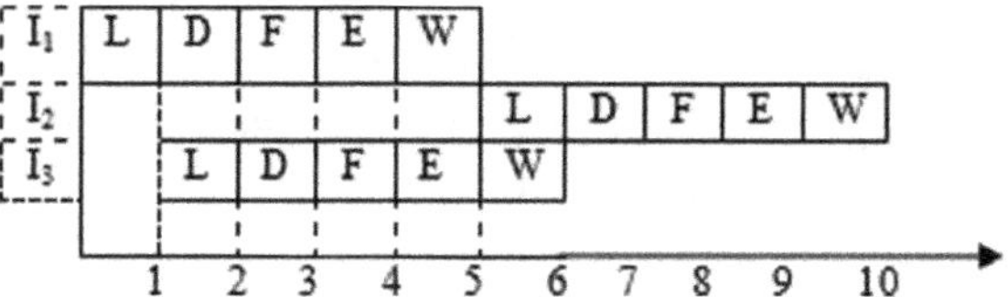

Fig. 4 Stall due to control hazard in super-pipeline

Until the time result of I_1 is written second instruction, I_2 will not able to start; because I_2 is control dependent on I_1, I3 is not dependent on any instructions; that is why I_3 can load on second clock cycle. This is shown in Fig. 4.

Case 3: Stall due to resource conflict

Let

$$I_1: x=27$$
$$I_2: a=x$$
$$I_3: b=23$$
$$I_4: \text{print } x$$

Instructions I_2 and I_4 both want to fetch the data during cycle 6.5. So, this is the clear example of fetch conflict during cycle 6.5, and to avoid this, 1 stall is given at cycle 6.5 for instruction I_4 which is shown in Fig. 5.

Case 4: Stall due to dependency on more than one instruction

Let

$$I_1: a=b-c$$
$$I_2: d=e*f$$
$$I_3: z=a+d$$

Here I_3 is dependent on I_1 and I_2 both; therefore, fetching of I_3 is possible only after completion of I_1 and I_2. This is shown in Fig. 6.

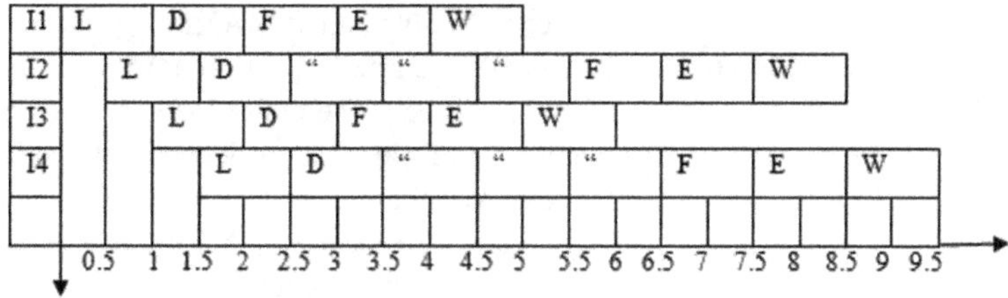

Fig. 5 Fetch conflicts in super-pipeline architecture

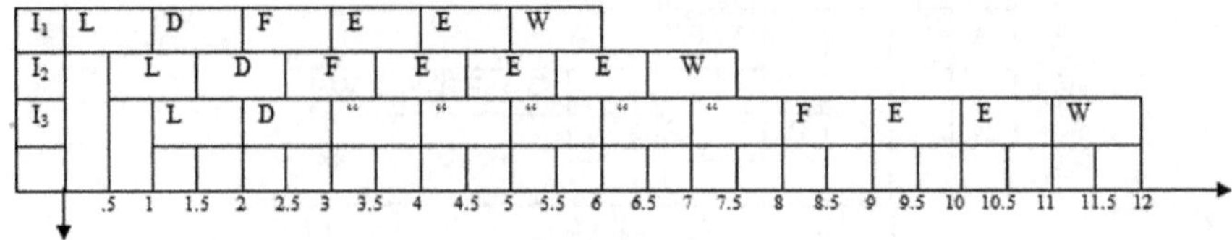

Fig. 6 Dependency on more than 1 instruction in super-pipeline architecture

3.2.2 Algorithm for Simulation

Step 1: START
Step 2: Declare 5 dynamic arrays for load, decode, fetch, execute, and write stage.
Step 3: Input—total number of instructions (n).
Step 4: For i=1 to n, repeat steps from 5 to 7.
Step 5: if this is first instruction, then

Step 5.1: Load will be started by i.
Step 5.2: Decode will be started after load and take one clock cycle.
Step 5.3: Fetch will be started after decode and take one clock cycle.
Step 5.4: For execute stage check type of instruction. For addition/subtraction
Step 5.5: Write will be started after the execute stage and take one clock cycle.
Take two clock cycles.
if type of instruction is multiplication/division
Take three clock cycles.
And if type of instruction is of else category
Take one clock cycle.

Step 6: If the instruction is not the first instruction, then

Step 6.1: Input—current instruction is dependent or not.
Step 6.2.1: If current instruction is dependent, then input dependent on how many instruction and type of dependency. Repeat Step 6.2.2 to Step 6.2.3
Step 6.2.2: Now find out maximum of all write cycles for which current instruction is dependent and stores into one variable called MAX.
Step 6.2.3: Load will be started by i-1 plus 0.5.
Decode will be started after load and plus dr.
Fetch will be started after decode with adding the value of MAX and fr.
For execution, repeat Step 5.4 with adding er.
Write will be started after execute with adding the value of wr.

Here, lr, dr, fr, er, and wr are the counter variables for load, decode, fetch, execute, and write stages, respectively, which are going to increment every time when there is a resource conflict. For finding resource conflict do for k=1 to <=i and check L[i]=L[k]; if yes, then increment lr by one, do L[i]=i+lr and again check, this process is repeated till L[i]!=L[k]. The same thing is checked for the rest of all stages.

Step 6.3: If current instruction is not dependent, then find out load, decode, fetch, execute, and write clock cycle the same as Step 6.2.3.

Step 7: Find out idle cycle: For finding out idle cycle check type of instructions & then find out total number of cycle to ideally execute that instruction and total number of actual cycle to execute that instruction. At last subtract both result will be idle cycle.
Step 8: Find out total number of idle cycle.
Step 9: Find out CPI and IPC from total cycle.
Step 10: Find out MIPS from IPC.
Step 11: STOP.

4 Results and Discussion

Simulation for super-pipelining is done, and the results of the simulation are provided in Table 1 for high dependency and in Table 2 for moderate dependency. Subsequently, graph for total cycle is plotted in Graph 1, for total stalls in Graph 2,

Table 1 High dependency

No. of instructions	Total clock cycles	Total stalls	CPI	IPC	Speedup
5	12.5	15	2.5	0.4	2.4
10	20.5	44	2.05	0.4878	2.68
15	25	99	1.66	0.6	3.6
20	29	132	1.45	0.689	4.034

Table 2 Moderate dependency

No. of instructions	Total clock cycles	Total stalls	CPI	IPC	Speedup
5	10	6	2.00	0.5000	3
10	14.5	17	1.4500	0.689655	3.93103
15	23	45	1.53333	0.65217	3.6956
20	23	52	1.15	0.8695	5

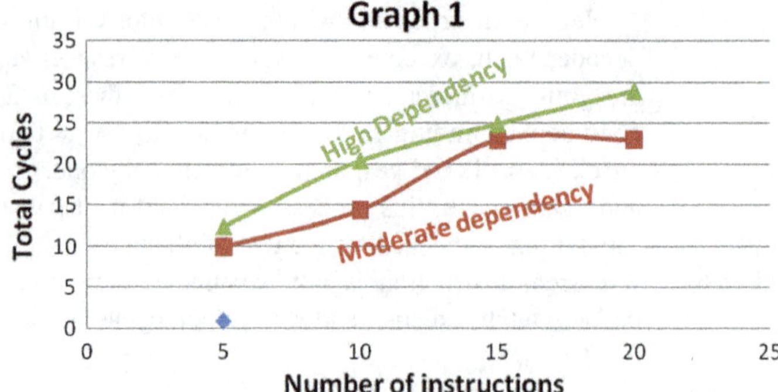

Graph. 1 Total cycle

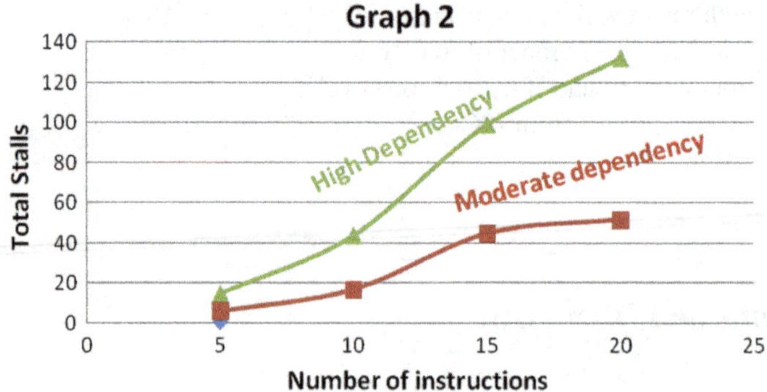

Graph. 2 Total stalls

and for speedup in Graph 3.

CPI and IPC are the derived parameters (calculated from total clock cycle); therefore, graphs for CPI and IPC are not drawn.

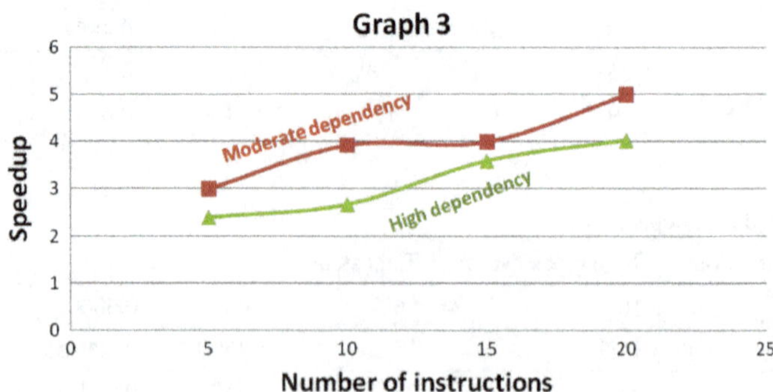

Graph. 3 Speedup

According to Tables 1, 2, and Graph 1, with increasing in the number of instructions, total clock cycles to execute the instruction are also increased, and total clock cycles are less in moderate dependency compared with high dependency as also shown in Graph 1. Because of dependency, total stalls are also more in high dependency and less in moderate dependency as shown in Graph 2. Speedup is calculated by using the formula of without pipelining upon with pipelining and speedup increases with the increase in number of instructions as shown in Graph 3. Speedup is more for moderate dependency compared with high dependency. CPI is more in high dependency, and IPC is more moderate dependency.

5 Conclusion and Future Work

This paper presented the designing and simulation of super-pipelining architecture for visualizing the effect of dependency. With the use of simulator presented in this paper, total clock cycles, total stalls, CPI, IPC, and speedup can be observed for any number of instructions. Results of simulation show that when dependencies among the instructions increase, then total clock cycles and total stalls also increase. In other words, the performance of super-pipeline architecture decreases with the increase in the level of dependency.

In the future, simulator can also be used for other pipelining architecture. The present simulator is based on 'out order issue and out order completion' scheduling algorithm, the performance of super-pipelining can be evaluated under other scheduling algorithms.

References

1. N.P. Jouppi, D.W. Wall, Available instruction level parallelism for superscalar and super-pipelined machines, in *ASPLOS-III Proceedings of Third International Conference on Architectural Support for Programming Languages and Operating Systems* (IEEE ACM, Boston, MA, USA, 1989), pp. 271–282
2. C.V. Ramamoorthy, H.F. Li, Pipeline architecture. ACM Comput. Surv. (CSUR) **9**(1), 61–102 (1977)
3. A. Briggs, K. Hwang, *Computer Architecture and Parallel Processing* (McGraw Hill, 2001), p. 148
4. M.J. Forsell, Minimal pipeline architecture—an alternative to Superscalar architecture. Microprocess. Microsyst. **20**, 277–284 (1996) (Elsevier)
5. J. Schneider, C. Ferdinand, Pipeline behavior prediction for superscalar processors by abstract interpretation, in *Proceedings of the ACM SIGPLAN 1999 Workshop on Languages, Compilers, and Tools for Embedded Systems*, vol. 34(7) (ACM, New York, NY, USA, 1999), pp. 35–44
6. A. Abnous, N. Bagherzadeh, Pipelining and bypassing in a VLIW processor. IEEE Trans. Parallel Distrib. Syst. **5**(6), 658–664 (1994)
7. V. Saravanan, D.P. Kothari, I. Woungang, An optimizing pipeline stall reduction algorithm for power and performance on multi-core CPUs. Hum. Cent. Comput. Inf. Sci. Springer Open J., 1–13 (2015)

8. R. Patel, S. Kumar, Visualizing effect of dependency in superscalar pipelining, in *Proceedings of the 4th IEEE International Conference on Recent Advances in Information Technology RAIT* (Indian Institute of Technology (ISM), Dhanbad, 2018), pp. 716–720. ISBN-978-1-5386-3038-9/18
9. L. Yizhen, L. Lin, W. Jun, The application of pipeline technology: an overview, in *2011 6th International Conference on Computer Science and Education (ICCSE)* (IEEE Xplore, Singapore, 2011), pp. 47–51
10. K. Hwang, *Advanced Computer Architecture* (Tata McGraw Hill, 2001), pp. 288–289
11. R. Patel, S. Kumar, Simulation and investigation on effect of dependency in under pipelining. Int. J. Comput. Appl. (IJCA) **152**(6), 12–15 (2016) (ISSN 0975-8887)
12. I.H. Unwala, E.E. Swartzlander, Superpipelined adder designs, in *IEEE International Symposium on Circuits and Systems (ISCAS'93)*, 3–6 May 1993, pp. 1841–1844
13. V. Lee, N. Lam, F. Xiao, A.K. Somani, Superscalar and superpipelined microprocessor design, and simulation: a senior project. IEEE Trans. Educ. **40**(1), 89–97 (1997)
14. R. Patel, S. Kumar, The effect of dependency on scalar pipeline architecture. IUP J. Comput. Sci. **11**(1), 38–50 (2017)
15. J.E. Thornton, The CDC 7600 project. IEEE Ann. Hist. Comput. **2**(4), 338–348 (1980)
16. S. Kumar, Mathematical modelling and simulation of a buffered fault tolerance double tree network, in *15th International Conference on Advanced Computing and Communications* (IEEE, 2007), pp. 422–426
17. P.P. Change, D.M. Lavery, S.A. Mahlke, W.Y. Chen, W.M.W. Hwu, The importance of prepass code scheduling for superscalar and superpipelined processors. IEEE Trans. Comput. **44**(3), 353–370 (1995)

Harmonic Minimization in Cascaded Multilevel Inverter Using PSO and BAT Algorithms

Md. Ekrama Arshad and Abrar Ahmad

Abstract AC-to-DC conversion is one of the basic requirements in the modern electric power system. Hence, inverters have become an essential part of the power system. The multilevel inverter (MLI) has emerged nowadays, especially for medium-voltage high-power application. The main problem with the power electronics inverter is that it injects harmonics in the system. Hence, a scheme to sort out the problem of harmonic distortion is suggested in this paper. The mathematical equation of the output voltage of the inverter involves nonlinear transcendental equation. Hence, the harmonic minimization in MLI is a complex problem. It is not possible to get a solution from these equations using conventional iterative method. In this thesis, particle swarm optimization (PSO) and bat algorithm are suggested to sort out this problem of harmonics present in the multilevel inverter. These two methods are used here to find the optimal value of the switching angle, so that THD can be decreased to the acceptable level. There are several topologies of the multilevel inverter, but due to various striking features, cascaded multilevel inverter (C-MLI) is widely accepted in the industry. Hence, the above-mentioned methods are applied to the 11-level cascaded H-bridge in MATLAB programming environment, and satisfactory results are found.

Keywords Multilevel inverter (MLI) · Cascaded multilevel inverter (C-MLI) · Selective harmonic elimination (SHE) · Particle swarm optimization (PSO) · Bat algorithm (BA) · Modulation index (*m*) · Total harmonic distortion (THD)

1 Introduction

The continuous rising demand of the electrical energy has motivated human being to look for the new alternatives, and huge population growth has almost depleted the stored fossil fuel reserves. Hence, we cannot afford to lose energy produced by the existing energy resources. To effectively utilize the energy produced, we need efficient devices in the system. Nowadays, power electronics systems with their

Md. E. Arshad · A. Ahmad (✉)
Department of Electrical Engineering, Jamia Millia Islamia, New Delhi, India
e-mail: aahmad15@jmi.ac.in

H. S. Saini et al. (eds.), *Innovations in Electronics and Communication Engineering*, Lecture Notes in Networks and Systems 107,
https://doi.org/10.1007/978-981-15-3172-9_71

proper control strategies are very efficient. They have the capability to minimize the energy wastage and improve the power quality. Most of the industrial application like adjustable speed drives demand medium voltage and high power. MLI has excellent performance in medium-voltage and high-power application because of its low switching loss, higher efficiency and more electromagnetic capability [1–2].

MLIs produce desired level of output voltage with very low harmonic distortion. To achieve desired level of voltage with reduced harmonics, different modulation strategies are available, e.g., space vector PWM and sinusoidal PWM, but these PWM techniques have their limitation that they cannot eliminate lower-order harmonics completely [3, 4]. There is another method known as SHE or programmable PWM which selects specific angles to eliminate the lower-order harmonics. The output voltage equation is very complex, i.e., they are nonlinear and transcendental in nature. It is very difficult to obtain arithmetic solution from these equations. To solve SHE problem, numerical analysis technique like Newton–Raphson can also be used [5–7]. But, it needs a very precise guess as initial value. It is very difficult to provide good initial value, especially in case of SHE problem as its search space is unknown and one cannot be sure about the existence of solution.

Resultant theory can be used to achieve the solution for SHE problem [9–9]. Here, at first transcendental equations are needed to be converted into polynomial equations only, and then, it provides a set of solutions, but if it is required to eliminate more number of harmonics, the polynomial equations also become complex as it attains very high degree. The increment in the degree of these is so rapid that if we only intend to remove up to sixth harmonics, then present computer tools will not be able to solve these equations [6]. Another approach is the use of stochastic techniques like genetic algorithm (GA) and PSO [10, 9]. These techniques are very efficient, but when the number of switching angle is increased, their search space becomes very complex and they may get trapped in the local optimal value and thus fails to provide the correct solution. In this project, PSO and BA are used to find out the optimal value of the switching angle and based on that THD is calculated.

2 Cascaded Multilevel Inverter

The typical structure of an 11-level C-MLI is shown in Fig. 1. It is clearly visible from the structure that cascaded MLI is made up of series connection of number of H-bridge units. It produces a staircase waveform, but an approximate sinusoid can be obtained if proper number of H-bridge is used. The waveform of the output voltage for 11-level C-MLI and a typical H-bridge unit is shown in Figs. 2 and 3, respectively. A H-bridge unit requires DC voltage source separately for each unit that can be provided by ultra-capacitor, fuel cell, battery, solar cell, etc. Each H-bridge contains four switches, and by proper combination of these switches, it can produce $+V_{\mathrm{dc}}$, 0 and $-V_{\mathrm{dc}}$. In order to achieve desired output voltage, a proper number of such units are connected in series. The number of DC sources decides the level of

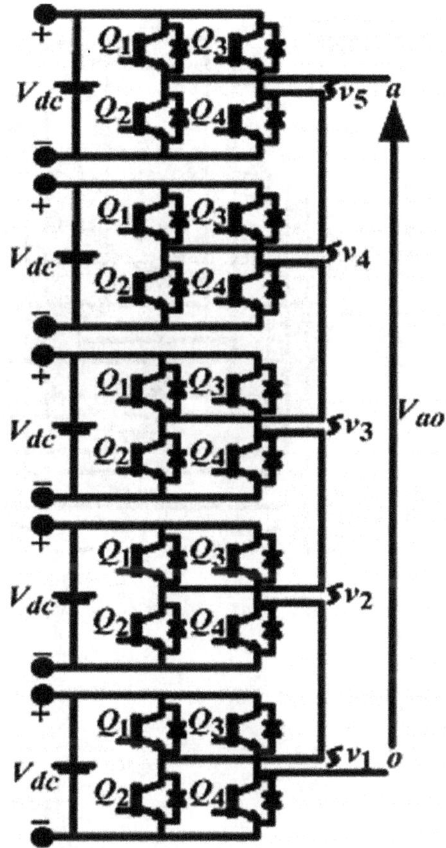

Fig. 1 Structure of 11-level C-MLI

output phase voltage which is given by the equation $m = 2\,s + 1$, where m = level of output phase voltage and s = separate DC voltage sources.

3 Problem of Harmonic Minimization in C-MLI

The waveform of output voltage of the C-MLI has a structure of staircase. This staircase waveform can be considered as sinusoidal waveform having harmonics. Hence, SHE technique is applied to remove harmonics from the staircase waveform. The number of separate DC source used poses a restriction on the number of switching in the quarter of cycle, thus limiting the number of harmonics that can be removed. If the number of DC source is given by 's', then only '$s - 1$' harmonics can be eliminated. Hence, in order to eliminate more harmonics without increasing the hardware requirement, SHE-PWM technique is used which is also known as virtual stage PWM. The switching frequency is lesser than other PWM technique. Figure 2 shows

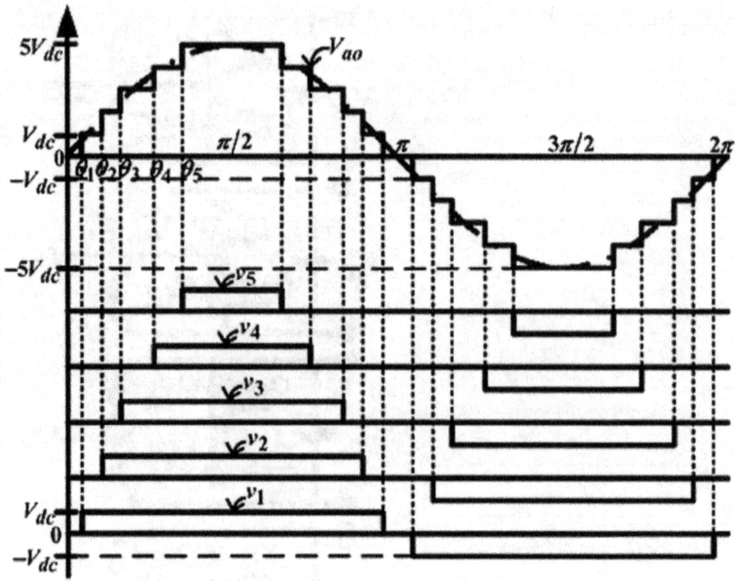

Fig. 2 Staircase output voltage of 11-level C-MLI

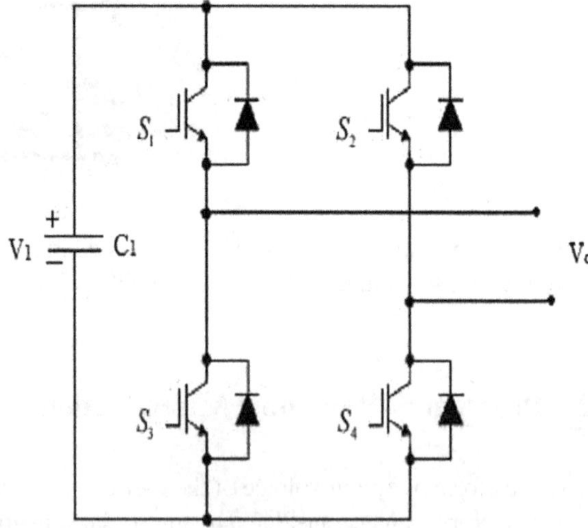

Fig. 3 A typical H-bridge unit

the output voltage with three times switching at each level. In SHE-PWM technique, the switching frequency is 'k' times the fundamental frequency, where k represents the number of switching at each level. So, $ks - 1$ harmonics will be eliminated in this case. For 11-level MLI, if $k = 3$, then 14 harmonics will be eliminated successfully. The expression for the output voltage is as follows:

$$V_{\text{an}}(\omega t) = \sum_{n=1,3,5}^{\infty} \frac{4V_{\text{dc}}}{n\pi}(\cos(n\alpha_1) + \cos(n\alpha_2) + \cdots + \cos(n\alpha_5))\sin(n\omega t) \quad (1)$$

where $0 \le \alpha_1 \le \alpha_2 \le \alpha_3 \le \alpha_4 \le \alpha_4, \ldots \le \alpha_{ks} \le \frac{\pi}{2}$

The objective is to choose the switching angle such that specified lower-order harmonics can be eliminated and desired fundamental value can be achieved.

$$\cos(\alpha_1) \pm \cos(\alpha_2) \pm \cdots \cos(\alpha_{ks-1}) \pm \cos(\alpha_{ks}) = \left(\frac{s\pi}{4}\right)M$$

$$\cos(5\alpha_1) \pm \cos(5\alpha_2) \pm \cdots \cos(5\alpha_{ks} - 1) \pm \cos(5\alpha_{ks}) = 0$$

$$\cos(7\alpha_1) \pm \cos(7\alpha_2) \pm \cdots \cos(7\alpha_{ks} - 1) \pm \cos(7\alpha_{ks}) = 0$$

$$\begin{aligned}&\cos(3(ks-2)\alpha_1) \pm \cos(3(ks-2)\alpha_2) \pm \cdots \cos(3(ks-2)\alpha_{ks-1}) \\ &\pm \cos(3(ks-2)\alpha_{ks}) = 0\end{aligned} \quad (2)$$

m is the modulation index and defined by $M = V_1/sV_{\text{dc}}$, V_1 = magnitude of fundamental frequency voltage and S = number of separate DC voltage source.

These are relevant equations, and by solving these ks equations, $3ks - 2$ is non-triplen harmonics if ks = odd and $3ks - 1$ if ks = even. In three-phase system in line–line voltage, triplen harmonics are canceled automatically.

Another approach is to minimize the harmonic to the acceptable level. Here, in this paper, 50 harmonics are considered for the calculation of THD. The objective is to minimize:

$$f(\alpha_1, \alpha_2, \ldots, \alpha_{ks})$$

where $0 \le \alpha_1 \le \alpha_2 \le \cdots \le \alpha_{ks} \le \frac{\pi}{2}$ and,

$$f(\alpha_1, \alpha_2, \ldots, \alpha_{ks}) = 100 * \left[\left|M - \frac{|V_1|}{sV_{\text{dc}}}\right| + \left(\frac{|V_5| + |V_7| + |V_{11}| + \cdots + |V_{49}|}{sV_{\text{dc}}}\right)\right] \quad (3)$$

K Ratio of switching frequency to the fundamental frequency and hence these equations can be used for fundamental harmonics if value of k is set to 1.

s Number of DC source.

4 PSO and BAT Algorithms

4.1 Particle Swarm Optimization

In 1995, Dr. Eberhert and Dr. Kennedy developed an optimization technique called PSO. It is a population-based optimization technique which is stochastic in nature. They took the inspiration from the social behavior of the swarms such as bird flocking and fish schooling. PSO is very much similar to the computational technique like genetic algorithm (GA), but unlike GA, it does not have any evolution operator such as mutation and crossover. PSO is very easy for implementation in most of the programming language. It is very effective for different types of optimization problem. In PSO, very few parameters are needed to be adjusted, and hence, with slight change, wide variation can be solved. An algorithm for PSO is shown in Fig. 4 to give an insight of the operation mechanism of this technique.

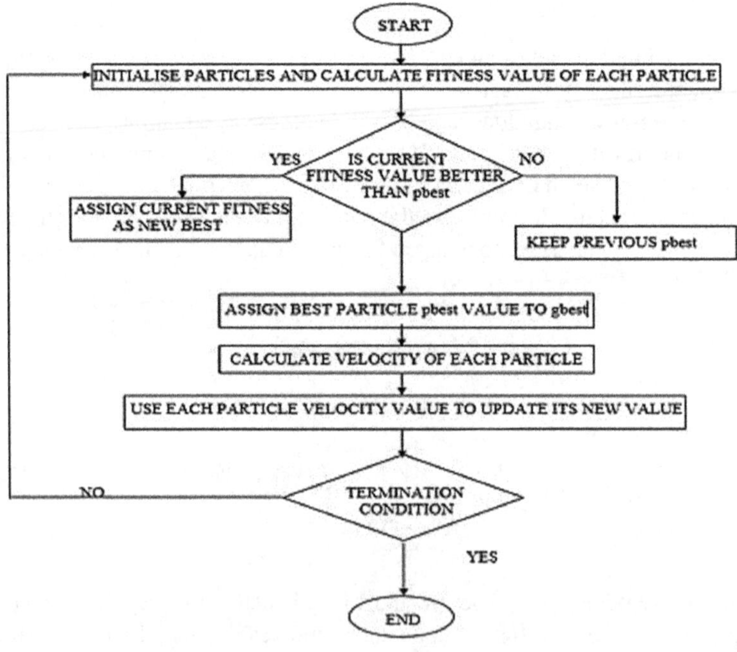

Fig. 4 PSO algorithm

4.2 BAT Algorithm

It is also a nature-inspired algorithm. In 2010, Xin-She Yang developed it, and since then it has emerged as a good alternative to solve the optimization problem. This algorithm is based upon the behavior of the bats and hence the name of the algorithm. The echolocation behavior of the bats is the basis of this algorithm. It is the only algorithm that uses frequency tuning to solve the dynamic behavior of bats. Each bat has certain velocity $v_{(i)}$ and location $x_{(i)}$ in the solution space. In the population of '*n*' bats, each bat updates its velocity and position with the best solution found so far, and thus, it reaches the optimum value. The position and the velocity can be updated as follows:

$$f_i = f_{\min} + (f\max - f_{\min})\beta \tag{4}$$

$$V_{i_{\text{new}}} = V_{i_{\text{old}}} + (X_{i_{\text{old}}} - X_*)f_i \tag{5}$$

$$X_{i_{\text{new}}} = X_{i_{\text{old}}} + V_{i_{\text{new}}} \tag{6}$$

where β is a random vector following uniform distribution and its value lies in [0, 1]. The loudness and pulse emission rate can also be changed during each iteration. This way selection of the best value is continued until stop criteria are met.

In MLIs, the equation of output voltage is nonlinear and transcendental in nature. It is not possible to analyze these equations using conventional method. Hence, in this paper, PSO and BAT algorithms are proposed to solve the minimization problem of harmonics. These algorithms are able to calculate the optimal switching angle even if the number of switching angle is increased. In 11-level C-MLI, these methods are applied and obtained results are presented in the next section. A comparative analysis of these two algorithms is also presented (Fig. 5).

5 Results

PSO and BAT algorithms are successfully applied to the problem of harmonic minimization in multilevel inverter. The voltage profile for different harmonics is shown in Figs. 6 and 7 for PSO and BAT algorithms, respectively. Here, value of modulation index is unity. Figure 8 shows the optimum switching angle found using the PSO algorithm, and using these angle THD is calculated which is shown in Fig. 9. In Figs. 10 and 11, switching angle for different modulation index varying from 0.1 to 0.9 is shown for both PSO and BAT algorithms, respectively. Figures 12 and 13 show the variation of THD versus *m* for PSO and BAT algorithms based on parameters

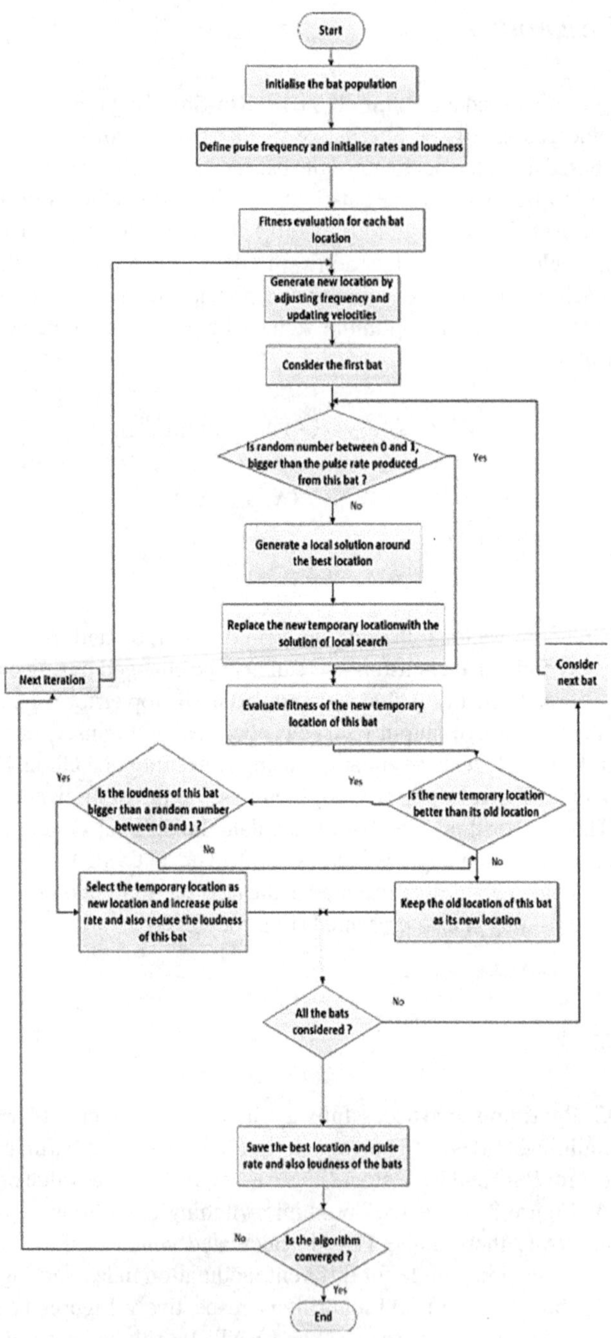

Fig. 5 BAT algorithm

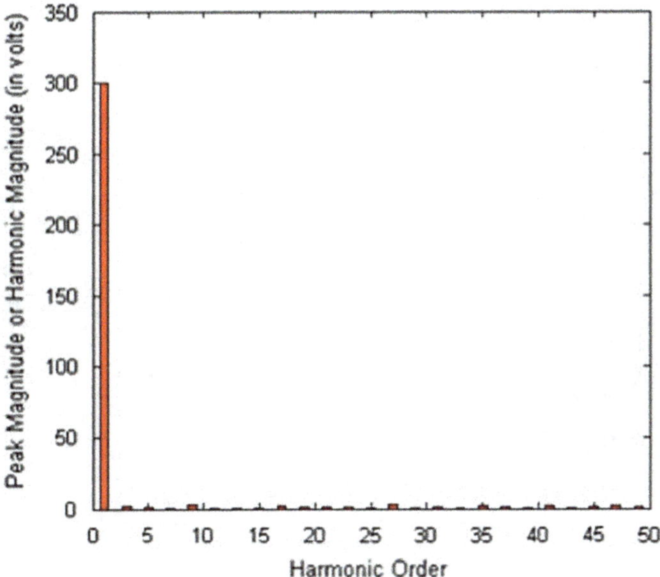

Fig. 6 Voltage magnitude of different harmonics using PSO algorithm

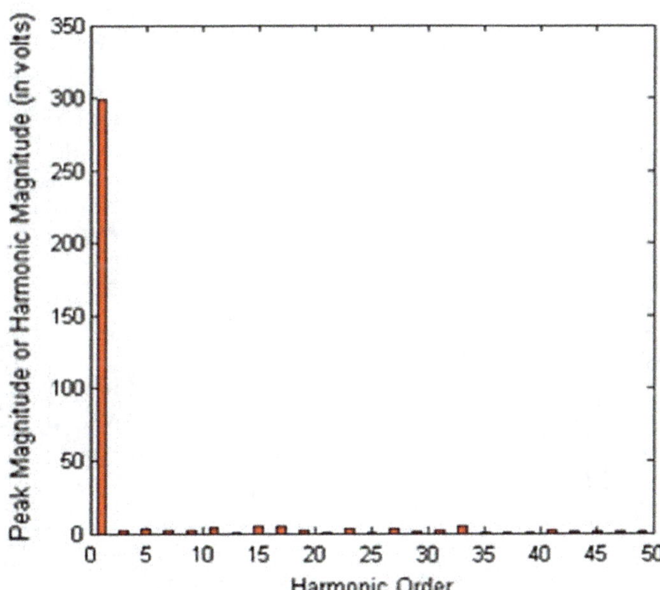

Fig. 7 Voltage magnitude of different harmonics using BAT algorithm

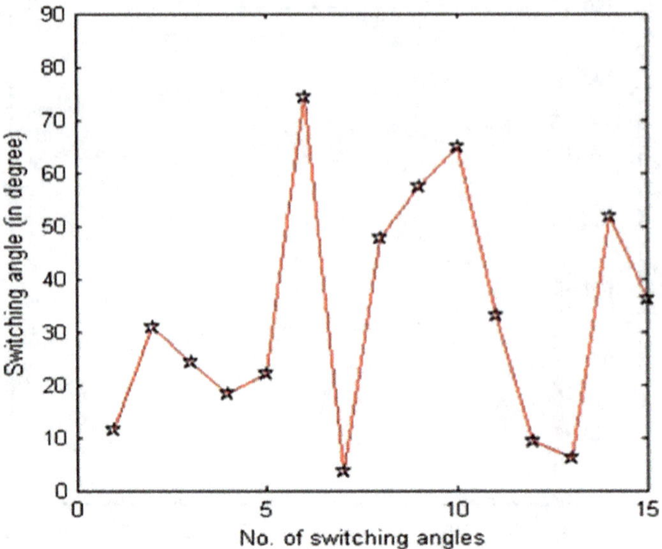

Fig. 8 Optimum switching angle with $m = 1$ for PSO

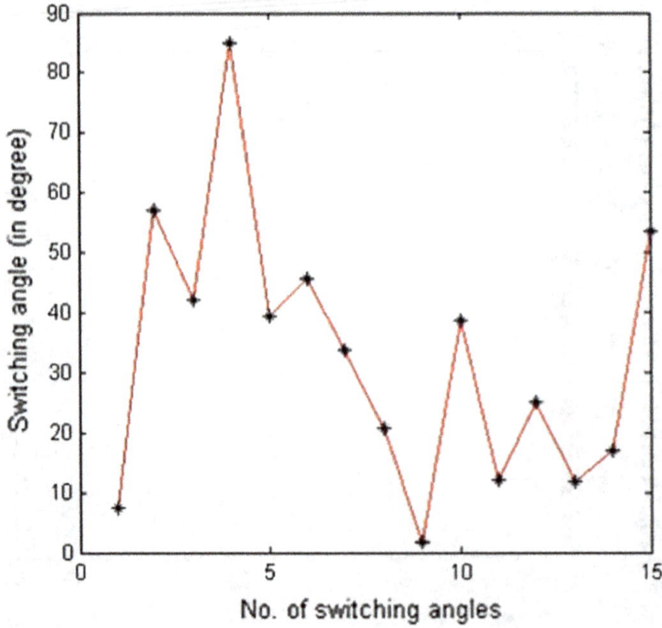

Fig. 9 Optimum switching angle with $m = 1$ for BAT

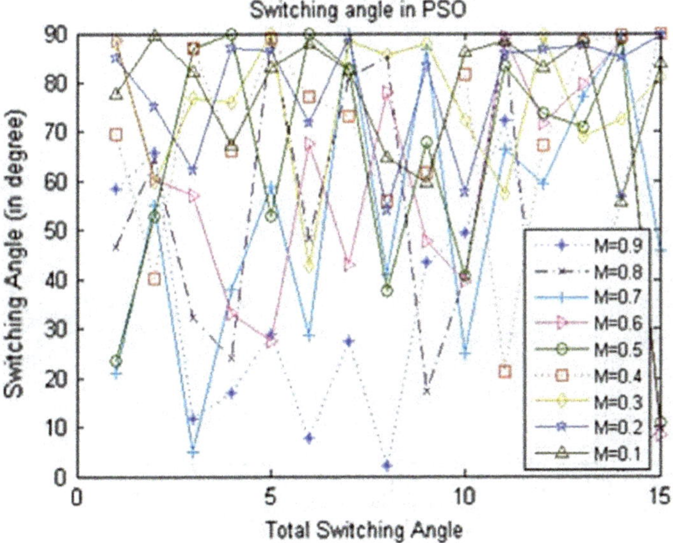

Fig. 10 Switching angle in PSO with different values of *m*

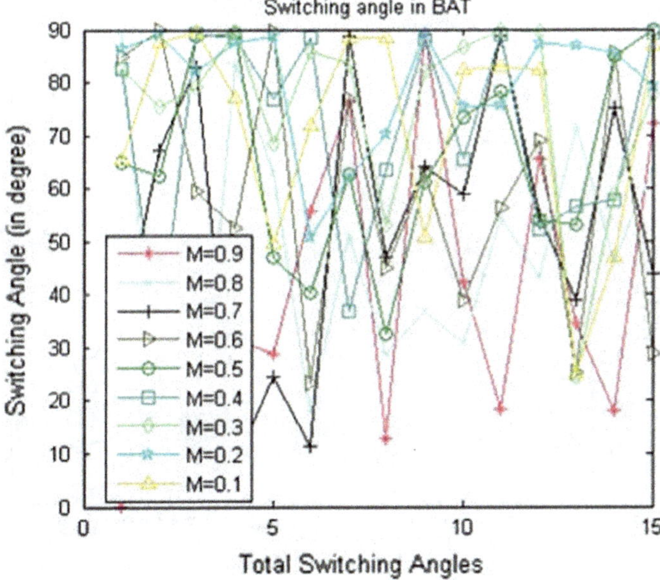

Fig. 11 Switching angle in BAT with different values of *m*

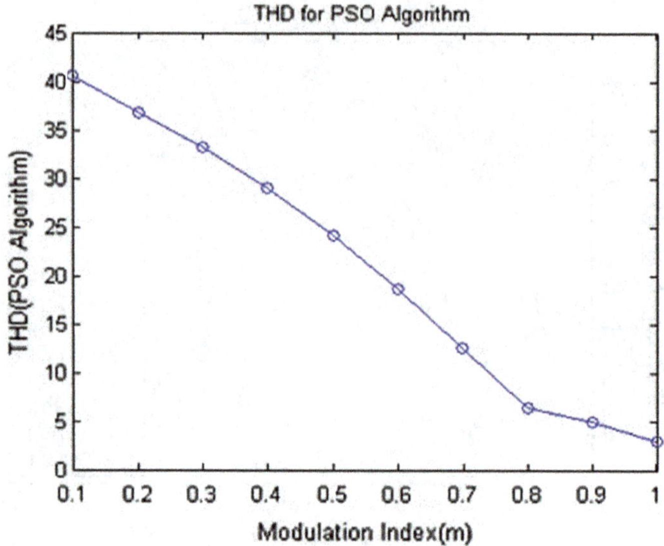

Fig. 12 THD versus *m* for PSO algorithm

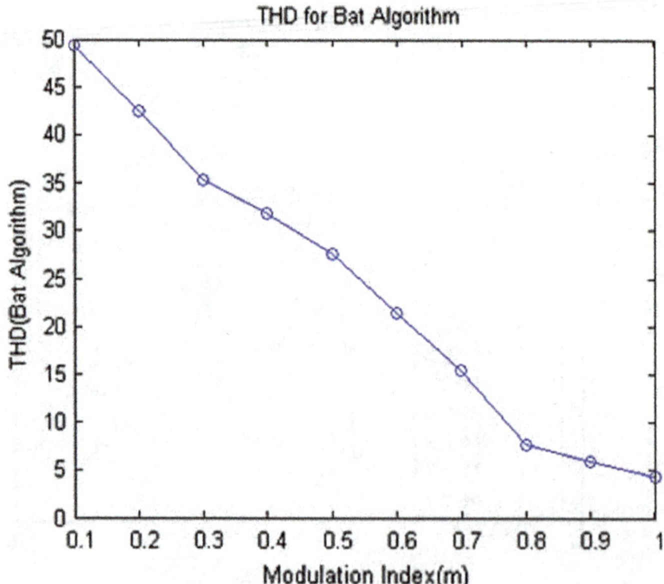

Fig. 13 THD versus *m* for BAT algorithm

Table 1 Simulation Parameters

No. of population	200
No. of iteration	1000
Magnitude of DC voltage used in each H-bridge unit	62 V
Velocity for PSO algorithm	−10 to +10
Frequency for BAT algorithm	0–2

Table 2 THD (PSO) & THD (BAT) Comparison

m	THD (BAT)	THD (PSO)
0.1	49.38	40.56
0.2	42.44	36.74
0.3	35.24	33.25
0.4	31.84	28.95
0.5	27.56	24.21
0.6	21.45	18.64
0.7	15.43	12.58
0.8	7.66	6.32
0.9	5.97	4.89
1	4.24	2.95

given in Table 1. Table 2 provides the value of THD for different values of modulation index in tabular form for both PSO and BAT algorithms, and a comparative graph for the same is shown in Fig. 14.

6 Conclusion

The PSO and BAT algorithms are successfully applied to the problem of harmonic minimization. The calculation of optimal switching angle is obtained for both the proposed method, and on the basis of these angles, THD is calculated. The harmonics up to the order of 50 is minimized, and it is shown in the results. The higher-order harmonics can be easily eliminated by simple filtering mechanism. The THD value using PSO algorithm is found to be lesser than that of BAT algorithm. Hence, PSO algorithm is more suitable than BAT algorithm for the minimization of harmonics content in the multilevel inverter.

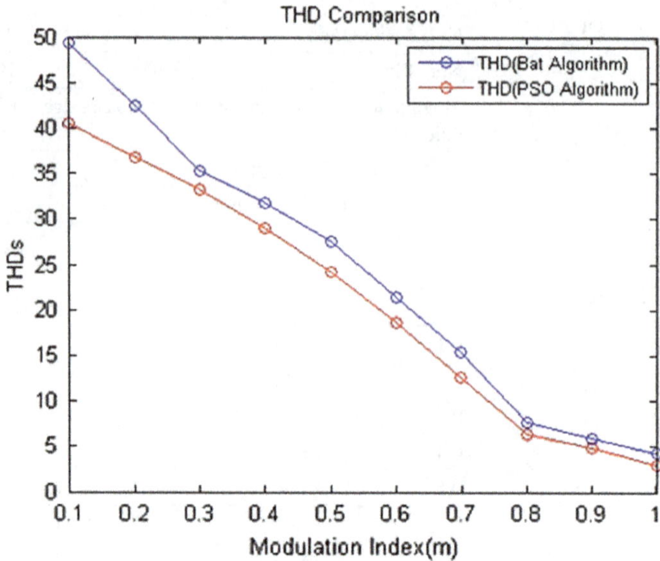

Fig. 14 THD comparison of PSO and BAT algorithms for different values of m

References

1. J.S. Lai, F.Z. Peng, Multilevel converters—a new breed of power converters. IEEE Trans. Ind. Appl. **32**(3), 509–517 (May/June 1996); J. Rodr ıguez, J. Lai, F. Z. Peng, Multilevel inverters: a survey of topologies, controls and applications. IEEE Trans. Ind. Electron. **49**(4), 724–738 (August 2002)
2. L.M. Tolbert, F.Z. Peng, T.G. Habetler, Multilevel converter for large electric drives. IEEE Trans. Ind. Appl. **35**(1), 36–44 (January/February 1999); J. Wang, F.Z. Peng, Unified power flow controller using the cascade multilevel inverter. IEEE Trans. Power Electron. **19**(4), 1077–1084 (July 2004); M. Manjrekar, G. Venkataramanan, Advanced topologies and modulation strategies for multilevel converters, in *Proceedings of IEEE Power Electron Special Conference* (Baveno, Italy, June 1996), pp. 1013–1018
3. D.G. Holmes, T.A. Lipo, *Pulse width modulation for power converters* (Piscataway, NJ: IEEE Press, 2003); H.S. Patel, R.G. Hoft, Generalized harmonic elimination and voltage control in thyristor inverters: part I—harmonic elimination. IEEE Trans. Ind. Appl. **IA-9**(3), 310–317 (May/June 1973)
4. H.S. Patel, R.G. Hoft, Generalized harmonic elimination and voltage control in thyristor inverters: part II—voltage control technique. IEEE Trans. Ind. Appl. **IA-10**(5), 666–673 (September/October 1974)
5. P.N. Enjeti, P.D. Ziogas, J.F. Lindsay, Programmed PWM techniques to eliminate harmonics: a critical evaluation. IEEE Trans. Ind. Appl. **26**(2), 302–316 (March/April 1990)
6. J.N. Chiasson, L.M. Tolbert, K.J. McKenzie, D. Zhong, Elimination of harmonics in a multilevel converter using the theory of symmetric polynomials and resultants. IEEE Trans. Control Syst. Technol. **13**(2), 216–223 (March 2005)
7. J.N. Chiasson, L.M. Tolbert, Z. Du, K.J. McKenzie, The use of power sums to solve the harmonic elimination equations for multilevel converters. Eur. Power Electron. Drives J. **15**(1), 19–27 (February 2005)

8. K.J. McKenzie, Eliminating harmonics in a cascaded H-bridges multilevel inverter using resultant theory, symmetric polynomials, and power sums. M.Sc. thesis, University Tennessee, Chattanooga (2004)
9. B. Ozpineci, L.M. Tolbert, J.N. Chiasson, Harmonic optimization of multilevel converters using genetic algorithms. IEEE Power Electron. Lett. **3**(3), 92–95 (September 2005)
10. H. Taghizadeh, M. Tarafdar Hagh, Harmonic elimination of multilevel inverters using particle swarm optimization, in *Proceedings of IEEE-ISIE* (Cambridge, U.K., 2008), pp. 393–397
11. B. Ozpineci, L.M. Tolbert, J.N. Chiasson, Harmonic optimization of multilevel converters using genetic algorithms. IEEE Power Electron. Lett. **3**(3), 92–95 (September 2005)
12. H. Taghizadeh, M. Tarafdar Hagh, Harmonic elimination of multilevel inverters using particle swarm optimization, in Proceedings of IEEE-ISIE(Cambridge, U.K., 2008), pp. 393–397

A Normalized Mean Algorithm for Imputation of Missing Data Values in Medical Databases

G. Madhu, B. Lalith Bharadwaj, K. Sai Vardhan and G. Naga Chandrika

Abstract Many medical research databases commonly consist of the missing value problems, and the presence of missing data value has a negative impact on machine learning models. However, the data with missing value can decrease the classifier performance and can lead to wrong insights by introducing biases. Imputation approaches are typically employed to impute the missing data value for data analysis. In addition, imputation helps us to build an effective classification model to discover hidden patterns which can provide insightful outcomes. In this paper, the normalized mean imputation approach is designed to fill the missing data value in numerical datasets. After normalizing the data, compute the mean and cube-root-of-cubic mean. Finally, impute the missing data value from the maximum value of these two methods which are the plausible data value in a given dataset. In addition, it is observed that after imputation some of the outliers are also eliminated in a dataset in this approach. The experiments are conducted on benchmark datasets and compared with mean imputation, median imputation, and mode imputation approaches. The experimental results show that the suggested imputation technique performed superior results compared with other state-of-the-art methods.

Keywords Datasets · Imputation · Multiple imputation · Normalized mean · Missing data value

1 Introduction

The concept of missing data values plays a significant role in data engineering and its applications for managing data in the machine learning process. In addition, the missing data values occur due to several reasons like [1]; (i) the data value might be omitted or ignored, (ii) data value not registered, (iii) improper measurements, and (iv) equipment malfunctions. However, many researchers developed and implemented imputation techniques for data analysis that includes classification and clustering techniques which require a complete dataset as the input value [2]. In this

G. Madhu (✉) · B. Lalith Bharadwaj · K. Sai Vardhan · G. Naga Chandrika
Department of Information Technology, VNRVJIET, Hyderabad, Telangana, India
e-mail: madhu_g@vnrvjiet.in

H. S. Saini et al. (eds.), *Innovations in Electronics and Communication Engineering*, Lecture Notes in Networks and Systems 107,
https://doi.org/10.1007/978-981-15-3172-9_72

view, imputation methods are required for handling this incomplete dataset [2]. In general, we have two types of approaches to handle missing data value problem, i.e., deleted or imputed with the plausible values [3]. The first one is the best way that concerns with the missing data value which is to be deleted in the dataset [2–4]. This is applicable only if a limited number of data records are missing, and it is a negative impact on classification problems [1, 3]. Other method is imputation i.e. replacing the missing data value with plausible data value [5]. However, traditional imputation approaches utilized statistical measures, namely mean and hot-deck imputation approaches [3]. These methodologies are single or multiple imputation methods; in a single imputation method, a missing data value is substituted by one probable data value. While in the multiple imputations, one or more data values are imputed; the multiple imputations perform better in terms of modeling the uncertainty [6]. In addition, few imputation methods are fixed values, random values, the nearest neighbor values, and mean values [7]. Using statistical methods for data imputation may reduce the bias with higher precision of the estimated data value that reduces the classification model performance [8]. The major challenge in the data analysis of missing data value is to determine the most plausible value [6]. However, the missing data value can create a serious defect for data analysis in the decision-making process. In general, imputation techniques are fully accounted for uncertainty while predicting the missing data value by imputing plausible variability into multiple imputed values [9].

In view of the aforementioned issues and challenges, we present the following salient features of our research:

- To determine the missing data values in a given dataset.
- Apply truncated Gaussian distribution on non-missing data values.
- Apply statistical method to generate a data value, and from this, truncate the data and then compute normalized mean on these data values.
- After normalizing the data, computed the cube-root-of-cubic mean.
- Next, imputed the missing data value from the maximum value of these two methods which are the plausible data value in the dataset.
- Finally, apply the de-normalization to obtain the original plausible data value in a given dataset.

During this process, we first make data to be continuous at a particular interval. Utilize the cube-root-of-cubic-mean and regular mean methodology to impute the missing values. When data is continuous at a certain interval, root-mean-square (RMS) and cube-root-of-cubic-mean produced good results. However, RMS gives a positive outcome instead of negative values in the datasets. Now by considering each attribute in the data sample, we apply our statistical method and impute missing values over that selected attribute. Finally, this procedure is repeated until all the missing data values are imputed. Furthermore, we compared its performance with state-of-the-art methods on benchmark dataset using various ensemble classifier methods.

2 Related Works

Recently, many researchers have been investigated in several imputation approaches for concerns with the missing data value. Lin and Tsai [10] presented an analysis of the imputation algorithms for missing value problems, and several issues are addressed during the missing value problem. Jerez et al. [11] presented imputation methods using statistical techniques (mean, hot-deck, and multiple imputations) on the breast cancer problem. Young et al. [12] presented a brief survey of a variety of imputation methods and highlighted their advantages and limitations. Karpievitch et al. [13] discussed various imputation methods to normalize and impute the missing values on the microarray data and mass spectrometry-based data. de França et al. [14] proposed a bi-clustering-based data imputation technique by using the mean squared residual metric which evaluates the degree of consistency between each object of the dataset. Yan et al. [15] proposed context-based linear mean with binary search and Gaussian mixture model technique for the imputation of missing data value in IoT data. Aljuaid and Sasi [16] presented a comparative study of numerous imputation procedures by using mean, median, mode, hot-deck, k-nearest neighbor, and expectation maximization. Myneni et al. [17] presented the imputation framework by using correlated-based clustering. The correlation between each data record in dataset with respect to missing data attributes and imputes the missing data value based on cluster mean value. Yelipe et al. [18] presented an efficient imputation approach for missing data value problem using IM-CBC method. Kabir et al. [19] discussed three types of imputation methods such as mean imputation, median imputation, linear regression-based imputation methods, and other three multiple types of imputation methods like an iterative model, multiple imputations of incomplete data, and the sequential-based imputation methods.

The aforementioned imputation methods are either impute or filled with plausible or approximate values based on various statistical techniques or clustering techniques. But these approaches do not yield an effective classification model.

3 Proposed Methods

Motivated by the aforementioned challenges, we proposed the normalized mean imputation approach to impute the missing data value in the datasets. After normalizing the data, compute the mean and cube-root-of-cubic mean. Then, impute the missing data value from the maximum value of these two methods which are the plausible data value in a given dataset. In addition, observed that after the imputation some of the outliers are also eliminated in a dataset using this approach. The comprehensive of the proposed imputation algorithm is presented in the subsequent steps (shown in Algorithm 1).

Algorithm 1: New Imputation Algorithm

Step-1: First, we consider non-missing data and apply Gaussian cumulative distribution function (GCDF) $\Phi(\mu, \sigma; x)$. Meanwhile, these data values are normalized; then, they have a floating point value, and they are frequently spread on a certain range by fitting into a bell-shaped data distribution.

$$\Phi(\mu, \sigma; x) = \frac{1}{2}[1 + \mathrm{erf}(\frac{\varepsilon}{\sqrt{2}})] \tag{1}$$

where erf is error function which is determined as

$$\mathrm{erf}(x) = \frac{2}{\sqrt{\pi}} \int_0^x e^{-t^2} \mathrm{d}t \tag{2}$$

and ε is called z-score which defined as below

$$\varepsilon = \frac{x - \mathrm{mean}(x)}{\sigma} \tag{3}$$

In Eq. (3), 'σ' is the standard deviation of the given dataset.

Step-2: In $\Phi(\mu, \sigma; x)$ is Gaussian CDF which converts the given dataset of values into continuous values by using the following equation.

$$\Psi(\mu, \sigma, a, b; x) = \frac{\Phi(\mu, \sigma; x) - \Phi(\mu, \sigma; a)}{\Phi(\mu, \sigma; b) - \Phi(\mu, \sigma; a)} \tag{4}$$

where $\Psi(\mu, \sigma, a, b; x)$ is truncated Gaussian cumulative distributive function for which 'a' is the lower truncated boundary and b is an upper truncated boundary.

Step-3: Now, with the help of the Step-2, we truncate or cut the required regions in a distribution which helps in removing the outliers when there is a humongous amount of data.

Step-4: Compute Inverse Gaussian CDF,

$$x = \mu + \sigma * [\sqrt{2}\mathrm{erf}^{-1}(2p - 1)] \tag{5}$$

where p is Gaussian CDF for that value x.

Step-5: After normalizing the data values, we apply the following two methodologies:

Mean: When there are fewer instances in data, mean this methodology works best.

$$\mu = \frac{1}{n} \Sigma_0^n x_i \quad (6)$$

Cube-root-of-cubic-mean: When there are large instances of continuous data, this method works best.

$$\omega = \frac{1}{n} \sqrt[3]{(\Sigma_0^n x_i^3)} \quad (7)$$

Step-6: After calculating these results [by using Eqs. (6) and (7)], we select the maximum value of these two methods

Step-7: Finally, the maximum data value is imputed on the missing data values of the original dataset, and a new dataset is generated without any missing values.

4 Experiments and Results

In this experiment study, we performed an overall analysis of the impact of our method with other three imputation methods on benchmark datasets. We have selected five datasets with missing values, which are downloaded from KEEL repository [20]. The complete description of the datasets is presented in Table 1.

Table 1 summarizes the properties and statistics of these datasets; the column categorized as '%MVs' indicates the percentages of missing values in the dataset. The datasets are collected from KEEL repositories, such as diabetes dataset (Pima), hepatitis dataset, house votes dataset, mammographic dataset, and Wisconsin dataset. The percentage of MVs varied from 48.39 to 2.29%, and we have used two-class datasets with prompted MVs as shown in Table 1.

We have employed various classification methods with these datasets from each data values imputed dataset. We used Adaboost (AdaBo), Extra Trees (ET), gradient boosting (GBC), random forest (RF), and XGboost classifiers for classification simulations. The classification accuracies are performed using k-fold cross-validation

Table 1 Missing value datasets used for the experiments

Datasets	# Instances	# Features	# Classes	# MV's
Diabetes dataset	8	768	2	15.39
Hepatitis	19	155	2	48.39
House votes	16	435	2	46.47
Mammographic	5	961	2	13.63
Wisconsin	9	699	2	2.29

Table 2 Test classifier metrics used normalized mean imputation with ETree classifier

Datasets	Normalized mean imputation + ETrees		
	Accuracy (%)	Variance (%)	AUC
Diabetes dataset	74.72	7.33	0.69
Hepatitis	80.30	12.40	0.80
House votes	95.40	3.05	0.98
Mammographic	77.20	4.30	0.81
Wisconsin	96.60	2.70	0.96

Table 3 Test classifier metrics used mean imputation with ETree classifier

Datasets	Mean imputation + ETrees		
	Accuracy (%)	Variance (%)	AUC
Diabetes dataset	69.4	6.62	0.75
Hepatitis	81.6	12.7	0.65
House votes	95.3	3.30	0.98
Mammographic	76.4	4.90	0.80
Wisconsin	96.4	3.5	0.94

test, variance scores, and area under the curve (AUC) values are presented in Tables 2, 3, 4 and 5 (where $k = 10$), variance scores and AUC curve (presented in Tables 2, 3, 4, and 5).

Table 4 Test classifier metrics used median imputation with ETree classifier

Datasets	Median imputation + ETrees		
	Accuracy (%)	Variance (%)	AUC
Diabetes dataset	69.50	6.60	0.70
Hepatitis	81.20	12.80	0.71
House votes	95.30	3.25	0.98
Mammographic	76.40	4.90	0.79
Wisconsin	96.40	3.50	0.94

Table 5 Test classifier metrics used mode imputation and ETree classifier

Datasets	Mode imputation + ETrees		
	Accuracy (%)	Variance (%)	AUC
Diabetes dataset	72.60	6.52	0.75
Hepatitis	81.20	14.40	0.70
House votes	95.40	3.32	0.98
Mammographic	76.80	4.87	0.79
Wisconsin	72.60	6.52	0.75

From the summary of Tables 2, 3, 4, and 5, we can say that the proposed algorithm (Algorithm 1) is superior to other imputation methods like mean, median, and mode with other classifiers (Fig. 1).

In addition, variance scores are computed on Wisconsin dataset with the proposed method and other imputation methods that are shown in Fig. 2.

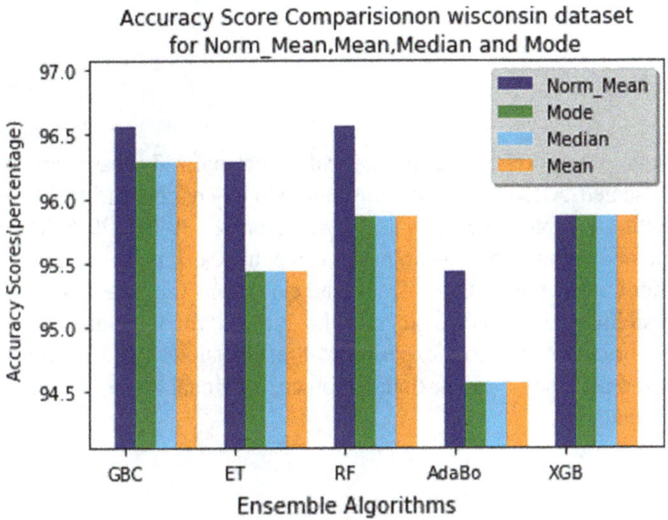

Fig. 1 Classifier accuracy compared with the proposed method versus other methods on Wisconsin dataset

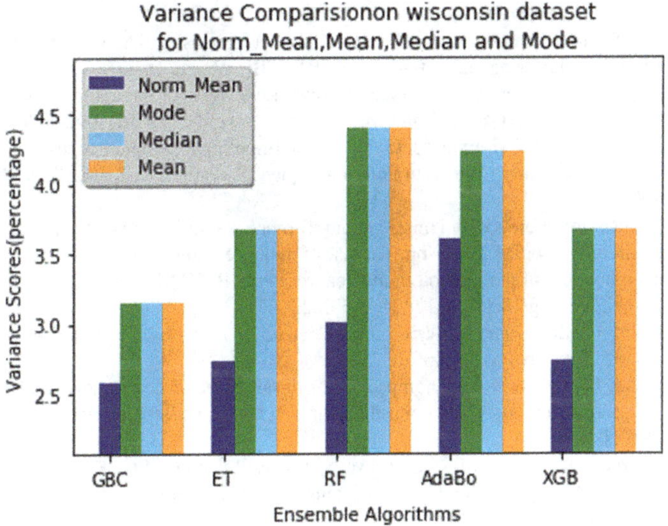

Fig. 2 Variance scores contrast with proposed method versus other methods on Wisconsin dataset

However, to impute the missing data value, various imputation approaches have been failed to deal with more than 50% of missing data values. Thus, we have successfully imputed the missing values by using the proposed Algorithm 1. The classifier accuracies are computed with Adaboost (AdaBo), Extra Trees (ET), gradient boosting (GBC), random forest (RF), and XGboost algorithms on Wisconsin dataset which is shown in Fig. 1.

5 Conclusions

In this paper, a comprehensive method called normalized mean imputation technique is presented. After imputation, this methodology has been tested on benchmark datasets with the percentage of MVs varying from 48.39 to 2.29%. The presented imputation method imputes plausible data value in the original dataset and evaluated the classifier accuracy with Extra Trees, and calculated variance scores, AUC curve values. In addition, we detected some of the outliers which are also eliminated in a dataset by our approach. From experimental results shows that the proposed imputation method accuracy is better than the other traditional mean, median, and mode imputation methods.

References

1. G. Madhu, et al., A novel index measure imputation algorithm for missing data values: a machine learning approach, in *IEEE International Conference on Computational Intelligence & Computing Research* (2012), pp. 1–7
2. M.C.P. De Souto, P.A. Jaskowiak, I.G. Costa, Impact of missing data imputation methods on gene expression clustering and classification. BMC Bioinform. **16**(1), 64 (2015)
3. A. Farhangfar, L.A. Kurgan, W. Pedrycz, A novel framework for imputation of missing values in databases. IEEE Trans. Syst. Man Cybern. Part A Syst. Hum. **37**(5), 692–709 (2007)
4. S. Chiewchanwattana, C. Lursinsap, C.-H.H. Chu, Imputing incomplete time-series data based on varied-window similarity measure of data sequences. Pattern Recognit. Lett. **28**(9), 1091–1103 (2007)
5. Donald B. Rubin, Inference and missing data. Biometrika **63**(3), 581–592 (1976)
6. P.H. Rezvan, K.J. Lee, J.A. Simpson, The rise of multiple imputation: a review of the reporting and implementation of the method in medical research. BMC Med. Res. Methodol. **15**(1), 30 (2015). https://doi.org/10.1186/s12874-015-0022-1
7. A. Gelman, J. Hill, *Data Analysis Using Regression and Multilevel/Hierarchical Models* (Cambridge University Press, 2006)
8. A.M. Wood, I.R. White, S.G. Thompson, Are missing outcome data adequately handled? a review of published randomized controlled trials in major medical journals. Clin. Trials **1**(4), 368–376 (2004)
9. J.A.C. Sterne, I.R. White, J.B. Carlin, M. Spratt, P. Royston, M.G. Kenward, A.M. Wood, J.R. Carpenter, Multiple imputation for missing data in epidemiological and clinical research: potential and pitfalls. BMJ **338**, b2393 (2009)
10. W.-C. Lin, C.-F. Tsai, Missing value imputation: a review and analysis of the literature (2006–2017). Artif. Intell. Rev.,1–23 (2019)

11. J.M. Jerez, I. Molina, P.J. García-Laencina, E. Alba, N. Ribelles, M. Martín, L. Franco, Missing data imputation using statistical and machine learning methods in a real breast cancer problem. Artif. Intell. Med. **50**(2), 105–115 (2010)
12. W. Young, G. Weckman, W. Holland, A survey of methodologies for the treatment of missing values within datasets: limitations and benefits. Theoret. Issues Ergon. Sci. **12**, 15–43 (2011)
13. Y.V. Karpievitch, A.R. Dabney, R.D. Smith, Normalization and missing value imputation for label-free LC-MS analysis. BMC Bioinform. **13**(16), S5 (2012)
14. F.O. de França, G.P. Coelho, F.J. Von Zuben, Predicting missing values with biclustering: a coherence-based approach. Pattern Recognit. **46**(5), 1255–1266 (2013)
15. X. Yan, W. Xiong, L. Hu, F. Wang, K. Zhao, Missing value imputation based on Gaussian mixture model for the internet of things, in *Mathematical Problems in Engineering 2015* (2015)
16. T. Aljuaid, S. Sasi, Proper imputation techniques for missing values in data sets, in *2016 International Conference on Data Science and Engineering* (*ICDSE*) (IEEE, 2016), pp 1–5
17. M.B. Myneni, Y. Srividya, A. Dandamudi, Correlated cluster-based imputation for treatment of missing values, in *Proceedings of the First International Conference on Computational Intelligence and Informatics* (Springer, Singapore, 2017), pp. 171–178
18. U.R. Yelipe, S. Porika, M. Golla, An efficient approach for imputation and classification of medical data values using class-based clustering of medical records. Comput. Electr. Eng. **66**, 487–504 (2018)
19. G. Kabir, S. Tesfamariam, J. Hemsing, R. Sadiq, Handling incomplete and missing data in water network database using imputation methods. Sustain. Resilient Infrastruct., 1–13 (2019)
20. J. Alcalá-Fdez, A. Fernandez, J. Luengo, J. Derrac, S. García, L. Sánchez, F. Herrera, KEEL data-mining software tool: data set repository, integration of algorithms and experimental analysis framework. J. Mult. Valued Log. Soft Comput. **17**(2–3), 255–287 (2011)

Impact of E-tools in Teaching and Learning for Undergraduate Students

Kiran Kumar Poloju and Vikas Rao Naidu

Abstract Education technology has reached the greater heights after the intervention of various e-tools and interactive media in order to enhance teaching and learning experience in higher education. Generally, students show interest to participate and involve in the activities rather than listening to faculty lecture during class and flipped teaching plays a vital role to achieve this. The newly joined undergraduate students feel bit difficult to understand the concept of core modules as they don't have any prior knowledge of those modules. This paper shows possible ways to encounter the problem using Socrative and PlayPosit to enhance student performance in their modules during the class. These methods create learning environment among students. The effectiveness of using these practices in teaching is measured through student feedback using Padlet, and it is concluded based on their responses that many advantages are found by using these practices like students enhanced their interest in their modules, and students are able to score good marks in their end semester examination, improved their understanding skills and relational abilities.

Keywords Flipped teaching · Socrative · PlayPosit · Padlet · E-learning · Active learning

1 Introduction

Most of the facilitators in higher colleges use traditional way of teaching to exhibit their knowledge while instructing also called teacher-centered learning. Fry [1] suggested, it is unfortunate, but true, that some academics teach students without having much formal knowledge of how students learn students. As per Biggs and Tang [2], students are not capable of listening to class for more than 15 min which says lectures are often viewed as ineffective learning experiences. To overcome this issue and to

K. K. Poloju (✉)
Department of Civil Engineering, Middle East College, Muscat, Oman
e-mail: kpoloju@mec.edu.om

V. R. Naidu
Vivekananda Global University, Jaipur, India
e-mail: vikasrn@gmail.com

H. S. Saini et al. (eds.), *Innovations in Electronics and Communication Engineering*, Lecture Notes in Networks and Systems 107,
https://doi.org/10.1007/978-981-15-3172-9_73

make students to engage for long time, the teacher has to conduct few activities during class by using various e-tools as a part of flipped teaching. Keppell et al. [3] explained that the technology-improved learning expanded student's cooperation in flipped classroom exercises. Smith [4] proposed that learning isn't only a component of individual endeavors to get data and information all things considered and assimilate it. Generally, students in a class are categorized into different groups As per VARK, [Visual (V), Auditory (A), Read/write (R), Kinesthetic (K)], they have their own learning styles. According to Kim et al. [5] research, the use of flipped learning improves student self-awareness and skills of undergraduate students of Korean university. Samantha Corcoran [6] suggested discussion in classrooms makes students to be more active and it is an instrument to be utilized as a part of teaching during class. When it is effectively utilized, it expands students' satisfaction in the class and enhances students' comprehension of ideas which help the greater part of the students in a classroom. Tsang [7] said that discussion "is a trade to exchange ideas where all individuals from the groups have a chance to take an interest and are required to do as such to some degree". As indicated by Ibrahim [8], a mix of various academic methodologies and learning methodologies could be considered by educators to enhance the way toward instructing and learning. This was reliable with Karlson and Janson [9] who found that various learning strategies and specifically in class gaming were helpful and added to the important learning knowledge. The incorporation of classroom addresses with innovation-based strategies was characterized by McCallum et al. [10] as mixed realizing where students have encountered learning in manners by which they are generally agreeable. As indicated by Garrison and Kanuka [11], students demonstrated a normal of 35% more grounded learning results for students educated in blended format in contrast to those learned by face to face teaching. Flipped learning method which can be called as flipped classroom means that student learning activities are flipped Slomanson [12]. In flipped learning method, students involve in the activity and understand the content at anywhere and discuss in which there are many studies revealed that flipped learning is very effective for improving student learning Bergmann and Sams [13]. Teacher can easily reach his/her expectation from student perspective, and students can build their own learning styles which lead to increase in motivation, interest and enthusiasm toward learning. Over the years, the use of Open-source software has grown in various areas including the field of teaching and learning. It is a better way to use software with a lower cost than available market tools, as well as being able to customize the same tool to suit an organization (or a specialization). By integrating the methods and techniques used in collaboration with the open-source tools [14]. Through combining the collaboration methods and techniques with the personalized open-source resources, the students and teachers would be able to achieve more efficient collaborative teaching and learning as well as the teacher would have a better way to deliver the learning outcomes through the software. It is also crucial that whatever learning style is adopted, it is necessary not only for teacher interaction with students to disseminate the information but also as a collaborative learning environment for enhanced learning experience among peers [15].

This paper describes teaching methodology using various e-tools during class and intended to achieve improvements in student consideration and concentration by redesigning approaches to teaching, learning and assessment of offered courses to accommodate fascinating method for learning styles and interesting way of learning styles. The importance of this paper is to determine the effectiveness of e-learning as a part of flipped teaching and active learning during class for undergraduate students. Moreover, benefits with using various e-tools, like PlayPosit, Socrative and Padlet, during class help to enhance student skills and understanding level.

2 Methodology

2.1 Problem Identification

During the implementation of flipped learning, there were few issues and hindrances faced by students. The most basic issue raised by the students were various reasons to class saying that don't have enough time to watch videos and come to class. This issue is considered as the most basic one and the practices specified beneath have acquired changes in student's state of mind and enthusiasm toward flipped teaching methodology. There are few e-tools like Socrative, and PlayPosit helps to assess student knowledge on topic. These practices help to overcome the problem in most of the modules and helps students to focus more on the topic and get interest on the module.

2.2 Teaching, Learning and Classroom Practice

Students don't have much knowledge of the module, and it is turned out to be more troublesome them to get it. In this strategy, facilitator utilizes different technologies and e-tools to encounter the threshold. To do so, various e-tools are implemented during class other than traditional way of teaching. Instructor shall upload relevant videos on learning management system (LMS) and conducting different activities using activity sheets, conducting live activities during class like quiz using e-tools like Socrative and PlayPosit, etc.

2.3 Details of Practice

Nowadays, flipped teaching and learning system has been executing in most of the colleges throughout the world. Various methods were implemented/implementing during the classes. This report is focused on few e-tools as a part of flipped teaching to enhance student's learning and understanding capabilities.

2.4 *Methodology Implemented*

Relevant self-recorded videos from YouTube are uploaded to students through Moodle, and facilitator can ask critical questions at any part of video as any mode of question like multiple choice, fill in the blank or short answer, etc. Students have access to view all the questions while watching the video and answer the questions. This makes students to understand the concept of topic, and finally, results are exported from these e-tools and facilitator can know how many number of students are watched and answered all questions.

Finally, during class a quiz is conducted using few e-tools like Socrative to assess student knowledge on topic. In order to reduce the paper usage to collect feedback on practice, Padlet is used to collect instant feedback.

2.5 *Enhancing Faculty and Student Learning Styles of Understanding*

Since the point of the instructing was to increase awareness of the decent variety of learning styles for students and educators.

For staff, it was vital that they understood the variations of learning styles that existed among students in their modules.

For students, it was vital that they knew and comprehended their own favored learning style, but it was also important to give them information on how they might work around differences between their learning style and the teaching they received.

3 Effectiveness of the Practice

Students' Anonymous Feedback: We trust that student feedback is vital to a specific end goal to assess the execution. It is essentially utilized as a reason for development. Besides, it was noticed that good relationship between student and teacher also enhances the nature of students' work and recognizes the regions of improvement. It is very useful technique. Most the student enjoyed these practices during class and they are completely involved in the activity. The feedback from students is noticed very positive and effective.

As seen from Figs. 1, 2, 3, 4, 5, 6, 7, feedback from students is measured using

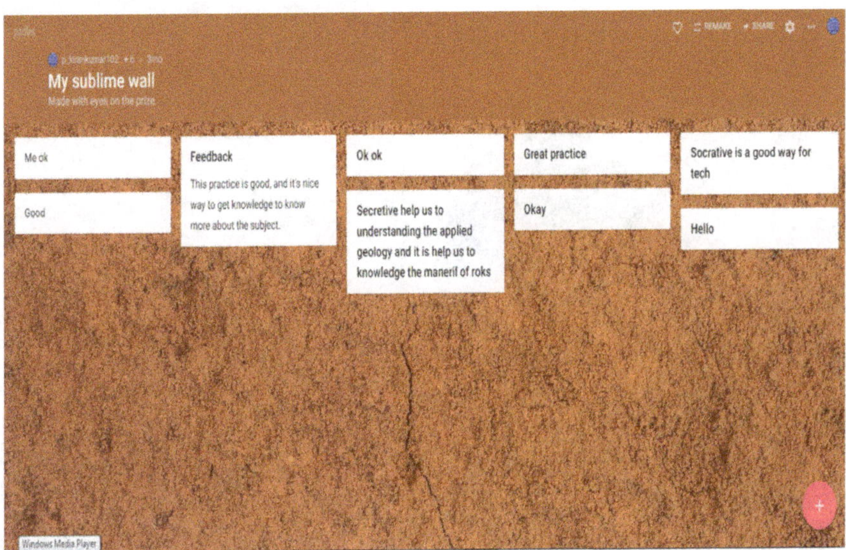

Fig. 1 Student feedback on Padlet

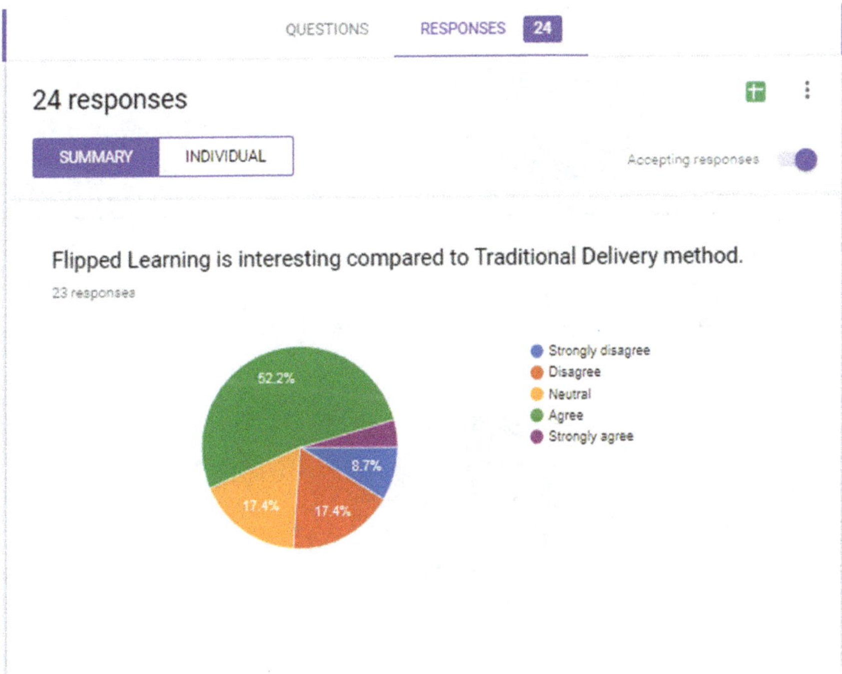

Fig. 2 Student response on PlayPosit

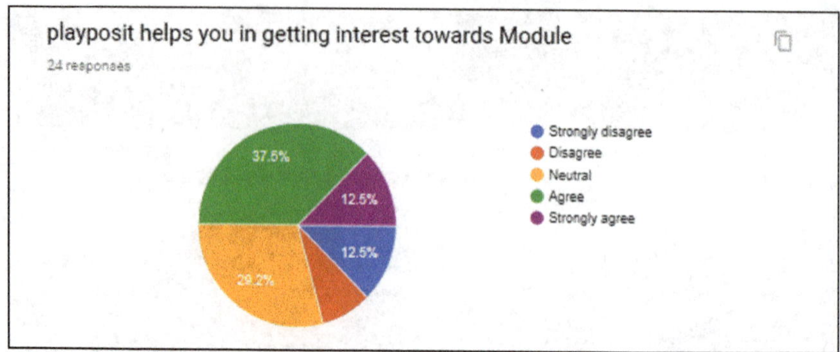

Fig. 3 Student response on PlayPosit

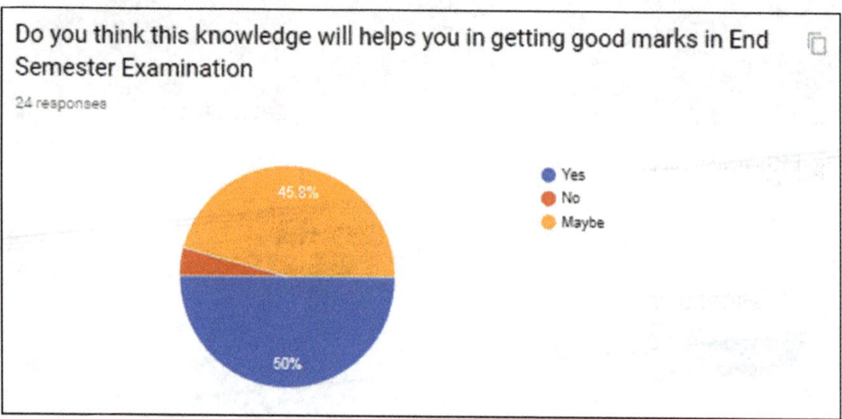

Fig. 4 Student response on PlayPosit

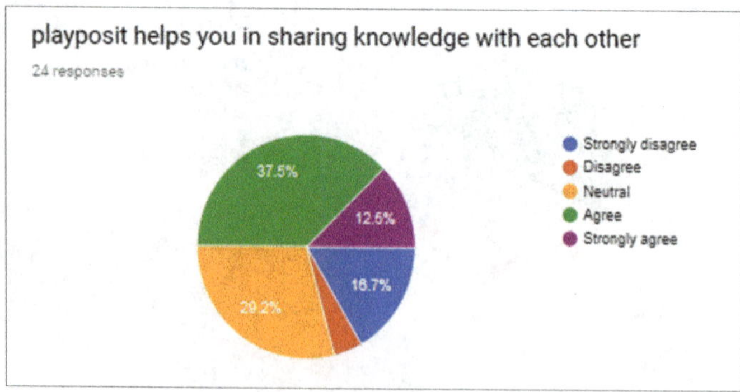

Fig. 5 Student response on PlayPosit

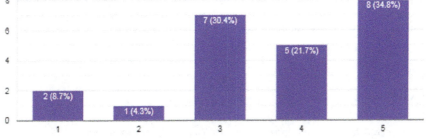

Fig. 6 Student response on PlayPosit

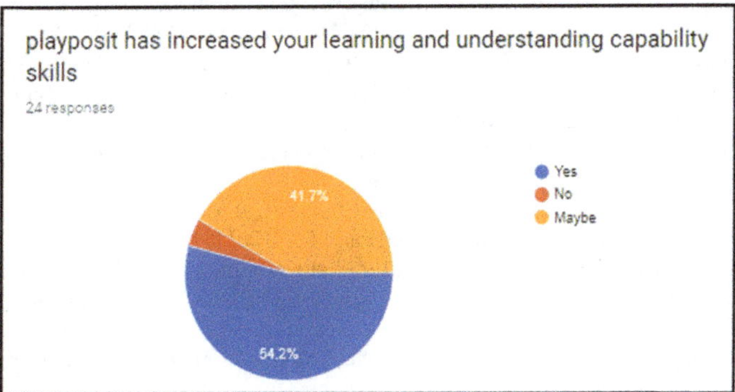

Fig. 7 Student response on PlayPosit

google forms and Padlet (instant feedback), which has resulted in a very positive output.

4 Conclusion

It is observed from the student's feedback that appropriate use of technologies made classes more fun filled and helped a lot in motivating them toward module. It helps in recalling as well as linking their lesson with their study. Students can easily correlate and understand the topic easier. Most of the students are happy with their improved speaking skills and their end semester results. Overall experience through the implementation of this flipped teaching technique was found to be commendable.

However, there are several tools available in the online platform for easy access and download in order to introduce various innovative methods in teaching and learning. Various free and open-source tools have opened up further opportunities in order to prepare teaching and learning content with an ease.

Acknowledgements I would like to acknowledge research ethical committee members Dr. Ahmed Nawaz Hakro and Dr. Nizar Al Bassam for their consent approval on student's anonymous feedback analysis on practices discussed in this paper.

References

1. H. Fry, *A Handbook for Teaching and Learning in Higher Education* (Routledge, London, 2009)
2. J. Biggs, C. Tang, *Teaching for Quality Learning at University* (Open University Press, Maidenhead, 2007)
3. M. Keppell, G. Suddaby, N. Hard, Assuming best practice in technology-enhanced learning environments. Res. Learn. Technol. **23** (2015). https://doi.org/10.3402/rlt.v23.25728
4. M.K. Smith, *Communities of Practice* (Encyclopaedia of Informal Education, 2009)
5. J.A. Kim, H.J. Heo, H. Lee, Effectiveness of flipped learning in project management class. Int. J. Softw. Eng. Appl. **9**(2), 41–46 (2015)
6. S. Corcoran, in *The Journal on Best Teaching Practices*. University of Wisconsin-River Falls. http://teachingonpurpose.org/
7. A. Tsang, In-class reflective Jigsaw method as a strategy for the development of students as evolving professionals. Int. J. Sch. Teach. Learn. **5**(1) (2011) (Article 7)
8. I. Ibrahim, Teaching project management for IT students methods and approach, in *2nd International Conference on Education and Management Technology* (IACSIT Press, Singapore, 2011)
9. G. Karlson, S. Janson, *The Flipped Classroom: A Model for Active Student Learning* (Portland Press Limited, 2016), pp. 127–136
10. S. McCallum, J. Schultz, K. Selke, J. Spartz, An examination of the flipped classroom approach on college student academic involvement. Int. J. Teach. Learn. High. Educ. **27**(1), 42–55 (2015). (**64**(1), 93–102)
11. D.R. Garrison, H. Kanuka, Blended learning: uncovering its transformative potential in higher education. Int. High. Educ. **7**, 95–105 (2004)
12. W.R. Slomanson, Blended learning: a flipped classroom experiment. J. Leg. Educ. **64**(1), 93–102 (2014)
13. J. Bergmann, A. Sams, Flip your students learning. Educ. Leadersh. **70**(6), 16–20 (2013)
14. Q.A. Mohammed, V.R. Naidu, R. Hasan, M. Mustafa, K.A. Jesrani, Re-defining the future of E-learning with free and open source E-tools for collaborative learning environment
15. M. Mustafa, V.R. Naidu, Q.A. Mohammed, K.A. Jesrani, R. Hasan, G. Al Hadrami, A customized framework to enhance students engagement in collaborative learning space in higher education

Policy Space Exploration for Linear Quadratic Regulator (LQR) Using Augmented Random Search (ARS) Algorithm

Sruthin Velamati and V. Padmaja

Abstract Considering the recent developments in embedded systems and automotive industry, it is quite evident that in very near future many application-based electronic devices will adapt the automation in its daily based activities. This automation will make the devices more powerful and will enhance its services. Currently, automation is the result of the algorithms which are pre-coded into the devices, but its future is the result of algorithms which enable devices to learn from environment in which it needs to work. It can be achieved utilizing the resources developed for a particular domain popularly known as reinforcement learning (RL). Main objective of this paper is to enable an agent to explore a policy for achieving a control of dynamic system such that it will be capable to find an optimal solution to solve the environment. It can be achieved using an algorithm known as augmented random search algorithm. To improve the training speed, we will use concept of multiprocessing and environment-specific customizations along with ARS algorithm.

Keywords Reinforcement learning · HalfCheetah · BipedalWalker · Machine learning · Parallelism

1 Introduction

1.1 Reinforcement Learning (RL)

Reinforcement learning (RL) is a learning mechanism in which an agent is trained to interact with dynamic environment in a multiple trial and error interactions. Reinforcement learning is a process which happens in multiple steps. At, each step, an

S. Velamati (✉) · V. Padmaja
Department of Electronics and Communication Engineering (ECE), VNR Vignana Jyothi Institute of Engineering and Technology (VNRVJIET), Bachupally, Hyderabad, Telangana, India
e-mail: sruthin.velamati@live.com

V. Padmaja
e-mail: padmaja_v@vnrvjiet.in

H. S. Saini et al. (eds.), *Innovations in Electronics and Communication Engineering*, Lecture Notes in Networks and Systems 107,
https://doi.org/10.1007/978-981-15-3172-9_74

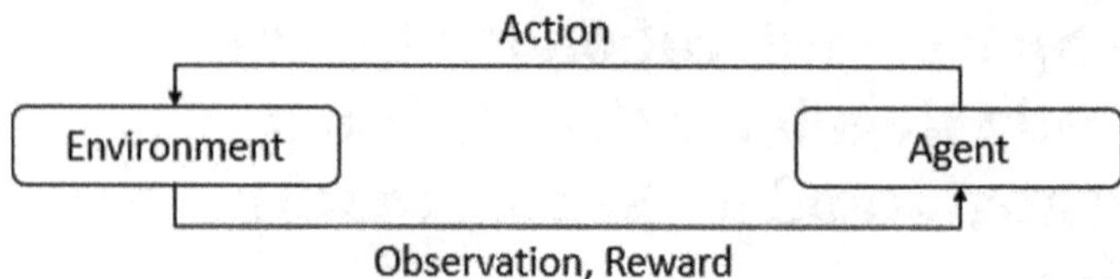

Fig. 1 Reinforcement learning model

agent tries to perform in the environment using the information it extracted from previous interactions with the environment.

Figure 1 explains the interactions between an environment and an agent. An agent communicates with agent as action while an environment communicates with agent as reward and observation or state information.

1.2 Strategies for Reinforcement Learning

Model-based and model-free learning are important strategies involved in solving reinforcement tasks. Model-based reinforcement learning fits a model to previously observed data and then uses this model in some fashion to approximate the solution. Model-free reinforcement learning will eliminate the need for a system's model to train a policy by directly seeking a map from observations/inputs to actions/outputs [1]. Model-free methods are primarily divided into two approaches: policy search or policy space exploration and approximate dynamic programming. In approximate dynamic programming, we will estimate a function that will best characterize the "cost to go" for experimentally observed states while in direct policy search, we attempt to find a policy that directly maximizes the optimal control problem using only input–output data(observation to action) mapping and directly searches for policies by utilizing data from earlier episodes [2].

1.3 Linear Quadratic Regulator

LQR is an optimal control regulator that can be used to keep a better track of a reference trajectory compared against existing traditional controllers. In LQR, future actions are predicted for every time step to minimize a global cost function which in our case is to maximize the reward with which we can better regulate offset in tracking. In robotics, LQR is very handy when nonlinear systems are linearized about specified equilibrium points. In extremely nonlinear systems, approximating higher-order terms can mean a drastic improvement to the degree at which we can dynamically control the interested system in the required operating regions. In [3], a

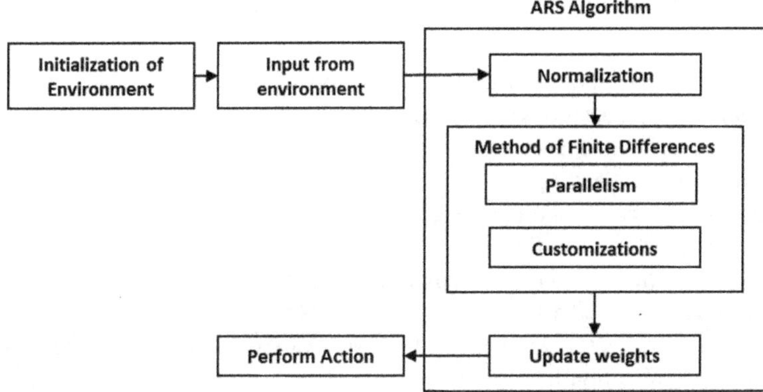

Fig. 2 Block diagram of the training process

high-dimensional robot is controlled using LQ methods in concomitance with reinforcement learning methods. In [4], LQ methods were used by authors to approximate a bat robots complex dynamics.

2 ARS Algorithm

The algorithm proposed in [1] is one of the state of the art model-free reinforcement learning using which an agent will learn to solve the environment using policy space exploration. In the algorithm, multiple perturbations are initiated which are applied on environment to be solved and required policy is updated based on the perturbation which yields the best reward. Section 3.1 will discuss the process of updating policy in deep as proposed in [1] along with parallelism and customization mentioned in Sect. 6 and 7. This algorithm creates and tunes a policy of dimensions Y outputs to X inputs, based on the environment chosen and provides a direct mapping of input to the outputs; it will predict the action needs to be taken by the agent at every particular time step, such that it can solve the agent (Fig. 2).

3 Finite Difference Method in ARS

Method of finite difference is used to update our policy and make it predict correct actions. It can be treated as replacement of traditional gradient descent algorithm which is used to update weights in network as gradient descent algorithm requires differential equation which can be computationally troublesome.

3.1 Steps in Method of Finite Differences

1. Choose hyperparameters (directions, best directions, learning rate, etc.).
2. Initialize N perturbations (δ_N). Let N be the number of directions chosen in hyperparameters. Performance of this process can be improved using multiprocessing as discussed in Sect. 6.
3. Consider M be the policy which need to be updated, add and subtract our perturbations to the policy.
4. Now, apply the policies to the environment and store the reward.
5. Continue step 3 and 4 for all perturbations (No. of directions).
6. Now, from all the rewards choose the best reward pairs. We can improve this step using customization from Sect. 7.
7. Update the policy using the equation [1].

$$M_{j+1} = M_j + \frac{\alpha}{b\sigma_R} \sum_{k=1}^{b} \left[r\left(\pi_{j,(k),+}\right) - r\left(\pi_{j,(k),-}\right) \right] \delta_{(k)}$$

4 Normalization

Normalization is a primary step which is similar to data whitening used in regression related tasks; this step will make sure that policies will apply same range of weights on different input states, for example, at one-time step, if state vector is in the range $[78, 93]$, and for other steps, state vector is in the range $[-2, 1]$. Here, actions with the first state may trigger larger changes while the second state may lead to small actions or vice versa. With this variation, our algorithm will lose uniformity over different states. Normalization will ensure that the abovementioned problem will not arise by reducing levels of all the input states to a similar range. To perform normalization, online mean and online variance calculation is used which will contain past states information for calculation of current states mean and variance.

5 Online Mean and Variance

If the amount of samples in the collection is small, normal mean and variance are sufficiently enough. However, as the amount of samples in the collection surge, a complication arises, when the summation of all the samples gets increases. If it gets too large, it might cause an overflow issue while processing huge collection of data. More subtly, loss of precision may occur if there is a huge difference in the magnitude of the sum when compared to the amount of samples in the list (especially while dealing with floating-point numbers). To help address these problems, there is a way of calculating the mean and variance using an incremental approach also known as online approach.

$$\text{Online mean: } \mu_t = \frac{X - \mu_{t-1}}{n}; \quad \text{Online variance : } \sigma^2 = \frac{(X - \mu_{t-1}) \times (X - \mu_t)}{n}$$

6 Parallelism

In the mentioned algorithm, it requires number of directions in which perturbations need to be considered; we can implement parallelism for these perturbations processing using multiprocessing. We will create few processes based on number of directions mentioned in step 1 on Sect. 3.1. We will create child process, and for communication between parent process and child process, we will use pipes. To implement parallelism, we will use a Python library, namely multiprocessing. To implement communication between parent and child processes, we will create pipes and exchange of communication can be done using command like send() and recv().

7 Customizations

In addition to the mentioned algorithm, customizations can be added to yield a faster training process. This includes the faulty behavior of the agent like solving in wrong directions, agent struck at single point. These scenarios can be identified by monitoring the reward of the perturbations against the number of steps, reward of the environment progress as agent tries to solve the environment, i.e., as time steps increase if time steps are more and reward is less or if reward of each time step is constant and negative for all the future steps, then these kinds of episodes can be excluded at step 6 of Sect. 3.1 (**Steps in Method of finite differences**).

8 Experimentation

For experimentation purpose, two Python libraries, OpenAI Gym and PyBullet [5, 6], have been used. Experimentation is done on two environments one from each, HalfCheetah (from PyBullet) to BipedalWalker (from OpenAI Gym—Box2D). Reward strategies used for evaluating performance of the algorithm in these environments are to attain a reward of 850 for HalfCheetah and reward of 300 for BipedalWalker. According to [1], reward of HalfCheetah can be achieved with 1000 steps, but using parallelism mentioned in Sect. 6 and customizations mentioned in Sect. 7,

Table 1 Table comprises the list of hyperparameters used for training the HalfCheetah and BipedalWalker environments

	Total steps	No. of directions	No. of best directions	Seed	Noise
HalfCheetah	600	16	16	1	0.03
BipedalWalker	600	50	40	1	0.03

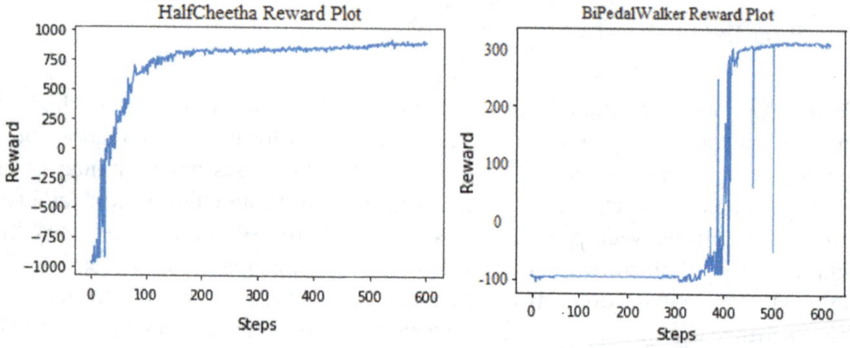

Fig. 3 HalfCheetah (Left) and BipedalWalker (Right) reward plot

```
Rendring HalfCheetha.....

Trial 1: Reward: 858.75368955
Trial 2: Reward: 862.75868955
Trial 3: Reward: 855.12596844

In [8]:
```

```
Rendring BiPedalwalker.....

Trial 1: Reward: 310.1353671627903
Trial 2: Reward: 308.3366909324753
Trial 3: Reward: 308.3097449766041

In [6]:
```

Fig. 4 Reward while rendering for HalfCheetah (Left) and BipedalWalker (Right)

we can achieve it in 600 steps. Table 1 shows the parameters and rewards for the environments (Fig. 3).

After training the policy for HalfCheetah and BipedalWalker and policy is stabilized for both environments, we will perform rendering for both environments in which both agents need to produce good rewards (Fig. 4).

9 Conclusion

Augmented random search algorithm is a policy space exploration technique which enables us to train an agent to find an optimal to solve the required environment by adapting itself to its surroundings. It can be implemented to improve the automation depth of many application-specific electronic devices. Reinforcement learning has huge amount of application in numerous domains such as in robotics and industrial automation; RL can be used to enable a robot to create an efficient adaptive control system for itself which learns from its own experience.

References

1. H. Mania, A. Guy, B. Recht, Simple random search provides a competitive approach to reinforcement learning. arXiv preprint arXiv:1803.07055 (2018)
2. B. Recht, A tour of reinforcement learning: the view from continuous control, in *Annual Review of Control, Robotics, and Autonomous Systems* (2018)
3. S. Levine, C. Finn, T. Darrell, P. Abbeel, End-to-end training of deep visuomotor policies. J. Mach. Learn. Res. **17**(1), 1334–1373 (2016)
4. A. Ramezani, X. Shi, S.-J. Chung, S. Hutchinson, Bat Bot (B2), a biologically inspired flying machine, in *2016 IEEE International Conference on Robotics and Automation* (*ICRA*) (2016), pp. 3219–3226
5. E. Coumans, Y. Bai, PyBullet, a Python module for physics simulation for games, robotics and machine learning (2016–2019). [Online]. Available: http://pybullet.org/
6. G. Brockman, V. Cheung, L. Pettersson, J. Schneider, J. Schulman, J. Tang, W. Zaremba, OpenAI Gym, arXiv:1606.01540 (2016)

Correction to: Innovations in Electronics and Communication Engineering

H. S. Saini, R. K. Singh, Mirza Tariq Beg, and J. S. Sahambi

Correction to: H. S. Saini et al. (eds.), *Innovations in Electronics and Communication Engineering*, Lecture Notes in Networks and Systems 107, https://doi.org/10.1007/978-981-15-3172-9

In the original version of the book, the following belated corrections have been incorporated as in below:

Chapter "Performance Evaluation of Various Modulation Techniques for Underwater Wireless Optical Communication System": The author affiliation "Department of Electrical and Electronics Engineering, Khulne University of Engineering and Technology, Khulna, Bangladesh" has been changed to "Department of Electrical and Electronic Engineering, Khulna University of Engineering & Technology, Khulna, Bangladesh".

Chapter "A Comparative Study on LSB Replacement Steganography". The author affiliation of Srinivas Bachu from "MLRIT, Hyderabad, India" to "Marri Laxman Reddy Institute of Technology and Management, Hyderabad, India".

The updated version of these chapters can be found at
https://doi.org/10.1007/978-981-15-3172-9_11
https://doi.org/10.1007/978-981-15-3172-9_57

H. S. Saini et al. (eds.), *Innovations in Electronics and Communication Engineering*, Lecture Notes in Networks and Systems 107,
https://doi.org/10.1007/978-981-15-3172-9_75

Author Index

H. S. Saini et al. (eds.), *Innovations in Electronics and Communication Engineering*,
Lecture Notes in Networks and Systems 107,
https://doi.org/10.1007/978-981-15-3172-9

CPSIA information can be obtained
at www.ICGtesting.com
Printed in the USA
LVHW081304010521
686188LV00001B/1

9 789811 531743